TOBACCO SMOKING AND NICOTINE

A Neurobiological Approach

ADVANCES IN BEHAVIORAL BIOLOGY

Recent Volumes in this Series

Volume 21 COCAINE AND OTHER STIMULANTS
Edited by Everett H. Ellinwood, Jr. and M. Marlyne Kilbey

Volume 22 DISCRIMINATIVE STIMULUS PROPERTIES OF DRUGS
Edited by Harbans Lal

Volume 23 THE AGING BRAIN AND SENILE DEMENTIA
Edited by Kalidas Nandy and Ira Sherwin

Volume 24 CHOLINERGIC MECHANISMS AND PSYCHOPHARMACOLOGY
Edited by Donald J. Jenden

Volume 25 CHOLINERGIC MECHANISMS: Phylogenetic Aspects, Central and Peripheral Synapses, and Clinical Significance
Edited by Giancarlo Pepeu and Herbert Ladinsky

Volume 26 CONDITIONING: Representation of Involved Neural Functions
Edited by Charles D. Woody

Volume 27 THE BASAL GANGLIA: Structure and Function
Edited by John S. McKenzie, Robert E. Kemm, and Lynette N. Wilcock,

Volume 28 BRAIN PLASTICITY, LEARNING, AND MEMORY
Edited by B. E. Will, P. Schmitt, and J. C. Dalrymple-Alford

Volume 29 ALZHEIMER'S AND PARKINSON'S DISEASES: Strategies for Research and Development
Edited by Abraham Fisher, Israel Hanin, and Chaim Lachman

Volume 30 DYNAMICS OF CHOLINERGIC FUNCTION
Edited by Israel Hanin

Volume 31 TOBACCO SMOKING AND NICOTINE: A Neurobiological Approach
Edited by William R. Martin, Glen R. Van Loon, Edgar T. Iwamoto, and Layten Davis

Volume 32 THE BASAL GANGLIA II: Structure and Function—Current Concepts
Edited by Malcolm B. Carpenter and A. Jayaraman

TOBACCO SMOKING AND NICOTINE

A Neurobiological Approach

Edited by

William R. Martin
Glen R. Van Loon
Edgar T. Iwamoto
and
Layten Davis

University of Kentucky
Lexington, Kentucky

PLENUM PRESS • NEW YORK AND LONDON

Library of Congress Cataloging in Publication Data

Tobacco smoking and nicotine.

Proceedings of an international symposium, held Dec. 2–4, 1985, in Lexington, Ky.

Includes bibliographies and index.

1. Nicotine—Physiological effect—Congresses. 2. Tobacco—Physiological effect—Congresses. 3. Neuropsychopharmacology—Congresses. I. Martin, William R. (William Robert), 1921– . [DNLM: 1. Behavior—drug effects—congresses. 2. Nervous System—drug effects—congresses. 3. Nicotine—adverse effects—congresses. 4. Nicotine—pharmacodynamics—congresses. QV 137 T6278 1985]

QP801.N48T63 1987 615′.782 87-18513

ISBN 0-306-42611-0

Proceedings of an international symposium on Tobacco Smoking and Health: A Neurobiological Approach, held December 2–4, 1985, in Lexington, Kentucky

A Division of Plenum Publishing Corporation
233 Spring Street, New York, N.Y. 10013

Printed in the United States of America

PREFACE

Nicotine is the major factor in the continuation of the smoking habit among humans. On December 2-4, 1985, under the sponsorship of the Tobacco and Health Research Institute in Lexington, Kentucky, leading scientists from around the world whose research efforts have focused on the role of nicotine in the tobacco habit participated in an International Symposium to provide the most comprehensive and extensive coverage of this topic so far. The material discussed was in the forefront of man's knowledge about nicotine, and both lectures and question and answer sessions were stimulating and enlightening. This publication contains the manuscripts presented at the Symposium, along with an Overview prepared by selected individuals. Lecture and poster session topics are delineated more fully in the Overview. This volume is expected to serve as a definitive reference on nicotine as it relates to the tobacco habit.

Symposium sessions were arranged under four headings: (1) behavioral effects of nicotine and nicotine dependence in humans and animals, (2) neurohumoral regulation of neuroendocrine and cardiovascular function by nicotine, (3) neuropharmacology of nicotine, and (4) neurochemistry.

This was the first assembly of this magnitude of scientists who had devoted years of research to nicotine and its effects. Indeed, it was a pleasure to have sponsored this important event.

Dr. Layten Davis, Director
Tobacco and Health Research Institute
Cooper and Alumni Drives
Lexington, KY 40546-0236

ACKNOWLEDGMENTS

The Editors of this volume express sincere appreciation to all of the contributors. Special acknowledgments are given to Ms. Elaine Fisher for the many hours spent typing the manuscripts and to Dr. Janice Humble and Ms. Carol Smith of the Scientific Information Section of the Tobacco and Health Research Institute for their valuable assistance in preparing the volume.

CONTENTS

Tobacco and Health Overview: A Neurobiological Approach 1
W.R. Martin

Tobacco Use and Associated Health Risks 15
D.L. Davis

BEHAVIORAL EFFECTS OF NICOTINE AND DEPENDENCE IN HUMANS AND ANIMALS

Nicotine Intake and its Regulation by Smokers 25
M.A.H. Russell

The Functions of Smoking 51
D.M. Warburton

Nicotine Increases Mental Efficiency: But How? 63
K. Wesnes

Nicotine: Abuse Liability, Dependence Potential and Pharmacologic Treatment of Dependence 81
J.E. Henningfield, S.R. Goldberg, and D.R. Jasinski

Nicotine Actions in Models of Learning/Memory and Reward 101
E.T. Iwamoto, P.J. Fudala, W.R. Mundy, and E.C. Williamson

Discriminative Stimulus Effects of Nicotine: Correlation with Binding Studies 113
I.P. Stolerman, R.K. Kumar, J.A. Pratt, and C. Reavill

Noncholinergic Mechanisms Involved in the Behavioral and Stimulus Effects of Nicotine! And Relationships to the Process of Nicotine Dependence 125
J.A. Rosecrans

Intracranial Self-Stimulation Studies with Nicotine 141
P.B.S. Clarke and R. Kumar

Autoradiographical Evidence of Nicotinic Receptors in Rat Brain 151
P.B.S. Clarke, and A. Pert

Nicotinic Modulation of Dopaminergic Neurotransmission: Functional Implications 169
A. Pert and P.B.S. Clarke

NEUROHUMORAL REGULATION OF NEUROENDOCRINE AND CARDIOVASCULAR FUNCTION BY NICOTINE

Current Concepts on the Effects of Nicotine on Neurotransmitter Release in the Central Nervous System 191
P. Rowell

Mechanisms of Nicotine Regulation of Dopamine Release in Neostriatum 209
T.C. Westfall, H. Perry, and L. Vickery

Effects of Nicotine and Exposure to Cigarette Smoke on Discrete Dopamine and Noradrenaline Nerve Terminal Systems of the Telencephalon and Diencephalon of the Rat: Relationship to Reward Mechanisms and Neuroendocrine Functions and Distribution of Nicotinic Binding Sites in Brain 225
K. Fuxe, K. Andersson, P. Eneroth, A. Harfstrand A. Nordberg, and L.F. Agnati

Nicotinic Regulation of Sympathoadrenal Catecholamine Secretion: Cross-Tolerance to Stress 263
G.R. Van Loon, J.A. Kiritsy-Roy, L.V. Brown and F.A. Bobbitt

Central Nicotinic Regulation of Arterial Blood Pressure 277
T. Kubo

Control of Blood Pressure by Muscarinic and Nicotinic Receptors in the Ventrolateral Medulla 287
H. Sapru

Stereoselectivity of Nicotine's Central Effects and its Relationship to Pain and Cardiovascular Function 301
B.R. Martin

Nicotinic Receptors in the Motor Endplate and in Sympathetic Ganglia 317
R.L. Volle

Nicotinic and Muscarinic Regulation of Adrenal Catecholamine Secretion 325
A.R. Wakade

Nicotinic Regulation of Adrenomedullary Opioid Peptide Synthesis and Secretion: A Model to Study Monoamine Neuropeptide Cotransmission 341
O.H. Viveros, C.D. Unsworth, T. Kanamatsu, J-S. Hong, and E.J. Diliberto, Jr.

NEUROPHARMACOLOGY OF NICOTINE

Metabolism, Pharmacokinetics and Pharmacodynamics of Nicotine in Man 357
N.L. Benowitz, and P. Jacob III

Metabolism of Nicotine in Smokers and Nonsmokers 375
D.W. Sepkovic and N.J. Haley

Does Smoking Decrease Eating and Eating Increase Smoking? 389
M.E. Jarvik

NEUROCHEMISTRY

The Biochemical and Functional Nature of the Nicotine Receptor in Rat Brain 401
L.G. Abood, A. Maiti, S. Grassi, and K.S. Salles

Brain and Muscle Nicotinic Receptors: Complex Homologous Proteins Carrying Multiple Binding Sites 413
B.M. Conti-Tronconi, and M.A. Raftery

The Effects of Chronic Nicotine Administration on Brain Nicotinic Receptor Numbers 439
A.C. Collins and M.J. Marks

Multiple Nicotinic Receptors: Nicotinic Ligands With Different Specificities 451
J.W. Sloan, W.R. Martin, and W.T. Smith

Nicotinic Cholinergic Receptor Recognition Sites in Brain 467
K.J. Kellar, R.D. Schwartz, and A.M. Martino

Nicotinic Binding Sites in the Brain: Properties, Regulation, and Putative Endogenous Ligands 481
H. Sershen and A. Lajtha

ABSTRACTS

Dose-related Nicotine Effects on CNS Binding and Behavior 493
C. Ksir and R.L. Hakan

[^{3}H]Acetylcholine Binding to Nicotinic Cholinergic Receptors in Brain: Localization and Regulation Visualized by Autoradiography 494
R.D. Schwartz, B.B. Wolfe, T.C. Rainbow, and K.J. Kellar

Acute Effects of Tobacco on Human Brain Stem Evoked Potentials 495
V.J. Knott

Effects of Nicotine on the Development of an Invertebrate Embryonic System 495
V. Martin, K. Kolberg, and W. Archer

Effect of Nicotine on the Turnover Rate of Catecholamine in the Rat Brain 496
T. Kubo, H. Amano, Y. Goshima, and Y. Misu

Cigarette Smoking and Parkinsonism 497
J.A. Baron, P. Adams, and H. Quinton

Stereospecific N-Methylation of Nicotine by Intact Guinea Pig Pulmonary Alveolar Macrophages 498
C. Gairola, A. Al Houdi, C.S. Godin, and P.A. Crooks

Biodegradable Microspheres for Sustained Delivery of α-Proteinase Inhibitor 498
F. Rypacek and P.P. DeLuca

In Vitro and *In Vivo* Biomethylation of Nicotine Enantiomers 499
C.S. Godin, W.F. Pool, C.G. Nwosu, and P.A. Crooks

Perturbation of S-Adenosylhomocysteine and Inhibition of Histamine Metabolism in the Guinea Pig, by Nicotine 500
C.S. Godin, A. Al Houdi, and P.A. Crooks

Cigarette Habituation in Nonhuman Primates 500
W.R. Rogers and M.L. Tuttle

Nausea, Cigarette Smoking, Nicotine, and Pituitary Hormone Release 501
M.J. Majchrzak, O.F. Pomerleau, and C.S. Pomerleau

The Role of Tobacco Smoking in Cutaneous Circulatory Changes Associated with Atherosclerosis 502
D. Richardson and G. Hyde

Effect of Cigarette Rod Length on Puff and Post-Puff Respiration Topography, Respiratory Co, and Subjective Ratings in Smokers 502
P.P. Woodson and R.R. Griffiths

Effects of Smoking Topography on Nicotine and Carbon Monoxide Exposure 503
J.P. Zacny, M.L. Stitzer, R.R. Griffiths, F.J. Brown, and J.E. Yingling

Chronic Nicotine Infusion in Rats: Evidence for some Noncholinergic Actions 504
M.D. Aceto, S.M. Tucker, J.R. Hinson, and G.S. Ferguson

Heterogeneity of Brain Nicotine Receptors: A Functional Approach 504
S. Wonnacott, C. Rapier, and G.G. Lunt

Attenuation of Rat Plasma Prolactin Response to Restraint Stress Following Acute and Chronic Dosing with Nicotine 505
B. Sharp, S. Beyer, A.S. Levine, J.E. Morley, and K. McAllen

Cognitive, Performance, and Electrophysiological Effects of Nicotine Deprivation and Nicotine Gum in Human Volunteers 506
W.B. Pickworth, R.I. Herning, F.R. Snyder, and J.E. Henningfield

Contributors 509

Index 517

TOBACCO AND HEALTH OVERVIEW

A NEUROBIOLOGIC APPROACH

William R. Martin

Department of Pharmacology
College of Medicine
University of Kentucky
Lexington, KY 40536

INTRODUCTION

A conference on the neurobiology of smoking, sponsored by the Tobacco and Health Research Institute of the University of Kentucky, summarized the role of the brain in tobacco dependence. This volume contains the reports and reviews of 34 investigators who addressed the conference. These scientific reports are technical in nature.

There are three purposes of this overview. The first purpose is to place substance dependencies in the context of current knowledge of brain function and to relate these findings to tobacco dependency. The second purpose is to provide a statement, in layman's terms, which relates these papers to the broad issue of tobacco and health. The third purpose is to provide a succinct summary of the conference.

Modern advances in the understanding of brain function and the effects of dependence-producing substances have revealed that almost all dependence-producing substances, in one way or another, alter normal chemical transmission in the brain. It is believed that most of the transactions of the brain are a consequence of nerve cells releasing chemicals (nerve transmitters) which affect the functioning of other nerve cells. Some of these important transmitters are: dopamine; the opioid enkephalins, endorphins and dynorphins; acetylcholine; and gamma-aminobutyric acid. These nerve transmitters exert their effects by interacting with a variety of chemical receptors which initiate changes in the functioning of nerve cells. Some drug types that alter or mimic these nerve transmitters are nicotine, opioids, and amphetamine-like drugs. All dependence-producing chemicals have been shown to produce feelings of well-being or euphoria. These feelings are largely characterized by an enhancement of self-image and associated feelings of being more effective and competitive. They do not produce exotic and ecstatic feelings. Some people may have an especially strong need for these types of mood changes and feelings. The question of whether a substance which enhances the effects of the naturally occurring nerve transmitters and fulfills a psychologic need should be viewed as an unnatural dependency is a major question. Many people have apparently minor disorders in feeling states that have profound influences upon behavior. These feelings can be improved or normalized by certain dependence-producing drugs and hence the drug may satisfy a psychologic need. Whether these needs are pathological or not is partly a social judgement and partly an issue of chronic toxicity. Many needs serve the purposes of both the individual and society. Whether the use of mood-enhancing chemicals serves either an individual or societal purpose is an issue deserving of investigation and discussion.

SHORT SUMMARY OF THE CONFERENCE

The world's consumption of cigarettes continues to increase in spite of established pulmonary and cardiovascular risks and published health warnings. In 1984 there were over 4.6 trillion cigarettes produced. Tobacco use may serve several useful purposes, an issue that was emphasized by several investigators. Drs. Warburton and Wesnes presented evidence that nicotine enhanced performance by facilitating the processing of information and sustaining arousal. A substantive body of data indicates that (-)-nicotine stimulates the reticular activating system, a part of the brain which sustains waking. Dr. Henningfield and his collaborators have shown that smoking and the administration of nicotine produce feelings of well-being in smokers. Dr. Jarvik has speculated that the high desirability of smoking after meals may be related to its ability to antagonize post-prandial sedation. All of these effects of nicotine can clearly serve a purpose for both the individual and society. They also may be closely related to the reasons why many individuals continue to smoke.

These considerations raise one of the most fundamental scientific issues discussed in this conference: the nature and the underlying neurobiology of pleasure. It seems clear that both smoking and (-)-nicotine produce pleasurable effects in the subjects. Dr. Henningfield described the effects of nicotine and smoking in subjects who were not only smokers but had also used heroin (37.5%), sedative (50%), and marihuana (62.5%). Although nicotine produces pleasant subjective effects which resemble those produced by modest doses of morphine, it is not known whether large doses of nicotine can produce the same subjective effects that are produced by larger doses of morphine. Animals will not work as hard to obtain (-)-nicotine as they will to obtain cocaine or amphetamines. Many alcoholics and narcotic addicts have hypophoric feelings and depressive symptoms. Some people smoke to relieve these hypophoric and depressive feelings and to decrease anxiety and fears. Others may smoke to obtain a more pleasurable state. Smoking can have both actions. Drs. Olds and Milner (1954) first observed that when electrodes were placed in certain regions of the brain, animals would work to administer electric shocks to these brain regions. This phenomenon is called intracranial self-stimulation (ICSS). These studies, which have been repeated and extended, suggest that specific regions of the brain are involved in pleasurable feelings and sensation. It is not clear at this time whether nicotine alters the activity of these pleasure centers, despite the fact that animals will work to administer nicotine to themselves. Rats will show a preference for a place where they have previously been injected with nicotine suggesting that nicotine's effects were pleasurable. Nicotine also produces aversive effects. Animals and humans can distinguish the effects of nicotine and to a degree generalize the effects of nicotine to other drugs which are presumed to be nicotine-like.

Nicotine has played a critical role in the understanding of how nerves affect the excitability of other nerves and tissues. Certain nerves make a chemical, acetylcholine (ACh), which is released by nerves, excites muscles causing them to contract, and stimulates other nerves. ACh exerts its actions by interacting with several receptors. These receptors are found in muscles, the adrenal medulla, glands, the autonomic nervous system, and the brain. Nicotine has been shown to release many types of nerve transmitters by interacting with these receptors. Some of these released chemicals are acetylcholine, dopamine, epinephrine, norepinephrine, opioid peptides (methionine-enkephalin), and serotonin. It may cause the release of other neurotransmitters. The nicotinic receptors are located on different parts of nerves: the dendrites, cell bodies, and axons. Hence, nicotine interacts with excitable cells in various parts of the brain and body.

It is not surprising that nicotine has many actions on the brain and other excitable tissues considering the ubiquity of its receptors: it alters blood pressure and pulse rate, both increasing and decreasing them; it increases and decreases respiration; it produces arousal and behavioral depression; it improves performance and enhances mood; it stimulates and paralyzes muscles; it diminishes pain-evoked responses; and it alters the release of many hormones. Some of these effects are beneficial, some effects are harmful, and some effects neither help nor harm.

Many of the effects of single doses of nicotine can be explained by its interacting with a cholinergic receptor (a receptor which is also activated by ACh). One of the best studied neurochemical responses to nicotine is its ability to release dopamine and catecholamines from nerve terminals. Dopamine is a nerve transmitter which has been convincingly demonstrated to be involved in the production of feelings of well-being and the antagonism of hypophoric feelings. When animals are exposed repeatedly to nicotine or tobacco smoke, tolerance appears to develop rapidly to nicotine's ability to release DA in some tissues and in some parts of the brain but not in others. Further, nicotine releases DA from nerves in several ways; one mechanism requires Ca^{++}, another does not. Two of the mechanisms release different pools of dopamine. Further, one mechanism of action requires micromolar concentration of nicotine for activation while another mechanism requires 1000 times greater concentrations.

A variety of types of information indicates nicotine probably acts in several different ways. In pharmacologic jargon, nicotine has several mechanisms of action because of its interacting with several different receptors. This is not a new concept. It has been known for many years that the receptor at the nicotinic synapse at the nerve-muscle junction is different from the receptor at autonomic ganglia. Further, it is known that nicotine can desensitize its own receptor. One of the newer concepts presented at this conference was that desensitization could be mediated through yet another type of nicotinic receptor. This receptor mechanism could relate to tolerance and to physical dependence on nicotine.

One of the major issues concerning the actions of nicotine on the brain is whether all of its effects are mediated through receptors which are also activated by a naturally occurring nicotinic ligand, acetylcholine, and whether there are other types of nicotinic nerve transmitters in the brain. These issues are at the forefront of nicotine research and cannot be answered at this time.

CONCLUSIONS

The logic of public health endeavors may not recognize the complexities of tobacco use in our society. The public health model is to identify the vector which is responsible for a disease and eradicate it.

One view of substance use is that the drug is the vector causing the disease. There is much to recommend this position. For example, several drugs of abuse, narcotic analgesics, and a variety of sedative-hypnotics give rise to a new disease "physical dependence" which not only can result in serious health problems when the dependence-producing drug is withheld but also may give rise to a strong need for the drug. Cigarette smokers have abstinence signs and symptoms; however, these are quite mild when compared with the abstinence syndromes associated with opioid and sedative hypnotic use. Further, the acute toxicity of tobacco use is far less than the acute toxicity of stimulants such as cocaine, opioids, and sedative hypnotics.

The issue of whether dependence-producing drugs serve a societal and mental health need must be considered drug by drug and issue by issue. Data have been presented indicating that nicotine and tobacco may have beneficial effects in enhancing performance, alertness, attention, and mood. These changes may have a physiologic basis. The study of Dr. Lee (1979) shows that the death incidence due to coronary heart disease in physicians who stopped smoking was less than physicians who continued to smoke; however, the incidence of deaths attributable to suicide and accidents was greater in abstaining physicians than in physicians who continued to smoke.

The public health approach assumes that the use of tobacco products is a disease vector and does not recognize that tobacco may serve tobacco users' purposes. Despite the many admonitions concerning the health risks of tobacco usage, over 4 trillion dosage units (e.g., cigarettes, cigars, etc.) are produced annually throughout the world.

An alternative solution to health problems associated with tobacco use is the evolution of safer substitutes. This has been implemented to the extent that low-tar, low-nicotine cigarettes, filtered cigarettes, and cigarettes designed to dilute the smoke have become increasingly available. To a degree these changes in cigarette design may increase the amount of tobacco products used by individuals in their efforts to obtain an adequate enhancement of feeling states. The use of smokeless tobacco or nicotine-containing gums may provide nicotine without the chronic adverse effects associated with the inhaling of tobacco smoke. However, ingesting nicotine by the oral route may decrease the ability of the subject to titrate themselves to their optimal subjective state and could affect the degree of tolerance and physical dependence produced by nicotine ingestion. Efforts are being made to allow the inhalation of nicotine as an aerosol and thus avoid substances which have a chronic toxicity but serve the tobacco user no purpose. Finally, the understanding of nicotine's mechanism of action may allow the development of safer reinforcing agents which may enhance the functioning of the nervous system.

All of these efforts may provide important information which will allow the amelioration of public health and mental health problems associated with tobacco use. However, a thorough understanding of nicotine's and related drugs' actions on brain and behavioral function will not resolve conflicting individual and social priorities. Thus, optimum health and safety comes at an economic cost. Economic and health considerations may have a major impact on the quality of life. The expectations of a rapidly evolving society may well stress ancient brain adaptive mechanisms which may provoke individuals to attempt to relieve these stresses through substance use. Thus, public health problems related to substance use are complex and are unlikely to be resolved by a desire to obtain universal abstinence.

TECHNICAL SUMMARY OF THE CONFERENCE

Tobacco use has been considered a type of substance dependency; however, as such it has many unique characteristics. Tobacco, regardless of the form, is used as a very complex substance which has, and yields, many chemicals. Further, different strains of tobacco plants differ in their chemical makeup. These chemicals have their own unique actions on biologic tissue and their own toxicity. The acute toxicity of tobacco use is quite low compared with most substances which are associated with dependency. Tobacco's chronic use is thought to give rise to cardiovascular and pulmonary toxicity, and these different toxicities are probably largely due to different constituents in tobacco, as well as the individual's response to the smoke components. Tobacco use is a volitional act which is exercised with the knowledge that its chronic use may cause health problems. This volume presents the works and thoughts of many investigators who have tried to understand the actions of tobacco and one of its constituents, nicotine, which is thought to be responsible for its chronic use. The reader will soon find that tobacco use is an extremely complex phenomenon which has ramifications at various levels of social functioning. Further, tobacco users commonly use other substances such as alcohol, as well as other drugs.

Dr. Davis reviews certain aspects of the history of tobacco use, cigarette production and consumption patterns, and the substances in tobacco smoke which may adversely effect health. Tobacco-related diseases are discussed. Changes in design of cigarettes that have led to cigarette smoke with greatly reduced content of tars, nicotine, and other chemicals are recounted. These modified cigarettes have now been on the market long enough for epidemiologic studies to soon reveal changes in tobacco use-related diseases. There is some indication that the incidence of lung cancer in adult males in the United States and parts of Europe is decreasing.

Dr. Russell's review deals with two major issues concerning smoking behavior: 1) the pharmacokinetics of tobacco ingestion and 2) the self-regulation of nicotine ingestion. The efficiency of deep inhalation as a method of delivering nicotine is emphasized. Even by this route, smokers employ different strategies. Smokers who inhale moderately may achieve repeated peaks and troughs in blood nicotine levels, whereas heavy frequent smokers tend to sustain a relatively constant level of blood nicotine. Dr. Russell postulates that moderate smokers obtain the reinforcing or pleasurable effects of nicotine with every puff, whereas the heavy smoker may smoke to avoid abstinence. The issue of the reinforcing effects of nicotine and nicotine dependence will be discussed subsequently. The sniffing of dry snuff is also an efficacious way of ingesting nicotine. On the other hand, smoking a pipe or cigar without inhaling, or using wet snuff or chewing tobacco depend largely on the absorption of nicotine across the oral mucosa. As a consequence, the absorbtion of nicotine is slower and no swings in blood nicotine levels are seen. Dr. Russell reviews his own work and that of others as well as several methodologic issues. If nicotine availability is decreased either by using low-tar, low-nicotine cigarettes or by increasing the excretion of nicotine, subjects tend to smoke more. If, on the other hand, the availability of nicotine is increased by the use of high-tar, high-nicotine cigarettes or by administering nicotine intravenously, subjects will decrease their nicotine aquisitive behavior. However, there appears to be a very large variation between individuals in the up-regulating or down-regulating of nicotine ingestion and in the blood levels that smokers achieve.

Dr. Warburton in his review first addresses the issue of why people smoke. A summary does not do justice to the subtleties of the different points of view of why people smoke, but all views have certain similarities which have also been proposed for other types of substance abuse. These include: (1) the induction of pleasurable states (euphoria) such as feeling relaxation or stimulation; (2) the suppression of unpleasant subjective states such as anger, anxiety, boredom, fatigue and hypophoric or depressive symptoms; (3) smoking out of habit; (4) social smoking; and (5) physical dependency. Dr. Warburton also examines the question of personality and smoking and cites several studies indicating that smokers exhibit both more extraversion and neuroticism than nonsmokers and that these personality characteristics are not only independent of each other but are present before smoking began. Several lines of evidence were presented which indicated that smoking may facilitate adjustment to stressful situations and that stressful situations may result in increased smoking.

Dr. Wesnes summarizes a large body of data strongly indicating that smoking enhances several dimensions of performance. Thus, tasks were performed more quickly, with fewer errors and with greater vigilance and persistence, when subjects were smoking. Similar improvements were observed following ingestion of nicotine tablets by both smokers and nonsmokers. Dr. Henningfield and his collaborators studied the effects of smoking and intravenous nicotine on subjective and physiologic parameters. Intravenous nicotine produced significant dose-related increases on the MBG (Morphine Benzedrine Group) scale and the "Liking" scale scores. The MBG scale measures feelings related to efficiency, pleasantness, and popularity. Smoking produced similar effects. Using a line analog scale to provide a global estimate of the intensity of subjective state, the changes in subjective state induced by nicotine and smoking had a very rapid onset and a very short duration (less than 5 minutes). The effects of nicotine on heart rate and blood pressure also showed a rapid onset, although somewhat slower than that observed for subjective states which persisted through the 30 minute observation period.

Both Drs. Russell and Benowitz have related the pharmacokinetics of nicotine to smoking behavior and to tolerance and physical dependence. Following the administration of a bolus of nicotine, blood levels show an initial rapid rise and decline (half-life of about 9 minutes) followed by a more prolonged decrease in blood levels (half-life of about 2 hrs). The early phase of nicotine elimination from the blood is due to a rapid redistribution into other body compartments while the more prolonged phase of elimination is due largely to its metabolism to an apparently inactive metabolite, cotinine. The important observation of Dr. Henningfield that the subjective effects of a single dose (bolus) of nicotine are largely over within minutes must indicate a rapid development and loss of tolerance to these effects of nicotine. Tolerance to other effects of nicotine develop more slowly. Thus, tolerance to the respiratory stimulant action of nicotine can be clearly demonstrated within 15 minutes in dogs infused intravenously with nicotine (Kamerling et al., 1983). Dr. Benowitz's work indicates that tachycardia is maximal following the first several cigarettes of the day. Heart rate remains elevated during the day and returns to abstinence levels only by the next morning. Dr. Benowitz stresses that the heavy smoker is exposed to nicotine 24 hours every day despite abstaining from smoking while asleep.

Drs. Sepkovic and Haley discuss their experiments concerning the *in vivo* oxidation of nicotine and the subsequent *in vivo* reduction of oxidized products. They also have found that nonsmokers passively exposed to smoke metabolize nicotine less rapidly than smokers; hence, nicotine and its metabolites may accumulate to a significant extent in nonsmokers. Further, sidestream smoke may be especially rich in toxic substances and carcinogens. Infants born of smoking mothers and infants exposed to sidestream smoke have elevated amounts of urinary cotinine, a major metabolite of nicotine.

The characterization of abstinence syndromes in dependent subjects has been an unexpectedly difficult problem both conceptually and phenomenologically for all dependence-producing drug types. Drs. Henningfield and Russell briefly discuss dependence on cigarettes and craving for tobacco. Bradycardia, increased food consumption, disrupted sleep, a decrement in cognitive function are commonly reported signs and symptoms associated with tobacco abstinence. The underlying problem in conceptualizing physical dependence relates to the identification of the apparent signs of abstinence which are the dissipation of the effects of smoking and the identification of alterations in brain plasticity. Since smoking is known to produce tachycardia and cognitive enhancement, the identification of parts of the associated apparent abstinence syndrome that can be attributed to disappearance of these pharmacologic effects of tobacco smoking cannot be determined. Thus, the question of which of nicotine's reinforcing actions are a consequence of its pharmacologic effects and which are a consequence of abstinence symptom suppression cannot be answered unambiguously.

Dr. Iwamoto, using a place preference paradigm, has shown that nicotine is reinforcing (rats tend to go to a specific place where nicotine has been previously administered). Pre- or post-treatment with conditioning doses of nicotine decreased the rate of acquisition of operant behavior for a food reward suggesting that nicotine may impair some types of learning.

Dr. Stolerman used (-)-nicotine as a discriminitive stimulus to perform a particular task to obtain a reward and found that rats generalize to cytisine, anabasine, and (+)-nicotine, but not to lobeline. He has related these actions to binding of (-)-nicotine to the high affinity site; however, the potencies of these drugs as discriminative stimuli are not well correlated with their binding affinities. This lack of correlation, however, is attributed to pharmacokinetic differences. The discriminative effects of (-)-nicotine can be antagonized by mecamylamine which does not bind to the high affinity site. It has thus been proposed that mecamylamine is a noncompetitive antagonist.

Dr. Rosecrans, who first applied discriminative techniques to the study of cholinergic drugs in the central nervous system, describes his studies showing that muscarinic and nicotinic drugs have different discriminative properties. He also found that when brain levels of acetylcholine (ACh) were increased by using an anticholinesterase, this change in brain chemistry produced changes which generalized as being muscarinic but not nicotinic-like. These observations raise the question of whether nicotinic discriminative properties involve a nicotinic cholinergic mechanism. To assess the role of norepinephrine (NE) and dopamine (DA) in the ability of rats to identify the discriminatory effects of nicotine, brain levels of NE and DA were reduced experimentally. The data suggest that central dopaminergic processes clearly play a role in nicotine's discriminatory action and NE a lesser role. Dr. Rosecrans points out that whereas tolerance to nicotine's disruptive effect on behavior develops rapidly, little tolerance develops to its discriminative properties. Data is presented that suggest tolerance is due to the effects of nicotine, not to the dosing schedule. Rats that have received nicotine chronically show a decrement in bar pressing behavior when nicotine is withdrawn. Tolerance and abstinence may be associated with a decrease in β-endorphin levels.

Dr. Clarke describes his work on the effects of (-)-nicotine on behavior directed toward intracranial self-stimulation (ICSS). In two series of experiments (-)-nicotine produced some early impairment of ICSS and performance, followed by an increased response rate. With repeated administration of (-)-nicotine, tolerance developed to the impairment of behavior but not to the increase in activity. (-)-Nicotine did not increase the accuracy of rats' behavior in going to places where they obtained ICSS; however, once tolerance had developed to (-)-nicotine's impairing effect, nicotine increased the time rats spent administering ICSS.

Dr. Jarvik also reviews a large body of data concerning (-)-nicotine's reinforcing actions as they relate to several hypotheses concerning nicotine's anorexic effect. Eating may increase craving for nicotine and enhance its pleasurable effects. This may be a consequence of nicotine's arousing action and its anti-vegetative effect.

Dr. B. E. Martin summarizes a number of studies comparing the (-)- and (+)-isomers of nicotine. The relative potencies of the two isomers have more than 50-fold range in producing a number of pharmacologic actions. Thus the (+)-isomer appears to be equivalent to the (-)-isomer in its actions in producing diaphragmatic blockade and its associated lethality in the rat and in releasing brain dopamine. (+)-Nicotine is approximately 1/50 as potent as (-)-nicotine in producing prostration, in serving as a discriminative stimulus, and in causing contracture of the guinea pig ileum. Many other potency estimates approximate (-)-nicotine being 10 times as potent as (+)-nicotine. Dr. B. E. Martin also summarizes experiments demonstrating nicotine's analgesic activity as well as experiments showing its mechanism and site of action.

Dr. Volle presents a succinct overview of older concepts of ganglionic transmission, as well as indicating some newer concepts and problems. (1) The first point concerns the ubiquity of nicotinic receptors in nervous tissue: they are found presynaptically and postsynaptically on several types of nerves. (2) Further nicotinic drugs have diverse actions, both exciting and blocking at the nicotinic receptor, desensitizing active receptors, and possibly blocking ion channels. (3) There is extensive transmitter pharmacologic redundancy in autonomic ganglion involving nicotinic, muscarinic, peptidergic, and possibly catecholaminergic processes. (4) The actions of nicotinic drugs involve the opening of Na^+ pores and increasing the intracellular levels of cGMP and cAMP. These latter changes do not appear to involve nicotinic, muscarinic or catecholaminergic processes. It is highly probable that synaptic processes in the brain are at least as complex as they are in the sympathetic ganglia.

Dr. Viveros reviews his own work relating to the release of catecholamines and enkephalins from the adrenal medulla and speculates about some of its implications. He and his co-workers have found that several secretagogues not only release catecholamines (CA) from the adrenal medulla but also release a variety of opioid peptides (OP) as well as other substances. Stressful stimuli which cause the release of CA and OP also increase the activity of mechanisms responsible for the synthesis of CAs and the OPs, an effect which persisted for several days after the stresses. The changes in CA and OP release mediated by nicotinic receptors are activated by micromolar concentrations of nicotine. Speculations are made about the physiologic implications of co-transmitters and about the recognition that the brain probably does not rely only on a binary logic.

Dr. Wakade also summarizes his experiments, which have shown that both nicotinic and muscarinic ligands release CAs from the rat adrenal medulla. Administration of both nicotine and muscarine increased the release of CAs; however, the nicotine dose-response line was biphasic and both the muscarine and ACh dose-response lines were monotonic. Further, tolerance or desensitization developed rapidly to the effects of infused nicotine but not to the effects of muscarine. The effects of nicotine but not muscarine were antagonized by mecamylamine, whereas the effects of muscarine but not nicotine were antagonized by atropine. Studies indicated that the rat adrenal muscarinic receptor is of the M_1 type.

The effects of (-)-nicotine on the regulation of blood pressure and heart rate have been discussed by Drs. Kubo and Sapru. Dr. Kubo has identified a site in the caudal part of the fourth ventricle of the anesthetized rat which produces both increases and decreases in blood pressure and a slowing of heart rate. Several lines of evidence indicate that (-)-nicotine is activating a physiologically active nicotinic cholinergic system, since other cholinergic nicotinic agonistic processes and antagonists produce appropriate responses. This system probably projects through the nucleus tractus solitarious. The predominant effect of nicotine injected into the lateral ventricle is an increase in blood pressure. Spontaneously hypertensive rats exhibit decreased hypotensive and bradycardic response to nicotine injected into the obex.

Dr. Sapru describes the effects of nicotine-related drugs on two sites in the rat ventrolateral medulla - an anterior site which yields vasopressor responses and tachycardia when stimulated by nicotine, glutamate, carbachol, and physostigmine, while a more posterior site gives vasodepressor responses and bradycardia when stimulated by (-)-nicotine and glutamate. The effects of nicotine at both sites were antagonized by hexamethonium.

Dr. Westfall and his collaborators summarize their studies of the ability of nicotine and dimethylphenylpiperazinium to release dopamine (DA) from the caudate nucleus. They have found two (-)-nicotine releasing mechanisms. (1) Nicotine in micromolar concentrations produces a Ca^{++} dependent release of newly synthesized DA. This release is antagonized by tetrodotoxin which inhibits secretion by nerve membranes and by the nicotinic antagonists d-tubocurarine and mecamylamine. (2) Nicotine in millimolar concentrations is required to release endogenous DA and newly taken up DA. This release does not require Ca^{++} and is not antagonized by tetroadotoxin nor by mecamylamine or d-tubocurarine.

Dr. Rowell reviews the literature on nicotine's effect on the release of nerve transmitter substances. (-)-Nicotine has been shown to release not only DA but also norepinephrine (NE), epinephrine (E), serotonin, and acetylcholine, as well as amino acid and peptide neurotransmitters. Dr. Rowell's work on the effects of micromolar concentrations of (-)-nicotine on DA release from the amygdala and nucleus accumbens, brain structures which may be involved in mood states, as well as from the striatum, is summarized. The DA-releasing effect of (-)-nicotine on the nucleus accumbens can be antagonized by hexamethonium and d-tubocurine. They have further shown that (-)-nicotine is less effective than K^+ and tyramine in releasing DA. It is more active than cotinine, ACh, and 2-methylpiperidine.

Dr. Fuxe reviews the findings of his laboratory and compares them with those of other laboratories. The distribution of the binding of nicotine and ACh in the brain are similar; however, it is quite different from α-bungarotoxin and d-tubocurarine. Thus [^{3}H](-)-nicotine produces a high degree of labeling of several thalamic nuclei, a moderate degree of labeling of the fronto-parietal and posterior cingulate cortex, and a low degree of labeling of the hippocampus, amygdala, and hypothalamus. In contrast, α-bungarotoxin and d-tubocurarine produce a high degree of labeling of the hypothalamus. Single doses of nicotine administered intravenously (iv) or into the lateral ventricles (icv) depleted DA in nerve fibers, in nerve terminals in the corpus striatum, and in nerve terminals of the diffuse type in the nucleus accumbens and tuberculum olfactorium. On the other hand, rats exposed to high levels of cigarette smoke (4 cigarettes) for 2 hours did not show changes in the DA content of the nerve terminals in these loci, but rather a decrease in the DA content of the dotted nerve terminals of the posterior nucleus accumbens. Lesser levels of smoke (1 or 2 cigarettes) produced changes similar to those produced by iv or icv administration of nicotine. Repeated high doses of (-)-nicotine deplete the catecholamines of various parts of the hypothalamus; changes are antagonized by mecamylamine. (-)-Nicotine enhances the release of several pituitary hormones including ACTH with an associated increase in plasma corticosterone, vasopressin, prolactin, lutienizing hormone, and β-endorphin. It decreases the release of thyrotropin (TSH) and growth hormone and does not change the release of somatotropin or leu-enkephalin. Mecamylamine partially antagonizes nicotine-induced changes in release of prolactin, lutienizing hormone, and TSH but not of growth hormone or corticosterone. Further, the effect of (-)-nicotine on lutienizing hormone is mediated through a D_1 receptor while its effect on prolactin is mediated through a D_2 receptor. The effects of smoking or repeated administration of (-)-nicotine on hypothalamic and pituitary function were somewhat different from single dose of (-)-nicotine administration iv or icv. Thus serum LH, prolactin, and TSH levels were decreased and corticosterone levels were increased. These effects on LH, prolactin, and TSH can be antagonized by mecamylamine. In contrast to acute exposure to nicotine, chronic exposure did not alter hypothalamic DA content but was associated with a lowering of NE levels. Nicotine also appears to alter the synthesis and release of neuropeptides such as cholecystokinin and substance P which may play roles in the release of pituitary hormones and appetite. Tolerance does not appear to develop to chronically administered nicotine's ability to suppress prolactin, LH, and FSH release. Tolerance does develop to its ability to enhance ACTH release and to decrease neuronal CA. In rats that had been exposed for 10 days to nicotine smoke and then withdrawn, only limited changes in hypothalamic CA levels and blood pituitary hormones levels were seen. Only a decrease in blood prolactin was observed and a loss of CA flourescence associated with increased utilization of DA in the medial palisade zone of the hypothalamus.

Dr. Van Loon describes experiments concerning tolerance to the sympathetic and cardiovascular effects of nicotine in unanesthetized rats. Intravenously administered (-)-nicotine produced a marked enhancement of plasma norepinephrine and epinephrine and an increase in blood pressure and bradycardia. Only frequently repeated intravenously administered doses induced tolerance to these actions of (-)-nicotine. When (-)-nicotine was administered to the brain, it preferentially increased plasma levels of E, and a single dose produced tolerance. There is cross-tolerance between stress and nicotine-induced release of CA.

Dr. Clarke describes and reviews binding and radioautography studies of (-)-nicotine, ACh, and α-bungarotoxin that strongly support the position that the pattern of acetylcholine and (-)-nicotine binding are quite similar; however, the binding pattern of both drugs is quite different from that of α-bungarotoxin. Thus the two classes of drugs exhibit high affinity binding to different layers of the cortex. The thalamus, habenula, and colliculi show the highest levels of (-)-nicotine binding, whereas the medulla, hippocampus, and hypothalamus show the most intense binding of α-bungarotoxin. The work of Dr. Clarke and his associates provide evidence that (-)-nicotine may exert some of its actions presynaptically via autoreceptors.

The neurochemistry and molecular biology of nicotinic receptors and the neuropsychopharmacology of nicotine have led to the confirmation of classic concepts about the mode of action of nicotine but have introduced new complexities. These new findings are by and large supportive of the idea that most of nicotine's actions are mediated through receptors. Further, great progress has been made in recognizing the intimate nature of nicotine and acetylcholine receptors and their complexities. An articulate summary of an enormous amount of work characterizing the chemical nature of the nicotinic cholinergic receptor (receptors in the electric torpedo), of the electric eel, of the calf myoneural junction, and of the chick optic lobe is presented by Dr. Conti-Tronconi. Several important generalizations can be made about the ACh (acetylcholine nicotinic) receptors. They are part of a large molecule which is one of the channels through which Na^+ enters and leaves nerve and muscle cells. The effect of the acetylcholine- or nicotine-mediated increases in Na^+ conductance is to excite certain nerve cells or muscle cells. This receptor is complicated and may be comprised of more than one receptor moiety. There are five different subunits which are long chain amino acids. Two of these have the same amino sequence but differ in other details. The other three subunits have different amino acid sequences. Dr. Conti-Tronconi speculates that because of the homology between the amino acid sequence, between the subunits, and between equivalent subunits of different species, these subunits probably originated from a common ancestoral gene. The extent of divergence of nicotinic receptors, however, is also substantive. Of particular importance is the observation that there is a difference in the amino acid sequence of the acetylcholine receptor from the brain and muscle of the chick that would be consistent with the idea that brain nicotinic receptors may differ in their specificity from peripheral nicotinic receptors. Further, Dr. Conti-Tronconi has presented evidence that cholinergic drugs can interact with two receptors, one that is a low affinity nicotinic agonist site that opens the Na^+ channel and another that is a high affinity nicotinic site that closes the Na^+ channels and produced desensitization. Dr. Abood has identified a nicotine receptor from the rat brain that is comprised of 4 subunits which differ significantly in their relative mass from the subunits seen in the torpedo nicotinic receptor. This suggests that brain nicotine receptors may differ not only from tissue to tissue but also from species to species. Dr. Kellar and his associates describe the similarities between ACh and nicotine binding in the rat brain, and they argue that the traditional view is that nicotine and ACh are acting on the same binding site and receptor.

Dr. Abood was the first investigator to show that (-)-nicotine had actions in the brain that were difficult to reconcile with the concept that it was acting only as a nicotinic cholinergic agonist. These observations have been extended in Dr. Abood's presentation. Research has focused on a prostration syndrome which is induced by nicotine. In a study of a series of nicotinic analogues and isomers there was excellent agreement between the ability of these agents to induce prostration and inhibit nicotine binding at a high affinity site. However, the nicotinic antagonists hexamethonium, mecamylamine, and α-bungaratoxin have a low affinity for the nicotinic high affinity site. Both mecamylamine and hexamethonium, but not α-bungaratoxin, antagonized the prostrating effect of nicotine in the rat.

The site of action of nicotine in producing prostration appears to be the vestibular nucleus and the X lobule (nodule) of the cerebellum. This conclusion is based on the observations that nicotine injected into these areas produces prostration. Other sites did not. Kainic acid lesions of this region abolished the prostrating effect of (-)-nicotine when administered icv. Lesioning of the vestibular nucleus and cerebellum produced an increase in both [^{3}H](-)-nicotine and [^{3}H]QNB binding. Dr. Abood speculates about the importance of the cerebellum and the vestibular nucleus in nicotine's reinforcing actions in view of their known ability to integrate sensory stimuli and motor activity. These observations may have even more significance in view of the observation that the vermis of the cerebellum is also involved in long-term habituation of the startle response.

Tolerance is comprised of elusive and complex phenomena despite the simplicity of its operational measures (Martin and Sloan, 1977). The nature of reduced effects associated with the chronic administration of nicotinic ligands is controversial. Collins and his collaborators administered nicotine by infusions for varying lengths of time and amounts in mice. Tolerance developed to nicotine's ability to inhibit Y maze crossings and rearings and to depress heart rate and body temperature. The ability of nicotine to slow and accelerate respiratory rate was not altered by the chronic administration of nicotine. These changes were associated with an increase in $[^3H]$(-)-nicotine and $[^{125}I]\alpha$-bungarotoxin binding but no changes in the binding of a muscarinic ligand, $[^3H]$QNB. Dr. Kellar, his collaborators, and others have observed that chronic administration of (-)-nicotine increased locomotor activity in the rat. The binding of $[^3H]$(-)-nicotine and $[^3H]$acetylcholine was increased by chronic nicotine treatment. However, in rats treated chronically with the anticholinesterase DFP, which increases the concentration of ACh in the brain and presumably in the milieu of the ACh receptors, a decrease in the amount of $[^3H]$ACh binding was observed.

Tolerance does not develop to all actions of nicotine when administered chronically. Dr. Abood described experiments in which rats, treated with nicotine administered chronically into the lateral ventricle, did not show any changes in responsivity to nicotine's effects when it was administered into the 4th ventricle. Further, chronic administration of nicotine did not alter either the dissociation constant or the amount of binding of nicotine. Drs. Sershen and Lajtha also address the issue of tolerance and binding site up-regulation and the fact that investigators who have used different dosing schedules have not always observed enhanced binding of nicotine.

The issue of tolerance, desensitization, and up-regulation are addressed in three contexts. Dr. Conti-Tronconi presented evidence that ACh and presumably nicotine act at two sites on the ACh receptor - a site which opens the Na^+ pore from the outside and a site which closes the pore on the inside. This action causes desensitization. Both Drs. Kellar and Collins and their associates have suggested that prolonged exposure to nicotine desensitizes the ACh receptor by converting it into a desensitized form. This, in turn, gives rise to enhanced receptor synthesis presumably through a negative feedback mechanism which increases the number of binding sites. Drs. Sloan and Martin have presented yet another proposal. They have identified 4 or 5 binding sites with which (-)-nicotine interact: a very high affinity site, a high affinity (-)-nicotine site, a high affinity (+)-nicotine site, a low affinity site, and a very low affinity site. There may even be subspecies of these receptors. The very high affinity binding site has been shown to up-regulate these high affinity sites and a stereospecific ligand has been identified for the very high affinity site. It has been proposed that the very high affinity site induces allosteric changes in the low affinity site, which may be the nicotinic cholinergic site, and converts it into a high affinity desensitized receptor. Hence, nicotine, by interacting with several receptors may produce diverse effects which include its ability to alter its own receptor through a receptor-mediated process.

Drs. Sershen and Lajtha have expanded on this point by noting that nicotine also binds to subcellular fractions and that it can modify protein metabolism. Drs. Sershen and Lajtha, as well as Drs. Sloan and

Martin, have conducted experiments with the goal of identifying endogenous brain ligands other than acetylcholine which interacts with brain nicotine receptors. Drs. Sershen and Lajtha have identified a substance, possibly a peptide, which inhibits both ACh and nicotine binding. Drs. Sloan and Martin found that both niacinamide and quinolinic acid could inhibit and enhance (-)-nicotine binding. Dr. Abood first studied piperidine as a possible noncholinergic ligand. Drs. Sloan and Martin have shown that piperidine, as well as several piperidine derivatives, bind to several nicotinic receptors.

There are many lines of evidence that suggest that (-)-nicotine has several modes of action which may all be receptor mediated and which produce diverse pharmacologic actions. It will be a major scientific accomplishment if the chemical nature of these receptors can be identified. These brain nicotine receptors may be of importance in mental health and brain function and indirectly in physical health. Although it is well established that acetylcholine is a nerve transmitter in the brain and that (-)-nicotine exerts some of its actions by mimicking acetylcholine, there is a real possibility that other brain chemicals could also interact with nicotinic receptors. Thus, continuing investigation of nicotine's actions on the brain may allow a better understanding of brain function and provide opportunities for improving mental health.

REFERENCES

Kamerling, S. G., W. R. Martin, K. M. Wu and J. G. Wettstein. Medullary kappa hyperalgesic mechanisms. II. The effects of ethylketazocine administered into the fourth cerebral ventricle of the conscious dog. Life Sciences 33(18):1839-1843, 1983.

Lee, P. N.: Has the mortality of male doctors improved with the reductions in their cigarette smoking? Brit. Med. J. 2:1538-1540, 1979.

Martin, W. R. and J. W. Sloan. Neuropharmacology and neurochemistry of subjective effects, analgesia, tolerance, and dependence, produced by narcotic analgesics. Drug Addiction I, Sect. II, Chap. 1, pp 43-158, 1977.

Olds, J. and P. M. Milner. Positive reinforcement produced by electrical stimulation of septal area and other regions of the rat brain. J. Comp. Physiol. Psychol. 47:419-427, 1954.

TOBACCO USE AND ASSOCIATED HEALTH RISKS

D. Layten Davis

University of Kentucky
Tobacco and Health Research Institute
Lexington, KY 40546-0236

In spite of extensive health warnings in many countries, the use of cigarettes on a worldwide basis continues to increase. However, there has been a decline in the per capita and even total consumption in some countries. The health risks associated with tobacco use have been well documented, especially for cancer, cardiovascular, and pulmonary diseases. The major reason for this continued use will be the primary focus of this Symposium.

In one form or another, tobacco has been used by humans for over 16 centuries. In the Bible, there is a plant referred to as mandrake that is said to have properties of a stimulating nature. This may represent the first record of tobacco use by humans. Tobacco seeds have been discovered during archeological excavations in both Mexico and Peru, and evidence has been gathered that indicates tobacco has been cultivated for many centuries. Perhaps the original form of tobacco usage was the chewing of the leaf. Archaeologists have located a stone carving left by the Mayans around 400 A.D. in Palenque, in the Mexican state of Shiapas which depicts a priest smoking. The art of "snuffing" may have originated with the Indians of Brazil. Jean Nicot, in whose honor the genus *Nicotiana* was named, grew tobacco and promoted the product in Europe for its magic cure-all. Over the years the use of tobacco has been controversial, with strong opinions being expressed by persons on different sides of the issue. As early as the 15th century, severe penalities were imposed on those found using tobacco. For example, it was a capital offense in Turkey to use tobacco in the early 1600s. Certain religious groups have promoted bans on the use of tobacco for many years. The health issue has been of concern for at least two hundred years. In 1761, John Hill, a London physician and botanist, concluded that nasal cancer could develop as a consequence of tobacco snuff use (Redmond, 1970). In 1789, a University of Cambridge scholar wrote a 63-page document on the danger of snuffing tobacco. With all of the attacks and counterattacks, tobacco has continued to be used by people in ever-increasing amounts. Beginning with the colonial days of this country, it has been an important commodity from an economic standpoint. Today, tobacco is a billion-dollar industry in Kentucky and remains economically important in many countries.

As indicated, tobacco has had a varied use over its history. It has been used in several forms including cigarettes, cigars, pipes, snuff taking and chewing. In this century the use of tobacco has been represented by an increase in commercial-type cigarettes. It is difficult to obtain reliable data from some countries in regard to actual usage. In some cases it is possible to obtain production information, but this may

Table 1. Recent Trends in Cigarette Production

Country	1980	1981	1982 (billions of pieces)	1983	1984*
Asia	1,620	1,770	1,870	1,879	1,996
Europe	1,085	1,070	1,091	1,035	1,038
N. America	883	907	864	830	829
USSR	364	364	359	369	385
S. America	248	235	229	231	235
Africa	148	150	154	165	169
Oceania	42	42	41	41	41
TOTAL	4,390	4,538	4,608	4,550	4,693

*Preliminary

Sources: USDA FT 7-85
Tobacco Journal International - 1985

be misleading due to trade among the different countries. In regard to cigarette production, it appears that there is a continued world-wide increase. In 1984, there were over 4.6 trillion cigarettes produced in the world (Table 1).

In some countries there has been a decline in cigarette sales, whereas in other's, an increase is noted. For example, in parts of Asia and Africa, there has been an increase in production since 1980 and in some countries of North America and Europe, there has been a decline. The total number of smokers in the United States has increased from 47.5 million in 1960 to over 55 million in 1983 (Table 2); however, for the first time, there may have been a decrease in 1984.

Smoking status can be further evaluated by an examination of the per capita consumption. Again, these data can be misleading since the population figures will vary over time. There is a wide variation in the per capita consumption of cigarettes (Table 3).

Table 2. Estimated Patterns of Smoking in the United States From 1960 - 1983

Year	Numbers of Smokers (18+) (millions)	Total Cigarettes Consumed (billions)	Cigarettes Per Day (number)
1960	47.5	484.4	27.95
1970	49.0	536.4	29.99
1980	53.1	631.5	32.60
1983	55.2	600.0	29.78

Sources: Statistical Abstracts of US - various issues
Tobacco Outlook and Situation
- various issues

Table 3. Cigarettes Use in Selected Countries in 1983

Country	Cigarettes Per Capita
USA	3494
Japan	3484
Australia	2965
England	2315
Venezuela	2193
Germany	2185
Norway	2070
France	2053
Argentina	1870
Italy	1818
Brazil	1590
Chile	1213
Mexico	1157

Source: Tobacco International 1985

In the United States, which still leads the world in per capita consumption, there has been a continued decline since the 1960's; and in 1984 the per capita consumption was still 3,454. In 1983 this was followed closely by Japan at 3,384, with Chile and Mexico having values of 1,157 and 1,213, respectively. For the United States, the per capita consumption increased from 54 in 1900 to a peak of 4,345 in 1963. In terms of total usage, in 1900, 2.5 billion cigarettes were smoked and this peaked at over 640 billion in 1981 (Table 4). In 1984, this figure was slightly over 604 billion. Therefore, in spite of all the health warnings it was not until the early to mid 1980s that the total use of cigarettes declined in the United States.

Smoking patterns differ among persons in various occupations (1985 Report of the United States Surgeon General). A higher percentage of male blue-collar workers compared with white collar workers currently smoke cigarettes. The same trend was found among female smokers; however, the differences were less pronounced.

Table 4. Ranges of Tar and Nicotine Yields For Commercial Cigarettes From Selected Locations

Location	Tar (mg/cig)	Nicotine (mg/cig)
Austria	6–36	0.40–2.71
China	21–33	0.72–1.24
France	17–44	0.57–2.24
Indonesia	18–75	0.85–2.81
Philippines	22–44	0.77–2.34

Source: Jenkins et al., 1983

Within the last decade, there has been an increased use of smokeless tobacco products by either snuff taking or chewing. Today, in certain parts of the world tobacco is chewed either alone or in a mixture with other ingredients (1985 IARC Monograph). Recently, there has been concern expressed about the health risks associated with the use of smokeless tobacco products in the United States. Three major factors have been responsible for this concern. First, there has been an increased usage in recent years especially among young adults. There were at least 12 million Americans who used these products in 1985 (1986 Report of the Advisory Committee to the United States Surgeon General, and approximately 50 percent of these were categorized as regular users). It appears that this usage may be constant at this point. The peak use of smokeless tobacco products in the United States occurred in the early 1930s. A second factor of concern has been the identification of nitrosamines and other carcinogens in these products (Brunnemann et al., 1985 and 1985 IARC Report). A third factor has been the implications for human health based on at least one epidemiological study in which it was found that the risk of developing oral cancer was increased among the users of certain forms of smokeless tobacco (Winn et al., 1981). Also, evidence has been presented that supports an association between smokeless tobacco use and gingival recession, as well as oral leukoplakia (1985 IARC Report).

There have been significant changes in cigarette design in terms of delivery over the last two decades (1981 Report of United States Surgeon General; 1984 Recent Advances in Tobacco Science). As a result, within the smoking population, there are several subgroups that may be defined. There are smokers who started smoking high-yield cigarettes and have remained in this category, although even in the high-yield classification, the upper yield limit has declined. Another group started on the high-yield brands and switched to lower yield products. A third group started on lower yield cigarettes and remained in that category. Many of the younger smokers would be in this group. A fourth group might be smokers who started on lower yield cigarettes and changed to higher yield products. Obviously, a fifth segment of the population would include exsmokers. In addition to the particular class of cigarettes being smoked, the number of cigarettes smoked by individuals within a specified time and the way in which the cigarettes are smoked are important considerations. Individual dosages, whether determined by cigarette design, smoker puffing and inhalation patterns, or a combination of the two, are important aspects that have not been addressed appropriately. Present and future epidemological studies should be designed to take these factors into account.

In many countries, during the past few decades there has been a gradual reduction in the tar and nicotine deliveries from cigarettes. However, in some parts of the world the yields remain relatively high (Jenkins et al., 1979, 1983). For example, in one country products with tar yields of 75 mg/cigarette are on the market.

In the United States, the market share of cigarettes with less than 15 mg of tar per cigarette increased dramatically from 1967 until 1981. Since that date there has been some decline in the sales of cigarettes in that category. To achieve the lower yields, there have been significant changes in the cigarette design over the past few decades. The major changes include more specific blend selection, variations in cigarette length and circumference, addition of filters, the use of reconstituted tobacco sheet, as well as expanded tobacco lamina, and development of ventilation techniques (Grise, 1984). The particular manufacturer may alter several of these parameters to achieve the desired final product in terms of delivery. The exact specifications are of proprietary nature; however, some general statements can be made about cigarette design in relation to smoke yield. Tobaccos produced under different production practices and from different genetic origins vary in leaf chemistry. In recent years there has been an increased use of imported tobaccos in American cigarettes. One of the changes in cigarette design has been the use of reconstituted sheet tobacco. In very general terms, reconstituted tobacco contains more cellulose, fiber and ash, and less alkaloids (nicotine), nitrogenous compounds, and reducing sugars (Norman, 1982). Another manufacturing technique used to reduce the yield of cigarettes is incorporation of expanded tobacco. Expanded tobacco made its appearance in the 1970s and is a significant component of modern cigarettes, especially those of ultra-low tar yield. This process involves "puffing up" or expanding the cured leaf, resulting in less tobacco being used per cigarette. The chemistry of the smoke is different primarily due to the effect of burn rate. Cigarettes that have large amounts of expanded tobacco have lower puff numbers per cigarette. Also, there have been changes in the cigarette paper that influence burn rate.

In some countries, especially the United States, filter cigarettes dominate the market. Cellulose acetate filters are the most widely used; however, charcoal filters continue to be placed on some brands. When compared with the nonfiltered version, the dry particulate matter and nicotine are significantly reduced for filter cigarettes; however, carbon monoxide delivery, compared with nicotine, may be higher especially for the higher efficiency filters (Norman, 1982). Phenols have been greatly reduced by plasticized cellulose acetate filters.

Another more recent change in the cigarette design has been air dilution or ventilation. Ventilation, which may be achieved by either increasing paper permeability or by providing ventilation pores at the filter, does more than dilute the smoke with air (Baker, 1984). As ventilation is increased, a smaller proportion of the puff volume is drawn through the pyrolysis zone. The level of most mainstream components is decreased. The outward diffusion of light gases such as CO is increased, thereby increasing these in the sidestream smoke. As the ventilation is increased, a larger proportion of the tobacco is consumed between puffs (smolder period) and the sidestream/mainstream ratios increase (Baker, 1984). The total ventilation of a ventilated cigarette will change as the cigarette is smoked, thereby affecting the per puff delivery of the cigarette. Obviously, the yields of mainstream smoke components are greatly affected by cigarette design, including ventilation (filter tip dilution and paper porosity), tobacco composition (blend, reconstituted sheet and expanded tobacco), and filtration. The yields of sidestream components can be affected by tobacco composition and to a lesser extent by ventilation.

For several years the Federal Trade Commission has published tar, nicotine, and now, carbon monoxide yields of commercial cigarettes. While it does provide a ranking under specified conditions, the standard method of determining yields (tar and nicotine) have little meaning in terms of human dosage. Individual smoking patterns are important. There is no absolute yield for any cigarette due to these factors.

Cigarette smoke is a complex mixture of particulate phase and gas phase components. A number of these have been described as being health related. A listing of selected components and ranges per cigarette is provided in Tables 5 and 6. These data were taken from a publication by Wynder and Goodman (1983). This listing should not be considered to be complete, but does represent some of the more important health-related components that have been quantified.

The health risks associated with tobacco use have been summarized in numerous reports, including a series of Surgeon General's Reports on Smoking and Health. The issues involving smoking and lung cancer were

Table 5. Health-Related Compounds in Particulate Phase of Cigarette Smoke

Particulate Phase	Ranges Per Cigarette
Nicotine	0.06–2.3 mg
Phenol	20–150 μg
Catechol	40–280 μg
Phytosteroids	130 μg
Naphthalene	2.8 μg
Pyrene	15–90 ng
Benzo(a)pyrene	8–40 ng
Harmane	1.1–3.1 μg
Nitroso-compounds	0.4.6 μg
Cadmium	100 ng
Polonium-210	.03–5pCi

Source: Hoffmann et al. (See Reference 5, page 185).

discussed in a publication by Wynder and Goodman (1983). Also, these authors discussed the low-yield cigarette as a risk factor in lung cancer as well as some aspects related to nutritional deficiencies or imbalances. Dose-response data were presented about smoking and lung cancer. For both men and women there was a steady increase in the risk for tobacco-related cancers with the number, duration and type of cigarettes smoked. Depending upon the type of cancer, the relative risk increased to about 30–40 cigarettes per day. Also, the most beneficial effect of filters in regard to decreased risk was in the 20–40 cigarettes per day range. These authors stated that smoking low-yield cigarettes may reduce the risks for lung cancer and cancer of the larynx but not for myocardial infarction. It is estimated that at least 30 percent of all cancer deaths in the United States can be attributed to tobacco use (1982 Report of the Surgeon General). In a recent IARC Monograph (1986), a working group reported that "there is sufficient evidence that the inhalation of tobacco smoke as well as topical application of tobacco

Table 6. Health Related Components in Vapor Phase of Cigarette Smoke

Vapor Phase	Ranges Per Cigarette
Carbon monoxide	0.5–26 mg
Carbon dioxide	10–80 mg
Nitrogen oxides	16–600 μg
Ammonia	10–130 μg
Hydrogen cyanide	180–550 μg
Hydrazine	32 μg
Formaldehyde	20–90 μg
Acrolein	10–140 μg
Nitroso-compounds	0.1–180 ng
Benzene	12–48 μg

Source: Hoffmann et al. (See Reference 5, page 185).

smoke condensate causes cancer in experimental animals," and "there is sufficient evidence that tobacco smoke is carcinogenic to humans." It was estimated that in populations with a long duration and heavy intensity of cigarette usage the proportion of lung cancer attributable to smoking is approximately 90 percent. It was stated that the occurrence of malignant tumors of the respiratory tract and upper digestive tract is causally related to the smoking of different forms of tobacco, and malignant tumors of the bladder, renal pelvis, and pancreas are causally related to cigarette smoking. In regard to the risk of cigarette smoke to the nonsmoker, it was concluded that the observations made to date are compatible with either an increased risk from "passive" smoking or an absence of risk. In most environments the cigarette smoke is rapidly diluted with air and the smoke is "aged," two factors which reduce the toxicity.

In addition to lung cancer, cigarette smoking is a risk factor in the development of other respiratory disorders and has been discussed in numerous publications, including a recent review by Jackson and Holle (1985). An early increase in airway resistance has been reported among persons who smoke. The changes in small airways may result in smooth muscle hypertrophy, goblet cell hyperplasia, and mild peribronchiolar fibrosis. There are increased numbers of inflammatory cells, including macrophages, found in the smokers' lungs. There is an impairment of the lung clearance mechanisms as a result of structural and functional abnormalities. The impairment of the mucociliary clearance system is associated with epithelial lesions, mucus hypersecretion and ciliary dysfunction. Chronic bronchitis is characterized by an impairment of the mucociliary clearance system. There is evidence that cigarette smoke alters the immune processes in the lung.

It has been estimated that 80 to 90 percent of the Chronic Obstructive Lung Diseases (COLD) in the United States is attributed to cigarette smoking (1984 Report of the Surgeon General). Several of the events that occur in COLD as a result of cigarette smoking are chronic mucus hypersecretion, chronic airflow obstruction, and emphysema (destruction of lung parenchyma). The current hypothesis regarding smoking and emphysema is one that includes an increase in protease and reduction or inactivation of the antiprotease inhibitor, which leads to the destruction of lung tissue (Laurell and Eriksson, 1963; Janoff et al., 1979; Gadek et al., 1981). A number of reports have been published on this hypothesis including a section of the Report of the Surgeon General (1984). The excess elastase activity may be due to increased concentrations of inflammatory cells in the smokers' lungs and decreased levels of antiprotease secondary to oxidation by cigarette smoke. In regard to the low-yield cigarette, most of the available data have not related COLD to cigarette yield.

Cigarette smoking has been established as a risk factor for a number of cardiovascular disorders including arteriosclerosis, which is the underlying cause of coronary heart disease, cerebrovascular disease, and atherosclerotic peripheral vascular disease (1983 Report of the Surgeon General). It has been estimated that approximately 30 percent of the deaths due to coronary heart disease in the United States are directly attributable to cigarette smoking. The events and mechanism responsible for the cardiovascular risk as it relates to smoking are for the most part unknown. The risk may be related to several deleterious cardiovascular effects, including production of an imbalance between myocardial oxygen supply and demand, a decrease in the threshold for ventricular fibrillation, and an increase in platelet aggregation. In a review article, McGill (1979) cited evidence that cigarette smoking may cause a slight to moderate elevation of total serum cholesterol concentration, a depression of HDL concentrations, and elevated LDL concentrations. This alteration of total serum cholesteral concentration and lipoprotein composition may lead to an increased development of atherosclerosis. It has been reported that cigarette smoke may be responsible for an unfavorable imbalance in prostacyclin and thromboxane ratios, which would favor platelet aggregation, vascular constriction, and pathogenesis of atherosclerosis (Hirsh et al., 1981). The cigarette smoke alkaloid nicotine has been shown to have acute and chronic effects on the cardiovascular system (Harrison, 1980). These include an increase in heart rate, systolic and diastolic blood pressures, myocardial contraction, oxygen consumption and excitability, coronary blood flow, and peripheral vasoconstriction. As with other smoke-related diseases, the cardiovascular problems appear to be dose related. The longer a person smokes, the greater the risk.

There are many factors to be considered in reducing the health risks associated with tobacco use including changes in cigarette design. It has been found that cytotoxicity of mainstream smoke decreases with smoke age, smoke dilution, and the addition of activated charcoal to the filter (Sonnenfeld et al., 1985). If a similar response occurs in humans, the risk in certain diseases may be reduced by the change in cigarette design. There may be some changes in pulmonary and/or cardiovascular incidences among smokers within the next few years as a result of the use of ventilated cigarettes. Obviously, it is a complex issue and one that requires objective research to reach accurate and beneficial conclusions.

REFERENCES

1. Baker, R.R. (1984). The effects of ventilation on cigarette combustion mechanisms. Recent Advances in Tobacco Science, Tobacco Chemists Research Conference. Vol. 10: 88–150.
2. Brunnemann, K.D., Genoble, L., and Hoffmann, D. (1985). N-nitrosamines in chewing tobacco: An international comparison. J. Agric. Food Chem. 33: 1178–1181.
3. Gadek, J.E., Fells, G.A., Zimmerman, R.L. Rennard, S.I., Crystal, R.G. (1981). Antielastases of the human alveolar structures: Implications for the protease-antiprotease theory of emphysema. J. Clin. Invest. 68(4): 889-898.
4. Grise, V.N. (1984). Market growth of reduced tar cigarettes. Recent Advances in Tobacco Science. Tobacco Chemistry Research Conference. Vol. 10: 4–14.
5. Hollmann D, Brunnemann, K.D., Klus H. Tobacco sidestream smoke and indoor air pollution (from Wynder and Goodman, 1983).
6. Harrison, D.C. (1980). Smoking, nicotine, and cardiovascular health. Prac. Cardiol 6: 53.
7. Hirsh, P.D., Campbell, W.B., Willerson, J.T. and Hillis, L.D. (1981). Am. J. Med. 71: 1009-1026.
8. IARC (1985). IARC monographs on the evaluation of the carcinogenic risk of chemicals to humans. Tobacco Habits Other Than Smoking; Betel-Quid and Areca-nut chewing; and some related nitrosamines. WHO-IARC: Vol. 37, p. 1-291.
9. IARC (1986). IARC monograph on the evaluation of the carcinogenic risk of chemicals to humans. Tobacco Smoking. WHO-IARC, Vol. 38 p. 314.
10. Jackson, F.N. and Holle, R.H.O. (1985). Smoking: Perspectives 1985, a symposium on pulmonary medicine. Primary Care, Vol. 12(2): 197-216.
11. Janoff, A., White, R., Carp, H., Harel, S., Dearing, R., Lee, D. (1979). Lung injury induced by leukocytic proteases. Am. J. Pathol. 97(1): 111-129.

12. Jenkins, R.A., Quincy, R.B., and Guerin, M.R. (1979). Selected constituents in the smokes of foreign commercial cigarettes: "tar," nicotine, carbon monoxide and carbon dioxide. Oak Ridge National Laboratory. TM-6869.
13. Jenkins, R.A., Holladay, S.K., Griest, W.H. & Guerin, M.R. (1983). Tar, nicotine, and carbon monoxide deliveries of commercial cigarettes from selected nations. In: Fifth World Conference on Smoking and Health, Winnipeg, Canada, 10-15 July 1983, Ottawa, Canadian Council on Smoking and Health.
14. Laurell, C.B., Eriksson, S. (1963). The electrophoretic α_1-globulin pattern of serum in α_1-antitrypsin deficiency. Scand. J. Clin. Lab. Invest. 15(2): 132-140.
15. McGill, H.C., Jr. (1979). Potential mechanisms for the augmentation of atherosclerosis and atherosclerotic disease by cigarette smoking. Prev. Med. 8(3): 390-403.
16. Norman, V. (1982). Changes in smoke chemistry of modern day cigarettes. Recent Advances in Tobacco Science, Tobacco Chemists Research Conference Vol. 8: 141-177.
17. Recent Advances in Tobacco Science (1984). Design of low tar cigarettes. Tobacco Chemists Research Conference. Vol. 10, pp. 1-159.
18. Redmond, D.E. (1970). Tobacco and cancer: The first clinical report, 1761. N. Engl. J. Med., 282: 18-23.
19. Sonnenfeld, G., Griffith, R.B., and Hudgens, R.W. (1985). The effect of smoke generation and manipulation variables on the cytotoxicity of mainstream and sidestream cigarette smoke to monolayer cultures of L-929 cells. Arch. Toxicol., Vol. 58: 120-122.
20. Statistical Abstracts of the United States, various issues from 1970-85.
21. Tobacco International, 1985. Good-sized Brazilian crop; cigarette sales still declining. Vol. 187 (13): 5-9.
22. Tobacco - Outlook and Situation, various issues from 1976-85. United States Department of Agriculture, Economic Research Service.
23. U.S. Department of Health and Human Services (1981). The health consequences of smoking, the changing cigarette, A Report of the Surgeon General, Rockville MD., Public Health Service Office on Smoking and Health pp. 1-237.
24. U.S. Department of Health and Human Services (1982). The health consequences of smoking, cancer, A Report of the Surgeon General, Rockville, MD., Public Health Service, Office on Smoking and Health pp. 1-301.
25. U.S. Department of Health and Human Services (1983). The health consequences of smoking, cardiovascular diseases, A Report of the Surgeon General, Rockville, MD., Public Health Service Office on Smoking and Health pp. 1-384.
26. U.S. Department of Health and Human Services (1984). The health consequences of smoking, chronic obstructive lung diseases. A Report of the Surgeon General, Rockville, MD., Public Health Service on Smoking and Health pp. 1-516.
27. U.S. Department of Health and Human Services (1985). The health consequences of smoking, cancer, and chronic lung disease in the workplace, A Report of the Surgeon General, Rockville, MD., Public Health Service, Office on Smoking and Health pp. 1-301.
28. U.S. Department of Health and Human Services (1986). The health consequences of using smokeless tobacco, A Report of the Surgeon General, Bethesda, MD., Public Health Service - NIH Publication No. 86-2874, p. 24.
29. Winn, D.M., Blot, W.J., and Fraumeni, J.F. (1981). Snuff dipping and oral cancer. N. Engl. J. Med. 305: 230-231.
30. Wynder, E.L. and Goodman, M.T. (1983). Smoking and lung cancer: Some unresolved issues. Epidemiol. Rev., Vol. 5: 177-207.

NICOTINE INTAKE AND ITS REGULATION BY SMOKERS

M.A.H. Russell

Addiction Research Unit
Institute of Psychiatry
101, Denmark Hill
London SE5 8AF
England

INTRODUCTION

People smoke in different ways. They vary in the number and type of cigarettes they smoke, and in the way they puff and inhale. It is therefore not surprising that nicotine intake also varies widely between individual smokers. It is my brief here to discuss self-regulation by smokers of the intake of nicotine into their blood and hence to their brain. It is the role of others to discuss what nicotine does when it gets there, and why so many smokers get so desperate when it doesn't.

Self-regulation of nicotine intake, or nicotine titration as it is sometimes called, has become a widely used approach for studying the role of nicotine in smoking. To what extent do smokers tend to maintain a constant intake over days or weeks or even years? To what extent do they modify their smoking pattern to maintain a constant intake when they switch to higher or lower yield cigarettes, or to a pipe or cigars? To what extent do they unwittingly puff harder and inhale more deeply when they smoke fewer cigarettes? Do they smoke and inhale less when given nicotine by another route, e.g., intravenously or by nicotine chewing gum? Above all, to what extent is any regulation of smoke intake mediated by a need to maintain a constant intake of nicotine, as opposed to tar, carbon monoxide, or some other component?

These are some of the questions I shall attempt to address. The answers are by no means completely clear. But, in summary, there is evidence that the intake of nicotine is regulated independently of other smoke components and that down-regulation to avoid excessive intake is more sensitive and precise than up-regulation to maintain usual intake. In general, smokers seem to tolerate a drop in blood nicotine to about two-thirds of their usual levels with some initial loss of satisfaction, and to adapt gradually to the lower level over two or three weeks. Greater reductions occur when fuller up-regulation is mechanically difficult or impossible (e.g., on ultra-low yield cigarettes) and generate increasing loss of satisfaction and craving. Other withdrawal symptoms become evident when blood nicotine levels fall below about one-third of the usual level.

Before considering the regulation of nicotine intake, I shall discuss the usual intake from cigarette smoking and other forms of tobacco use as well as certain theoretical and practical methodological issues which affect understanding of the intake regulation behaviour of smokers. The implications for less harmful forms of tobacco use, and less harmful cigarettes in particular, will be discussed in a final section.

NICOTINE INTAKE FROM CIGARETTE SMOKING

The modern cigarette is a highly effective device for getting nicotine into the brain. The smoke is mild enough to be inhaled deeply into the alveoli of the lungs from where it is rapidly absorbed. It takes about 7 seconds for nicotine absorbed through the lungs to reach the brain compared with the 14 seconds it takes for blood to flow from arm to brain after an intravenous injection. Thus, after each inhaled puff, the smoker gets an intravenous-like 'shot' or bolus of blood containing a high concentration of nicotine which reaches the brain more rapidly than from an intravenous injection. The uptake of nicotine by the brain is also extremely rapid as was clearly demonstrated by the autoradiographic studies of Appelgren et al. more than 20 years ago (1).

This puff-by-puff bolus form of nicotine intake has a number of implications which might explain the high dependence-producing potential of cigarette smoking. Firstly, each puff is pharmacologically reinforced so that the number of reinforcements is far greater than with drug intake from other routes; 20 cigarettes per day is 7,300 a year, each puffed about 10 times, making a total of more than 70,000 intravenous-like shots of nicotine per year. Secondly, the nicotine concentrations in the post-inhalation boli are unknown but must be many times higher than those measured in mixed venous blood following dispersion of the boli. Uptake of nicotine by the brain may be sufficiently rapid for brain concentrations to reflect those in the boli rather than the levels in mixed venous blood. There is some evidence for this from the work of Stalhandske (Table 1) (2). Thirdly, unlike those of mixed venous blood, nicotine concentrations in the arterial boli are unaffected by any increase in the rate of nicotine metabolism that may result from acquired pharmacokinetic tolerance.

Cigarette smokers who do not inhale absorb little nicotine, even if they are heavy smokers of 40 cigarettes or more per day. But, in smokers who inhale, up to 92% of the inhaled dose of nicotine is absorbed depending on the depth of inhalation (3). On average, heavy smokers

Table 1. Ratio of brain-to-blood nicotine levels after intravenous and intraperitoneal injection in mice. The higher ratios after intravenous injection suggest that the brain cells equilibrate to some extent with the higher levels of the intravenous bolus.

Minutes after Injection	Intravenous	Intraperitoneal
1	5.6	1.6
2.5	5.2	1.9
5	5.1	3.0
10	-	2.9
20	5.8	2.2
60	4.0	2.7

Abstracted from Stalhandske (1970). (2)

absorb about 1 mg nicotine from each cigarette (4,5) although their intake from the first cigarette of the day may be somewhat higher, about 1.4 mg (6). Blood nicotine levels of smokers vary widely, from below 5 ng/ml to over 70 ng/ml, with an average level for heavy smokers of about 35 ng/ml. The distribution of peak blood nicotine concentrations just after a cigarette is shown in Fig. 1 for a sample of heavy smokers (7). Although the curve for smokers in the general population would be somewhat to the left, measurable pharmacological effects are produced with blood nicotine levels of 10 ng/ml or less. It is thus apparent that most regular smokers inhale and absorb sufficient nicotine to produce pharmacological effects.

PHARMACOKINETICS AND BLOOD NICOTINE PROFILES

In two early studies the half-life of nicotine in plasma was reported to be about 30 minutes (3) or less (8). Both these studies based their estimates on blood concentrations in both initial and terminal phases which were not differentiated. We have shown the decline in blood nicotine concentration to be biexponential with the half-lives of the initial and terminal phases averaging 9 minutes (SD 1.2 min) and 133 minutes (SD 20 min) respectively (5). These values were essentially the same before and after a loading period of smoking six cigarettes indicating the absence of saturation kinetics at the concentrations that build up during smoking. In an earlier study, Benowitz et al. (9) found a similarly long terminal half-life averaging 119 minutes (SD 44 min). Since they used a slow infusion of nicotine, they were unable to estimate the initial half-life. These pharmacokinetic data indicate that

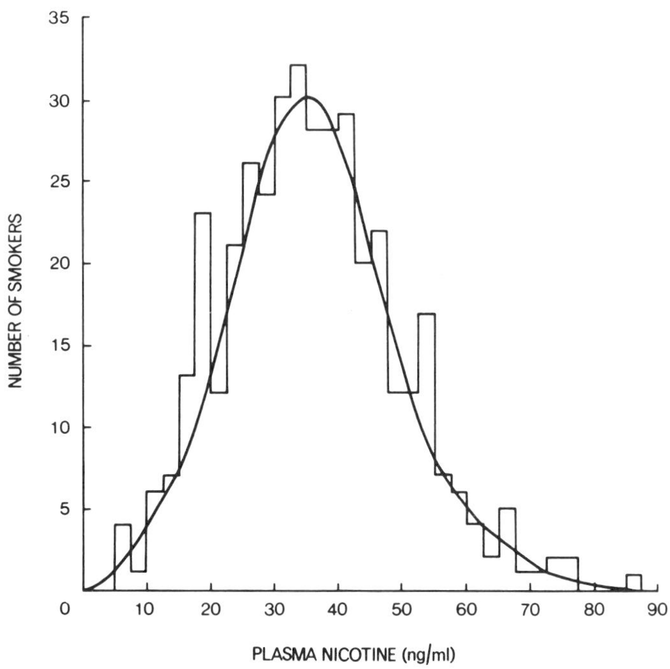

Fig. 1. Distribution of peak blood nicotine concentrations of 393 cigarette smokers two minutes after completing a cigarette during the afternoon of a day of usual smoking. From Russell and Jarvis (1985) (7).

appreciable accumulation of nicotine would occur even in relatively moderate smokers of 15 per day (about one cigarette per daytime hour) and that the problem would be exacerbated in heavy smokers whose rate of smoking approached one cigarette in two to three initial phase half-lives (e.g., 2–3 cigarettes per hour). This situation is shown in Figs. 2 and 3.

Among smokers who inhale, two types of plasma nicotine profile are evident (Fig. 2 and 3). An inhaler who smokes at a rate of one cigarette an hour (Fig. 2) can obtain rapid plasma nicotine peaks following each cigarette and these are no doubt accompanied by positive pharmacological effects. Some accumulation of nicotine is evident but even at steady-state there are prominent peaks after each cigarette. In contrast, at higher rates of smoking the peaks are negligible (Fig. 3) and would have little positive pharmacological effect. It is likely, therefore, that heavy smokers of this type smoke mainly for negative reasons, i.e., to avoid the discomfort of withdrawal effects resulting from a drop in blood nicotine level, rather than for the positive effects of periodic peaks. Indeed, it would be pharmacokinetically impossible for heavy smokers to obtain prominent peaks without going up to very high levels which appear to be aversive even to heavy smokers who have developed considerable tolerance to nicotine (see literature on forced rapid smoking aversion therapy) (12). It is also noteworthy that both the positive and the negative reinforcement type of smokers appear to be subject to an aversive barrier set at a similar level. In these two examples, the peak levels of the lower rate smoker are slightly higher than those of the heavier smoker. The general picture is that among light smokers of less than 15 per day there is a positive correlation between cigarette consumption and blood nicotine concentration, but among heavier smokers it is minimal; r=0.13 in a sample of 330 heavy smokers we studied (13), with similar findings in the large sample reported recently by Gori and Lynch (14).

Finally, there is evidence that smokers tend to achieve their individual steady-state blood nicotine levels by mid-morning and to then

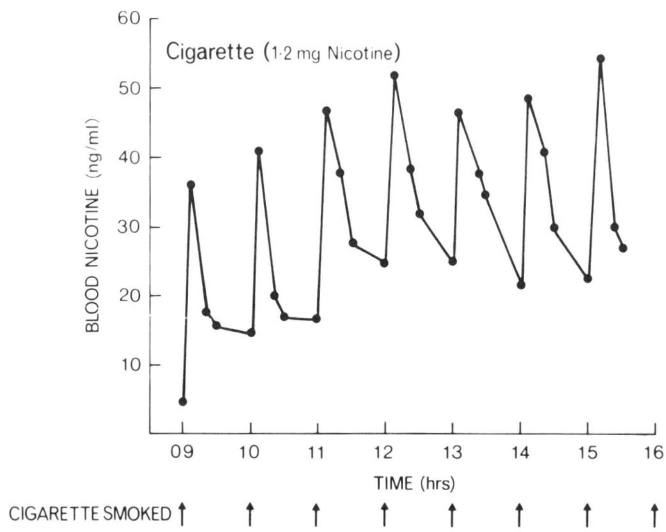

Fig. 2. Blood nicotine levels of an inhaling smoker, smoking one cigarette an hour. From Russell et al. (1976) (10).

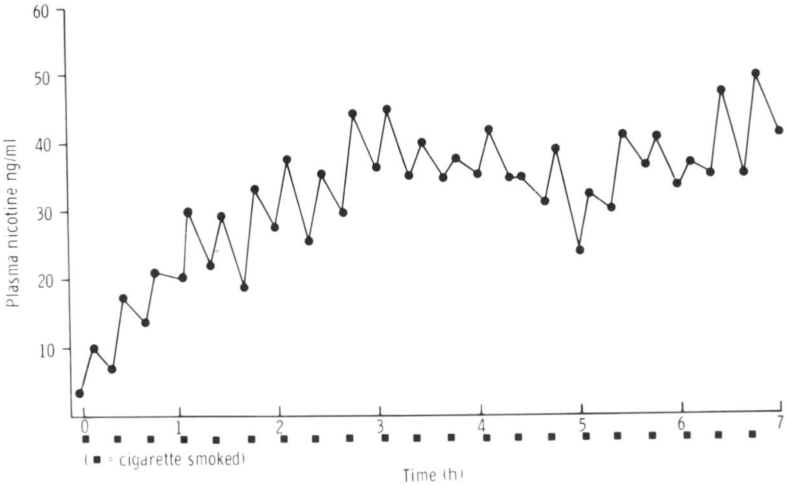

Fig. 3. Blood nicotine levels of a heavy smoker, smoking three cigarettes an hour. From Russell & Feyerabend (1978) (11).

maintain those levels throughout the rest of the day and from one day to the next. The correlation within smokers across days is about 0.9 (15,16). The average blood nicotine boost from a single cigarette in regular smokers at steady-state during the afternoon is about 10 ng/ml (17). At present these data are derived from small samples of heavier than average smokers over relatively short periods of time. There is a great need for data on the consistency of intake in large samples of normal smokers extending over time periods of several years.

NICOTINE INTAKE FROM OTHER TOBACCO PRODUCTS

Tobacco use has fluctuated between chewing, snuffing and smoking for more than 400 years. In this section the nicotine intake from other forms of tobacco use is considered together with the implications for the regulation of nicotine intake and for the potential of the nasal and buccal routes for the administration of nicotine as a substitute for smoking for either short-term use as an aid to cessation, or for long-term use as a less harmful source of nicotine.

Pipe and Cigar Smoking

Unlike cigarette smoking, pipe and cigar smoking can produce pharmacological doses of nicotine even if the smoke is not inhaled. This is achieved by absorption through the buccal mucosa and occurs for two reasons. Firstly, the smoke is more alkaline (about pH 8.5) than the smoke of flue-cured cigarette tobaccos (about pH 5.5) and the higher pH facilitates the absorption of nicotine. Secondly, the dose taken into the mouth is larger. The smoke is unfiltered and has a higher nicotine concentration, and, in the case of a fully packed pipe or a large cigar, there is more of it. Nicotine is also absorbed through the buccal mucosa when taken in the form of nicotine chewing gum. However, as can be seen in Fig. 4, the buccal route is far less efficient than the lungs and absorption is slow. Although high levels of blood nicotine can be built up with repeated doses of nicotine gum or non-inhaled pipe or cigar smoking, there are no rapid peaks and no bolus effect without inhalation.

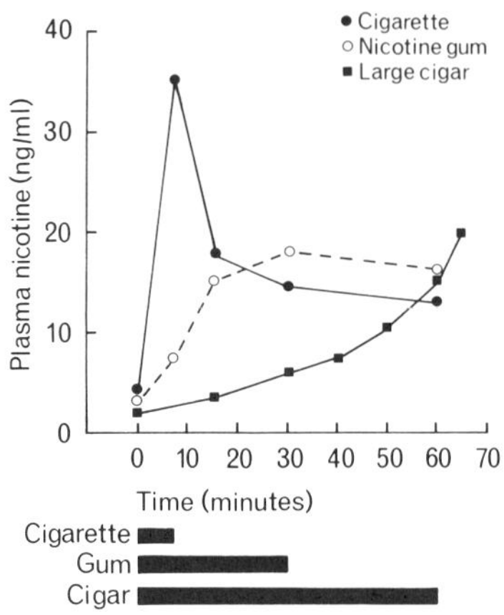

Fig. 4. Blood nicotine levels before, during and after non-inhaled smoking of a large cigar compared with inhaled smoking of one cigarette (1.2 mg nicotine) and chewing one piece of nicotine gum (4 mg nicotine) in a smoker who had abstained for at least 12 hours. The rate of absorption from the cigar increases as the pH and nicotine concentration in the smoke rise as it shortens during smoking. The greater efficiency of buccal absorption from the gum is due to buffers which keep the pH in the mouth at about 8.5, whereas the buffering capacity of cigar smoke is rather limited. From Russell et al. (1980) (18).

Despite these limitations, most so-called primary pipe and cigar smokers, who have never been regular cigarette smokers, appear to be non-inhalers and to derive sufficient satisfaction from this slow form of nicotine intake. On the other hand, when cigarette smokers switch to become so-called secondary pipe and cigar smokers, they almost inevitably continue to inhale. Having become addicted to rapid bolus doses of nicotine, it seems that they are no longer satisfied by the slower and less intense effects of non-inhaled smoking. They therefore regulate their nicotine intake by inhaling pipe or cigar smoke and thereby maintain cigarette-like blood nicotine levels (19,20). When such regulation is impossible, as in switching to nicotine gum, satisfaction is incomplete. Although it alleviates many of the effects of cigarette withdrawal, nicotine gum does not give the same pleasure or eliminate fully the craving for cigarettes (21-23).

Snuff Use

There are two forms of snuff, known as 'wet' and 'dry' snuff. Use of wet snuff is quite common in the USA and Scandinavia. It consists of moist ground tobacco which is placed in the vestibule of the mouth behind the lower lip enabling nicotine to be absorbed through the buccal mucosa. Dry snuff is finely powdered tobacco which is sniffed into the nose from where nicotine is absorbed through the nasal mucosa. It is still used by a few people in Britain, but was widely used in Europe in the 18th century.

It is clear from Fig. 5, which refers to dry snuff, that the nose is a fairly effective route for nicotine administration (24). Absorption is far more rapid than from a non-inhaled cigar or nicotine gum. Indeed, it takes no longer for the blood nicotine concentration to reach a peak than the time taken to smoke a cigarette. However, there are no high-nicotine boli in the case of snuff. As shown in Fig. 6, both the trough blood

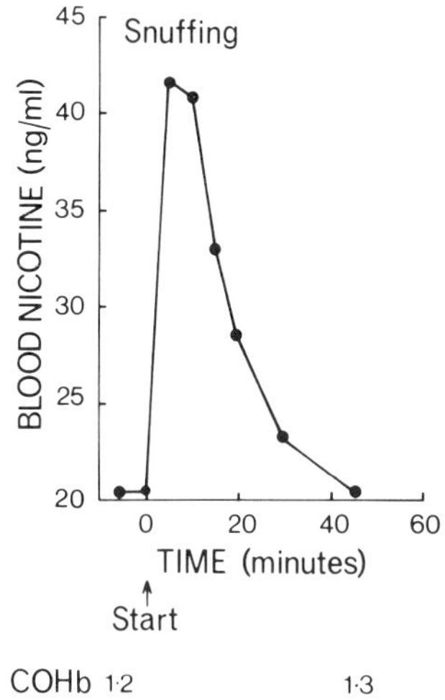

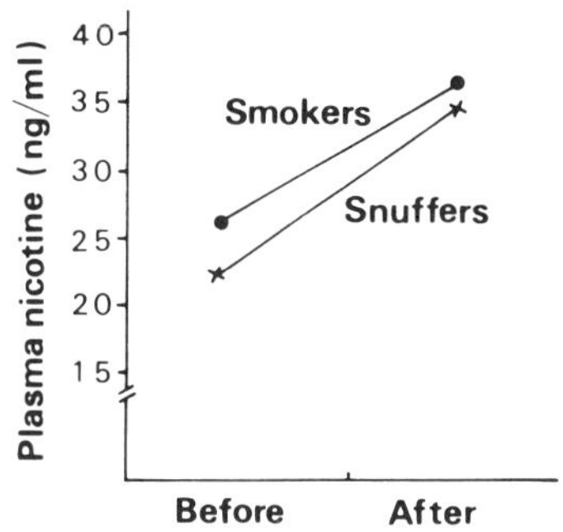

Fig. 5. Blood nicotine levels of a regular snuffer before, during and after one pinch of snuff during the afternoon of a day of usual snuffing. From Russell et al. (1980) (24).

Fig. 6. Comparison of average trough and peak blood nicotine levels of regular snuffers (n = 7) and cigarette smokers (n = 13) before and after dosage during the afternoon of a day of usual usage. From Russell et al. (1981) (17).

nicotine level and the boost received by a further dose in regular snuff takers are very similar to those of cigarette smokers (17). It is unlikely that such similarities between such very different rituals are mere coincidence. The most plausible explanation is that the rituals are determined by the nicotine concentrations they produce and that snuff-takers and smokers regulate their nicotine intake to achieve those levels.

As with non-inhaled pipe or cigar smoking and use of nicotine gum, the rate of nicotine absorption from wet snuff is also much slower than from a cigarette (Fig. 7) (25,26). However, with repeated use the blood nicotine levels of regular wet snuff takers build up to levels similar to those of smokers (27). A direct comparison of the rate of nicotine absorption from the nasal and buccal routes is illustrated in Fig. 7 by the concentration-time curves of 2 mg nicotine doses of nicotine chewing gum and nasal nicotine solution (25). Absorption is clearly more rapid via the nasal route suggesting that smokers would find it easier to switch to nasal snuff or nasal nicotine solution than to wet snuff, chewing tobacco or nicotine chewing gum. Theoretically, inhalation of a nicotine aerosol would mimic cigarette smoking more closely, but a satisfactory one has yet to be developed.

METHODOLOGICAL ISSUES IN INTAKE REGULATION

Self-regulation of nicotine intake by smokers has been studied in several different ways and many indirect as well as direct measures of intake have been used. The problems of studying this area are great, but are not always fully recognized. For example, up-regulation, or compensatory smoking as it is often called, can be affected by various

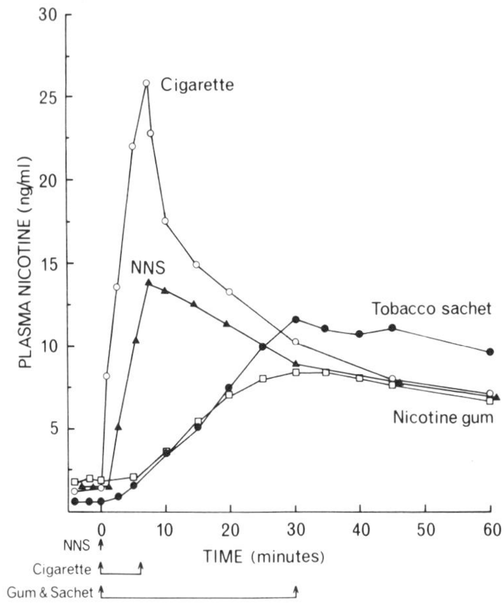

Fig. 7. Comparison of blood nicotine levels following nicotine dosage by three different routes: Inhalation from smoking one middle-tar cigarette (1.4 mg nicotine yield), nasal absorption from one drop of nasal nicotine solution (NNS) containing 2 mg nicotine, and absorption through the buccal mucosa from nicotine chewing gum (Nicorette, 2 mg) and sucking one sachet of wet snuff (Skoal Bandit). The values shown are the means of three subjects who had abstained from tobacco use for at least 12 hrs (25,26).

mechanical problems. It is important that some of these methodological issues are appreciated before going on to discuss intake regulation.

Methods of Study

Nicotine intake regulation has been studied by measuring the changes in smoking behavior and nicotine intake in response to: (i) variations in the nicotine yield of cigarettes; (ii) nicotine administration by injection, infusion, chewing gum, etc.; (iii) pharmacological blockade of nicotine actions; and (iv) altering nicotine excretion by inducing changes in urinary pH.

A major problem with the first approach is the lack of experimental cigarettes of commercial quality and flavor that have deliveries of nicotine which are varied independently of those of other components such as tar, and which are also similar in taste, draw resistance, etc. One or two studies have attempted to overcome this problem (although with only limited success) and the results have suggested that the smoking pattern is affected more by the delivery of nicotine than that of tar (28,29). There have also been several studies using shortened cigarettes, but smoke differs greatly from distal to proximal end and gets progressively more irritant and difficult to inhale deeply.

Of the other more direct approaches involving the manipulation of nicotine alone, (ii) and (iii) are limited in that they only test down-regulation to avoid the aversive toxic barrier and do not inform us on the need for smokers to keep nicotine intake above a lower limit for acceptability. These issues have been fully discussed some years ago (30,31).

Methods of Measurement

Another issue is how the regulation of nicotine intake is measured. Apart from the direct measurement of nicotine and cotinine in blood, urine or saliva, many indirect measures have been used: crude consumption of cigarettes, butt length, weight of tobacco burned, puff analysis (rate, volume etc), chest strain gauges and more elaborate measures of inspiration, butt analysis for nicotine, measures of CO intake and thiocyanate levels. The value and validity of any one measure or combination of measures depends on the particular purpose for which it is used. Isolated indirect measures of nicotine intake are not adequate. Puffing and inhalation patterns are different when a holder is interposed between cigarette and lips. Cigarettes tend to be smoked more intensively in a laboratory than in natural settings. Mouth-level intake (total volume puffed) correlates no more than 0.49 with inhaled dose as measured by blood nicotine concentration (32). Measures of inspiration are suspect in that smoke can be held in the mouth while inspiration takes place through the nose and to clip the nose makes the situation unnatural. Some of these issues have been discussed elsewhere (33,34). There is no doubt, however, that at the present time, if it is nicotine intake that is being studied, there is no substitute for the direct measurement of nicotine in blood, backed by cotinine as a longer term measure. The deluxe model for such studies is that of Benowitz in which the estimation of nicotine intake is based on complete pharmacokinetic data from each individual subject (35).

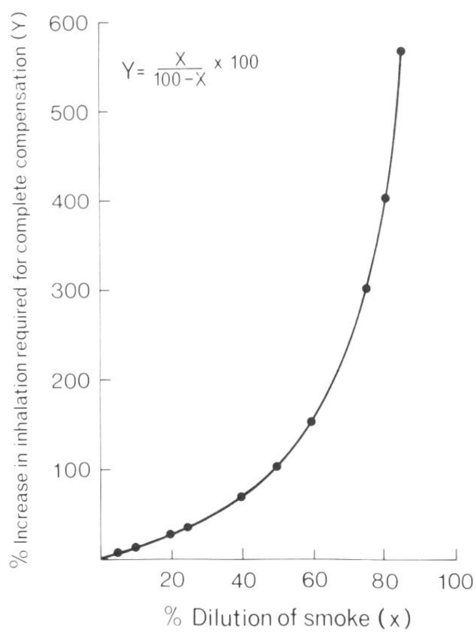

Fig. 8. Theoretical relationship between smoke dilution and the increase in the volume of smoke that it is necessary to inhale to maintain intake and compensate completely. (Reprinted with permission, from Sutton et al. (1978) (36).

Mechanical Problems in Compensatory Smoking

A final methodological issue concerns the ease with which it is possible for the smoker to make a regulatory adjustment to nicotine intake. There is clearly little difficulty about inhaling a stronger cigarette less deeply. Not so, with lower yields. Puff volume and inhalation can be increased to maintain intake and nicotine concentrations in post-inhalation boli, but only up to a point. After this, compensatory changes in smoking pattern become intrusive, then aversive and eventually, with very low yields, complete compensation becomes impossible. Several factors may influence this.

(i) The mathematical relationship between the nicotine concentration in the smoke and the volume it is necessary to inhale to obtain a given dose is hyperbolic (Fig. 4) so that things soon get difficult as the smoker reaches the steep part of the curve (36).

(ii) A smoker who seeks a given dose per puff will have fewer compensatory strategies than one who seeks a given dose per cigarette. In the latter, this can also be achieved by increasing puff rate, taking more puffs and smoking to a shorter butt length. A smoker who smokes to maintain a minimal plasma nicotine trough without rapid peaks after each cigarette can also compensate by smoking more cigarettes.

(iii) The manufacturing techniques used to reduce deliveries affect the extent and ease with which it is possible for a smoker to smoke so as to exceed the standard machine-smoked delivery. Differences in the effects on nicotine deliveries of 'oversmoking' cigarettes with ventilated vs unventilated filters are unknown to me, although it is well-known that deliberate or inadvertent blocking of ventilation holes is a simple way to enhance nicotine deliveries (37). Cigarettes in which low yields are achieved mainly by lowering the nicotine content (less tobacco and/or less nicotine in the tobacco) are to my knowledge not available on the market. Presumably they would be unpopular because it would be impossible for the smoker to get out more nicotine than was present in the cigarette. Indeed, it seems as if the tobacco industry in the USA, but not yet in the UK, has worked in the opposite direction. By increasing the length of filter overwraps, and hence reducing the number of standard machine-smoked puffs, the FTC standard yields have been artifactually reduced to provide 'low' and 'ultra-low' yield cigarettes which smokers find it relatively easy to smoke so as to exceed the standard yield (38).

SELF-REGULATION OF NICOTINE INTAKE

The discussion so far has been concerned with the wide variation in the way different people smoke and how this results in a similarly wide range of individual blood nicotine profiles which tend to be fairly constant from one day to the next. The present section examines the extent to which smokers actively regulate the way they smoke to maintain a constant intake of nicotine. Is it, in other words, the nicotine intake and blood nicotine level that control the smoking behavior? Or is the constant nicotine intake merely incidental to the regularity of a habit controlled by other factors? As mentioned above, the controlling influence of nicotine intake is suggested by the similarity of the blood nicotine levels of smokers and snuff takers, and by the tendency for cigarette smokers to continue inhalation on switching to a pipe or cigars. There is now extensive literature and several reviews on self-regulation of smoke intake (30,33,39,40). This section is concerned with self-regulation of nicotine intake and focuses on studies which have used more direct measures or which have a historical role in the development of our knowledge of the subject.

Historical Background

The first demonstration of intake regulation, or nicotine titration as it is sometimes called, was the study of Finnegan, Larson and Haag, published in Science in 1945, although the authors seem to have been unaware of this aspect (41). Their aim was to look for the development of withdrawal symptoms after 'blind' substitution of low-nicotine cigarettes. This they found. But, in addition, as is clear in Table 2, about half the subjects increased their cigarette consumption on the low yield cigarette and had few withdrawal symptoms. Those subjects who did not up-regulate by smoking more cigarettes suffered severe withdrawal symptoms which persisted throughout the month that they were smoking the low-nicotine cigarettes. Another interesting feature of this study is that the experimental cigarettes were prepared by using a tobacco naturally low in nicotine and then adding nicotine to produce the higher yielding cigarette. The two cigarettes therefore differed only in their nicotine yields indicating that it was lack of nicotine rather than other components which induced withdrawal symptoms in half the subjects and an increase in cigarette consumption in the others.

The next landmark was the classic study of Lucchesi et al. (42) some 20 years later in 1967. They showed, under double-blind conditions, that slow intravenous infusion of nicotine at a rate of 4 mg per hour had an inhibitory effect on ad libitum smoking compared with saline control.

Table 2. Effect of nicotine yield on cigarette consumption

Subject group	Average daily cigarette consumption per group			
	usual brand	1.96 mg nicotine	0.34 mg nicotine	1.96 mg nicotine
No withdrawal symptoms (n=6)	26.9	26.6	30.9	26.8
Transient mild withdrawal symptoms (n=6)	22.4	22.0	26.5	23.9
Transient moderate withdrawal symptoms (n=3)	23.6	28.3	28.6	27.6
Severe withdrawal symptoms persisting for 1 month (n=9)	25.0	24.7	24.6	24.9

Abstracted for Finnegan et al. (41). The nicotine yield of the subjects' usual brand was not given but an average yield for cigarettes of that period would have been around 2.0 mg. The experimental cigarettes were made from tobacco naturally low in nicotine (i.e., 0.34 mg yield), but nicotine was added to this tobacco to provide the cigarettes with the higher yield (1.96 mg). The order of administration was as shown from left to right. The exact period on each brand was not clearly stated but was about 4 weeks on the usual and low nicotine brands and 2 weeks for each period on the 1.96 mg brand. Withdrawal symptoms on switching to the low nicotine cigarettes were graded as follows: mild = vague lack of satisfaction initially; moderate = missed their usual brand but eventually adapted to the low nicotine cigarettes; severe = missed their usual brand even after 1 month and experienced irritability, poor concentration and/or hunger-like feelings.

There was a 27% reduction in the number of cigarettes smoked per session and subjects also took fewer puffs and discarded their cigarettes earlier. Although the amount of suppression was small in relation to the dose of nicotine given, the infusion would not have simulated intermittent bolus dosage from smoking. Also any decrease in inhalation would have been undetected. This study is well known and has been much cited, yet there was no attempt at replication for another 10 years until the study of Kumar et al. in 1977, which failed to show an inhibitory effect on puffing during and after intravenous nicotine (43). It seems that the issue has rested in this unsatisfactory state for almost 10 more years, but we have just completed a replication and our results appear to support Lucchesi et al. rather than Kumar et al.

It was thus apparent by 1967, from the two key studies of Finnegan et al. and Lucchesi et al., that smokers tend to alter their smoking behavior to up-regulate their nicotine intake on switching to a lower yielding brand and to down-regulate their intake from smoking when receiving intravenous nicotine. Furthermore, these two studies also showed that the intake regulation was controlled by nicotine rather than other smoke components -- an issue which few subsequent studies have been designed to address. However, the matter was opened to doubt by subsequent studies in which cigarette consumption was used as the measure of smoke intake. One example which strongly influenced thought at that time was a study by Waingrow and Horn in 1968. They analyzed self-report data from a national sample of 1,466 American cigarette smokers interviewed in 1964 and 1966 and found no tendency for smokers who had switched to a lower yield brand to increase their daily cigarette consumption (44). With hindsight, we of course realize that cigarette consumption is a poor measure of nicotine intake and that intake and its regulation are more strongly related to the way smokers puff and inhale their cigarettes than to the number they smoke. Thus, while a change in consumption may serve as positive evidence of intake regulation, its absence cannot be construed as negative evidence.

Brand-switching Studies

The issue of intake regulation remained unclear until 1970, when Frith (45) and Ashton and Watson (46) independently showed that smokers modified their puffing behavior according to the tar and nicotine yields of their cigarettes. When smoking higher yielding cigarettes, they took smaller, less frequent puffs and when smoking lower yield cigarettes, puff volume and puff rate were increased. In the case of Ashton and Watson, the finding was serendipitous and was noted during studies on the effect of smoking on perceptual-motor tasks. Frith, however, set out to test the hypothesis, having been inspired by the Lucchesi et al. study. Numerous studies have since confirmed these findings that smoke intake (to the mouth) is partially regulated by changing puffing behavior and there are several reviews on the topic (e.g., 33).

As mentioned before, the amount of smoke puffed into the mouth is an indirect and crude index of inhalation into the lungs and hence of absorption of nicotine into the bloodstream. The first reports of changes in COHb and blood nicotine concentrations after switching to high and ultra-low yield cigarettes appeared in 1973 and 1975 (16,47). They were derived from the same study but the plasma samples had to be kept frozen for more than a year until Feyerabend's assay method was sufficiently refined to undertake the analyses for nicotine. The subjects were ten office staff who were studied before and after 5-hour periods of smoking under natural conditions at their place of work. The results, in Table 3, show that cigarette consumption over the 5-hour period was decreased on switching to the high-nicotine brand, CO intake was also reduced and blood nicotine levels were unchanged compared with the usual brand condition,

Table 3. Effect on cigarette consumption, carbon monoxide and nicotine intake of switching to high and ultra-low nicotine cigarettes (means of 10 subjects).

Brand of cigarette	Mean nicotine yield mg	Mean carbon monoxide yield mg	Mean number smoked in 5-hour test period	Weight smoked in 5-hour test period g	Mean COHb after 5-hour test period %	Plasma nicotine after 5-hour test period ng/ml
Usual	1.34	17.2	10.7	6.0	8.16	30.1
Ultra-low nicotine	0.14	5.0	12.5	6.5	6.59	8.5
High nicotine	3.2	16.7	6.7	4.2	5.09	29.2

Abstracted from Russell et al. (16,47). The mean CO yields are derived from Russell et al. (48). The order of taking the high and low nicotine cigarettes was balanced within the group. The data for the usual brand are the means of two 5-hour test periods. COHb% is a unit of measurement not a % change.

indicating complete down-regulation of nicotine intake to avoid excessive levels. Despite increasing cigarette consumption on the ultra-low yield brand, COHb and blood nicotine concentrations were lower than on the usual brand. However, the drop in COHb and blood nicotine was proportionately much less than the drop in cigarette yields indicating some compensatory increase in inhalation, albeit insufficient to maintain blood nicotine at usual levels (i.e., partial compensation or incomplete up-regulation). Similar results have been found in a number of subsequent studies of short-term and more prolonged experimental switching to lower or higher yielding brands. The results of some of these have been summarized in Table 4 and confirm the general picture of accurate down-regulation but incomplete up-regulation of nicotine intake (15,49,50).

Nicotine Intake from Self-selected Brands

Some support for the controlled brand-switching studies comes from several large-scale studies of the nicotine intake from self-selected brands. In 1980, we reported a study of the blood nicotine levels of 206 women and 124 men who had been smoking their usual brand in their usual way (13). Some were smokers of low-yield cigarettes, others of higher yield brands. Most had been smoking their own particular brand for a year or more. Blood was sampled during the afternoon or evening by which time steady-state conditions would have occurred in most cases. The results are shown in Fig. 9. The most striking feature is the wide variation between individuals in blood nicotine levels (4-72 ng/ml in this sample). It is also evident that the blood nicotine levels of men and women were similar and that there was very little relation to the nicotine yield of the cigarettes. It is very obvious that it was possible for high blood nicotine levels to be obtained from low-nicotine cigarettes delivering (according to official smoking-machine estimates) only 0.6 mg nicotine. The correlation between blood nicotine and nicotine yield of cigarette,

Table 4. Changes in plasma nicotine concentrations after switching to high, low or ultra-low yield cigarettes (values shown are group means).

	Brand of cigarette	Tar yield (mg/cig)	Nicotine yield (mg/cig)	Plasma nicotine (ng/ml)
Ashton et al. (1979)				
(n=12)	Usual	18.0	1.4	39.6
	Low tar	6.5	0.6	30.9
	High tar	26.5	1.84	42.8
% reduction	(low vs usual)	64%	57%	22%
Russell et al. (1982)				
(n=12)	Usual	17.4	1.33	32.4
	Middle tar	19.0	1.3	35.2
	Usual	17.4	1.33	33.2
	Low tar (2 days)	10.9	0.7	20.8
	" (8 wks)	"	"	22.7
	" (10 wks)	"	"	22.8
% reduction	(low vs usual)	39%	47%	34%
West et al. (1984)				
(n=11)	Usual	14.2	1.3	22.0
	Ultra-low tar	1.0	0.1	9.4
	% reduction	93%	92%	57%

Abstracted from Ashton et al., Russell et al., and West et al. (49,15,50). On switching to lower yield cigarettes the blood nicotine concentrations are significantly less than on the usual brands. However, the reductions in blood levels are proportionately less than the reductions in nicotine yields indicating that some compensatory increase in puffing and inhalation has occurred although it was not sufficient to compensate completely for the drop in yield (i.e., partial compensation). Cigarette consumption did not change significantly in any of the three studies. On switching to a high-yield brand (Ashton et al.) or another middle-tar brand (Russell et al.) the plasma nicotine levels did not change significantly. The drop at 8 and 10 weeks after switching to a low-yield brand was similar to that at 2 days after switching (Russell et al.). The plasma nicotine concentrations in the West et al. study were 'trough' levels, whereas the other two studies used 'peak' levels two minutes after a cigarette.

though statistically significant, was low (0.21, p<.001) showing that the nicotine yield of the cigarettes accounted for only 4.4% of the variation in blood nicotine levels. The number of cigarettes smoked on the day accounted for even less (r = 0.13).

These findings have been widely replicated (14,51,52) and are essentially the same when the range of nicotine yields is extended below 0.6 mg which was the lower limit in our sample. The data from all these studies suggest that up-regulation is more complete in 'natural switchers' who regularly smoke lower yield cigarettes than in those who switch temporarily during an experimental study. Many smokers try low-yield cigarettes and it is possible that the self-selected minority who continue to smoke them are those who compensate more completely. The low correlations between cigarette consumption and blood nicotine

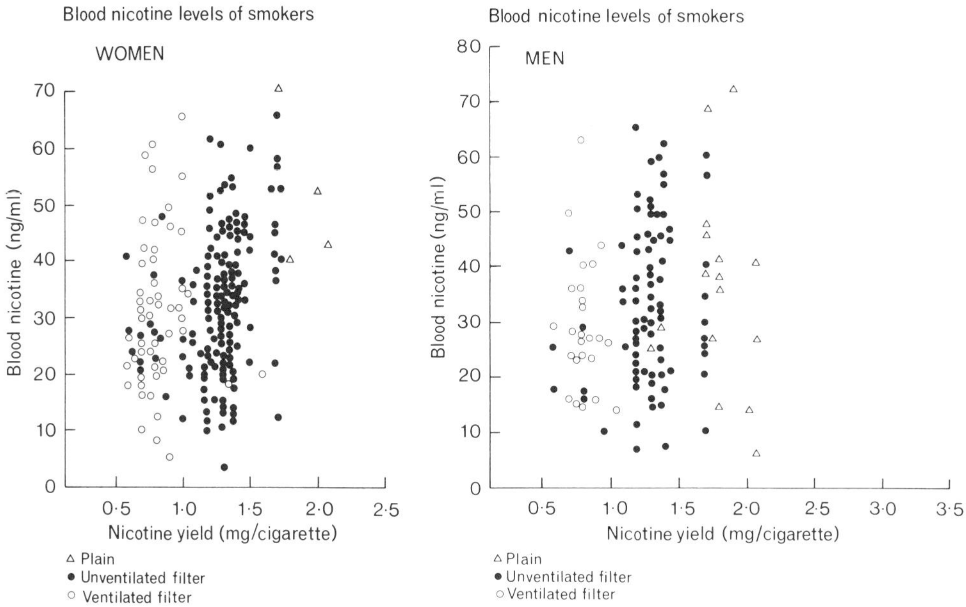

Fig. 9. Blood-nicotine levels of smokers plotted against the nicotine yields of their cigarettes. All smokers had been smoking their own self-selected brand in their usual way. (△) Plain; (●) unventilated filter, (○) ventilated filter. (Reprinted, with permission from Russell et al. (1980) (13).

concentrations indicate that when smoking under natural conditions, nicotine intake is regulated by modifying the pattern of puffing and inhalation rather than the number of cigarettes smoked.

Factors Controlling Down-regulation

Since tar and nicotine yields are highly correlated it is not clear whether the precise down-regulation demonstrated in the brand-switching studies was in fact determined by the nicotine level. Tar or irritants in the gas phase could be implicated. There have been relatively few direct studies which avoid confounding with tar intake and other factors. All of these are consistent with the view that down-regulation of smoke intake is controlled by nicotine and that excessively high blood nicotine levels are avoided because they are aversive. Thus the modest inhibitory effect on smoking behavior of intravenous nicotine in the Lucchesi et al. study (42) and that produced by nicotine gum (53,54) are consistent with the avoidance of excessive nicotine levels. Also consistent with this is the fact that people smoke more cigarettes and take more puffs when the pharmacological effects of nicotine are blocked by mecamylamine (55) and that they vary their cigarette consumption (56) and nicotine intake from cigarettes (35) inversely to the rate of nicotine excretion when this is manipulated by tablets which alter urinary pH.

All these studies show clearly that nicotine is involved in the down-regulation of smoking behavior to avoid excessive intake. The precision of the down-regulation was shown in the two studies which included measures of blood nicotine (54,35). In one of these (54), when subjects were taking additional nicotine in the form of ten pieces of 2 mg nicotine gum per day while continuing to smoke freely, there was no overshoot in peak blood nicotine concentrations after a cigarette. The average level was 27.4 ng/ml when smoking and taking nicotine gum, compared with 30.1 ng/ml when smoking without gum. In a recent study, Benowitz and Jacob (35) manipulated urinary pH and found that blood nicotine concentrations were not increased when urinary nicotine excretion was reduced by sodium bicarbonate (i.e., perfect down-regulation) but were 15% lower during urinary acidification with ammonium chloride (i.e., incomplete up-regulation).

In summary, there is good evidence that nicotine intake is down-regulated with great precision and that it is nicotine rather than other smoke components that determines the down-regulation of smoke intake. The sensitivity of down-regulation is no doubt partly due to the lack of any mechanical problems posed by puffing and inhaling less vigorously, but it also suggests an upper toxic barrier at which the effects of nicotine become aversive and are avoided. It would seem, furthermore, that most smokers of the type taking part in experimental studies keep their peak nicotine levels bouncing against their individual toxic barriers with most of the cigarettes they smoke.

Factors Controlling Up-regulation

Although down-regulation seems to be controlled mainly by the aversive effects of too much nicotine, it is less clear to what extent up-regulation is determined by a need to obtain optimal doses of nicotine, to avoid insufficient doses, or indeed by other factors. It is beyond my brief here to discuss these issues fully since they involve the whole field of smoking motivation, the nature of the desire or craving to smoke, the determinants of the withdrawal syndrome and the complex interaction of nicotine and other factors in all this.

To date there are only two studies, one published in 1945, the other in 1985, which avoid the confounding factors of other smoke components and so provide direct evidence that the compensatory increases in smoke intake are determined by a need to up-regulate nicotine. The first is the study of Finnegan et al. (41) described at the start of this section (Table 2). The other is the recent study of Benowitz and Jacob (35) mentioned briefly in the previous section on down-regulation. They manipulated urinary pH, and hence the urinary excretion of nicotine, in smokers who were free to smoke their usual brand in their usual way. Daily nicotine intake from smoking was measured using metabolic clearance data together with repeated measures of blood and urine nicotine concentrations over periods of 24 hours. When urinary nicotine excretion was increased by acid loading with ammonium chloride, the smokers up-regulated their nicotine intake from smoking by 18% compared with the placebo condition. The up-regulation in this case can only have been determined by a desire or need to raise the blood nicotine concentrations towards the accustomed levels. As with brand-switching studies, up-regulation was incomplete and blood nicotine levels were 15% lower than those on placebo*.

---

*Had the up-regulation been complete it could not have been inferred that the smokers had a need to raise their nicotine levels. The hypothesis that they were inhaling for some other motive but down-regulating to avoid the aversive effects of excessive nicotine could not have been rejected. This is why the earlier study of Schachter et al. (56) which did not include blood nicotine measures, while demonstrating the role of nicotine in down-regulation of smoke intake, does not inform on its role in up-regulation. This has been discussed more fully elsewhere (31).

To return to brand-switching, it has been discussed how, despite increasing the intensity of their puffing and inhalation, smokers tend to compensate only partially for a given drop in yield. On average, the drop in blood nicotine level is about two-thirds as much as would be expected if puffing and inhalation remained unchanged. In other words, smokers by their extra effort in puffing harder and inhaling more deeply raise their blood nicotine levels about one-third of the difference between the expected level on the lower yield brand and their former level on their usual brand (i.e., about 30-35% compensation). The reasons why they do this much and yet do not do more are unknown.

One problem is that the studies to date have been designed mainly to demonstrate the occurrence and extent of up-regulation rather than to ascertain its determinants. The data are based on averages of small samples which, though statistically significant, conceal wide individual variation between smokers. Some smokers up-regulate almost completely, others hardly at all. But there is no evidence, in the small numbers involved, that the extent of up-regulation is related to baseline cigarette consumption or blood nicotine level, or that those who up-regulate completely are less likely to experience loss of satisfaction after switching to lower-yield cigarettes. Similarly, there is no suggestive evidence from comparisons across studies that the degree of up-regulation is affected by the size of the reduction in yield of the test cigarette or by the absolute value of its yield. Again, the number of such studies is small and no thorough meta-analysis has been done.

It is well known that low-yield cigarettes provide less satisfaction and are unacceptable to most smokers (see below). How much this is due to lack of taste, to incomplete up-regulation of nicotine intake, or to aversiveness of the extra puffing and inhalation necessary to achieve partial up-regulation is unknown. A change in taste and smoking characteristics of a different brand can generate loss of satisfaction even when tar and nicotine yields are unchanged (15). It is evident from the few studies that have sampled blood on more than one occasion after switching (15,36) that any up-regulation that occurred took place immediately and that there was no subsequent change in the degree of up-regulation for periods up to 10 weeks. In contrast, loss of satisfaction and the degree to which the usual brand was missed decreased progressively over time. However, it is not clear whether this represents an adjustment to a lower nicotine intake, to harder puffing and inhalation, to the change in taste, or to all three factors.

It appears then that, despite partial up-regulation, the blood nicotine levels of smokers drop to roughly two-thirds of their usual level when they switch to low-yield cigarettes (about 0.6 mg nicotine). They also experience a loss of satisfaction which is probably due to a combination of the change in taste, the harder puffing and inhalation, and the drop in nicotine intake. The incompleteness of up-regulation probably reflects a compromise between the aversiveness of a further drop in nicotine intake and that of more intensive puffing and inhalation. Under natural conditions most smokers return to a higher yielding brand, but in complying with a study protocol they appear to adapt gradually over a few weeks and to regain satisfaction despite the lower level of nicotine intake.

On switching to ultra-low nicotine cigarettes (about 0.1 mg), full up-regulation is mechanically difficult or impossible (see Fig. 8 above). In the recent study of West et al. (50), despite a 57% drop in blood nicotine levels down to an average of 9.4 ng/ml, the subjects reported only slight craving for their usual brand. Withdrawal symptoms were not prominent, and among a variety of measures the only significant effects were an increase in hunger and a drop in heart rate. This suggests that withdrawal effects are allayed by relatively low blood nicotine levels, i.e., about one-third of those produced by smoking. This is in keeping with recent findings that many of the withdrawal symptoms after quitting smoking are alleviated by 2-mg nicotine gum (21-23), although blood nicotine levels produced by the 2-mg gum averaged only 9.3 ng/ml (21). The fact that placebo gum did not relieve withdrawal symptoms indicates that they were nicotine-related.

In conclusion, there is clear evidence that smokers up-regulate their nicotine intake and that it is the drop in nicotine rather than other smoke components that determines compensatory changes in smoking behavior on switching to lower yield brands. A drop in blood nicotine to about two thirds of the usual level generates some loss of satisfaction, but more intense craving and other withdrawal symptoms become evident only when levels fall below about one third of the usual smoking level. Up-regulation tends to be incomplete due to the aversiveness of intensive puffing and inhalation.

HEALTH IMPLICATIONS OF NICOTINE REGULATION

For the most part smokers regulate their smoke intake to regulate their nicotine intake, and they regulate their nicotine intake because they have become dependent on the psychopharmacological effects they get from it or because they would suffer withdrawal effects if they did not. How much these withdrawal effects are elicited by secondary conditioning to other stimuli (sensory and environmental), to simple absence of the positive pharmacological effects of nicotine, or to acquired physiological changes over and above those attributable to simple absence of the positive pharmacological effects is not fully understood and requires much further study.

A state of dependence on nicotine underlies the smoking behavior of most smokers. Besides making it difficult for them to stop smoking, it also determines how they smoke and how they would respond to other forms of tobacco use and other purer forms of nicotine substitution. It is not my purpose here to consider nicotine dependence and the problems of smoking cessation. These have been covered in a recent review (57). This section is confined to consideration of the implications of nicotine intake regulation for the health of those seeking a feasible but less harmful way to continue tobacco use.

Cutting Down

Epidemiological studies have shown that the risks of smoking are related to the number of cigarettes smoked per day. But a person who

Table 5. Blood nicotine and carbon monoxide levels in smokers whose cigarette consumption is reduced by 50% (means of 16 subjects)

	Cigarettes per day	COHb (%)	Plasma nicotine (ng/ml)
Usual smoking	30	7.7	17.0
50% reduction	15	7.9	24.8

Abstracted from Ho-yen et al. (1982) (58).

smokes 15 cigarettes per day and who once smoked 30 per day is different from one who smokes 15 per day and has never smoked more. As shown in Table 5, heavy smokers who cut down the number of cigarettes they smoke do not cut down their smoke intake (58). They tend to regulate their intake by getting more out of each cigarette.

Switching to a Pipe or Cigars

It has been mentioned above how primary pipe and cigar smokers, who have never been regular cigarette smokers, tend to be non-inhalers. Their blood nicotine levels rise gradually during smoking and they seem to derive sufficient satisfaction from this slower form of nicotine intake which is devoid of post-inhalation boli. Because they do not inhale, they absorb little carbon monoxide and take little tar into their lungs. It is this kind of pipe and cigar smoking to which epidemiologists have attributed only minimal risks. In contrast, most cigarette smokers who switch to become secondary pipe or cigar smokers continue to inhale, taking tar, CO, and other harmful products into their lungs, and so do not reduce the risks to their health. Having become dependent on the rapid bolus form of nicotine intake, former cigarette smokers are not satisfied by the slower rate of absorption through the buccal mucosa so that the inhalation habit is carried over to pipe and cigar smoking. The tobacco industry has made it easier for them to inhale by providing a wide choice of small mild cigars which are little different from the strong cigarettes of the past. Secondary cigar and pipe smokers cannot, therefore, be regarded as ex-smokers. It is this mistake which has so distorted the conclusions of the MRFIT* study. By regarding pipe and cigar smokers as nonsmokers, the cut-off point for thiocyanate validation was set too high and intervention success was therefore overestimated (59). The failure of the trial to find an effect of intervention on mortality may therefore not be as puzzling as has been thought.

Switching to Smokeless Tobacco

There is no doubt that switching to smokeless tobacco would be far less harmful than inhaled smoking. The small risk of local cancers would be many times offset by the absence of risk of lung cancer and other lung diseases, and a probable reduction in risk of cardiovascular disease. Buccal absorption of nicotine from smokeless tobacco is more efficient than from a non-inhaled pipe or cigar due to a higher pH. Switching should therefore be easier, and "cheating" by inhalation would of course be impossible. Nicotine absorption through the nose appears to be even more rapid than through the buccal mucosa, but the effects of periodic high-nicotine boli would still be lacking so that brain nicotine levels would be lower than with cigarette smoking. The main disadvantage of smokeless tobacco is that it is messy, which suggests a potential for purer and safer substitutes like nicotine gum and the nasal solution.

Low-tar Cigarettes

For many years there have been policies in most developed countries for lowering the tar yields of cigarettes. Because health authorities have responded more to the advice of epidemiologists than to behavioral scientists, these policies have invariably advocated lowering tar and nicotine yields rather than adopting a medium-nicotine low-tar approach. For more than a decade it has been evident that the health benefits of the traditional low-tar low-nicotine approach would be undermined by compensatory smoking to maintain nicotine intake, and that at some point down the nicotine scale acceptability to smokers would be insuperably

*MRFIT, Multiple Risk Factor Intervention Trial

impaired. Both these problems could be met by adopting a medium-nicotine low-tar approach. Since people smoke mainly for nicotine but die mainly from the tar, CO, and other harmful gases, this approach is so exquisitely logical (60). However, while policy makers have dillied the tobacco industry has not dallied, and this approach is clearly being gradually implemented.

Sales-weighted tar and nicotine yields of cigarettes declined rapidly in most countries during the 1960's, coinciding with the switch to filter-tipped cigarettes. This was a switch which the smoking population made rapidly and easily. Since the nicotine intake of smokers probably averages little more than 1 mg/cigarette (4,5) it is hardly surprising that they found it easy to switch down from nicotine yields above 2.5 mg to yields of about 1.3 mg/cigarette. But since then things have not been quite so easy. As Fig. 10 shows, while sales-weighted tar yields have continued to decline steadily in the UK, the sales-weighted mean nicotine yield has remained static at about 1.3 mg since 1972 (61). This suggests an acceptability barrier linked more closely to nicotine than to tar yield. Supportive evidence of a barrier to the acceptance of low-tar low-nicotine cigarettes comes from the fact that only 15-17% of British smokers smoke a low-tar brand (<10.5 mg). As shown in Table 6, this figure too has remained static at about this level since 1977, despite widespread availability and promotion of low-tar brands (61).

The trends in the US are somewhat different. Sales-weighted nicotine yields have continued to decline, albeit slowly, and have not yet bottomed out as in the UK. Moreover, they have gone through the UK barrier down to a mean of about 1.0 mg nicotine per cigarette. The most likely explanation is the trend in the US to longer filter overwraps, as mentioned above (38), which makes it easier for smokers to get above the standard machine-smoked yield.

Finally, the data in Fig. 10 indicate that the sales-weighted tar/nicotine yield ratio has been declining steadily in the UK. This has been especially marked among the lower yield brands (61). But the overall picture evident from the figure is that the 29% decline in sales-weighted tar yields from 21 mg in 1972 to 15 mg in 1983 occurred without accompanying changes in sales-weighted nicotine yields. If we assume that

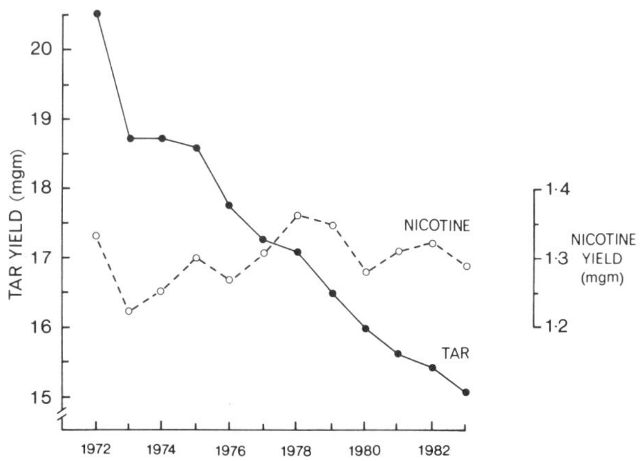

Fig. 10. Average sales-weighted tar and nicotine yields of cigarettes smoked in the UK from 1972 to 1983. From Jarvis and Russell (1985) (61).

Table 6. Market share (%) of low tar cigarettes (<10.5 mg tar) in UK, 1972-1983, from two different sources.

	Source of estimate	
	NOP	Maxwell
1972	<1	-
1973	2	<1
1974	5	3
1975	6	5
1976	8	7
1977	15	12
1978	15	13
1979	16	13
1980	17	16
1981	17	15
1982	15	16
1983	16	15

Source, Jarvis and Russell (1985) (61).

average blood nicotine levels of the population of smokers were similarly unchanged, it follows that the tar intake into the lungs of British smokers has decreased by 29% over this period. This is a clear demonstration of the importance of focussing on lowering tar/nicotine ratios rather than overall yields -- in other words, the potential of the medium-nicotine low-tar approach to less harmful cigarettes. Table 7 shows that low-yield smokers take in less nicotine (about 15% less, if the blood nicotine and cotinine data are averaged). They also take in about 10% less CO and, due partly to the improved tar/nicotine yield ratio, about 25% less tar (62). The health implications of all this for present-day low-tar smokers and for the future development of less harmful cigarettes are rather obvious.

SUMMARY AND CONCLUSIONS

The development of reliable methods for measuring nicotine in blood has enabled research over the past decade to demonstrate the extent to which tobacco use, in all its forms, is a drug-taking activity. Cigarette smokers vary widely in their individual smoking styles and this is reflected in their blood nicotine profiles. The tenfold variation in blood nicotine concentrations between individual smokers bears little relation to their self-reports of cigarette consumption and inhalation or to the machine-smoked yields of their cigarettes, but individual smokers tend to regulate their intake to maintain fairly constant levels from one day to the next. Thus, when they switch to higher yield brands, they puff and inhale less. Conversely, when they switch to lower yield brands or smoke fewer cigarettes, they compensate by puffing harder and inhaling more deeply, and when they switch to a pipe or cigars they tend to continue to inhale to maintain the rapid absorption and post-inhalation bolus form of nicotine intake characteristic of inhaled cigarette smoking.

There is now sound evidence that self-regulation of smoke intake is controlled by nicotine independently of other factors such as tar, CO or other gas phase components of the smoke. Down-regulation of nicotine

Table 7. Average percentage reductions in tar, nicotine and carbon monoxide intake by low-tar smokers in comparison with smokers of higher yield brands.

	Non-low-tar Smokers (n=241)	Low-tar Smokers (n=151)	% Reduction
Cigarette data			
Tar yield (mg/cig)	17.3	9.2	46.8
Nicotine yield (mg/cig)	1.42	0.86	39.3
CO yield (mg/cig)	16.6	11.0	34.1
T/N ratio	12.3 (.08)	10.8 (.10)	12.1***
T/CO ratio	1.06 (.01)	0.85 (.01)	19.8***
Intake measures			
Plasma nicotine (ng/ml)	38.3 (.91)	31.8 (.99)	17.0***
Plasma cotinine (ng/ml)	379 (13.7)	333 (16.2)	12.1*
COHb (%)	7.81 (.18)	7.06 (.20)	10.6**
Index of tar intake			
TI (Nic)	469 (11.2)	341 (10.4)	27.3***
TI (Cot)	4601 (166)	3543 (164)	23.0***
TI (CO)	8.13 (.18)	5.93 (.16)	27.1***

Note: The plasma continine data are based on smaller samples of 146 non-low-tar smokers and 95 low-tar smokers, but these subsamples did not differ from the remainder of the subjects in any of the other measures used. The percentage reduction in COHb was calculated after subtracting 0.7 to correct for the background level in non-smokers. Standard errors are shown in parentheses. The index of tar intake was derived from the measured blood level of a marker and the ratio of the tar to marker yields of the cigarette. E.g., using nicotine as the marker, TI (Nic) = Plasma nicotine X T/N yield ratio. The cigarette yield ratios and indices of tar intake were computed for individual smokers before averaging them to obtain the group means. Statistical significance of differences are based on t tests between non-low-tar and low-tar smokers: ***p<.001, **p<.01, *p<.05.
From Russell et al. (62).

intake to avoid a toxic aversive barrier is very precise and in this respect nicotine is unlike other addictive drugs. But up-regulation is less sensitive and is usually incomplete. This may be partly due to the aversiveness of intensive puffing and inhalation. On balance, smokers seem to tolerate a drop in blood nicotine to roughly two-thirds of their usual levels with some initial loss of satisfaction but they adapt gradually to the lower level over two or three weeks. Greater reductions occur when fuller up-regulation is mechanically difficult or impossible (e.g., on ultra-low yield cigarettes) and generate increasing loss of satisfaction and craving. However, other withdrawal symptoms only become prominent when blood nicotine levels fall below about one-third of the usual smoking level.

A state of dependence on nicotine underlies the smoking behavior of most smokers and they regulate their intake because they have become dependent on the psychopharmacological effects of nicotine. Nicotine dependence is the major obstacle to the various strategies for reducing smoking-related disease. It makes it difficult for smokers to stop smoking and limits the acceptability of low yield cigarettes. The compensatory changes in smoking pattern involved in the up-regulation of nicotine intake undermine the health benefits of smoking fewer cigarettes and of switching to lower yield cigarettes or to a pipe or cigars. However, some progress is being made by dealing more specifically with these problems. The use of nicotine chewing gum as a temporary aid to cessation increases the chances of success. A nasal nicotine solution is being developed for the same purpose. For those who continue smoking, the tobacco industry is moving slowly towards providing medium-nicotine low-tar cigarettes and there is evidence that the tar intake of present-day low-tar smokers is substantially reduced.

Some time in the 21st Century we may see the demise of tobacco use, but it is doubtful whether nicotine use will be abandoned. It is the impurities in tobacco and its smoke which kill, while nicotine provides most of the pleasure, stimulation, and relief from stress. It is not so much the potential of purer forms of nicotine as temporary aids to smoking cessation, but their potential use for long-term self-administration which merits the most serious consideration. Conventional tobacco products may in future be as archaic as the unrefined use of alkaloids in folk medicine appear now in comparison with the modern products of the pharmaceutical industry. The principle for all drugs has been to purify them as much as possible. If the tobacco industry does not do this with its drug, the pharmaceutical industry will. It is beginning to do so already.

ACKNOWLEDGEMENTS

I thank the Medical Research Council for financial support, Pamela Hancox for secretarial help and my colleagues Martin Jarvis and Ann McNeill for their comments on the manuscript.

REFERENCES

1. Appelgren LE, Hansson E and Schmiterlow LG. (1962) The accumulation and metabolism of C14-labelled nicotine in the brain of mice and cats. Acta Physiol Scand. 56, 249-257.
2. Stalhandske T. (1970) Effects of increased liver metabolism of nicotine on its uptake, elimination and toxicity in mice. Acta Physiol Scand. 80, 222-234.
3. Armitage AK, Dollery CT, George CF, Houseman TH, Lewis PJ and Turner DM. (1975) Absorption and metabolism of nicotine from cigarettes. Brit. Med. J. 4, 313-316.
4. Benowitz NL and Jacob P. (1984) Daily intake of nicotine during cigarette smoking. Clin. Pharmac. Ther. 35, 499-504.
5. Feyerabend C, Ings RMJ and Russell MAH. (1985) Nicotine pharmacokinetics and its application to intake from smoking. Brit. J. Clin. Pharmac. 19, 239-247.
6. Russell MAH, West RJ and Jarvis MJ. (1985) Intravenous nicotine simulation of passive smoking to estimate dosage to exposed non-smokers. Brit. J. Addict. 80, 201-206.
7. Russell MAH and Jarvis MJ. (1985) Theoretical background and clinical use of nicotine chewing gum. Pages 110-130, In "Pharmacological Adjuncts in Smoking Cessation." Eds. Grabowski J. and Hall SM. National Institute on Drug Abuse Research Monograph 53. U.S. Government Printing Office, Washington, D.C.
8. Isaac PF and Rand MJ. (1972) Cigarette smoking and plasma levels of nicotine. Nature, 236, 308-310.

9. Benowitz NL, Jacob P, Jones RT and Rosenberg J. (1982) Interindividual variability in the metabolism and cardiovascular effects of nicotine in man. J. Pharmacol Exp Ther. 21, 368-372.
10. Russell MAH and Feyerabend C and Cole PV. (1976) Plasma nicotine levels after cigarette smoking and chewing nicotine gum. Brit. Med. J. 1, 1043-1046.
11. Russell MAH and Feyerabend C. (1978) Cigarette smoking: a dependence on high-nicotine boli. Drug Metabolism Reviews, 8, 29-57.
12. Russell MAH, Raw M, Taylor C, Feyerabend C and Saloojee, Y. (1978) Blood nicotine and carboxyhaemoglobin levels after rapid-smoking aversion therapy. J. Consult. and Clin. Psychol. 46, 1423-1431.
13. Russell MAH, Jarvis MJ, Iyer R and Feyerabend C. (1980) Relation of nicotine yield of cigarettes to blood nicotine concentrations in smokers. Brit. Med. J. 280, 972-975.
14. Gori GB and Lynch CJ. (1985) Analytical cigarette yields as predictors of smoke bioavailability. Regulatory Toxicol & Pharmacol. 5, 314-326.
15. Russell MAH, Sutton SR, Iyer R, Feyerabend C and Vesey CJ. (1982) Long-term switching to low-tar low-nicotine cigarettes. Brit. J. Addict. 77, 145-158.
16. Russell MAH, Wilson C, Patel UA, Feyerabend C and Cole PV. (1975) Plasma nicotine levels after smoking cigarettes with high, medium and low nicotine yields. Brit. Med. J. 2, 414-416.
17. Russell MAH, Jarvis MJ, Devitt G and Feyerabend C. (1981) Nicotine intake by snuff users. Brit. Med. J. 283, 814-817.
18. Russell MAH, Raw M and Jarvis MJ. (1980) Clinical use of nicotine chewing gum. Brit. Med. J. 280, 1599-1602.
19. Turner JAM, Sillett RW and McNichol MW. (1977) Effect of cigar smoking on carboxyhaemoglobin and plasma nicotine concentrations in primary pipe and cigar smokers and ex-cigarette smokers. Brit. Med. J. 2, 1387-1389.
20. Armitage AK, Dollery CT, Houseman TH, Kohner E, Lewis PJ and Turner DM. (1978) Absorption of nicotine from small cigars. Clin Pharmac Ther. 23, 143-151.
21. West RJ, Jarvis MJ, Russell MAH, Carruthers ME and Feyerabend C. (1984) Effect of nicotine replacement on the cigarette withdrawal syndrome. Brit. J. Addict. 79, 215-219.
22. Schneider NG, Jarvik ME and Forsythe AB. (1984) Nicotine versus placebo gum in the alleviation of withdrawal during smoking cessation. Addict. Behav. 9 149-156.
23. Hughes JR, Hatsukami DK, Pickens RW, Krahn D, Malin S and Luknic A. (1984) Effect of nicotine on the tobacco withdrawal syndrome. Psychopharmacology. 83, 82-87.
24. Russell MAH, Jarvis MJ and Feyerabend C. (1980) A new age for snuff? Lancet, 1, 474-475.
25. Russell MAH, Jarvis MJ, Feyerabend C and Ferno O. (1983) Nasal nicotine solution: a potential aid to giving up smoking? Brit. Med. J. 286, 683-684.
26. Russell MAH, Jarvis MJ, West RJ and Feyerabend C. (1985) Buccal absorption of nicotine from smokeless tobacco sachets. Lancet, 2, 1370.
27. Gritz ER, Baier-Weiss V, Benowitz NL, Van Vunakis H and Jarvik ME. (1981) Plasma nicotine and cotinine concentrations in habitual smokeless tobacco users. Clin. Pharmacol. Ther. 30, 201-209.
28. Goldfarb TL, Gritz ER, Jarvik ME and Stolerman IP. (1976) Reactions to cigarettes as a function of nicotine and tar. Clin. Pharmacol. Ther. 19, 762-772.
29. Herning RI, Jones RT, Bachman J and Mines AH. (1981) Puff volume increases when low-nicotine cigarettes are smoked. Brit. Med. J. 283, 187-189.

30. Russell MAH. (1978) Self-regulation of nicotine intake by smokers, pages 108-122 in "Behavioural Effects of Nicotine," Edited by K. Battig, Karger, Basel.
31. Russell MAH. (1979) Tobacco dependence: Is nicotine rewarding or aversive? pages 100-122. In "Cigarette Smoking as a Dependence Process." Ed. N.A. Krasnegor. National Institute on Drug Abuse Research Monograph 23. U.S. Government Printing Office, Washington, D.C.
32. Sutton SR, Russell MAH, Iyer R, Feyerabend C and Saloojee Y. (1982) Relationship between cigarette yields, puffing patterns and smoke intake: evidence for tar compensation? Brit. Med. J. 285, 600-603.
33. McMorrow MJ and Foxx RM. (1983) Nicotine's role in smoking: an analysis of nicotine regulation. Psychological Bulletin, 93, 302-327.
34. Benowitz NL. (1983) Use of biologic fluid samples in assessing tobacco smoke consumption. Pages 6-26, In "Measurement in the analysis and treatment of smoking behavior." (Eds) Grabowski J and Bell CS. National Institute on Drug Abuse Research Monograph 48. U.S. Government Printing Office, Washington, D.C.
35. Benowitz NL and Jacob P. (1985) Nicotine renal excretion rate influences nicotine intake during cigarette smoking. J. Pharmacol. Exp. Ther. 234, 153-155.
36. Sutton SR, Feyerabend C, Cole PV and Russell MAH. (1978) Adjustment of smokers to dilution of tobacco smoke by ventilated cigarette holders. Clin. Pharmacol. Ther. 24, 395-405.
37. Kozlowski LT, Frecker RC, Khouw V and Pope MA. (1980) The misuse of 'less hazardous' cigarettes and its detection: hole-blocking of ventilated filters. Am. J. Public Health, 70, 1202-1203.
38. Grunberg NE, Morse DE, Maycock VA and Kozlowski LT. (1985) Changes in overwrap and butt length of American filter cigarettes: An influence on reported tar yields. New York State J. Med. 85, 310-312.
39. Gritz ER. (1980) Smoking behavior and tobacco abuse. Pages 91-158, In "Advances in Substance Abuse." Ed. Mello NK, JAI Press, Greenwhich, Connecticut.
40. Moss RA and Prue DM. (1982) Research on nicotine regulation. Behavior Therapy, 13, 31-46.
41. Finnegan JK, Larson PS and Haag HB. (1945) The role of nicotine in the cigarette habit. Science, 102, 94-96.
42. Lucchesi BR, Schuster CR and Emley GS. (1967) The role of nicotine as a determinant of cigarette smoking frequency in man with observations of certain cardiovascular effects associated with the tobacco alkaloid. Clin. Pharmacol. Ther. 8, 789-796.
43. Kumar R, Cooke EC, Lader MH and Russell MAH. (1977) Is nicotine important in tobacco smoking? Clin. Pharmacol. Ther. 21, 520-529.
44. Waingrow S and Horn D. (1968) Relationship of number of cigarettes smoked to tar rating. National Cancer Institute Monograph, No. 28, 29-33.
45. Frith CD. (1971) The effect of varying the nicotine content of cigarettes on human smoking behavior. Psychopharmacologia, 19, 188-192.
46. Ashton H and Watson DW. (1970) Puffing frequency and nicotine intake in cigarette smokers. Br. Med. J. 3, 679-681.
47. Russell MAH, Wilson C, Patel UA, Cole PV and Feyerabend C. (1973) Comparison of the effect on tobacco consumption and carbon monoxide absorption of changing to high and low nicotine cigarettes. Brit. Med. J. 4, 512-516.
48. Russell MAH, Cole PV, Idle MS and Adams L. (1975) Carbon monoxide yields of cigarettes and their relation to nicotine yield and type of filter. Brit. Med. J. 3, 71-73.

49. Ashton H, Stepney R and Thompson JW. (1979) Self-titration by cigarette smokers. Brit. Med. J. 2, 357-360.
50. West RJ, Russell MAH, Jarvis MJ and Feyerabend C. (1984) Does switching to an ultra-low nicotine cigarette induce nicotine withdrawal effects? Psychopharmacology, 84, 120-123.
51. Ebert RV, McNabb ME, McCusker KT and Snow SL. (1983) Amount of nicotine and carbon monoxide inhaled by smokers of low-tar low-nicotine cigarettes. JAMA, 250, 2840-2842.
52. Benowitz NL, Hall SM, Herning RI, Jacob P, Jones RT and Osman AL. (1983) Smokers of low-yield cigarettes do not consume less nicotine. N. Engl. J. Med. 309, 139-142.
53. Kozlowski LT, Jarvik ME and Gritz ER. (1975) Nicotine regulation and cigarette smoking. Clin. Pharmacol. Ther. 17, 93-97.
54. Russell MAH, Wilson C, Feyerabend C and Cole PV. (1976) Effect of nicotine chewing gum on smoking behaviour and as an aid to cigarette withdrawal. Brit. Med. J. 2, 391-393.
55. Stolerman IP, Goldfarb T, Fink R and Jarvik ME. (1973) Influencing cigarette smoking with nicotine antagonists. Psychopharmacologia, 28, 247-259.
56. Schachter S, Kozlowski LT and Silverstein B. (1977) Effects of urinary pH on cigarette smoking. J. Exp. Psychol. (Gen). 106, 13-19.
57. Russell MAH. (1985) Smoking, nicotine dependence and its treatment with nicotine chewing gum, pages 7-42, In "Nicorette in Smoking Cessation." Excerpta Medica Asia Pacific Congress Series No 39, Excerpta Medica, Hong Kong.
58. Ho-yen DO, Spence VA, Moody JP and Walker WF. (1982) Why smoke fewer cigarettes? Brit. Med. J. 284, 1905-1907.
59. Jarvis MJ, West RJ, Tunstall-Pedoe H and Vesey C. (1984) An evaluation of the intervention against smoking in the Multiple Risk Factor Intervention Trial. Preventive Medicine, 13, 501-509.
60. Russell MAH. (1976) Low-tar medium-nicotine cigarettes: a new approach to safer smoking. Brit. Med. J. 1, 1430-1433.
61. Jarvis MJ and Russell MAH. (1985) Tar and nicotine yields of UK cigarettes 1972-83: Sales-weighted estimates from non-industry sources. Brit. J. Addict. 80, 429-434.
62. Russell MAH, Jarvis MJ, Feyerabend C and Saloojee Y. (1986) Reduction of tar, nicotine and carbon monoxide intake in low-tar smokers. J. Epidemiol & Community Health, 40, 80-85.

THE FUNCTIONS OF SMOKING

David M. Warburton

Department of Psychology
Reading University
Reading RG6 2AL England

INTRODUCTION

This chapter will outline the functional view of smoking. Smoking in this paper will be taken to refer to cigarette smoking. This functional approach regards smoking as a person's use of nicotine to control his/her psychological state. This view is a formalization of ideas that have been discussed previously in a series of papers from our laboratory (1, 2, 3, 4, 5) and other laboratories (6, 7). In our papers smoking was described as a form of "self-medication," a coping strategy for everyday problems and the functional model can be seen as a more general form of this view without the quasi-medical connotations. The work of Ashton and Stepney (7) described smoking as a "psychological tool," i.e., in functional terms.

The functional model that I have adopted has similarities with the work of Alexander and Hadaway (8). In their paper on opiate use, they distinguish between the "exposure" orientation and the "adaptive" orientation. The adaptive orientation refers to opiate use for acute distress rather than for help with chronic distress. The functional model is more general than the adaptive orientation because it refers to smoking for any purpose including help with distress and the cause can be acute as well as chronic. In the next section I will outline some evidence that points towards a functional interpretation of smoking and show that smoking is not an irrational habit but is adopted because it enables smokers to effectively control their psychological state.

EVIDENCE FOR THE FUNCTIONAL MODEL

The functional interpretation of smoking behavior as an attempt by the smoker to control his psychological state does not fit the conventional medical view but is implicit in the literature on smoking. Evidence for it comes from studies of smoking motives, smoking situations, laboratory studies of smoking, and the personality of smokers.

Smoking Motives and Smoking Situations

Support for the model has come from the assessment of stated smoking motives and a second related body of evidence is from the study of smoking situations.

One of the first studies of smoking motives was done by Tomkins (9, 10). Tomkins proposed a theory of smoking which is based on his theory of affect. He suggested that there were four types of smokers: (a) smokers who smoked to produce or increase pleasant feelings (Positive Affect Smokers); (b) smokers who smoked to reduce unpleasant feelings of anger, anxiety, etc. (Negative Affect Smokers); (c) smokers who smoked automatically but not affectively (Habitual Smokers); (d) smokers who always were aware of not smoking and this awareness generates negative affect which is reduced only by smoking (Addictive Smokers).

On the basis of Tomkin's theoretical model, Horn and his colleagues (11) devised an affective smoking questionnaire to look at the affective conditions that could lead to smoking. Horn found three factors which represented various aspects of positive affect smoking: (a) pleasurable relaxation smoking; (b) stimulation smoking; and (c) sensorimotor manipulation smoking. The three other remaining factors were: (d) negative affect smoking; (e) habitual smoking; and (f) addictive smoking. The Tomkin's model has been substantiated, although his concept of the habitual smoking (automatic) has received only weak support (12).

In a study of smoking situations, McKennell (13) classified smokers according to the occasions in which they thought that they were likely to smoke. The answers to the questions were factor-analyzed and seven smoking situations were found: (a) nervous irritation smoking; (b) relaxation smoking; (c) smoking alone; (d) activity accompanying smoking; (e) food substitution smoking; (f) social smoking; and (g) social confidence smoking. The first five types of smoking were correlated and they formed an "inner need factor." The other two factors were associated and could be identified as "social factor." The "nervous irritation" factor included "smokes when anxious or worried" and "smokes when nervous" and also "smokes when angry." When cigarette consumption was equated, female smokers had higher scores for "nervous irritation" smoking.

Another study of situational smoking was done by Frith (14). He asked smokers to rate their desire for a cigarette in a set of imagined situations. A principal components analysis of their answers had a positive-loading on desire to smoke in each situation, as well as on cigarette consumption, while the second component contrasted desire to smoke in high and low arousal-inducing situations, i.e., sedative and stimulant smoking. Frith (14) subdivided his proposed high arousal-inducing situations into (a) those involving emotional stressors and anxiety and (b) those in which the stressors could be due to mental activity. Women were more likely than men to smoke in stressful, high-arousal situations. The low-arousal situations focussed on relaxation, boredom, repetitive work, and fatigue. This finding raised the possibility of making more detailed classifications of smoking situations.

The Tomkins-Horn and the McKennell typologies provided descriptions of relatively restricted psychological conditions, although they were both based on large samples and their factor structures have been largely replicated (15, 16). Equivalences exist between the Tomkins-Horn factors of "pleasurable relaxation," "stimulation," and "negative affect" and McKennell's factors of "relaxation," "activity accompaniment," and "nervous irritation." However, these typologies do not explain how smoking modifies affect, why smoking occurs in certain situations, nor do these typologies account for the role of nicotine in smoking (17).

In order to improve the understanding of smoking motives, Russell et al. (17) combined some of McKennell's items and some of the Tomkins-Horn items into a single questionnaire in order to see whether there was a similarity between pleasurable relaxation, stimulation, and negative affect smoking of the Tomkins-Horn typology, and the relaxation, activity accompaniment, and nervous irritation of McKennell. Another aim was to expand the relationship between smoking motives by using items that had psychosocial connotations for which the work of McKennell, Mausner, and Platt (15) had argued. Thus the questionnaire covered the social and psychological aspects of smoking of the Reinforcement Model of Smoking proposed by Russell (6).

The results of the factor analysis were six factors which Russell et al. (17) labeled as psychosocial, indulgent, sensorimotor, stimulation, addictive, and automatic. However, Russell identified a major pharmacological dependence factor which he called "pharmacological addiction" that combined his intercorrelated stimulation, automatic and addictive factors. On an item analysis of their Smoking Motives Questionnaire, Russell et al. (17) found that 93 percent of smokers who attended a smoking clinic and 74 percent of nonclinic smokers said that they smoked when worried, while Warburton and Wesnes (2) found that 88 percent of students answered in the same way. It was interesting that there are differences in female and male motives for smoking; it is said that women smoke more to relieve anxiety and anger than men (17). This data fitted very neatly with the findings of McKennell and Frith on smoking motives and smoking situations.

Combination of the questionnaires was carried a stage further by a study (12) which used all the items of Russell's Smoking Questionnaire with all the items of Frith's Situational Smoking Questionnaire, together with data on cigarette consumption, the nicotine yield of their preferred cigarette brand, their reported depth of inhalation, and the age at which they started to smoke. In a two-factor solution, one factor grouped the "high-stress" factor from Frith's Situational Smoking Questionnaire with the pharmacological dependence factor from Russell's Smoking Questionnaire. The other factor combined Frith's "low-stress" factor with items from the nonpharmacological factor from Russell's questionnaire, the pleasurable relaxation items and some of the sensory items. It appeared that the pharmacological motive for smoking was much more closely related to smoking in stressful than in relaxing situations. Thus the data from the Smoking Situations questionnaire support the Russell differentiation of pharmacological and nonpharmacological motives for smoking.

Another item that made up the "nervous irritation" factor of smoking in the McKennell (13) questionnaire was anger. Thomas (18) reported that cigarette smokers and former smokers had higher levels of irritability and anger in comparison with lifetime nonsmokers. When the smokers were divided into light (less than 20 a day) and heavy groups, the heavy smokers showed greater anger under stress than the light smokers. Smoking when angry was one major reason given for smoking by smokers attending a smoking-cessation clinic (83%), nonclinic smokers (47%), and student smokers (65%) (17, 2).

Laboratory Studies

A third set of supporting evidence comes from experimental studies of smoking. Item analyses of the questionnaire surveys (e.g., 12, 16, 17) have shown that desires for stimulation and for sedation are major smoking motives and the specific item on the smoking motive questionnaires (e.g., 17) indicates clearly that smokers believe that smoking helps them to think and to concentrate. Laboratory studies have provided evidence for more efficient performance after nicotine and for the effects of smoking on mood states.

An extensive review of the effects of smoking and nicotine on performance has been published (19) and our recent work is in the paper by

Wesnes in this volume. In brief, these studies of smoking and nicotine tablets and rapid information processing (20, 21, and 22) provide unequivocal evidence for more efficient processing in terms of improved detection of triplets of three odd or three even digits in a string of single numbers flashed on a computer screen. However, nicotine from cigarettes and tablets also reduced reaction time. This improvement was around seven percent and between 65 and 77 percent of the subjects were improved depending on the cigarette's nicotine delivery. This finding is very important because it shows that there is no speed and accuracy trade-off and so there was an overall improvement in processing efficiency.

Studies of the effects of nicotine on learning and memory have shown that nicotine was facilitating the input of information to storage but had no direct effect on storage or retrieval (23). This behavioral data fits with the study of Edwards, Wesnes, Warburton, and Gale (24) which showed that smoking decreased the latency of the P300 wave in the rapid information processing task. Nicotine increases electrocortical arousal, (25) and so this work fits neatly with research demonstrating that compounds which modify electrocortical arousal change attentional performance (see review 26). Thus both behavioral and electrocortical measures indicate that nicotine is enabling more efficient processing of information in the brain and provides support for the subjective experience of smokers that smoking helps them to concentrate.

Experimental studies of mood are difficult to conduct and no satisfactory experiments on nicotine and anger in humans have been done. Schachter (27) reports a study by Perlick which compared irritability in "unrestrained" smokers and "restrained" smokers who were trying to cut down their smoking. Each group rated the annoyance caused by aircraft noise after smoking either a 1.3 mg nicotine cigarette, a 0.3 mg nicotine cigarette, or not smoking. Unrestrained smokers, when deprived or smoking the 0.3 mg nicotine cigarette, were more annoyed than when smoking the 1.3 mg cigarette. However, unrestrained smokers, when smoking the 1.3 mg cigarette, were no better than nonsmokers. The restrained group were just as irritated in all three conditions as the deprived or low nicotine, unrestrained smokers. There are at least two explanations for these results; either irritability is an abstinence symptom which is the consequence of nicotine dependence, or smokers are constitutionally more irritable and nicotine from smoking helps to reduce this feeling.

Work on the use of smoking to control appetite is outlined in the chapter by Jarvik in this volume and will not be considered here.

The Personality Characteristics of Smokers

A fourth area of research has considered the personality characteristics of smokers. One influential theory of personality has been based on the traditional dichotomy between introverts and extraverts. This has been formalized in a personality questionnaire, the Eysenck Personality Inventory (28). It can be used to assess the degree of extraversion based on questions about behavior and attitudes. The same questionnaire has items that are designed to assess the degree of neuroticism, which is an independent dimension of personality. The stereotype of a person who scores high on neuroticism is one who displays much more labile emotional behavior. They are more anxious, irritable, moody, restless, excitable, changeable -- an unstable person. The contrasting type is the stable person who is calm, even-tempered, and reliable.

The largest surveys of smokers were made by Eysenck (29, 30). From these investigations a very highly significant positive correlation was found between cigarette smoking and level of extraversion. Heavy smokers had higher degrees of extraversion, medium smokers had lower degrees of extroversion, while light smokers were lower still and nonsmokers were the lowest. Ex-smokers scored between light and medium on the extroversion scale. Subsequent studies on both sexes have confirmed this statistical relationship between smoking and extraversion including our own (2).

Eysenck's studies of 6,000 men gave no evidence for a significant correlation of smoking with neuroticism (30), and Rae's (31) survey of 253 female students revealed no differences in the degree of neuroticism between nonsmokers, ex-smokers, light smokers, or medium smokers (less than 15 per day). However, at least seven studies have suggested a positive relationship between smoking and neuroticism, and this association is particularly strong for women smokers (32, 33, 34, 35, 36, and 4).

Although this evidence suggests that smokers are constitutionally more anxious, it could be argued that repeated exposure to nicotine had caused anxiety and a higher neuroticism score (27). Evidence against this argument comes from a crucial study done by Cherry and Kiernan (37) which has followed a cohort of 2,853 young people for 25 years. At the age of 16 years, they completed the Maudsley Personality Inventory before most of them had begun to smoke. At 20 years and 25 years of age they completed a smoking habits questionnaire and it was found that the cigarette smokers, as a group, scored more highly on extraversion and neuroticism. The two personality dimensions were independent and additive in their effect on the likelihood of becoming a habitual smoker. This finding argues strongly that constitutional factors underlying neuroticism determine whether some people smoke and not that smoking induces anxiety.

From this evidence comes the functional view that smoking is the outcome of the interaction of the person and the situation. However, the personality characteristics of the smoker are critical in this interaction and also in the resulting level of smoking. The situation x individual x smoking interaction was examined by studying the amount of nicotine taken by subjects in the vigilance test mentioned earlier (2). A positive correlation was found between neuroticism and the difference in smoke generation between the first and last cigarette for both the low-nicotine cigarette and the high-nicotine cigarette, indicating that the more neurotic subjects smoke more intensely at the beginning of the session than at the end of the test. This result provides further evidence that one aspect of smoking behavior is the outcome of the interaction of the situation and the individual and increases in stressful situations - smokers use more cigarettes, smoke them more intensely, and inhale the smoke more deeply, with those who are more anxious and susceptible to stress showing the most marked effects.

An interesting question is whether smoking does help to cope with stress and improve behavioral efficiency. We have analyzed performance increments as a function of neuroticism and found that there was a direct association between the degree of neuroticism and the amount of performance improvement after smoking (2). The latter association is consistent with data obtained by Kucek (38) in an experiment where subjects were tested under conditions of information overload. A comparison of neurotic smokers who were allowed to smoke and neurotic smokers who were deprived of smoking showed that smoking had a beneficial effect on the performance of neurotic subjects allowed to smoke.

Nicotine is the most likely constituent of cigarettes for producing this improvement in performance, and we found that for both females and males the performance improvements after nicotine tablets were directly correlated with neuroticism scores (2). This significant association with neuroticism gives some evidence for the hypothesis that nicotine is the tranquillizing ingredient of cigarette smoke.

There is clear evidence of variation of smoking behavior with stressful events. We have studied the smoking patterns of students during examination periods (4). They smoked more throughout the day during the examination period than during the equivalent times in the nonexamination period. During the examination period they also used more cigarettes on those mornings preceding an afternoon examination. In fact, only 2 of the 48 subjects did not increase their morning smoking during the examination period. Analysis of the cigarette butts showed a slight decrease which would result in higher nicotine absorption during the examination period but the product of the mouth nicotine concentration and number of cigarettes indicated an increased intake of nicotine. Throughout the examination period the students reported that they inhaled more strongly.

It is sensible to ask whether smoking had any effect on their performance that might be expected from the laboratory work studies. While we have not found any effects on gross measures like percentage pass rates and finals degree class, we have found significant differences in examination marks and a set of tutorial marks between smokers and nonsmokers (39) which give some evidence for an association between smoking and academic success.

Of course, these data do not allow us to infer that smoking produces higher achievement. However, we have already cited some studies from our laboratory that have demonstrated that cigarette smoking improves mental efficiency in tasks involving sustained concentration (see review in 19). It would seem to be a valid extrapolation that academic work performance like studying and essay writing, which depend on sustained concentration, would be facilitated by smoking as we have found in this study. In addition, we have presented evidence that smokers smoke more while studying during a stressful period of their lives, and smokers claim that smoking helps relax them.

However, there is also some evidence that one of the personality characteristics of smokers is a need for achievement, so a third hypothesis can be devised by combining the other two hypotheses. People with a higher need for achievement adopt many coping strategies to obtain their goals. If experience with cigarettes shows that smoking helps them think and concentrate, then they continue to smoke for these benefits. This hypothesis would fit with other evidence that smokers adjust their smoking behavior to match their needs.

As we have pointed out elsewhere (3), there seem to be several motives for smoking. A functional model does not require a single motive for all smokers. Casual use of cigarettes is not at all uncommon and many people have an occasional cigar on social occasions. This sort of casual use is completely consistent with the functional model. People may smoke on particular occasions but not in other situations. The level of smoking will reflect the personal function of smoking for the individual, and since most lives are stable, use will be relatively stable and increase only to meet extra needs. Smokers seem to reach a stable level of consumption and it remains constant throughout life provided their life is stable.

From the point of view of the functional model, the positive reinforcing effects do not result from euphoria but from the functions that smoking serves for the person. The acquisition and maintenance of smoking can be seen as being reinforced by the personal control that the smoker has over his psychological state. This control is very precise because of a very effective delivery process for nicotine whereby nicotine from cigarette smoke passes from the mouth to the brain via the lungs and heart in about 10 seconds (40, 5). Thus smoking is experienced as providing rapid control of the psychological state, and the level of smoking reflects the need for control and varies accordingly.

Smoking Cessation

Smoking cessation studies provide evidence for the functional model because they show that after abstinence, smokers report an increase in the effects that they claim they smoke to avoid. It is consistent with the functional model that ex-smokers could suffer some consequences of abstaining. On smoking cessation, clearly defined abstinence symptoms only occur in ex-smokers for whom smoking could still serve some function and alternative coping strategies have not been developed. It is significant that two major abstinence symptoms are anxiety and lack of concentration, which is consistent with self-reports of smoking and experimental studies of the habit (17, 5). However, 25 percent of smokers give up their habit without any perceived symptoms, and for others the symptoms are extremely variable (41). Reductions in smoking without withdrawal symptoms occur for many smokers during holidays. Orthodox Jews abstain on the Sabbath without experiencing abstinence symptoms (27).

Studies of withdrawal symptoms that follow cessation of smoking show that one common symptom is anxiety (42) and it is more likely to occur in women, among whom there is a greater proportion of neurotic smokers (32, 41). An analysis of the situations that resulted in a return to smoking, by Marlatt (43) and Shiffman (41), indicated that 80 percent of these situations fell into three categories: coping with anxiety and other negative emotional states (43%), social pressure (25%), and coping with social stress (12%).

In the Shiffman (41) study, two-thirds of the subjects were under stress at the time of relapse and anxiety was particularly common among ex-smokers who relapsed at work, which suggests that work-related anxiety was a contributory factor. Smokers who have high degrees of neuroticism are more anxiety-prone, and so it is not surprising that they find it difficult to stop smoking or relapse if they abstain (37). These data on smoking to cope with stress fit neatly with studies showing that deprived smokers are more likely to feel anxious in stressful situations (27). Studies of abstinence have shown hostility and agression increase markedly during abstinence (44, 42). Anger commonly triggers relapse among ex-smokers, and these relapses occurred in the absence of smoking-related cues such as "people around were smoking" (41). Clearly, the smoker who is deprived of nicotine is less calm. Evidence from a study by Thomas (18), which showed similar scores for anger when stressed for smokers and ex-smokers, suggests that smokers are constitutionally more angry and irritable. In summary, cessation studies are consistent with the view that smoking serves the function of reducing anger and anxiety.

The functional model predicts that if smoking is serving some function for the person, then some other chemical agent could be substituted for smoking or an increase in psychological problems could result in some ex-smokers. A study by Lee (45) of British doctors who gave up cigarettes after 1964 shows that there is not the mortality savings that might have been expected. British doctors have reduced their cigarette smoking over 20 years, and there has been some of the expected improvement in mortality rate from "smoke-related" diseases, but the overall death rate did not fall as much as would have been predicted from other groups of similar socioeconomic status.

Mortality from other causes had increased, particularly from accidents, poisonings, suicide, and cirrhosis of the liver (stress-related deaths). Lee concludes that the increased mortality from stress-related diseases may have been due in part to the loss of the stress-reducing effects of smoking. It is always very hazardous to argue from correlation to causality, but of relevance to the functional argument was the increase in alcoholism that suggests that these ex-smokers had switched to alcohol, a functionally similar but chemically different substance, for the reduction of distress. This substitution is completely consistent with the functional model.

THE FUNCTIONAL MODEL

If smoking behavior is an attempt by smokers to control their psychological state, then both internal and external factors will be determiners. Thus smoking is the outcome of the interaction of the person and the situation. Depending on the situation, smokers will adjust their smoking behavior in terms of the number of cigarettes smoked, smoke generation, and amount of smoke inhaled to control the nicotine levels reaching the brain; and in this way they control their psychological state. Deprivation of cigarettes results in abstinence symptoms, which are manifested as a less efficient function in certain situations, and so relapse is likely to occur.

Differences in smoking behavior will be a function of the intensity of the individual-situation interaction, its duration, and more probably, its density (the product of intensity and duration). In other cases smoking may not even be initiated until the intensity, duration, or density of the interaction is at a critical level. The occurence of smoking and the pattern of smoking behavior controlling nicotine intake depend on the individual-situation interaction.

In addition, smoking behavior can be the outcome of not only the characteristics of the situation but also the personality of the individual. Personality characteristics will determine the way in which the individual selects and interprets information within the situation, i.e., there is an interaction going on between individual factors and situational factors. Many smokers smoke in anticipation of a future need, before the situation has occurred or the critical stress level has been reached and they are still coping. Here the smoking behavior is very clearly a response to the individual's interpretation of the situation, his expectations about what may occur, rather than the objective characteristics of the situation. Smokers have learned to control their mood by smoking, enabling them to function more efficiently. In addition, smoking may help them avoid the undesired consequences of other situations and, consequently, anticipatory smoking can be seen as a rational coping strategy on the basis of their past experience.

There seem to be several motives for smoking, but a functional model does not require a single motive for all smokers. It only requires that a function of smoking can be identified for each smoker. An individual smoker may have more than one motive and smoking may have different functions on different occasions. Studies of self-medication that were discussed in Warburton (1) suggest that people use those classes of agents that fulfill their own individual needs. From smoking research there is evidence that people titrate nicotine to obtain a specific dose to fit their needs, i.e., precise personal control over nicotine intake and psychological state. Nicotine seems to be a unique substance because it combines both performance enhancement with anti-anxiety and anti-anger action. Thus smoking can be seen as providing rapid control of the psychological state and the level of smoking reflects the amount of need for control.

CHARACTERISTICS OF THE FUNCTIONAL MODEL IN SUMMARY

(1) In the functional model, the reasons for smoking predate its initiation and smoking is maintained because it satisfied these needs.

(2) The causes in the functional model can be due to both exogenous and endogenous causes.

(3) The functional model is purposive; it views smoking as a coping activity for everyday needs.

(4) In the functional model, smokers are a heterogeneous group who are smoking for different reasons. Thus smoking is a multidimensional behavior which must be conceptualized by a multifactorial model.

(5) The functional model does not view smoking per se as a problem, although problems may arise from excessive use. The functional model does conceive the possibility of nonproblem use.

REFERENCES

1. Warburton, D.M.: Self medication. In (eds.) Oborne, D.J., Gruneberg, M.M. and Eiser, J.R. Psychology and Medicine, pp 196-204, Academic Press, London, 1979.
2. Warburton, D.M. and Wesnes, K.: Individual differences in smoking and attentional performance. In (ed.) Thornton, R.E. Smoking Behaviour, pp 19-43, Churchill-Livingstone, Edinburgh, 1978.
3. Warburton, D.M. and Wesnes, K.: Mechanisms of habitual substance use: food, alcohol and cigarettes. In (eds.) Gale, A. and Edwards, J. Physiological Correlates of Human Behavior. 1: London: Academic Press, pp 277-298, 1983.
4. Warburton, D.M., Wesnes, K. and Revell, A.: Personality factors in self-medication by smoking. In (ed.) Janke, W., Response Variability to Psychotropic Drugs, in press, Pergamon Press, London, 1983.
5. Warburton, D.M.: Nicotine and the smoker. Rev. Environ. Health, in press, 1985.
6. Russell, M.A.H.: Cigarette smoking: Natural history of a dependence disorder. Brit. J. Med. Psychol., 44: 1-16, 1971.
7. Ashton, H. and Stepney, R.: Smoking: Psychology and Pharmacology. University Press, Cambridge, 1982.
8. Alexander, B.K. and Hadaway, P.F.: Opiate Addiction: The case for an adaptive orientation. Psychol. Bull., 92: 367-81, 1982.
9. Tomkins, S.S.: Psychological model for smoking behavior. Amer. J. Public Health, 56: 17-20, 1966.
10. Tomkins, S.S.: A modified model of smoking behavior. In (eds.) Borgatta, E. and Evans, R. Smoking, Health and Behavior, pp 165-86. Chicago: Aldine, 1968.
11. Ikard, F.F., Green, D.E. and Horn, D.A.: A scale to differentiate between types of smoking as related to management of affect. Int. J. Addict., 4: 649-59, 1969.
12. Stanaway, R.G. and Watson, D.W.: Smoking Motivation: A factor-analytical study. Pers. Ind. Diff. 1: pp 371-380, 1980.
13. McKennell, A.C.: Smoking motivation factors. Brit. J. Soc. Clin. Psychol. 9: 8-22, 1970.
14. Frith, C.D.: Smoking behaviour and its relationship to the smoker's immediate experience. Brit. J. Soc. Clin. Psychol., 10: 73-8, 1971.
15. Mausner, B. and Platt, E.S.: Smoking: A Behavioral Analysis. New York: Pergamon Press, 1971.

16. McKennell, A.C.: A comparison of two smoking typologies. Research paper No. 12. London: Tobacco Research Council, 1973.
17. Russell, M.A.H., Peto, J. and Patel, U.A.: The classification of smoking by factorial structure of motives. J. Roy. Stat. Soc. A. 137: 313-333, 1974.
18. Thomas, C.B.: The relationship of smoking and habits of nervous tension. In (ed.) Dunn, W.L., Smoking Behavior: Motives and Incentives, pp 157-170, Winston (J. Wiley), New York, 1973.
19. Wesnes, K. and Warburton, D.M.: Smoking, nicotine and human performance. Pharmacol. Therap., 21: 189-208, 1983a.
20. Wesnes, K. and Warburton, D.M.: The effects of smoking on rapid information processing performance. Neuropsychobiol., 9: 223-229, 1983b.
21. Wesnes, K. and Warburton, D.M.: Effects of scopolamine and nicotine on human rapid information processing performance. Psychopharmacol., 82: 147-150, 1984a.
22. Wesnes, K. and Warburton, D.M.: The effects of cigarettes of varying yield on rapid information processing performance. Psychopharmacol., 82: 338-342, 1984.
23. Warburton, D.M., Wesnes, K., Shergold, K. and James, M.: Facilitation of learning and state dependency with nicotine. Psychopharmacology, in press, 1985.
24. Edwards, J.A., Wesnes, K., Warburton, D.M. and Gale, A.: Evidence of more rapid stimulus evaluation following cigarette smoking. Addictive Behaviours, 10: pp. 113-126, 1985.
25. Edwards, J. and Warburton, D.M.: Smoking, nicotine and electrocortical activity. Pharmacol. Therap. 19: 147-164, 1983.
26. Warburton, D.M.: Neurochemical bases of behaviour. Brit. Med. Bull. 37: 121-125, 1981.
27. Schachter, S.: Pharmacological and psychological determinants of smoking. In (ed.) Thornton, R.E., Smoking Behaviour, pp 208-228, Churchill-Livingstone, Edinburgh, 1978.
28. Eysenck, H.J. and Eysenck, S.B.G.: Manual of the Eysenck Personality Inventory. University of London Press, London, 1964.
29. Eysenck, H.J., Tarrant, M., Woolf, M. and England, L.: Smoking and Personality. Brit. Med. J., 1: 1456-460, 1960.
30. Eysenck, H.J.: Personality and cigarette smoking. Life Sci. 3: 777-792, 1963.
31. Rae, G. Extraversion, neuroticism and cigarette smoking. Brit. J. Soc. Clin. Psychol. 14: 429-430, 1975.
32. Guilford, J.S.: Factors Related to Successful Abstinence from Smoking. 31 Pittsburgh, American Institutes for Research, 1966.
33. Meares, R., Grimwade, J., Bickley, M. and Wood, C.: Smoking and neuroticism. Lancet 2: 770, 1971.
34. Waters, W.E.: Smoking and neuroticism. Brit. J. Prevent. Soc. Med. 25: 162-164, 1971.
35. Dunnell, K. and Cartwright, A.: Medicine Takers, Prescribers and Hoarders. Routledge & Kegan Paul, London, 1972.
36. Shiffman, S.M.: Analysis of relapse episodes following smoking cessation. Paper presented at the 4th World Congress on Smoking and Health, 1979b.
37. Cherry, N. and Kiernan, K.: A longitudinal study of smoking and personality. In (ed.) Thornton, R.E. Smoking Behaviour, pp 12-18, Churchill-Livingstone, Edinburgh, 1978.
38. Kucek, P.: Effect of smoking on performance under load. Studia Psychologia 17: 204-212, 1975.
39. Warburton, D.M., Wesnes, K. and Revell, A.: Smoking and academic performance. Current Psychological Research and Reviews, 3: 25-31, 1984.

40. Russell, M.A.H.: Tobacco smoking and nicotine dependence. *In* (eds.) Gibbins, R.J., Israel, Y., Kalant, H., Popham, R.E., Schmidt, W. and Smith, R.G.: *Research Advances in Alcohol and Drug Problems*, Vol. 3: pp 1-48, Wiley, New York, 1976.
41. Shiffman, S.M.: The tobacco withdrawal syndrome. *In* (ed.) Krasnegor, N.A., *Cigarette Smoking as a Dependence Process*, National Institute for Drug Abuse, Washington, DC, 1979a.
42. Shiffman, S.M. and Jarvik, M.E.: Smoking withdrawal symptoms in two weeks of abstinence. *Psychopharmacol.*, *50*: 35-39, 1976.
43. Marlatt, A.: A cognitive-behavioral model of the relapse process. *In* (ed.) Krasnegor, J., *Behavioral Analysis and Treatment of Substance Abuse*, pp 191-199, National Institute for Drug Abuse, Washington DC, 1979.
44. Schechter, M.D. and Rand, M.J.: Effect of acute deprivation of smoking on aggression and hostility. *Psychopharmacol.*, *35*: 19-28, 1974.
45. Lee, P.N.: Has the mortality of male doctors improved with the reductions in their cigarette smoking? *Brit. Med. J.* *2*: 1538-1540, 1979.

NICOTINE INCREASES MENTAL EFFICIENCY: BUT HOW?

Keith Wesnes

Smoking Research Group
Department of Psychology
University of Reading
Reading, RG6 2AL
England

INTRODUCTION

Over the last 12 years our research group has carried out an extensive investigation into the effects of cigarette smoking and nicotine tablets on human mental efficiency. The consistent finding from this program has been that both cigarettes and nicotine favorably affect human information processing. The purpose of this paper is to consider the findings from this research in relation to, firstly, the ways in which smoking and nicotine improve mental efficiency and, secondly, how these effects might be produced.

THE WAYS IN WHICH CIGARETTES AND NICOTINE IMPROVE MENTAL EFFICIENCY

Having answered the basic question of whether or not smoking and nicotine improve human performance (for reviews see 1, 2, 3), more interesting questions emerge such as "which aspects of performance," "under what conditions," "in what ways," "to what extent," and "in whom." The following sections will consider the light the research from this laboratory has shed on these questions.

Aspects of Performance Improved by Smoking and Nicotine Tablets

The efficient processing of incoming sensory information is an essential requirement for most aspects of human performance. Attention is the common term for many aspects of such information processing. In general terms the work from our laboratory has concerned the effects of smoking and nicotine on human attention. As has been pointed out previously (1), in human psychopharmacology, substance-induced changes in the efficiency of attention are inferred from changes in the efficiency of the performance of a particular task. The strength of the inference is determined by the demands of the task. For example, if a substance was found to improve performance on a task involving complex skilled motor behavior, it may have done so by enhancing the attention paid to the task or it may have had a direct effect on the responses necessary for the execution of the task. For this reason we have generally used tasks which require efficient and sustained information processing but which require no skilled motor movements and do not require a high memory load for efficient performance.

In early experiments the effect of smoking on the performance of 80-minute vigilance tasks was studied. In both visual and auditory tasks, smoking facilitated information processing by preventing the decline in the signal detection theory measure of stimulus sensitivity to the experimental targets which occurred over time in nonsmoking or nicotine-free smoking conditions (2). Nicotine tablets also reduced the decline in stimulus sensitivity to the targets in a visual vigilance task (4). In another study, nicotine tablets reduced the magnitude of the Stroop Effect, which can be interpreted to represent enhanced processing of relevant information despite the presence of distracting irrelevant information (2). Furthermore, in a learning task, nicotine was found to facilitate the acquisition of information (5).

An experimental task which has been used in a large number of studies involves the processing of rapidly presented information. The task measures both the ability to sustain attention and the speed of information processing. In this task, subjects search for target sequences from a continuous series of digits presented singly at the rate of 100 per minute on a video screen. The target sequences are any three consecutive odd or even digits and subjects signal their detections by pressing a response key. The task is computer controlled allowing the measurement of correct detections (Hits), the speed of such detections, and incorrect responses (False Alarms). The typical finding has been that smoking improves both the speed of response and the number of targets detected while either having no effects on false alarm rates or actually decreasing them (6, 7). Furthermore, nicotine tablets have comparable effects on this task (8). In order to determine whether smoking would enable smokers to perform two concurrent tasks more efficiently or whether smoking would simply benefit one task at the expense of another, we carried out a study in which the smokers performed an auditory version of the rapid visual information processing task together with the standard visual task. The subjects were instructed to press a response button with their left hand if they detected targets in the auditory task and a second response button with the right hand if they detected targets in the visual task. Smoking was found to increase efficiency on both tasks, which suggests that smoking increases overall information processing as opposed to concentrating processing resources on a particular task.

The studies described in this section provide evidence consistent with the hypothesis that smoking facilitates information processing. In vigilance tasks, this effect is seen as a prevention of the 'vigilance decrement' which occurs over long sessions. Smoking and nicotine thus enable individuals to sustain their attention to infrequent stimuli over long vigils. In another type of task which lasts only a few minutes, information processing was facilitated in another way - by the increased ability to ignore distracting stimuli. In a learning task, nicotine improved the aquisition of new information, possibly by enhanced processing. On the rapid information processing task, smoking and nicotine improve both the speed and the accuracy of target detection. This task requires sustained attention together with efficient processing of rapidly presented information, and these effects are therefore interpreted as improvements in both the ability to sustain attention and the speed with which the information is processed. Finally, such improvements were also found when the subjects had to divide their attention between two concurrent tasks.

In summary, the aspects of information processing which are improved by smoking and nicotine in the smoker include the abilities to: sustain attention, ignore distraction, acquire information, process information quickly and accurately, and divide attention between two tasks.

The Conditions under which Smoking Improves Information Processing

The theme of this section will be that smoking improves performance on the rapid information processing task despite a number of methodological variations. There have been three major types of variations, the first being the use of different cigarettes, the second being changes in task design, and the third being alterations to the environmental conditions. The only products which have not improved performance have been nicotine-free herbal cigarettes (2, 6). When commercially available products have been used, the improvements are related to the overall yields of the cigarettes, higher deliveries of tar and nicotine producing greater improvements than lower deliveries (6, 7, 9, 10).

For studies in which the design of the rapid visual information processing task has been varied, smoking has still been found to improve performance. Work stress was manipulated in one experiment by both decreasing and increasing the speed with which the digits were presented. During the 'high stress' condition the digits were presented at the rate of 150 per minute and subjects were rewarded financially for the number of correct detections made. In the low stress condition, the speed of presentation of the digits was reduced to 50 per minute. Despite these alterations, smoking improved performance relative to not smoking in both 'high' and 'low' work stress conditions. In the Divided Attention study mentioned earlier, the postsmoking performance period was reduced from the conventional 20 to 10 minutes, but neither this nor the performance of two tasks at once prevented smoking from improving performance. During all of the studies described so far, performance has been measured before and after smoking. In the latest series of studies, which will be considered in more detail in a later section, performance has been measured while subjects have been smoking. This does not prevent smoking from improving performance; if anything, performance improvements are larger. Furthermore, in one study, the average target density was increased by 50%, although this did not affect the improvements produced by smoking. Finally, in another study which will be described in more detail in a later section, there was no pre-smoking baseline; instead, subjects performed the task for a continuous 30-minute session following smoking. Once again, this variation in design did not prevent smoking from improving information processing performance.

In a number of studies the environmental conditions under which the testing was carried out have been varied. 'Social stress' was induced in one experiment by filming the subjects while they performed the task and informing them that the films would be analyzed by a panel of trained psychologists. The subjects were not filmed in the "no-stress" condition. Both conditions were carried out with and without smoking. It can be seen from Fig. 1 that in the high stress nonsmoking condition, performance was elevated during the first 10 minutes compared to the low stress nonsmoking condition. However, in both smoking conditions, performance was increased relative to not smoking. Thus, the effects of smoking were not obscured by filming the subjects; instead, the effects of smoking and stress appeared to be additive. In two studies carried out in collaboration with Prof. T. Gale and Dr. J. Edwards at Southampton University, electrocortical activity was recorded while subjects performed the task (12, 10). In both studies smoking improved performance despite carrying out the study in a different laboratory environment and concomitantly measuring electrocortical activity. Finally, as the vast majority of the studies described have been carried out during the

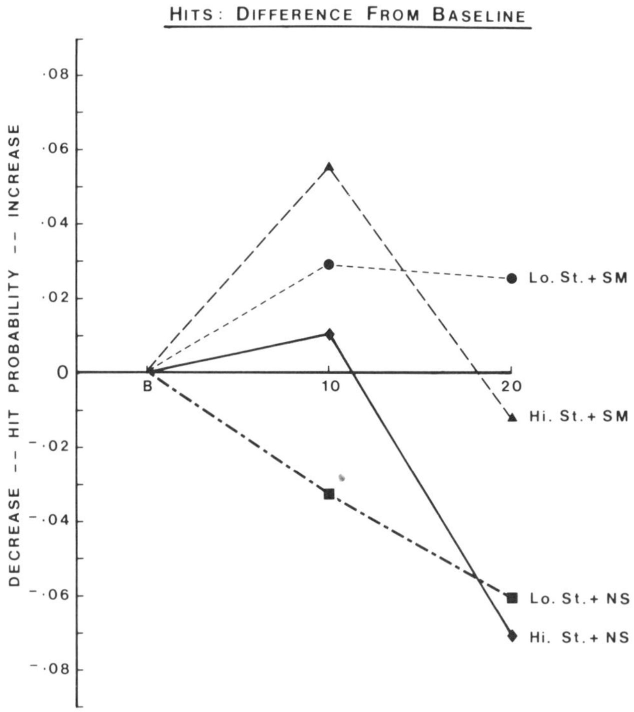

Fig. 1. The accuracy of information processing performance when subjects smoke and do not smoke in "high stress" and "low stress" situations (see text for explanation).

morning, a study has recently been carried out in which subjects ate a heavy lunch prior to performing the task in the early afternoon. Smoking was again found to improve performance despite the lunch and the afternoon testing.

To summarize, the improvements in information processing produced by smoking occur under a wide range of conditions. Every nicotine-containing cigarette we have studied improves performance. Improvements occur irrespective of the duration of testing, the speed of the presentation of the digits, the density of the targets, whether or not subjects smoke while performing, whether or not they are filmed, whether or not electrocortical activity is measured in another laboratory, and whether testing is carried out in the morning or afternoon.

The Nature of the Improvements Produced by Smoking and Nicotine

In this section the nature of the improvements in information processing which are produced by smoking and nicotine will be considered. In the vigilance studies, these improvements were seen as a prevention of the decrement in efficiency which occurred as a result of performing the task over an extended period. No evidence was obtained from these studies that smoking or nicotine could improve performance above resting levels.

As described earlier, three measures of performance are made from the rapid visual information processing task. Two of these measures enable the determination of the sensitivity to the experimental targets: the number of correct detections and the number of False Alarms. Smoking generally increases correct detections and has either no effect on, or reduces, False Alarms. Thus the subjects are not simply increasing their overall responding rates, but are better able to detect the targets. However, over a 10-minute period on this task, subjects correctly detect on average 65% of the 80 targets, while they make on average between two and three False Alarms. Clearly, False Alarms play a minor role in overall efficiency, but as they are either decreased or unaffected by smoking, do substantiate the interpretation that increases in hits represent improved sensitivity. Thus the two major measures of performance are firstly the percentage or probability of correct detections, and secondly the speed with which these detections are made. This latter measure provides a second way of evaluating improvements. By measuring both the speed and the accuracy of performance, we can consider not only the quantitative level of efficiency but also the qualitative level of efficiency. This is illustrated in Fig. 2. Here, not only is it evident that performance is quantitatively greater after smoking than after not smoking, but, as both speed and accuracy are improved, performance can also be seen to be qualitatively better. Furthermore, nicotine tablets have been found to produce comparable effects (8). Considering now the quantitative level of improvements, not only does smoking prevent the decrement which occurs over time in not-smoking conditions but it also improves performance about the pre-smoking level. This improvement has been interpreted as an absolute improvement in efficiency (7). However, as will be seen from studies in which performance is resolved over smaller time periods, this interpretation needs to be qualified.

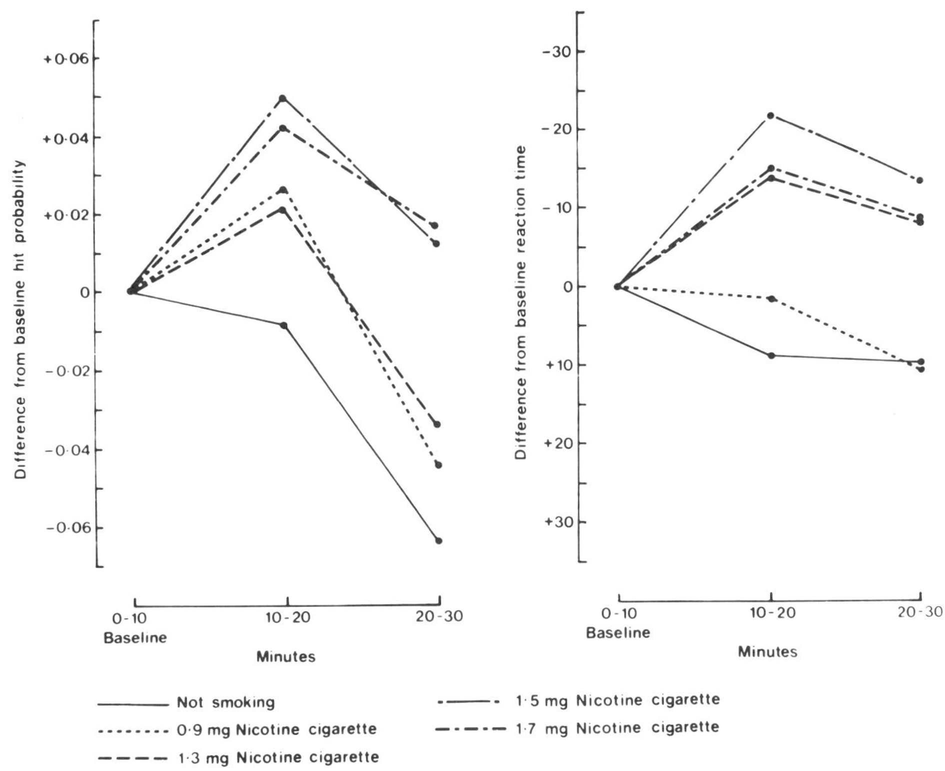

Fig. 2. The effects of smoking four different cigarettes on accuracy and speed in the rapid visual information processing task.

Over the first five years during which the rapid visual information processing task was used, performance was resolved over 10-minute periods (one prior to smoking, and two following). However, as body nicotine is known to drop rapidly following smoking it was considered worthwhile to attempt to resolve performance over shorter time periods. To do this, in one study the number of targets in each 5-minute period was standardized, which when using the conventional design gave two successive measures of performance prior to smoking and four after smoking. The results of this trial are presented in Fig. 3. The first thing which is immediately apparent from the data is that the 10-minute period prior to smoking is not homogenous, instead it comprises a marked drop in efficiency from the first 5 minutes to the second. Smoking clearly improves performance but does not take it to a higher level than that during the first 5 minutes of the pre-smoking period. Thus, the improvement is only absolute if we average over 10-minute periods. However, if we look at the nonsmoking condition, it is clear that the 10-minute rest from the task does not restore performance to its original level. Therefore, there is a residual decrement in performance from the first 10-minute period of performance to the 20-minute postsmoking/nonsmoking period. This decrement may prevent smoking from enhancing performance above the initial level, and thus the possibility remained that performance could be absolutely enhanced above performance if there were no pre-smoking baseline.

To study this, an experiment was carried out in which performance was measured for 30 minutes following either smoking or not smoking. These results are presented in Fig. 4 from which it can be seen that performance immediately following smoking was greater than that following not smoking. Statistical analysis revealed this improvement to be significant for hits during the first 5 minutes, and thereafter to be significant for both hits and reaction times during each successive 5-minute period. These data indicate that even when measuring performance over short periods of time, smoking produces absolute improvements compared to not smoking.

In a further series of studies, the effects of performing the task while smoking were studied. In order to prevent smoking from distracting subjects from the task, short gaps were introduced into the task to allow them to light the cigarettes and take puffs. The subjects were required to perform the task for a 20-minute period. After the first 5 minutes, the task stopped for 12 seconds to enable the subjects to light the cigarettes. At the beginning of each of the 7th to 14th minutes the task was halted for 7 seconds to allow the subjects to puff on the cigarettes. The task thus comprised a 5-minute pre-smoking baseline, 10 minutes of smoking while performing, and 5-minutes postsmoking. The subjects were given the same brief rest periods in the nonsmoking condition. The results of this study are presented in Fig. 5 for each successive 5-minute period. Here a clear enhancement above the baseline level is evident during the first 5 minutes of smoking.

In summary, smoking and nicotine enable subjects to maintain their level of vigilance over extended periods. Using the rapid visual information processing task, smoking and nicotine improve both the speed and accuracy of performance. Furthermore, during the 10 minutes after smoking, performance is significantly better than during the 10 minutes prior to smoking. However, if performance is measured over 5-minute periods, smoking is seen to restore performance to its level during the first 5 minutes of the task, and to reduce the decline which occurs over the next 5 minutes. Nonetheless, if there is no presmoking baseline, performance after the first 5 minutes of smoking is significantly better than during the first 5 minutes after not smoking. Performance is also found to be improved during smoking. Thus, smoking produces both

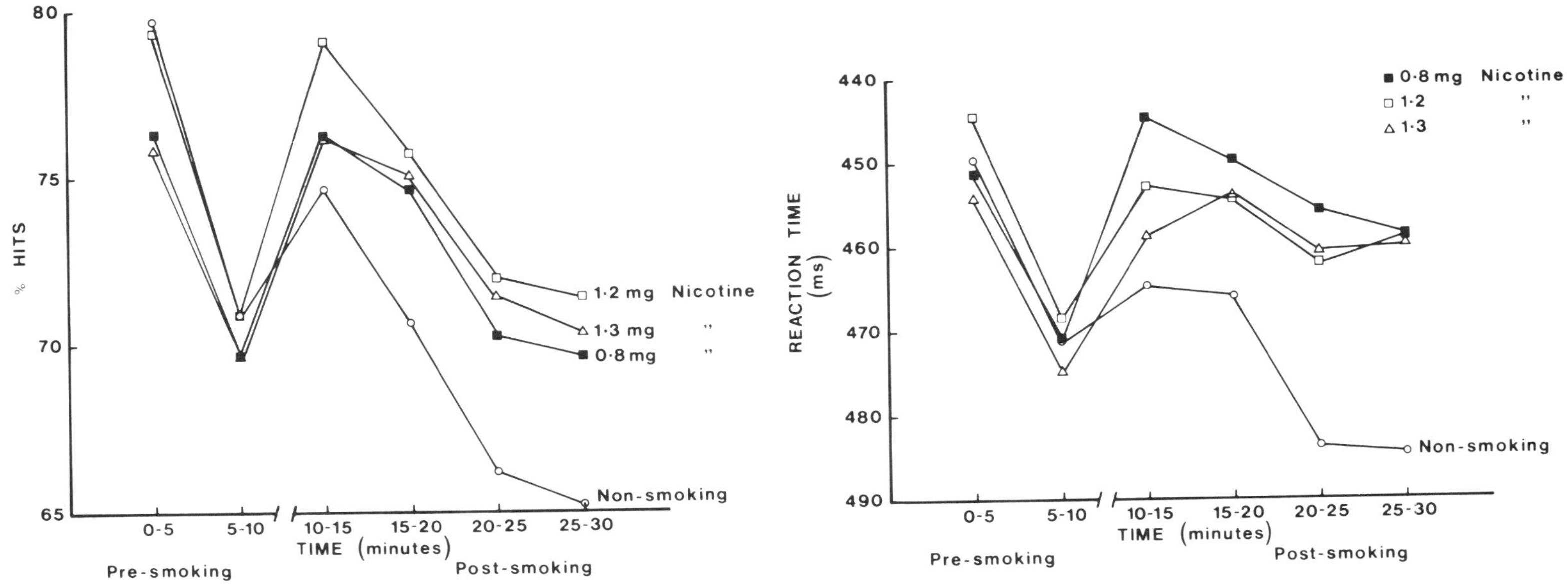

Fig. 3. The effects of smoking on accuracy and speed in the rapid visual information processing task when performance is resolved over successive five minute periods.

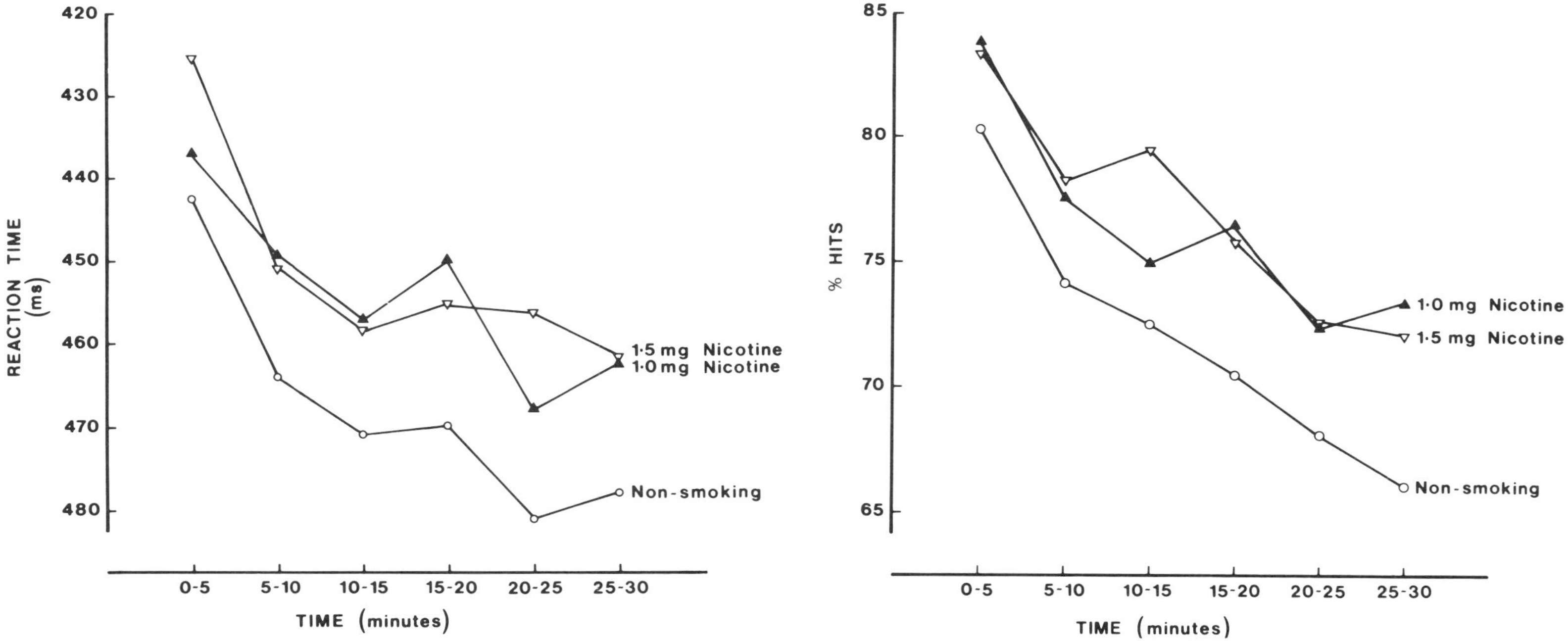

Fig. 4. The speed and accuracy of performance on the rapid visual information processing task for thirty minutes following either smoking or not smoking.

qualitative and quantitative improvements on information processing, and if baseline testing is omitted, then the level of efficiency is raised absolutely.

The Extent of the Improvements in Information Processing

Having considered what aspects of information processing are improved under which conditions and in which ways by smoking, it is now appropriate to consider the extent of these effects, both in terms of the magnitudes of the improvements and the durations for which they persist.

The magnitudes of the effects of smoking on performance can be evaluated in two ways. Firstly, the performance after smoking can be compared to the level prior to smoking, and secondly, performance during or after smoking can be compared to performance at equivalent times when not smoking. The magnitudes of the effects using three different smoking designs will now be presented. Considering first studies in which performance has been measured over 10-minute periods, experiments have been carried out with groups of heavy smokers in which the primary difference between the studies was the type of cigarette. One study has been described earlier (7); others are being prepared for publication. Averaging over these studies, the mean percentage improvement for the first 10 minutes after smoking compared to the 10 minutes prior to smoking are 6.4% and 3.4% for hits and speed of reaction, respectively. Compared to not smoking, these products produced an average improvement during the first 10 minutes after smoking of 7.3% and 4.6% for hits and speed of reaction, respectively, and during the second 10 minutes after smoking of 7.9% and 3.0%, respectively. Looking secondly at the study described earlier (see Fig. 5) in which subjects performed the task while smoking, the average improvements from the first 5 minutes of the task to the first 5 minutes of smoking were 5% and 3.3% for hits and speed of reaction, respectively. Compared to not smoking, these cigarettes produced average improvements of 13.6% and 7.4% for hits and speed, respectively, during the first 5 minutes of smoking, 13.3% and 7.6% during the second 5 minutes, and 14% and 6.7% during the 5 minutes after smoking. Taking thirdly, the study described earlier (see Fig. 4) in which performance was measured for 30 minutes after smoking, the percentage improvements over not smoking for each successive 5-minute periods were 4.2%, 5.0%, 6.4%, 8.1%, 6.3% and 10.1% for accuracy and 2.5%, 3.0%, 2.8%, 3.7%, 3.9% and 3.3% for speed of reaction.

From these analyses it is clear that the magnitudes of the improvements depend on the experimental design used. In the early studies in which the subjects rested from the task for ten minutes while smoking or not smoking, the improvements over not smoking were approximately half of those obtained when the subjects smoked while performing. This suggests that not resting the subjects allows more scope for improvement. Nonetheless, improvements occur immediately following smoking, even when the subjects have not performed baseline testing, and thus time on task prior to smoking is not necessary for improvements to occur, but can modulate the size of the improvements if no rest periods are present.

It is also evident that the improvements relative to not smoking persist throughout the periods during which subjects are tested, even up to 30 minutes following smoking. Further, there is little evidence that the improvements tail-off over time; quite the opposite occurred for accuracy in the study in which performance was measured 30 minutes after smoking.

The subjects who performed these studies were well trained on the experimental task and, being young university students, can reasonably be

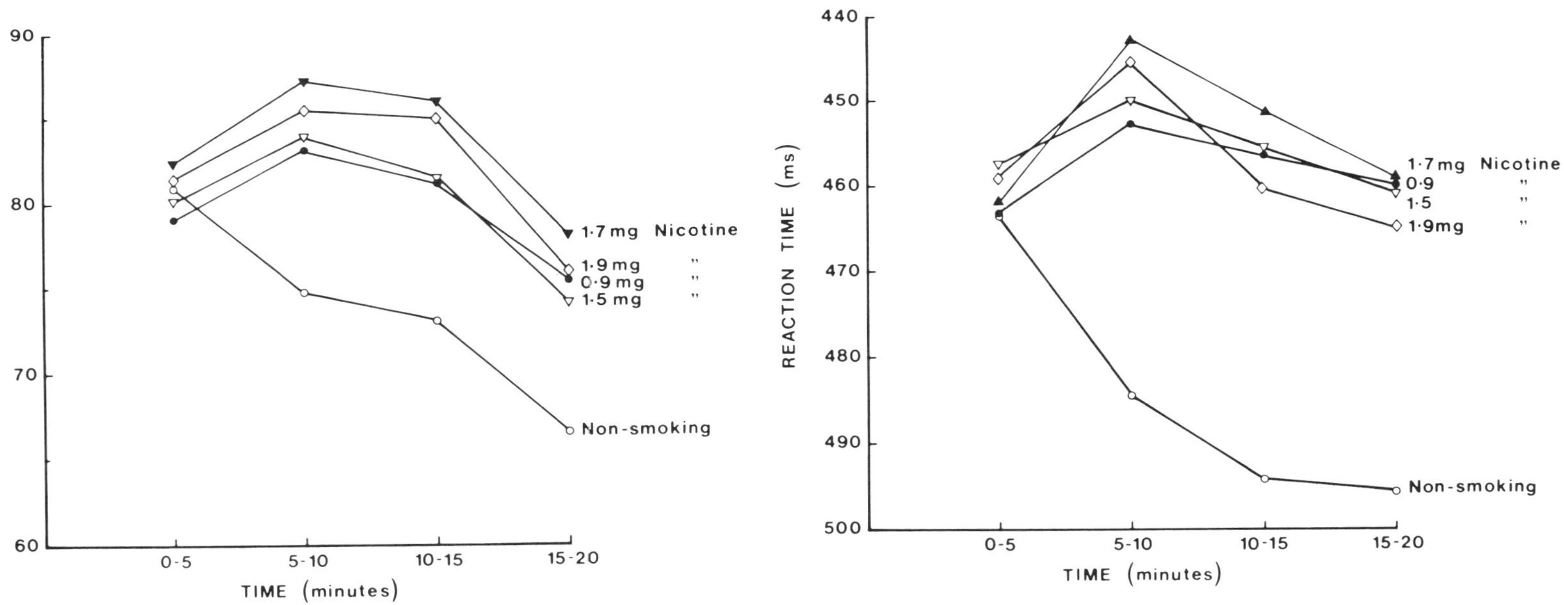

Fig. 5. The accuracy and speed of information processing both during and after smoking.

expected to be at peak levels of cognitive efficiency. Thus, to improve their performance by even a few percentage points is quite an achievement and to increase accuracy by as much as 14% is remarkable. Laties and Weiss (13) in considering the benefits amphetamines produce on athletic performance, conclude "the amount of change induced by the amphetamines is usually small, of the order of a few percent. Nevertheless, since a fraction of a percent improvement can make the difference between fame and oblivion, the margin conferred by these drugs can be quite important." Although of course we are measuring cognitive, not athletic, efficiency, the same arguments could be applied. If it is further considered that from the studies described here two independent aspects of performance are both being simultaneously improved, the compound improvement in task efficiency is even greater.

In summary, smoking can simultaneously increase accuracy by up to 14% and speed by up to 7.6%, and the improvements detected persist up to 30 minutes after smoking. The combined benefit of faster and more accurate information processing in healthy young students which persists over long periods indicates that smoking can improve efficiency to a considerable extent.

Who is Improved by Nicotine and Smoking?

Nicotine tablets have comparable effects on the performance of both smokers and nonsmokers. In the study described earlier of the effects of nicotine on visual vigilance, three groups of 12 subjects were studied: heavy smokers (15 or more per day), light smokers (5 or less per day) and nonsmokers. When performance was analyzed using the nonparametric index of stimulus sensitivity, no differences between the groups were found in their response to nicotine (4). Similarly, in the study of nicotine and the Stroop Effect when the six nonsmokers and the six smokers were analyzed separately, no differences were found between them in the effects of nicotine (14). The subjects in the study of nicotine on rapid visual information processing were all nonsmokers (8), as were the subjects in a second study in which nicotine reversed the effects of scopolamine (15).

In the majority of the studies on smoking, heavy smokers (15 or more per day) have been studied. No differences have been found between male and female heavy smokers in the effects of smoking. However, in studies in which male and female light smokers performed the task before and after smoking a range of commercially available cigarettes, consistent improvements were not found. In subsequent interviews with these subjects, it was found that they generally did not smoke in the morning when the testing actually took place. A second study was then carried out in which light smokers performed the task during the afternoon. In contrast to the studies carried out in the morning, the findings of this study were that the performance of both male and female light smokers was improved by the cigarettes. This finding emphasizes the need to study the effects of smoking on behavior at times when smokers usually smoke.

In the United Kingdom, the Health Department subdivides brands of cigarettes into 'tar' groups, two of the groups being called Low Tar (0-10 mg/cig) and Low to Middle Tar (11-16 mg/cig), although very recently these 'tar' brand definitions have been modified. For a recent study, smokers were recruited whose preferred brands fell within these two groups as defined above (9). Both Low Tar (LT) smokers and Low to Middle Tar (LMT) smokers performed the task once while not smoking and once while smoking a cigarette from the LT group. The LMT smokers also performed the task while smoking a cigarette from the LMT group. The task design was the smoking while performing version described earlier in which subjects perform the task for 5 minutes prior to smoking, 10 minutes while smoking, and 5 minutes after smoking. For both LT and LMT smokers, the LT cigarette did not improve accuracy until the second 5 minutes of smoking, whereas the LMT cigarette improved the LMT smokers throughout the 10 minutes of smoking (see Fig. 6). It is interesting to note from the figure that when not smoking, the performance of the LT smokers was stable throughout the first 10 minutes of the task whereas that of the LMT smokers declined over this period. Low tar products are also low in nicotine, and this raises the intriguing possibility that LT smokers will smoke LT cigarettes because they are able to maintain their concentration over longer periods and thus do not need smoking to aid their concentration until the later stages of smoking. LMT smokers, on the other hand, show faster decrements and thus their preferred brand is one which improves performance during the early stages of smoking. The vast majority of smokers in previous studies have not been Low Tar smokers, and performance decrements have consistently been found in nonsmoking sessions from the first 5 minutes to the second (e.g., Figs. 4 & 5). However, less than 15% of our University subject pool smoke Low Tar products, and this compares well with near 15% market share for such low tar brands in the United Kingdom, and the 15% of the marketplace captured by similar products (termed ultralow-tar brands, 0–10 mg tar/cig) in the U.S.A. (16). If this finding were replicated on a larger population of LT smokers (19 were used in this study) it might provide a possible explanation of why only a small proportion of smokers smoke such low tar, low nicotine cigarettes.

The majority of the smokers in the studies described have been undergraduates between the ages of 18 and 25 years. A large comparative study has just been completed in which a cohort of 60 undergraduates were

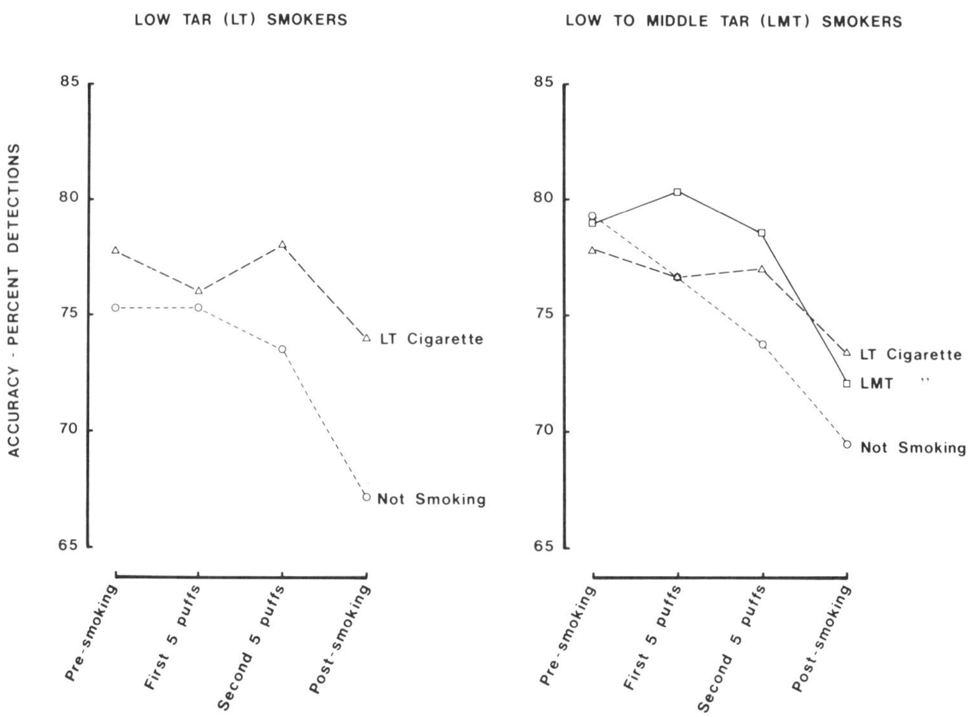

Fig. 6. The effects of smoking on accuracy of information processing in a study using both Low Tar smokers and Middle Tar smokers (see text for explanation).

compared on a large number of measures to a cohort of 60 smokers of all ages recruited from outside the university. Although all the results have not yet been analyzed, it was found that both groups performed the rapid visual information processing task more efficiently when smoking cigarettes than when not smoking. This result strongly suggests that the general findings of the studies described here can be applied to the smoking population as a whole.

In summary, both nonsmokers and smokers are favorably affected by nicotine. Male and female heavy smokers respond similarly to a range of products but light smokers have to be tested in the afternoon for these effects to emerge. Low tar smokers do not show decrements in accuracy over the first 10 minutes of performance, which might possibly explain why they smoke cigarettes that do not affect performance until later stages of testing. Finally, university undergraduates resemble non-university smokers of all ages in their response to smoking.

Future Research

The study of the cognitive actions of cigarette smoking and nicotine is still in its infancy. Despite the large numbers of trials described here, many of the conclusions which can be drawn are inclusive as opposed to exclusive. Thus, there is good evidence that smoking improves various aspects of information processing, although this does not exclude the possibility that nicotine may directly affect other psychological mechanisms involved in cognition and performance. Similarly, smoking affects performance up to 30 minutes following smoking, although this does not exclude the possibility that the effects of smoking may persist far longer than this. Clearly, we know considerably more about the effects of smoking and nicotine than we did 10 years ago but there are still vast areas to explore in the future. To spell out the areas for future research would cover more space than has already been allocated to discussing current knowledge. All of the avenues of research described above have proven fruitful and should be studied further.

One important question concerns the subjective experiences which are reflected in improved information processing. Many smokers may not smoke to help them work but may still enjoy subjective conscious experiences such as increased awareness, even at times of relaxation. More detailed studies of the relationships between subjective levels of consciousness and improved performance should further our understanding of a habit which is practiced in such a wide variety of situations.

Another important area worthy of fuller future investigation is individual differences. In the early stages of research into any psychotropic compound, individual differences are a nuisance variable, controlled for by using large groups of volunteers, baseline measures, and within-subject designs. However, when the general effects of a compound have been established, the individual differences in response to the compound become the major interest. In all of the research described above, the results described were the effects of smoking and nicotine on groups of smokers. In these studies the great majority of subjects were improved. For those subjects who were not, the nature of the experimental design does not reveal whether such subjects never benefit from smoking or nicotine or whether, due to chance variations, these effects were not revealed. If different subpopulations could be identified, perhaps they could be further differentiated on other behavioral variables such as personality, life-style, smoking behavior, preferred brands, etc., or other constitutional differences, such as metabolic rates, endogenous rhythms etc. Such research should provide much information on the neurobiologic mechanisms which underly the smoking habit.

To summarize, many of the major effects of smoking and nicotine described above still require further detailed investigation. Many others await to be discovered. Furthermore, work should now proceed in parallel on the reliability and extent of the individual differences to the known behavioral responses to nicotine and smoking.

THE NEURAL MECHANISMS BY WHICH SMOKING AND NICOTINE IMPROVE PERFORMANCE

Although the smoking studies described above stand alone as investigations into the performance effects of a major social habit, and the nicotine studies stand alone as basic investigations into the effects of a psychotropic drug on human performance, our initial interest in nicotine stemmed from its cholinergic effects in the CNS. Warburton (17, 18) proposed that the ascending cholinergic pathways from the ventral tegmental area to the cortex, by modulating electrocortical arousal, play a major role in the process of information processing. The evidence for this cholinergic theory of information processing has been reviewed extensively elsewhere (17, 18, 19, 20), but as the initial rationale for the work described above was to test a hypothesis from this theory in man, some of the major steps in the reasoning will be briefly presented.

In both animals and man, electrocortical arousal is directly correlated with behavioral efficiency (see reviews in 19, 20, 21). Cholinergic systems in the brain have long been implicated in the control of electrocortical arousal (22). Injections of the cholinergic agonist, carbamylcholine chloride, into the mesencephalic area close to the ventral tegmental area induce electrocortical arousal (23). In animals, nicotine injections induce electrocortical arousal, which can be prevented by lesions rostral to the reticular formation (24, 25, 26), by muscarinic and nicotinic cholinergic blockade (24, 25) but not by lesions caudal to the reticular formation nor by adrenergic blockade (24). Furthermore, the electrocortical arousal produced by nicotine injections is accompanied by increased release of acetylcholine from the cortex, whereas when applied directly to the cerebral cortex, nicotine affects neither acetylcholine output nor electrocortical arousal (27). Such findings indicate that nicotine must produce these effects by acting subcortically, that the ascending pathways from the reticular formation to the cortex must be intact, and that cholinergic neurones must be involved, although adrenergic neurones do not appear to be.

In animals, cholinergic blockers impair both electrocortical arousal and information processing, whereas cholinergic agonists have the opposite effects (17). In man, nicotine injections and cigarette smoking both produce electrocortical arousal (21) whereas the muscarinic cholinergic antagonist, scopolamine, lowers electrocortical arousal (28). As was demonstrated in the first section both nicotine and smoking improve information processing, and in parallel with these findings, we have found increased electrocortical arousal (12) as well as shifts in components of the evoked potential indicative of enhanced stimulus evaluation (10). Furthermore, scopolamine impairs performance in the vigilance and rapid information processing tasks described earlier (8, 29) and the latter effect can be reversed by nicotine (15).

From the evidence presented above, the following mechanism for the actions of nicotine on mental efficiency is therefore proposed. Nicotine increases electrocortical arousal by stimulating neural pathways ascending from the classical reticular formation to the cortex. Cholinergic neurones are involved in this effect and this cortical activation results in improved information processing, which is the basis for the effects of smoking and nicotine on mental efficiency described earlier. What is unclear is the precise locus of action of nicotine in the CNS and the extent to which other neurotransmitters besides acetylcholine could be involved. Nicotine may activate the ascending reticular pathways via a secondary mechanism; certainly there is good evidence that some of the behavioral effects of nicotine cannot be explained on the basis of traditional nicotinic cholinergic mechanisms (30). Undoubtedly, the wealth of contributions from other workers in this volume will shed further light on the precise neural mechanisms responsible for the effects of nicotine on human information processing.

IMPLICATIONS FOR SMOKING MOTIVATION

Smokers say that smoking helps them to think and concentrate, and a large proportion claim that this is part of the motivation to smoke (7). The effects of smoking on cognitive efficiency are entirely consistent with such subjective impressions, and questionnaire data from studies which are being prepared for publication indicate firstly that the beneficial effects of smoking on efficiency are accompanied by increased impressions of alertness, secondly that smokers are aware that they have performed more efficiently while smoking, and thirdly that these effects are attributed to the cigarettes. Nicotine itself has comparable effects on performance, strongly suggesting that the nicotine absorbed while smoking plays a major role in the effects measured.

Such findings strongly support the idea that the effects of nicotine on mental efficiency play an important role in smoking. As was suggested earlier, it is also possible that improved information processing is a reflection of other subjective states which the smoker may enjoy. A smoker may for example enjoy increased alertness, even at times when he is relaxing. The purpose of much of the research described here is to help understand one of the mechanisms which motivate the smoking habit. A cigarette which delivers sufficient nicotine for the smoker's needs, but with reduced levels of other substances would be a first step towards minimizing the risk from smoke-related diseases. We are beginning to understand some of the complex psychological motives which underlie the smoking habit. This knowledge should assist future smoking and health research.

SUMMARY

Research from this laboratory into the effects of cigarette smoking and nicotine tablets is considered in relation firstly, to the ways in which smoking and nicotine improve mental efficiency and, secondly, to the mechanisms by which these effects might be produced. Both smoking and nicotine favorably affect information processing on a number of tasks. These effects are obtained with a wide range of cigarettes, on different populations, under a wide variety of task and environmental conditions. These actions are interpreted in terms of the action of nicotine on cholinergic pathways controlling electrocortical arousal. It is argued that these actions of nicotine on mental efficiency play an improtant role and are relevant to the development of cigarettes with minimum risk from smoke-related diseases.

REFERENCES

1. Wesnes, K.: The effects of psychotropic drugs on human behavior. _Mod. Probl. Pharmacopsychiatry_ _12_:37-58, 1977.

2. Wesnes, K. and Warburton, D.M.: The effects of cigarette smoking and nicotine tablets upon human attention. In (ed.) Thornton, R.E., Smoking Behavior: Physiological and Psychological Influences, pp. 131-147, Churchill Livingstone, Edinburgh, 1978.
3. Wesnes, K. and Warburton, D.M.: Smoking, nicotine and human performance. Pharmacol. Ther. 21:189-208, 1983.
4. Wesnes, K., Warburton, D.M. and Matz, B.: Effects of nicotine on stimulus sensitivity and response bias in a visual vigilance task. Neuropsychobiology 9:41-44, 1983.
5. Warburton, D.M., Wesnes, K., Shergold, K. and James, M. Facilitation of learning and state dependency with nicotine. Psychopharmacology 89:55-59, 1986.
6. Wesnes, K., and Warburton, D.M.: Effects of smoking on rapid visual information processing performance. Neuropsychobiology 9:223-229, 1983.
7. Wesnes, K., and Warburton, D.M.: The effects of cigarettes of varying yield on rapid information processing performance. Psychopharmacology 82:338-342, 1984.
8. Wesnes, K., and Warburton, D.M.: Effects of scopolamine and nicotine on human rapid information processing performance. Psychopharmacology 82:147-150, 1984.
9. Wesnes, K., and Warburton, D.M.: The differential effects of cigarettes varying yield on rapid information processing performance. Paper presented at IVth World Congress of Biological Psychiatry, Philadelphia, 1985.
10. Edwards, J.A., Wesnes, K., Warburton, D.M., and Gale, A: Evidence of more rapid stimulus evaluation following cigarette smoking. Addict. Behav. 10:113-126, 1985.
11. Wesnes, K., Warburton, D.M. and Revell, A.: Work and stress as motives for smoking. In (eds.) Cumming, G. and Bonsignore, G. Smoking and The Lung. pp. 233-249, New York: Plenum, 1984.
12. Warburton, D.M. and Wesnes, K.: The role of electrocortical arousal in the smoking habit. In (eds.) Remond, A. and Izard, C. Electrophysiological Effects of Nicotine, pp. 183-200, Amsterdam: Elsevier, 1979.
13. Laties, V.G. and Weiss, B.: The amphetamine margin in sports. Fed. Pro. 40:2689-2692, 1981.
14. Wesnes, K.: The effects of nicotine and scopolamine on human attention. Unpublished Doctoral Thesis, Reading University, 1979.
15. Wesnes, K. and Revell, A.: The separate and combined effects of scopolamine and nicotine on human information processing. Psychopharmacology 84:5-11, 1984.
16. Fielding, J.E.: Smoking: Health effects and control. New Engl. J. Med. 313:555-561, 1985.
17. Warburton, D.M.: Brain, Behavior and Drugs. London: Wiley, 1975.
18. Warburton, D.M.: Neurochemical bases of behavior. Br. Med. Bull. 37:121-125, 1981.
19. Warburton, D.M.: Stimulus selection and behavioral inhibition. In (eds.) Iversen, LL, Iversen, SD and Snyder SH. Handbook of Psychopharmacology, Vol. 8: Drugs, Neurotransmitters and Behavior, pp. 385-432, New York: Plenum, 1977.
20. Warburton, D.M.: Neurochemical bases of consciousness. In (eds.) Brown, K. and Cooper, S. Chemical Influences in Behavior, pp. 421-462, 1979.
21. Edwards, J.A. and Warburton, D.M.: Smoking, nicotine and electrocortical activity. Pharmacol. Ther. 19:147-164, 1982.
22. Rinaldi, F. and Himwich H.E.: Alerting responses and actions of atropine and cholinergic drugs. AMA Arch. Neurol. Psychiat. 73:387-395, 1955.
23. Endroczi, E., Hartmann, G. and Lissak, K.: Acta Physiol. Hung. 24:200-209, 1963.

24. Il'yuchenok, R. Yu, and Ostrovskaya, R.U.: The role of mesencephalic cholinergic systems in the mechanism of nicotine activation of the electroencephalogram. Bull. Exp. Biol. Med. 54:753-757, 1962.
25. Domino, E.F.: Electroencephalographic and behavioral arousal effects of small doses of nicotine: A neuropsychopharmacological study. Ann. N.Y. Acad. Sci. 142:216-244, 1967.
26. Kawamura, M. and Domino E.F.: Differential actions of m and n cholinergic agonists on the brain stem activating system. Int. J. Neuropharm. 8:105-115, 1969.
27. Armitage, A.K., Hall, G.H. and Sellers, C.M.: Effects of nicotine on electrocortical activity and acetylcholine release from the cat cerebral cortex. Br. J. Pharmacol. 35:152-160, 1969.
28. Ostfeld, A.M. and Aruguette, A.: Central nervous system effects of hyoscine in man. J. Pharmacol. Exp. Ther. 137:133-139, 1962.
29. Wesnes, K. and Warburton, D.M.: Effects of scopolamine on stimulus sensitivity and response bias in a visual vigilance task. Neuropsychobiology 9:154-157, 1983.
30. Abood, L.G.: Sites and mechanisms for nicotine's action in the brain. Neurosci. and Biobehav. Rev. 5:479-486, 1981.

NICOTINE: ABUSE LIABILITY, DEPENDENCE POTENTIAL AND PHARMACOLOGIC TREATMENT OF DEPENDENCE

Jack E. Henningfield[1,2], Steven R. Goldberg[1,3] and Donald R. Jasinski[2]

NIDA Addiction Research Center
P.O. Box 5180/4940 Eastern Avenue
Baltimore, MD 21224
[1]National Institute on Drug Abuse Addiction Research Center; [2]The Johns Hopkins University School of Medicine; [3]University of Maryland, School of Medicine

INTRODUCTION

Clinical observations, historical anecdotes, and testimony, suggest that at least some users of tobacco are unable to permanently abstain even when ill health is evident (e.g., Connolly, Winn, Hecht, Henningfield, Hoffman, and Walker, 1986). However, whereas such observations suggest that tobacco self-administration (S-A) can become an addictive behavior, such observations are not scientifically definitive statements that provide the basis for evaluating the abuse liability and dependence potential of tobacco products. Specifically, such observations do not provide a basis for differentiating tobacco S-A from other commonly accepted "addictive behaviors" such as television watching, exercise, or excessive eating; furthermore, such observations do not provide scientifically valid evidence that an addictive drug is critical to the maintenance of the addictive behavior. The distinction between nonspecific addictive behaviors and those involving certain drugs is not just semantic; the implications are critical to issues of public health, legislative jurisdiction, and treatment of the behavior. The purpose of the present report is to summarize evidence that tobacco is a dependence-producing substance according to the same scientific criteria by which other substances are evaluated.

Since many critical terms in this paper have been variably defined and used, the following usage will be adhered to in this chapter. Whereas the term "addiction" is often used interchangeably with "abuse" and "dependence," the term addiction is often used so generally and variably that it has lost much of its technical usefulness. For instance, addiction is widely used to describe compulsive behaviors in which no specific drug is involved, and the term is also used when a drug produces physiological changes but when no compulsive drug taking was engendered. Therefore, the term "drug dependence" will be used in accord with the definition of the World Health Organization Expert Committee on Drug Dependence (WHO, 1969) and other scientifically accepted sources (e.g., Jaffe, 1985). In brief, the definition of drug dependence as so used is as follows: substance-seeking behavior that is critically determined by the activity of a constituent drug in the central nervous system;

tolerance and physiologic dependence may or may not be present (World Health Organization, 1969; Jaffe, 1985). The potential of a substance to produce drug dependence is determined by evaluating its "abuse liability" and "physiologic dependence potential" (Brady and Lukas, 1984; Jasinski, Johnson and Henningfield, 1984). Abuse liability refers to those effects of a drug which result in its compulsive self-administration, often in the face of mounting cost, physical and social dysfunction, damage, and exclusion of more socially acceptable behavior (see also, Jasinski, 1977). Physiologic dependence potential refers to the direct physiologic effects, or neuroadaptation produced by repeated administration of the drug (see also Jaffe, 1985). Neuroadaptation is assessed by the demonstration of tolerance to the effects of the drug and physiologic rebound phenomena ("abstinence" or "withdrawal") that occur following termination of drug administration. Such neuroadaptation is a frequent, but not invariant, correlate of abuse liability and is often relevant to the treatment of drug dependence.*

The literature relevant to the assessment of abuse liability and dependence potential of tobacco may be usefully divided into two general categories. The first is a comparative evaluation in which tobacco is systematically compared to the prototypic dependence-producing drugs. Such a comparison shows that patterns of use and effects of tobacco share critical points of commonality with prototypic dependence-producing drugs such as morphine and cocaine. The points of commonality not only cluster tobacco with drugs of abuse but they also distinguish tobacco (the substance) from substances such as food, and tobacco use (the behavior) from habitual behaviors such as television watching and exercise. The second category of literature to be reviewed is recent studies in which nicotine itself, in the absence of tobacco smoke or use, was evaluated using the same methods and criteria used to evaluate any substance suspected of causing abuse and dependence. This evaluation shows that nicotine meets rigorous experimental criteria as a drug that has a liability for abuse and the potential to produce dependence.

TOBACCO USE AND DRUG DEPENDENCE: COMMONALITIES

The rational basis for the categorization of tobacco as a dependence-producing drug and for the compulsive behavior of tobacco self-administration as a form a drug dependence is that tobacco and tobacco use resemble dependence-producing drugs and drug dependence, respectively. That is to say, when systematically compared to prototypic drugs of abuse, tobacco is similar along most usual and all critical points of comparison. Systematic reviews of these commonalities have been published (Jarvik, 1970, 1973; Russell, 1971; Jaffe and Kanzler, 1979; Henningfield, Griffiths and Jasinski, 1981). Table 1 provides an overview of the previously described commonalities.

As shown in the table, despite the commonly held differences in the use of tobacco and other substances of abuse -- tobacco use is legal, widely accepted, and occurs with relatively little disruption to cognitive

*An additional concept central to many discussions of drug dependence is that the substance produces "damage" or "debilitation." This aspect of tobacco dependence will not be addressed in this report as there are extensive data indicating the actual toxicity of tobacco and the widespread perception by tobacco users that the substance is harmful (see the Reports to the Surgeon General on the Health Consequences of Tobacco Use).

Table 1. Specific Commonalities Between Tobacco Use and Drug Dependence

Spread is socially mediated and is persistent
Patterns of relapse are similar following treatment
Use persists in the face of damage (individual & social)
Personality types overlap
Centrally (CNS) acting substance (drug) is delivered
The drug is a reinforcer for animals
Deprivation increases drug-seeking behavior
Tolerance develops with repeated use
"Therapeutic effects" may be produced
Patterns of self-administration and dose-response functions are orderly

and behavioral performance -- a systematic comparison reveals many points of commonality. The fundamental commonality is that the substance-taking behaviors are orderly, compulsive, and occur in the face of damage. These facts do not provide the basis for distinguishing drug dependence as a subset of habitual behaviors, however. Such a distinction is made on the basis that use of the substance delivers a specific drug to the central nervous system which produces behavioral and physiological effects characteristic of those of prototypic dependence-producing drugs. Thus, use of opium, coca derivatives, and alcoholic beverages results in the delivery of morphine, cocaine, and ethanol, respectively, and all of these drugs produce effects that share critical points of commonality. The analogous substance in tobacco is nicotine. Use of all common forms of tobacco results in the delivery of nicotine to the central nervous system (Schmiterlaw, Hansson, Andersson, Applegren, and Hoffman, 1967). Since most of the points of commonality listed in Table 1 have been previously reviewed (Henningfield, 1984a; Henningfield, Griffiths and Jasinski, 1981) they will be only briefly summarized here.

Discriminative Effects are Centrally Mediated

Nicotine, like other drugs of abuse, produces dose-related discriminable effects in animals which are attenuated by centrally acting antagonists; these effects partially generalize to another drug of abuse, amphetamine (Stolerman, 1987; Rosecrans, 1979, 1987; Rosecrans and Meltzer, 1981). These and more recent data are reviewed elsewhere in this volume (see chapters by Stolerman and by Rosecrans). The findings from animal studies are also consistent with those of a human study in which the dose-related discriminable effects of intravenously-administered nicotine are attenuated by mecamylamine pretreatment (Henningfield, Miyasato, Johnson, and Jasinski, 1983).

Nicotine as a Reinforcer for Animals

Most drugs that are abused by humans are voluntarily self-administered when available in laboratory studies; that is, the drug serves as a reinforcer or reward (Griffiths and Balster, 1979; Griffiths, Bigelow and Henningfield, 1980). When such studies are properly conducted and controlled, the findings confirm that the biologic effects of the drug in the central nervous system are sufficient for the substance to control behavior by virtue of its reinforcing effects. Until the early 1980s, it was not clear whether or not nicotine was such a substance. However, nicotine now has been shown to function as a reinforcer for several nonhuman animal species and under a variety of conditions (see review, Henningfield and Goldberg, 1985, and the chapter by Goldberg and Henningfield elsewhere in this volume). Furthermore, its functional behavioral effects are similar to those engendered when other drugs of abuse (e.g., cocaine) serve as reinforcers.

Patterns of Tobacco Self-Administration are Orderly

Within-day patterns of smoking are also orderly. Dependent smokers tend to smoke their first cigarette within 30 minutes of waking, after a night of sleep, and find it difficult to abstain for more than a few hours (Fagerstrom, 1978). If smoking behavior is relatively unconstrained, regular patterns develop which more closely resemble those of psychomotor stimulant self-administration in animals (Griffiths, Bigelow and Henningfield, 1980). Even within the smoking of cigarettes, patterns of tobacco self-administration are orderly. Several studies have now shown that across successive puffs on a cigarette, puff duration tends to decrease (Griffiths and Henningfield, 1982; Chait and Griffiths, 1982; Nemeth-Coslett and Griffiths, 1984a, 1984b), and these changes may be due to factors other than simply the nicotine in the cigarette (Nemeth-Coslett and Griffiths, 1985).

Tobacco Self-Administration Varies as a Function of Nicotine Dose

The effective dose of a substance may be varied by changing the quantity of drug per unit, by pretreating the organism with either an agonist or antagonist, or by altering the rate of elimination of the substance. Studies involving the first manipulations have been done extensively with other drugs and, more recently, with nicotine. Some studies of nicotine self-administration (often in the form of tobacco smoking) have also been conducted in which subjects were pretreated with alternate forms of nicotine, such as nicotine gum, or were pretreated with nicotinic antagonists such as mecamylamine. Following a major review of the literature published before 1980, Gritz came to a conclusion that still accurately describes the collected literature on the effects of nicotine dose manipulations on cigarette smoking: "Almost all of the studies demonstrate some increase in smoking as cigarette nicotine content falls below accustomed levels, and a decrease in smoking when cigarette nicotine content is unusually high" (Gritz, 1980). Similarly, pretreatment of smokers with nicotine gum or mecamylamine has effects similar to those of increasing or decreasing, respectively, the available dose of nicotine.

Neuroadaptation to Nicotine: Tolerance and Physiologic Dependence

Administration of most drugs of abuse results in neuroadaptation which is measured by tolerance to the repeated administration of the drug and a subsequent rebound (withdrawal) when drug administration is terminated (Jaffe, 1985). Tolerance to drug effects is determined by either the diminished response to repeated doses of a drug or to the requirement of increasing doses to achieve the same response. Tolerance to the behavioral and physiologic effects of nicotine have been studied for decades (cf. review by Henningfield, 1984a). As is the case with other drugs of abuse, a variety of mechanisms account for tolerance to many of nicotine's effects. These include metabolic tolerance (Beckett and Triggs, 1967), behavioral tolerance (Clarke and Kumar, 1983; Stitzer, Morrison and Domino, 1970; Stolerman, Bunker and Jarvik, 1974), and physiologic tolerance (Faulkeborn, Larsson, and Nordberg, 1981; Domino, 1979; Fagerstrom and Gotestam, 1977). More recently, studies have shown that subjective effects of nicotine also show tolerance with repeated dosing (Jones, Farrell and Herning, 1978; Henningfield, 1984b).

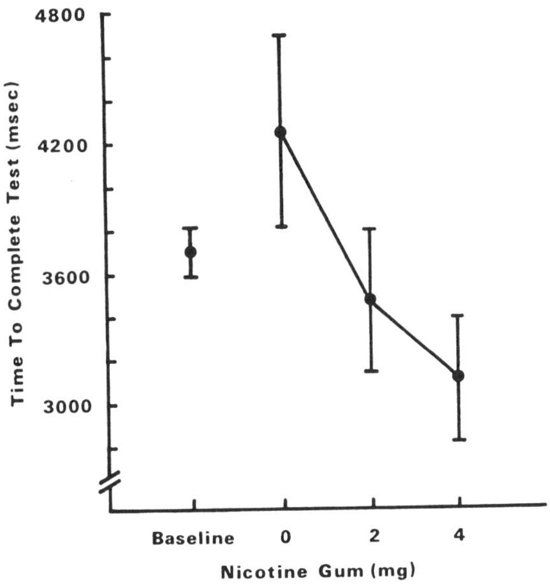

Fig. 1. Data points represent the average change from baseline in each of the treatment groups after 12 hours of cigarette abstinence as assessed on one of the measures (Logical Reasoning) of the computerized performance assessment battery. During the baseline sessions, subjects were allowed to smoke normally up to the time of the test. The 2 and 4 mg nicotine gum reversed the effect of abstinence, whereas placebo gum shows the detrimental effects of 12 hours of tobacco deprivation in smokers. (Unpublished data by F. Snyder and J.E. Henningfield)

Physiologic dependence to drugs is determined by showing that termination of drug administration produces a syndrome of effects, generally opposite in direction of those produced by drug administration, and that this syndrome is reversible (at least in its early stages) by administration of the drug; prolonged drug abstinence ("detoxification") results in ultimate return to baseline ("normal") values of behavioral and physiological functions. Fagerstrom (1978, 1981) showed that a psychometric instrument (Fagerstrom Tolerance Questionnaire) could be used to predict which tobacco users would show signs of physiologic dependence and which would also be most efficaciously treated using a nicotine substitution procedure (nicotine gum). These were critical data showing the specificity of the tobacco withdrawal syndrome to the administration and withdrawal of nicotine. Figure 1 shows data which confirm the specificity of the tobacco abstinence syndrome to nicotine. As shown in the figure, behavioral performance deficits associated with tobacco abstinence may be reversed when subjects are treated with nicotine in the form of a chewing gum (Snyder and Henningfield, 1986). The generality and the conditions under which physiologic dependence occurs have yet to be resolved. For instance, the specific role of preabstinence nicotine intake, the possible ancillary role of other tobacco constituents (e.g., the CO and CO_2 in tobacco smoke), and species differences remain to be resolved. However, recent studies now confirm the earlier findings, and some of these recent findings will be summarized later in this report.

Effects Considered Desirable or "Therapeutic" to Tobacco Users

Most drugs of abuse have had specific therapeutic applications; nicotine is no exception (e.g., Austin, 1978; Brecker, 1972). The degree to which the therapeutic effects of nicotine depend upon the individual having had a history of use of nicotine, as opposed to the possibility that nicotine is efficacious for preexisting conditions, remains to be thoroughly investigated; similar issues remain for other drugs of abuse. Pomerleau and his co-workers have studied a variety of mechanisms by which possibly weak, initial reinforcing effects of nicotine can be greatly strengthened by its array of somewhat more subtle (or at least more difficult to measure) effects on mood, cognition, and normal physiological and behavioral functioning; these effects appear to be mediated by the diverse effects of nicotine on hormonal release and regulation (Pomerleau and Pomerleau, 1984). Other effects of nicotine considered therapeutic by many smokers are reviewed elsewhere in this volume in chapters by Warburton, Wesnes, and by Jarvik (see also, Wesnes and Warburton, 1982, 1983, 1984).

Although the euphoriant properties of drugs can stand apart from collateral therapeutic actions (as is the case with morphine, amphetamine, and alcohol), attention to such drug effects may enhance the understanding of the underlying addictive process, as well as strengthen the efficacy of treatment. Since nicotine, in the form of tobacco, is widely available, relatively inexpensive, and in a convenient form for precise dose regulation, it provides a means of "self-medication" for the cigarette smoker, e.g., as an anorectant or anxiolytic. These effects may contribute to the abuse liability of tobacco and are of demonstrable significance in the treatment of tobacco dependence.

Implications of Commonalities Among Tobacco Use and Drug Dependence

The preceding review has shown that tobacco shares many points in common with prototypic dependence-producing drugs. These similarities provide a strong conceptual basis for the categorization of tobacco as a dependence-producing drug. The behavioral process is orderly, tobacco self-administration results in the delivery of a centrally active drug (nicotine), and that drug appears critical in the control of the compulsive behavior of tobacco self-administration. These findings are consistent with those expected if tobacco use is an orderly form of drug self-administration, similar to that observed when other drugs of abuse are self-administered by animal and human subjects (Griffiths, Bigelow and Henningfield, 1980). They provide the conceptual basis for the categorization of tobacco as a dependence-producing drug.

RECENT STUDIES OF THE ABUSE LIABILITY AND DEPENDENCE POTENTIAL OF NICOTINE

The foregoing comparison of tobacco to prototypic dependence-producing drugs is a summary of the conceptual basis for the categorization of compulsive tobacco use as a form of drug dependence in which nicotine plays a critical role in the behavior. However, if nicotine's role is like that of the CNS-affecting agents present in other substances of abuse (e.g., the cocaine in coca leaves), then nicotine, in the absence of the multitude of stimuli associated with cigarette smoking, should be an abusable substance as well. Objective methods for abuse liability assessment were available before the recent interest in nicotine (Brady, and Lukas, 1984; Jasinski, Johnson, and Henningfield, 1984). With consideration given to the fact that nicotine has more rapid effects than many other drugs of abuse, these methods were readily adapted to studies of the abuse liability of nicotine. Essentially, the approach is to test the hypothesis that the drug in question is psychoactive and will serve as a euphoriant and a reinforcer, as assessed by methods used to assess other dependence-producing drugs. Psychoactivity and euphoria are determined by assessing the pharmacodynamic subjective effects of single doses of the drug ("single-dose" or "abuse liability" studies), and are validated by observed behavioral and physiologic responses (Jasinski, 1977). Demonstrating that a drug will serve as a reinforcer is determined by assessing the ability of the drug to strengthen and maintain orderly patterns of behavior leading to subsequent administration of the drug (Henningfield, Lukas, and Bigelow, 1986; Young and Herling, 1986). Physiologic dependence potential may also be evaluated according to standardized procedures, such as the substitution approach, in which administration of the subject's maintenance drug is terminated and replaced with either placebo or another drug and the possible ensuing signs and symptoms of withdrawal are evaluated (Martin, 1977; Jasinski, Johnson and Henningfield, 1984). Nicotine has now been assessed using these approaches; the results are summarized below.

Abuse Liability Studies of Nicotine: Psychoactivity and Euphoria

The basic method utilized in these experiments was developed for assessing the abuse liability of morphine-like compounds in humans (Jasinski, 1977). Volunteers are given a range of doses of the test compound and placebo under double-blind conditions. Individuals with histories of drug abuse are used as subjects because they are able to accurately discriminate compounds with a potential for abuse and can compare the effects of the compounds to those of abused drugs (Jasinski, Johnson, and Henningfield, 1984). In a study at the Addiction Research Center, nicotine was given both intravenously and in the form of tobacco smoke over a range of doses to eight subjects with histories of drug abuse (Henningfield, Miyasato, and Jasinski, 1985). Three doses of nicotine and placebo were given intravenously, and three doses of research cigarettes (controlled nicotine delivery) and an unlit cigarette were inhaled according to a standardized puffing procedure. Each dose was presented to each subject on four different occasions. Self-reported (subjective), observer-reported (behavioral), and physiologic variables were measured before, during, and after drug administration.

The functional equivalence of intravenous nicotine to tobacco smoke was shown by the findings that nicotine produced a similar profile of effects across a variety of measures when given by both routes of administration (Fig. 2). Furthermore, nicotine was shown to be psychoactive, as evidenced by the reliable discrimination of nicotine from placebo. Its self-reported effects peaked within one minute after administration (by either route) and dissipated within a few minutes: peak and duration of response were directly related to the dose. The two hallmark indicators of euphoria in such studies are the Liking Scale (Single Dose Questionnaire) and the Morphine Benzedrine Group (MBG) Scale (Addiction Research Center Inventory, or ARCI) (Jasinski, Johnson and Henningfield, 1984). Figure 3 shows responses on the 5-point Liking Scale, which asked how much the drug was liked (0=not at all, 4=an awful lot). As shown in the figure, nicotine produced responses on the Liking Scale similar to those of more commonly studied drugs of abuse such as morphine and _d_-amphetamine. Scores on the MBG scale of the ARCI were consistent with the Liking Scale data, confirming that nicotine, given by both routes of administration, was a euphoriant. When asked to identify the injections from a list of commonly used and abused drugs, subjects most frequently identified nicotine injections as cocaine.

Similar results for intravenous and inhaled nicotine were also obtained on several physiologic measures, including pupil diameter, blood pressure, and skin temperature. These similarities in subjective and

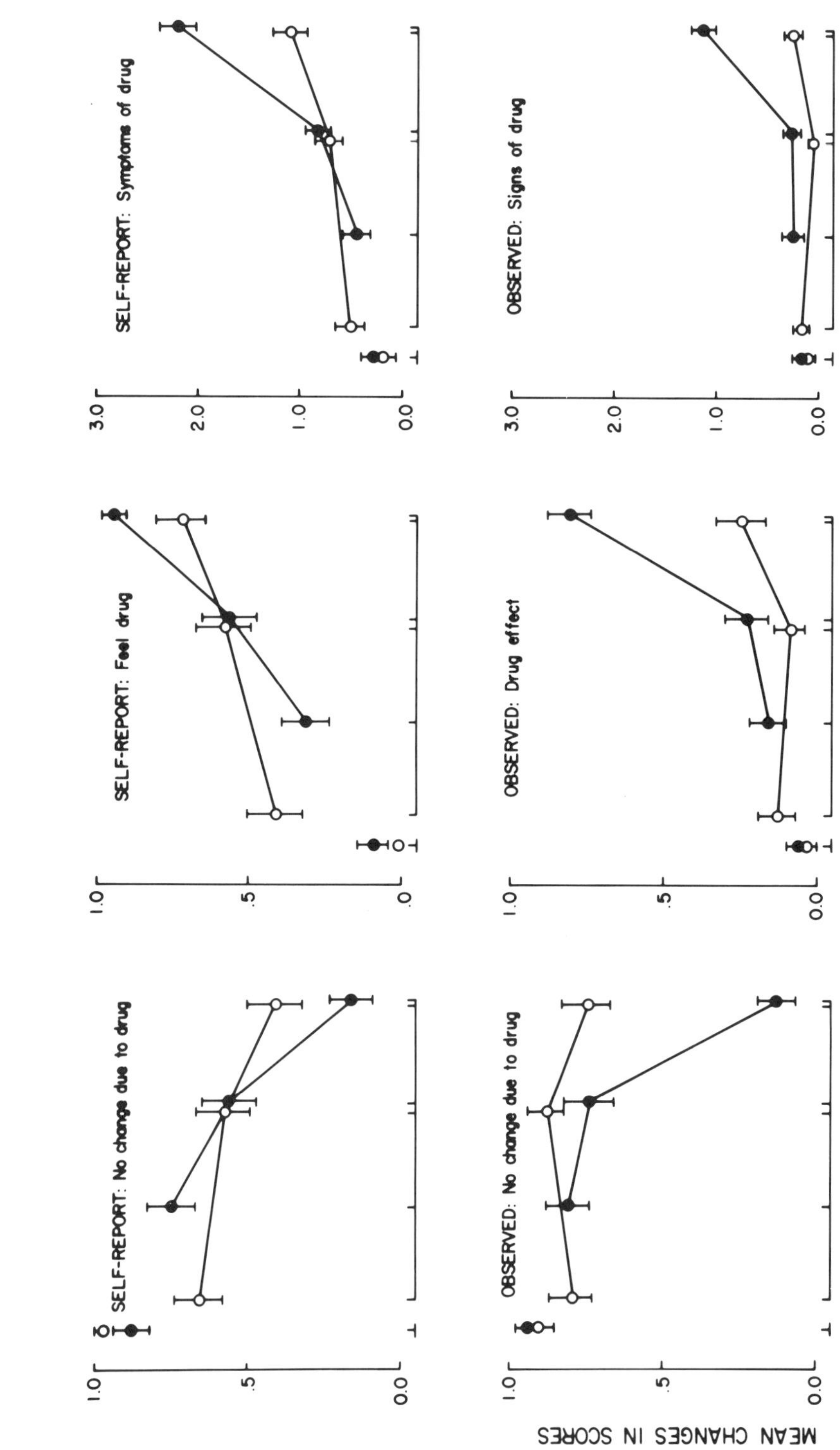
SELF-REPORT: No change due to drug
SELF-REPORT: Feel drug
SELF-REPORT: Symptoms of drug
OBSERVED: No change due to drug
OBSERVED: Drug effect
OBSERVED: Signs of drug
MEAN CHANGES IN SCORES
1.0
.5
0.0
.0
3.0
2.0
1.0
0.0

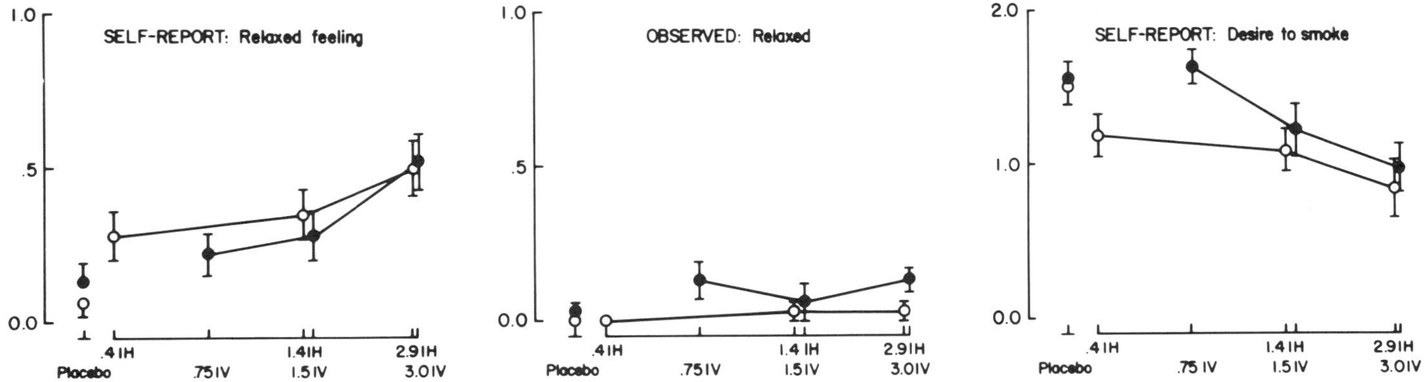

Fig. 2. Summary of change from baseline scores on instruments for measuring self-reported and observer-reported changes. Measures were taken immediately before and about 3 minutes following drug administration (n = 32, 8 subjects x 4 sessions). Open circles = inhaled (IH) nicotine, closed circles = intravenous (IV) nicotine. (Henningfield, Miyasato and Jasinski, 1985).

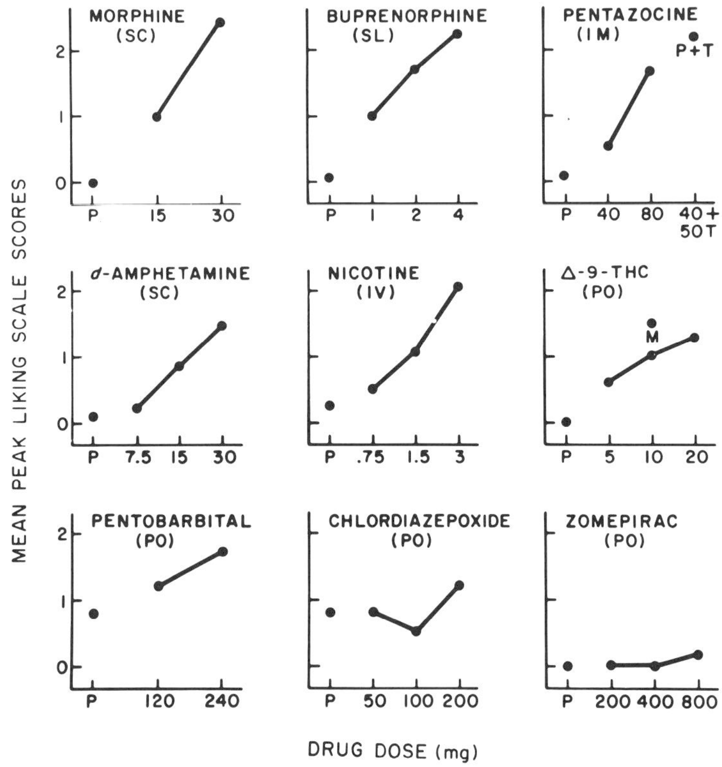

Fig 3. Mean scores on the "liking" scale of the Single Dose Questionnaire from subjects tested at the Addiction Research Center show that nicotine is similar to other drugs of abuse on this measure. The number of subjects in each group range from 6 (pentobarbital and chlordiazepoxide) to 13 (d-amphetamine). The high dose of each drug except zomepirac produced significant (P< 0.05) increases in scores above placebo data. The responses are peak responses which occurred after the drug had been given. The time of the peak response ranged from about 1 min (nicotine) to 5 h (buprenorphine). Morphine and zomepirac data are from the same group of subjects as are the pentobarbital and chlordiazepoxide data. The "P+T" point on the pentazocine graph is the score, by the same subjects, to 40 mg pentazocine given in combination with 50 mg tripelennamine (an antihistamine that produced a "liking" scale score of 0.9): the street combination called "T's and Blue's." The "M" point on the delta-9-THC graph is the score, from the same subjects, obtained after smoking a marijuana cigarette that contained 10 mg (1% by weight) delta-9-THC (from Jasinski, Johnson and Henningfield, 1984).

physiologic responses to nicotine given as either tobacco smoke or intravenous nicotine confirmed that nicotine was the critical pharmacologic compound that accounted for these effects of tobacco smoke. A subsequent study showed that nicotine's subjective and physiologic effects could be partially blocked by pretreating the subjects with mecamylamine (Henningfield, Miyasato, Johnson, and Jasinski, 1983); these results are analogous to those obtained in animal discrimination studies (see chapters in this volume by Rosecrans and by Stolerman).

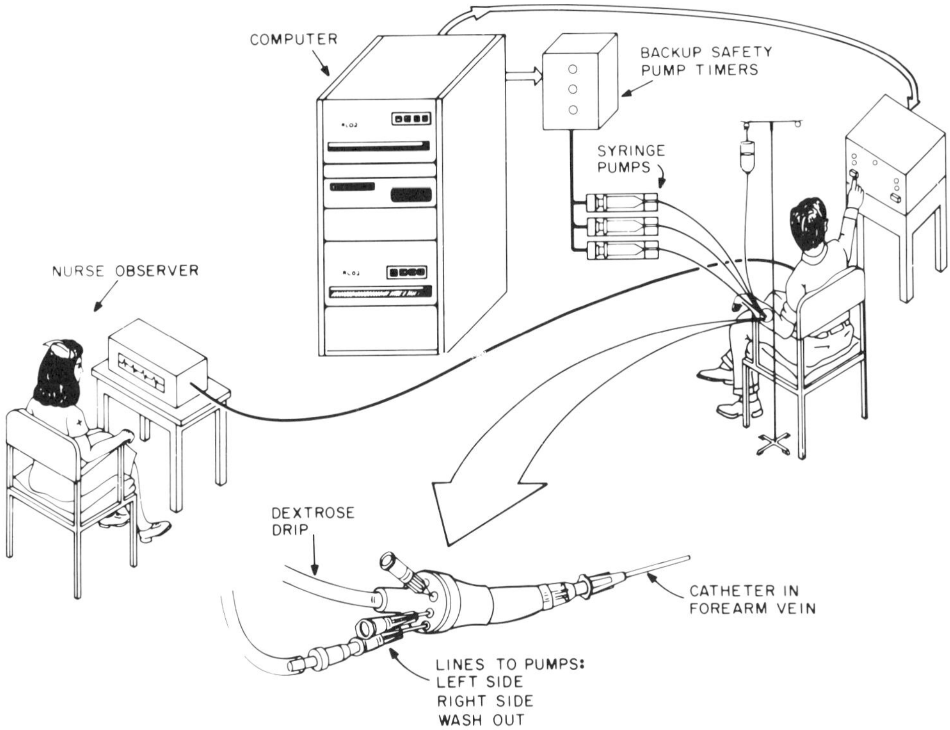

Fig. 4. A schematic diagram of the intravenous drug self-administration preparation used in human abuse liability studies. Before test sessions, the subject is equipped with a catheter in a forearm vein. Up to three pumps can be used to give the subject a choice among test solutions (e.g., placebo versus nicotine) to be delivered when the subject presses the lever.

Abuse Liability Studies of Nicotine: Self-Administration

A second approach to studying abuse liability involves measuring the conditions under which a subject will voluntarily take the substance (see review of human self-administration studies by Henningfield, Lukas, and Bigelow, 1986). Self-administration studies determine whether the drug serves as a positive reinforcer. They can be conducted in both animals and human subjects, thereby providing a means of establishing the biologic generality of the phenomena while controlling the possible confounding influence of personality, social, or cultural variables. A high degree of concordance between findings from animal and human studies has been established over a wide range of drugs (Griffiths and Balster, 1979; Griffiths, Bigelow, and Henningfield, 1980). This section will focus on the results from studies using human volunteers. Figure 4 shows a schematic diagram of the preparation used to study intravenous nicotine self-administration.

In one such study, volunteers were tested during three-hour sessions in which 10 presses on a lever resulted in the intravenous (iv) injection of either nicotine or placebo (Henningfield, Miyasato, and Jasinski, 1983). Subjects were not permitted to smoke cigarettes for one hour before or during the study. A variety of safeguards ensured the safety of the subjects. The main finding was that all six subjects voluntarily self-administered nicotine. Patterns of self-administration (injections)

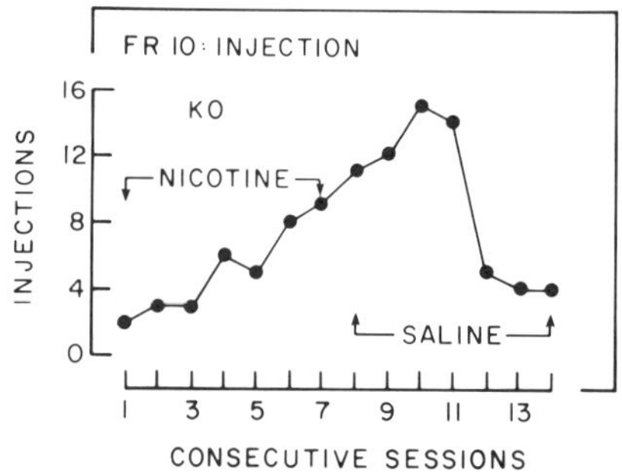

Fig. 5. Subject KO was a cigarette smoker without a history of drug abuse. Ten lever presses on one lever resulted in delivery of either saline or nicotine; pressing the other lever had no programmed consequence. Nicotine was available (1.5 mg nicotine per injection) during seven consecutive sessions; then saline was substituted for an additional seven sessions. Number of injections per session are shown on the x-axis. (From Henningfield and Goldberg, 1983B).

were similar to those observed when human subjects smoke cigarettes and when rhesus monkeys take intravenous amphetamine injections in comparable experimental situations (Griffiths, Bigelow, and Henningfield, 1980).

One of the six subjects described above was not a drug abuser. As shown in Fig. 5, the pattern of acquistion of nicotine self-administration developed gradually, over several sessions; double-blind substitution of saline for nicotine resulted in extinction of the self-injection behavior. Similarly, when subjects were given simultaneous access to both nicotine and placebo at the same time (by pressing alternate levers), they chose nicotine, confirming that nicotine had come to serve as a positive reinforcer (Henningfield and Goldberg, 1983a). These data indicate that the pharmacologic activity of nicotine itself was necessary for the maintenance of the behavior.

Nicotine self-administration has also been studied in a variety of nonhuman species, under a variety of experimental conditions (see reviews, Goldberg and Spealman, 1982; Goldberg, Spealman, Risner and Henningfield, 1983; Henningfield and Goldberg, 1983b). These studies confirm that nicotine may function as a positive reinforcer in five nonhuman species, under a variety of conditions; furthermore, the resulting findings are similar to those observed when other dependence producing drugs are similarly tested. It should also be noted, however, that the conditions under which nicotine serves as a reinforcer for animals are more restricted than with morphine or cocaine. It is plausible that nicotine self-administration, in the form of tobacco smoke inhalation or smokeless tobacco use, provides ideal confluences of conditions for the establishment and maintenance of nicotine dependence in humans (Henningfield, 1984a), since peripheral taste and olfactory stimuli are immediate and abundant. Consistent with this hypothesis are data from studies conducted using squirrel monkeys in which the highest rates of responding leading to iv nicotine self-administration were maintained under second-order schedules in which responding by the subjects produced frequent presentations of brief light stimuli that were intermittantly associated with nicotine injection (Goldberg, Spealman and Goldberg, 1981; Spealman and Goldberg, 1982).

Implications of Abuse Liability Studies

The results of these studies provide direct evidence that nicotine itself, apart from its being presented in combination with all of the sensory properties of tobacco smoke, is an abusable drug. That is, nicotine meets the critical criteria of being psychoactive, producing euphoriant effects, and serving as a reinforcer. These findings confirm that the role of nicotine in cigarette smoking is similar to the roles played by other drugs in the maintenance of other kinds of substance self-administration, e.g., morphine in opium use, cocaine in coca leaf use, and ethanol in alcoholic beverage consumption (see public policy implications in Pollin, 1984; US DHHS 1983; US DHHS, 1984).

Physiologic Dependence Potential of Nicotine

As reviewed earlier, many studies have examined various aspects of the physiologic dependence potential of nicotine. Until recently, however, no single study had been conducted in which the same general strategies used to assess physiologic dependence potential of opioids and sedatives had been used to study these effects in nicotine (see a review and critique in Henningfield, 1984a). For instance, previous research examining tobacco withdrawal symptoms has been either retrospective, employed only a limited number of measures, employed no baseline period, or failed to include a control group. An important series of studies, recently completed at the University of Minnesota, has addressed many of these issues. The Minnesota studies determined a reliable battery of indices of tobacco withdrawal by examining physical, subjective, and behavioral effects of tobacco deprivation. The results of three of these studies are summarized below.

The first such study involved 27 smokers who resided for seven days on a research ward (Hatsukami, Hughes, Pickens, and Svikis, 1984). Following baseline, they were assigned to either abstain from smoking or to continue smoking for four days. A battery of physiologic, subjective, and behavioral measures were obtained and analyzed. The second study was conducted on a nonresidential basis to assess tobacco withdrawal in the nonlaboratory environment (Hughes and Hatsukami, 1986). In this study, signs and symptoms of tobacco withdrawal were measured in 100 smokers. Following baseline determination, subjects were randomly assigned to either nicotine or placebo gum which was to be chewed at their own rate. They returned on three different occasions for assessment. The third study assessed the reliability of the tobacco withdrawal syndrome within subjects (Hatsukami, Hughes, and Pickens, 1985). This study employed a modified within-subject experimental design (baseline smoking, tobacco deprivation, return to baseline smoking, and tobacco deprivation assessed in each subject).

The results of these studies demonstrated that a syndrome of withdrawal occurs reliably and consistently in chronic smokers after tobacco deprivation which includes the following: (1) decreased heart rate, (2) increased caloric intake/eating, (3) increased number of awakenings during sleep, (4) increased desire to smoke cigarettes, and (5) increased confusion. Other changes which were found to occur in some but not all studies included increased irritability and decreased vigor. A prospective examination of data from both residential and nonresidential studies revealed no statistically significant differences between men and women in either number or severity of tobacco withdrawal symptoms (Svikis, Hatsukami, Hughes, Carroll and Pickens, 1986).

Another series of studies in which the physiologic dependence potential of nicotine is being more systematically explored is nearing completion at the Addiction Research Center in Baltimore, by Henningfield and his colleagues (see one result in Figure 1). Preliminary results of the ARC studies confirm and extend those of the Minnesota studies by the use of cognitive performance measures, electrophysiologic measures of brain function, and evaluation of nicotine gum efficacy in attenuating the tobacco abstinence syndrome.

Implications of Dependence Potential Studies

These recent studies confirm and extend the findings of earlier investigations demonstrating that tobacco had the potential to produce physiologic dependence. It is now clear that the syndrome is orderly, is due to the administration and withdrawal of nicotine, and may be attenuated by administration of other forms of nicotine administration (e.g., nicotine gum). The signs of nicotine withdrawal are more subtle than those which mark opioid and sedative withdrawal, but they are not necessarily less important to the individual; for instance, withdrawal effects such as mood changes, performance deficits, and weight gain are of considerable potential disruption to the normal functioning of the individual. It is anticipated that just as detoxification and treatment of opioid and sedative dependence have benefited from improved understanding of these syndromes of withdrawal, so also may detoxification and treatment of tobacco withdrawal benefit.

PHARMACOLOGIC TREATMENT STRATEGIES FOR TOBACCO DEPENDENCE

Since tobacco use may occur as a form of drug dependence, strategies for treatment of other forms of drug dependence should be applicable. Most extant information and strategies for treatment of tobacco use are almost exclusively behaviorally and psychologically oriented. Experience in the treatment of these other kinds of drug dependence has made clear the importance of also addressing the pharmacologic components of dependence (Graboswki, Stitzer and Henningfield, 1984). This conclusion is strengthened by the observations that persons being treated for opioid dependence regard tobacco to be as necessary to them as methadone (Blumberg, Cohen, Dronfield, Mordecai, Roberts and Hawks, 1974), and that persons successfully treated for other kinds of drug dependence are often unable to give up tobacco (*Journal of Clinical Psychiatry*, Supplement 1984).

Pharmacologic treatment of opioid dependence is basically of three types: substitution therapy (e.g., methadone for opioid dependence), in which a more manageable form of the drug is provided according to a prearranged maintenance protocol; blockade therapy (e.g., naltrexone for opioid dependence), in which the effects of the abused drug are blocked by pretreatment with an antagonist; and less specific supportive therapy, in which the patient is treated symptomatically, as by the temporary use of clonidine and other agents during opioid detoxification (Grabowski, Stitzer and Henningfield, 1984). All three approaches have been used in the treatment of cigarette smoking with varying degrees of success (see Henningfield, 1984b; Tennant, Tarver and Rawson, 1983; Glassman, Jackson, Walsh, Roose and Rosenfeld, 1984). The newest widely used treatment, and the first recognized as efficacious by the FDA, is directly modeled after treatment of opioid dependence by methadone substitution; namely, substitution of tobacco with nicotine delivered via polacrilex (gum) (American Hospital Formulary Service, 1984). Nicotine gum-based treatment of cigarette smoking is potentially efficacious though success rates are limited by a variety of factors (see Grabowski and Hall, 1985). Such findings are analogous to those obtained when intravenous opioid users are treated with other opioids given via other routes (Grabowski, Stitzer and Henningfield, 1984). For instance, methadone administration may reverse objective signs and symptoms of opioid withdrawal while still leaving the patient feeling partially treated and likely to relapse if not combined with a behavioral treatment component (Grabowski, Stitzer and Henningfield, 1984). Tobacco dependence does not appear any less resistant to modification than dependence to opioids, sedatives, and alcohol, but the experience gained through prior studies of treatment of these forms of drug dependence has clearly facilitated the development of treatment strategies for tobacco dependence.

SUMMARY

Historical observations suggested that tobacco use leads to dependence in much the same way as opium and alcohol use leads to dependence. Despite the complexities of the process of tobacco smoke self-administration, systematic analysis has confirmed that the resulting dependence is similar to that produced and maintained by other dependence-producing drugs in both human and animal subjects. Animals can learn to discriminate nicotine from other substances due to its effects in the central nervous system. These effects are related to dose as is also the case with other drugs of abuse. Other studies have shown that nicotine functions as a reinforcer under a variety of conditions. Such data confirm that nicotine can function in all of the capacities that characterize a drug with a liability to widespread abuse. Additionally, as is the case with most other drugs of abuse, nicotine produces effects that the user considers desirable. These effects are due to the nicotine and not simply the vehicle of delivery (tobacco or tobacco smoke).

The methods and criteria used to establish these similarities are identical to those used for other drugs suspected of having the potential to produce abuse and physiologic dependence. Specifically, the abuse liability studies (and to a lesser degree, the physiologic dependence potential studies) were conducted according to methods used to evaluate the abuse liability of morphine-like drugs in humans (Jasinski, 1977; Brady and Lukas, 1984; Jasinski, Johnson and Henningfield, 1984). The following conclusions may be drawn from these studies: (1) nicotine is psychoactive, producing transient dose-related changes in mood and feeling; (2) it is a euphoriant, producing dose-related increases in scores on standards measures of euphoria; (3) it is a reinforcer in both human and animal intravenous self-administration paradigms, functioning as do other drugs of abuse; additionally, (4) nicotine through smoking produces the same effects; and (5) nicotine causes neuroadaptation that leads to tolerance and physiological dependence. Taken together, these results confirm the hypothesis that the role of nicotine in the compulsive use of tobacco is similar to the role of morphine in the compulsive use of opium derivatives or of cocaine in the compulsive use of coca derivatives. These data are consistent with the fact that more than one-quarter of adult Americans use tobacco while most of them acknowledge that the effects are harmful.

The commonalities shared by tobacco and other dependence-producing substances, the pharmacologic role of nicotine dose in regulating tobacco intake, the dependence liability of nicotine, and finally, the direct evidence that orally delivered nicotine retains critical characteristics of a dependence-producing drug, provides evidence that smoked and smokeless tobacco products are dependence producing. More specifically, these nicotine-delivering substances produce the same functional relations between drug administration and measures of abuse liability and dependence potential as do the prototypic dependence-producing drugs, morphine and cocaine. A few other characteristics of tobacco products in general may operate to further enhance the number of individuals afflicted by nicotine dependence: (1) nicotine-delivering products are widely available and relatively inexpensive; (2) the self-administration of such products is relatively well tolerated by society; (3) nicotine produces a variety of individual-specific therapeutic actions such as mood and performance enhancement; (4) the brief effects of nicotine ensure many opportunities for conditioning to occur as the behavior is frequently associated with the effects of nicotine and concomitant environmental stimuli.

REFERENCES

American Hospital Formulary Services. Drug Information Supplement B. Bethesda, Maryland: American Society of Hospital Pharmaceuticals, pp 39B-44B, 1984.

Austin, G.A.: Perspectives on the History of Psychoactive Substance Use. National Institute on Drug Abuse Research Monograph 24. U.S. Gov't. Printing Office, Washington, D.C., 1978.

Beckett, A.H. and Triggs, E.J.: Enzyme induction in man caused by smoking. Nature 216:587, 1967.

Blumberg, H.H., Cohen, S.D., Dronfield, B.E., Mordecai, E.A., Roberts, J.C. and Hawks, D.: British opiate users: I. People approaching London drug treatment centers. Int J Addict 9:1-23, 1974.

Brady, J.V. and Lukas, S.E. (Eds): Testing Drugs for Physical Dependence Potential and Abuse Liability. National Institute on Drug Abuse Research Monograph 52, U.S. Gov't Printing Office, Washington, D.C., 1984.

Brecker, E.M. and the Editors of Consumers Union Reports: Licit and Illicit Drugs. Boston: Little, Brown and Company, 1972.

Chait, L.D. and Griffiths, R.R.: Smoking behavior and tobacco smoke intake: Response of smokers to shortened cigarettes. Clin Pharmacol Ther 32:90-97, 1982.

Clarke, P.B.S. and Kumar, R.: The effects of nicotine on locomotor activity in non tolerant and tolerant rats. Br J Pharmacol 78:329-337, 1983.

Connolly, G.N., Winn, D.M., Hecht, S.S., Henningfield, J.E., Hoffman, D. and Walker, B.: The re-emergence of smokeless tobacco. N Engl J of Med 314: 1020-1027, 1986.

Domino, E.F.: Behavioral, electrophysiological, endocrine and skeletal muscle actions of nicotine and tobacco smoking. In: A. Remond and C. Izard (Eds), Electrophysiological Effects of Nicotine. Amsterdam: Elsevier, pp. 133-146, 1979.

Fagerstrom, K.: Measuring degree of physical dependence to tobacco smoking with reference to individualization to treatment. Addict Behav 3:235-241, 1978.

Fagerstrom, K.: Tobacco smoking, nicotine dependence and smoking cessation. Doctoral thesis, Uppsala University, 1981.

Fagerstrom, K.O., and Gotestam, K.G.: Increase in muscle tonus after tobacco smoking. Addict Behav 2:203-206, 1977.

Faulkerborn, Y., Larsson, C. and Nordberg, A.: Chronic nicotine exposure in rat: A behavioral and biochemical study of tolerance. Drug Alcohol Depend 8:51-60, 1981.

Glassman, A.H., Jackson, W.K., Walsh, T., Roose, S.P. and Rosenfeld, B.: Cigarette craving, smoking withdrawal, and clonidine. Science 226:864-866, 1984.
Goldberg, S.R. and Spealman, R.D.: Maintenance and suppression of behavior by intravenous nicotine injections in squirrel monkeys. Fed Proc 41:216-220, 1982.
Goldberg, S.R., Spealman, R.D. and Goldberg, D.M.: Persistent behavior at high rates maintained by intravenous self-administration of nicotine. Science 214:573-575, 1981.
Goldberg, S.R., Spealman, R.D., Risner, M.E. and Henningfield, J.E.: Control of behavior by intravenous nicotine injections in laboratory animals. Pharmacol Biochem Behav 19:1011-1020, 1983.
Grabowski, J. and Hall, S.M. (Eds.): Pharmacological Adjuncts in Smoking Cessation. National Institute on Drug Abuse Research Monograph 53, U.S. Gov't Printing Office, Washington, D.C., 1985.
Grabowski, J., Stitzer, M.L. and Henningfield, J.E. (Eds.): Behavioral Intervention Techniques in Drug Abuse Treatment. National Institute on Drug Research Monograph 46, U.S. Gov't Printing Office, Washington, D.C., 1984.
Griffiths, R.R. and Balster, R.L.: Opioids: Similarity between evaluations of subjective effects and animal self-administration results. Clin Pharmacol Ther 25:611-617, 1979.
Griffiths, R.R., Bigelow, G.E. and Henningfield, J.E.: Similarities in animal and human drug taking behavior. In N.K. Mello (Ed), Advances in Substance Abuse: Behavioral and Biological Research. Greenwich, Connecticut:JAI Press, pp. 1-90, 1980.
Griffiths, R.R. and Henningfield, J.E.: Pharmacology of cigarette smoking behavior. Trends Pharmacol Sci 3:260-263, 1982.
Gritz, E.R.: Smoking behavior and tobacco abuse. In N.K. Mello (Ed). Advances in Substance Abuse. Greenwich, Connecticut:JAI Press, pp. 91-158, 1980.
Hatsukami, D.K., Hughes, J.R. and Pickens R.W.: Characteristics of tobacco abstinence: physiological and subjective effects. In: Pharmacological Adjuncts in Smoking Cessation (Eds), J. Grabowski and S.M. Hall. National Institute on Drug Abuse Research Monograph 53. U.S. Gov't Printing Office, Washington, D.C., 1985.
Hatsukami, D.K., Hughes, J.R., Pickens, R.W. and Svikis, D.: Tobacco withdrawal symptoms: An experimental analysis. Psychopharmacology 84:231-236, 1984.
Henningfield, J.E.: Behavioral Pharmacology of Cigarette Smoking. In T. Thompson, P.B. Dews, and J.E. Barrett (Eds) Advances in Behavioral Pharmacology, Vol. 4, 131-210., 1984(a).
Henningfield, J.E.: Pharmacologic basis and treatment of cigarette smoking. J Clin Psychiatry, 45:24-34, 1984(b).
Henningfield, J.E. and Goldberg, S.R.: Nicotine as a reinforcer in human subjects and laboratory animals. Pharmacol Biochem Behav 19:989-992, 1983(a).
Henningfield, J.E. and Goldberg, S.R.: Control of behavior by intravenous nicotine injections in human subjects. Pharmacol Biochem Behav 19:1021-1026, 1983(b).
Henningfield, J.E. and Goldberg, S.R.: Stimulus properties of nicotine in animals and human volunteers: A review. Behavioral Pharmacology: The Current Status. pp. 433-449, 1985.
Henningfield, J.E., Griffiths, R.R. and Jasinski, D.R.: Human dependence on tobacco and opioids: Common factors. In T. Thompson and C.E. Johanson (Eds), Behavioral Pharmacology of Human Drug Dependence National Institute on Drug Abuse Research Monograph 37. Washington, D.C.: U.S. Govt. Printing Office, 1981.
Henningfield, J.E., Lukas, S.E. and Bigelow, G.E. Human studies of drugs as reinforcers. In: S.R. Goldberg and I.P. Stolerman (Eds),

Behavioral Analysis of Drug Dependence. New York: Academic Press, pp. 69-122, 1986.

Henningfield, J.E., Miyasato, K. and Jasinski, D.R.: Cigarette smokers self-administer intravenous nicotine. Pharmacol Biochem Behav 19:887-890, 1983.

Henningfield, J.E., Miyasato, K., Johnson, R.E. and Jasinski, D.R.: Rapid physiologic effects of nicotine in humans and selective blockade of behavioral effects by mecamylamine. Problems of Drug Dependence National Institute on Drug Abuse Research Monograph 43. Washington, D.C.: U.S. Govt. Printing Office, 1983.

Henningfield, J.E., Miyasato, K. and Jasinski, D.R.: Abuse liability and pharmacodynamic characteristics of intravenous and inhaled nicotine J Pharmacol Exp Ther 234:1-12, 1985.

Hughes, J.R. and Hatsukami, D.: Signs and symptoms of tobacco withdrawal Arch Gen Psychiatry 43:289-294, 1986.

Jaffe, J.H.: Drug addiction and drug abuse. In: A.G. Gilman, L.S. Goodman, T.W. Rall and F. Murad (Eds), Goodman and Gillman's The Pharmacological Basis of Therapeutics. New York: Macmillan, 532-581, 1985.

Jaffe, J.H. and Kanzler, M.: Smoking as an addictive disorder. In: Cigarette Smoking as a Dependence Process National Institute on Drug Abuse Research Monograph 23, Washington, D.C.: U.S. Govt. Printing Office, pp. 4-23, 1979.

Jarvik, M.: The role of nicotine in the smoking habit. In: W.A. Hunt (Ed.). Learning Mechanisms in Smoking. Chicago: Aldine, pp. 155-190, 1970.

Jarvik, M.: Further observations on nicotine as the reinforcing agent in smoking. In: W.L. Dunn (Ed). Smoking Behavior: Motives and Incentives. Washington, D.C.: Winston, pp. 33-49, 1973.

Jasinski, D.R.: Assessment of the abuse potentiality of morphine-like drugs (methods used in man) In: W.R. Martin (Ed.) Handbook of Experimental Pharmacology. Heidelberg, West Germany: Springer-Verlag, 197-258, 1977.

Jasinski, D.R., Johnson, R.E. and Henningfield, J.E.: Abuse liability assessment in human subjects. Trends Pharmacol Sci 5:196-200, 1984.

Jones, R.T., Farrell, T.R. and Herning R.I.: Tobacco smoking and nicotine tolerance. In: Self-administration of Abused Substances: Methods for Study. National Institute on Drug Abuse Research Monograph 20, Washington, D.C.: U.S. Govt. Printing Office, 202-208, 1978.

Journal of Clinical Psychiatry, 45:12(2) Double diagnosis: Double delemma. The polyaddictions, alcoholism, substance abuse, smoking, gambling, 1984.

Martin, W.R., W.R. Martin (Ed.), In: Handbook of Experimental Pharmacology: Vol. 45. Drug Addiction I. pp. 3-42. Berlin Heidelberg: Springer-Verlag, 1977.

Nemeth-Coslett, R. and Griffiths, R.R.: Determinants of puff duration in cigarette smokers: I. Pharmacol Biochem Behav 20:965-971. 1984(a).

Nemeth-Coslett, R. and Griffiths, R.R.: Determinants of puff duration in cigarette smokers: II. Pharmacol Biochem Behav 21:903-912, 1984(b).

Nemeth-Coslett, R. and Griffiths, R.R.: Effects of cigarette rod length on puff volume and carbon monoxide delivery in cigarette smokers. Drug Alcohol Depend 15:1-13, 1985.

Pollin, W.: The role of the addictive process as a key step in causation of all tobacco-related diseases. JAMA 252:2874, 1984.

Pomerleau, O.F. and Pomerleau, C.S.: Neuroregulators and the reinforcement of smoking: Towards a biobehavioral explanation. Neurosci Biobehav Rev 8:503-513, 1984.

Rosecrans, J.A.: Nicotine as a discriminative stimulus to behavior: Its characterization and relevance to smoking behavior. In: Cigarette Smoking as a Dependence Process. National Institute on Drug Abuse Research Monograph 23, Washington, D.C.: U.S. Govt. Printing Office, pp. 58-69, 1979.

Rosecrans, J.A.: Brain area cites of nicotine action: Evidence for non-cholinergic mechanisms. This volume, 1987.

Rosecrans, J.A. and Meltzer, L.T.: Central sites and mechanisms of action of nicotine. Neurosci Biobehav Rev 5:497-501, 1981.

Russell, M.A.H.: Cigarette smoking: national history of a dependence disorder. Br J Med Psychol 44:1-16, 1971.

Schmiterlaw, C.G., Hansson, E., Andersson G., Applegren, L.E. and Hoffman, P.C.: Distribution of nicotine in the central nervous system. New York Academy of Science 143:2-14, 1967.

Snyder, F. and Henningfield, J.E. Nicotine gum reverses cognitive performance deficits caused by tobacco abstinence. Manuscript in preparation, 1986.

Spealman, R.D. and Goldberg, S.R.: Maintenance of schedule-controlled behavior by intravenous injections of nicotine in squirrel monkeys. J Pharmacol Exp Ther 223:402-408, 1982.

Stitzer, M., Morrison, J. and Domino, E.F.: Effects of nicotine on fixed-interval behavior and their modification by cholinergic antagonists. J Pharmacol Exp Ther 171:166-177, 1970.

Stolerman, I.P. Discriminative stimulus properties in nicotine: Correlations with nicotine binding. This volume, 1987.

Stolerman, I.P., Bunker, P. and Jarvik, M.E.: Nicotine tolerance in rats: Role of dose and dose interval. Psychopharmacology 34:317-324, 1974.

Svikis, D.S., Hatsukami, D.K., Hughes, J.R., Carroll, K.M. and Pickens, R.W.: Sex differences in tobacco withdrawal syndrome. Addict Behav in press, 1986.

Tennant, F.S., Tarver, A.L. and Rawson, R.A.: Clinical evaluations of mecamylamine for withdrawal from nicotine dependence. In: Problems of Drug Dependence, National Institute on Drug Abuse Research Monograph, U.S. Gov't Printing Office, 1983.

U.S. Department of Health and Human Services: Drug Abuse and Drug Abuse Research. The First in a Series of Triennial Reports to Congress., 1984, pp. 85-104.

U.S. Department of Health and Human Services (HHS), Public Health Service. Why People Smoke Cigarettes (PHS Publication No. PHS 83-50195). Washington, D.C.: U.S. Gov't Printing Office, 1983.

Wesnes, K. and Warburton, D.M.: The effects of cigarettes of varying yield on rapid information processing performance. Psychopharmacology 82:338-342, 1984.

Wesnes, K. and Warburton, D.M.: Smoking, nicotine and human performance. Pharmacol Ther 21:189-208, 1983.

World Health Organization Technical Report Series, No. 407, Geneva, Switzerland, 1969.

Young, A.M. and Herling, S. Drugs as reinforcers: Studies in laboratory animals. In: S.R. Goldberg and I.P. Stolerman (Eds), Behavioral Analysis of Drug Dependence. New York: Academic Press, pp. 9-68, 1986.

NICOTINE ACTIONS IN MODELS OF LEARNING/MEMORY AND REWARD

Edgar T. Iwamoto[+], Paul J. Fudala[*],
William R. Mundy[*], and Edwin C. Williamson[+]

Dept. of Pharmacology, College of Medicine[+] Graduate Center for Toxicology, The Graduate School[*] and the Tobacco and Health Research Institute, University of Kentucky
Lexington, Kentucky 40536 USA

INTRODUCTION

Nicotine appears to be the primary substance responsible for maintaining the use by humans of tobacco products. Although it is readily demonstrated that nicotine is capable of inducing psychologic, physiologic, and pharmacologic changes in both animals and humans, these effects may not necessarily relate directly to the phenomenon of tobacco use in humans. Although many past experiments in both animals and humans have given rise to many isolated hypotheses as to why people use tobacco, there has been little integration of the results of these many, isolated experiments. This laboratory has been using the methods of strong inference and of multiple working hypotheses to address the question, "Why do people smoke?" This report will integrate the results from two aspects of the effects of nicotine: those effects of nicotine on learning and memory, and those effects of nicotine on the subjective state. We will use this integration to formulate a new hypothesis based upon the commonalities shared by learning/memory and reinforcement of behavior. This approach using multiple hypotheses will lead to a better understanding of the phenomenon of tobacco use in humans.

Perhaps the simplest answer for the question, "Why do people smoke?", is that nicotine induces a positive change of the affective state. It is important here to emphasize that the nicotine-subjective state does not have to be the massive euphoria or the stupendous sense of well-being that have been associated with heroin or cocaine. In contrast, the nicotine-subjective state may just be a very subtle difference in affect from one's usual demeanor. Whatever the subtle changes induced by nicotine, the behavioral effects are desirable enough to be sought after by tobacco users via long-term self-administration.

A recent model being used to estimate the reinforcing properties of drugs is the place-conditioning paradigm. The administration of a drug or saline is repeatedly paired with distinctive cues. It is hypothesized that pairing environmental cues with rewarding drugs should produce an approach to the cues; pairing cues with aversive substances should produce cue avoidance. Rats will spend more time in the compartment of a three-compartment alley that was previously paired with food, morphine, cocaine, amphetamine, or with subcortical injections of opioids and opioid peptides. Rats avoid compartments previously paired with lithium chloride, naloxone, ethanol, cholecystokinin, vasopressin, phencyclidine, or pentobarbital. This laboratory was the first to report the rewarding effects of nicotine in the place-conditioning paradigm (Fudala et al., 1985).

It is apparent that the basis of the place-conditioning method derives from associative learning theory. The subjective state induced by the drug, the unconditioned stimulus, becomes associated with neutral, environmental cues (conditioned stimuli) of the place-conditioning apparatus. When the animal is reexposed to the conditioned environmental stimuli during the testing situation, either an approach or an avoidance response is evoked by the conditioned stimuli; these responses are interpreted as place-preference or place-aversion responses, respectively.

It is also apparent that in the place-conditioning paradigm, the association of the drug-induced subjective state with environmental cues must be remembered. A recent work has suggested that the processes underlying place-conditioning depend on the memory-improving properties in addition to the affective properties of the particular reinforcer being tested (White and Carr, 1985). In particular, these workers posit that in order for a drug treatment to cause either place-preference or place-aversion, it must also facilitate memory for retention of the associations made during the conditioning sessions. Thus, current data indicate that drugs that are reinforcers of behavior must alter at least two different processes, affective state and memory.

We have been interested also in the effects of nicotine on learning and memory from the standpoint that nicotine is known to release vasopressin and that vasopressin is thought to be involved in the consolidation of memory (Mundy and Iwamoto, 1985). There have been very few experiments with nicotine using positively motivated tasks, such as food or water reward, to assess nicotine's effect on learning and memory. Since one of our multiple working hypotheses on why people smoke has been that nicotine improves learning and/or memory, we have conducted experiments testing the pharmacologic effects of nicotine in models of learning and memory using a positively reinforced task in rats.

In this chapter we will present some of our evidence regarding the rewarding nature of nicotine according to the place-conditioning paradigm and our unexpected finding that nicotine impairs learning/memory, but not task performance, after acute administration. We will then relate our findings to past and present knowledge and forward an hypothesis on how nicotine could cause both reward and impairment of learning memory. Finally, we will relate our findings to the original question of, "Why do people smoke?" and pose further questions to be answered by future preclinical and clinical experimentation.

METHODS FOR PLACE-CONDITIONING

Adult male Sprague-Dawley rats (Harlan Industries, Indianapolis, IN) weighing approximately 250-430 g at the time of use were initially quarantined for 10 days before being housed in groups of two for one to two weeks. The animals were then housed individually 1 to 2 days before the beginning of and throughout each experiment. Food and water were freely available in the home cages, and the animals were maintained on a 12 hr light/dark cycle.

The place-conditioning apparatus has been described previously (Fudala et al., 1985). The wooden apparatus had one three-walled chamber painted white with a triangular mesh floor 25 cm on each side, a smaller middle passageway (10 x 10 cm floor) painted gray, and a cuboidal chamber with black walls and a bar-grid floor (25 x 25 cm). All walls were 20 cm high, and the three chambers could be closed-off with sliding doors when necessary. The walls could be lifted and separated from the floor panel to allow the floors to be thoroughly washed after each animal-conditioning session. On testing day, interruptions of infrared beams, which passed through each of the white, gray and black chambers, were used to monitor animal location. Cumulative timing began when beams in the white or black chambers were interrupted. Immediate re-interruption of the white-chamber beam, for example, did not alter timing. However, subsequent breakage of both the white- and gray-chamber beams or of the gray-chamber beam alone stopped timing. Cumulative timing for the white and black chambers restarted when their respective beams were interrupted anew.

All experiments were carried out between 0800 and 1700 hrs in a small room (4 sq. m. floor area) illuminated by one shaded 60-watt light bulb. The reflected light intensity from the walls in the white chamber was approximately 16 lux and 4 lux in the black chamber. On conditioning days, the floor panel of the place-conditioning apparatus was washed and dried, and the mesh floor was swabbed with 0.06% acetic acid. After recording the body weights of 2 rats, 1 was injected subcutaneously (SC) with saline and placed in the closed (doors in place) white chamber, and the other was injected SC with drug solution and placed in the closed black chamber. In some experiments the first injection was followed 1 min later by a second injection. Total conditioning time was 20 min per day. The next day, these 2 rats switched positions and treatments. Over 6 days of conditioning, a given rat in the drug/saline-treatment group received a total of 3 saline/white-chamber pairings and 3 drug/black-chamber pairings. In contrast, a rat in the saline/saline-treatment group received 3 saline/white-chamber pairings and 3 saline/black-chamber pairings. The daily order of injecting the 32 rats of a typical experiment was randomized over the 6 days of conditioning. On day 7, preference testing day, one animal was placed in the closed gray chamber, the sliding doors were removed, and the amount of time (in sec) spent in the white, gray, and black chambers during the 900-sec testing period was recorded automatically.

Nicotine base (> 98% purity, Eastman Kodak) was dissolved in phosphate-buffered saline and dose was expressed in terms of the free base assuming a density of 1 mg/microliter. All injections were made subcutaneously (SC).

<u>Treatment of the Data</u>

The statistic used in all of the experiments was the residence ratio (RR), where RR = (W-B)/(W+B). For each rat, W equals the time in seconds spent in the white chamber, and B equals the time in seconds spent in the black chamber on test day. For the calculation of the NIC dose-response, an estimate of the interval containing 95% of the control values (95% CI) was obtained according to the formula: 95% CI = mean RR controls ± 1.96 x SD (standard deviation from the mean). Using the RRs calculated for each rat in the NIC treatment group, the animals at each dose level were categorized as showing reward, aversion, or no difference from vehicle control according to the following assignments:

REWARD: Observed RR < lower limit of the 95% CI
AVERSION: Observed RR > upper limit of the 95% CI
NO RESPONSE: Observed RR within the 95% CI
The % REWARD values obtained were analyzed by linear regression.

The data from the other experiments were initially analyzed by ANOVA. Subsequent pairwise comparisons were then performed as necessary.

PLACE-CONDITIONING RESULTS

Dose-Response of Nicotine in the Place-Conditioning Paradigm

Nicotine, when administered just prior to placement into the conditioning boxes, produced a dose-dependent place-preference but not place-aversion within the dose range tested (0.001-1.5 mg/kg). In rats which exhibited a place preference, the conditioned response was a function of dose between 0.1 and 0.8 mg/kg (Table 1). The percent reward values (a measure of the proportion of animals at each dose level exhibiting a significant place preference) ranged from 6 to 31%. No animals at the 0.001, 0.025 or 0.05 mg/kg dose level were classified as showing reward or aversion. The 1.2 and 1.5 mg/kg dose levels were not used to estimate the regression due to the apparent decreasing response at these doses. The regression equation for the predicted percent response is given by: % REWARD = 0.04 + 0.38 (DOSE). The test for significant regression yielded: $F(1,3) = 47.4$, $p<0.01$. The coefficient of determination of the regression line $R^2 = 0.93$.

Effects of Time of Nicotine or Saline Administration Prior to Conditioning Sessions

The nicotine cue which conditions place-preference occurs within 60 min after its administration (Table 2). The data in Table 2 represent the mean RRs (± SE) from groups of rats (N=16) administered NIC 0.8 mg/kg or SAL (1 ml/kg) at zero, 60 or 120 min prior to conditioning sessions. The data were analysed using a 2-way ANOVA. No significance was found for the factor associated with the treatment prior to conditioning (NIC or SAL). Significance was found for the factor associated with the week of conditioning, $F(4,150) = 2.91$, $p<0.025$ and for the treatment x week interaction, $F(4,150) = 2.65$, $p<0.04$. To interpret the interaction, the Least Squares Means analysis was used to compare the NIC treatment groups to their corresponding controls. The only difference found was at zero min prior to conditioning, $t(150) = 2.86$, $p<0.005$.

Effects of Single Daily Nicotine Injections between Repeated Conditioning and Testing Cycles

The data in Table 3 represent the mean RRs (± SE) from groups of rats

Table 1. Dose-Response of Nicotine in the Place-Conditioning Paradigm

Nicotine-Dose (mg/kg)	N	Percent Reward
0.001	16	0%
0.025	32	0%
0.050	16	0%
0.1[a]	16	6%
0.2[a]	16	12%
0.4[a]	16	18%
0.6[a]	16	31%
0.8[a]	16	31%
1.2[a]	16	18%
1.5	16	6%

[a]Percent reward was a function of dose between 0.1 and 0.8 mg/kg. There were two control groups of 64 rats each. The two 95% CI's used in these studies were: (-0.25 to 0.64) and (-0.42 to 0.54).

Table 2. Effect of Time of Nicotine or Saline Administration Prior to Conditioning (N = 16 per value).

Time Before Conditioning	Treatment	Mean Residence Ratio
0 min	Saline	0.10 ± 0.06
	Nicotine	-0.14 ± 0.08[a]
60 min	Saline	0.11 ± 0.04
	Nicotine	0.22 ± 0.05
60 min	Saline	0.11 ± 0.05
	Nicotine	0.08 ± 0.07
60 min	Saline	0.13 ± 0.07
	Nicotine	0.11 ± 0.06
120 min	Saline	0.10 ± 0.05
	Nicotine	0.17 ± 0.06

[a]Negative RR = "REWARD", $p<0.005$. Animals in this zero min group were injected with either saline or nicotine and then placed immediately into the conditioning chambers.

(N=15 or 16) administered NIC 0.6 mg/kg or SAL, respectively, over two conditioning and testing cycles. Between cycles, rats were administered single daily injections of nicotine 0.6 mg/kg or received no treatments. The data were analyzed using a 2-way repeated measures ANOVA with the factor treatment (NIC or SAL) repeated over the factor, cycle of conditioning. For those rats that received daily nicotine injections between conditioning cycles, there was a significant difference between NIC and SAL groups, $F(1,30) = 13.87$, $p<0.0009$. There were no differences between the first and second cycle measurements and there was no treatment x cycle interaction. For those rats that received no treatment between conditioning cycles, there was also a significant difference between NIC and SAL groups, $F(1,29) = 9.93$, $p<0.005$. There were no differences between the first and second cycle measurements and there was no treatment x cycle interaction. For all groups tested, rats conditioned with nicotine showed a significant place-preference when compared to their corresponding controls regardless of conditioning cycle or treatment between cycles.

Nicotine Administrations Immediately After Place-Conditioning Sessions

Delay conditioning with nicotine induced a place-avoiding response in the place-conditioning paradigm in rats (Table 4). Instead of being injected immediately before placement into the place-conditioning apparatus (as in Tables 1, 2, and 3), rats first spend 20 min inside either the white or black conditioning chambers and are then injected SC with vehicle, 0.2, 0.4, 0.8 mg/kg of nicotine 5 min after removal from the apparatus. Table 4 indicates that the vehicle control group had a slight baseline preference for the black chamber; negative Residence Ratio's larger than control denote aversion in this experiment. The results also indicate that delay conditioning with the higher doses of nicotine, 0.4 and 0.8 mg/kg, induced a significant place aversion on test day. We are currently characterizing this very interesting discovery (Fudala and Iwamoto, 1986).

Table 3. Effect of Single, Daily Nicotine Injections Between Two Conditioning and Testing Cycles.

Conditioning Group	Mean Residence Ratio After First Conditioning and Testing Cycle	Intervening Two-week Treatment	Mean Residence Ratio After Second Conditioning and Testing Cycle
Saline	0.12 ± 0.05	No Treatments	0.13 ± 0.06
Nicotine[a]	-0.08 ± 0.04[b]		-0.05 ± 0.06[b]
Saline	0.07 ± 0.05	Nicotine, 0.6 mg/kg SC, Daily for Two Weeks	0.11 ± 0.05
Nicotine	-0.15 ± 0.07[b]		-0.08 ± 0.05[b]

[a]N = 15; all other groups, N = 16. The conditioning dose of nicotine was also 0.6 mg/kg SC.
[b]Significantly different from the saline group immediately above as indicated in text.

Experiments were carried out in modular operant test cages (E10-10, Coulbourn Instruments, Lehigh Valley, PA) housed in sound-attenuating enclosures. The test cages were equipped with one pellet delivery trough located in the center and 2 cm above the grid floor, one retractable lever 5 cm to the right and 3 cm above the floor, and triple cue lamps 12 cm above the lever. A centrally located house light 2 cm from the ceiling remained on during the session. Animal contact with the lever in the extended ("lever touch") or in the retracted position ("nose poke") was monitored via a high resistance contact input circuit. Two photoelectric detectors used to assess animal location were positioned with the beam axes parallel to the modular wall containing the retractable lever. The two beams, located 2 cm above the grid floor, were 6 ("front") and 27 cm ("rear") from the modular wall. The reflected light intensity within the test cages was approximately 10 lux.

METHODS FOR LEARNING/MEMORY EXPERIMENTS

Animals, Materials, and Procedure

Adult, male Sprague-Dawley rats, weighing 265 to 345 g at the time of the experiments, were obtained from Harlan Industries (Indianapolis, IN) and held in a quarantine room for 10 days before experiments began. The animals were food-deprived and maintained at 80% of their initial body weight with free access to water. Rats were housed under automatically controlled conditions with consistent temperature (21°C) and humidity (35-55%) on a 12h/12h light cycle with lights on at 0700 h. Experiments were done between 0730 and 1530 h. During the course of an experiment, body weights were maintained with the appropriate amount of Purina Rat Chow. Complete daily records of body weights were maintained.

Procedure for Acquisition of an Autoshaped, Lever-Touch Response

Solutions, prepared as previously described, were administered SC either 15 min before the autoshaping session or immediately after the session. An animal was placed into the test cage and given 10 trials - each trial was a lever presentation with retraction of the lever and

Table 4. Effect of Nicotine Delay Conditioning in the Place-Conditioning Paradigm.

Delay Conditioned Group [a]	Mean Residence Ratio	No. Rats in Aversion Category
Vehicle	-0.05 ± 0.06	0
Nicotine, 0.2 mg/kg	-0.04 ± 0.08	1
Nicotine, 0.4 mg/kg	-0.22 ± 0.08	3
Nicotine, 0.8 mg/kg	-0.26 ± 0.08[b]	5

[a]N = 8 rats per group. Vehicle or nicotine was administered SC 5 min after 20 min of conditioning in the white chamber; vehicle was administered to all rats 5 min after the 20 min conditioning sessions in the black chamber on alternating days. Conditioning continued for 8 days with a 15 min test session on day 9.
[b]Significantly different from vehicle control, $P<0.05$, Newman-Keuls multiple comparisons test.

pellet delivery after the rat had touched the lever or after 15 sec had elapsed. The lever was presented on a random interval 48 sec schedule (RI 48). Ten trials comprised a daily session and experiments were continued until a criterion of 10 lever touches per 10 lever presentations was attained (100% Maximum Response) or until 10 sessions (10 days) had elapsed. The data are presented as Mean Number of Lever Touches versus Session Days. Significant effects of dose were assessed by repeated measures analysis of variance.

RESULTS OF NICOTINE EFFECTS ON LEARNING/MEMORY

Presession nicotine significantly decreased acquisition of the autoshaped lever-touch response (Fig. 1). Groups of 10 animals were injected SC with saline or nicotine (0.45 and 0.8 mg/kg) 15 min before each of the 10 daily sessions. Acquisition training continued until animals attained the criterion of 10/10 correct lever touches in one session, or until 10 days had elapsed. The saline data represent pooled data from two experiments. The rates of acquisition of the autoshaped lever-touch response in the group of animals administered 0.45 and 0.8 mg/kg of nicotine were significantly less than saline controls (repeated measures analysis of variance).

Postsession nicotine also significantly decreased the acquisition of the lever-touch response (Fig. 2). Groups of 10 animals were injected SC with saline or nicotine (0.45 and 0.8 mg/kg) immediately after completing their daily sessions of 10 discrete trials. Acquisition training continued until animals attained the criterion of 10/10 correct lever touches in one session (100% responding), or 10 days had elapsed. Repeated measures analysis of variance revealed that postsession nicotine (0.45 and 0.8 mg/kg) significantly decreased lever-touch acquisition compared to controls.

In animals with previous autoshaped behavior training, nicotine did not alter lever-touch performance (data not shown). Animals with a previous history of autoshaped behavior training were used after a period of 3 weeks during which animals received no training but remained

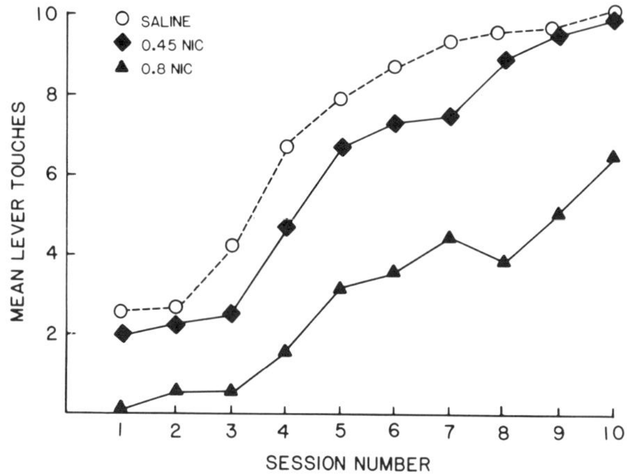

Fig. 1. Effect of Presession Nicotine or Saline Injections on Acquisition of Lever-Touch Responding in Rats.

food-deprived. All animals were then retrained to a criterion of 10/10 correct lever touches in one session for 3 consecutive days. The next day, animals were randomly divided into groups of 5 animals and injected SC with saline or nicotine (0.1, 0.25, 0.45, and 0.8 mg/kg) 15 min before a single session of 10 trials. The data indicated that in rats which already had acquired lever-touch responding, 0.1, 0.25, 0.45, and 0.8 mg/kg SC of nicotine failed to alter lever-touch performance.

DISCUSSION

Our data clearly demonstrate that nicotine administration just prior to place conditioning induces dose-related place preferences after doses

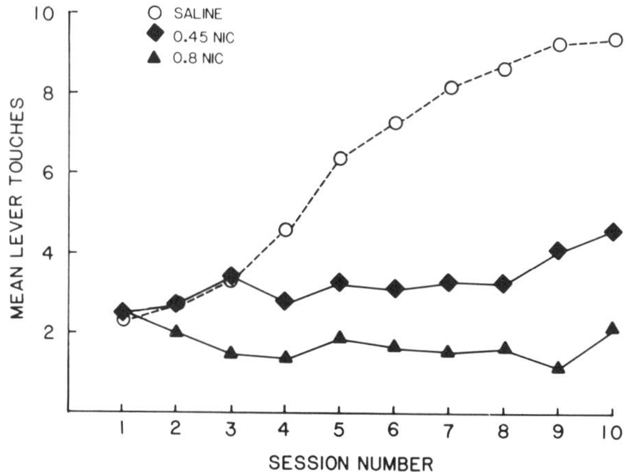

Fig. 2. Effect of Postsession Nicotine or Saline Injections on Acquisition of Lever-Touch Responding in Rats.

of 0.4 to 0.8 mg/kg SC. The nicotine-induced subjective state occurred within the first 60 min after SC administration. Tolerance to nicotine-induced place preference did not develop during the dosage regimens used. In contrast, nicotine induced a dose-related place aversion when it was administered 5 min after the drug/environment conditioning sessions. From our learning/memory experiments, it was discovered that acute nicotine administration (0.45 and 0.8 mg/kg) depresses the rate of acquisition of an autoshaped-learning task. The acute administration of nicotine, however, does not alter performance of tasks that have already been acquired. Finally, since postsession nicotine injections impaired learning/memory to a greater extent than presession nicotine injections, our data show that nicotine appears to have greater retroactive effects than proactive effects on the consequences of autoshaped-learning trials. We will now address the significance of our findings that nicotine induces "reward" and impairs learning/memory in relation to previously published work.

In 1974, German and Bowden reported that activation of central catecholamine systems was essential for the maintenance of intracranial self-stimulation (ICSS). One of the brain sites that support ICSS was the nigrostriatal dopaminergic system, a major component of the medial forebrain bundle. In 1969, Steiner et al. reported that rats escaped from previously self-produced rates of ICSS. Rats implanted with electrodes in the medial forebrain bundle were trained to self-stimulate by pressing a manipulandum; concurrently, their response patterns were recorded on magnetic tape. In subsequent brain-stimulation sessions in which the delivery of electrical stimulation was driven by the prerecorded tape, animals learned to terminate the stimulation. It appeared that the prerecorded, noncontingent stimulation was aversive, in contrast to the self-initiated, response-contingent ICSS which was rewarding.

In 1976, Lichtensteiger et al. reported that 1 mg/kg of nicotine SC in male rats induced a transient (1 min) decrease in the rate of firing of dopaminergic cells of the substantia nigra which was followed by an intense increase in the rate of firing which lasted at least 30 min. This increase in firing rate was accompanied by a marked increase in the cellular fluorescence intensity of the dopaminergic cells of the substantia nigra. It was postulated that nicotine increases the neuronal activity of the nigrostriatal pathway resulting in the release of dopamine from nerve terminals in the caudate nucleus; in fact, nicotine increases the levels of HVA in the caudate of nicotine-treated rats (Lichtensteiger et al., 1976).

In 1976, Fibiger and Phillips reported that electrical stimulation of the nigrostriatal bundle disrupted learning/memory in rats engaged in acquiring a passive avoidance task. In contrast, Major and White (1978) reported that electrical stimulation of the nigrostriatal pathway facilitated memory consolidation of rats engaged in acquiring a one-trial appetitive learning task. Although these two papers appear to contradict each other, there was a crucial difference in methodology. Memory was disrupted if electrical stimulation was presented noncontingently; that is, if the intracranial stimulation was delivered via the electrodes in a pattern chosen by the experimenter (Fibiger and Phillips, 1976). In contrast, memory and retention were strengthened if ICSS was self-initiated by the animal after it had finished participating in the learning task (Major and White, 1978). It may be that contingent activation of the nigrostriatal pathway facilitates memory consolidation.

From the past findings enumerated above, and from our present data (Tables 1-4 and Figs. 1-2), we now forward a hypothesis integrating several sources of information: Nicotine-induced reward as evidenced by conditioned place preference and nicotine-induced impairment of learning/memory as evidenced by our autoshaped learning experiments are both mediated by the release of dopamine from nerve terminals of the nigrostriatal pathway.

It has long been known that the release of dopamine was associated with drugs that either support their own self-administration or are rewarding according to the place-conditioning paradigm. A corollary to our proposal is that dopamine-receptor blockers are expected to antagonize the rewarding effects of such drugs. In fact, the results of a pilot experiment indicate that spiperone, a dopamine-receptor antagonist, at doses of 0.03, 0.06 and 0.12 mg/kg co-administered with 0.8 mg/kg presession conditioning doses of nicotine, is able to block nicotine-induced place-preference in a dose-related manner (significant slope of the graph of spiperone-antagonism, $F(1,46) = 4.32$, $P = 0.043$).

A major paradox has been uncovered by our experiments: Nicotine can be "rewarding" if administered just prior to the place-conditioning sessions, but it also can be aversive when it is given just after place-conditioning. According to the reports cited above, we speculate that nicotine induces a sudden increase in dopamine release in the CNS within 5 to 10 min, and that this release may be responsible for the "reward" response as evidenced by presession, nicotine-induced place-preference. However, as nicotine continues to release dopamine (for a period of time greater than 30 min, see Lichtensteiger et al., 1976), the nicotine cue now becomes punishing, or "aversive." The sustained release of CNS dopamine caused by nicotine then may be the predominant cue responsible for the place-aversion conditioned by postsession injections of nicotine (Table 4). This nicotine-induced, noncontingent activation of the nigrostriatal path would also result in a disruption of memory according to the previous literature cited above. As an extension of this proposal and of our findings, it may be predicted that if nicotine were contingently self-administered after the learning session instead of noncontingently injected, memory may be enhanced instead of being impaired. Experiments investigating these possibilities are currently in progress.

How do our data relate to the smoking phenomenon? First, it must be realized that the doses of nicotine used in our studies are high; the dose of nicotine in rats which yields plasma levels commonly observed in human smokers of 35 mg/ml is approximately 0.1 mg/kg administered subcutaneously. With respect to the place-conditioning paradigm, 0.1 mg/kg SC of nicotine had little or no effect; however, it must be remembered that place-preference is induced after only 3 drug/environment pairings using our usual methods (Fudala et al., 1985). Perhaps more drug-conditioning sessions are required with the lower doses of nicotine. It is also possible that nicotine-induced place-preferences may be magnified by prior chronic exposure to nicotine; we are currently examining this hypothesis to see if chronic nicotine enhances or diminishes the reward value of the lower doses of nicotine. Secondly, our data on nicotine-induced reward can explain the long-term nature of human smoking behavior. Our data indicate that tolerance to the rewarding effects of nicotine does not develop. This is an important finding in that it explains the chronic nature of smoking behavior: people will still feel the positive effects of nicotine even after long-term use. Finally, it is tempting to speculate that the subjective effects of nicotine may change during acquisition, maintenance, and extinction of human smoking behavior. For example, the rewarding effects of nicotine may be moderate in intensity during acquisition (during the time individuals first experiment with smoking), diminish in intensity slightly and then plateau during maintenance (during which time smokers gradually increase cigarette consumption to a fixed level), and then suddenly increase in intensity during the initial stages of abstinence with a subsequent diminution of reward intensity when a smoker has "quit." In parallel with these phenomena, our data suggest that nicotine may initially disrupt learning/memory during acquisition of the smoking habit. Tolerance and dependence to the disruptive effects of nicotine on memory may develop such that the smoking individual now depends upon the continued presence of nicotine in order to concentrate, read, think, speak, and comprehend. During extinction from smoking behavior, learning/memory is initially disrupted until, at later stages of cigarette abstinence, a new equilibrium is reached because the former smoker has readjusted and relearned how to cope with day-to-day events without nicotine. These hypotheses can and should be tested in the clinic.

REFERENCES

Fibiger, H.C. and A.G. Phillips: Retrograde amnesia after electrical stimulation of the substantia nigra: Mediation by the dopaminergic nigroneostriatal bundle. Brain Research 116:23-33, 1976.

Fudala, P.J., K.W. Teoh and E.T. Iwamoto: Pharmacologic characterization of nicotine-induced conditioned place preference. Pharmacology Biochemistry and Behavior 22:237-241, 1985.

Fudala, P.J. and E.T. Iwamoto: Nicotine-induced conditioned place preference: Dose-response, temporal requirements and lack of tolerance development. Society for Neuroscience (Abst.) 11:1294, 1985.

Fudala, P.J. and E.T. Iwamoto: A novel effect of nicotine: Delay-conditioned aversion in the place-preference paradigm. Pharmacologist (Abst.) 28:190, 1986.

German, D.C. and D.M. Bowden: Catecholamine systems as the neural substrate for intracranial self-stimulation: a hypothesis. Brain Research 73:381-419, 1974.

Iwamoto, E.T. and W.R. Mundy: A study of nicotine, desglycinamide arginine vasopressin, and scopolamine in an autoshaped lever-touch model of learning in rats. Society for Neuroscience (Abst.) 11:381, 1985.

Lichtensteiger, W., D. Felix, R. Lienhart and F. Hefti: A quantitative correlation between single unit activity and fluorescence intensity of dopamine neurones in zona compacta of substantia nigra, as demonstrated under the influence of nicotine and physostigmine. Brain Research 117:85-103, 1976.

Major, R. and N. White: Memory facilitation by self-stimulation reinforcement mediated by the nigroneostriatal bundle. Physiology and Behavior 20:723-733, 1978.

Mundy, W.R. and E.T. Iwamoto: The actions of desglycinamide arginine vasopressin (DGAVP), scopolamine (SCOP), and nicotine (NIC) in an autoshaping model of "learning" in rats. Pharmacologist 27:235 (Abst.), 1985.

Steiner, S.S., B. Beer and M.M. Shaffer: Escape from self-produced rates of brain stimulation. Science 163:90-91, 1969.

White, N.M. and G.D. Carr: The conditioned place preference is affected by two independent reinforcement processes. Pharmacology Biochemistry and Behavior 23:37-42, 1985.

DISCRIMINATIVE STIMULUS EFFECTS OF NICOTINE: CORRELATION WITH BINDING STUDIES

I.P. Stolerman, R. Kumar, J.A. Pratt* and C. Reavill

Departments of Pharmacology and Psychiatry
Institute of Psychiatry
De Crespigny Park
London SE5 8AF, England
*Department of Pharmacology
University of Strathclyde, Scotland

INTRODUCTION

The main reason for smoking tobacco is to obtain the effects of nicotine; an improved way of dealing with the pharmacological component of tobacco addiction will only come from advances in understanding both the behavioral mechanisms involved and the mode of action of nicotine in the CNS. This article reviews recent work which has been aimed at identifying underlying neuropharmacological mechanisms.

Drug discrimination experiments require animals trained to make different behavioral responses depending on whether or not they have been injected with a particular drug. Correct discriminations are reinforced by presenting either food or another of the conventional stimuli much used in behavioral research. The observation that nicotine is well discriminated from the nondrug condition does not itself indicate whether it is "addictive" or reinforcing. However, the nicotine discriminative stimulus or cue has been very useful as a behavioral assay in studies primarily concerned with nicotine's neuropharmacological mode of action. There is another reason why these studies may be of interest - with many classes of drugs, there is a remarkable correlation between discriminative effects in animals and subjective effects in humans. The proper interpretation of this relationship remains controversial, but a reasonable case can be made that discriminative effects of drugs represent the closest known analogue in animals of subjective effects in humans.

Discriminative effects of drugs are particularly suitable for use as "behavioral assays" because they can be assessed in an objective, quantitative manner, they may be very specific in a pharmacological sense, and they may be extremely robust and reproducible. Morrison and Stephenson (1) trained rats to discriminate the effects of nicotine (0.4 mg/kg sc) from saline. This was done in the now conventional manner: presses on one of two bars were reinforced with food in sessions after nicotine injections, whereas presses on the other bar were reinforced in sessions after saline injections. Generalization tests and pretreatment experiments with a range of drugs were then carried out to determine the characteristics of the nicotine cue. These and all subsequent experiments have used the naturally occurring (-)-nicotine isomer for training.

Rosecrans and his co-workers carried out a lengthy series of studies using both maze and two-bar operant techniques (2,3). This work established the nicotine cue as an exceptionally reliable and valid method for assessing the sensitivity of subjects to central effects of nicotine. One major achievement was the demonstration that central muscarinic and nicotinic receptors were clearly distinguishable; drugs acting upon them were not generalized with each other, and muscarinic and nicotinic antagonists had the expected selective blocking effects.

The two main questions concerning the neuropharmacology of nicotine have been, firstly, does nicotine act primarily through cholinoceptive mechanisms in the CNS and, secondly, how and where in the brain does nicotine act to bring about the effects sought by tobacco users? In the peripheral nervous system, it seems that all the major pharmacological effects of nicotine are achieved by mimicking the effects of acetylcholine at cholinoceptive sites. Both stimulatory and blocking effects at cholinoceptors are well documented, but only the former are thought to occur with concentrations of nicotine obtained from tobacco smoke. It would be unwise to extrapolate from this knowledge to the CNS.

In recent years, attempts have been made to identify and characterize putative nicotinic receptors in the CNS by means of ligand-binding studies. This strategy, which was so successfully used with other classes of drugs, has only recently been applied to nicotine. When the aim is to find out how and where nicotine acts in the CNS, it is necessary to use nicotine itself as the ligand. The earliest attempts to do so were hampered by several problems, not the least of which was the low specific activity of [^{3}H]-nicotine available at the time (4,5). Not until 1980 was a binding site for nicotine identified with characteristics resembling those expected for a receptor, i.e., saturable binding displaceable by unlabelled nicotine and with some evidence for stereospecificity (6). The ligand used in this work was [^{3}H]-(±)-nicotine, whereas the product in tobacco is (-)-nicotine.

Qualitative aspects of the binding found by Romano and Goldstein have been confirmed by several groups of workers (7,8). The binding is displaced by other nicotinic-cholinergic agonists (e.g., anabasine, carbachol, cytisine, dimethyl-phenylpiperazinium, and lobeline). (-)-Nicotine is 10 - 15 times more potent than (+)-nicotine in displacing [^{3}H]-(±)-nicotine. Drugs acting at other neurotransmitter receptors (including muscarinic-cholinergic compounds) are ineffective except at very large concentrations. For most purposes, it is advisable to use [^{3}H]-(-)-nicotine as the ligand instead of racemic nicotine.

Current results from binding experiments indicate that the well-established nicotinic blocking drugs (mecamylamine, pempidine, hexamethonium, chlorisondamine, tubocurarine and alpha-bungarotoxin) do not inhibit nicotine binding. Similar findings have been obtained in several laboratories including some in which autoradiographic techniques were used to provide a very detailed atlas of nicotine binding sites in rat brain (9). Binding sites were spread widely through many areas of the brain, with particularly high densities in the thalamus, interpeduncular nucleus, superior colliculus, medial habenula, cerebral cortex, substantia nigra and dentate gyrus of the hippocampus. Some structures were almost devoid of binding sites (e.g., globus pallidus). This information will be of great value as a guide for localizing nicotine's sites of action in the brain.

The ability of nicotinic-cholinergic agonists to displace the binding of nicotine suggests that if the binding site is a receptor and not merely an acceptor site with no functional significance, then the endogenous ligand may well be acetylcholine. High affinity binding sites for acetylcholine have been identified in the presence of atropine (which inhibits binding to muscarinic sites). This remaining acetylcholine binding is inhibited by nicotine and the characteristics of the site including its regional distribution are almost identical with those of sites for nicotine itself (9-12).

The present paper summarizes experiments designed to characterize the nature of mechanisms mediating the nicotine discriminative stimulus. Studies with both nicotinic agonists and antagonists are presented. Stolerman (13) has provided a detailed review of work on the discriminative effects of nicotine.

METHODS

Male, hooded rats were housed individually in rooms maintained at about 22°C with a regular light-dark cycle (light from 8 a.m. to 8 p.m.). Throughout the experiments the rats were fed restricted amounts of food so as to maintain their weights at about 80% of those under free-feeding conditions. Water was available in the living cages at all times.

Standard experimental chambers (Campden Instruments) were contained in sound-insulated, ventilated enclosures. The chambers were fitted with two response bars separated by a recess in which 45 mg pellets of food could be presented. White noise at 78 db above a reference level of 0.0002 dyn/cm was present at all times to mask external sounds.

The procedure for establishing nicotine discrimination has been described in detail (14) and only the main features are presented here. Rats were first trained to press bars for food without receiving any injections. Then, in 15-min sessions beginning 15 min after subcutaneous administration of nicotine, the animals were reinforced with food for pressing one of the two bars: presses on the other bar were reinforced in sessions after saline injections. The final schedule of food reinforcement was a tandem variable interval 1-min fixed ratio 10; under this schedule, food was presented following the tenth consecutive response on the correct bar after a randomly-determined interval (mean = 1 min). Responses during the intervals were not reinforced. All systemic injections were given subcutaneously in a volume of 1 ml/kg and all doses were calculated as those of the base. Discriminative effects of drugs were determined with groups of 6-8 trained rats in 5-min extinction tests during which no responses were reinforced. Such tests took place twice weekly, with training sessions continuing on the intervening days. For each drug, a range of doses was tested; all rats received each dose once and the sequence of treatments was determined separately for each rat by a randomization procedure. The largest doses used of most drugs markedly suppressed responding and it was not possible to assess discriminative effects at these doses.

Intraventricular injections were given in a volume of 2 µl of saline, while rats were in a stereotaxic apparatus, under halothane anaesthesia. Injections aimed at brain regions took place through previously implanted bilateral, guide cannulae. The sites of intracerebral injections were verified histologically.

In order to test putative blocking drugs, the rats received two injections before each test session. The first injection consisted of a dose of the blocking drug or its vehicle, whereas the second injection was nicotine, cytisine or saline.

The main index used to assess discriminative effects was the number of responses on the bar appropriate for the training drug expressed as a percentage of the total number of responses on both bars. This quantitative index was calculated separately for each rat and means were then taken. A supplementary (quantal) index was obtained by calculating the percentage of rats selecting the drug-appropriate bar; the selected bar was defined as the one on which the rat first totalled 10 responses. The total number of responses on both bars in a 5-min extinction test was used as an index of overall response rate.

RESULTS AND DISCUSSION

Generalization Tests: Nicotinic Drugs

Fig. 1 shows typical dose-response functions for rats trained to discriminate nicotine in the two-bar task. It can be seen from Fig. 1a that the numbers of responses on the bar appropriate for nicotine were strongly related to dose, reaching a maximum of 88% after the 0.4 mg/kg dose of nicotine used for training, as compared with 3% after saline. Doses of nicotine up to 0.4 mg/kg had little effect on the total numbers of responses on both bars (Fig. 1b), showing the superior sensitivity of the discriminative index of response. The mean plasma nicotine concentrations associated with nicotine administrations were determined in other rats of the same sex, strain and age. Fig. 1c shows that at the 0.4 mg/kg training dose, the peak plasma nicotine concentration was 146 ng/ml. In other experiments, nicotine-appropriate responding increased as early as 2.5 min after sc injections; it became maximal between 2.5 – 20 min and disappeared at 80 – 160 min. These results also correlated well with the time-course for nicotine in plasma (14).

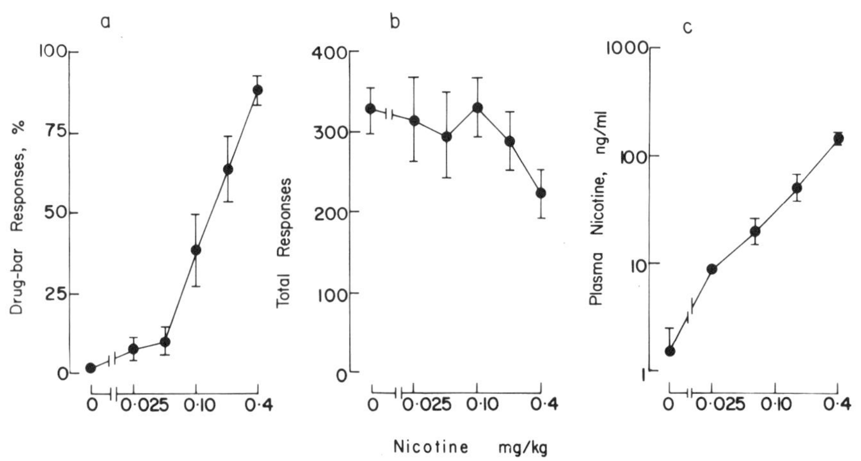

Fig. 1. Nicotine dose-response determinations in rats trained to discriminate 0.4 mg/kg nicotine from saline (n = 8). The discriminative stimulus (cue) effect of nicotine is shown in (a) by dose-related increases in percentages of responding on the drug-appropriate bar (means ± S.E.M.). The weak reduction in overall response rate is shown in (b) by total numbers of responses on both bars. Corresponding mean plasma nicotine concentrations determined in groups of 6 – 7 untrained rats are shown in (c). All injections were s.c. 15 min before sampling behavior or plasma (from Pratt et al. (14)). These data show the powerful discriminative effect of nicotine, much used as a behavioral assay in studies of its mode of action.

Drugs which produce nicotine-like responding in trained rats are said to be generalized with nicotine, and they may help to elucidate the ways in which nicotine acts in the brain. Using rats trained in a T-maze shock-escape task, Romano et al. (15) presented preliminary evidence of generalization with the nicotine analogue anabasine; this was an important observation because anabasine inhibited the binding of nicotine to a recently discovered site on rat brain membranes. The analogue cytisine was behaviorally inactive, possibly because of its poor penetration into the CNS. Subsequently, clear dose-related generalization to both anabasine and cytisine has been obtained in the two-bar procedure. Fig. 2 illustrates these findings which were clearest in rats trained with a small, 0.1 mg/kg dose of nicotine (16). It can be seen that virtually complete generalization was obtained with both quantitative (Fig. 2a) and

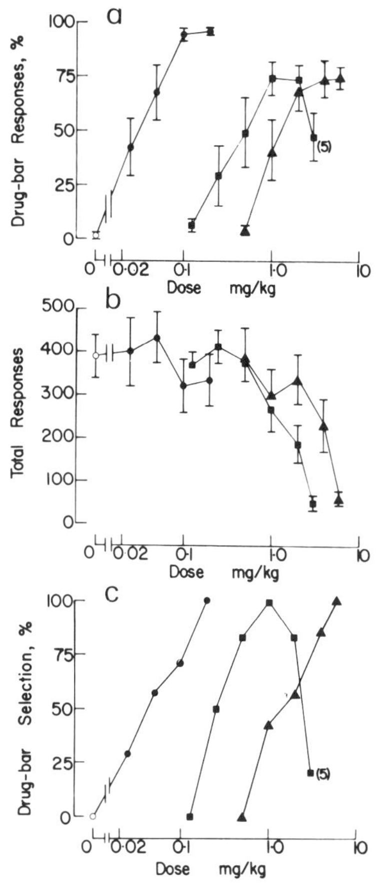

Fig. 2. Dose-response generalization tests with nicotine (●), and the nicotine analogues cytisine (■) and anabasine (▲), in rats trained to discriminate 0.1 mg/kg of nicotine from saline (n = 6 - 8). Generalization to anabasine and cytisine is shown by a quantitative index, percentage of responses on the nicotine-appropriate bar (a), and by a quantal index, percentage of rats selecting the nicotine-appropriate bar (c). Generalization occurred with doses of the analogues producing only minimal reductions in overall rates of responding (b). All injections s.c. 15 min before tests (from Stolerman et al. (16)). These data indicate that the analogues produce specific, nicotine-like behavioral effects.

quantal (Fig. 2c) response indices, at doses of the analogues which had minimal effects on overall numbers of responses (Fig. 2b). The stereoisomer (+)-nicotine also generalizes with (-)-nicotine (15,17).

In recent studies carried out in collaboration with Dr. M. Risner and Dr. S.R. Goldberg (NIDA Addiction Research Center, Baltimore), we have found that doses of (+)-nicotine and of the nicotine analogue nornicotine can be fully generalized with nicotine (unpublished results). A small, 0.1 mg/kg training dose of nicotine was used in all these experiments.

The findings with (+)-nicotine, anabasine, and cytisine provide a very strong indication that the site at which these compounds inhibit the binding of nicotine is a functional receptor mediating at least the discriminative effect of nicotine. The relative potencies of the two isomers of nicotine and of cytisine in the behavioral and biochemical procedures correlate reasonably well when allowances are made for pharmacokinetic factors (15,18). The possibility of a similar correlation for anabasine cannot be tested because pharmacokinetic data are not available. The clearest generalization to anabasine and cytisine was obtained in rats trained with a 0.1 mg/kg dose of nicotine, which produced a peak plasma nicotine concentration of 35 ng/ml (Fig. 1c); this was similar to plasma nicotine concentrations found in cigarette smokers who inhaled (range = 4 - 72 ng/ml (19)). Thus, only nicotinic-cholinergic agonists can both inhibit nicotine binding and generalize with nicotine. It follows that there is an important cholinoceptive link in the central mediation of nicotine's effects in the dose range relevant to the human use of tobacco.

In some experiments, nicotine or saline has been injected into the region of the dorsal hippocampus or the globus pallidus. There were marked increases in drug-appropriate responding after injections of nicotine into either site (10 μg bilaterally). Saline injections did not produce this effect. The similarity in effect across the two regions was unexpected since the globus pallidus contains few binding sites for nicotine (9). One explanation would be based on the extremely rapid removal of nicotine from brain sites into which it is micro-injected (20). Generalization effects such as those described above may be the result of drug actions at relatively remote sites. Rosecrans and Meltzer (21) have reported some increases in drug-appropriate responding when unilateral injections of nicotine are aimed at the hippocampal or reticular formations.

Some generalization experiments have failed to support the cholinoceptor hypothesis. Lobeline inhibits nicotine binding but is not generalized (15); however, there is no evidence that lobeline penetrates to the CNS, nor has it been tested in rats trained to discriminate small (0.1 mg/kg) doses of nicotine. Mixtures of physostigmine, atropine and hexamethonium or of physostigmine and atropine also failed to generalize although they would be expected to produce selective increases in cholinergic stimulation at central nicotinic-cholinergic sites (13,21). The results of these complex manipulations have led Rosecrans and Meltzer (21) to speculate that the nicotine cue is mediated through noncholinergic mechanisms.

Generalization Tests: Non-nicotinic Drugs

Tests of generalization to nicotine analogues would be practically uninterpretable in terms of receptor mechanisms if the nicotine cue was pharmacologically nonspecific. Fig. 3 shows results of some generalization tests carried out over a range of doses with two non-nicotinic drugs. It can be seen that neither the convulsant picrotoxin nor the benzodiazepine chlordiazepoxide produced any increases in nicotine-appropriate responding. The doses of picrotoxin and chlordiazepoxide included some which may themselves be discriminable when used to train animals and their behavioral activity was confirmed by reductions in overall numbers of responses (Fig. 3).

Many other studies have shown that rats trained to discriminate nicotine do not identify non-nicotinic drugs as nicotine-like. Drugs from many different pharmacological classes either failed to increase nicotine-appropriate responding at all, or produced small increases which were easily distinguishable from those produced by nicotine itself. Among the drugs tested were compounds acting at receptors for catecholamines, opioids, 5-hydroxytryptamine, benzodiazepines and muscarinic-cholinergic agonists (13). Similar negative results were obtained with nicotine antagonists and certain nicotine analogues. There was no evidence for an overall loss of specificity when very small doses of nicotine were used for training (16). Amphetamine administered systemically increased nicotine-appropriate responding, but not to the same extent as nicotine itself (16). Such an effect of amphetamine also occurs in rats trained to discriminate small doses of some non-nicotinic drugs and its interpretation remains unclear.

Pretreatment Experiments

Pretreatment studies with drugs thought to act specifically on particular neurotransmitter receptors are as important as

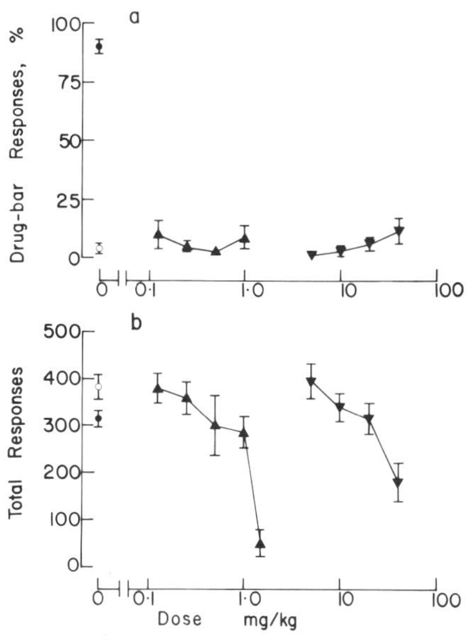

Fig. 3. Generalization tests with two non-nicotinic drugs in rats trained to discriminate 0.4 mg/kg of nicotine from saline (n = 8). Neither picrotoxin (▲) nor chlordiazepoxide (▼) produced any increase in nicotine-appropriate responding at any of the doses tested (Pratt and Stolerman, unpublished data). Both drugs reduced overall rates of responding, as shown in Fig. 3b. Injections were either 15 min (picrotoxin) or 30 min (chlordiazepoxide) before tests. These data illustrate the type of evidence used to establish the specificity of the nicotine cue.

generalization tests in characterizing a drug-produced cue. The only drugs which to date have been found to block the nicotine cue are compounds which block nicotinic-cholinergic receptors in autonomic ganglia. This has been an extremely reliable effect which has been obtained consistently in several different laboratories (1,2,3,15). The active drugs are those ganglion-blockers which penetrate well into the CNS, i.e., mecamylamine and pempidine. Mecamylamine also blocks the generalization response to cytisine (16,22). Such studies provide evidence that the receptor mediating the discriminative effect of nicotine may resemble the cholinoceptive site in autonomic ganglia. A variety of drugs which act on a range of other neurotransmitter systems have failed to block the nicotine cue (13).

These observations are all the more interesting because of the failure of mecamylamine and pempidine to inhibit binding of nicotine to rat brain membranes. This suggests that the drugs may act through noncompetitive mechanisms. One would expect the block of nicotine's effects to be reversed by increases in the dose of nicotine if mecamylamine acted competitively. Fig. 4a shows that the block of the nicotine cue by mecamylamine could not be fully reversed by increases in the doses of nicotine. The block of the response rate-reducing effect of nicotine was

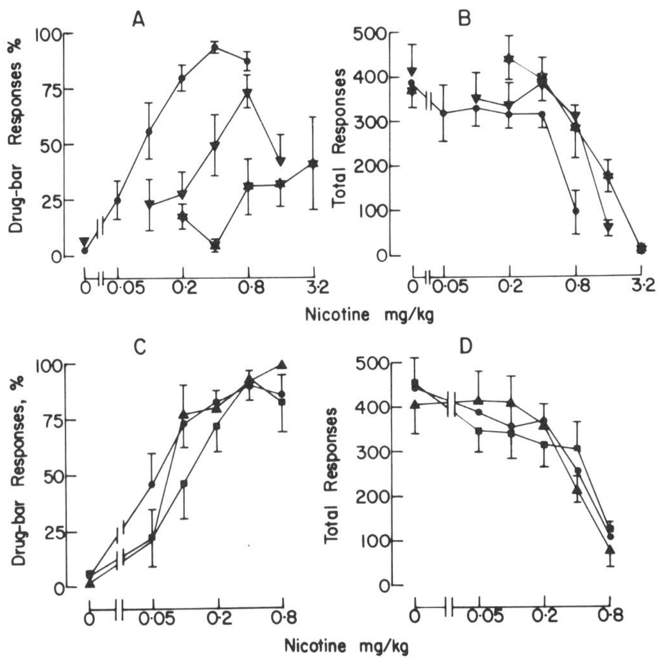

Fig. 4. Dose-response curves for the discriminative and response-rate reducing effects of nicotine in rats trained to discriminate 0.4 mg/kg of nicotine from saline (n = 7–8). Mecamylamine blocked both effects, but only the block of response-rate decreases could be reversed by increasing the dose of nicotine: (a) shows discriminative effect of nicotine after pretreatment with saline (●) or mecamylamine in doses of 0.25 (▼) or 0.75 mg/kg (✶), whereas (b) shows total numbers of responses in the same rats. Hexamethonium did not affect either response to nicotine, (c) shows discriminative effects after pretreatment with saline (●) or hexamethonium in doses of 2.5 (▲) or 7.5 mg/kg (■), whereas (d) shows total numbers of responses (from Stolerman et al. (22)).

reversible (Fig. 4b), but response rate decreases do not indicate specific nicotine-like effects. Effects of nicotine which do not reflect nonspecific behavioral depression (e.g., increases in response rates) seem irreversible, whereas response rate decreases seem reversible (22,23). These studies are, therefore, consistent with the possibility that mecamylamine acts through noncompetitive mechanisms and generally fit in with the results of the ligand-binding work.

Quaternary ganglion-blocking drugs such as hexamethonium and chlorisondamine do not penetrate well into the CNS and when injected systemically, they do not block the nicotine cue except at very large doses (15). This suggests that the nicotine cue is primarily of central origin, thus confirming the conclusion also reached from generalization tests with nicotine methiodide, which also acts mainly in the periphery (13). However, a single intraventricular injection of a small (5 μg) dose of chlorisondamine blocked the nicotine cue for a period of several weeks (24). In contrast (22), hexamethonium was not found to block the nicotine cue, even when large doses were injected ivt (Reavill and Stolerman, unpublished results). Further studies of how antagonists such as mecamylamine and chlorisondamine block the discriminative effects of nicotine may make useful contributions to understanding central nicotinic mechanisms.

CONCLUSIONS: MODELS OF THE CNS NICOTINIC RECEPTOR

It is now possible to develop hypotheses about central nicotinic receptors by pooling results of neurochemical, behavioral, and other studies (13). The overall structure of the receptor may resemble those in autonomic ganglia because most of the central effects of nicotine can be prevented by ganglion-blocking drugs. This argument would be strengthened if drugs blocking nicotinic receptors at the neuromuscular junction could be shown not to block behavioral effects of nicotine. Such information is not available, although the convulsant effect of nicotine is prevented by ganglion but not neuromuscular blockers (25). In contrast, Wonnacott et al. (26) have reported 5% immunological cross-reactivity between a partly purified alpha-bungarotoxin-binding component of rat brain and antibodies for rat muscle. Immunological studies with monoclonal antibodies to nicotinic receptors from chick muscle also suggest some similarity of CNS nicotinic receptors to those at the neuromuscular junction (27). The immunological studies reopen the question of whether central nicotinic receptors resemble most closely the ganglionic or the neuromuscular receptors, and the possibility that both types of receptor have their analogues in brain cannot be discounted (9).

Several pieces of evidence suggest that the central effects of nicotine involve more than simply a recognition site for nicotine. Highly effective nicotine antagonists such as mecamylamine, pempidine and chlorisondamine appear to act through noncompetitive mechanisms and do not inhibit nicotine binding. There is strong evidence that peripheral nicotinic receptors are linked to ion channels which play a crucial role in the chain of events leading from a drug effect at the nicotine recognition site to an observed pharmacological effect. The possibility that nicotine acts centrally, not simply through a single site, but via a receptor complex must be considered. The different elements in the complex may be in close physical proximity to each other and pharmacological responses would be influenced by events at any one of them.

The primary site in a proposed receptor complex is that at which nicotine and other cholinergic agonists bind. The widely used nicotine antagonists probably act elsewhere because they do not inhibit nicotine binding and because the antagonism exhibits noncompetitive characteristics. Only some recently identified toxins may possibly act as antagonists at the nicotine site (28,29,30). Presumably, therefore, there is a second site through which ganglion-blockers such as mecamylamine act. Ligand-binding studies with such drugs have not been published. Identification of such a site could introduce a substantial new element into knowledge of central nicotinic mechanisms. Presumably any endogenous agonist for such a site would not be acetylcholine, hence the idea may help to resolve the controversy about mediation of nicotine effects by showing the involvement of both cholinoceptive and other mechanisms. There is the matter of possible linkage to an ion channel. This idea is firmly established in studies of nicotinic mechanisms at the neuromuscular junction, and various channel blockers, including phencyclidine and perhydro-histrionicotoxin, have been identified (31). The relevance of the idea to the CNS can only be assessed if more information becomes available.

The simplest model that seems feasible comprises only two sites, the nicotine recognition site and the antagonist site. In this case, it is assumed either that there is no ion channel involved in the central nicotinic receptor complex or that the antagonist site is actually the ion channel. The latter assumption gains credibility from studies in peripheral tissues suggesting that some nicotinic antagonists, including mecamylamine and chlorisondamine, act as channel blockers (32,33). However, it cannot be assumed that this situation necessarily applies to the CNS, and the problem will only be solved by studies ranging from *in vitro* experiments to tests of functional significance in intact, behaving subjects.

SUMMARY

The nicotine discriminative stimulus has been used as a behavioral assay for analyzing central nervous system (CNS) mechanisms mediating the psychological effects of nicotine. Rats were trained to discriminate the effects of a small, 0.1 mg/kg dose of nicotine from saline in a standard, two-bar, operant conditioning procedure. Nicotine yielded clear dose-response curves in doses associated with plasma nicotine concentrations in the smoking range. Only nicotinic-cholinergic agonists that inhibited the binding of nicotine to rat brain membranes *in vitro* produced nicotine-like behavioral effects. Compounds from a wide range of other classes (including psychomotor stimulants, tranquilizers, narcotic analgesics and muscarinic-cholinergic agonists) failed to produce nicotine-like discriminative effects. These results support the view that behavioral and psychological effects of nicotine are mediated mainly through central cholinoceptors. Actions at only a single site can hardly account for all the available information and a receptor complex is proposed, with separate sites for nicotinic-cholinergic agonists and antagonists, and perhaps linked to an ion-channel.

ACKNOWLEDGEMENTS

We thank the Medical Research Council for financial support.

REFERENCES

1. Morrison, CF and Stephenson, JA: Nicotine injections as the conditioned stimulus in discrimination learning, *Psychopharmacologia* *15*: 351-360, 1969.

2. Rosecrans, JA and Chance, WT: Cholinergic and non-cholinergic aspects of the discriminative stimulus properties of nicotine. In (ed.) Lal, H, Discriminative Stimulus Properties of Drugs, 1977, 155-185, Plenum, New York.
3. Rosecrans, JA and Chance, WT: The discriminative stimulus properties of N- and M-cholinergic receptor stimulants. In (eds.) Ho, BT, Richards, DW and Chute, DL, Drug Discrimination and State Dependent Learning, 1978, 119-130, Academic Press, New York.
4. Schleifer, LS and Eldefrawi, ME: Identification of the nicotinic and muscarinic acetylcholine receptors in subcellular fractions of mouse brain, Neuropharmacology 13: 53-63, 1974.
5. Yoshida, K and Imura, H: Nicotinic cholinergic receptors in brain synaptosomes, Brain Res. 172: 453-459, 1979.
6. Romano, C and Goldstein, A: Stereospecific nicotine receptors on rat brain membranes, 1980, Science 210: 647-650, 1980.
7. Marks, MJ and Collins, AC: Characterization of nicotine binding in mouse brain and comparison with the binding of α-bungarotoxin and quinuclidinyl benzilate, Mol. Pharmac. 22: 554-564, 1983.
8. Repond, C, Pratt, JA, Stolerman, IP, Mayer, JM, Jenner, P, Marsden, CD and Testa, B: Affinity of pyridylalkylamines for nicotinic, muscarinic and histaminic recognition sites in brain tissue preparations, Drug Res., in press.
9. Clarke, PBS, Schwartz, RD, Paul, SM, Pert, CB and Pert, A: Nicotinic binding in rat brain; autoradiographic comparison of [^{3}H]-acetylcholine, [^{3}H]-nicotine and [^{125}I]-alpha-bungarotoxin, J. Neurosci. 5: 1307-1315, 1985.
10. Rainbow, TC, Schwartz, RD, Parsons, B and Kellar, KJ: Quantitative autoradiography of nicotinic [^{3}H]acetylcholine binding sites in rat brain, Neurosci. Letters 50: 193-196-1983.
11. Clarke, PBS, Pert, CB and Pert, A: Autoradiographic distribution of nicotine receptors in rat brain, Brain Res. 323: 390-395, 1984.
12. Schwartz, RD, McGee, R and Kellar, KJ: Nicotinic cholinergic receptors labelled by [^{3}H]acetylcholine in rat brain, Mol. Pharmac. 22: 56-62, 1982.
13. Stolerman, IP: Psychopharmacology of nicotine: stimulus effects and receptor mechanisms. In (eds) Iversen, LL, Iversen, SD and Snyder SH, Handbook of Psychopharmacology, 19, in press, Plenum, New York.
14. Pratt, JA, Stolerman, IP, Garcha, HS, Giardini, V, and Feyerabend, C: Discriminative stimulus properties of nicotine: further evidence for mediation at a cholinergic receptor, Psychopharmacology 81: 54-60, 1983.
15. Romano, C, Goldstein, A and Jewell, NP: Characterization of the receptor mediating the nicotine discriminative stimulus, Psychopharmacology 74: 310-315, 1981.
16. Stolerman, IP, Garcha, HS, Pratt, JA and Kumar, R: Role of training dose in discrimination of nicotine and related compounds by rats, Psychopharmacology 84: 413-419, 1984.
17. Meltzer, LT, Rosecrans, JA, Aceto, MD and Harris, LS: Discriminative stimulus properties of the optical isomers of nicotine, Psychopharmacology 68: 283-286, 1980.
18. Martin, BR, Tripathi, HL, Aceto, MD and May, EL: Relationship of the biodisposition of the stereoisomers of nicotine in the central nervous system to their pharmacological actions. J. Pharmac. exp. Ther. 226: 157-163, 1983.
19. Russell, MAH, Jarvis, M, Iyer, R, and Feyerabend, C: Relation of nicotine yield of cigarettes to blood nicotine concentrations in smokers. Br. Med. J. 280: 972-976, 1980.
20. Iwamoto, ET, Williamson, EC, Walsh, C and Hancock, R: Improved drug infusion pump for injecting nanoliter volumes subcortically in awake rats, Pharmac. Biochem. Behav. 20: 959-963, 1984.

21. Rosecrans, JA and Meltzer, LT: Central sites and mechanisms of action of nicotine, Neurosci. Biobehav. Rev. 5: 497-501, 1981.
22. Stolerman, IP, Pratt, JA, Garcha, HS, Giardini, V and Kumar, R: Nicotine cue in rats analysed with drugs acting on cholinergic and 5-hydroxytryptamine mechanisms, Neuropharmacology 22: 1029-1037, 1983.
23. Spealman, RD, Goldberg, SR and Gardner, ML: Behavioral effects of nicotine: schedule-controlled responding by squirrel monkeys. J. Pharmac. exp. Ther. 216: 484-491, 1981.
24. Garcha, HS, Kumar, R, Norris, EA, Reavill, C and Stolerman, IP: Long-term blockade of nicotine cue by chlorisondamine in rats, Br. J. Pharmac. 85: 245P, 1985.
25. Caulfield, MP, and Higgins GA: Mediation of nicotine-induced convulsions by central nicotinic receptors of the 'C_6' type, Neuropharmacology 22: 347-351, 1983.
26. Wonnacott, S, Harrison, R and Lunt, GG: Immunological cross-reactivity between the α-bungarotoxin binding component from rat brain and nicotinic acetylcholine receptor, J. Neuroimmunology 3: 1-13, 1982.
27. Mehraban, F. Kemshead, JT and Dolly, JO: Properties of monoclonal antibodies to nicotinic acetylcholine receptor from chick muscle, Eur. J. Biochem. 138: 53-61, 1984.
28. Dryer, SE and Chiappinelli, VA, Kappa-bungarotoxin: an intracellular study demonstrating blockade of neuronal nicotinic receptors by a snake neurotoxin, Brain Res. 289: 317-321, 1983.
29. Hayashi, E. Isogai, M, Kagawa, Y, Takayanagi, N, and Yamada, S: Neosurugatoxin, a specific antagonist of nicotinic acetylcholine receptors, J. Neurochem. 42: 1491-1494, 1984.
30. Rapier, C, Harrison, Lunt, GG and Wonnacott, S: Neosurugatoxin blocks nicotinic acetylcholine receptors in the brain. Neurochem. Int. 7: 389-396, 1985.
31. Albuquerque, EX, Tsai, M-C, Aronstam, RS, Witkop, B, Eldefrawi, AT and Eldefrawi, ME.: Phencyclidine interactions with the ionic channel of the acetylcholine receptor and electrogenic membrane, Proc. Natl. Acad. Sci. USA 77: 1224-1228, 1980.
32. Ascher, P, Large, WA, and Rang, HP: Studies on the mechanism of action of acetylcholine antagonists on rat parasympathetic ganglion cells, J. Physiol. 295: 139-170, 1979.
33. Lingle, C: Blockade of cholinergic channels by chlorisondamine on a crustacean muscle, J. Physiol. 339: 395-417, 1983.

NONCHOLINERGIC MECHANISMS INVOLVED IN THE BEHAVIORAL AND STIMULUS EFFECTS OF NICOTINE, AND RELATIONSHIPS TO THE PROCESS OF NICOTINE DEPENDENCE

John A. Rosecrans

Department of Pharmacology
and Toxicology
School of Basic Health Sciences
Virginia Commonwealth University
Richmond, VA 23298

INTRODUCTION

A major objective of research conducted by this investigator over the last 18 years has been to evaluate the pharmacological mechanisms by which nicotine can alter behavior, and the role that such mechanisms may have in the process of nicotine-induced dependence. The overall goals of this research have been to: (1) determine potential brain area sites of nicotine action important to its pharmacological effects; (2) determine whether nicotine is acting at cholinergic receptors (pre- or postsynaptic) excited (or inhibited) by the neurotransmitter acetylcholine (ACh); and (3) to determine the potential role that noncholinergic neurons play in the CNS pharmacology of nicotine. The overall approaches utilized in this research evolved from earlier research conducted by Domino (1) in which he very clearly showed that brain cholinergic neurons were not interdependent, as in the periphery, but existed as two independent receptor populations, muscarinic (M-Cholinergic; M-Ch) and nicotinic (N-Cholinergic; N-Ch).

This review will discuss research conducted in this laboratory which has contributed to the view that several of nicotine's effects are the result of an interaction of nicotine at receptors not sensitive to ACh. In addition, the impact of this research on the development of concepts dealing with nicotine-induced behavioral and pharmacological dependence will also be discussed.

BRAIN AREA SITES AND MODE OF ACTION OF THE DISCRIMINATIVE STIMULUS PROPERTIES OF NICOTINE AT CHOLINERGIC RECEPTORS

Because of nicotine's variable effects on operant and simple behaviors (2), it was necessary to develop an *in vivo* CNS-mediated pharmacological model of nicotine which was specific and reliable, and not dependent upon its behavioral effects. With Drs. Schechter and Hirschhorn (3,4,5), a model was developed which utilized the discriminative stimulus (DS) properties of nicotine. As described by Stolerman and co-workers (6,7), a two-lever operant procedure was used in which correct-lever responding is under the stimulus control of nicotine. The nicotine DS (or cue) has been found to be extremely specific and selective (8); it is antagonized by mecamylamine and generalizes to a relatively small group of structurally related compounds (Table 1). Once the DS properties of nicotine were characterized to the point where we were satisfied that we were working with a specific and sensitive *in vivo* centrally-mediated pharmacological model, a research program was designed to elucidate the psychopharmacology of nicotine with respect to brain area sites and cholinergic mechanisms of action.

Central Sites of Nicotine Action

Our basic approach was influenced by Domino (1) and others (9,10) who have viewed both the reticular activating system and hippocampus as essential brain area sites at which nicotine might be acting. Thus, the first study conducted involved an experiment in which rats trained to discriminate subcutaneous (s.c.) injections of 200 μg/kg of nicotine from saline were implanted with unilateral cannulae directed towards the dorsal hippocampus (Fig. 1). This experiment was quite instructive from several points of view. First, rats trained to discriminate nicotine 10 min after injection had difficulty generalizing to the hippocampus if tested at the same peripheral training time. To demonstrate generalization to the hippocampus, rats had to be tested immediately after injection into the hippocampus, indicating the rapid spread of nicotine to surrounding brain tissue. Under these conditions, an excellent dose-related generalization

Table 1. Summary of the Current Status of the Mechanisms of Action and Specificity of the Discriminative Stimulus Properties of Both Arecoline and Nicotine

	Nicotine	Arecoline
Stimulus-Dose Related	Yes	Yes
Antagonized	Mecamylamine	Atropine Sulfate
Not Antagonized	Hexamethonium Atropine Sulfate Naltrexone Propranolol Dibenamine BC-105	Atropine Methylnitrate Mecamylamine
Generalized[a]	3-PMP (+)-nicotine Amphetamine (partial) Nornicotine (partial)	Oxotremorine Pilocarpine (partial)
No Generalization	Arecoline Dizaepam Caffeine[b] Cotinine Lobeline	Nicotine Amphetamine

[a]For additional information see Stolerman et al. (this volume)
[b]Generalization was observed i.c.v. at low training doses

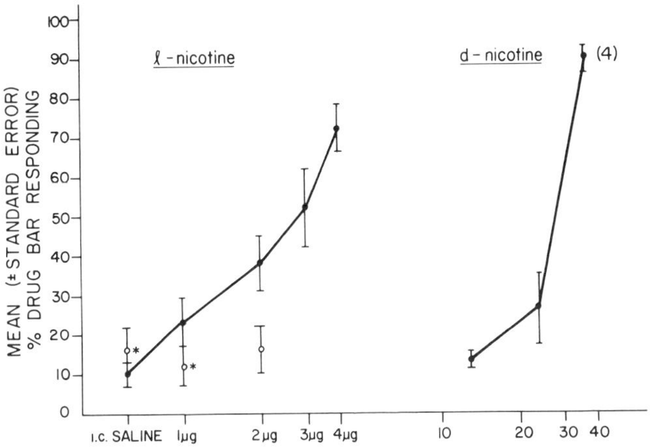

Fig. 1. Generalization of systemically administered (–)-nicotine (200 μg/kg, s.c., N=12) to nicotine injected unilaterally into the dorsal hippocampus (Hp). Sterioisomers of nicotine were administered into the Hp just prior to being placed into the operant chamber during a 2.5-min test session. Open circles represent data obtained when nicotine or vehicle was administered 10 min prior to testing.

was evident up to 4 μg (approximately 8 μg/kg) when the (–)-isomer of nicotine was injected into the hippocampus. As observed from peripheral studies (11), the fact that the (+)nicotine isomer was also several times less active than the (–)isomer established the stereoselective nature of the DS effects of nicotine in the hippocampus.

A second study (12) was conducted in two groups of rats trained to discriminate either nicotine (400 μg/kg, s.c.) or the M-Ch agonist arecoline (1.14 mg/kg, s.c.). Half of the rats in each drug group were implanted with unilateral cannulae aimed at either the dorsal hippocampus or midbrain reticular formation. This study provided additional information indicating that the hippocampus is an important site at which nicotine may be producing its effects, but also showed the reticular formation to be more sensitive to nicotine. Peripheral mecamylamine administration was also observed to antagonize the generalization of the peripherally generated nicotine DS to either the reticular formation or hippocampus, and these data provided additional evidence that the peripheral DS was acting at some central Ch receptor. It should also be noted that the generalization at hippocampal sites was greater in rats trained at lower nicotine doses, indicating the greater sensitivity of the lower training dose (Fig. 1). In contrast to the demonstration of nicotine's central sites of action, arecoline administration into either site did not generalize to the peripherally generated arecoline DS. In addition, the nicotine DS (s.c.) did not generalize to arecoline administered into either site; these data again indicated the specificity and differential sites of action of these two cholinergic drugs.

Cholinergic Nature of the Nicotine and Arecoline DS in the Control of Behavior

A major question which has always intrigued our laboratory was the question of how these select cholinergic receptors (M-Ch and N-Ch) could be linked together, and whether the endogenous ligand at both receptors was acetylcholine (ACh). The task was to test whether nicotine and/or arecoline were acting at ACh-sensitive receptor sites. In the following experiments rats, were first trained to discriminate either nicotine (400 μg/kg, s.c.) or arecoline (1.14 mg/kg, s.c.). The approach was to block all receptors that are nonessential to having a rat learn to discriminate either drug. The following approach was used:

(1) Central and peripheral M-receptors were blocked with 4 mg/kg of atropine (s.c.) and peripheral N-receptors were blocked with 1 mg/kg of hexamethonium (s.c.) in nicotine-trained rats. As observed in Fig. 2, rats administered 400 μg/kg of nicotine were able to discriminate nicotine, which again demonstrated receptor specificity of the DS.

(2) In arecoline-trained rats, on the other hand, central N-receptors were antagonized with mecamylamine (1 mg/kg, s.c.), while peripheral M-receptors were blocked by methylatropine (2.0 mg/kg, s.c.). Again, rats were able to discriminate arecoline (Fig. 2). This experiment further substantiated the existence of two separate cholinergic receptors, and also demonstrated that when all receptors were blocked (mecamylamine + atropine), neither drug was discriminable and animals responded as if they had been given saline.

(3) With nonessential receptors blocked in different groups of rats, the next task was to attempt to mimic both M- and N- cues by increasing ACh levels at respective M-Ch and N-Ch receptors. To

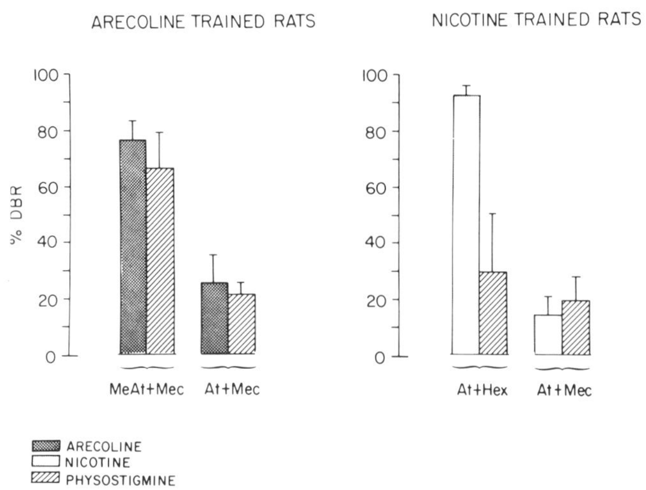

Fig. 2. An attempt to mimic the DS effects produced by either arecoline (1.14 mg/kg) or nicotine (400 μg/kg) by increasing brain acetylcholine (ACh) via the inhibition of cholinesterase with physostigmine (250 μg/kg, s.c., 20 prior to testing); all drugs were administered as the free base. The approach utilized to block specific central cholinergic receptors with mecamylamine (Mec; 1 mg/kg), hexamethonium (Hex; 1 mg/kg), atropine (AT; 2 mg/kg), or methyl atropine (MeAt; 2 mg/kg). Data is presented as % arecoline- or % nicotine-correct responding (%DBR). Redrawn from Hendry and Rosecrans, In: (ed.) Balfour, Nicotine and the Tobacco Smoking Habit, Int. Encyclopedia Pharmacol. Theraps, Sec 14, Pergamon Press, Oxford, pp. 75-97, 1983 (with permission).

accomplish this goal, physostigmine, an acetylcholinesterase (AChE) inhibitor, was utilized to increase ACh levels at cholinergic receptors which were not inhibited (13). In this study, rats were first administered specific antagonists and then administered 250 μg/kg of physostigmine to determine if the M- and N- cues could be mimicked.

As seen in Fig. 2, we were able to mimic the arecoline DS but not that of nicotine. In addition to these studies, we have been able to shift the arecoline dose-response curve to the left when animals were pretreated with physostigmine, providing further evidence that arecoline may be acting at an ACh-sensitive receptor (Meltzer and Rosecrans, unpublished observations). In contrast, physostigmine was unable to facilitate or mimic the nicotine cue. Thus, is nicotine acting at a noncholinergic receptor, or is it acting via interneurons, which are ACh-sensitive? These questions are not yet answerable. In terms of the DS effects of nicotine, the lack of ability of nicotine to generalize to physostigmine has also been observed by Stolerman (7), but no one has attempted to replicate these findings using the approach just described.

Schwartz and co-workers (14) have provided similar findings to ours using an _in vitro_ receptor approach. Their studies indicate that nicotine will compete with ACh binding sites in the presence of atropine and physostigmine, and that mecamylamine will not compete with ACh under similar binding conditions. These data suggest that nicotine may be producing its effects via some noncholinergic site. These same workers have shown that arecoline and oxotremorine (a drug which generalizes to arecoline, Table 1) appear to compete with ACh at a muscarinic binding site, but unlike mecamylamine vs. ACh, atropine competes with ACh at almost the same molar concentrations as arecoline (15). That is, atropine appears to be blocking M-Ch receptors sensitive to ACh. Their data fits our drug discrimination data quite well, and suggests that nicotine may be acting on at least two receptor sites, one cholinergic and one noncholinergic.

While these data suggest that nicotine may not be acting at an ACh sensitive receptor, additional research needs to be conducted before any final conclusions can be made. For example, it should be determined whether nicotine will generalize to ACh at some nicotine-sensitive brain site. Studies such as these would greatly clarify the role that ACh plays in the action of nicotine.

ROLE OF DOPAMINE-CONTAINING NEURONS IN THE PSYCHOPHARMACOLOGICAL EFFECTS OF NICOTINE

Several researchers have observed that Parkinson's disease, a neurological syndrome characterized by a loss of brain dopamine (DA) neurons, is less prevalent in smokers when compared with nonsmokers (16,17). These initial observations have more recently been verified by Baron et al. (this volume), suggesting a possible relationship between nicotine and DA function. Westfall and co-workers (this volume) have also observed nicotine to alter DA release from central neurons, while Singer et al. (18) have shown that the acquisition of the schedule-induced self-administration of nicotine was attenuated by 6-hydroxydopamine (which is neurotoxic to DA neurons) injected into DA-rich neurons of the nucleus accumbens, providing additional information suggesting that nicotine may interact with the DA neurons. Thus, does nicotine alter DA neurons in some manner, or is nicotine's pharmacological effect contingent upon the integrity of this catecholamine projection system?

To begin to determine whether such a relationship exists, rats depleted of DA neonatally were studied as adults as to their ability to discriminate nicotine from saline using the procedures described above. To accomplish this goal, neonate male rats were depleted of brain DA or norepinephrine (NE) via the intracisternal injection of 6-hydroxydopamine (18). NE was depleted by injecting 50 μg of 6-hydroxydopamine/rat on postnatal days 1 and 3, while dopamine was depleted by injecting 150 μg of 6-hydroxydopamine/rat intracisternally 30 min following an intraperitoneal dose of desmethylimipramine (25 mg/kg) on neonatal day 14. 6-Hydroxydopamine produces long-term destruction of both amines as shown when animals are evaluated 60-90 days later. In our hands, DA is usually depleted by 70-90%, while NE levels are reduced on the average 40-60% of control levels. These procedures are unusually specific, as only the amines in question were depleted, while the other important biogenic amine, serotonin, or peripheral catecholamines were unaffected.

Important to an understanding of the relationship between these catecholamines and nicotine is the fact that such amine-depleted rats were able to discriminate nicotine even though its discriminability (% nicotine-correct responding - % saline-correct responding) was slightly reduced in the DA-depleted rat. Thus, neither amine system is essential to an animal learning to discriminate nicotine. However, the sensitivity of the nicotine cue was altered as the ED50 generalization dose of nicotine was increased by 65% in norepinephrine-depleted rats, while the antagonist dose of mecamylamine needed to reduce nicotine-correct responding was attenuated in both groups of amine-depleted rats. The last experiment conducted in these animals involved a hippocampal generalization study similar to that conducted above (Fig. 1). Rats trained to discriminate nicotine peripherally (400 μg/kg) generalized to 500 ng/rat administered bilaterally into the hippocampus of control and NE-depleted rats, whereas generalization in the DA-depleted rat was completely antagonized (19,20). This latter finding is extremely important, as Robinson et al. (21) have shown that the depletion of septal DA-containing neurons (via 6-hydroxydopamine) will facilitate hippocampal acetylcholine turnover, providing evidence that nicotine-sensitive cholinergic receptors may be under control of a DA-projecting neuron to the hippocampus. More recent evidence of such a relationship has been provided by the work of Schwartz et al. (22), in which nicotinic-cholinergic receptors were significantly reduced in the rat by the intraventricular administration of either 6-hydroxydopamine or 5,7-dihydroxytryptamine (neuro-toxic to serotonin neurons). The data suggests that nicotinic receptors are also located presynaptically on both catecholamine as well as indoleamine-containing neurons. Therefore, these findings suggest that noncholinergic neurons may be involved in the sensitivity to the nicotine-induced DS, but are not essential to a rat's learning the nicotine cue. However, we have yet to evaluate the nicotine cue in the serotonin-depleted rat.

In addition to evaluating the nicotine DS in amine-depleted rats, response rates on both levers were also evaluated under all conditions (Fig. 3). In this evaluation, the response rates were pooled across all doses of nicotine and nicotine + mecamylamine; data also represents response rates on both levers. This presentation of the data can provide us with a single measure of response rates and is a good way to analyze the overall trend of single and/or combined treatments. The most important finding observed was that DA-depletion significantly facilitated response rates when nicotine was administered, which was antagonized by mecamylamine. Thus, while DA-depletion did not alter the ability of rats to learn the nicotine cue, their response rates were independently increased. The fact that mecamylamine was able to antagonize the rate increase also suggests that this effect is related to an interaction

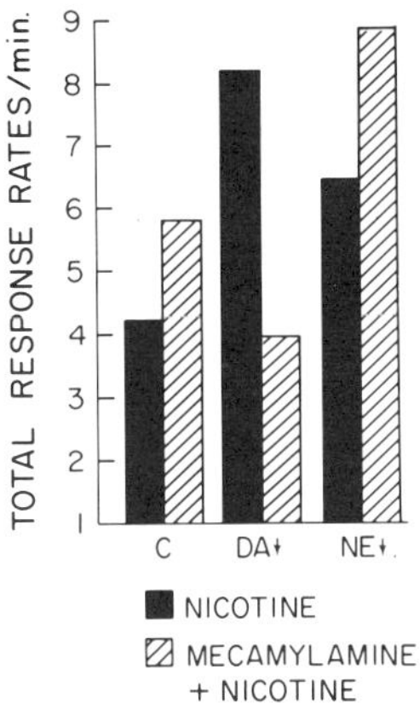

Fig. 3 Effects of biogenic amine depletion on nicotine-induced rates of responding in a drug discrimination paradigm. Rats depleted of either DA or NE were trained to discriminate nicotine (400 μg/kg, s.c.). These data represent pooled response rates derived from nicotine dose-response generalization (100, 200, and 400 μg/kg) studies involving amine depletion and mecamylamine challanges. Discrimination and amine-depletion data appear.

between cholinergic and dopaminergic neurons. NE-depletion, on the other hand, induced a somewhat different profile of effects; response rates during nicotine dose-response generalization were facilitated, but unlike what was observed in DA-depleted rats, mecamylamine facilitated response rates when administered prior to nicotine. The fact that both depleted groups increased the AD50 dose of mecamylamine suggests that percent discrimination and response rates may be under the control of a different amine projection system. Interpretation of these findings are not possible at this time, but the results emphasize the potential role that these catecholamine-containing neurons have in relation to nicotine's effects and to the potential interactions that DA and NE neurons have at sites presynaptic to the nicotinic receptor.

To further evaluate the potential role that DA-containing neurons might have on nicotine-induced alterations in rates of responding, a second study was conducted in neonatal DA-depleted rats (23). Rats depleted of DA were initially trained to bar-press for food using a variable interval 15-sec (VI-15) schedule of reinforcement. Once behavior stabilized, rats were administered nicotine (750 μg/kg, s.c. daily) for 36 days. The results of this experiment confirmed our initial findings that DA-depletion facilitated response rates after nicotine administration. These results suggest that presynaptic DA neurons do appear to play a role in the psychopharmacological effects of nicotine. Whether this finding is due to an effect at a presynaptic nicotinic receptor (22) or whether such an effect could be the result of a postsynaptic supersensitivity needs to be resolved. Thus, DA neurons may have a prominent role on nicotine's behavioral effects, and may help us better understand the mechanism by which nicotine can serve as either a primary or secondary reinforcer.

DEVELOPMENT OF PHARMACOLOGICAL TOLERANCE TO NICOTINE: EVIDENCE FOR A WITHDRAWAL SYNDROME

Development of Nicotine-Induced Pharmacological Tolerance

The major objective of this study was to determine the nature of tolerance development to nicotine's disruptive effects on schedule-controlled behavior. As demonstrated in drug discrimination studies, tolerance to nicotine's disruptive effects on behavior develop very rapidly, but tolerance to the cue does not. The question asked by this research concerned whether tolerance in these behavioral situations was a function of behavioral adaptation, or whether tolerance occurred because of some neuronal adjustment by the organism. To begin to answer these questions, mice were initially trained to bar-press for food reward using a Fixed Ratio-25 (FR-25) schedule of reinforcement during 30-min behavioral sessions (24). The mice were then divided into two groups: one group was administered nicotine immediately prior to each daily behavioral session, and the other group was administered the same daily dose of nicotine immediately after each behavioral session. An initial dose-disruption curve was determined prior to chronic nicotine dosing for each of the experimental groups (pre- or postsession nicotine treatment). The approximate 80% disruptive dose (1.2 mg/kg) was then administered daily until the development of tolerance was evident in the pretreatment chronic nicotine group; rates of responding were not significantly reduced. A dose-disruption curve was redetermined for each treatment group prior to the behavioral session during this tolerance phase.

The results of this investigation are presented in Fig. 4. Mice were administered 1.2 mg/kg (s.c.) of nicotine daily until the pre-nicotine group became tolerant as evidenced by a return of behavioral responding to the pre-chronic nicotine baseline levels; tolerance developed within 30 days. The most important finding in this study was the observation that mice from both the chronic pre- and postsession treatment groups were equally tolerant when the nicotine was administered prior to the behavioral test session. Thus, animals did not become tolerant simply because of behavioral adaptation; these data suggested that tolerance development appeared to be related to some neuronal compensatory mechanism. At this point, it would be interesting to speculate that this effect was related to the up-regulation of N-receptors, but not M-receptors, induced by chronic nicotine administration observed by other investigators (25). These data again emphasize the potential differences between the sites of nicotine action and ACh-sensitive cholinergic receptors. In support of these differential relationships between M-Ch and N-Ch agonists, Meltzer and Rosecrans (26) observed that tolerance to the behaviorally disruptive effects of arecoline was devoid of any pharmacological component because only the pre-session arecoline group exhibited any tolerance to this M-Ch agonist.

Evidence of a Behavioral and Neurochemical Withdrawal Syndrome After Chronic Nicotine Administration in the Mouse

In addition to evaluating the development of tolerance in the study described above, we also attempted to determine the rate of tolerance disappearance to a behaviorally disruptive dose of nicotine (24). Pharmacological tolerance disappeared within 21 days after the last dose of the chronic nicotine dosage regimen. In addition to the rapid loss of tolerance, response rates over this period following saline injections also declined to levels below initial baseline rates suggesting a behavioral withdrawal; this has also been observed by Collins (this volume). It should be added that these effects were independent of

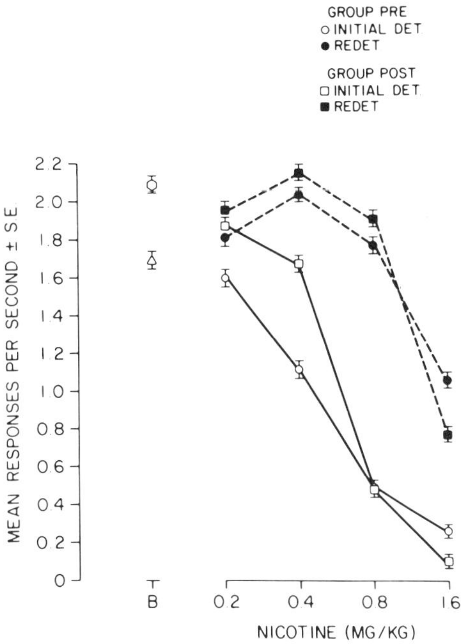

Fig. 4 The effects of single injections of nicotine on response rates per sec to mice in group pre (circles) and group post (squares) prior to (open symbols) and during (closed symbols) chronic treatment with 1.2 mg/kg nicotine. B=Baseline responding prior to the initiation of chronic nicotine administration. Reproduced from Hendry and Rosecrans, Psychopharmacology 77:339-343, 1982 (24) (with permission).

whether rats received chronic nicotine pre- or postsession, and suggest that these effects were related to the pharmacological nature of tolerance development.

To obtain more information concerning potential neurochemical mechanisms involved in postnicotine withdrawal, an analogous study was conducted in mice using the same nicotine-dosing regimen. Mice were administered nicotine (1.2 mg/kg s.c.) once a day for 30 days and killed at 1, 7 and 14 days after the last dose of nicotine. The results of this study indicated that nicotine had no significant acute or chronic effects on receptor binding (QNB, diazepam, or DA), nor were any peptides affected acutely (27,28). Chronic nicotine administration did induce a significant decrease (36%) of hypothalamic immunoreactive beta-endorphin (BE), but not met-enkephalin (Fig. 5), 24 hrs after the last nicotine dose. The reduction of BE was less 7 days later (25%), and there was evidence of an overshoot of BE levels 7 days later. This finding was most unexpected, and should be studied in greater detail. Our initial interpretation of these results was that either nicotine is inducing a very specific interaction with endorphin-containing neurons that may be related to its up-regulation of nicotinic receptors, or that this effect could be the result of a nonspecific stress-induced activation of BE-containing neurons. These data, especially the overshoot of BE levels at day 14, suggest that BE-containing neurons were either chronically inhibited by nicotine, or that BE synthesis was unable to keep up with a nicotine-induced release (directly or indirectly via a neuronal feedback), or both. Thus, the increase in BE during the postwithdrawal period could

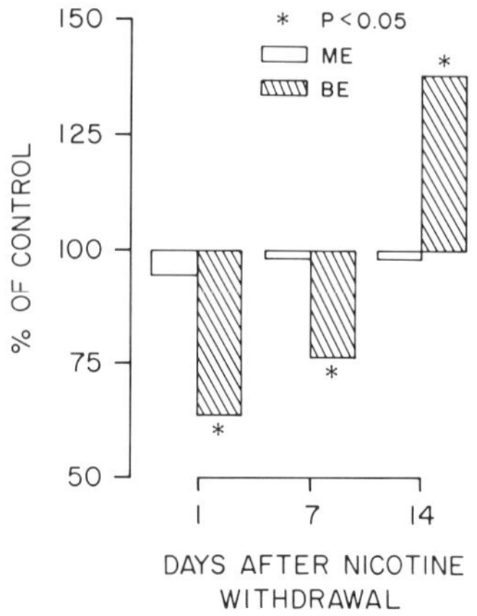

Fig. 5 Hypothalamic immunoreactive met-enkephalin (ME) and beta-endorphin (BE) levels 1, 7, and 14 days after the cessation of chronic nicotine (1.2 mg/kg, s.c.) administration in the mouse. Reproduced from Rosecrans et al., Pharmacol. Biochem. Behav. 23:141-143, 1985 (with permission).

be a reflection of an increase in rate of BE synthesis resulting from these chronic nicotine-induced effects, or a diminished degradation.

CONCLUSIONS

Mechanisms of Action of Nicotine

Research conducted in this laboratory has provided a wealth of information concerning the psychopharmacologic effects of nicotine. However, many questions remain unanswered. Our initial view of how nicotine can alter neuronal function stems from our earlier concepts that this cholinergic drug acts at specific receptors sensitive to ACh (1). However, our perceptions have been considerably altered by the observations that mecamylamine does not appear to bind at nicotinic-cholinergic receptors, and that we were unable to mimic the nicotine cue by increasing brain ACh levels (Fig. 2). In addition, whether nicotinic receptors are pre- and/or postsynaptic, and whether these receptors are under the influence of other neurotransmitters, has complicated our perceptions as to how nicotine's CNS effects are mediated. The observations that nicotine and/or nicotinic receptors may be important to the development of Parkinsons's (Baron, this volume) and Alzheimer's diseases (Kellar, this volume) also supports the important role that N-Ch receptors may play in brain function.

Thus, the neurochemical substrates upon which nicotine can act may be diverse; and conversely, it is clear that ACh-containing neurons, N-Ch or M-Ch, can be regulated by other neurotransmitter systems such as DA. If we look at one major cholinergic site, the hippocampus, the puzzle becomes even more intriguing (Fig. 6). As can be observed, hippocampal cholinergic function is under the control of several biogenic amine systems as well as the inhibitory neurotransmitter gamma-aminobutyric acid. In fact, there is additional evidence to suggest that ACh turnover may also be under the control of selected neuropeptides (29). In addition

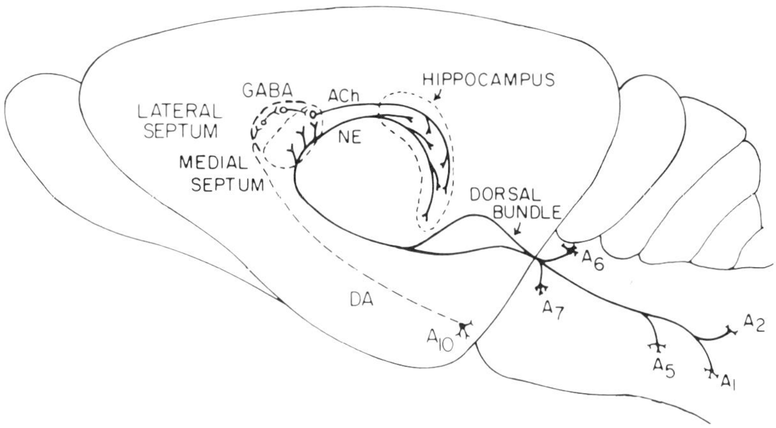

Fig. 6 Interrelationships between hippocampal acetycholine (ACh) turnover with other neuronal systems projecting to this organ. Redrawn from Robinson, SE: In: (ed.) Singh, MM, Warburton, DM and Lal, H, Central Cholinergic Mechanisms and Adaptive Dysfunctions, Plenum Publishing Co., NY, pp. 376, 1984 (with permission).

to the apparent control of ACh via select amine pathways, there have been some reports on the potential interaction between N- and M-Ch receptors within the hippocampus that suggest that these neurons may be mutually antagonistic (30,31,32). Thus, the ability of DA depletion to alter the DS properties of nicotine (Fig. 3), for example, could be related to an alteration of the M-Ch/N-Ch neuronal balance caused by the destruction of DA neuronal pathways innervating the hippocampus that exert a tonic inhibitory influence on ACh turnover (21). In addition to the role that catecholamine projecting systems may play in controlling ACh-containing neurons, Robinson and co-workers (33) have also observed that raphe nucleus lesions can facilitate ACh turnover in either the hippocampus or frontal cortex, depending upon which forebrain serotonin-projection system was destroyed. In all cases these amine pathways appear to exert a tonic inhibitory control over ACh turnover. Thus, the neuronal mechanisms by which nicotine can affect behavior are complex, and the task at this point will be to separate out which systems are involved with specific behavioral and/or physiological functions affected by nicotine.

The Nicotine Dependence Process

There are two aspects of nicotine-induced dependence which need to be discussed. First, does nicotine produce physical dependence? The preliminary data obtained in the mouse suggests that it does (Fig. 5). There appears to be evidence of a withdrawal syndrome in the mouse following cessation of chronic nicotine doses, especially in relation to hypothalamic beta-endorphin levels. In support of these animal data, preliminary observations with the drug clonidine, which has been used to reduce withdrawal symptoms in heroin-dependent individuals, indicate that this compound may attenuate withdrawal symptoms in smokers as well; these findings suggest some relationship between noradrenergic and opiate systems and chronic nicotine exposure (34). However, these studies need to be replicated and extended before any conclusions can be drawn with respect to the relationship of nicotine's chronic effects to other drug dependencies.

The second issue which needs to be examined concerns the behavioral properties which lead to, and maintain, smoking behavior. A major contention of several researchers is that nicotine can serve as primary reinforcer and this is why humans use tobacco (35,36). While this is an attractive hypothesis, it may not be as well founded as one would like. First of all, the doses used to test this hypothesis ranged from 1-2 mg administered intravenously. Such doses are rather large when considering the actual levels of nicotine taken via tobacco products and it should be realized that these intravenous nicotine doses are quite aversive to many of the subjects studied (36). Studies conducted in our laboratories (37) indicate that humans can detect a 1-mg difference in nicotine between different cigarettes when smoked in the usual manner; smoking is quite different from a single bolus of 1-2 mg intravenously. Subjects in our study provided us with very few clues as to how they were able to detect a 1-mg difference in nicotine concentration. Throat irritation or some dizziness were the only clues volunteered. Even more interesting was the fact that more than half of the subjects studied were unable to detect any subjective effect from smoking cigarettes containing 1.3 mg of nicotine, even though they were, in fact, able to detect differences between high and low nicotine concentrations as measured by nicotine-correct lever responding. Thus, none of our subjects experienced the positive subjective effects that others have reported (37).

The last aspect to be discussed is the behavioral factors which may be involved in the dependence process. Drug dependence as defined by this author is related to the ability of one to extinguish chronic drug use. As evident from many reports, smokers have great difficulty in stopping smoking, and thus by definition tobacco is a dependence-producing substance; tobacco dependence is probably due to the presence of nicotine. We are especially interested in two aspects of the dependence process: conditioning factors, and nicotine-induced, state-dependent learning. Regardless of the reasons for the initiation of tobacco use, one is exposed to some level of nicotine, which, as a stimulus, can be paired with a variety of environmental stimuli and form a number of conditioning complexes. The significance of this interaction becomes even more evident when one realizes that the average smoker will be exposed to over 7000 (20 cigarettes/day) opportunities to produce a conditioned relationship with his or her environment in one year. The specificity of its pharmacology and ability to enter the brain rapidly via smoking (7 sec; 40) makes nicotine an ideal stimulus. In addition to the conditioning element, the state-dependent properties of nicotine probably play a major role in the maintenance of this behavior. That is, the more one continues to function under the nicotine state, the more one needs to smoke in order to continue to function in specific situations associated with tobacco use. This was well documented by Warburton and Wesnes (this volume).

What has been attempted in this review is to demonstrate the behavioral and neurochemical complexities of nicotine's effects in rats and mice. What has also been attempted was to try to convince the reader that nicotine use and dependency is not a simple matter of nicotine producing a pharmacologically-induced positive reinforcing effect, but may be the result of a constellation of behaviors including its ability to exert stimulus control of behavior. Furthermore, because of its ability to serve as stimulus, it is further hypothesized that the use of nicotine can be conditioned by environmental stimuli as well. In addition, chronic nicotine exposure may also alter cholinergic up-regulation of nicotinic ACh-sensitive binding sites, as well as brain neuropeptide function. Thus, in addition to all of the behavioral and sociological contingencies which make it difficult to stop smoking, one may have to also contend with possible affective behavioral fluctuations induced by alterations in brain neurotransmitter function. Finally, it should be realized that this discussion is not unique to nicotine, but may be applicable to all dependence-producing drugs. However, nicotine may provide us with a better example of these processes because of its unique and subtle properties.

SUMMARY

The major objective of research conducted in this laboratory has been to determine central sites and mechanisms of action of the psychopharmacological effects of nicotine which may be important to its dependence-producing effects. We have utilized a variety of different approaches to study nicotine, and find it to be a complex drug in relation to how it affects both behavior and a variety of neurotransmitter substances in the brain. While it seems to be somewhat clear to some that nicotine may be acting at cholinergic receptors sensitive to acetylcholine, evidence obtained from this laboratory suggests that nicotine may have very prominent effects on noncholinergic receptors as well, especially in relation to its discriminative stimulus (DS) effects. Furthermore, it seems that catecholamine-containing neurons, primarily dopamine neurons, may play a prominent role in nicotine's psychopharmacological effects, which may be helpful to understanding nicotine's apparent positive reinforcing effects.

In addition to utilizing the DS effects of nicotine as a means of determining how nicotine is acting at specific brain sites, we have also observed that chronic nicotine administration in the mouse produces biphasic effects on the brain neuropeptide beta-endorphin (BE) that are suggestive of a physiological/psychological withdrawal syndrome. These observations, combined with nicotine's ability to induce stimulus control of behavior, has led us to develop the hypothesis that nicotine is capable of producing physical as well as psychological dependence. The basic hypothesis, however, emphasizes the role that the stimulus properties of nicotine play in the dependence process. Nicotine as a discriminative stimulus is capable of producing conditioned relationships with other environmental stimuli such that smoking behavior may be controlled to various degrees by a variety of environmental stimuli. Therefore, smokers with a history of many conditionings would find this behavior extremely difficult to extinguish. It should be remembered that a typical smoker (20 cigarettes per day) is exposed to at least 7000 conditioning possibilities per year. In addition, the ability of nicotine to induce state-dependent behavioral effects and the potential for a physical withdrawal syndrome, as subtle as they may be, produce an added burden on those individuals attempting to stop smoking. Thus, from a public health point of view, nicotine should no longer be considered only habit-producing, but should be classified with other drugs exhibiting pronounced dependence-producing effects such as the CNS stimulants, depressants, and opiates.

ACKNOWLEDGEMENTS

The author would like to thank the AMA Education Research Foundation and the Tobacco Research Council, USA, for their generous support of this research. The author would also like to acknowledge the research efforts of the following: Drs. MD Schechter, ID Hirschhorn, MJ Kallman, ME Elchisak, WT Chance, JS Hendry, J-S Hong, and LT Meltzer.

REFERENCES

1. Domino, EF: Neurophysiology of nicotine and tobacco smoking. In: (ed.) Dunn, WL, Jr, Smoking Behavior: Motives and Incentives. H. Winston and Sons, Washington, DC, pp. 5-31, 1973.
2. Hendry, JS and Rosecrans, JA: Effects of nicotine on conditioned and unconditioned behaviors in experimental animals. Pharmacol. Therap. 17:431-454, 1982.
3. Rosecrans, JA, Kallman, MJ and Glennon, R: The nicotine cue: An overview. In (ed.) Colpaert, FC and Rosecrans, JA, Stimulus Properties of Drugs: Ten Years of Progress. Elsevier Biomedical Press, Amsterdam, pp. 69-82, 1978.
4. Schechter, MD and Rosecrans, JA: CNS effect of nicotine as the discriminative stimulus for the rat in a T-maze. Life Sci. 10:821-832, 1971.
5. Hirschhorn, ID and Rosecrans, JA: Studies on the time course and the effect of cholinergic and adrenergic receptor blockers on the stimulus effect of nicotine. Psychopharmacology 40:109-120, 1974.
6. Pratt, JA, Stolerman IP, Garcha, HS, Giardini, V, and Feyerabend, C: Discriminative stimulus properties of nicotine: further evidence for mediation at a cholinergic receptor. Psychopharmacology 81:54-60, 1983.
7. Stolerman, IP, Garcha, HS, Pratt, JA and Kumar, R: Role of training dose in discrimination of nicotine and related compounds by rats. Psychopharmacology 84:413-419, 1984.
8. Rosecrans, JA and Meltzer, LT: Central sites and mechanisms of action of nicotine. Neurosci. Biobehav. Rev. 5:489-497, 1982.
9. Abood, LG, Lowy, K, Tometsko, A and Booth, H: Electrophysiological, behavioral and chemical evidence for a noncholinergic, stereospecific site for nicotine in rat brain. J. Neurosci. 3:327-333, 1978.
10. Nelsen, JM, Pelley, K and Goldstein, L: Protection by nicotine from behavioral disruption caused by reticular formation stimulation in the rat. Pharmacol. Biochem. Behav. 3:749-754, 1975.
11. Meltzer, LT, Rosecrans, JA, Aceto, MD and Harris, LS: Discriminative stimulus properties of the optical isomers of nicotine. Psychopharmacology 68:283-286, 1980.
12. Meltzer, LT and Rosecrans, JA: Investigations on the CNS sites of action of the discriminative stimulus effects of arecoline and nicotine. Pharmacol. Biochem. Behav. 15:21-26, 1981.
13. Rosecrans, JA, Dren, AT and Domino, EF: Effects of physostigmine on rat brain acetylcholine, acetylcholinesterase and conditioned pole jumping. Neuropharmacology 7:127-134, 1968.
14. Schwartz, RD, McGee, R and Kellar, KJ: Nicotinic cholinergic receptors labelled by [^{3}H]acetylcholine in rat brain. Mol. Pharmacol. 22:56-62, 1982.
15. Kellar, KJ, Martino, AM, Hall, DP, Schwartz, RD and Taylor, RL: High-affinity binding of [^{3}H]acetylcholine to muscarinic cholinergic receptors. J. Neurosci. 5:1577-1582, 1985.
16. Kessler, II, Parkinson's disease. Perspectives on epidemiology and pathogenesis, In: Perspectives on Parkinson's Disease, Academic Press, NY, pp. 88-105, 1973.
17. Kessler, II and Diamond, EL: Epidemiologic studies of Parkinson's disease. I. Smoking and Parkinson's disease: A survey and explanatory hypothesis. Amer. J. Epidemiol. 94:16-25, 1971.
18. Singer, G, Wallace, M and Hall, R: Effects of dopaminergic nucleus accumbens lesions on the acquisition of schedule induced self injection of nicotine in the rat. Pharmacol. Biochem. Behav. 17:579-581, 1982.

19. Rosecrans, JA, Chance, WT, Schechter, MD: The discriminative stimulus properties of nicotine, d-amphetamine and morphine in dopamine depleted rats. Psychopharmacol. Commun. 2:349-356, 1976.
20. Rosecrans, JA, Spencer, RM, Krynock, GM and Chance, WT: Discriminative properties of nicotine and nicotine-related compounds. In: (ed.) Battig, K, Behavioral Effects of Nicotine, Karger Press, Basel, pp. 70-82, 1978.
21. Robinson, SE, Malthe-Sorenssen, D, Wood, PL and Commissiong, J: Dopaminergic control of the septal-hippocampal cholinergic pathway. J. Pharmacol. Exp. Therap. 208:476-479, 1979.
22. Schwartz, RD, Lehmann, J and Kellar, KJ: Presynaptic nicotinic cholinergic receptors labeled by [^{3}H]acetylcholine on catecholamine and serotonin axons in brain. J. Neurochem. 42:1495-1498, 1984.
23. Kallman, MJ, Harry, GJ, Elchisak, MA and Rosecrans, JA: Behavioral effects of chronic nicotine in adult rats depleted of dopamine. Fed. Proc. 45:428, 1986.
24. Hendry, JS and Rosecrans, JA: The development of pharmacological tolerance to the effect of nicotine on schedule-controlled responding in mice. Psychopharmacology 77:339-343, 1982.
25. Kellar, KJ and Schwartz, RD: In vivo regulation [^{3}H]acetylcholine recognition sites in brain by nicotinic cholinergic drugs. J. Neurochem. 45:427-433, 1985.
26. Meltzer, LT and Rosecrans, JA: Tolerance to the disruptive effects of arecoline on schedule-controlled behavior. Psychopharmacology 77:85-93, 1982.
27. Hendry, JS, Rosecrans, JA, Hong, J-S and Bondy, SC: Chronic nicotine reduces hypothalamic beta-endorphin in the mouse. Fed. Proc. 41:1077, 1982.
28. Rosecrans, JA., Hendry, JS and Hong, J-S: Biphasic effects of chronic nicotine treatment on hypothalamic immunoreactive beta-endorphin in the mouse. Pharmacol. Biochem. Behav. 23:141-143, 1985.
29. Botticelli, LJ and Wurtman, RJ: Septohippocampal cholinergic neurons are regulated trans-synaptically by endorphin and corticotrophin neuropeptides. J. Neurosci. 2:1316-1321, 1982.
30. Ross, JF and Grossman, SP: Intrahippocampal application of cholinergic agents and blockers. J. Comp. Physiol. Psychol. 86:590-600, 1974.
31. Segal, M: The acetylcholine receptor in the hippocampus: nicotinic, muscarinic or both? Neuropharmacology. 17:619-623, 1978.
32. Rovira, C, Cherubini, E and Ben-Ari, Y: Opposite actions of muscarinic and nicotinic agents on hippocampal dendritic negative fields recorded in rats. Neuropharmacology 22:239-243, 1983.
33. Robinson, SE: Effect of specific serotonergic lesions on cholinergic neurons in the hippocampus, cortex and striatum. Life Sci 32:345-353, 1983.
34. Glassman, AH, Jackson, WK, Walsh, BT and Roose, SP: Cigarette craving, smoking withdrawal, and clonidine. Science 226:864-866, 1984.
35. Goldberg, SR, Spealman, RD, Risner, ME and Hennigfield, JE: Control of behavior by intravenous nicotine injections in laboratory animals. Pharmacol. Biochem. Behav. 19:1011-1020, 1983.
36. Henningfield, JE and Goldberg, JR: Control of behavior by intravenous nicotine injections in human subjects. Pharmacol. Biochem. Behav. 19:1021-1026, 1983.
37. Kallman, WM, Kallman, MJ, Harry, GJ, Woodson, PP and Rosecrans, JA: Nicotine as a discriminative stimulus in human subjects. In: (ed.) Colpaert, FC and Slangen JL, Drug Discrimination: Application in CNS Pharmacology, Elsevier Biomedical Press, Amsterdam, pp. 211-218, 1982.

INTRACRANIAL SELF-STIMULATION STUDIES WITH NICOTINE

Paul B.S. Clarke[1] and R. Kumar[2]

[1]Kinsmen Laboratory of Neurological Research
Dept. of Psychiatry
Univ. of British Columbia
Vancouver, B.C.
Canada V6T 1W5
[2]Dept. of Psychiatry
Institute of Psychiatry
Univ. of London
London SE5 8AF
England

INTRODUCTION

More than thirty years ago James Olds and Peter Milner (1954) fortuitously observed that a rat would return to the place in an open field where it had previously received electrical stimulation of the brain (ESB) delivered via a septal electrode. Subsequently, it was found that rats and many other species including man would learn to perform arbitrarily assigned tasks in order to obtain electrical stimulation at certain sites in the brain (see Olds, 1977). This behavior, termed "self-stimulation" or "intracranial self-stimulation" (ICSS), appeared to offer a direct handle on putative systems in the brain responsible for pleasure or reward.

Although ESB often seems to be highly rewarding, it has proven difficult to quantify these rewarding properties, which is presumably a necessary step in the elucidation of putative central reward substrates. Initially, a variety of behavioral methods were employed in animals but problems of measurement and interpretation were soon recognized (Valenstein, 1964). Despite the burgeoning ICSS literature, we still do not know of the precise neurochemical basis for brain stimulation reward (BSR). A lack of reliable behavioral indices of BSR may be partly responsible. In recent years the neurotransmitter dopamine has been recognized to play a key role both in the mediation of BSR and in the reinforcing properties of psychostimulant and opiate drugs (Fibiger, 1984; Phillips, 1984).

Nicotine is consumed in a variety of ways by man but the basis of this behavior is not clear. The alkaloid exerts several pharmacological actions which could conceivably be reinforcing. Central stimulant properties are found both in man and animals, and questionnaire studies suggest that some people smoke tobacco in order to obtain a stimulant effect (Kumar and Lader, 1981). Nicotine stimulates the activity of ascending dopaminergic pathways in the brain (see Pert and Clark, this volume), raising the possibility that the drug acts on the same neural substrate as do both rewarding brain stimulation and psychomotor stimulants such as amphetamine. In order to explore possible interactions between nicotine and rewarding brain stimulation, it was first necessary to devise a method for quantifying BSR in rats.

Theoretical Aspects

Animals will work for electrical stimulation of the brain, and, following the terminology of Skinner (1938), ESB which increases the frequency of behaviors upon which it is made contingent is regarded as a positive reinforcer. Since an animal can generally be trained to respond in a variety of topographically different ways for ESB, ICSS is not simply a reflex behavior (Skinner, 1938). Indeed, rats will cross an electrified grid in spaced trials in order to obtain ESB (Olds, 1958), indicating that in some situations, at least, ESB is predominantly rewarding. In other situations a second process may be envisaged: responding may be maintained by an immediate effect of the ESB which is neither rewarding nor aversive. For example, Justeson et al. (1963) reported that in cats which had received no training for ICSS, brain stimulation, which was not rewarding, nevertheless maintained responding above control rates through a nonspecific acceleration of the operant response rate. Electrical stimulation at certain brain sites increases behavioral signs of arousal (e.g., Rolls and Kelly, 1972) which may be channeled into a transient increase in responding. Most probably, rewarding ESB exerts additional effects unrelated to reward. Self-stimulating rats have been variously reported to experience hyperthermia, autonomic effects, hypothalamo-pituitary responses, turning movements, analgesia, perioral stereotypies and other motor effects, seizures, and "stimulus-elicited" behaviors (see Valenstein, 1980).

In order to analyze ICSS, we assume that ESB has two types of consequence: a reward component (i.e., BSR) and a nonreward component, which includes any effects of the stimulation not related to reward which may affect responding for ESB. In this scheme a drug may alter ICSS performance in three ways: by enhancing or reducing the reward component of ESB, by interacting with the nonreward component of ESB so as to alter its influence on ICSS performance, and by exerting effects of its own on performance which are independent of either ESB component. The purpose of our experiments, described below, was to examine whether nicotine altered the brain's sensitivity to brain stimulation reward.

Studies of Nicotine and ICSS

In contrast to several other drugs which are self-administered by man, nicotine has attracted scant attention in studies of ICSS. In the few such studies which preceded our own (Clarke and Kumar, 1983a, 1984), the principal dependent measure was the rate of lever pressing for rewarding ESB. Thus, the question of whether nicotine alters brain stimulation reward magnitude had not been seriously addressed. In addition, the chronic effects of nicotine had not been systematically investigated.

Nicotine can stimulate or depress ICSS lever responding in nontolerant animals. Typically, response rates are reduced for several minutes after systemic administration, especially at higher doses and if predrug response rates are high (Olds and Domino 1969a; Pradhan and Bowling, 1971), whereas responding is stimulated at longer intervals after injection and if the baseline rate is low (Wanner and Battig, 1966; Olds and Domino, 1969a & b; Pradhan and Bowling, 1971; Newman, 1972). Broadly similar effects of nicotine had been reported across a range of conditioned and unconditioned behaviors, including responding for food or water (Morrison, 1967; Stitzer et al. 1970), avoidance responding (Bignami and Michalek, 1978), and locomotor activity (Morrison and Lee, 1968; Clarke and Kumar, 1983b). Since nicotine appeared to produce unselective alterations of responding, it was particularly necessary to devise a behaviorally specific measure of BSR.

For our own ICSS experiments in rats, electrodes were positioned in the medial forebrain bundle at the level of the lateral hypothalamus so as to impinge upon dopaminergic fibres ascending from the pons. Stimulation at this brain site carries little, if any, aversive component (Olds and Olds, 1963), and the possibility of aversive effects was further reduced by employing short pulse trains of electrical stimulation (Shizgal and Matthews, 1977). In an attempt to reduce possible drug-induced disruption of performance, the conventional lever press response was rejected in favor of photobeam interruption. In order to minimize the possibility of motoric side effects of the ESB, it was necessary to keep deliveries of ESB well spaced temporally. In both procedures described below, the animal was presented with a choice between two comparable responses, only one of which resulted in the delivery of ESB.

Y Maze ICSS Experiments

Full details are given elsewhere (Clarke and Kumar, 1983a). In brief, rats with chronically indwelling lateral hypothalamus/medial forebrain bundle (LH/MFB) electrodes were trained in a symmetrical Y maze. During most of the 80-minute session, two arms of the maze provided BSR. The rat received a brief (0.5 ms) pulse train when it reached the end one of these two "ON" arms and interrupted a photobeam. Repetitive entries into the same photobeam were not rewarded - the next available pulse train was made available in the other ON arm, and hence the rat received BSR by shuttling between the two ON arms. The third arm did not provide ESB and served to monitor unrewarded entries. Approximately every minute, the two ON arms were randomly reselected, and when this occurred, the rat redetermined where the brain stimulation was available. The first ten responses made after the arms had been reallocated were used to determine the rate-free measure of BSR. This measure, termed "accuracy," was calculated as the percentage of responses which were rewarded. Chance performance yielded a value of around 67%. In addition, the overall rate of rewarded responding (i.e., ON arm entries per minute) was determined. Within a session, both measures were assessed as a function of current intensity. Intermittently throughout the session there were signaled periods of "time-out" when ESB was unavailable in any arm.

Following training, there were three consecutive stages of testing: (1) acute-dose response study of nicotine, (2) tests with nicotine and mecamylamine, and (3) chronic effects of nicotine. Test doses of nicotine bitartrate were administered subcutaneously immediately before the 80-minute session. Doses of nicotine and mecamylamine refer to the base of each compound; those of d-amphetamine sulphate refer to the salt.

Acute-Dose Response Study

Rats were initially drug-naive and were each tested with saline and with five doses of nicotine (0.025 - 0.4 mg/kg) in a random order. In the first 20 minutes nicotine markedly depressed both rewarded and unrewarded responding in a dose-related way; ataxia and prostration occurred, especially at the higher doses. Too few responses were made under drug influence to determine accuracy of responding. From 40 to 80 minutes the predominant drug action was stimulant, and this was dose-dependent at certain current intensities and also in periods of time-out. Nicotine increased response rates without altering accuracy.

Pretreatment with Mecamylamine

When given alone, mecamylamine (2.0 mg/kg sc) produced a modest decrease in response rates, whether ESB was available or not, but failed to alter accuracy significantly. The antagonist completely prevented the initial depressant actions of nicotine (0.4 mg/kg) described above. Antagonism of nicotine's later stimulant actions was not statistically significant.

Chronic Effects of Nicotine

Rats were tested for 24 consecutive days. After several baseline sessions, subjects were given seven consecutive pairs of tests. Each pair consisted of a test with saline and with nicotine (0.4 mg/kg) given in a counterbalanced order across subjects. On a day when a subject was tested with saline, it received nicotine (0.4 mg/kg sc) after the session. As shown in Fig. 1, the initial depression of high response rates disappeared with chronic administration, and the stimulant action of nicotine, seen at lower control rates of responding, increased with chronic treatment.

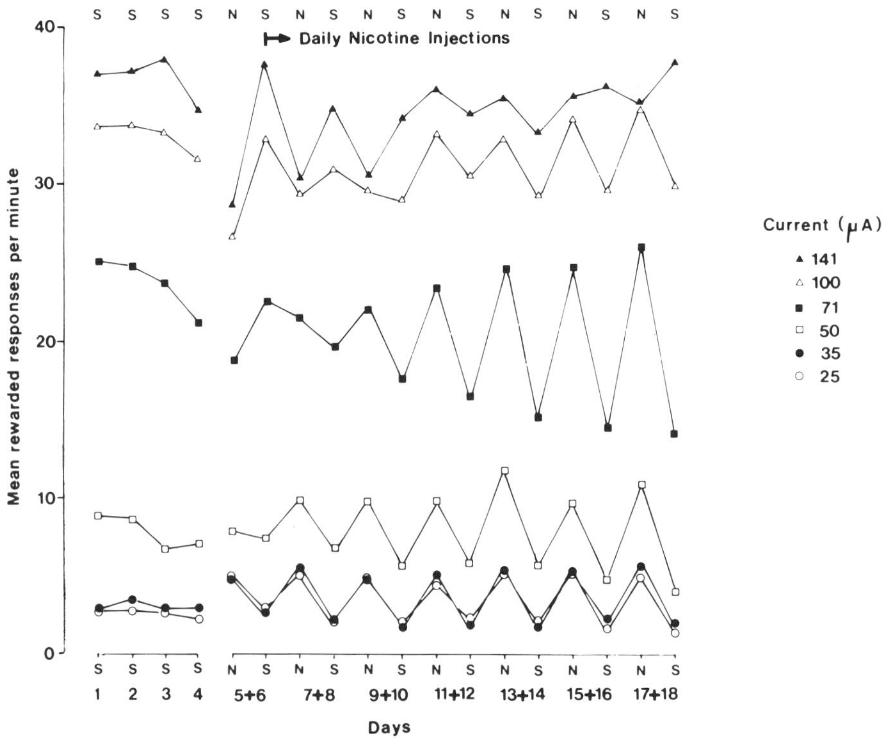

Fig. 1. The effects of chronic nicotine on rates of rewarded responding (Y maze). After four saline tests, rats (n = 7) were tested daily, alternately just before or just after an injection of nicotine (0.4 mg/kg sc). At first (days 5 and 6), nicotine enhanced low rates of responding and depressed high rates. Over subsequent days, the depressant action waned and a stimulant action emerged or became more pronounced. Saline response rates declined over successive tests. From Clarke and Kumar, 1983a.

Subsequently, the chronic nicotine treatment regime was continued, and the drug was found to increase rates of rewarded responding in a dose-related fashion (0.05 - 0.4 mg/kg) without significantly affecting response accuracy (Fig. 2). As further evidence of a general stimulant action, nicotine also increased time-out responding in a dose-dependent way.

The Y maze ICSS experiments therefore indicated that nicotine could stimulate or depress responding in a behaviorally nonspecific manner, depending not only on dose, interval after injection, and baseline response rate, but also on previous administration of the drug. In particular, robust dose-dependent stimulant effects were observed in rats that had become tolerant to the depressant action of the drug. Most importantly, nicotine failed to alter the rate-free measure of BSR under conditions where it reliably increased rates of responding. Parallel increases in time-out responding provide further evidence for a broad behavioral spectrum of action.

Shuttle Box ICSS Experiments

Confirmation of these conclusions was then sought, using a second "rate-free" procedure for measuring BSR, refined from that of Valenstein and Meyers (1964). Full details are given elsewhere (Clarke and Kumar, 1984). Rats were implanted with chronic LH/MFB electrodes as before, and trained to self-stimulate in a shuttle box. The box was equipped with two photobeams running along opposite walls. By breaking one of these

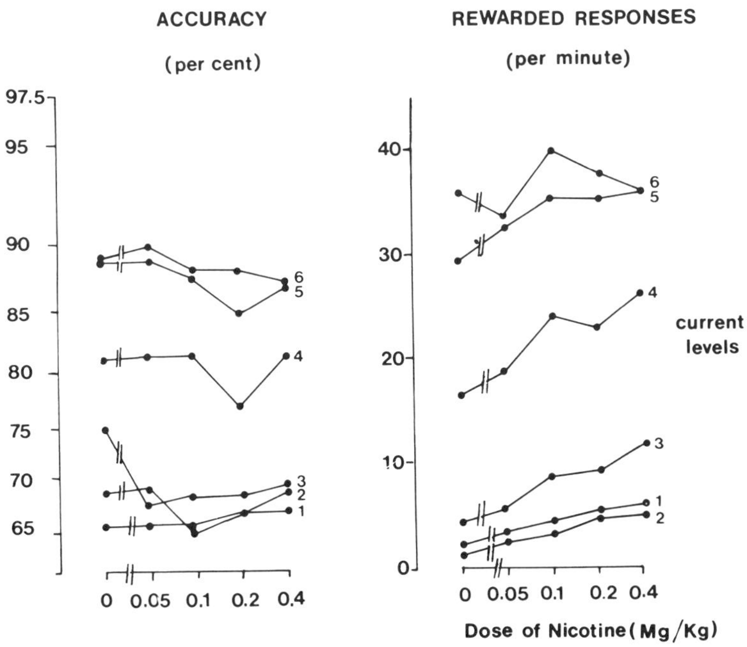

Fig. 2. Dose response study in tolerant rats (Y maze). Accuracy data are plotted as the mean (± SEM) of the transformed individual scores (n = 6), in equal arcsin square root units. Current levels 1 - 6 refer to logarithmically spaced ascending current intensities. Nicotine increased rewarded responding in a dose-related way without significantly altering accuracy. From Clarke and Kumar, 1983a.

photobeams (the "ON" beam), a subject could initiate the delivery of intermittent but regularly occurring trains of ESB. Brain stimulation could be terminated by interrupting the other "OFF" beam, and reinitiated and terminated *ad libitum*. The ON and OFF beams reversed every 30 sec, and each time this occurred, the rat was permitted 10 sec to reorient before data was collected. Brain stimulation reward was measured by the amount of time that the rat spent receiving current over four successive intervals of 20 sec. This measure, termed "self-stimulation time" (SST), had a maximum value of 80 sec and a chance value of 40 sec. The number of initiating or terminating responses was counted concurrently with SST.

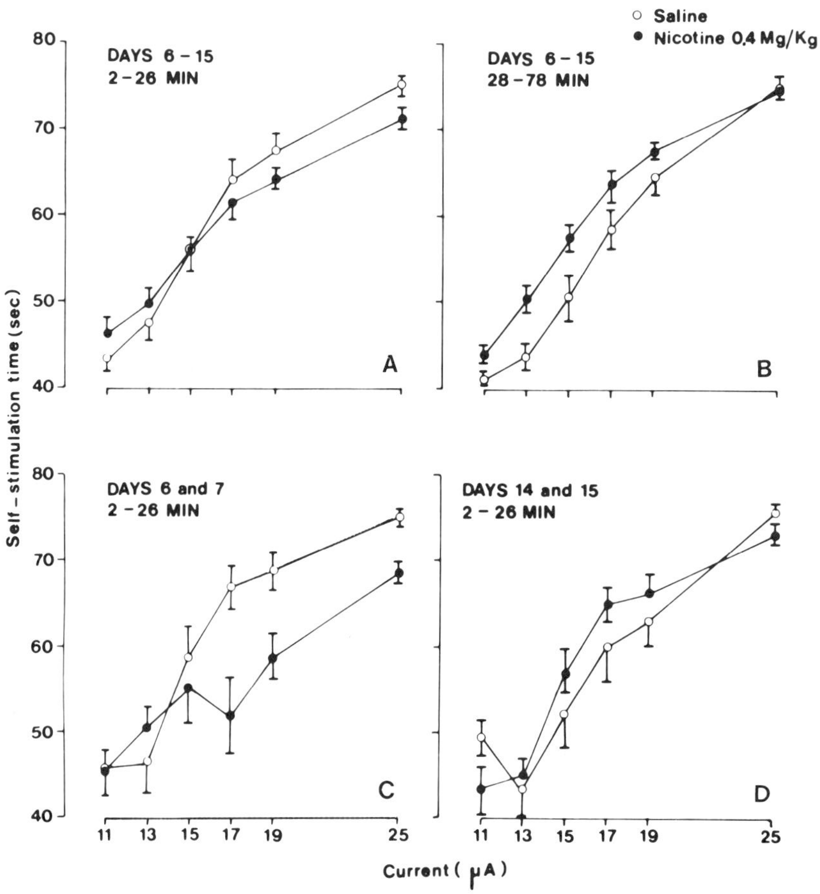

Fig. 3. (A-D): Effects of nicotine on self-stimulation time (SST) in a shuttle box. Subjects were tested on ten consecutive days, saline tests alternating with nicotine (0.4 mg/kg sc) tests. The top two figures show data averaged across all five saline tests (open circles) and all five nicotine tests (closed circles). The amount of time receiving brain stimulation (SST) increased with increasing current from around chance (40 s) to near maximum (80 s). In the first third of the session (A), nicotine's actions varied across successive tests; in drug-naive rats (C), nicotine reduced SST and also impaired motor performance, but tolerance developed to these effects by the final drug test (D). In the last two-thirds of the session, nicotine enhanced SST (B). Bars represent one SEM either side of the group mean (n = 10). From Clarke and Kumar, 1984.

Both measures were obtained as a function of current within each session. Rats were tested daily, and once baseline (saline) performance had stabilized, rats received several pairs of saline and nicotine (0.4 mg/kg) tests while receiving a constant daily dose of the drug (0.4 mg/kg), in similar fashion to the Y maze experiment.

As expected, undrugged subjects spent more time receiving the brain stimulation the higher the current (Fig. 3). When first tested with nicotine (days 6 and 7), they became motorically impaired, which probably accounts for the reduced SST scores early in the session. By the last nicotine test (days 14 and 15), neither effect was observed. In the latter part of the session (26 - 78 min), nicotine increased SST and this effect did not change across successive tests. Drug-induced enhancement of SST was accompanied by a marked increase in responding, even during periods when current was not available. Subsequently, closely similar results were obtained with d-amphetamine sulphate (0.25, 0.75 mg/kg), which has been reported to enhance BSR in a variety of ICSS paradigms.

GENERAL DISCUSSION

Thus, using two rather different behavioral procedures, both providing rate-independent measures of BSR, we obtained two different results: in the Y maze, nicotine appeared not to enhance the magnitude of BSR, whereas in the shuttle box precisely this effect was observed. It is difficult to reconcile these results. It seems highly unlikely that the drug-induced increase in responding contributed to the enhancement of BSR that was seen in the shuttle box; the animal was required to withhold movements into the OFF photobeam in order to obtain maximal brain stimulation. Moreover, at the lowest current level both nicotine and d-amphetamine produced a preference for brain stimulation where none existed in the undrugged state. Such a result cannot be attributed to hyperactivity. In the Y maze, nicotine tended to impair the accuracy measure at high current intensities, and although this trend was not statistically significant, it may indicate that the drug impaired discrimination ability per se. Such an impairment could have masked a drug-induced enhancement of brain stimulation reward. Conceivably, different putative neural systems underlie BSR, depending on the operant response (White, Brown and Yachim, 1978). Whatever the basis of our conflicting results, it appears on balance that nicotine can enhance brain stimulation reward, at least under certain conditions.

How may our studies of ICSS be related to recent demonstrations of nicotine-induced place preference conditioning (see Iwamoto, this volume) and intravenous self-administration of this drug (see Goldberg, this volume) in elucidating nicotine's reinforcing actions? In the rat, responding for intravenous nicotine administration is very slow and is accompanied by marked noncontingent drug effects which may complicate interpretation of results (Cox, Goldstein and Nelson, 1984). Nevertheless, the intravenous self-administration paradigm offers the most direct approach to studying the primary reinforcing actions of nicotine, which presumably account for much of the tobacco smoking habit. Quite how conditioned place preference, a measure of secondary reinforcement produced by single spaced injections of nicotine, relates to intravenous self-administration is not clear yet.

Finally, is it possible that a more simple animal model exists for elucidating the neural mechanisms underlying nicotine reinforcement? Nicotine, especially in chronically treated animals, exerts a broad stimulant action which is readily quantified in locomotor activity cages (Morrison and Stephenson, 1972; Clarke and Kumar, 1983b and c; Clarke,

1984). In addition, nicotine appears to activate ascending dopaminergic systems (see Pert and Clark, this volume), an action shared by other psychomotor stimulants such as amphetamine. The reinforcing and stimulant actions of psychomotor stimulants appear to share a common dopaminergic substrate (Fibiger, 1984). Possibly, then, the reinforcing actions of nicotine are reflected most simply in the drug's locomotor stimulant action. Humans are said to jump for joy; the rat, a more prosaic creature, seems to run around more.

SUMMARY

Animals and man will perform intracranial self-stimulation; that is, they will work in order to obtain rewarding electrical stimulation delivered at certain brain sites. Several drugs of abuse, including amphetamine and morphine, appear to enhance the rewarding properties of such stimulation, and it has been proposed that interactions of this kind represent convergent actions on a central reward system. We therefore investigated the possible effects of nicotine on rewarding brain stimulation. Two "rate-free" methods were devised in order to measure the magnitude of the reward produced by electrical stimulation of the medial forebrain bundle in behaving rats. In a Y maze choice procedure, nicotine affected response rates without measurably altering the magnitude of brain stimulation reward. However, in a second choice procedure which employed a shuttle box, nicotine enhanced the rewarding nature of the electrical stimulation, as did d-amphetamine. In neurochemical terms, the release of mesolimbic dopamine is strongly implicated in drug- and electrically-induced reward, and it is of interest that nicotine appears to activate this neurotransmitter pathway.

REFERENCES

Bignami, G and Michalek, H: Cholinergic mechanisms and aversively motivated behaviors. In (eds.) Anisman, H and Bignami, G, Psychopharmacology of Aversively Motivated Behavior, Plenum, New York, 1978.

Clarke, PBS: Chronic central nicotinic blockade after a single injection of the bisquaternary ganglion blocking drug chlorisondamine. Br. J. Pharmac. 83:527-535, 1984.

Clarke, PBS and Kumar, R: Nicotine does not improve discrimination of brain stimulation reward by rats. Psychopharmac. 79:271-277, 1983a.

Clarke, PBS and Kumar, R: Effects of nicotine on locomotor activity in non-tolerant and tolerant rats. Br. J. Pharmac. 78:329-337, 1983b.

Clarke, PBS and Kumar, R: Characterization of the locomotor stimulant action of nicotine in tolerant rats. Br. J. Pharmac. 80:587-594, 1983c.

Clarke, PBS and Kumar, R: Effects of nicotine and amphetamine on intracranial self-stimulation in a shuttle box in rats. Psychopharmac. 84:109-114, 1984.

Cox, BM, Goldstein, A and Nelson, WT: Nicotine self-administration in rats. Br. J. Pharmac. 83:49-55, 1984.

Fibiger, HC: The neurobiological substrates of depression in Parkinson's Disease: a hypothesis. Can. J. Neurol. Sci. 11:105-107, 1984.

Goldberg, SR: this volume, 1986.

Iwamoto, ET: this volume, 1986.

Justeson, DR, Sharp, JC, and Porter, PB: Self-stimulation of the caudate nucleus by instrumentally naive cats. J. Comp. Physiol. Psychol. 56:371-374, 1963.

Kumar, R and Lader, MH: Nicotine and smoking. In (eds.) Essman, WB and Valzelli, L, Current Developments in Psychopharmacology, Vol. 6, pp 127-164, Spectrum, New York, 1981.
Morrison, CF: Effects of nicotine on operant behavior of rats. Int. J. Neuropharmac. 6:229-240, 1967.
Morrison, CF and Lee, PN: A comparison of the effects of nicotine and physostigmine on a measure of activity in the rat. Psychopharmac. 13:210-221, 1968.
Morrison, CF and Stephenson, JA: The occurrence of tolerance to a central depressant effect of nicotine. Br. J. Pharmac. 46:151-156, 1972.
Newman, LM: Effects of cholinergic agonists and antagonists on intracranial self-stimulation behavior in the rat. J. Comp. Physiol. Psychol. 79:394-413, 1972.
Olds, J: Self-stimulation of the brain. Science 127:315-324, 1958.
Olds, J and Milner, PM: Positive reinforcement produced by electrical stimulation of septal area and other regions of the rat brain. J. Comp. Physiol. Psychol. 47:419-427, 1954.
Olds, ME and Olds, J: Approach-avoidance analysis of rat diencephalon. J. Comp. Neurol. 120:259-295, 1963.
Olds, J and Domino EF: Comparison of muscarinic and nicotinic cholinergic agonists on self-stimulation behavior. J. Pharm. Exp. Ther. 166:189-204, 1969a.
Olds, J and Domino, EF: Differential effects of cholinergic agonists on intracranial self-stimulation and escape behavior. J. Pharm. Exp. Ther. 170:157-167, 1969b.
Olds, J: Drives and Reinforcements. Raven Press, New York, 1977.
Pert, A and Clark, P: this volume, 1986.
Phillips, AG: Brain reward circuitry: a case for separate neural systems. Brain Res. Bull. 12:195-201, 1984.
Pradhan, SN and Bowling, C: Effects of nicotine on self-stimulation in rats. J. Pharmacol. Exp. Ther. 176:229-243, 1971.
Rolls, ET and Kelly, PH: Neural basis of stimulus-bound locomotor activity in the rat. J. Comp. Physiol. Psychol. 81:173-182, 1972.
Shizgal, P and Matthews, G: Electrical stimulation of the rat diencephalon: Differential effects of interrupted stimulation on on- and off-responding. Brain Res. 129:319-333, 1977.
Skinner, BF: The Behavior of Organisms. Octavo, New York, 1938.
Stitzer, M, Morrison, J, and Domino, EF: Effects of nicotine on fixed interval behavior and modification by cholinergic antagonists. J. Pharm. Exp. Ther. 171:165-177, 1970.
Valenstein, ES: Problems of measurement and interpretation with reinforcing brain stimulation. Psychol. Rev. 71:415-437, 1964.
Valenstein, ES and Meyers, W: A rate-independent test of reinforcing consequences of brain stimulation. J. Comp. Physiol. Psychol. 57:52-60, 1964.
Valenstein, ES: Stereotypy and sensorimotor changes evoked by hypothalamic stimulation: possible relation to schizophrenic behavior patterns. In (ed.) Routtenberg, A, Biology of Reinforcent: Facets of Brain Stimulation Reward. Academic Press, New York, 1980.
Wanner, HU and Battig, K: Wirkung von Nikotin und Amphetamin auf die Selbstreizung bei der Ratte. Helv. Physiol. et Pharm. Acta. 24: C122-124, 1966.
White, NZ, Brown, M and Yachim, P: Effects of catecholamine manipulations on three different self-stimulation behaviors. Pharmac. Biochem. Behav. 9:273-278, 1978.

AUTORADIOGRAPHICAL EVIDENCE OF NICOTINIC RECEPTORS IN RAT BRAIN

P.B.S. Clarke[1] and A. Pert[2]

Kinsmen Laboratory of Neurological Research[1]
Dept. of Psychiatry
Univ. of British Columbia
Vancouver, B.C.
Canada V6T 1W5
and
Biological Psychiatry Branch[2]
NIMH, Bldg. 10, Room 3N212
Bethesda, MD 20205

INTRODUCTION

In the peripheral nervous system, nicotine and muscarine mimic different actions of acetylcholine (ACh) and act through different receptors. Peripheral nicotinic receptors fall into two classes (1): C6 type cholinoceptors, found principally at autonomic ganglia, where cholinergic neurotransmission is selectively blocked by hexamethonium; and C10 type cholinoceptors, occurring at the muscle endplate, where decamethonium is a much more potent antagonist. There is less consensus as to the nature of central nicotinic receptors; cholinoceptors with mixed nicotinic and muscarinic properties have been invoked, while others have argued for a noncholinergic receptor for nicotine. What can be said with confidence, however, is that nicotine generally acts through central receptors of some sort. Thus, many biochemical (2), electrophysiological (3), and behavioral (4,5) studies have demonstrated central actions of nicotine that are blocked by nicotinic antagonists, and some of these actions are known to be stereoselective (6,7,8).

RADIOLIGANDS FOR LABELING PUTATIVE NICOTINIC RECEPTORS

Various compounds with known peripheral actions have been employed as radioligands for labeling brain nicotinic receptors. Until recently, the most frequently used was the nicotinic antagonist [^{125}I]alpha-bungarotoxin (BTX), which binds to a subunit of the well-characterized nicotinic receptor/ionophore macromolecule of Torpedo californica and also at the mammalian neuromuscular junction. In brain homogenates, BTX binds in a manner consistent with a nicotinic receptor label, insofar as binding is saturable, of high affinity, and selectively inhibited by nicotinic compounds (for review, see 9). Among other peripheral antagonists, nicotinic binding to brain has been reported with radiolabeled d-tubocurarine, naja naja siamensis alpha-toxin, and dihydro-beta-erythroidine, but the extent to which these ligands label the same nicotinic site has not been determined.

In recent years, attention has focussed on nicotinic agonists. In 1980, Romano and Goldstein (10) reported the stereospecific and saturable binding of tritiated nicotine to rodent brain homogenates. As subsequently confirmed by other groups (11,12,13), binding was of high affinity (as reflected by a nanomolar dissociation constant), and was potently inhibited by nicotinic agonists including ACh, but not by C6- or C10-selective antagonists. A second, lower-affinity site has also been detected, but this is of doubtful pharmacological significance, since it is of high capacity and possesses little or no regional distribution (10,11,13,14). In contrast, Sloan, Todd, and Martin (15) reported the possible existence of five sites at which l-nicotine may act, including a stereoselective site (kD approx. 5 nM) which may correspond to the high-affinity site previously reported. In another report (16), two sites were described in mouse brain, both with much lower affinities, and the higher-affinity site, which lacked stereoselectivity, was not saturable by l-nicotine in concentrations as high as 100 uM. Schwartz, McGee, and Kellar (17) described the use of [^{3}H]ACh as a ligand for putative nicotinic cholinoceptors in brain; in the presence of excess unlabeled displacer and an inhibitor of acetylcholinesterase, [^{3}H]ACh bound to homogenates with characteristics similar to those of the high-affinity [^{3}H]nicotine binding site.

ANATOMICAL DISTRIBUTION OF NICOTINIC BINDING SITES IN RAT BRAIN

There is little agreement as to the regional distribution of high-affinity [^{3}H]nicotine binding determined in tissue homogenates of microdissected brain regions. In particular, variable amounts of binding are attributed to the hypothalamus and hippocampus relative to other brain areas (11,12,14,18,19). In order to obtain a much more detailed map of binding sites, we turned to autoradiography (for details, see ref. 20). For direct comparison with autoradiography, the binding of [^{3}H]d,l-nicotine was characterized and optimized using unfixed slide-mounted sections of rat brain which were transferred into vials for liquid scintillation counting. Displaceable binding, assessed by the addition of excess unlabeled l-nicotine, was enhanced by the addition of calcium chloride and reached equilibrium after 10 to 20 minutes at room temperature. Following incubation with radiolabeled nicotine, sections were washed in ice cold buffer which selectively removed nondisplaceable binding with little loss of displaceable binding. The amount bound was proportional to section thickness (8 - 48 μm). Scatchard analysis of coronal sections taken through the midstriatal forebrain revealed a single-affinity site (Fig. 1). The dissociation constant (3.5 nM) obtained was lower than in most reports employing homogenates, whereas the binding capacity (B_{max}) was comparable. As originally described in homogenates (10), binding to tissue sections was selectively inhibited by nicotinic agonists; antagonists selective for either ganglia or muscle endplate were weak inhibitors. The biologically more potent isomer, l-nicotine, was seventeen times more potent than d-nicotine in inhibiting binding of the racemic radioligand; and at the low concentration used for autoradiographic labeling, it is likely that [^{3}H]d-nicotine was binding to a negligible extent. Consistent with this suggestion, we have since obtained kD values of around 1 nM using pure [^{3}H]l-nicotine instead of the racemic label, and we have found that the two ligands yield autoradiographic patterns which are qualitatively indistinguishable (unpublished observations). Displaceable binding accounted for 90% of total binding across a range of radioligand concentrations (Fig. 1), and the high signal-to-noise ratio was reflected in the autoradiographs which were obtained at 3.5 nM [^{3}H]nicotine. The autoradiographic distribution was discrete and respected anatomical demarcations. Dense labeling was observed in the medial habenula and interpeduncular nucleus, which appear to belong to a common cholinergic system; in the so-called specific motor

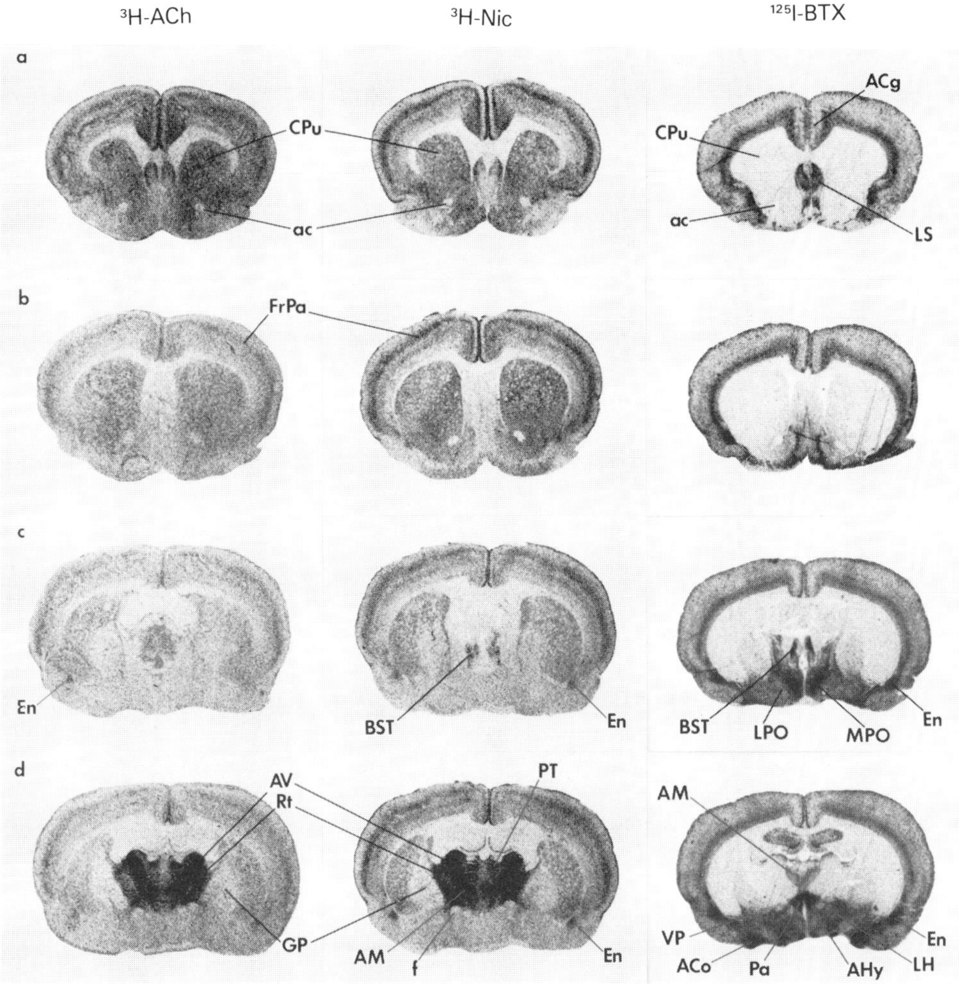

Fig. 1. Autoradiographic images showing the distribution of [^{3}H]acetylcholine (left), [^{3}H]nicotine (middle), and [^{125}I]alpha-bungarotoxin (right) labeling in rat brain. Near-adjacent brain sections were prepared unfixed and incubated with radioligand (see ref. 20 for details). Labeling with [^{3}H]acetylcholine was performed in the presence of excess atropine in order to occlude muscarinic sites. Images a to n represent coronal sections taken consecutively along a rostrocaudal axis. Images o and p represent parasagittal sections. The tritiated agonists, which may label a central nicotinic cholinergic receptor of the ganglionic type (see text), bound densely in specific nuclei of thalamus (e.g., anteroventral - AV and ventroposterior - VPM), in the lateral and medial geniculate nucei (DLG, VLG, MG), and in laminae III/IV of cerebral cortex, suggesting an involvement in sensory processing. The dense agonist labeling in the mesolimbic dopaminergic cell body (VTA) and terminal (nucleus accumbens) regions provides a clue to nicotine's reinforcing actions. Alpha-bungarotoxin labeling has a markedly different pattern, possibly representing central nicotinic receptors of the neuromuscular type. Reproduced from Ref. 20.

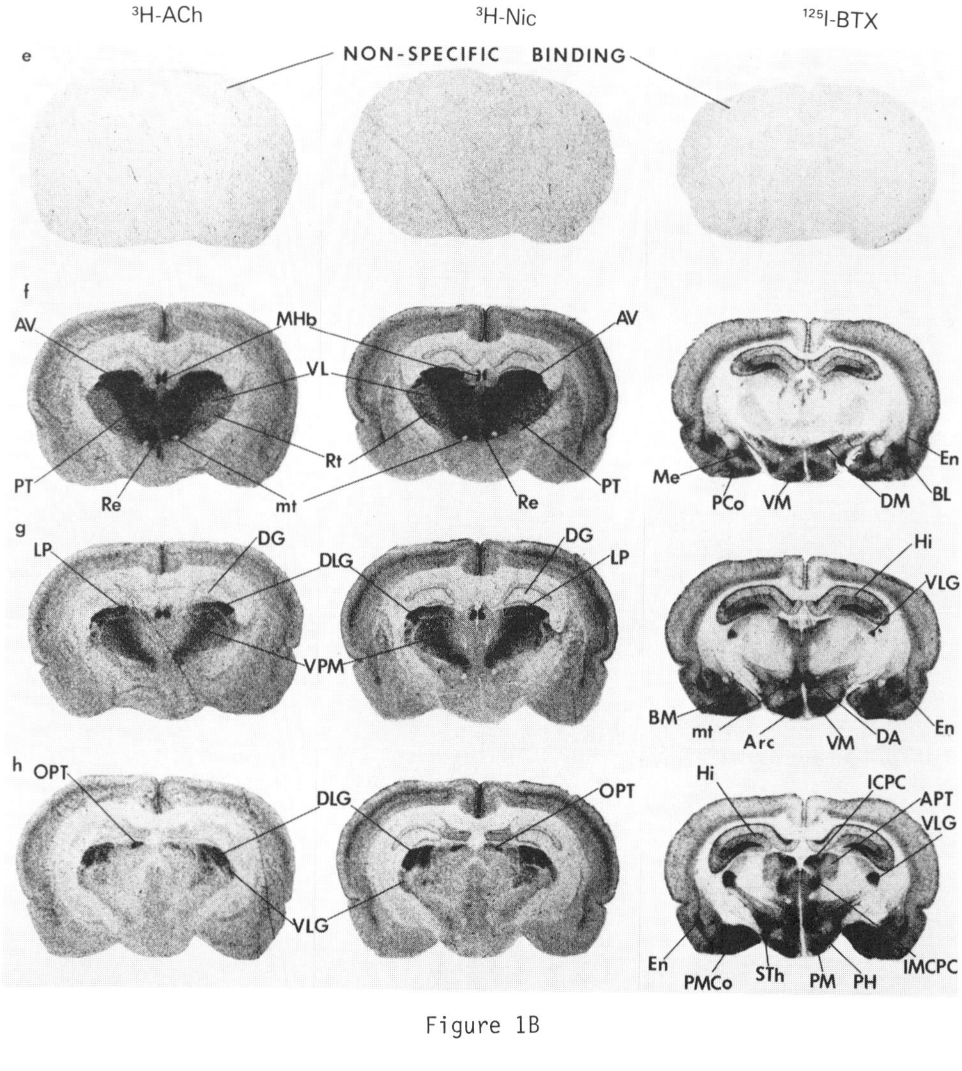

^{3}H-ACh
^{3}H-Nic
^{125}I-BTX
e
NON-SPECIFIC BINDING
f
AV
MHb
VL
Rt
PT
Re
mt
En
Me
PCo
VM
DM
BL
g
LP
DG
DLG
VPM
Hi
VLG
BM
Arc
DA
h
OPT
ICPC
APT
PMCo
STh
PM
PH
IMCPC

Figure 1B

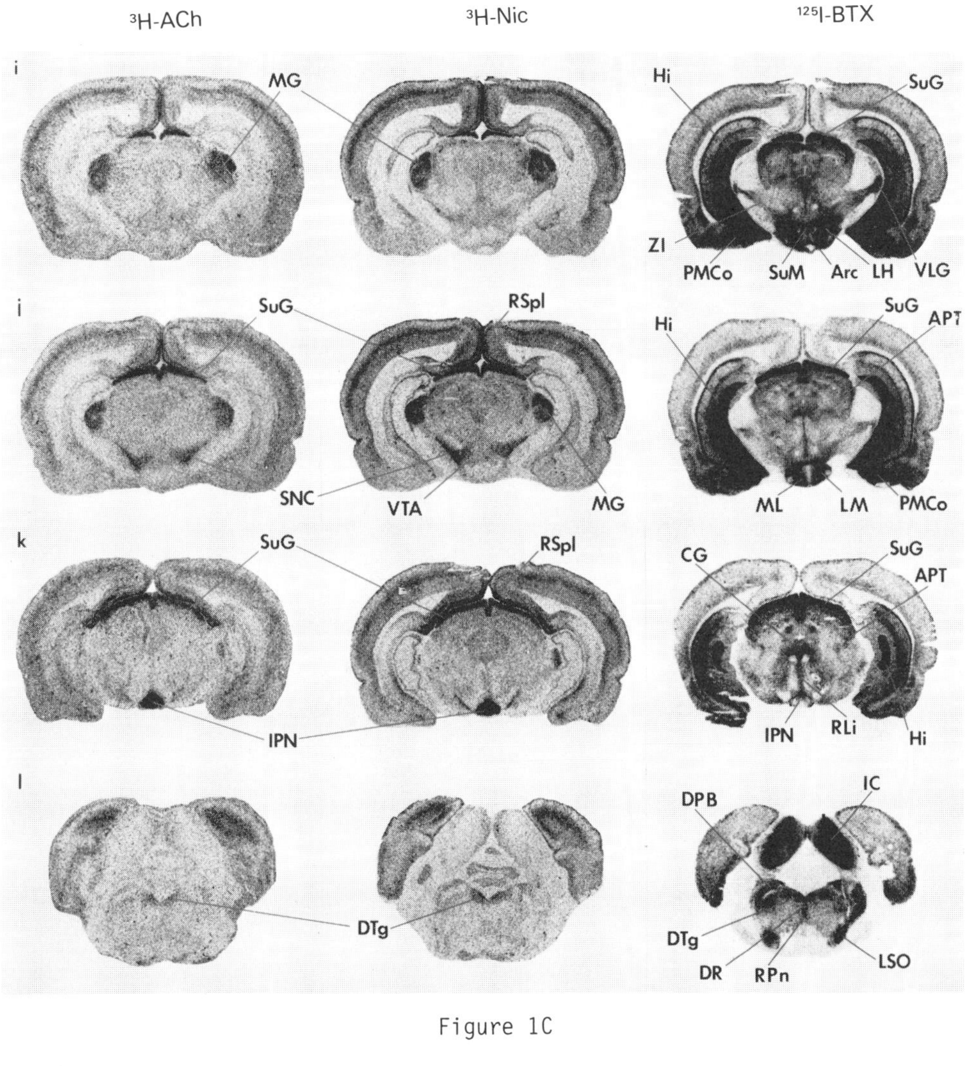

³H-ACh
³H-Nic
¹²⁵I-BTX
i
MG
Hi
SuG
ZI
PMCo
SuM
Arc
LH
VLG
j
SuG
RSpl
Hi
SuG
APT
SNC
VTA
MG
ML
LM
PMCo
k
SuG
RSpl
CG
SuG
APT
IPN
IPN
RLi
Hi
l
DPB
IC
DTg
DTg
DR
RPn
LSO

Figure 1C

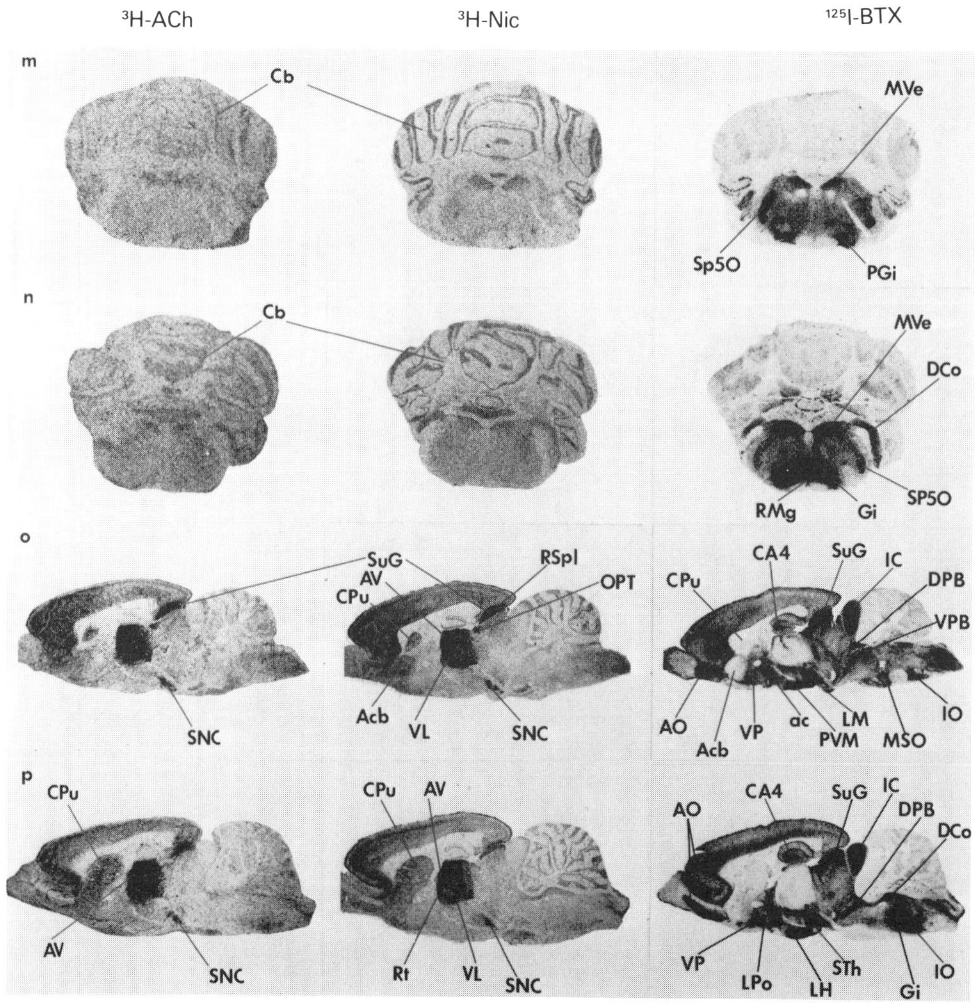

^{3}H-ACh
^{3}H-Nic
^{125}I-BTX
m
Cb
MVe
Sp5O
PGi
n
Cb
MVe
DCo
RMg
Gi
SP5O
o
SuG
AV
CPu
RSpl
OPT
Acb
VL
SNC
CA4
SuG
IC
CPu
DPB
VPB
AO
Acb
VP
ac
LM
PVM
MSO
IO
p
CPu
AV
SNC
CPu
AV
Rt
VL
SNC
AO
CA4
SuG
IC
DPB
DCo
VP
LPo
LH
STh
Gi
IO

Figure 1D

and sensory nuclei of the thalamus, and layers III and/or IV of cerebral cortex with which they communicate; in the substantia nigra pars compacta and ventral tegmental area, containing the perikarya of ascending dopaminergic pathways; and in the molecular layer of the dentate gyrus, the presubiculum, and superficial layers of the superior colliculus. Labeling was conspicuously absent in the hypothalamus and hippocampus.

A quantitative analysis of autoradiographic [^{3}H]nicotine labeling has been provided by London and colleagues (21). Binding densities were determined within many brain areas, but they probably do not correspond to B_{max} values, since autoradiographic labeling was measured in brain sections incubated at a single, probably nonsaturating, concentration of radioligand. Moreover, the preliminary biochemical characterization was performed in sonicates whose binding properties may differ from those of brain sections; indeed, the reported kD value resembled that obtained in homogenates and was much higher than our own determination. In this study, nonradioactive hexamethonium completely inhibited displaceable binding, and this appears to be the only report of appreciable inhibition by this compound. Nevertheless, the densitometric values reported are in striking accord with our qualitative description. The only major difference was the dense labeling reported in layer V of cerebral cortex.

Parallel work using [^{3}H]ACh (in the presence of atropine) yielded similar results (22). Binding to brain sections was of high (nanomolar) affinity and was potently inhibited by nicotine and carbachol. The autoradiographs showed a rather diffuse distribution, but densitometry (performed at a single nonsaturating concentration of radioligand) revealed dense labeling in certain areas of thalamus, in superior colliculus, and in medial habenula, in agreement with [^{3}H]nicotine autoradiography. However, low levels were described throughout the pons, which contains the substantia nigra, ventral tegmental area, and also the interpeduncular nucleus, the structure most heavily labeled by [^{3}H]nicotine (21).

Thus, the available evidence suggested an overlap in the regional distributions of [^{3}H]nicotine and [^{3}H]ACh. There seemed to be little correspondance with the autoradiographic distribution of [^{125}I]BTX binding sites as initially described by Polj-Tejera et al. (23) and subsequently elaborated (24,25). In similar vein, Schwartz et al. (17) had noted a lack of correlation between the regional distributions of nicotinic [^{3}H]ACh and [^{125}I]BTX binding in rat brain homogenates, and mismatches of [^{3}H]nicotine- and [^{125}I]BTX-binding properties had also been found in mouse brain (11). Other observations suggested that BTX labeling overlapped with a subset of mammalian nicotinic cholinoceptors: the toxin bound to autonomic ganglia, but failed to block nicotinic cholinergic neurotransmission (1); in the spinal cord, extrasynaptic binding was observed (26); the well-known nicotinic cholinergic excitation of spinal Renshaw cells was unaffected by BTX (27), and binding to these neurons was at best equivocal (24, 28).

Next, we compared the autoradiographic distributions obtained with the three putative nicotinic receptor radioligands in adjacent sections of rat brain (29). The exact labeling conditions were optimized for each ligand, and the procedures for [^{3}H]nicotine and [^{3}H]ACh labeling were similar or identical to those used previously (20, 22). In preliminary experiments, we characterized binding of [^{125}I]BTX to brain sections processed as for autoradiography. This confirmed the existence of a single high-affinity site having a selectively nicotinic profile, as observed in membrane preparations.

The resulting film images revealed a striking concordance of [^{3}H]nicotine and [^{3}H]ACh labeling. No clear differences were seen in any part of the neuraxis as far caudal as the medulla. The pattern observed appeared to be identical to that previously obtained using [^{3}H]nicotine (20). The distribution produced by [^{125}I]BTX was quite different, and almost resembled a photographic negative of tritiated agonist labeling. Only in the superior colliculus was there a convergence of dense labeling. Neither of the two distributions represented a subset of the other (Fig. 1). This may seem surprising, since unlabeled nicotine inhibits [^{125}I]BTX binding. However, nicotine inhibits toxin binding in the micromolar range, whereas [^{3}H]nicotine binds at nanomolar concentrations. Biochemical similarities between [^{3}H]nicotine and [^{3}H]ACH binding strongly support a common recognition site (30).

Recently, the presence of high-affinity [^{3}H]nicotine binding sites was briefly reported in human brain homogenates (30). In brain slices from rhesus monkey cortex, we have found that [^{3}H]l-nicotine binds to a single class of site that closely resembles that found in the rat with respect to dissociation constant and nicotinic displacement profile (unpublished results). The autoradiographic labeling pattern is also quite similar, with considerable labeling in cerebral cortex and thalamus (31, 32). However, the medial habenula, so densely bound in the rat, appeared devoid of binding, and the laminar distribution in cerebral cortex differed too.

TRITIATED NICOTINE BINDING SITES: PRE- OR POSTSYNAPTIC?

Tritiated agonists bind primarily in the synaptosomal fraction of rat brain (13,17). Since microiontophoresed nicotine appears to excite neurons directly (3), some nicotinic receptors, at least, are probably located on the postsynaptic membrane of somata or dendrites. In the substantia nigra pars compacta, dopaminergic neurons are sensitive to microiontophoretic nicotine (33) and possess high-affinity tritiated nicotine binding sites (34,35). Biochemical experiments (2) indicate that nicotine acts at receptors located directly on striatal dopaminergic terminals to control release of dopamine. Relative to the dopaminergic terminal, the receptors can be termed "presynaptic." However, they are probably located on postsynaptic membranes since nicotine-induced release occurs even in the absence of transneuronal excitation. Consistent with this conclusion, neurochemically selective lesions of dopaminergic systems reduce tritiated agonist labeling in striatum (34), apparently through a loss of receptor number rather than affinity (36). Similar results have been reported in hypothalamus following serotoninergic lesions (36).

Many neurotransmitters control their own release by acting at autoreceptors located on the presynaptic membrane (2). In the peripheral myenteric plexus, nicotinic agonists enhance ACh release via presynaptic autoreceptors, and inhibitory control occurs through a muscarinic mechanism (37). Rowell and Winkler (38) have reported a similar nicotinic action in mouse cerebral cortical synaptosomes. These findings are timely: in Alzheimer's disease, presynaptic cholinergic markers are reduced in cerebral cortex, with little or no change in the density of muscarinic sites but with a marked loss of [^{3}H]nicotine binding sites (30). Denervation is not usually associated with a down-regulation of postsynaptic receptors, and it seems likely that presynaptic autoreceptors are lost.

In several neurotransmitter systems, autoreceptors and postsynaptic receptors differ pharmacologically and are selectively labeled by different radioligands. In order to determine whether [^{3}H]nicotine might label nicotinic autoreceptors, we examined the interpeduncular nucleus (IPN). This brain stem nucleus, together with its main afferent pathway, the fasciculus retroflexus, FR, provides an ideal model system for detecting autoreceptors: the nucleus lacks intrinsic cholinergic neurons and derives all or almost all of its cholinergic innervation from the fasciculus; bilateral lesions of this pathway result in a near-total loss of presynaptic cholinergic markers within the nucleus as early as three days. At this time interval, transneuronal degeneration cannot be seen, and denervation-induced changes in postsynaptic binding would not be expected. Fasciculus retroflexus afferents to IPN do not form axoaxonic contacts; and most importantly, the existence of nicotinic cholinergic autoreceptors has been indicated on FR terminals (39). These characteristics imply that a lesion-induced loss of nicotinic binding sites in IPN, occurring shortly after bilateral FR lesions, could not be ascribed to a loss of receptors located either postsynaptically on FR terminals or postsynaptically on neurons intrinsic to IPN; most probably, the change would reflect a loss of receptors located presynaptically on FR afferents. Our first analysis (39a) confirmed a subnuclear distribution of [^{125}I]BTX labeling within IPN and also revealed a differential distribution for [^{3}H]nicotine. Bilateral FR lesions duly reduced [^{3}H]nicotine labeling in certain subnuclei, and subsequent Scatchard analysis of microdissected tissue in homogenates indicated a loss of receptor number and not affinity (39b). Thus, it appears that [^{3}H]nicotine binding sites can be located on presynaptic elements which are presumably cholinergic terminals, although this has yet to be directly demonstrated.

FUNCTIONAL SIGNIFICANCE OF NICOTINIC BINDING SITES IN BRAIN

Two anatomically distinct populations of nicotinic binding sites are identified autoradiographically and biochemically (11,17) in rodent brain. Tritiated nicotine and ACh not only bind with the same autoradiographic distribution (29) but also appear to label the same molecular recognition site (30). We have previously proposed that tritiated agonists and BTX label two different populations of nicotinic cholinergic receptor, probably resembling the ganglionic and neuromuscular receptors, respectively (29). This hypothesis presumes that central nicotinic receptors exist that can be distinguished from central muscarinic cholinoceptors, are activated endogenously by ACh, and resemble peripheral nicotinic cholinoceptors. None of these assumptions is universally accepted.

The existence of cholinoceptors with mixed nicotinic/muscarinic characteristics has been proposed on the basis of microiontophoretic experiments in hippocampus and spinal cord (3). Mixed receptors have been inferred when both nicotinic and muscarinic antagonists are found to block an effect of ACh on neuronal firing (40). This conclusion must be treated with caution in view of the interpretational difficulties associated with the microiontophoretic technique (41). Interestingly, Bradley and Lucy (42) noted a similar lack of antagonist specificity in relation to ACh-induced neuronal excitation, but not in relation to excitation induced by nicotine or muscarine. Possibly, then, intraneuronal crosstalk between nicotinic and muscarinic receptors accounts for the reported lack of antagonist selectivity. Nevertheless, the possibility of a hippocampal cholinoceptor with mixed nicotinic and muscarinic properties should be reevaluated at the appropriate level of analysis by recording from single channels.

In other brain areas, a clear distinction exists between central muscarinic and nicotinic actions. For example, nicotinic but not muscarinic antagonists block some central effects of nicotine on transmitter release (2), gastric secretion (43), cardiovascular function (44), and on temperature regulation (45). In the behavioral domain,

Rosecrans and colleagues have elegantly demonstrated the independence of nicotinic and muscarinic discriminative stimuli (4). Finally, the binding characteristics of nicotinic and muscarinic radioligands are different in several respects, not least in their regional distribution and *in vivo* regulation by cholinergic agonists (46).

The term "cholinergic," as classically defined, denotes the involvement of endogenous ACh. Several groups have suggested that nicotine may act through "noncholinergic" sites in the brain, but there is little evidence in favor of this view. Abood et al. (18,47) have presented arguments for such sites, and the basis of this claim appears to be three-fold: (1) intracerebroventricular administration of nicotine in rats produced a prostration syndrome which was not blocked by prior central injection of several known nicotinic cholinergic antagonists (decamethonium, BTX, and d-tubocurarine); (2) various synthetic analogues of nornicotine and piperidine blocked nicotine-induced prostration but failed to inhibit the binding of radiolabeled tubocurarine or BTX to brain membranes; (3) the prostration syndrome was not reproduced by ventricular injection of ACh (given in the absence of a hydrolysis inhibitor) or carbachol. Recent data, however, show that ganglion-selective antagonists, which were not originally evaluated, do indeed prevent prostration (18,48). Moreover, by no means do all pharmacological anatagonists bind at the agonist recognition site in order to block the effects of agonists (49). Finally, the inefficacy of carbachol may be due to poor drug penetration following intraventricular administration. In another behavioral paradigm, Rosecrans and Meltzer (50) reported that a drug combination (physostigmine, atropine, hexamethonium) which would be expected to selectively enhance central nicotinic cholinergic transmission, failed to mimic the nicotine discriminative stimulus. Since only one combination of doses was tested, the failure of generalization may have reflected a quantitative, rather than a qualitative difference in stimulus properties. Hence, to conclude that the nicotine cue is mediated by a noncholinergic mechanism seems premature. A noncholinergic site of action has also been suggested on the basis of binding experiments, in which several cholinergic agents (decamethonium, d-tubocurarine, ACh) failed to inhibit the binding of tritiated nicotine to mouse brain membranes (16). However, two affinity sites were found, and binding inhibition was apparently only studied in relation to a site of low affinity, high capacity, and little stereoselectivity.

Do tritiated agonist binding sites represent ganglionic-type nicotinic cholinoceptors? Following systemic administration, nicotine exerts dose-dependent central actions which typically occur at doses (e.g., 50-400 ug/kg sc: 51,52) that in the periphery would be expected to stimulate ganglionic transmission selectively. It is perhaps unsurprising, therefore, that most if not all of these central actions can be prevented by centrally acting C6-selective antagonists such as mecamylamine. Moreover, tachyphylaxis, which occurs readily at ganglia (53), is a common feature of central C6-type actions of nicotine (see 29).

The high (nanomolar) affinities of [^{3}H]nicotine and [^{3}H]ACh binding to brain do not match their pharmacological potencies either *in vitro* or *in vivo*, where micromolar concentrations produce dose-related effects. It has been suggested that tritiated nicotinic agonists bind to a desensitized form of the receptor which has a higher affinity for agonists than has the undesensitized form (10,17); alternatively, in the low concentrations used in most binding assays, radiolabeled agonists may bind to a site of higher affinity which controls desensitization rather than channel opening (54). Either mechanism would help to explain why few nicotinic antagonists displace tritiated agonist binding with appreciable potency; it is suggestive that the only ones that do are dihydro-beta-erythroidine (DHBE) and neosurugatoxin (55), of which the latter is ganglion-specific. Receptor desensitization may account for the paradoxical up-regulation of tritiated agonist binding that may occur following chronic nicotine treatment (56,57,58) and, indeed, animals receiving chronic nicotine are subsensitive to acute challenges with the drug (58). In the periphery, higher doses are required to induce tachyphylaxis at the neuromuscular junction than at ganglia (53), and interestingly, BTX binding sites up-regulate less readily than tritiated nicotine sites upon exposure to chronic nicotine (46).

Recent experiments employing the 2-deoxyglucose technique have elegantly indicated that central actions of nicotine that are blocked by mecamylamine are associated with high-affinity agonist binding sites (59). After systemic injection of nicotine in rats, a pattern of neuronal activation was observed that closely resembled the autoradiographic distribution of tritiated nicotine labeling. Thus, marked increases occurred in medial habenula, superior colliculus, and interpeduncular nucleus, and were blocked by mecamylamine, a ganglion-selective antagonist.

We would predict that central nicotinic actions on ganglionic-type receptors would be seen in areas rich in tritiated agonist binding but devoid of BTX binding. One example is the striatum, where a nicotinic enhancement of dopamine release is blocked by low concentrations of mecamylamine and pempidine (2) but not of BTX (60). The substantia nigra pars compacta possesses little, if any, BTX labeling, but dopaminergic neurons in this cell layer are densely labeled with tritiated nicotine (34). These neurons are excited by microiontophoretic nicotine and ACh (33), and nicotine-induced excitation is prevented by mecamylamine (61). Conversely, several brain areas possess appreciable BTX labeling but appear devoid of tritiated agonist binding. For example, in rat brain medulla and in the CA1 area of hippocampus, microiontophoretic mecamylamine failed to block the direct actions of nicotinic agonists on single units, whereas the antagonist DHBE, which acts at both ganglia and muscle endplate, was effective (42,62).

It seems likely that endogenous ACh acts on neuronal elements which bind tritiated agonists with high affinity. Tritiated agonist binding appears to fall within known cholinergic systems in the brain. Most of the main cholinergic projection areas contain tritiated agonist binding, including cerebral cortex, thalamus, interpeduncular nucleus, medial habenular and dentate gyrus. Some regions containing cholinergic cell bodies are also labeled (e.g., laterodorsal tegmental nucleus, medial habenula, septum). The loss of tritiated agonist binding following chronic _in vivo_ treatment with acetylcholinesterase inhibitors (12,56,57) may reflect cholinergicity, but cytotoxic effects must also be considered. Recently we have obtained neuroanatomical and electrophysiological evidence for an excitatory cholinergic input to dopaminergic neurons of the substantia nigra, which appears to synapse onto nicotinic receptors labeled by tritiated nicotine (Clarke, Pert, and Skirboll, submitted).

In mammals, BTX blocks nicotinic cholinergic transmission at the muscle endplate but not at autonomic ganglia (1). If BTX binding sites in brain resemble peripheral neuromuscular nicotinic receptors, they will be activated only at high doses of nicotine sufficient to affect the muscle endplate. Thus, they are unlikely to be activated in most _in vivo_ studies. However, local application of nicotinic agonists directly onto brain areas possessing dense BTX labeling should produce effects at BTX-labeled sites. Indeed, this seems to be the case: in the inferior colliculus, microiontophoretic BTX blocked the effects of cholinergic agonists on single unit activity (63); microinjection of carbachol into the suprachiasmatic nucleus of the hypothalamus mimicked the effects of light on pineal enzyme activity, and local application of BXT blocked this effect (64); susceptibility to nicotine-induced seizures is correlated with BTX-binding capacity in hippocampus (65). In addition, there is some evidence from experiments employing chronic esterase treatment that some BTX sites are cholinergic (A.C. Collins, personal communication); nevertheless, dense BTX labeling is encountered in certain areas of the hippocampus that lack cholinergic innervation (66). Recently, the nicotinic antagonist, DHBE, was reported to bind to a "neuromuscular" nicotinic recognition site in brain (67). This interpretation, however, rested on the mistaken claim that DHBE is a selective neuromuscular blocker (see 1).

Published autoradiographic studies reveal very little tritiated agonist labeling in hypothalamus and hippocampus, whereas appreciable binding is found in membrane preparations derived from microdissected tissue of these same areas. Inaccurate dissection may partly account for this discrepancy. However, it seems quite possible that, with the autoradiographic labeling conditions employed to date, only a subpopulation of tritiated agonist binding sites has been detected. In general, the issue of receptor heterogeneity has not been adequately addressed, although in rat brain homogenates, tritiated nicotine has been found to bind to thalamus with an apparent affinity different from that in hypothalamus and hippocampus (68). The presence of nicotinic receptors in the hypothalamus is indicated by several pharmacological and electrophysiological experiments, and while nicotine may exert certain local actions by interacting with BTX-labeled sites (see above), ganglionic-type cholinoceptors may also be present (69). In autoradiographic and membrane-binding experiments, negative results should be treated with caution, not the least because receptor densities which are below our limits of detection may yet be pharmacologically significant.

NICOTINE BINDING AND SMOKING

Clinical and experimental data suggest that many people smoke primarily in order to activate central nicotinic receptors of the ganglionic type (70,71). If such an action is localized in the brain, tritiated agonist autoradiographs may provide some clues to the reinforcing mechanism. Animal studies suggest that several self-administered drugs are reinforcing because they enhance the release of transmitter from the ascending mesolimbic dopaminergic system (72). Of particular interest, therefore, mesolimbic dopaminergic neurons possess tritiated nicotine binding sites (34) and appear to be activated by nicotine (35). In addition, nicotine produced electrocortical desynchrony and enhanced release of cortical ACh, which are frequent concommitants of behavioral arousal (73). Transection studies suggest that nicotine produces desynchrony through a primary action in the pons (73); one candidate is the laterodorsal tegmental nucleus, which is labelled by tritiated nicotinic agonists and sends cholinergic afferents to neocortex. The release of ACh from cortical afferents, activated at source, may be further promoted by a nicotinic action on autoreceptors (38). Nicotine and smoking can enhance mental performance (74,75), and improved attentiveness is a common experience during tobacco smoking. The dense autoradiographic labeling found in sensory relay nuclei (thalamus, geniculate nuclei) and in the corresponding laminae of cerebral cortex, suggest a nicotinic role in the modulation of sensory information, possibly to enhance selective attention. Finally, not all central actions of nicotine occur in the brain: the muscle-relaxant action of nicotine, mediated indirectly by spinal Renshaw cells (73), provides yet another possible source of reinforcement.

SUMMARY

In the peripheral nervous system, nicotine mimics the effects of acetylcholine by acting at two kinds of cholinergic receptors, the one located principally at autonomic ganglia, the other found at the neuromuscular junction. At low doses, nicotine selectively affects ganglionic transmission in the periphery, and also exerts central effects which are blocked by ganglion-selective antagonists. To date, two populations of nicotinic binding sites have been distinguished in brain, each with a characteristic and discrete anatomical distribution, as demonstrated by receptor autoradiography: one population is labeled with high affinity by the tritiated agonists nicotine and acetylcholine, the other population by the neuromuscular blocker alpha-bungarotoxin. Tritiated agonist binding sites appear to represent nicotinic cholinergic receptors of the ganglionic type, and it is through these central receptors that nicotine produces its reinforcing actions in tobacco smokers. Alpha-bungarotoxin binding sites may represent central nicotinic receptors of the neuromuscular type and are probably activated by high concentrations of nicotine, possibly above the range encountered in smokers. The autoradiographic visualization of tritiated nicotinic agonist binding sites suggests ways in which nicotine may affect central nervous system functioning. In particular, an association with the ascending dopaminergic pathways of the brain indicates how nicotine may gain access to putative central mechanisms of reinforcement.

ACKNOWLEDGEMENT

It is a pleasure to acknowledge the collaboration of Rochelle Schwartz, Candace Pert, Geoffrey Hamill, David Jacobowitz, David Freidman, and Blanche O'Neill in many of the studies described above.

REFERENCES

1. Brown, DA: Neurotoxins and the ganglionic (C6) type of nicotinic receptor. In (eds.) Ceccarelli, B and Clementi, F, Advances in Cytopharmacology, vol. 3, pp 225-230, 1979.
2. Chesselet, MF: Presynaptic regulation of neurotransmitter release in the brain: facts and a hypothesis. Neuroscience 12:347-375, 1984.
3. Krnjevic, K: Acetylcholine receptors in vertebrate central nervous system. In (eds.) Iversen, LL, Iversen, SD and Snyder, SH, Handbook of Psychopharmacology, vol. 6, 97-126, Plenum, New York, 1975.
4. Rosecrans, JA and Chance, WT: Cholinergic and noncholinergic aspects of the discriminitive stimulus properties of nicotine. Adv. Behav. Biol. 22:155-185, 1977.
5. Clarke, PBS: Chronic central nicotinic blockade after a single administration of the bisquaternary ganglion blocking drug chlorisondamine. Br. J. Pharmac. 83:527-535, 1984.
6. Romano, C, Goldstein, A, and Jewell, NP: Characterization of the receptor mediating the nicotine discriminative stimulus. Psychopharmac. 74:310-315, 1981.
7. Clarke, PBS and Kumar, R: Characterization of the locomotor stimulant action of nicotine in tolerant rats. Br. J. Pharmac. 80:587-594, 1983.
8. Kumar, R, Pratt, JA, and Stolerman, IP: Characteristics of the conditioned taste aversion produced by nicotine in rats. Br. J. Pharmac. 79:245-253, 1983.
9. Oswald, RE and Freeman, JA: Alpha-bungarotoxin binding and central nervous system nicotinic acetylcholine receptors. Neuroscience 6: 1-44, 1981.

10. Romano, C and Goldstein, A: Stereospecific nicotine receptors on rat brain membranes. Science 210:647-650, 1980.
11. Marks, MJ and Collins, AC: Characterization of nicotine binding in mouse brain and comparison with the binding of alpha-bungarotoxin and quinuclidinyl benzilate. Mol. Pharmac. 22:554-564, 1982.
12. Costa, LG and Murphy, SD: [^{3}H]nicotine binding in rat brain: Alteration after chronic acetylcholinesterase inhibition. J. Pharmac. Exp. Ther. 226:392-297, 1983.
13. Benwell, MEM and Balfour, DJK: Nicotine binding to brain tissue from drug-naive and nicotine-treated rats. J. Pharm. Pharmac. 37:405-409, 1985.
14. Martin, BR and Aceto, MD: Nicotine binding sites and their localization in the central nervous system. Neurosci. Biobehav. Rev. 5:473-478, 1981.
15. Sloan, JW, Todd, GD, and Martin, WR: Nature of nicotine binding to rat brain P2 fraction. Pharmac. Biochem. Behav. 20:899-909, 1984.
16. Sershen, H, Reith, MEA, Lajtha, A, and Gennaro, J: Noncholinergic saturable binding of (±)[^{3}H]nicotine to mouse brain. J. Recept. Res. 2:1-15, 1981.
17. Schwartz, RD, McGee, R, and Kellar, K: Nicotinic cholinergic receptors labelled by [^{3}H]acetylcholine in rat brain. Mol. Pharmac. 22:55-62, 1982.
18. Abood, LG, Reynolds, DT, Booth, H, and Bidlack, JM: Sites and mechanisms for nicotine's actions in the brain. Neurosci. Biobehav. Rev. 5:479-486, 1981.
19. Larsson, C and Nordberg, A: Comparative analysis of nicotine-like receptor-ligand interactions in the rodent brain. J. Neurochem. 45:24-31, 1985.
20. Clarke, PBS, Pert, CB, Pert, A: Autoradiographic distribution of nicotine receptors in rat brain. Brain Res. 323:390-395, 1984.
21. London, ED, Waller, SB, and Wamsley, JK: Autoradiographic localization of [^{3}H]nicotine binding sites in the rat brain. Neurosci. Lett. 53:179-184, 1985.
22. Rainbow, TC, Schwartz, RD, Parsons, B, and Kellar, KJ: Quantitative autoradiography of [^{3}H]acetylcholine binding sites in rat brain. Neurosci. Lett. 50:193-196-1984.
23. Polj-Tejera, G, Schmidt, J, and Karten, HJ: Autoradiographic localization of alpha-bungarotoxin binding sites in central nervous system. Nature 258:349-351, 1975.
24. Hunt, SP and Schmidt, J: Some observations on the binding patterns of alpha-bungarotoxin binding sites within the central nervous system of the rat. Brain Res. 157:213-232, 1978.
25. Segal, M, Dudai, Y and Amsterdam, A: Distribution of alpha-bungarotoxin binding and cholinergic nicotinic receptor in rat brain. Brain Res. 148:105-119, 1978.
26. Ninkovic, M and Hunt, SP: Alpha-bungarotoxin binding sites on sensory neurones and their axonal transport in sensory afferents. Brain Res. 272:57-69, 1983.
27. Duggan, AW, Hall, JG, and Lee, CY: Alpha-bungarotoxin, cobra neurotoxin, and excitation of Renshaw cells by acetylcholine. Brain Res. 107:166-170, 1976.
28. Duggan, AW, Hall, JC, Headley, PM, Hendry, IA and Minchin, MCW: Absence of binding of alpha-bungarotoxin and cobra neurotoxin to central acetylcholine receptors - an autoradiographic study. Neurosci. Lett. 3:123-127, 1976.
29. Clarke, PBS, Schwartz, RD, Paul, SM, Pert, CB, and Pert, A: Nicotinic binding in rat brain: Autoradiographic comparison of [^{3}H]acetylcholine, [^{3}H]nicotine, and [^{125}I]alpha-bungarotoxin. J. Neurosci. 5:1307-1315, 1985.
30. Kellar, KJ: see this volume.

31. Friedman, DP, Clarke, PBS, O'Neill, JB, and Pert, A: Distributions of nicotinic and muscarinic receptors in monkey thalamus. Neurosci. Abstr. 11:307.3, 1985.
32. O'Neill, JB, Clarke, PBS, Friedman, DP, and Pert, A: Distributions of nicotinic and muscarinic receptors in monkey cerebral cortex. Neurosci. Abstr. 11:307.2, 1985.
33. Lichtensteiger, W, Hefti, D, Felix, T, Huwyler, E, Melamed, E, and Schlumpf, M: Stimulation of nigrostriatal dopamine neurones by nicotine. Neuropharmac. 21:963-968, 1982.
34. Clarke, PBS and Pert, A: Autoradiographic evidence for nicotine receptors on nigrostriatal and mesolimbic dopaminergic neurons. Brain Res., in press, 1985.
35. Pert, A: see this volume.
36. Schwartz, RD, Lehmann, J and Kellar, K: Presynaptic nicotinic cholinergic receptors labelled by [^{3}H]acetylcholine on catecholamine and serotonin axons in brain. J. Neurochem. 42:1495-1498, 1984.
37. Briggs, CA and Cooper, JR: Cholinergic modulation of the release of [^{3}H]acetylcholine from synaptosomes of myenteric plexus. J. Neurochem. 38:501-508, 1982.
38. Rowell, P and Winkler, DL: Nicotinic stimulation of [^{3}H]acetylcholine release from mouse cerebral cortical synaptosomes. J. Neurochem. 43:1593-1598, 1984.
39. Brown, DA, Dougherty, RJ, and Halliwell, JV: The action of cholinomimetic substances on impulse condition in the habenulo-interpeduncular pathway of the rat in vivo. J. Physiol. 353:101-109, 1984.
39a. Hamill, GS, Clarke, PBS, Pert, A, Jacobowitz, DM. ^{3}H-nicotine and ^{125}I-alpha-bungarotoxin receptors in interpeduncular nucleus of rats. I. Subnuclear distribution. J Comp Neurol 251:398-406, 1986.
39b. Clarke, PBS, Hamill, GS, Nadi, NS, Jacobowitz, DM, and Pert, A: ^{3}H-nicotine and ^{125}I-alpha-bungarotoxin receptors in the interpeduncular nucleus of rats. II. Effects of habenular deafferentation. J Comp Neurol 251:407-413, 1986.
40. Bird, SJ and Aghajanian, GK: Denervation supersensitivity in the cholinergic septo-hippocampal pathway: a microiontophoretic study. Brain Res. 100:355-370, 1975.
41. Bloom, FE: To spritz or not to spritz: the doubtful value of aimless iontophoresis. Life Sci. 14:1819-1834, 1974.
42. Bradley, PB and Lucy, AP: Cholinoceptive properties of respiratory neurones in the rat medulla. Neuropharmac. 22:853-858, 1983.
43. Osumi, Y, Ishikawa, T, Nagasaka, Y, and Fujiwara, T: Central effects of nicotine on gastric acid secretion in rats. Eur. J. Pharmac. 68:409-415, 1980.
44. Kubo, T and Misu, Y: Changes in arterial blood pressure after microinjections of nicotine into the dorsal area of the medulla oblongata in rats. Neuropharmac. 20:521-524, 1981.
45. Hall, GH: Changes in body temperature produced by cholinomimetic substances injected into the cerebral ventricles of unanaesthetized cats. Br. J. Pharmac. 44:634-641, 1972.
46. Marks, J and Collins, AC: Tolerance, cross-tolerance, and receptors after chronic nicotine or oxotremorine. Pharmac. Biochem. Behav. 22:283-291, 1985.
47. Abood, LG, Lowy, K, and Booth, H: Acute and chronic effects of nicotine in rats and evidence for a noncholinergic site of action. NIDA Res. Monog. 23:136-149, 1979.
48. Schwab, LS and Kritzer, MF: The effect of cholinergic antagonists on a central response to nicotine. Experentia 38:119-120, 1982.
49. Conti-Tronconi, BM and Raftery, MA: The nicotinic-cholinergic receptor: correlation of molecular structure with functional properties. Annu. Rev. Biochem. 51:491-530, 1982.

50. Rosecrans, JA and Meltzer, LT: Central sites and mechanisms of action of nicotine. Neurosci. Behav. Rev. 5:497-501, 1981.
51. Clarke, PBS and Kumar, R: The effects of nicotine on locomotor activity in non-tolerant and tolerant rats. Br. J. Pharmac. 78:329-337, 1983.
52. Clarke, PBS and Kumar, R: Some effects of nicotine on food and water intake in undeprived rats. Br. J. Pharmac. 82:233-239, 1984.
53. Paton, WDM and Savini, EC: The action of nicotine on the motor endplate in the cat. Br. J. Pharmac. 32:360-380, 1968.
54. Conti-Tronconi, BM, Dunn, SMJ, and Raftery, MA: Independent sites of low and high affinity for agonists on torpedo californica acetylcholine receptor. Biochem. Biophys. Res. Commun. 107:23-29, 1982.
55. Hayashi, E, Isogai, M, Kagawa, Y, Takayanagi, N, and Yamada, S: Neosurugatoxin, a specific antagonist of nicotinic acetylcholine receptors. J. Neurochem. 42:1491-1494, 1984.
56. Schwartz, RD and Kellar, KJ: Nicotinic cholinergic receptor binding sites in the brain: regulation in vivo. Science 220:214-216, 1983.
57. Schwartz, RD and Kellar, KJ: In vivo regulation of [^{3}H]acetylcholine recognition sites in brain by nicotinic cholinergic drugs. J. Neurochem., in press, 1985.
58. Marks, MJ, Burch, JB, and Collins, AC: Effects of chronic nicotine infusion on tolerance development and nicotine receptors. J. Pharmac. Exp. Ther. 226:817-825, 1983.
59. London, ED, Connolly, RJ, Szikszay, M and Wamsley, JK: Distribution of cerebral metabolic effects of nicotine in the rat. Eur. J. Pharmac. 110:391-392, 1985.
60. Belleroche, J and Bradford, HF: Biochemical evidence for the presence of presynaptic receptors on dopaminergic nerve terminals. Brain Res. 142:53-68, 1978.
61. Clarke, PBS, Hommer, DW, Pert, A, and Skirboll, LR: Electrophysiological actions of nicotine on substantia nigra single units. Br. J. Pharmac. 85:827-835, 1985.
62. Rovira, C, Ben-Air, Y, Cherubini, E, Krnjevic, K, and Ropert, N: Pharmacology of the dendritic action of acetylcholine and further observations on the somatic disinhibition in the rat hippocampus in situ. Neuroscience 8:97-106, 1983.
63. Farley, GR, Morley, BJ, Javel, E, and Gorga, MP: Single unit responses to cholinergic agents in the rat inferior colliculus. Hearing Res. 11:73-91, 1983.
64. Zatz, M and Brownstein, MJ: Injection of alpha-bungarotoxin near the suprachiasmatic nucleus blocks the effects of light on pineal enzyme activity. Brain Res. 213:438-442, 1981.
65. Miner, LL, Marks, MJ, and Collins, AC: Classical genetic analysis of nicotine-induced seizures and nicotinic receptors. J. Pharmac. Exp. Ther. 231:545-554, 1984.
66. Hunt, SP and Schmidt, J: the relationship of bungarotoxin binding activity and cholinergic termination within the rat hippocampus. Neuroscience 4:585-592, 1979.
67. Williams, M and Robinson, JL: Binding of the nicotinic cholinergic antagonist dihydro-beta-erythroidine to rat brain tissue. J. Neuroscience 4:2906-2911, 1984.
68. Yoshida, K, Engel, J, and Lindquist, S: The effect of chronic ethanol administration on high affinity [^{3}H]nicotine binding in rat brain. Naunyn-Schied. Arch. Pharmac. 321:74-76, 1982.
69. Hery, F, Bourgion, S, Hamon, M, Ternaux, JP, and Glowinski, J: Control of the release of newly synthesised [^{3}H]5-hydroxytryptamine by nicotinic muscarinic receptors in rat hypothalamic slices. Naunyn-Schied. Arch. Pharmac. 296:91-97, 1977.

70. Stolerman, IP, Goldfarb, T, Fink, R, and Jarvik, ME: Influencing cigarette smoking with nicotine antagonists. Psychopharmac. 28:247-259, 1973.
71. Tennant, FS, Tarver, AL, and Rawson, RA: Clinical evaluation of mecamylamine for withdrawal from nicotine dependence. NIDA Research Monograph 49:239-246, 1983.
72. Fibiger, HC: The neurobiological substrates of depression in Parkinson's disease: a hypothesis. Can. J. Neurol. Sci. 11:105-107, 1984.
73. Domino, EF: Neuropsychopharmacology of nicotine and tobacco smoking. In (ed.) Dunn, WL, Smoking Behavior: motives and incentives, 5-31, VH Winston, Washington, DC, 1973.
74. Warburton, DM: see this volume.
75. Wesnes, K: see this volume.

NICOTINIC MODULATION OF DOPAMINERGIC NEUROTRANSMISSION: FUNCTIONAL IMPLICATIONS

Agu Pert and Paul B.S. Clarke

Biological Psychiatry Branch
NIMH, Bethesda, Maryland

There is substantial evidence from diverse sources to indicate an important interactive relationship between the cholinergic and dopaminergic systems in brain. It has been known for some time that dopaminergic agonists and antagonists exert a profound influence on the release, turnover, or concentration of striatal acetylcholine (ACh) (1-5). In general, the findings from such studies have suggested that the function of dopaminergic nigrostriatal fibers is to exert a tonic inhibitory influence on cholinergic interneurons. Likewise, cholinergic agents also seem to exert an effect on dopaminergic function. Systemic injections of muscarinic cholinergic agonists have been found to enhance the synthesis (6) and utilization (7-10) of striatal as well as mesolimbic dopamine (DA). Muscarinic antagonists, on the other hand, appear to reduce the rate of utilization of dopamine (6, 11-15).

Systemic injections of nicotinic compounds also appear to exert a modest but nevertheless significant effect on dopaminergic activity. Lichtensteiger et al. (16), for example, have reported that 1 mg/kg of nicotine (s.c.) increases the striatal concentration of HVA in rats. In the same study nicotine injections also increased the fluorescence intensity of DA neurons in the substantia nigra when assessed with histochemical microfluorimetry. Such increases in intensity have been related previously to increases in neuronal activation (17). Similar increases in the fluorescence of nigral neurons following nicotine administration were confirmed also in a subsequent study by the same authors (18). Fuxe et al. (19) on the other hand, reported a decrease in catecholamine (presumably DA) fluorescence in the medial palisade zone of the medial eminence following systemic injections of nicotine. Anderson et al. (20), using the same technique, also reported that a single injection of nicotine (1 mg/kg i.p.) produced a significant reduction of dopamine stores in the medial caudate and in the diffuse dopamine nerve terminal systems of the anterior portion of the nucleus accumbens. Following inhibition of tyrosine hydroxylase, nicotine was found to produce a preferential decrease in DA (increase in DA turnover) in the medial caudate and within the posterolateral portion of the olfactory tubercle. Lichtensteiger et al. (21) also have reported more recently that systemic nicotine produces a significant, although moderate, increase in DA turnover in the striatum that was quantitatively similar to that produced by electrical stimulation of the nigrostriatal pathway. Nicotinic antagonists appear to produce opposite effects to nicotine on striatal and mesolimbic DA function. Ahtee and Kaakkola (22), for example, have reported that systemic injections of mecamylamine slowed the rate of depletion of DA from the striatal and mesolimbic DA regions following pretreatment with a tyrosine hydroxylase inhibitor.

Originally, the effects of cholinergic agents on DA neurotransmission were thought to be mediated indirectly through striato-nigral feedback pathways. It is apparent now that at least some of the effects of these compounds are mediated directly through cholinergic receptors located on both terminals as well as cell bodies of DA neurons.

Are Dopaminergic Neurons Cholinoceptive?

There is considerable evidence to suggest that dopaminergic neurons are cholinoceptive in nature. Receptor binding studies, for example, have revealed that dopaminergic neurons possess both nicotinic as well as muscarinic receptors. Kato et al. (23) reported that unilateral 6-hydroxydopamine lesions of the substantia nigra in rats decreased specific [^{3}H]atropine binding in the ipsilateral striatum by 20% when assessed 4 to 15 days following surgery. During this time a concomitant increase in atropine binding was seen in the contralateral striatum. However, neither an increase nor a decrease in [^{3}H]atropine binding was found in the striatum 21 or 40 days following surgery even though dopamine levels were still depressed significantly. Two alternate explanations were entertained to account for these findings. First, the decreased binding of ^{3}H-atropine might suggest that muscarinic cholinergic receptors are localized on dopamine neurons in the striatum. This explanation, however, does not readily explain the transitory nature of the decreased binding in the ipsilateral striatum nor the increased binding initially observed on the contralateral side. The authors suggest that both of these effects could be accounted for by a trans-synaptic action which is mediated by a long lasting activation induced by removal of tonic inhibitory dopaminergic influences (24-27). In addition, since unilateral interruption of nigrostriatal DA neurons results in activation of the controlateral pathway (28), it was further proposed that such increased inhibitory influences by DA on cholinergic neurons on the contralateral side is translated into a compensatory up-regulation of post-synaptic muscarinic receptors. De Belleroche et al. (29) also observed an 8% decrease in the specific binding of N-methyl-atropine in striatal homogenates of rats three days following intraventricular injections of 6-OHDA at which time tyrosine hydroxylase activity of striatal synaptosomes was reduced by 70%. More persistent alterations in the binding of muscarinic ligands in striatum have been reported by others. Suga (30) found a 14% decrease in specific binding of ^{3}H-quinuclidinyl benzilate (QNB) ipsilateral to a 6-OHDA lesion of the nigrostriatal pathway 45 days following surgery. Interestingly, when animals were treated chronically with L-DOPA the difference between the lesioned and unlesioned side (11%) no longer reached statistical significance. Based on the apparent ability of L-DOPA to reverse the lesion effect, this author also suggested that the decreased binding of QNB could be accounted for by assuming that the removal of DA inhibitory input to cholinergic striatal interneurons resulted in up-regulation of the post-synaptic cholinergic muscarinic receptor. This interpretation however is questionable. Although the author concludes that treatment with L-DOPA had restored the decrease in muscarinic receptor binding induced by the lesion, it is possible that this manipulation simply increased the variability of the data thus masking the small difference between the lesioned and unlesioned side. There was certainly no indication that the binding of [^{3}H]QNB was quantitatively different in the striatum ipsilateral to the lesion between control rats and those treated with L-DOPA. Findings from several other studies are also inconsistent with the notion that decreased muscarinic ligand binding is secondary to a disinhibitory effect of 6-OHDA lesions on cholinergic neurons. Namura et al. (31), for example, reported that rats treated neonatally with 6-hydroxydopa had decreased binding of [^{3}H]QNB in mesolimbic and striatal membrane homogenates when tested 2-3 months later. Such long term effects would be difficult to reconcile with the disinhibition hypothesis. Gurwitz et al. (32) have also observed decreased binding of a muscarinic ligand to striatal homogenates of mice up to 60 days following an intraventricular injection of 6-OHDA. In this study, chronic treatment of the animals with reserpine (which depletes brain catecholamines) failed to alter the binding of [^{3}H]4-methyl-N-piperidyl-benzilate, a muscarinic antagonist. This finding also appears to be inconsistent with the notion that decreased muscarinic binding following destruction of dopaminergic neurons is determined by removal of tonic inhibitory influences on cholinergic neurons.

To add further confusion to the issue, McGeer et al. (33) did not find decreases in [^{3}H]QNB binding in the rat striatum 10-15 days following intraventricular injections of 6-OHDA. Reisine et al. (34) also failed to find alterations in [^{3}H]QNB binding in this structure or the substantia nigra following 6-OHDA lesions of the nigrostriatal pathway. McGeer et al. (33) on the other hand, did find a strong correlation between QNB binding and GAD levels following kainic acid injections into the striatum suggesting that muscarinic receptors are localized on neurons intrinsic to this structure. Hruska et al. (35) have also reported that kainic acid lesions of the striatum reduce binding of [^{3}H]QNB by 40% in the rat. If nonspecific damage to striatal afferent axon terminals can be ruled out, these findings would indicate that muscarinic receptors are localized primarily on striatal perikarya or at least on striatal interneurons. It is also likely that a small percentage of such receptors are localized on dopaminergic terminals.

While there have been a considerable number of attempts to define the localization of muscarinic receptors on striatal neurons, very little research has been directed at evaluating the presence of nicotine receptors on striatal neuronal constituents. Recently, Schwartz et al. (36) have reported a decrease in striatal nicotinic receptor binding sites labeled by [^{3}H]acetylcholine following intraventricular injections of 6-OHDA. We have also evaluated the presence of nicotine receptors on mesolimbic as well as nigrostriatal DA neurons with autoradiographic procedures (37). Rats were lesioned unilaterally in the medial forebrain bundle with 6-OHDA which produced an 89% decrease in striatal dopamine content ipsilateral to the lesion. Five weeks following surgery the animals were prepared for autoradiography as previously described (38). Following unilateral 6-OHDA lesions, [^{3}H]nicotine labeling was reduced in the striatum, as well as the zona compacta of the substantia nigra (Fig. 1) ipsilateral to the lesion in every animal (Table 1). More significantly, the loss of labeling in the striatum was related to the degree of striatal DA depletion ($r = 0.71$, $p < .05$). Within the mesolimbic DA system, a visible reduction of labeling was observed in the ventral tegmental area and olfactory tubercle (Fig. 1), as well as in the nucleus accumbens (Table 1). Labeling was unaltered in the medial fronto-parietal cortex (laminae II-V) which lacks a dopaminergic innervation. Previous analyses by Schwartz et al. (36) suggest that the changes in binding of nicotine ligands following such dopaminergic neuronal destruction reflect a loss of receptors rather than an attenuation of receptor affinity. These results strongly suggest that nicotinic cholinoceptors are located on DA forebrain terminals as well as on mesencephalic DA perikarya.

Studies employing alpha-bungarotoxin as a nicotine ligand have also provided evidence for the localization of nicotinic receptors on DA neurons. McGeer et al. (33), for example, found a strong correlation between decreases in tyrosine hydroxylase activity and alpha-BTX binding in the striatum following intraventricular 6-OHDA. Scatchard analyses

Table 1. [^{3}H]Nicotine Labeling and Striatal DA Content Following 6-OHDA Lesion

Site	Displaceable OD x 1000		Ratio lesioned unoperated (%)
	Unoperated Side	Lesioned Side	
Anterior striatum	140.3 ± 11.3	95.7 ± 12.6	67.3 ± 4.3
Midstriatum	118.3 ± 7.7	86.8 ± 11.4	70.6 ± 5.4
Posterior striatum	107.2 ± 5.0	81.3 ± 9.2	75.0 ± 5.6
Nucleus accumbens	117.9 ± 12.5	72.3 ± 8.8	61.4 ± 3.9
Frontoparietal cortex	144.4 ± 7.9	144.0 ± 8.3	99.8 ± 2.6
Anterior SN	168.1 ± 23.2	75.3 ± 17.2	45.0 ± 8.3
Posterior SN	119.8 ± 18.8	82.4 ± 10.6	72.6 ± 7.6
pg DA/μg protein	166.2 ± 13.9	32.8 ± 18.5	19.6 ± 11.5

DA content was measured in micropunches (ca 0.6 mg tissue) of striatum taken between the anterior and midstriatal levels, by HPLC combined with electrochemical detection. OD values quoted above correspond to displaceable binding and were calculated in individual rat (n = 6) by subtracting the mean OD of 4 sections labeled in the presence of non-radioactive nicotine (10 μM) from the mean OD of 4 adjacent sections labeled in the absence of excess non-radioactive nicotine. Lesion-induced changes in [^{3}H]nicotine binding were assessed by comparison with the unoperated side in each rat. Experimental values are mean ± S.E.M.

suggested that the decreased binding was due to a loss of receptors and not a change in affinity. De Belleroche et al. (29) have also reported decreases in binding of alpha-BTX in the striatum following i.c.v. injections of 6-OHDA which decreased tyrosine hydroxylase activity by 70%.

The relationship between findings from studies that have employed [^{125}I]alpha-bungarotoxin and [^{3}H]acetylcholine (in the presence of atropine) or [^{3}H]nicotine as ligands is not clear however. We have recently compared the binding of [^{3}H]nicotine, [^{125}I]alpha-BTX and [^{3}H]acetylcholine in the presence of atropine to rat brain sections using autoradiographic procedures (39). The autoradiographic distributions of [^{3}H]Ach and [^{3}H]nicotine were clearly concordant with the highest densities in the interpeduncular nucleus, most thalamic nuclei, superior colliculus, medial habenula, presubiculum and molecular layer of the dentate gyrus. Binding was also prominent in the substantia nigra of the zona compacta, ventral tegmental area, and certain laminae of the cerebral cortex. Moderate densities of labeling were found in the neostriatum, ventral striatum, dorsal tegmental nucleus and cerebellum.

The pattern of [^{125}I]BTX on the the other hand was strikingly different, the only significant overlap with the agonist binding being the cerebral cortex and superior colliculus. [^{125}I]BTX binding was also dense in the inferior colliculus, hypothalamus, and hippocampus. Most notably, the striatum, nucleus accumbens and mesencephalic regions encompassing the substantia nigra and ventral tegmental area were strikingly devoid of [^{125}I]BTX labeling. Clearly there are at least two distinct populations of nicotinic binding sites in rat brain. Various lines of evidence (39) suggest that the high affinity agonist binding sites in brain labeled by [^{3}H]nicotine and [^{3}H]ACh correspond to

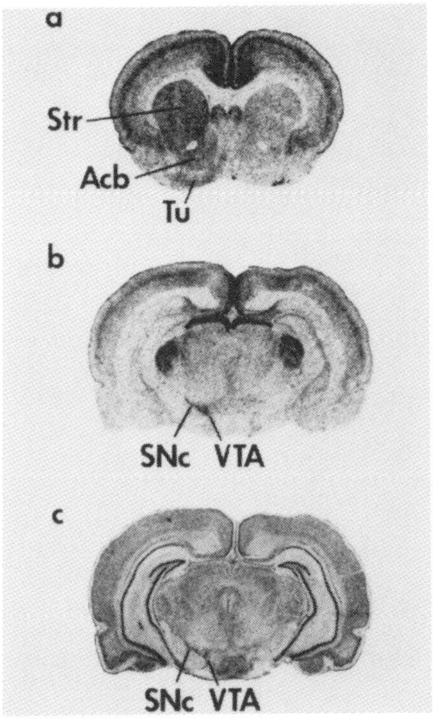

Fig. 1. Photomicrographs (a,b) of [^{3}H]nicotine receptor distribution in coronal sections of rat brain, at the level of the anterior striatum (a) and anterior SN (b). The lower photograph (c) shows the Nissl-stained section which provided the nigral autoradiograph (b). Five weeks prior to sacrifice, the animal received a unilateral injection of 6-OHDA into the ascending DA fibre bundle. This resulted in a near-total depletion of DA in the ipsilateral striatum and the disappearance of DA cell bodies in the SN and adjacent ventral tegmental area (VTA), as shown in c (right side). The unilateral degeneration of DA neurons was accompanied by a reduction of [^{3}H]nicotine labeling in terminal areas of the nigrostriatal system (Str, striatum) and the mesolimbic system (Acb, nucleus accumbens; Tu, olfactory tubercle); and a virtual disappearance of specific labeling in the corresponding DA cell body regions (SN and VTA).

nicotinic cholinergic receptors similar to those found at autonomic ganglia. BTX binding sites in brain may also serve as receptors for nicotine and are possibly related to neuromuscular nicotinic receptors. Considering the absence of BTX binding sites in regions populated by dopaminergic neurons (39-41), it is not likely that these receptors play an important role in regulating dopaminergic function.

Cholinergic Modulation of Dopamine Release: Striatal Actions

Considering the presence of cholinergic receptors on dopamine neurons, it is not surprising that cholinergic compounds have been found to modulate dopaminergic functions. Acetylcholine has been known to regulate striatal DA release for some time. Besson et al. (42) were the first to describe the ability of ACh in the presence of physostigmine to evoke the release of newly synthesized DA from striatal tissue. Westfall (43), on the other hand, reported that ACh significantly inhibited the release of [^{3}H]DA from striatal slices evoked by either potassium or electrical stimulation. De Belleroche and Bradford (44) also reported an inhibitory role for ACh in regulating DA functions. There investigators found that ACh, in the presence of the nicotinic antagonists hexamethonium or alpha-bungarotoxin, inhibited the potassium-evoked release of dopamine from rat striatal synaptosomes. Acetylcholine by itself or in the presence of the muscarinic antagonist atropine stimulated DA release in the same preparation. These findings prompted the hypothesis that nicotinic receptors on striatal dopamine terminals mediate the excitatory actions of ACh on DA release while muscarinic receptors mediate an inhibitory action. Such a dual cholinergic regulation of dopamine release would be analogous to the modulation of noradrenergic release in peripheral neurons (45). While this was a rather attractive hypothesis, it has not received substantiation from other reports which have shown that both muscarinic, as well as nicotinic agonists are capable of enhancing the release of dopamine in the forebrain. Bartholini and Stadler (46), for example, have shown that perfusion of the cat caudate nucleus with ACh in the presence of physostigmine produces a significant increase of DA in the perfusate. Giorguieff-Chesselet et al. (47), using push-pull perfusion techniques, have also found that ACh (10^{-5} M), in the presence of physostigmine, stimulated the release of [^{3}H]DA continuously formed from [^{3}H]tyrosine in the cat caudate nucleus. Interestingly, this effect was no longer seen during concurrent activation of DA neurons by Substance P infused into the ipsilateral substantia nigra. This finding suggests that the presynaptic regulation of DA release in the striatum by ACh is dependent on the rate of firing of nigrostriatal DA neurons. Muscarinic receptor activation has also been shown to enhance the release of DA _in vitro_. Giorguieff et al. (48) have reported that carbachol and ACh enhance the release of [^{3}H]DA from rat striatal slices. In a subsequent report from the same laboratory (49) the effects of ACh on striatal DA release were shown to be abolished in the absence of calcium and partially blocked by pempidine, atropine or scopolamine. Oxotremorine also enhanced DA release, an effect that was reversed by atropine but not pempidine. Since the addition of tetrodotoxin did not alter the stimulatory effect of either ACh or oxotremorine, it seems likely that the excitatory actions of nicotinic and muscarinic agonists on DA release are mediated through cholinoceptors located on DA terminals in the striatum. Several more recent studies have characterized the actions of ACh on striatal as well as mesolimbic DA release somewhat further. Raiteri et al. (50) for example, have shown that it is possible to enhance the potassium-evoked release of [^{3}H]DA from rat striatal synaptosomes with ACh by using lower concentrations of KCl than those used in the earlier studies. This action was blocked by atropine, indicating a muscarinic receptor mediated event. Since the characteristics of muscarinic receptor-mediated modulation of DA release may depend on the state of depolarization of DA nerve terminals, the failure of Westfall (43), as well as de Belleroche and Bradford (44) to observe an enhanced release with ACh may be due to the excessively high concentrations of KCl that were used in their studies. In a more recent study, de Belleroche and Gardiner (51), however, have found that both oxotremorine and ACh enhance the release of DA evoked by 34 mM K^+ in slices from the nucleus accumbens. The actions of oxotremorine on DA release were calcium-dependent and blocked by atropine but not mecamylamine.

Recently, Lehman and Longer (52) have attempted to study the regulation of DA release by ACh under more physiologically meaningful conditions. Essentially, they examined the effects of exogenously administered ACh and cholinomimetic agents on the release of DA from striatal slices evoked presumably by electrically induced release of endogenous ACh. They found that extremely low concentrations (1.8 – 10 μM) of the muscarinic receptor agonist oxotremorine, as well as the anticholinesterase physostigmine, produced a significant increase in electrically evoked release of [^{3}H]DA from slices of cat caudate. The effects of both compounds were antagonized by atropine. These findings suggest that in the electrically depolarized caudate, endogenous ACh may interact with muscarinic receptors, facilitating depolarization-evoked release of [^{3}H]DA.

Nicotinic receptors localized on DA terminals in the striatum also appear to regulate DA release. Westfall (53) first demonstrated that high concentrations of nicotine (10^{-3} M) produced a significant release of [^{3}H]DA from rat striatal slices which was antagonized by the addition of hexamethonium. Interestingly, this effect was also reduced by the addition of ACh and methacholine, which were thought to exert their effects predominantly through the inhibitory muscarinic receptors as described above. Giorguieff et al. (48) have shown that ACh enhances the release of newly synthesized [^{3}H]DA from rat striatal slices as well as from the cat caudate nucleus _in vivo_. Both of these effects were blocked by the addition of mecamylamine or hexamethonium, suggesting the involvement of nicotinic receptors. Release of [^{3}H]DA from striatal slices by nicotinic activation has been confirmed by a number of subsequent reports. Giorguieff et al. (49) also found that ACh stimulated the release of [^{3}H]DA from striatal slices, an effect that was partially blocked by pempidine, a ganglionic blocker. Sakurai et al. (54) have reported recently that high concentrations of ACh (5×10^{-4} M), in the presence of physostigmine enhanced the release of [^{3}H]DA from both striatal punches as well as synaptosomes. This effect was partially inhibited by the nicotinic antagonist, D-tubocurarine but not by atropine. Nicotine at $2\text{-}5 \times 10^{-4}$ M as well as lobeline, continine and sparteine also enhanced the spontaneous release of [^{3}H]DA from both striatal punches and synaptosomes. Neither D-tubocurarine, hexamethonium, tetraethylammonium nor alpha-BTX were found to inhibit [^{3}H]DA release elicited by lobeline or nicotine however. Recently, Marien et al. (55), also using relatively high concentrations of nicotine, have demonstrated a concentration-related release of [^{3}H]DA from rat caudate slices as well. This effect was not significantly influenced by the absence of Ca^{2+}, Mg^{2+}, nor by the addition of hexamethonium to the superfusion medium. Arqueros et al. (56) have also reported a calcium independent release of [^{3}H]DA from rat striatal slices by high concentrations of nicotine (5 mM). There investigators concluded that nicotine enhances the release of DA by displacing the monoamine from the vesicle storage complex which is not dependent on extracellular calcium. Both of these studies, however, employed exceedingly high concentrations of nicotine and nicotinic agonists. The ability of nicotine to enhance DA release in a calcium independent manner and the failure to block such nicotinic actions with appropriate antagonists may be related to the high concentrations of nicotine employed. Giorguieff et al. (57) have shown that lower concentrations of nicotine (10^{-6} M) will induce a calcium-dependent DA release from rat striatal slices which is blocked by pempidine and d-tubocurarine. These effects were probably mediated through receptors localized on DA terminals since the addition of tetrodotoxin (which should interrupt interneuronal processes) did not modify the actions of nicotine.

Cholinergic-Dopaminergic Interactions in the Substantia Nigra

In addition to the presence of nicotinic receptors on DA perikarya in the zona compacta of the substantia nigra (SN) (37), there is considerable evidence for additional cholinergic markers in this structure as well. For example, the SN contains a high-affinity choline uptake system (58) and appreciable amounts of acetylcholine and choline acetyltransferase (ChAT) (59). Within the zona compacta, DA neurons do not themselves contain ChAT, but do contain acetylcholinesterase (AChE), which appears to be localized in the soma as well as dendrites (59). It has been proposed that the AChE in dendrites of nigral zona compacta neurons may serve to inactivate a cholinergic input upon the processes (59).

There are a number of studies that have demonstrated cholinergic-dopaminergic interactions in the substantia nigra. Intranigral injections of muscarinic cholinergic agents, for example, seem to produce behaviors that are dopamine dependent. James and Massey (60) have reported strong contralateral rotational behavior following unilateral injections of hemicholinium into the rat substantia nigra, an effect that was accompanied by an increase in DA utilization within the ipsilateral striatum. Carbachol injections into the same site induced strong ipsilateral rotations with a decrease in ipsilateral DA utilization. Since animals presumably rotate contralateral to the striatum with the preponderance of DA activity, the findings above suggest that muscarinic activation of the SN leads to inhibition of nigrostriatal DA transmission while inhibition produces an opposite effect. Similar findings have been reported also by others (61-64).

While considerable effort has been directed at studying the behavioral and neurochemical effects following direct applications of muscarinic agents into the SN, there are no comparable studies that have used nicotinic compounds. There is evidence, however, that nicotinic agonists (unlike muscarinic agonists) activate DA neurons in the midbrain, presumably through a direct action on the cells.

Lichtensteiger et al. (16, 18, 21), using extracellular single unit recording procedures, have demonstrated that nicotine administered subcutaneously (1 mg/kg) increases the firing rate of zona compacta cells in the substantia nigra. The excitation induced by nicotine was antagonized by iontophoretic application of dihyro-B-erythroidine but not by atropine (21). Recently, we also have evaluated the effects of systemic nicotine on the firing rate of both zona compacta and zona reticulata neurons in the substantia nigra of chloral hydrate-anaesthetized rats (65). Either subcutaneously or intravenously administered nicotine increased the spontaneous firing rate of zona compacta neurons (Fig. 2). The excitatory actions of nicotine were prevented by mecamylamine but not by a ganglionic-blocking dose of the bisquaternary compound chlorisondamine (Fig. 3). Mecamylamine by itself was found to decrease the spontaneous activity of dopaminergic neurons, suggesting that the cholinergic input may be tonically active. Besides stimulating the dopaminergic neurons in the zona compacta, nicotine also increased the spontaneous activity of non-dopaminergic neurons in the zona reticulata. This effect was probably mediated through peripheral actions since chlorisondamine prevented the marked excitation. The excitatory actions of nicotine on DA neurons appears to be direct. Lichtensteiger et al. (16, 18, 21) have reported that the firing of DA neurons in the zona compacta of the substantia nigra was accelerated following iontophoretic administration of nicotine. Iontophoretic application of ACh also produced a similar but smaller increase in firing rate (16, 21). The excitatory effects of both nicotine and ACh were antagonized by dihydro-B-erythroidine, confirming the nicotinic nature of the response (18, 21).

Although findings from studies that have examined the direct effects of nicotine on DA zona compacta neurons are relatively consistent, findings from studies that have evaluated the direct actions of ACh are more equivocal. While Dray and Straughan (66) have also reported that zona compacta neurons are excited by iontophoretic ACh, a number of other investigators have failed to report such actions. Zona reticulata

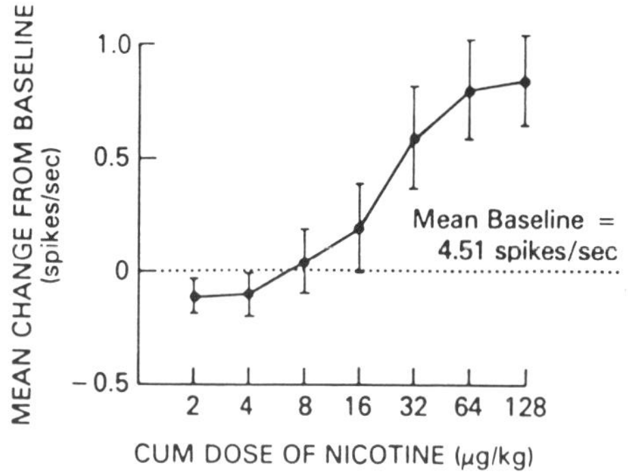

Fig. 2. Change in firing rate of SNC DA neurons following intravenous administration of nicotine. Each rat received a series of increasing doses of nicotine (2, 2,4 .....64 μg/kg). Injections were spaced 3 min. apart. The effect of nicotine was assessed by comparing the mean firing rate (i.e., before any nicotine was given). The data are expressed as the group mean (± s.e. mean) of the corresponding difference scores (n = 11). The mean firing rate was 4.51 spikes/sec.

neurons, on the other hand, always seem to be excited by iontophoretic ACh (66–69). It is likely that the two populations of cells are differentially sensitive to ACh. Even Lichtensteiger et al. (16) reported that ACh produced only a very modest increase in firing rate of zona compacta neurons while the effects of nicotine were much more apparent.

If zona compacta neurons in the SN are indeed cholinoceptive, what is the precise origin of the cholinergic input? It is not likely that the cholinergic neurons innervating either zona compacta or zona reticulata cells are intrinsic to the SN. Firstly, the SN appears to contain few if any interneurons (60). Secondly, the absence of immunohistochemically identified cholinergic perikarya in the SN (70–72) and the results of lesion experiments (73) suggest an extrinsically derived cholinergic innervation. On the basis of histochemical studies on AChE, Oliver et al. (74) originally suggested that the SN receives a descending cholinergic input from the caudate nucleus. The existence of such an input seems unlikely, however, since a variety of striatal lesions have failed to alter ChAT in the SN (75–76). A remaining possibility is that the cholinergic innervation of the SN is derived from more caudally situated cholinergic cell bodies. One possible source is the population of cholinergic cells which traverse the pontine tegmentum, starting just caudal to the nigra and extending in a dorsocaudal direction to include the peribrachial region. Mesulam et al. (71) have divided this population into two groups, one located primarily within the pedunculopontine nucleus and the other in the region of the dorsolateral tegmental nucleus. Retrograde tracing studies have in fact identified these nuclei as a potential source of projections to the SN (77–78, Clarke et al., unpublished findings). We have preliminary evidence (Clarke et al., unpublished findings) suggesting that cholinergic neurons in the peribrachial region provide an excitatory input to zona compacta neurons in the SN. Briefly, we have found that injections of kainic acid (a perikaryal excitant) into the peribrachial region of the rat brain produce an elevation in the firing rate of zona compacta neurons that is antagonized by an intravenous injection of mecamylamine. Although these effects may be polysynaptically mediated, when considered in the context of the studies reviewed above, it seems likely that the SN receives an excitatory nicotinic innervation from neurons located in the peribrachial region.

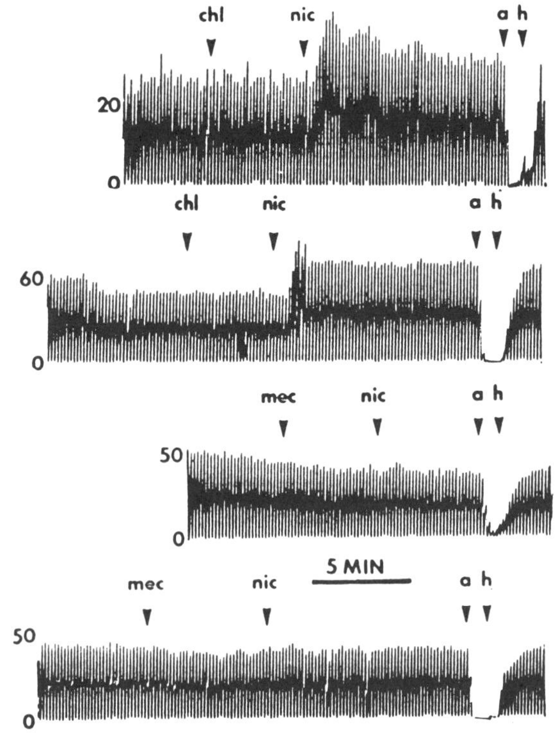

Fig. 3. Ratemeter records showing the responses of single SNC neurons to nicotine, in rats pretreated with mecamylamine or chlorisondamine. Extracellular spikes were aggregated over periods of 10s (vertical axes). Following a baseline period of 5 min, the ganglion blockers chlorisondamine (chl - 0.1 mg/kg) or mecamylamine (mec - 2 mg/kg) were given intravenously. Of these two compounds, only mecamylamine readily penetrates the CNS and only mecamylamine prevented the excitatory effect of nicotine (1.0 mg/kg sc). Each experiment was concluded with i.v. injections of apomorphine (a - 0.1 mg/kg) and haloperidol (h - 0.1 mg/kg) to show that the cells were dopaminergic.

Behavioral Effects of Nicotine Involving Dopaminergic Activation

One of the most noticeable behavioral actions of nicotine is its ability to alter locomotor output. Acute injections of nicotine produce a dose-dependent depression of locomotor behavior (79-83). Such initial depressant actions are often followed by locomotor ecitation (79-80, 82-83). Following repeated injections of nicotine, the initial depressant actions of the drug are replaced rapidly by a dose-dependent excitatory effect (79-81). Since mecamylamine but not hexamethonium antagonizes the depressant, as well as excitatory effects of nicotine, it appears that both actions are mediated through the central nervous system (79-80, 84-85).

It is apparent that both the nigrostriatal, as well as mesolimbic dopamine systems play an important role in locomotor activity (86-88). Drugs that enhance transmission at DA synapses appear to increase locomotor activity while drugs that block DA receptors depress locomotor output (86). We have recently completed a series of studies which suggest that nicotinic agonists also enhance locomotor output by activating the ascending DA mesolimbic pathways. In the initial studies, rats were implanted with chronic stainless steel cannulae guides aimed for an area 1.5 mm dorsal to the left lateral ventricle. Following recovery from surgery, the animals were injected intraventricularly with 1-100 n moles of cytisine, a potent nicotinic agonist (89-90) that is considerably less lipophilic than nicotine (89), thereby making it a more suitable agent to use for intracerebral studies. Intraventricular injections of cytisine were found to produce a characteristic dose-dependent initial depression of locomotor output that was followed by an excitatory action which persisted over an hour following the largest dose (Fig. 4). In order to ascertain whether the excitatory effects of cytisine were mediated through dopaminergic pathways, rats were implanted with bilateral cannulae guides aimed for either the caudate nucleus, nucleus accumbens, substantia nigra or ventral tegmental area. Injections of cytisine (5 n moles bilaterally) into the VTA (the origin of the mesolimbic pathway) produced a significant increase in locomotor output (Fig. 5) while injections into the other three regions had no such effect. This excitatory effect of cytisine was blocked by systemic injections of mecamylamine (Fig. 6) or by 6-OHDA lesions of the nucleus accumbens (Fig. 7). These preliminary findings suggest that the excitatory effects of nicotinic agonists are mediated through the activation of mesolimbic DA systems at the level of the VTA.

While nicotine agonists appear to produce their locomotor excitatory effects through activation of mesolimbic DA pathways, there is some evidence that nicotine may also exert behaviorally relevant actions through activation of the nigrostriatal DA systems. One behavioral paradigm that has been useful in evaluating the functional activity of the

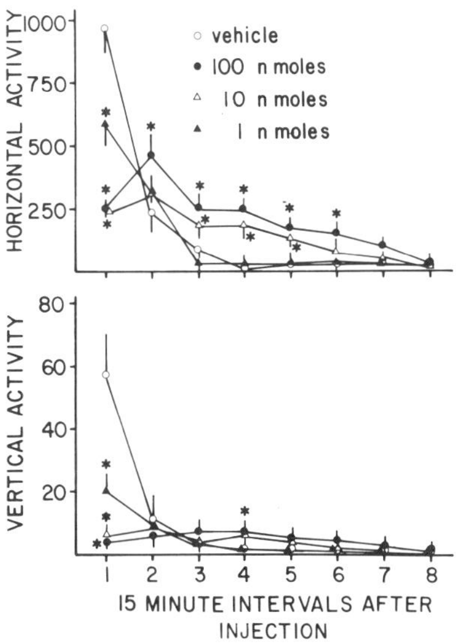

Fig. 4. Effects of intraventricular cytisine on locomotor activity in rats. All animals (n = 12) were injected with 5 μl of saline and 1, 10, 100 n moles of cytisine in 5 μl of ml of saline in a counterbalanced order separated by at least three days. Locomotor activity was assessed in Columbus Instrument activity monitors (model opto-varimex minor). Repeated measures analysis of variance revealed a statistically significant drug effect (P< .01), whereas *p< .05 for comparisons of drug treatment effects with the saline control using the Dunnett t-statistic.

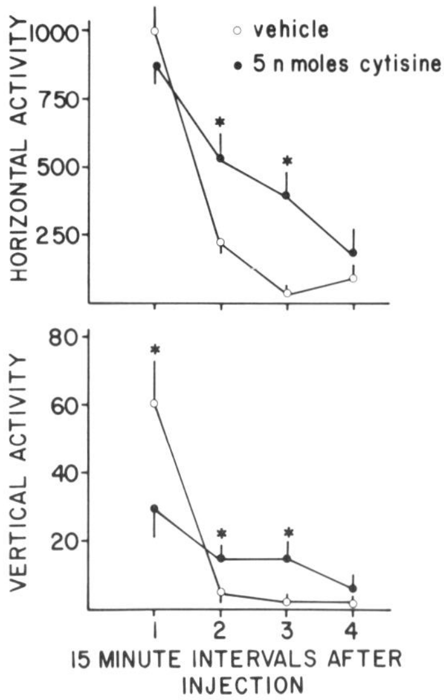

Fig. 5. Effects of cytisine on locomotor activity following injections into the ventral tegmental area (VTA). All rats (n = 8) were injected bilaterally in the VTA with 1 μl saline and 5 n moles of cytisine/ 1 μl saline in a counterbalanced order separated by at least three days. A repeated measures analysis of variance revealed a statistically significant drug effect ($p < .05$); *p=< .05 for individual comparisons of treatment effects with the saline control.

nigrostriatal DA system is the assessment of rotational behavior in animals lesioned unilaterally in the nigrostriatal pathway (91-92). As already noted, this model is based on the principle that animals will rotate contralaterally to the striatum with the preponderance of DA activity. Thus, when the ascending DA nigrostriatal pathway is lesioned unilaterally with 6-OHDA, rats will rotate contralaterally to the lesion following amphetamine and ipsilaterally following apomorphine (presumably due to the development of denervation supersensitivity in the ipsilateral striatum).

Using this paradigm Lichtensteiger et al. (18) have found evidence that nicotine activates the nigrostriatal dopamine pathways. They reported that systemic nicotine produced weak rotational behavior ipsilateral to the 6-OHDA lesion during the first 15 minutes after injection. The effect was no longer present when the rats were tested 35 minutes following administration. Kaakkola (93) has also evaluated the effects of nicotine and its interactive effects with amphetamine on rotational behavior in rats lesioned unilaterally in the nigrostriatal pathway with 6-OHDA. While nicotine itself had little effect on rotational output it potentiated significantly rotational behavior induced by systemic injections of amphetamine. Conversely, systemic injections of mecamylamine (but not hexamethonium) decreased the actions of amphetamine. In addition, the muscarinic agonist pilocarpine antagonized amphetamine while the muscarinic antagonist, atropine, seemed to potentiate the effects of this sympathomimetic. These findings prompted the authors to suggest that stimulation of nicotinic receptors on nigrostriatal neurons increases the release of DA while stimulation of muscarinic receptors inhibits the nigrostriatal DA pathways either directly or indirectly. While these studies suggest that nicotine

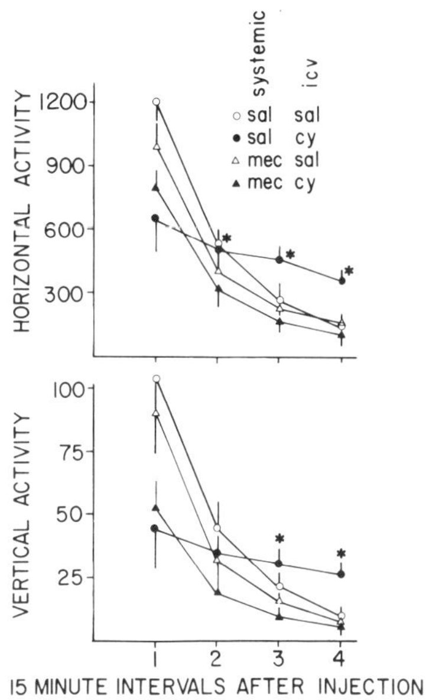

Fig. 6. Effects of mecamylamine pretreatment on cytisine induced alterations in locomotor output. All rats (n = 12) received the following drugs and drug combinations: systemic saline + i.c.v. cytisine (100 n moles), systemic saline + i.c.v. saline (5 μl), systemic mecamylamine (1 mg/kg i.p.) + i.c.v. cytisine (100 n moles), systemic mecamylamine (1 mg/kg i.p.) + i.c.v. saline (5 μl). Systemically administered drugs were given 20 minutes prior to the i.c.v. injection. All animals received all drug combinations in a counterbalanced order separated by at least 5 days. A repeated measures analysis of variance revealed a statistically significant drug effect ($p < .05$); * = $p < .05$ for comparisons of systemic saline + i.c.v. cytisine with systemic mecamylamine + i.c.v. cytisine.

produces behaviorally relevant effects through the nigrostriatal DA system, Fuxe et al. (19) have failed to find that either nicotine or mecamylamine had any significant effect on amphetamine-induced rotational behavior in lesioned rats. An examination of the data, however, does reveal a trend in the predicted direction which may have failed to reach significance due to rather large variability. There are several factors which may make it difficult to measure significant rotational behavior following systemic injections of nicotine. Firstly, since acute injections of nicotine produce an initial depression of locomotor output, it is conceivable that such locomotor depressant actions may mask the expression of nigrostriatal activation induced by this drug. The behavioral effects of nigrostriatal activation by nicotine may be studied best in rats that are initially made tolerant to the depressant actions of nicotine. Secondly, the effect may be somewhat transitory or may appear only briefly after the initial depressant actions of the drug are replaced by excitation. Undoubtedly, it is important to assess the effects of nicotinic agonists in this model over a sufficiently long time interval following injection. Thirdly, behavioral effects of nicotine mediated through the nigrostriatal system are probably at best rather subtle, thus attempts should be made to maximize the expression of rotational output.

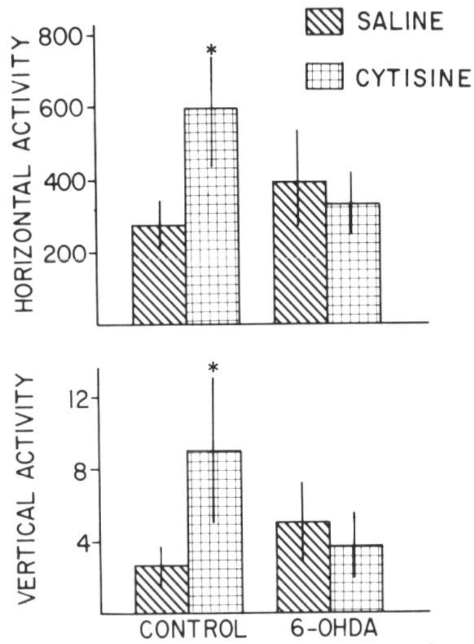

Fig. 7. Effects of 6-OHDA lesions of the nucleus accumbens on enhanced locomotor output induced by injections of cytisine into the ventral tegmental area. One group of rats (n = 11) was lesioned in the n. accumbens with 12 μg of 6-OHDA. The other group remained unlesioned (n = 12). Both groups were implanted with ventricular cannulae. Twelve days later half of the animals in each group were injected i.c.v. with saline (5 μl). The other half were injected with cytisine (100 n moles). Five days later the i.c.v. drug conditions were reversed for all rats. Activity counts in figure represent total activity during the first 30 minutes after injection; *p=< .05 for comparisons of the drug effects with saline effects using a t-test for dependent samples.

Significance of Dopaminergic Activation in Understanding the Addictive Liability of Nicotine

Over the past several years, considerable evidence has accumulated in support of reinforcing properties for nicotine in both animals and man. Nicotine has been found to be self-administered by rats (94), dogs (95), monkeys (96-97), and baboons (98). There is also strong evidence suggesting that nicotine is a euphorogenic and serves as a positive reinforcer to maintain cigarette smoking in man. Following the early observation by Johnston (99) that nicotine produced pleasant sensations in cigarette smokers, there have been a number of recent studies that have described and evaluated the subjective effects produced by both tobacco smoke as well as nicotine. Jasinski et al. (100) reported that nicotine produced dose-dependent increases in scores on scales of euphoria and consequent decreases in the desire to smoke in human subjects. When compared to other drugs of abuse, nicotine has been found to produce a profile of subjective reactions in man that is very similar to that produced by morphine and cocaine (101). In addition, some subjects with a history of drug dependence have identified nicotine injections as cocaine from a list of commonly used drugs (102). Consistent with these observations is the finding that smokers will self-inject nicotine intravenously when cigarettes are not available (102), an effect that appears to be blocked by pretreatment with mecamylamine.

Although it is clear that nicotine has reinforcing properties and produces pleasureable sensations in man, similar to other abused substances, the neuroanatomical and neurochemical substrates underlying these actions are not understood. One possibility is that the reinforcing properties of all drugs of abuse, including nicotine, are mediated through a common neurohumoral substrate. Sympathomimetics, like cocaine and amphetamine, appear to produce their rewarding effects by activating the dopamine systems in brain. For example, self-administration of amphetamine and cocaine are attenuated or abolished following inhibition of catecholamine synthesis with alpha-methyl-p-tyrosine (104) or blockade of dopamine receptors with dopamine and antagonists (105-109). The mesolimbic dopamine pathways originating from the ventral tegmentum and projecting to the nucleus accumbens appear to be critical since 6-OHDA lesions of the nucleus accumbens also decrease the rewarding effects of cocaine as well as amphetamine (112). The rewarding effects of opiates also appear to involve the mesolimbic dopamine pathways. Rats, for example, will self-administer morphine into the ventral tegmental area. Conditioned place preference has also been reported following injections of morphine into this structure (114). Morphine injections into the VTA have also been found to increase locomotor activity (115-117), an effect that appears to be related to mesolimbic dopamine activation. Considering the ability of nicotine to increase the activity of both nigrostriatal and mesolimbic neurons, it is quite likely the rewarding effects of this substance are mediated also (at least partially) through this neurohumoral system.

Acknowledgment

We gratefully acknowledge the expert assistance of Drs. C.C. Chiueh, D.W. Hommer, L.R. Skirboll, H.D. Everist, and Mr. T. Sullivan during various phases of the research described above. We also wish to thank Ms. S. Rucker for her excellent secretarial support in the preparation of this manuscript.

REFERENCES

1. Bartholini, G, Lloyd, KG and Stadler, H: The effect of drugs on the release of striatal neurotransmitters. CINP IX Congress. J. Pharmacol. (Paris), 5, suppl I: 60, 1974.
2. Guyenet, PG, Agid, Y, Javoy, F, Beaujouan, JC, Rossier, J and Glowinski, J: Effects of dopaminergic receptor agonists and antagonists on the activity of the neostriatal cholinergic system. Brain Res. 84: 227-244, 1975.
3. Ladinsky, H, Consalo, S, Bianchi, S, Samanim, R and Ghezgi, D: Cholinergic dopaminergic interaction in the striatum: the effect of 6-hydroxydopamine or pimozide treatment on the increased striatal acetylcholine, piribedil and d-amphetamine. Brain Res. 84: 221-226, 1975.
4. Stadler, H, Lloyd, KG, Gadea Ciria, M and Bartholini, G: Enhanced striatal acetylcholine release by chlorpromazine and its reversal by apomorphine. Brain Res. 55: 476-480, 1973.
5. Trabucchi, M, Cheney, D, Racogni, G and Costa, E: Involvement of brain cholinergic mechanisms in the action of chlorpromazine. Nature (London), 219: 664-666, 1974.
6. Javoy, F, Agid, Y and Glowinski, J: Oxotremorine and atropine induced changes of dopamine metabolism in the rat striatum. J. Pharm. Pharmacol. 27: 677-681, 1975.
7. Carrodi, H, Fuxe, K, Hammer, W, Sjoqvist, F and Ungerstedt, U: Oxotremorine and central monoamine neurons. Life Sci. 6: 2557-2566, 1967.

8. Laverty, R and Sharman, DF: Modification by drugs of the metabolism of 3-4-dihydroxyphenylethylamine, noradrenaline and 5-hydroxytryptamine in the brain. Brit. J. Pharmacol. 24: 759-763, 1965.
9. Nose, T and Takemato, H: Effect of oxotremorine on homovanillic acid concentration in the striatum of the rat. Eur. J. Pharmacol. 25: 51-55, 1976.
10. Westerink, BC and Korf, J: Regional rat brain levels of 3,4-dihydroxyphenylacetic acid and homovanillic acid: concurrent fluorometric measurement and influence of drugs. Eur. J. Pharmacol. 38: 281-291, 1976.
11. Anden, NE and Bedard, P: Influences of cholinergic mechanisms on the function and turnover of brain dopamine. J. Pharm. Pharmacol. 22: 460-462, 1971.
12. Bartholini, G and Pletscher, A: Atropine-induced changes of cerebral dopamine turnover. Experientia, 27: 1302, 1971.
13. Carrodi, H, Fuxe, K and Lidbrink, K: Interaction between cholinergic and catecholaminergic neurons in rat brain. Brain Res. 43: 397-416, 1972.
14. Hitzeman, RJ, Loh, HH and Domino, E: Effect of scopolamine on the cerebral accumulation of ^{14}C-catecholamines for ^{14}C-tyrosine. Pharmacology, 8: 291-299, 1972.
15. O'Keefe, R, Sharman, DF and Vogt, M: Effect of drugs used in psychoses on cerebral dopamine metabolism. Brit. J. Pharmacol. 382: 287-304, 1970.
16. Lichtensteiger, W, Felix, D, Lienhart, R and Hefti, F: A quantitative correlation between single unit activity and fluorescence intensity of dopamine neurons in zona compacta of substantia nigra, as demonstrated under the influence of nicotine and physostigmine. Brain Res. 117: 85-103, 1976.
17. Lichtensteiger, W: Effect of electrical stimulation on the fluorescence intensity of catecholamine-containing tuberal nerve cells. J. Physiol. (London), 218: 63-84, 1971.
18. Lichtensteiger, W, Felix F, Hefti, F and Schlumpf: Effects of nicotine on dopamine neurons of adult and prenatal mammals and of invertebrates. In (eds) Remond, A and Izard C, Electrophysiological Effects of Nicotine, 15-30, Elsevier/North-Holland Biomedical Press, 1979.
19. Fuxe, K, Agnati, L, Enerath, P, Gustafsean, J-A, Hakfelt, T, Lofstrom, A, Skett, B and Skett, P: The effect of nicotine on central catecholamine neurons and gonadotropin secretion. I. Studies in the male rat. Medical Biology, 55: 148-157, 1977.
20. Anderson, K, Fuxe, K and Agnati, LF: Effects of single injections of nicotine in the rat. Acta Physiol. Scand. 112: 345-347, 1981.
21. Lichtensteiger, W, Hefti, F, Felix, D, Huwyler, T, Melamed, E and Schlumpf, M: Stimulation of nigrostriatal dopamine neurons by nicotine. Neuropharmacol. 21: 963-968, 1982.
22. Ahtee, L and Kaakkola, S: Effects of mecamylamine on the fate of dopamine in striatal and mesolimbic areas of rat brain; interaction with morphine and haloperidol. Br. J. Pharmacol. 62: 213-218, 1978.
23. Kato, G, Carson, S, Kemel, ML, Glowinski, J and Giorguieff, MF: Changes in striatal specific ^{3}H-atropine binding after unilateral 6-hydroxydopamine lesions of the nigrostriatal dopaminergic neurons. Life Sci. 22: 1607-1614, 1978.
24. McGeer, PC, Grewald, DS and McGeer, EG: Influence of noncholinergic drugs on rat striatal acetylcholine levels. Brain Res. 80: 211-217, 1974.
25. Guyenet, PG, Agid, Y, Javoy, F, Beaurjauan, SC, Rauier, J and Glowinski, J: Effect of dopaminergic receptor agonists and antagonists on the activity of the neo-striatal cholinergic system. Brain Res. 84: 227-244, 1975.

26. Ladinsky, H, Consolo, S, Bianchi, S, Samanin, R and Ghezzi, D: Cholinergic-dopaminergic interaction in the striatum: The effect of 6-hydroxydopamine on pimozide treatment on the increased striatal acetylcholine levels induced by amorphine, piribedil and d-amphetamine. Brain Res. 84: 221-226, 1975.
27. Racogni, G, Cheney, DL, Zsilla, G and Costa, E: The measurement of acetylcholine turnover in the rat brain structures. Neuropharmacol. 15: 723-726, 1976.
28. Nieollon, A, Cheramy, A and Glowinski, J: Interdependence of the nigrostriatal dopaminergic systems on the two sides of the brain in the cat. Science 198: 416-418, 1977.
29. de Belleroche, J, Lugmani, Y and Bradford, HF: Evidence for presynaptic cholinergic receptors on dopaminergic terminals: degeneration studies with 6-hydroxydopamine. Neuroscience Letters II: 209-213, 1979.
30. Suga, M: Effects of long-term L-dopa administration on the dopaminergic and cholinergic (muscarinic) receptors of striatum in 6-hydroxydopamine lesioned rats. Life Sci. 27: 877-882, 1980.
31. Namura, Y, Kajiyama, H, Nakata, Y and Segua, T: Muscarinic cholinergic binding in striatal and mesolimbic areas of the rat: reduction by 6-hydroxydopa. Europ. J. Pharmacol. 18: 125-131, 1979.
32. Gurwitz, D, Kloog, Y, Egazi, Y and Sokalovsky, M: Central muscarinic receptor degeneration following 6-hydroxydopamine lesion in mice. Life Sci. 26: 79-84, 1980.
33. McGeer, PL, McGeer, EG and Innanen, VT: Dendro axonic transmission. I. Evidence from receptor binding of dopaminergic and cholinergic agents. Brain Res. 169: 433-441, 1979.
34. Reisine, TD, Nagy, JI, Beaumont, K, Fibiger, HC and Yamamura, HI: The localization of receptor binding sites in the substantia nigra and striatum of the rat. Brain Res. 177: 241-252, 1979.
35. Hruska, RE, Schwarcz, R, Coyle, JT and Yamamura, HI: Alterations of muscarinic cholinergic receptors in the rat striatum after kainic acid injections. Brain Res. 152: 620-625, 1978.
36. Schwartz, RD, Lehmann, J and Kellar, KJ: Presynaptic nicotinic cholinergic receptors labeled by [^{3}H]acetylcholine on catecholamine and serotonin axons in brain. J. Neurochem. 42: 1495-1498, 1984.
37. Clarke, PBS and Pert, A: Autoradiographic evidence for nicotine receptors on nigrostriatal and mesolimbic dopaminergic neurons. Brain Res. 348: 355-358, 1985.
38. Clarke, PBS, Pert, CB and Pert, A: Autoradiographic distribution of nicotine receptors in rat brain. Brain Res. 323: 390-395, 1984.
39. Clarke, PBS, Schwartz, RD, Paul, SM, Pert, CB and Pert, A: Nicotine binding in rat brain: autoradiographic comparison of ^{3}H-acetylcholine, ^{3}H-nicotine and ^{125}I-alpha-bungarotoxin. J. Neursci. 5: 1307-1315, 1985.
40. Amenta, F, Bernardi, G, Floris, V and Marciani, MG: Localization of alpha-bungarotoxin binding sites within the rat corpus striatum. Neuropharmacol. 18: 319-322, 1979.
41. Segal, M, Dudai, Y and Amsterdam, A: Distribution of an alpha-bungarotoxin-binding cholinergic nicotinic receptor in rat brain. Brain Res. 148: 105-109, 1978.
42. Besson, MI, Cheramy, A, Feltz, P and Glowinski, J. Release of newly synthesized dopamine from dopamine-containing terminals in the striatum of the rat. Proc. Nat. Acad. Sci. 62: 741-748, 1969.
43. Westfall, TC: Effect of muscarinic agonists on the release of ^{3}H-norepinephrine and ^{3}H-dopamine by potassium and electrical stimulation from rat brain slices. Life Sci. 14: 1641-1652, 1974.
44. de Belleroche, JS and Bradford, HF: Presynaptic control of the synthesis and release of dopamine from striatal synaptosomes: a comparison between the effects of 5-hydroxytryptamine, acetylcholine and glutamate. J. Neurochem. 35: 1227-1234, 1980.

45. Muscholl, E: Presynaptic muscarinic receptors and inhibition of release. In (ed.) Paton, DM, The release of catecholamines from adrenergic neurons, 87-110, Pergamon Press, Oxford, 1979.
46. Bartholini, G and Stadler, H: Cholinergic and GABA-ergic influence on the dopamine release in extrapyramidal centers. In (eds.) Almgren, O, Carlsson, A and Engel, J, Chemical tools in catecholamine research Vol. II, 235-241, North-Holland Publishing Co., 1975.
47. Giorguieff-Chesselet, MF, Kernel, ML and Glowinski, J: The presynaptic stimulating effect of acetylcholine on dopamine release is suppressed during activation of nigro-striatal dopaminergic neurons in the cat. Neurosci. Letters 14: 177-182, 1979.
48. Giorguieff, MF, Le Floc'h, ML, Westfall, TC, Glowinski, J and Besson, MJ: Nicotinic effect of acetylcholine on release of newly synthesized [^{3}H]dopamine in rat striatal slices and cat caudate nucleus. Brain Res. 106: 117-131, 1976.
49. Giorguieff, MF, Le Floc'h ML, Glowinski, J and Besson, MJ: Involvement of cholinergic presynaptic receptors of nicotinic and muscarinic types in the control of the spontaneous release of dopamine from striatal dopaminergic terminals in the rat. J. Pharmacol. Exp. Ther. 200: 535-544, 1977.
50. Raiteri, M, Marchi, M and Maura, G: Presynaptic muscarinic receptors increase striatal dopamine release evoked by quasi-physiological depolarization. Europ. J. Pharmacol. 83 127-129, 1982.
51. de Belleroche, JS and Gardiner, IM: Cholinergic action in the nucleus accumbens: modulation of dopamine and acetylcholine release. Br. J. Pharmacol. 75: 359-365, 1982.
52. Lehmann, J and Langer, SZ: Muscarinic receptors on dopamine terminals in the cat caudate nucleus: neuromodulation of [^{3}H]dopamine release in vitro by endogenous acetylcholine. Brain Res. 248: 61-69, 1982.
53. Westfall, TC: Effect of nicotine and other drugs on the release of ^{3}H-norepinephrine and ^{3}H-dopamine from rat brain slices. Neuropharmacol. 13: 693-700, 1974.
54. Sakurai, Y, Takano, Y, Kahjimoto, Y, Honda, K and Kamiya, H-O: Enhancement of [^{3}H]dopamine release and its [^{3}H] metabolites in rat striatum by nicotinic drugs. Brain Res. 242: 99-106, 1982.
55. Marien, M, Brien, J and Jhamandas, K: Regional release of [^{3}H]dopamine from rat brain in vitro: effects of opioids on release induced by potassium, nicotine, and L-glutamic acid. Can. J. Physiol. Pharmacol. 61: 43-60, 1983.
56. Argueros, L, Naquira, D and Zunino, E: Nicotine-induced release of catecholamines from rat hippocampus and striatum. Biochem. Pharmacol. 27: 2267-2674, 1978.
57. Giorguieff-Chesselet, MF, Kennel, ML, Wandscheer, D and Glowinski, J: Regulation of dopamine release by presynaptic nicotinic receptors in rat striatal slices: effect of nicotine in a low concentration. Life Sci. 25: 1257-1262, 1979.
58. Massey, SC and James, TA: The uptake of tritiated choline and tritiated acetylcholine in the rat substantia nigra. Life Sci. 23: 345-350, 1978.
59. Butcher, LL and Talbot, K: Acetylcholinesterase in rat nigro-striatal neurons. In (ed.) Butcher, LL, Cholinergic-monoaminergic interactions in the brain, 25-95, Academic Press, New York, 1978.
60. James, TA and Massey, S: Evidence for a possible dopaminergic link in the action of acetylcholine in the rat substantia nigra. Neuropharmacology 17: 687-690, 1978.
61. Costall, B, Naylor, RJ and Olley, JE: Catalepsy and circling behavior after intracerebral injections of neuroleptic, cholinergic and anticholinergic agents into the caudate-putamen, globus pallidus and substantia nigra of rat brain. Neuropharmacol. 11: 645-663, 1972.

62. De Montis, GM, Olianos, MC, Serra, G, Tagliamonte, A and Scheel-Kruger, J: Evidence that a nigra GABAergic-cholinergic balance controls posture. Eur. J. Pharmacol. 53: 181-190, 1979.
63. Javoy, F, Agido, Y, Bouvet, D and Glowinski, J: Changes in neostriatal DA metabolism after carbachol or atropine microinjections into the substantia nigra. Brain Res. 68: 253-260, 1974.
64. Nashold, BS, Urbaniak, JR and Hatcher, MA: Chemical stimulation of red nucleus, substantia nigra and basis pedunculi in alert cats. Neurology 15: 604-612, 1965.
65. Clarke, PBS, Pert, A, Hommer, DW and Skirboll, LR: Electrophysiological actions of nicotine on substantia nigra single units. Brit. J. Pharmacol. 85: 527-533, 1985.
66. Dray, A and Straughan, DW: Synaptic mechanisms in the substantia nigra. J. Pharm. Pharmacol. 28: 400-405, 1976.
67. Aghajanian, GK and Bunney, BS: Dopaminergic and non-dopaminergic neurons in the substantia nigra: different responses to putative transmitters. In (eds.) Boissier, Jr, Hippius, H and Pichat, P, Proceedings 9th International Congress of the CINP, 444-452, Amsterdam, Excerpta Medica, 1974.
68. Collinridge, GL and Davies, J: The influence of striatal stimulation and putative neurotransmitters on identified neurons in the rat substantia nigra. Brain Res. 212, 345-359, 1981.
69. Pinnock, RD and Dray, A: Differential sensitivity of presumed dopaminergic and non-dopaminergic neurons in rat substantia nigra to electrophoretically applied Substance P. Neurosci. Letters 29: 153-158, 1982.
70. Armstrong, DM, Saper, CB, Levey, AI, Wainer, BH and Terry, RD: Distribution of cholinergic neurons in rat brain: demonstrated by immunocytochemical localization of choline acetyltransferase. J. Comp. Neurol. 21: 53-68, 1983.
71. Mesulam, M-M, Mufson, EJ, Wainer, BH and Levey, AI: Central cholinergic pathways in the rat: an overview based on an alternative nomenclature (Chl to Ch6). Neuroscience 10: 1185-1201, 1983.
72. Satoh, K, Armstrong, DM and Fibiger, HC: A comparison of the distribution of central cholinergic neurons as demonstrated by acetylcholinesterase pharmacohistochemistry and choline acetyltranferase immunohistochemistry. Brain Res. Bull. 11: 693-720, 1983.
73. Nagy, JI, Vincent, SR, Lehmann, J, Fibiger, HC and McGeer, EG: The use of kainic acid in the localization of enzymes in the substantia nigra. Brain Res. 149: 431-441, 1978.
74. Olivier, A, Parent, A, Simard, H and Poirier, LI: Cholinesterasic striatopollidal and striatonigral efferents in the cat and the monkey. Brain Res. 18: 273-282, 1970.
75. Fonnum, F, Grofova, I, Rinvik, E, Storm-Mathisen, J and Walberg, F: Origin and distribution of glutamate decarboxylase in substantia nigra of the cat. Brain Res. 71: 77-92, 1974.
76. McGeer, PL, McGeer, EG, Fibiger, HC and Wickson, V: Neostriatal choline acetylase and cholinesterase following selective brain lesions. Brain Res. 35: 308-314, 1971.
77. Gerfen, CR, Staines, WA, Arbuthnott, GW and Fibiger, HC: Crossed connections of the substantia nigra in the rat. J. Comp. Neurol. 207: 283-303, 1982.
78. Jackson, A and Crossman, AR: Nucleus tegmental pedunculopontinus: efferent connections with special reference to the basal ganglia, studied in the rat by anterograde and retrograde transport of horseradish peroxidase. Neurosci. 10: 725-765, 1983.
79. Clarke, PBS and Kumar, R: Characterization of the locomotor stimulant action of nicotine in tolerant rats. Br. J. Pharmacol. 80: 587-594, 1983.

80. Clarke, PBS and Kumar, R: The effects of nicotine on locomotor activity in non-tolerant and tolerant rats. Br. J. Pharmacol. 78: 329-337, 1983.
81. Morrison, CF and Stephenson, JA: The occurrence of tolerance to the central depressant effect of nicotine. Br. J. Pharmacol. 46: 151-156, 1972.
82. Stolerman, IP, Fink, R and Jarvick, ME: Acute and chronic tolerance to nicotine measured by activity in rats. Psychopharmacol. 30: 329-342, 1973.
83. Battig, K, Driscoll, P, Schlatter, J and Uster, HI: Effects of nicotine on the exploratory locomotion patterns of female Roman high- and low-avoidance rats. Pharmacol. Biochem. Behav. 4: 435-439, 1976.
84. Barthelemy, C, Tremblay, E and Jacob, I: Comparison de divers antagonistes de l'eserine et de la nicotine chez la souris. J. Pharmacol. (Paris) I: 369-382, 1970.
85. Morrison, CF, Goodyear, JM and Sellors, CM: Antagonism of anti-muscarinic and ganglion-blocking drugs of some of the behavioral effects of nicotine. Psychopharmacol. 15: 341-350, 1969.
86. Beninger, RJ: The role of dopamine in locomotor activity and learning. Brain Res. 6: 173-196, 1983.
87. Costall, B and Naylor, RJ: Behavioral aspects of dopamine agonists and antagonists. In (eds.) Horn, AS, Karf, J and Westink, BHC, The Neurobiology of Dopamine, 555-576, Academic Press, London, 1979.
88. Ungerstedt, U: Central dopamine mechanisms and behavior. In (eds.) Horn, AS, Korf, J and Westerink, BHC, The Neurobiology of Dopamine, 577-596, Academic Press, London, 1979.
89. Romano, C, Goldstein, A and Jewell, NP: Characterization of the receptor mediating the nicotine discriminative stimulus. Psychopharmacol. 74: 315-320, 1981.
90. Romano, C and Goldstein, A: Stereospecific nicotine receptors on rat brain membranes. Science 210: 647-649, 1980.
91. Anden, NE, Dahlstrom, A, Fuxe, K and Larsson, K: Functional role of the nigrostriatal dopamine neurons. Acta Pharmac. Tox. 24: 262-274, 1966.
92. Ungerstedt, U: Striatal dopamine release after amphetamine or nerve degeneration revealed by rotational behavior. Acta Physiol. Scand. 32, Suppl. 367: 49-68, 1971.
93. Kaakkola, S: Effect of nicotinic and muscarinic drugs on amphetamine- and apomorphine-induced circling behavior in rats. Acta Pharmacol. et Toxicol 48: 162-167, 1981.
94. Latiff, AA, Smith, LA and Lang, WJ: Effects of changing dosage and urinary pH in rats self-administering nicotine on a food delivery schedule. Pharm. Biochem. Behav. 13: 209-213, 1980.
95. Risner, ME and Goldberg, SR: A comparison of nicotine and cocaine self-administration in the dog. Fixed-ratio and progressive-ratio schedules of intravenous drug infusion. J. Pharmacol. Exp. Ther. 224: 319-326, 1983.
96. Goldberg, SR, Spealman, RD and Goldberg, DM: Persistent high-rate behavior maintained by intravenous self-administration of nicotine. Science 214: 573-575, 1981.
97. Dougherty, I, Miller, D, Todd, G and Kostenbauder, HB: Reinforcing and other behavioral effects of nicotine. Neurosci. Biobehav. Res. 5: 487-495, 1981.
98. Ator, N and Griffiths, RR: Nicotine self-administration in baboons. Pharm. Biochem. Behav. 19: 993-1003, 1983.
99. Johnston, LM: Tobacco smoking and nicotine. Lancet 2: 742, 1942.
100. Jasinski, DR, Johnson, RE and Henningfield, JE: Abuse liability assessment in human subjects. Trends Pharmacol. Sci. May: 196-200, 1984.

101. Jasinski, DR, Haertzen, CA, Henningfield, JE, Johnson, RE, Makhazaumi, HM and Miyasato, K. Progress report of the NIDA Addiction Research Center, 45-52, 1982.
102. Henningfield, JE, Miyasato, K and Jasinski, DR: Cigarette smokers self-administer intravenous nicotine. Pharm. Biochem. Behav. 19: 887-890, 1983.
103. Henningfield, JE and Goldberg, SR: Control of behavior by intravenous nicotine injections in human subjects. Pharmacol. Biochem. Behav. 19: 1021-1026, 1983.
104. Davis, WM and Smith, SG: Blocking effect of alpha-methyltyrosine on amphetamine based reinforcement. J. Pharm. Pharmacol. 25: 174-177, 1973.
105. Davis, WM and Smith, SG: Effect of haloperidol on (+)-amphetamine self-administration. J. Pharm. Pharmacol. 27: 540-542, 1975.
106. De Wit, H and Wise, RA: Blockade of cocaine reinforcement in rats with the dopamine receptor blocker pimozide, but not with the noradrenergic blockers phentolamine or phenoxybenzamine. Canad. J. Psychol. 31: 195-203, 1977.
107. Yokel, RA and Wise, RA: Attenuation of intravenous amphetamine reinforcement by central dopamine blockade in rats. Psychopharmacology 48: 311-318, 1976.
108. Yokel, RA and Wise, RA: Increased lever pressing for amphetamine after pimozide in rats: implications for a dopamine theory of reward. Science 187: 547-549, 1975.
109. Risner, ME and James, BE: Role or noradrenergic and dopaminergic processes in amphetamine self-administration. Pharmacol. Biochem. Behav. 5: 477-482, 1976.
110. Lyness, WH, Friedle, NM and Moore, KE: Destruction of dopaminergic nerve terminals in nucleus accumbens: effect on d-amphetamine self-administration. Pharm. Biochem. Behav. 11: 553-556, 1979.
111. Roberts, DCS, Corcorn, ME and Fibiger, HC: On the role of ascending catecholaminergic systems in intravenous self-administration of cocaine. Pharm. Biochem. Behav. 6: 615-620, 1977.
112. Roberts, DCS, Koob, GF, Klonoff, P and Fibiger, HC: Extinction and recovery of cocaine self-administration following 6-hydroxydopamine lesions of the nucleus accumbens. Pharm. Biochem. Behav. 12: 781-787, 1980.
113. Bozarth, MA and Wise, RA: Intracranial self-administration of morphine into the ventral tegmental area in rats. Life Sci. 28: 551-555, 1981.
114. Phillips, AG and Le Piane, FG: Reinforcing effects of morphine microinjected into the ventral tegmental area. Pharmacol. Biochem. Behav. 12: 965-968, 1980.
115. Joyce, EM and Iversen, SD: The effect of morphine applied locally to mesencephalic dopamine cell bodies on spontaneous motor activity in the rat. Neurosci. Lett. 14: 207-212, 1979.
116. Joyce, EM and Koob, GF, Strecker, R, Iversen, SD and Bloom, FE: The behavioral effects of enkephalin analogues injected into the ventral tegmental area and globus pallidus. Brain Res. 221: 359-370, 1981.
117. Kelly, AE, Stinus, L and Iversen, SD: Interaction between D-Ala-Met-enkephalin, A-10 dopaminergic neurons and spontaneous behavior in the rat. Behav. Brain Res. 1: 3-24, 1980.

CURRENT CONCEPTS ON THE EFFECTS OF NICOTINE ON NEUROTRANSMITTER RELEASE IN THE CENTRAL NERVOUS SYSTEM

Peter P. Rowell

Department of Pharmacology and Toxicology
University of Louisville School of Medicine
Louisville, KY 40292

ABSTRACT

In this overview, the present knowledge about the effects of nicotine on neurotransmitter release in the central nervous system has been evaluated. Five characteristics relating to nicotine mechanisms and sites of action in the brain are presented. These are the following: the site of action of nicotine at the neuronal level; the areas of the brain in which nicotine has been shown to affect neurotransmitter release; the neurotransmitter affected by nicotine; the mechanistic characteristics of release such as receptor mediation, calcium dependence, and subcellular pools affected; and, the pharmacological characteristics of nicotine such as potency, efficacy, time course, stereoselectivity, and structure-activity relationships. Experimental information is presented concerning the effects of nicotine on [^{3}H]dopamine release from rat striatal synaptosomes and tissue punches from mesolimbic nuclei. These include concentration-response relationships, the effects of agonists and antagonists, and the time-course and efficacy of nicotine on the stimulation of [^{3}H]dopamine release.

An assimilation of the results indicates that nicotine stimulates the release of neurotransmitters, particularly the catecholamines, from selective brain areas, notably the hypothalamus and striatum. Nicotine is able to release these neurotransmitters at concentrations likely to be present in the brains of cigarette smokers. The effect appears to be mediated by presynaptic receptors which resemble, but are not identical to, the N_1-type cholinergic receptors. This indicates that there are presynaptic neural connections at selective nerve terminals in the brain where an endogenous nicotinic neuromodulator acts to facilitate neurotransmitter release. Whether the neuromodulator is acetylcholine or another unidentified nicotinic compound is unknown. It is quite likely that the ability of nicotine to stimulate catecholamine release in selective brain areas such as the mesolimbic system is involved in the pleasurable aspects associated with smoking. The discovery of new agents that mimic or block the effects of nicotine on neurotransmitter release may lead to the identification of endogenous reinforcing compounds as well as permit the design of successful low-tar, low-nicotine cigarettes with a diminished risk associated with the pharmacological effects of nicotine.

Investigations into the reasons why people smoke and the mechanisms accounting for the pleasurable aspects of smoking have continued to be of major interest to neurobiologists for obvious reasons. The identification and characterization of why people smoke could have practical implications such as the design of more effective smoking cessation programs and the production of safer cigarettes.

Although a number of factors appear to be involved in the reason for cigarette smoking, such as subjective pleasurable sensations, decreased anxiety, improvement of memory, repetitive motor activity, social elements, and learned responses (1), most investigators would agree that the underlying pharmacological basis for smoking is due to the presence of nicotine. Not only is nicotine the most abundant and pharmacologically active component of cigarette smoke (2), but also it has been shown that nicotine is a positively reinforcing agent which is associated with the pleasurable experiences associated with cigarette smoking (3-5). Further, nicotine appears to be the component of cigarette smoke that correlates with the frequency of smoking (6-9), and substitution of nicotine with other compounds results in a decrease in cigarette smoking (7,10-12). It is also thought that the dependence, tolerance, and withdrawal symptoms that are characteristic of chronic cigarette smoking are due to the actions of nicotine (9,13-15). From all these studies it would appear that the rewarding stimuli associated with cigarette smoking are centered on the pharmacological effects of nicotine.

It is also generally accepted by neuroscientists that the site of nicotine's action is the central nervous system. Although it is not inconceivable that at least part of nicotine's pleasurable stimuli is mediated by its autonomic effects in the peripheral nervous system, it is far more likely that nicotine's actions are due to its effects in the CNS. Unlike most neurons in the peripheral nervous system, however, central neurons have no effector organ but instead function to integrate and transfer information to adjacent neurons. The process by which this takes place involves the release of neurotransmitters from the nerve terminal. Therefore, it is evident that any drug which has CNS activity must, either directly or indirectly, influence the release of some neurotransmitter in the brain. Even drugs with a purely postsynaptic site of action must eventually affect neurotransmitter release to have CNS activity. If neurotransmitter release were not affected by a substance, it is difficult to see how CNS activity could be influenced. Therefore, from the premises that (a) the pleasurable aspects of smoking are due to nicotine, (b) the site of nicotine's action is the CNS, and (c) centrally neuroactive substances have effects on neurotransmitter release, it is evident that one of the most fruitful areas of investigation into why people smoke is the study of the effects of nicotine on neurotransmitter release in the central nervous system.

In studies of neurotransmitter release, it is obviously not sufficient to simply measure the effects of nicotine administration *in vivo* or *in vitro* in order to define nicotine's mechanism and site of action in the brain. Instead, a number of characteristics should be considered which, taken together, will provide useful information about nicotine's central neuroactive effects. In this overview, five of these characteristics are presented and the studies that address these considerations are evaluated. The five major considerations are (1) the neuronal site of action of nicotine; (2) the areas of the brain in which nicotine has been shown to affect release and the relationship of these structures to CNS function; (3) the neurotransmitters affected by nicotine; (4) the mechanistic characteristics of release such as receptor mediation, calcium dependency, and subcellular pools affected; and, (5) the pharmacological characteristics of the drug such as potency, relative activity, time course, stereoselectivity, and structure-activity relationships. It is hoped that this analysis will aid in defining what is currently known about nicotine's effects on neurotransmitter release in the CNS and aid in elucidating the sites and mechanisms of nicotine as the neuroactive component of tobacco.

NEURONAL SITE OF ACTION

The premise that nicotine releases neurotransmitters is not a novel consideration. In fact, the best-characterized and most widely recognized action of nicotine is based on this effect. Nicotine's primary activity in the peripheral nervous system is its agonist activity at N_1-type cholinergic receptors in the autonomic ganglia which give rise to the release of norepinephrine and acetylcholine at the postganglionic sympathetic and parasympathetic nerve terminals, respectively. It is therefore well-documented that nicotine releases norepinephrine as well as acetylcholine in the peripheral nervous system. An analogous situation could be responsible for nicotine's activity in the central nervous system as well. Nicotine could be acting as a postsynaptic receptor agonist or antagonist, and this activity would then be manifested by changes in neurotransmitter release at the nerve terminal affected. Alternatively, nicotine could release a neurotransmitter by a direct action on the presynaptic nerve terminal. Therefore, it is clear that a distinction must be made between an indirect postsynaptically-mediated dendrosomatic type of release and a direct presynaptically-mediated release of neurotransmitter. The neuronal site of nicotine's activity is certainly an important characteristic of release which must be considered in order to elucidate the true mechanism of action of the drug.

By far the majority of studies designed to investigate the effects of nicotine on neurotransmitter release have been conducted with intact tissue, usually slices, but also brain mince or punches (16-24). Nicotine has been shown to produce an enhancement of neurotransmitter release in most of these studies, but the site of action at the level of the neuron is unknown. Since many areas of the central nervous system contain short cholinergic interneurons, such as the striatum, hippocampus, and nucleus accumbens, even the use of small minced tissue does not rule out a postsynaptic dendrosomatic site of action. Since acetylcholine acts via postsynaptic nicotinic receptors in many areas of the brain to produce excitatory or inhibitory potentials (25,26), it would be surprising if nicotine did not release neurotransmitters in the brain via this dendrosomatic mechanism.

In investigations using intact tissue, any nicotine-induced release of neurotransmitters which is observed could be taking place either pre- or postsynaptically. In order to make the distinction, one must determine if release can still be produced when neuroconduction is inhibited or when isolated nerve terminals are used. Experiments conducted using these techniques have shown that nicotine is able to release the catecholamine neurotransmitters by a direct action on the nerve terminal. Giorguieff and co-workers (27) found that the acetylcholine-induced release of dopamine from striatal slices still occurred in the presence of tetrodotoxin, and later demonstrated that nicotine's ability to release dopamine was also largely tetrodotoxin insensitive (28). The experiments of Balfour (29) were some of the earliest to demonstrate nicotine-induced release of norepinephrine from isolated nerve endings. Many recent studies using synaptosomal preparations have also shown that nicotine is able to release norepinephrine (30) as well as dopamine (31-34) by a presynaptic mechanism. Our studies on the nicotine-induced release of acetylcholine from cerebral cortical synaptasomes have also demonstrated a presynaptic mechanism (35). Studies by Yoshida and Imura (36), de Belleroche and Bradford (37), and Schwartz et al. (38) demonstrating the presence of nicotinic cholinergic receptors in synaptosomes supports the concept that nicotine can release neurotransmitters by a direct action on the nerve terminal. Whether this effect is taking place by intraneuronal transmitter displacement, direct vesicular exocytosis, presynaptic cholinergic receptors, or some other mechanism requires further characterization of the effects as presented below.

AREAS OF THE BRAIN

Although it would certainly be an oversimplification to assign specific activities to any one area of the brain, it is clear that discrete areas of the central nervous system are related to various functions. The hypothalamus is the most important neuroendocrine center of the brain, the striatum is primarily a motor center, the hippocampus is involved in learning and memory, the mesolimbic system is intimately involved in pleasurable and aversive stimuli; the cerebral cortex is the center of cognitive function, etc. Therefore, the characterization of the areas of the brain where nicotine has its effects could provide useful information about nicotine's physiological activities in the body.

The vast majority of experiments designed to study the central effects of nicotine on neurotransmitter release have been conducted in two brain areas, the hypothalamus and the striatum. In the hypothalamus, nicotine has been shown to release norepinephrine, dopamine, and serotonin (17-19,29,30,39-42). It is not surprising, then, that nicotine has effects on the hypothalamic hormones as well (42-45). It is quite plausible that the ability of nicotine to produce changes in neuroendocrine function and hormone levels plays a role in the stimulus to smoke.

The striatum is an area of the brain in which there is a close functional relationship between the cholinergic and dopaminergic systems. It is logical then, that many investigators have studied the effects of nicotine in the striatum. Again, it has been found that nicotine is able to release dopamine from striatal tissue (16,17,21,22), an effect which appears to be due to a direct action on the nerve terminal (28,31,33,34). Although nicotine has pronounced effects on motor activity, it is not clear how the effects of nicotine in the striatum are related to why people smoke.

It has been well established that the hippocampal centers of the brain are related to learning and memory function (46,47) and that cholinergic pathways are involved (48-52). Although relatively little work has been done on investigating the effects of nicotine in the hippocampus, a few studies have shown that nicotine also releases dopamine (16) and norepinephrine (29). In studies of electrical activity in the hippocampus, however, Benardo and Prince (53) found that nicotine had little effect. In any event, nicotine has been reported to improve learning and memory (54,55), and it is possible that nicotine's effects in the hippocampus could have important implications to the neuroactive effects of nicotine.

The mesolimbic system is an important brain center involved in reinforcement and reward (56-58). Like the striatum, there also appears to be a close functional relationship between the cholinergic and dopaminergic systems in this area of the brain (59-62). It is therefore surprising that no studies have been reported on the effects of nicotine on neurotransmitter release in this area. Marien (21) conducted

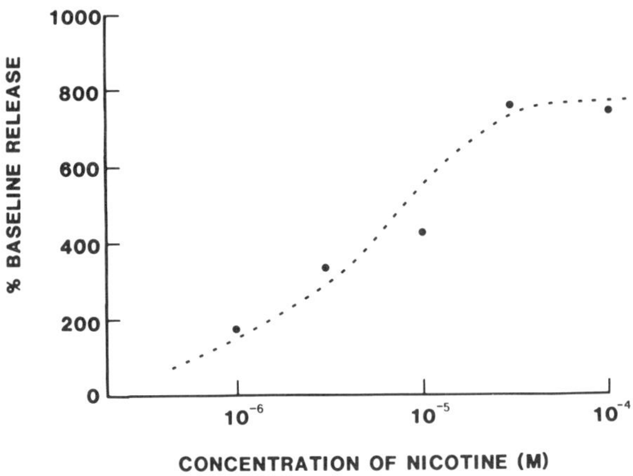

Fig. 1. Concentration-response effect of nicotine on the release of [^{3}H]dopamine from the amygdala. Values represent the cumulative increase in [^{3}H]dopamine release during a 15 min treatment period expressed as percent of spontaneous baseline release.

experiments on the effects of opioids and nicotine in several brain areas including the nucleus accumbens, but the direct effects of nicotine in this area were not studied.

We have recently conducted studies in our laboratory examining the effects of nicotine on the release of [^{3}H]dopamine from tissue punches of the amygdala. These results are presented in Fig. 1. It is seen that nicotine causes a dose-dependent release of [^{3}H]dopamine from this brain area with an EC50 value of 5 x 10^{-6} M. It is very likely that the increase in dopamine release produced by nicotine in mesolimbic nuclei is involved in the pleasurable effects associated with cigarette smoking.

NEUROTRANSMITTERS AFFECTED

At the appropriate concentration and in the appropriate tissue, nicotine has been shown to release all five major biogenic amine neurotransmitters (norepinephrine, epinephrine, dopamine, serotonin, and acetylcholine) as well as a number of putative amino acid and peptide transmitters and neuromodulators. The effects of nicotine on these substances have recently been reviewed by Balfour (43). Most early neurochemical studies centered on nicotine's ability to release norepinephrine, probably as an analogy to amphetamine's neuroactive effects. Most notably, studies by Arqueros and co-workers (16), Balfour (29), Hall and Turner (39), Goodman (17), and Yoshida (30) have all conclusively shown that nicotine can release norepinephrine in specific areas of the brain. Andersson and co-workers (42) have demonstrated that the chronic administration of nicotine releases norepinephrine and leads to decreased levels of this neurotransmitter in the brain.

In the last five years, more investigators have been directing their research toward the study of nicotine's effect on dopamine release. In particular, the experiments of Giorguieff (28), Sukurai et al. (31), Andersson et al. (42), Marien et al. (21), Connelly and Littleton (32), Takano et al. (33), Westfall et al. (22), and Mills and Wonnacott (34) have demonstrated that nicotine releases dopamine, principally from the striatum and caudate nucleus. Serotonin has also been shown to be released by nicotine *in vitro* as documented by Balfour (29), Fuxe et al. (19), and Westfall *et al.* (22). The characteristics of the release and the concentrations of nicotine used in many of these studies of the monoamine neurotransmitters has lead to some doubt about the physiological significance of some of these findings. This will be discussed in the following sections.

In contrast to the monoamines, the effect of nicotine on acetylcholine release in the CNS has received very little attention, in spite of the fact that an indirect action of nicotine to release acetylcholine has been hypothesized as part of its cholinergic agonist activity in the peripheral nervous system for over 20 years (63-66). A number of investigators have shown that nicotine administered to animals can release acetylcholine and decrease the levels of acetylcholine in the brain (67-69). Studies in our laboratory (35) have shown for the first time that nicotine releases acetylcholine *in vitro* in the CNS, an effect which has now been well documented for the monoamine neurotransmitters.

Nicotine has been shown to release not only the biogenic amine transmitters but a number of other substances as well (43). The importance of the effect on each neurotransmitter system and the physiological significance of these effects to nicotine's presence in cigarette smoke must be evaluated on the basis of the additional considerations presented in this overview.

CHARACTERISTICS OF RELEASE

Some of the most important considerations in assessing nicotine's effects in the brain are the mechanistic characteristics of the release process. These characteristics include the involvement of calcium ions in nicotine-induced release, the identification and classification of receptors that may be mediating the effects of nicotine, and the subcellular locations from which the neurotransmitter is released.

If neurotransmitter release is shown to be calcium dependent, this provides an indication that nicotine is acting via the normal physiological process by which neurotransmitters are released in response to depolarization. Alternatively, a calcium-independent process would indicate that nicotine could be acting to release the neurotransmitter directly from its vesicular sites, acting as a neurotransmitter channel from the cytoplasm, or liberating the neurotransmitter in some other fashion unrelated to physiological depolarization-induced release. Although the characterization of the involvement of calcium is important in elucidating the mechanism of nicotine's effect, it is not of practical importance to nicotine's neuroactive effects. Once released, neurotransmitters would obviously produce the same responses regardless of how the release took place.

Most studies have shown that the release of neurotransmitters by nicotine is at least partially dependent on the presence of calcium ions. Both Westfall (18) and Giorguieff and co-workers (27,28) have reported that nicotinically induced release of catecholamines is almost totally calcium dependent. Connelly and Littleton's (32) studies as well as our own studies with acetylcholine (35) have shown nicotine-induced neurotransmitter release to be largely calcium dependent in symaptosomes. In contrast, other studies (16,21) have found that the omission of calcium ions from the medium had no effect on nicotine's response. In these latter studies, however, the concentration of nicotine used was extremely large (3 and 5 mM) actually being higher than that of calcium itself.

Many investigators have also studied the involvement of receptors in the nicotine-induced release of neurotransmitters. Using hexamethonium as the prototypic N_1 cholinergic antagonist, Westfall (18), Rowell and Winkler (35), and Kaji and coworkers (70) have shown that the nicotinically induced release is largely or completely eliminated in the presence of the antagonist. Other investigators have used other N_1-type antagonists such as pempidine (28,32), mecamylamine (30,34,41,42,71,72), and dihydro-β-erythroidine (73) with similar results. Interestingly, Connelly and Littleton (32) found that only pempidine, and not hexamethonium, inhibited the release of dopamine in whole-brain synaptosomes. Also, Sakurai and co-workers (31) have reported that neither N_1 (hexamethonium, tetraethylammonium) nor N_2 (d-tubocurarine) cholinergic antagonists block the nicotine-induced release of dopamine. Similarly, Marien and co-workers (21) found that hexamethonium did not attenuate the response to nicotine; however, the concentration of nicotine used was quite large (3 mM).

Although in most studies the effects of nicotine on neurotransmitter release are attenuated by classical N_1-type cholinergic antagonists, it has certainly not been established that the nicotinic receptors that appear to be involved are the same as those which have been well-characterized in the peripheral nervous system. For example, studies in our laboratory on the release of $[^3H]$dopamine from the nucleus accumbens have shown that both N_1- as well as N_2-type antagonists reduce, but do not abolish, the nicotine-induced release of $[^3H]$dopamine from the mesolimbic system. These results are presented in Fig. 2. It is important to note that the concentrations of the antagonists used in these studies (3×10^{-4} M hexamethonium and 1×10^{-4} M d-tubocurarine) have been shown to completely block the effects of nicotine at the peripheral autonomic ganglia or the somatic neuromuscular junction. The incomplete

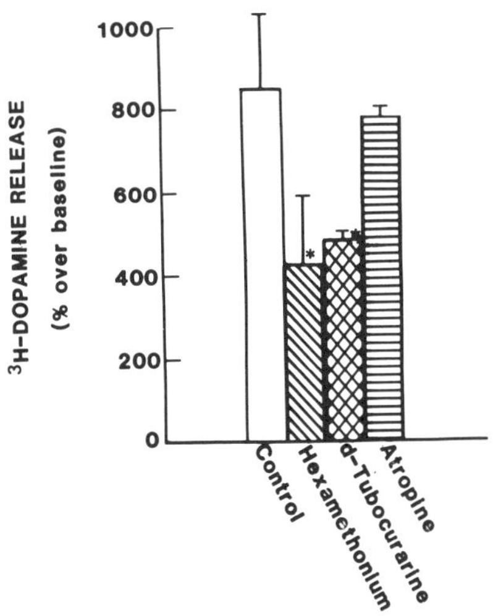

Fig. 2. Effect of cholinergic receptor blockers on the nicotine-induced release of $[^3H]$dopamine from rat nucleus accumbens. Responses represent the cumulative effect of 1×10^{-4} M nicotine during a 15 min exposure expressed as the percent of spontaneous baseline release. Nicotine was added alone (open bar) or in the presence of 3×10^{-4} M hexamethonium (diagonal stripes), 1×10^{-4} M d-tubocurarine (cross hatched) or 1×10^{-6} M atropine (horizontal stripes). Values are means of 4-6 experiments ± S.E.M. * = different from control, $P < 0.05$.

nature of the blockade could indicate that receptors are only partially involved in the release process. In addition, peripheral responses are usually abolished by either N_1 or N_2 antagonists but not by both at these concentrations. As in our studies, Hery and co-workers (24) found that both N_1 and N_2, but not muscarinic antagonists, blocked the release of serotonin from hypothalamic slices induced by carbachol. The fact that the effects of nicotine on neurotransmitter release are blocked by both types of nicotinic antagonists indicates that the nicotinic receptors involved in this response are different than those in the periphery. This conclusion has also been reported by Sakurai et al. (31) and Connelly and Littleton (32).

In the case of the effect of nicotine on acetylcholine release in the CNS, the possibility exists that the presynaptic nicotinic receptors are actually autoreceptors for acetylcholine. Recent studies in the peripheral nervous system have provided evidence for the existence of nicotinic excitatory autoreceptors (74,75) analogous to the muscarinic inhibitory autoreceptors which are fairly well documented (76,77). Our studies in cerebral cortical synaptosomes (35) provide evidence that excitatory autoreceptors of the nicotinic type exist at central cholinergic nerve terminals as well.

Studies by Abood and co-workers (78,79) have shown that some of the central effects of nicotine probably do not involve cholinergic receptors at all. It is not clear that the motor responses investigated in these studies are involved in the behavioral effects of nicotine that relate to cigarette smoking. The ongoing investigations of Sloan and Martin (80,81) have helped to elucidate the subtypes of nicotinic receptors present in the brain. The investigations of Clarke and Pert (82,83) as well as Mills and Wonnacott (34) have already shown that nicotine and acetylcholine interact with discrete binding sites in the CNS which are not the same as the α-bungarotoxin binding sites.

Beyond investigating the effects of calcium and receptor antagonists on the nicotine-induced release of neurotransmitters, a number of other important considerations have received very little attention. Among these is the intracellular pools from which the neurotransmitter is released. Studies by Sakurai and Takano (31,33) investigating the differences in dopamine and its metabolites resulting from potassium depolarization and nicotine suggest that nicotine releases the transmitter from the cytoplasmic pool rather than the synaptic vesicles. Early studies by Chiou (65,66) found that nicotine was able to release acetylcholine directly from isolated synaptic vesicles. It is certainly important to establish that nicotine does not cause neurotransmitter release by producing leaky membranes or cell damage (32), an experiment usually overlooked in most studies. It is clear that the elucidation of nicotine's mechanism of action will require more definitive neurochemical characterizations of nicotine's actions at the cellular and subcellular level.

PHARMACOLOGICAL CHARACTERISTICS

The final consideration in examining nicotine's effects on central neurotransmitter release is the analysis of the drug effect itself with regard to the potency, efficacy, time course, stereoselectivity, and structure-activity relationships. Of these, the most obvious consideration is the concentration of nicotine required to produce its effect. In many studies (16-18,21,22,31,32,39) the concentration of nicotine used has been quite high, from 0.5 to 20 mM. It is usually not clear whether lower concentrations were tested and found to be ineffective, although concentration-effect relationships have been presented in some studies (21,32). It is important to note that the concentration of nicotine found in the blood of cigarette smokers is in the range of 0.1 to 0.5 μM (84-86). To be sure, the concentration of nicotine at its site of action may be higher than this since nicotine reaches the brain quite rapidly during cigarette smoking and the blood levels of nicotine fall rather quickly due to its rapid metabolism (87-89). It is also conceivable that nicotine could be sequestered in some brain locations or concentrated intracellularly in central neurons. Nevertheless, it is difficult to see how effects produced by nicotine *in vitro* at concentrations thousands of times higher than blood levels *in vivo* are physiologically relevant.

There have been a number of studies in which the effects of nicotine at micromolar concentrations have been investigated (20,28-30,34,35). These studies demonstrate that the release of neurotransmitters in a number of brain areas can be produced by nicotine at concentrations likely to be present in the brains of cigarette smokers. For example, the results presented in Fig. 1 show that nicotine produces a measurable, though not maximal, release of $[^3H]$dopamine from the amygdala at a concentration of 1.0 μM. A maximal effect was observed at 30 μM, a concentration still quite reasonable in view of the rapid bioavailability of the compound from cigarette smoke.

In addition to the potency, the relative activity of nicotine is an important consideration as well. Unfortunately, this property of the drug has been given very little attention. Arqueros and co-workers (16) have reported information relating the release of $[^3H]$dopamine by potassium depolarization with that of nicotine. Nicotine was found to have approximately 30% of the activity of potassium depolarization. Similarly, our studies (35) demonstrated that nicotine's maximal effect to release acetylcholine from synaptosomes is only about 20% that of depolarization by 50 mM potassium ion.

Figure 3 compares the maximally effective concentration of nicotine to release $[^3H]$dopamine from rat striatal synaptosomes with the effects produced by 50 mM potassium depolarization as well as 1×10^{-4} M tyramine. Again it is seen that nicotine is considerably less efficacious than the other two. Upon reflection, one might expect that nicotine would have lower activity than potassium or tyramine. As already mentioned, most studies have indicated that nicotine acts to release neurotransmitters via a receptor-mediated process. This is a completely different mechanism than direct depolarization or direct release by phenylethylamine agents. It is not altogether surprising that the nicotine-induced release, being more physiological, would produce considerably less release than the more direct treatments. More importantly, the release produced by potassium depolarization and tyramine would be relatively nonselective. That is, they would act to release dopamine from all dopaminergic terminals in the tissue. Nicotine, on the other hand, would act only on those dopaminergic terminals possessing presynaptic nicotinic receptors. These selective dopaminergic terminals may be a small subpopulation of the total number of dopaminergic terminals in the brain. Nicotine may be quite, perhaps maximally, effective on a much smaller number of terminals and therefore would only appear to be less efficacious than the nonselective treatments.

The relative time course of the nicotine-induced effect is also an important characteristic. The use of the superfusion technique allows one to make dynamic determinations of the effects of drugs or other treatments on neurotransmitter release with a time resolution dependent only upon the sampling interval and sensitivity of detection. An analysis of the time course presented by a number of investigators (18,19,21,28,30,31,35)

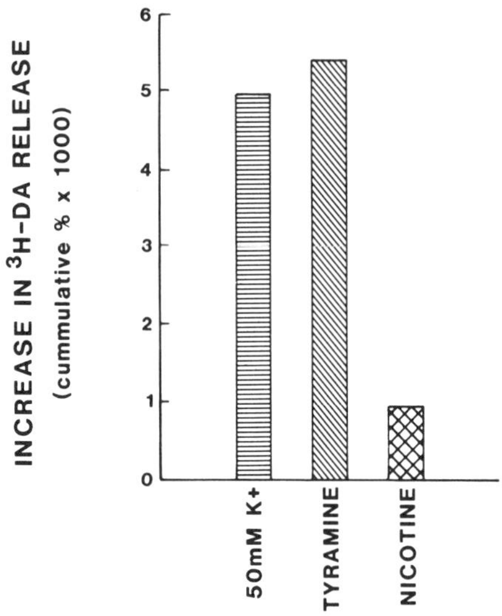

Fig. 3. Maximal depolarization and drug-induced release of [^{3}H]dopamine from striatal synaptosomes. Values represent the cumulative increase in [^{3}H]dopamine release during a 10 min treatment period expressed as percent of spontaneous baseline release. 50 mM k^+ was substituted for an iso-osmotically equivalent amount of Na^+, tyramine and nicotine were each present at a concentration of 100 μM.

demonstrates that the nicotine-induced release of transmitter has a rapid onset, but the time to reach maximum effect is rather slow, about five minutes, relative to potassium depolarization or a phenylethylamine-type release.

An example of the time course of both 1 x 10^{-4} M nicotine and 15 mM potassium ion on the release of [^{3}H]dopamine from rat striatal synaptosomes is shown in Fig. 4. Potassium depolarization produces a much more rapid release which reaches a maximum within two minutes and rapidly returns to baseline. Nicotine, on the other hand, causes a more gradual effect which plateaus in about eight minutes and returns to baseline much more gradually. The relatively slow response produced by nicotine in the CNS is further evidence that the central nicotinic receptors that stimulate neurotransmitter release are different from those in the periphery. Both N_1- and N_2-type cholinergic receptors in the peripheral nervous system give rise to rapid excitatory responses that trigger action potentials at excitable membranes. The slow modulatory responses seen in the CNS are more analogous to those produced at muscarinic receptors.

Finally, a number of investigators have compared the effects of nicotine with other compounds in order to determine whether cholinergic agonists or structural analogues of nicotine can substitute for the nicotine-induced release of neurotransmitter. Studies by Sakurai et al. (31) demonstrated that lobeline, cotinine and sparteine all stimulate the release of [^{3}H]dopamine from striatal synaptasomes, and the release is actually more sensitive to these three agents than to nicotine or acetylcholine. Nicotine, lobeline and sparteine all seemed to be equally as active, however. Yoshida et al. (30) found that carbachol was also similar to nicotine in its effect to release norepinephrine from hypothalamic synaptosomes, whereas arecoline had little effect. Giorguieff et al. (90) also found that carbachol was effective in

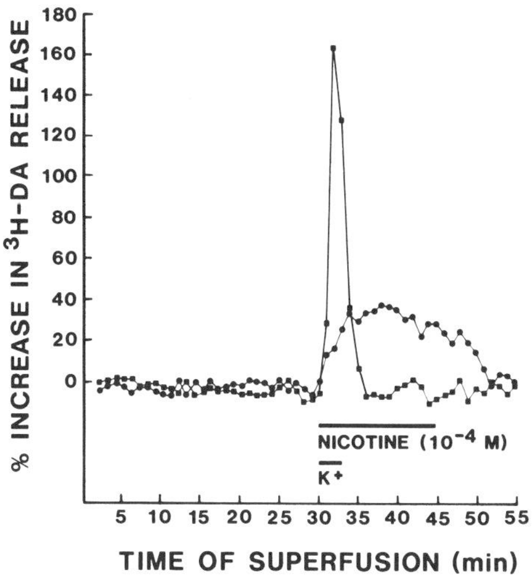

Fig. 4. Time course of evoked release of [^{3}H]dopamine from rat striatal synaptosomes. Values represent the percent change in [^{3}H]dopamine release compared to spontaneous baseline release in two representative experiments. Squares represent the effect of 15 mM K^+ added to the superfusate for 3 min; circles represent the effect of 100 μM nicotine added for 15 min.

releasing [^{3}H]dopamine from rat striatal slices. Westfall et al. (22) have shown that DMPP has similar activity to nicotine in releasing dopamine and serotonin from the striatum. Similarly, we have shown that DMPP and nicotine are equally potent and active in releasing acetylcholine from cerebral cortical synaptosomes (35).

In recent studies of the release of [^{3}H]dopamine from the rat mesolimbic nuclei, we have examined the relative activity of a number of agents compared to nicotine. These results are presented in Fig. 5. In this system we found that at equimolar concentrations (10 μM), nicotine was the most active compound with DMPP, a nicotinic agonist, and arecoline, a compound with both nicotinic and muscarinic activity (91), having about half the activity of nicotine. It was interesting that acetylcholine had little activity compared to nicotine in this system.

An important contribution has been made by the studies of Connelly and Littleton (32) examining the stereoselectivity of the nicotine-induced release of transmitter. They have shown that (+)-nicotine has the same potency and activity as the naturally occurring (-)-nicotine in releasing dopamine from whole-brain synaptosomes. However, the concentration of nicotine necessary to produce the effect (1 mM and higher) does not necessarily indicate that the demonstrated effects of nicotine at micromolar concentrations in other systems are not stereospecific. Nevertheless, these studies are important, particularly in view of the finding that a significant amount of (+)-nicotine is produced pyrolytically in cigarette smoke (92) and that the *in vivo* metabolism of (+)-nicotine leads to compounds which may have unique pharmacological properties (93,94). The studies of Rosecrans and Meltzer (95) as well as those of Stolerman and co-workers (96,97) using drug discrimination have yielded important results about the ability of other agents to substitute for the psychological effects of nicotine. It is clear that considerably more information must be obtained about the structure-activity relationships of nicotine's neuroactive effects in order to elucidate its mechanism of action in the brain.

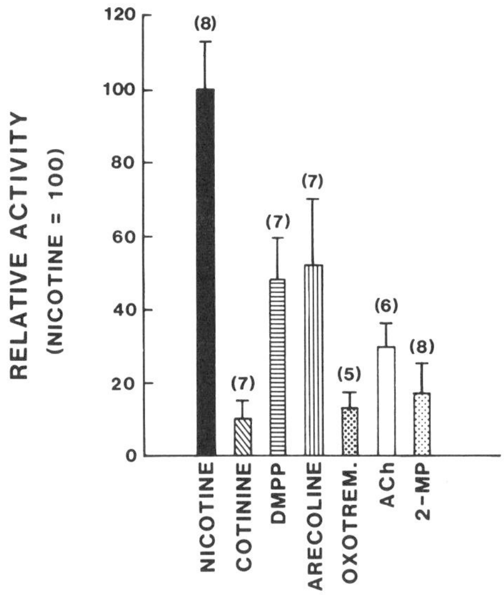

Fig. 5. Comparison of the effects of nicotine with other compounds on the release of [^{3}H]dopamine from rat nucleus accumbens. Values represent relative equimolar activity of cotinine, dimethyl-phenylpiperazinium (DMPP), arecoline, oxotremorine, acetylcholine (ACh) or 2-methylpiperidine (2-MP) compared to nicotine. All compounds were added to the superfusate for 10 min at a concentration of 1×10^{-5} M. Values represent means of 5-8 experiments ± S.E.M. Experiments with ACh were conducted in the presence of 10 μM neostigmine.

SUMMARY

In the above overview, an attempt has been made to categorize our present knowledge about the effects of nicotine on neurotransmitter release in the brain. The wide variety of procedures, tissues, and concentrations employed and the diversity of neurotransmitters studied and approaches used has provided a very extensive amount of information for assessing nicotine's effects. This has also made it difficult to make definitive statements about the sites and mechanisms of nicotine in the brain. Nevertheless, there appear to be a few generalizations that can be made and that are supported by a significant number of studies.

It seems clear that nicotine causes the release, rather than inhibits the release, of primarily the catecholamine neurotransmitters in the brain. This effect is quite well documented in the hypothalamus and striatum, and can be produced at the micromolar concentrations likely to be present in the brains of cigarette smokers. Further, the effect is apparently mediated by nicotinic receptors which resemble, but are probably not identical to, the N_1-type cholinergic receptors in the peripheral nervous system. These receptors appear to be located at the presynaptic nerve terminal of these catecholaminergic nerves in the hypothalamus and striatum. This indicates that there are presynaptic neural connections at selective terminals in the CNS where an endogenous "nicotinic" neuromodulator acts to stimulate the release of catecholamines from the adjacent nerve terminal. There is presently no agreement as to whether the endogenous neuroregulator is acetylcholine or an, as yet, unidentified substance such as a pyridine-like compound analogous to nicotine.

From this information, it would seem that a very promising goal of future investigations would be the discovery of potent and selective agonist and antagonist compounds that mimic or block the effects of nicotine to release the catecholamine neurotransmitters in the hypothalamus, striatum, and perhaps mesolimbic nuclei of the brain. This might lead not only to the identification of new endogenous positively reinforcing compounds, but permit the design of successful low-tar, low-nicotine cigarettes with a diminished risk associated with the autonomic, cardiovascular, and pulmonary effects of nicotine.

REFERENCES

1. Pomerleau, OF and Pomerleau, GA: Neuroregulators and the reinforcement of smoking: Towards a biobehavioral explanation. Neurosci. Biobehav. Rev. 8:503-513, 1984.
2. Brunnemann, KD, Adams, JD, Ho, DPS and Hoffmann, D: The influence of tobacco smoke on indoor atmospheres. II. Volatile and tobacco specific nitrosamines in main- and side-stream smoke and their contribution to indoor pollution. Proceedings 4th Joint Conference on Sensing of Environmental Pollutants, Am. Chem. Soc., 876-880, 1978.
3. Ejrup, B: The role of nicotine in smoking pleasure, nicotinism, and treatment. In (ed.) Von Euler, VS, Tobacco Alkaloids and Related Compounds, Pergamon, Oxford, 1965, pp. 333-346.
4. Jarvik, ME: Further observations on nicotine as the reinforcing agent in smoking. In (ed.) Dunn, WL, Smoking Behavior: Motives and Incentives, Winston and Sons, Washington, D.C., 1973, pp. 33-49.
5. Larson, PS and Silvette, H: Tobacco: Experimental and Clinical Studies. Supp. 3. William and Wilkens, Baltimore, 1975.
6. Lucchesi, BR, Schuster, CR and Emley, GS: The role of nicotine as a determinant of cigarette smoking frequency in man with observations of certain cardiovascular effects associated with the tobacco alkaloid. Clin. Pharmacol. Therap. 8:789-796, 1967.
7. Russell, MAH, Wilson, C, Feyerabend, C and Cole, PV: Effect of nicotine chewing-gum on smoking behavior and as an aid to cigarette withdrawal. Brit. Med. J. 2:391-393, 1976.
8. Goldfarb, T, Gritz, ER, Jarvik, MR and Stolerman, IP: Reactions to cigarettes as a function of nicotine and "tar." Clin. Pharmacol. 19:767-772, 1976.
9. Schachter, S: Regulation: withdrawal and nicotine addiction. In (ed.) Kranegor, NA, Cigarette Smoking as a Dependent Process, NIDA Research Monograph 23, U.S. Dept. Health, Education and Welfare, 1979, pp. 123-133.
10. Jarvik, ME, Glick, SD and Nakamura, RK: Inhibition of cigarette smoking by orally administered nicotine. Clin. Pharmacol. Therap. 11:574-576, 1970.
11. Kozlowski, LT, Jarvik, ME and Gritz, ER: Nicotine regulation and cigarette smoking. Clin. Pharmacol. Therap. 17:93-97, 1975.
12. Brentmark, B, Ohlin, P and Westling H: Nicotine-containing chewing gum as an anti-smoking aid. Psychopharmacol 31:191-200, 1973.
13. Hanson, HM, Ivester, CA and Morton, BR: Nicotine self-administration in rats. In (ed.) Kranegor, NA, Cigarette Smoking as a Dependent Process, NIDA Research Monograph 23, U.S. Dept. Health Education and Welfare, 1979, pp. 70-90.
14. Stolerman, IP, Fink, R and Jarvik, ME: Acute and chronic tolerance to nicotine measured by activity in rats. Psychopharmacol. 30:329-342, 1973.
15. Hubbard, JE and Gohd, RSL Tolerance development to the arousal effects of nicotine. Pharmacol. Biochem. Behav. 3:471-476, 1975.
16. Arqueros, L, Naquira, D and Zunino, E: Nicotine-induced release of catecholamine from rat hippocampus and striatum. Biochem. Pharmacol. 27:2667-2674, 1978.

17. Goodman, FR: Effects of nicotine on distribution and release of ^{14}C-norepinephrine and ^{14}C-dopamine in rat brain striatum and hypothalamus slices. Neuropharmacol. 13:1025-1032, 1974.

18. Westfall, TC: Effect of nicotine and other drugs on the release of ^{3}H-norepinephrine and [^{3}H]dopamine from rat brain slices. Neuropharmacol. 13:693-700, 1974.

19. Fuxe, K, Everett, BJ and Holsfelt, T: On the action of nicotine and cotinine on central 5-hydroxytryptamine neurons. Pharmacol. Biochem. Behav 10:671-677, 1979.

20. Torrens, Y, Beaujouan, JC, Besson, MJ, Michelot, R and Glowinski, J: Inhibitory effects of GABA, L-glutamic acid and nicotine on the potassium-evoked release of substance P in substantia nigra slices of the rat. Eur. J. Pharmacol. 71:383-392, 1981.

21. Marien, M, Brien, J and Thamandas, K: Regional release of [^{3}H]dopamine from rat brain in vitro: effects of opioids on release induced by potassium, nicotine, and L-glutamic acid. Can. J. Physiol. Pharmacol. 61:43-60, 1983.

22. Westfall, TC, Grant, H and Perry, H: Release of dopamine and 5-hydroxytryptamine from rat striatal slices following activation of nicotinic cholinergic receptors. Gen. Pharmacol. 14:321-325, 1983.

23. Weiler, MH, Misgeld, V and Cheong, DK: Presynaptic muscarinic modulation of nicotinic excitation in the rat neostriatum. Brain Res. 296:111-120, 1984.

24. Hery, F, Bourgoin, S, Hamon, M, Ternaux, JP and Glowinski, J: Control of the release of newly synthetized ^{3}H-5-hydroxytrytamine by nicotinic and muscarinic receptors in rat hypothalamic slices. Naunyn. Schmiedebergs. Arch. Pharmacol 296:91-97, 1977.

25. Krnjevils, K: Chemical nature of synaptic transmission in vertebrates. Physiol. Rev. 54:418-540, 1974.

26. Phillis, JW: Acetylcholine and synaptic transmission in the central nervous system. In (eds.) Hockman and Bieger, Chemical Transmission in the Mammalian Central Nervous System, University Park Press, Baltimore, 1976, pp. 159-213.

27. Giorguieff, MF, LeFloch, ML, Glowinski, J and Besson, MJ: Involvement of cholinergic presynaptic receptors of nicotinic and muscarinic types in the control of the spontaneous release of dopamine from striatal dopaminergic terminals in the rat. J. Pharmacol. Exp. Therap. 200:535-544, 1977.

28. Giorguieff-Chesselet, MF, Kemel, ML, Wandscheer, D and Glowinski, J: Regulation of dopamine release by presynaptic nicotinic receptors in rat striatal slices: effects of nicotine in a low concentration. Life Sci. 25:1257-1262, 1979.

29. Balfour, DJK: Effects of nicotine on the uptake and retention of ^{14}C-noradrenaline and ^{14}C-5-hydroxytryptamine by rat brain homogenates. Eur. J. Pharmacol. 23:19-26, 1973.

30. Yoshida, K, Kato, Y and Imura, H: Nicotine-induced release of noradrenaline from hypothalamic synaptosomes. Brain Res. 182:361-368, 1980.

31. Sakurai, Y, Takano, Y, Kohjimoto, Y, Honda, K and Kamiya, HO: Enhancement of [^{3}H]dopamine release and its [^{3}H]metabolites in rat striatum by nicotinic drugs. Brain Res. 242:99-106, 1982.

32. Connelly, MS and Littleton, JM: Lack of stereoselectivity in ability of nicotine to release dopamine from rat synaptosomal preparations. J. Neurochem. 41:1297-1302, 1983.

33. Takano, Y, Sakurai, Y, Kohjimoto, Y, Honda, K and Kamiya, HO: Presynaptic modulation of the release of dopamine from striatal synaptasomes: differences in the effects of high K^{+} stimulation, methamphetamine and nicotinic drugs. Brain Res. 279:330-334, 1983.

34. Mills, A and Wonnacott, S: Antibodies to nicotine acetylcholine receptors used to probe the structural and functional relationships between brain α-bungarotoxin binding sites and nicotine receptors. Neurochem. Int. 6:249-257, 1984.

35. Rowell, PP and Winkler, DL: Nicotinic stimulation of [^{3}H]acetylcholine release from mouse cerebral cortical synaptosomes. J. Neurochem. 43:1593-1598, 1984.
36. Yoshida, K and Imura, H: Nicotinic cholinergic receptors in brain synaptosomes. Brain Res. 172:453-459, 1979.
37. de Belleroche and Bradford, HF: Biochemical evidence for the presence of presynaptic receptors on dopaminergic nerve terminals. Brain Res. 142:53-68, 1978.
38. Schwartz, RD, Lehmann, J and Keller KJ: Presynaptic nicotinic cholinergic receptors labeled by [^{3}H]acetylcholine on catecholamine and serotonin axons in brain. J. Neurochem. 42:1495-1498, 1984.
39. Hall, GH and Turner, DM: Effects of nicotine on the release of ^{3}H-noradrenaline from the hypothalamus. Biochem. Pharmacol. 4:1829-1836, 1972.
40. Andersson, K, Eneroth, P and Agnati, LF: Nicotine-induced increases of noradrenaline turnover in discrete noradrenaline nerve terminals systems of the hypothalamus and the median eminence of the rat and their relationship to changes in the secretion of adrenohypophyseal hormones. Acta Physiol. Scand. 113:227-231, 1981.
41. Andersson, K, Fuxe, K, Eneroth, P, Gustafsson, JA and Agnati, LF: Mecamylamine induced blockade of nicotine induced inhibition of gonadotropin and TSH secretion and of nicotine induced increases of catecholamine turnover in the rat hypothalamus. Acta. Physiol. Scand. Suppl. 479:27-29, 1980.
42. Andersson, K, Fuxe, K, Eneroth, P and Agnati, LF: Involvement of cholinergic nicotine-like receptors as modulators of amine turnover in various types of hypothalamic dopamine and noradrenaline nerve terminal systems and of prolactin, LH, FSH and TSH secretion in the castrated male rat. Acta Physiol. Scand. 116:41-50, 1982.
43. Balfour, DJK: The effects of nicotine on brain neurotransmitter systems. Pharmacol. Ther. 16:269-282, 1982.
44. Eneroth, P, Fuxe, K, Gustafsson, JA, Hokfelt, T, Lofstrom, A, Skett, P and Agnati, L: The effect of nicotine on central catecholamine neurons and gonadotropin secretion. II. Inhibitory influence of nicotine on LH, PSH and prolactin secretion in the ovariectomized female rat and its relation to regional changes in dopamine and noradrenaline levels and turnover. Med. Biol. 55:158-166, 1977.
45. Andersson, K, Fuxe, K, Eneroth, P and Agnati, LF: Effects of acute central and peripheral administration of nicotine on hypothalamic catecholamine nerve terminal systems on the secretion of adenohypophyseal hormones in the male rat. Med. Biol. 60:98-111, 1982.
46. Douglas, RJ: The hippocampus and behavior. Psychol. Bull. 67:416-442, 1967.
47. Isaacron, RL and Pribam, KH: The Hippocampus. Plenum Press, New York, 1975.
48. Sitaram, N, Weingartner, H and Gillin, J: Human serial learning: Enrichment with arecoline and choline and impairment with scopolamine. Science 201:274-276, 1978.
49. Bartus, RT, Dean, RL, Goar, JA and Lippa, AS: Age-related changes in passive avoidance retention: Modulation with dietary choline. Science 209:301-303, 1980.
50. Davis, K, Mohs, R, Tinklenberg, J, Pfefferbaum, A, Hollister, L and Kopell, B: Physostigmine: Improvement of long-term memory processes in normal humans. Science 201:272-274, 1978
51. Drachman, DA and Sahakian, BJ: Effects of cholinergic agents on human learning and memory. In (eds.) Barbeau, A, Growdon, JH and Wurtman, R, Choline and Lecithin in Brain Disorders, Raven Press, New York, 1979.

52. Greene, EG and Lomax, P: Impairment of alternation learning in rats following microinjection of carbachol into the hippocampus. Brain Res. 18:355-359, 1970.
53. Benardo, LS and Prince, DA: Cholinergic pharmacology of mammalian hippocampal pyramidal cells. Neuroscience 7:1703-1712, 1982.
54. Anderson, K: Effects of cigarette smoking on learning and retention. Psychopharmacol. 41:1-5, 1975.
55. Nelson, JM, Pelley, K and Goldstein, L: Protection by nicotine from behavioral description caused by reticular formation stimulation in the rat. Pharmacol. Biochem. Behav. 3:749-754, 1975.
56. Van Rossum, JM, Broekkamp, CLE and Pijnenburg, AJJ: Behavioral correlates of dopaminergic function in the nucleus accumbens. Adv. Biochem. Psychopharmacol. 16:201-207, 1977.
57. Iversen, SD and Koob, GF: Behavioral implications of dopaminergic neurons in the mesolimbic system. Adv. Biochem. Psychopharmacol. 16:209-214, 1977.
58. Wise, RA: The dopamine synapse and the notion of pleasure centers in the brain. Trends. Neurosci. 3:91-95, 1980.
59. Costa, E, Cheney, DL, Mao, CC and Moroni, F: Action of antischizophrenic drugs on the metabolism of aminobutyric acid and acetylcholine in globus pallidus, striatum and nucleus accumbens. Fed. Proc. 37:2408-2414, 1978.
60. Langnickel, R, Bluth, R and Oelszner, W: Various dose-dependent influences of apomorphine on the acetylcholine turnover in striatum and mesolimbic areas of rat brain. Biomed. Biochem. Acta 42:937-946, 1983.
61. de Belleroche, JS and Gardiner, IM: Cholinergic action in the nucleus accumbens: Modulation of dopamine and acetylcholine release. Brit. J. Pharmacol. 75:359-365, 1982.
62. Gold, R and Bluth, R: Dopaminergic modulation of acetylcholine release from slices of rat nucleus accumbens. Biogenic Amines 2:211-218, 1985.
63. Day, M and Vane, JR: An analysis of the direct and indirect actions of drugs on the isolated guinea-pig ileum. Brit. J. Pharmacol. 20:150-170, 1963.
64. Henderson, PT, Ariens, EJ and Simonis, AM: Differentiation of various types of cholinergic and other spasmogenic actions on the isolated guinea-pig ileum. Eur. J. Pharmacol. 4:62-70, 1968.
65. Chiou, CY, Long, JP, Potrepha, R and Spratt, JL: The ability of various nicotinic agents to release acetylcholine from synaptic vesicles. Arch. Int. Pharmacodyn. Ther. 187:88-96, 1970.
66. Chiou, CY: Mechanism of acetylcholine release by drugs and its blockade. Arch. Int. Pharmacodyn. Ther. 201:170-181, 1973.
67. Pepeu, G: Nicotinic e acetilcolina cerebrate. Arch. Ital. Sci. Farmacol. 15:93-94, 1965.
68. Essman, WB: Metabolic and behavioral consequences of nicotine. In (ed.) Smith, WL, Drugs and Cerebral Function, Charles C. Thomas Publishers, Springfield, 1971.
69. Armitage, AK, Hall, GH and Seller, CM: Effects of nicotine on electrocortical activity and acetylcholine release from the cat cerebral cortex. Brit. J. Pharmacol. 35:152-160, 1969.
70. Kaji, M, Chihara, K, Minamitarsi, N, Kodama, H, Yanaihara, N and Fujita, T: Release of vasoactive intestinal polypeptide into the cerebrospinal fluid of the fourth ventricle of the rat: involvement of cholinergic mechanism. Brain Res. 269:303-310, 1983.
71. Bhagat, B: Influence of chronic administration of nicotine on the turnover and metabolism of noradrenaline in the rat brain. Psychopharmacologia 18:325-332, 1970.
72. Morgan, WW and Pfeil, KA: Mecamylamine blockade of nicotine enhanced noradrenaline turnover in rat brain. Life Sci. 24:417-420, 1979.

73. Lichtensteiger, W, Heft, F, Felix, D, Huwyler, T, Medamed, E and Schlumpf, M: Stimulation of nigrostriatal dopamine neurons by nicotine. Neuropharmacol. 21:963-968, 1982.
74. Fosbraey, P and Johnson, ES: Release-modulating acetylcholine receptors on cholinergic neurons of the guinea-pig ileum. Brit. J. Pharmacol. 68:289-300, 1980.
75. Briggs, CA and Cooper, JR: Cholinergic modulation of the release of [^{3}H]acetylcholine from synaptosomes of the myenteric plexus. J. Neurochem. 38:501-508, 1982.
76. Molenaar, PC and Polak, RL: Stimulation by atropine of acetylcholine release and synthesis in cortical slices from rat brain. Brit. J. Pharmacol. 40:406-417, 1970.
77. Nordstrom, O and Bartfai, T: Muscarinic autoreceptor regulates acetylcholine release in rat hippocampus: In vitro evidence. Acta Physiol. 108:347-353, 1980.
78. Abood, LG, Lowey, K, Tometsko, A and Booth, H: Electrophysiological, behavioral and chemical evidence for a noncholinergic, stereospecific site for nicotine in rat brain. J. Neurochem. 3:327-333, 1978.
79. Abood, LG, Lowey, K, Tometsko, A and MacNeil, M: Evidence for a noncholinergic site for nicotine's action in brain: psychopharmacological, electrophysiological and receptor binding studies. Arch. Int. Pharmacodyn. Therap. 237:213-229, 1979.
80. Sloan, JW, Martin, WR, Hoods, R and Hernandey, J: Structure-activity relationships of some pyridine, piperidiene, and pyrolidine analogues for enhancing and inhibiting the binding of (+)-(-)[^{3}H]nicotine to the rat brain P2 preparation. J. Med. Chem. 28:1245-1251, 1985.
81. Sloan, JW, Todd, GD and Martin, WR: Nature of nicotine binding to rat brain P2 fraction. Pharmacol. Biochem. Behav. 20:899-909, 1984.
82. Clarke, PB, Schwartz, RD, Paul, SM, Pert, CB and Pert A: Nicotinic binding in rat brain: autoradiographic comparison of [^{3}H]acetylcholine, [^{3}H]nicotine, and [^{125}I]α-bungarotoxin. J. Neurosci. 5:1307-1315, 1985.
83. Clarke, PB, Pert, CB and Pert A: Autoradiographic distribution of nicotine receptors in rat brain. Brain Res. 323:390-395, 1984.
84. Armitage, AK, Dollery, CT, Houseman, TH, Lewis, PJ and Turner, DM: Absorption and metabolism of nicotine from cigarettes. Brit. Med. J. 4:313-316, 1975.
85. Russell, MA, Jarvis, M. Iyer, R and Feyerabend, C: Relation of nicotine yield of cigarettes to blood nicotine concentrations in smokers. Brit. Med. J. 280:972-976, 1980.
86. Kogan, MJ, Verebey, K, Jaffee, JH and Mule, SJ: Simultaneous determination of nicotine and cotinine in human plasma by nitrogen detection gas-liquid chromatography. J. Forensic Sci. 26:6-11, 1981.
87. Adir, J, Miller RP and Rotenberg, KS: Disposition of nicotine in the rat after intravenous administration. Res. Comm. Chem. Path. Pharmacol. 13:173-183, 1976.
88. Miller, RP, Rotenberg, KS and Adir, J: Effect of dose on the pharmacokinetics of intravenous nicotine in the rat. Drug Metab. Dispo. 5:436-443, 1977.
89. Taylor, P: Ganglionic stimulants and blocking agents. (eds.) Gilman, AG, Goodman, LS and Gilman, A. In The Pharmacological Basis of Therapeutics, 6th Edition. New York, Macmillan, 1980, pp. 211-219.
90. Giorguieff, MF, LeFloch, ML, Westfall, TC, Glowinski, J and Besson, MJ: Nicotinic effect of acetylcholine on the release of newly synthesized [^{3}H]dopamine in rat striatal slices and cat caudate nucleus. Brain Res. 106:117-131, 1976.
91. Tripathi, ON: Arecoline induced nicotinic and muscarinic stimulation of the superior cervical ganglion of cat. Biomed. Biochem. Acta. 42:275-282, 1983.

92. Klus, H and Kuhn, H: A study of the optical activity of smoke nicotines. Fachliche Mitt. Oesterr. Tabakregie 17:331-336, 1977.
93. Sato, M and Crooks, PA: N-methylnornicotinium ion, a new in vivo metabolite of R-(+)-nicotine. Drug Metab. Dispos. 13:348-352, 1985.
94. Cundy, KC, Godin, CS and Crooks, PA: Stereospecific in vitro N-methylation of nicotine in guinea pig tissues by an S-adenosylmethionine-dependent N-methyltransferase. Biochem. Pharmacol. 34:281-284, 1985.
95. Rosecrans, JA and Meltzer, LT: Central sites and mechanisms of action of nicotine. Neurosci. Biobehav. Rev. 5:497-501, 1981.
96. Stolerman, IP, Pratt, JA, Garcha, HS, Giardini, V and Kumar, R: Nicotine cue in rats analyzed with drugs acting on cholinergic and 5-hydroxytryptamine mechanisms. Neuropharmacol. 22:1029-1037, 1983.
97. Pratt, JA, Stolerman, IP, Garcha, HS, Giardini, V and Feyerabend, C: Discrimination stimulus properties of nicotine: further evidence for mediation at a cholinergic receptor. Psychopharmacol. 81:54-60, 1983.

MECHANISMS OF NICOTINE REGULATION OF DOPAMINE RELEASE IN NEOSTRIATUM

Thomas C. Westfall, Holly Perry, and Lillian Vickery

Department of Pharmacology
St. Louis University School of Medicine
St. Louis, MO 63104

ABSTRACT

The mechanism(s) of the evoked release of dopamine from rat striatal slices by nicotine and the nicotinic agonist dimethylphenylpiperazinium (DMPP) has been studied. Release of dopamine was assessed by three separate procedures: 1) release of [^{3}H]dopamine from slices preincubated with the labeled amine (newly taken up dopamine); 2) release of endogenous transmitter measured by high pressure liquid chromatography coupled to electrochemical detection (endogenous dopamine), and 3) release of [^{3}H]dopamine continuously formed by superfusion of slices with [^{3}H]tyrosine (newly synthesized dopamine). Both nicotinic agonists were observed to cause a release of dopamine from rat striatal slices. The results obtained are consistent with the interpretation that these agents release dopamine by two mechanisms, one seen at low and the other seen at high concentrations of the drugs.

The effect of high concentrations of nicotine (100 - 10,000 μM) does not appear to be dependent upon extracellular calcium, is not sensitive to block by tetrodotoxin, and is not easily antagonized by classical antagonists such as d-tubocurarine or mecamylamine. High concentrations of nicotine and DMPP appear to release dopamine by displacement of the amine in a manner similar to indirectly acting adrenergic and dopaminergic drugs or to activation of an atypical nicotinic receptor. In contrast, the releasing effect of low concentrations of nicotine (1 - 100 μM) appears to involve activation of nicotinic cholinergic receptors as this effect is antagonized by d-tubocurarine and mecamylamine, requiring extracellular calcium in the incubation buffer and antagonized by tetrodotoxin.

Although both mechanisms may come into play in explaining some of the central pharmacological actions of smoking, it is likely that the low dose effect of nicotine is most relevant since this would better correlate with concentrations of nicotine reaching the central nervous system in humans during smoking.

INTRODUCTION

Nicotine, the active ingredient in tobacco smoke, is well known to produce multiple psychopharmacological, behavioral, and neurochemical effects in the central nervous system (1,2,3,4,5). The mechanism or mechanisms underlying these effects is unclear but nicotine and nicotinic agonists are known to affect catecholamine systems in the brain (6,7). Nicotine has been shown to cause a release of dopamine from slices of synaptosomes of the caudate nucleus. This was first demonstrated in 1974 (8) and has since been verified by numerous investigators (9,10,11,12,13). The mechanism(s) involved in the release of dopamine is less clear. Although it was first thought that nicotine released dopamine by activation of classical nicotinic-cholinergic receptors, there are several inconsistencies that cast some doubt on this interpretation. First, although some investigators have reported that nicotinic antagonists block the nicotine-induced release of dopamine (4,8,10,12,14,15), others have failed to see this effect (16,17). Secondly, in the same preparation some nicotinic antagonists reduced the effect while others did not; and finally, in all cases where nicotinic antagonists have been observed to work, the agents only reduced but never completely antagonized the response. This situation caused us to further probe the mechanism(s) by which nicotinic agents release dopamine in the striatum of the rat. The present report summarizes results obtained with two nicotinic agonists, nicotine itself and dimethylphenylpiperazinium (DMPP). Our results are consistent with the interpretation that nicotinic agents act by two mechanisms, one seen at low and the other seen at high concentrations of the drug.

MATERIALS AND METHODS

Male Sprague-Dawley rats weighing between 160-300 g obtained from SASCO, (St. Louis, MO) were used in all experiments. Animals were maintained under standard conditions with food and water ad libitum. Animals were sacrificed by decapitation and striata were rapidly removed in a cold room (4°C) using glass manipulators. Striata were chopped into slices 0.5 mm thick with a McIlwain tissue chopper and placed in vials containing ice cold Krebs-Ringer bicarbonate buffer, containing in mM: NaCl, 118: KCl, 4.85; $CaCl_2$, 2.5; $MgSO_4$, 1.15; KH_2PO_4, 1.15; glucose, 11.1; and $NaHCO_3$, 25.0. Release of dopamine in response to DMPP was determined by measuring the release of endogenous dopamine by high pressure liquid chromatography coupled to electrochemical detection (HPLC-EC). In experiments to assess whether release was from the cytoplasmic or the vesicular pool, the release of newly taken up [^{3}H]dopamine was measured. Release of dopamine in response to nicotine was determined by three separate procedures and experiments: 1) release of [^{3}H]dopamine from slices preincubated for 30 min with the labelled transmitter (newly taken up dopamine); 2) release of endogenous transmitter as measured by high pressure liquid chromatography coupled to electrochemical detection (endogenous dopamine); 3) release of [^{3}H]dopamine continuously formed by superfusion of slices with [^{3}H]tyrosine (newly synthesized dopamine).

Release of Newly Taken Up [^{3}H]Dopamine

Slices were prepared and placed in glass minivials containing 750 µl of Krebs-Ringer bicarbonate buffer, bubbled with 95% O_2 - 5% CO_2, and maintained at 37°C in a water bath. The slices were preincubated for 10 min and then transferred to a vial and incubated for 30 min in the presence of [^{3}H]dopamine final concentration 5 x 10^{-8} M. The slices were placed in vials containing normal buffer at 15 or 30 min intervals. They were then exposed to various concentrations of nicotine for 5 min.

In experiments examining the effect of various perturbations or manipulations on the nicotine-induced release of dopamine, the slices were placed in a modified buffer for 15 or 30 min prior to exposure to nicotine. Following the 5 min stimulation period, the slices were placed in another vial containing normal buffer and washed for 1 min before analysis. In addition, the effects of various concentrations of nicotine ranging from 10 to 1000 μM, the effects of removing extracellular calcium, and the effects of mecamylamine, d-tubocurarine, or tetrodotoxin on the release of [^{3}H]dopamine produced by 0.5 and 1.0 mM nicotine were examined. The amount of [^{3}H]dopamine released into the media was determined by liquid scintillation spectrometry following absorption and elution from alumina.

[^{3}H]dopamine was determined by counting 1 ml of sample in 10 ml of Scintiverse E (Fisher Scientific) in a Packard 300 C liquid scintillation counter. The nicotine-induced [^{3}H]dopamine release was calculated as the difference between stimulated [^{3}H]dopamine release and an approximation of spontaneous release during the period of stimulation based on prestimulation levels of basal release. This difference was divided by the total [^{3}H]dopamine released in superfusates plus the [^{3}H]dopamine remaining in the tissue at the end of the experiment and expressed as a percent fractional release. The amount of [^{3}H]dopamine in the tissue at the start of any time interval was calculated by adding cumulatively the amount of [^{3}H]dopamine released into the medium to the amount of [^{3}H]dopamine in the tissue at the end of the experiment.

Endogenous Dopamine

Striata were placed into glass vials and were vigorously vortexed to separate the slices. Two slices from each striatum were selected, placed in a wire-mesh basket in a mini-vial containing 750 μl buffer, incubated at 37°C in a shaking water bath, and bubbled with 95% O_2 - 5% CO_2. The slices were incubated for at least 15 min before being challenged with various concentrations of nicotine or DMPP for 5 min. The slices were then transferred to vials containing either a modified or normal buffer and incubated for another 15 or 30 min. The slices were then placed in a buffer containing nicotine or DMPP for a second 5 min stimulation period. In some experiments following the second 5 min stimulation period, the slices were placed in another vial containing normal buffer for an additional 15 or 30 min followed by exposure for a third time to nicotine or DMPP. Following the third 5 min stimulation period, the slices were placed in a vial containing normal buffer and washed for 1 min before analysis. Release was studied under several sets of conditions. The release of dopamine was determined by measuring the amount in the media and tissue. Four fractions from each experiment - the two 5 min stimulating fractions and the two 5 min fractions immediately preceding them - were assayed for dopamine. The media fractions were concentrated by an alumina batch technique using [^{3}H]dopamine as an internal standard. The eluate from the alumina was then analyzed for released dopamine by HPLC-EC. The solvent used routinely for analysis was 10% methanol, 125 mM acetic acid, 2 mM EDTA, and 2 mM HSA. The pH was adjusted to 3.8 with 13.3 M NaOH. Dopamine was quantified by comparison of peak heights with a known amount of authentic dopamine.

The release of endogenous dopamine was expressed as the nanograms released per 5 min or as the percent fractional release. The nicotine-induced dopamine release or percent fractional release were calculated as the difference between stimulated dopamine release and an approximation of spontaneous release during the period of stimulation based on prestimulation levels of basal release in a fashion analogous to [^{3}H]dopamine release already mentioned.

Release of Newly Synthesized [^{3}H]Dopamine

Slices were placed in individual superfusion chambers and continuously superfused at a flow rate of 200 µl/min with [^{3}H]tyrosine. [^{3}H]tyrosine was purified by successive passage over alumina and Dowex columns and maintained under an atmosphere of nitrogen. The [^{3}H]tyrosine was reconstituted in buffer before each experiment and used at a final concentration of 40–50 µCi/ml. A steady state formation of [^{3}H]dopamine occurs following 25–30 min of superfusion. Once a steady state of [^{3}H]dopamine was obtained, the effect of nicotine in the absence or presence of d-tubocurarine, mecamylamine, the removal of extracellular Ca^{2+} or the presence of tetrodotoxin was assessed. Superfusate effluents were continuously collected in 2.5 min fractions by means of a fraction collector. [^{3}H]dopamine in the superfusate effluents was determined following passage over amberlite followed by alumina columns according to previously published techniques (18). In experiments in which pertubations were assessed, the drugs were present 15 min prior to stimulation by nicotine. Calcium was removed from the superfusion buffer 15 or 30 min prior to stimulation with nicotine. The manner in which the data were calculated was based on the method described by Nieoullan (19) and utilized by us previously (20). The amount of [^{3}H]dopamine released during the steady state obviously varies from animal to animal. Therefore, each animal served as his own control. The control (100%) value for each animal was calculated by taking the mean value of dopamine released from the three fractions preceding the drug treatment. Release of dopamine in each successive fraction was expressed as a percentage of the control value. Data from various animals were pooled and the mean ± standard error of the mean was calculated.

Release of Dopamine from Cytoplasmic or Vesicular Pool

Experiments carried out with DMPP examined the question of whether or not DMPP released dopamine from a vesicular or cytoplasmic pool. In these experiments, rats were first treated with reserpine at a dose of 1 mg/kg/24 hours on two consecutive days. This was shown to produce greater than 98% depletion of dopamine stores. One hour before killing, the animals were also treated with the monoamine oxidase inhibitor, pargyline (10 mg/kg). Animals were killed and striatal slices prepared. The slices were incubated in the presence of [^{3}H]dopamine (10^{-8} M) plus pargyline. The slices were placed in three consecutive vials containing normal buffer at 15 min intervals. They were then exposed to tyramine (1.7 mM), amphetamine (1 µM), and DMPP (500 µM) for 5 min. The slices were placed in another vial containing normal buffer and washed for 1 min prior to analysis. The amount of [^{3}H]dopamine released into the media was determined by liquid scintillation spectrometry following absorption and elution from alumina columns.

Statistics and Chemicals

The number of animals used to generate the data is indicated in each figure. Data are expressed as mean ± S.E.M. Depending on the design and type of experiment, statistical significance was determined by either Students t-test using two-tailed probabilities where one control response was compared to one perturbation or analysis of variance when multiple variables were compared to a single control response.

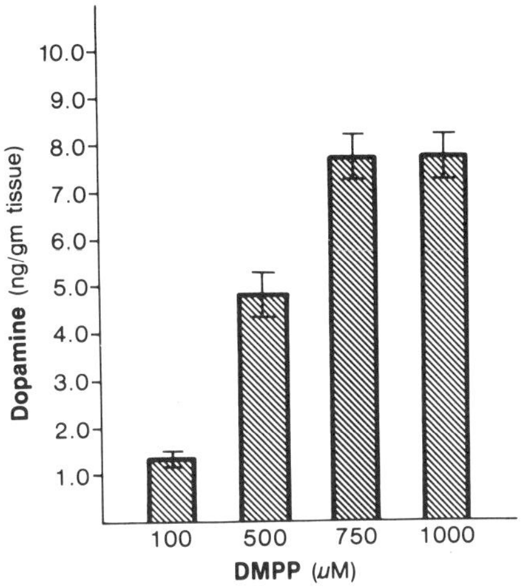

Fig. 1. The effect of DMPP on the release of endogenous dopamine from incubated rat striatal slices. Data are plotted as dopamine released in ng/gm/5 min to increasing concentrations of DMPP (in μM). Each bar represents the mean ± S.E.M. of 5-7 experiments. Data are corrected by subtracting an approximation of spontaneous release (prestimulation value) from that released during the 5 min period of stimulation.

RESULTS

Release of Dopamine by DMPP

DMPP was observed to produce a concentration-dependent release of dopamine from incubated striatal slices (Fig. 1). A minimal effective concentration was 10 μM DMPP with a maximal effect detected at a concentration of 750 μM DMPP. Tetrodotoxin at a concentration of 10^{-5}M, which was shown to block the field-stimulation induced release of dopamine (data not shown), failed to alter the DMPP-induced release of dopamine (Fig. 2). The removal of extracellular Ca^{2+} from the incubation medium (Fig. 3) or the presence of low Na^+ buffer also failed to alter the DMPP-induced release of dopamine (Fig. 4).

Experiments to examine the source of the dopamine pool (cytoplasmic or vesicular) are depicted in Fig. 5. A dose of reserpine (1 mg/kg/day twice daily) produced a greater than 98% depletion of dopamine measured in striatal tissue. The animals were then treated with pargyline (10 mg/kg) to inhibit monoamine oxidase. Animals were killed and striatal slices incubated with [^{3}H]dopamine. Following the treatment regimen (reserpine plus pargyline), [^{3}H]dopamine is limited to cytoplasmic sites.

It was observed that tyramine, amphetamine, and DMPP produced the same degree of release of [^{3}H]dopamine from slices of reserpine plus pargyline-treated animals as that from nontreated animals. In contrast, field-stimulation or potassium depolarization failed to induce release of [^{3}H]dopamine from slices of reserpinized and pargyline-treated animals.

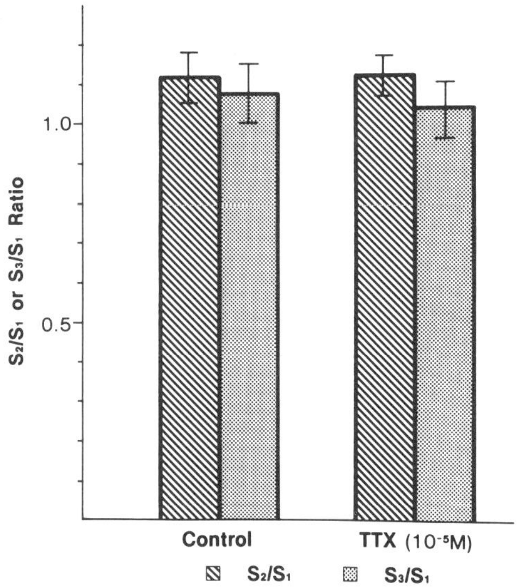

Fig. 2. The effect of tetrodotoxin on the DMPP (500 μm) induced release of endogenous dopamine from rat striatal slices. Data are plotted as S_2/S_1 or S_3/S_1 ratios. The slices were exposed three times to DMPP (500 μM for 5 min) S_1, S_2 and S_3. Tetrodotoxin (10^{-5} M) was present in the incubation medium prior to and during the second stimulation (S_2). Each bar is the mean ± S.E.M. of 5-7 experiments.

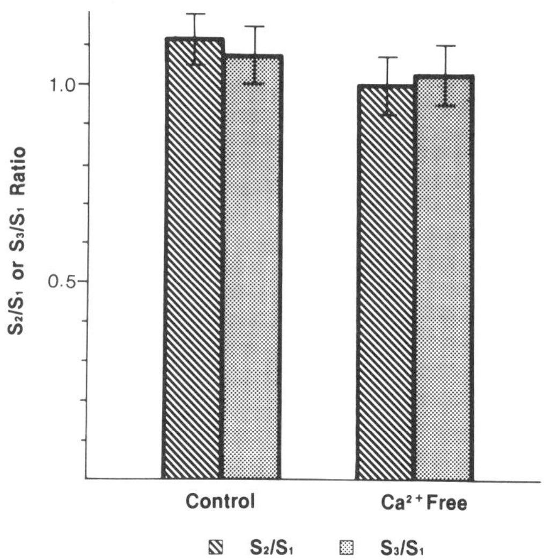

Fig. 3. The effect of removing extracellular Ca^{2+} from the incubation medium or the DMPP (500 μM) induced release of endogenous dopamine from rat striatal slices. Data are plotted in an identical fashion as in Fig. 2. Each bar is the mean ± S.E.M. of 5-7 experiments.

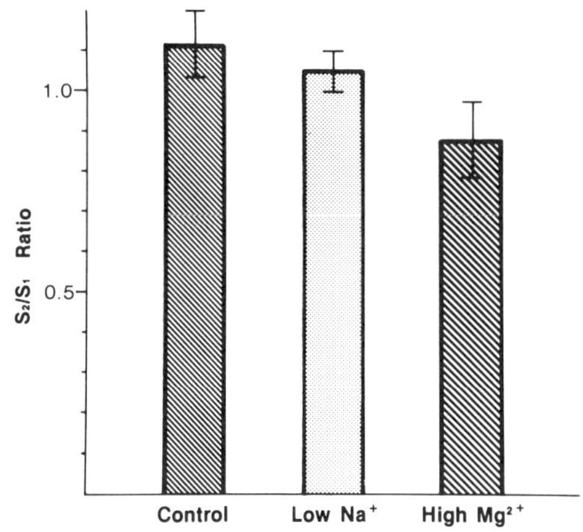

Fig. 4. The effect of a low sodium (23 mM) or high magnesium (24 mM) buffer on the DMPP (500 μM) induced release of endogenous dopamine from rat striatal slices. Data are plotted as S_2/S_1 ratios (normal (S_1) or altered (S_2) buffers). Each bar is the mean ± S.E.M. of 5-7 experiments.

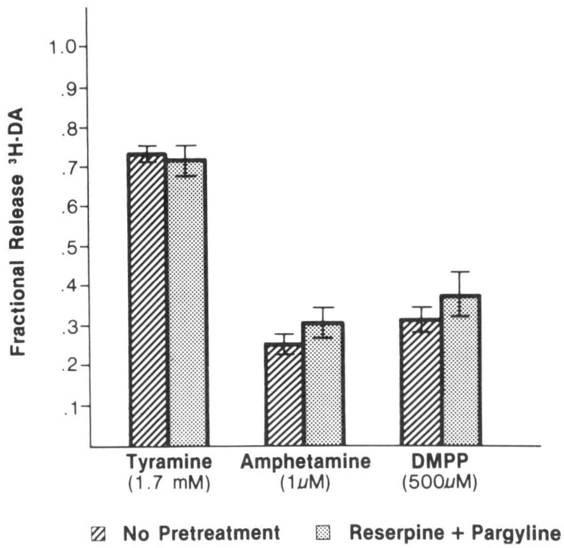

Fig. 5. The effect of reserpine pretreatment (see Methods) on the evoked release of [^{3}H]dopamine from striatal slices produced by tyramine, amphetamine or DMPP. Data are plotted as fractional release of [^{3}H]dopamine. Each bar is the mean ± S.E.M. of 5-7 experiments.

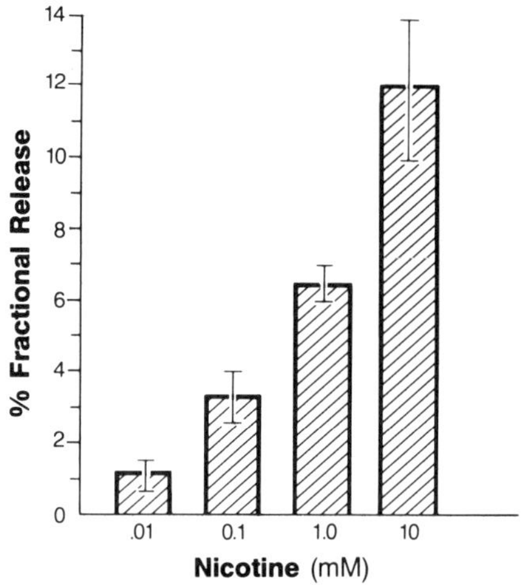

Fig. 6. The effect of nicotine on the release of dopamine from striatal slices previously incubated with [^{3}H]-dopamine to label the endogenous stores. Data are plotted as percent fractional release of [^{3}H]-dopamine against the concentration of nicotine in mice. Each bar is the mean ± S.E.M. of 5-7 experiments. A significant release was observed by > 0.1, 1.0 and 10 mM ($p < .05$, .01, .001), respectively.

Release of Dopamine by Nicotine

The effect of increasing concentrations of nicotine on the release of [^{3}H]dopamine from slices previously incubated with the labelled transmitter is depicted in Fig. 6. A concentration of 0.01 mM just barely resulted in a detectable release of dopamine, whereas a concentration-dependent release could be seen with concentrations above 0.1 mM. Using 1 mM nicotine as a test concentration, the effect of the nicotinic antagonists mecamylamine and d-tubocurarine on the nicotine-evoked release was examined. Neither antagonist resulted in a significant reduction in the evoked release of [^{3}H]dopamine. The effect of d-tubocurarine on the release induced by nicotine or DMPP is depicted in Fig. 7. No attenuation by the antagonist was observed. Tetrodotoxin or the removal of extracellular Ca^{2+} from the incubation medium also failed to alter the nicotine-induced release of [^{3}H]dopamine (data not shown).

Similar results were obtained when the effect of nicotine on the release of endogenous dopamine from incubated slices was examined. The evoked release of dopamine by 1 mM nicotine was not significantly altered by removal of extracellular Ca^{2+} or the presence of tetrodotoxin or mecamylamine (Fig. 8).

The use of the third technique to analyze the release of dopamine, namely, [^{3}H]dopamine newly synthesized from [^{3}H]tyrosine in superfused slices, resulted in dramatically different results. First, it was possible to see a readily detectable increase in [^{3}H]dopamine release with concentrations of 1 μM and less. This was not possible with either of the other two ways of assessing dopamine release. A concentration

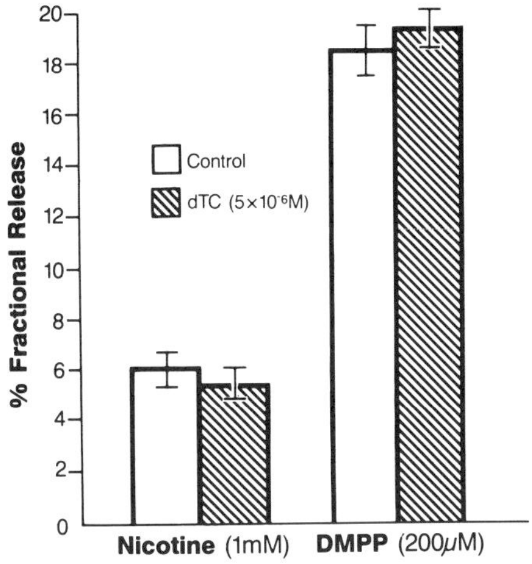

Fig. 7. The effect of d-tubocurarine (5 x 10^{-6} M) on the nicotine (1 mM) or DMPP (200 μM) induced release of [^{3}H]dopamine from rat striatal slices. Data are plotted as percent fractional release. Each bar is the mean ± S.E.M. of 5-8 experiments. d-TC did not alter the evoked release of [^{3}H]DA to either nicotine or DMPP.

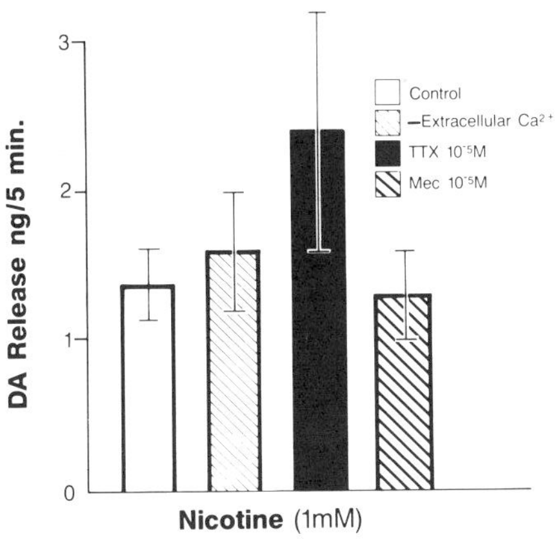

Fig. 8. The effect of nicotine (1 mM) to release endogenous dopamine from incubated striatal slices in the absence or presence of various perturbations. Removal of extracellular calcium or presence of tetrodotoxin (10^{-5} M) or mecamylamine (10^{-5}M). Data are plotted as dopamine released in nanograms/5 min and each bar represents the mean ± S.E.M. of 5-8 experiments. None of the perturbations altered the nicotine-induced response.

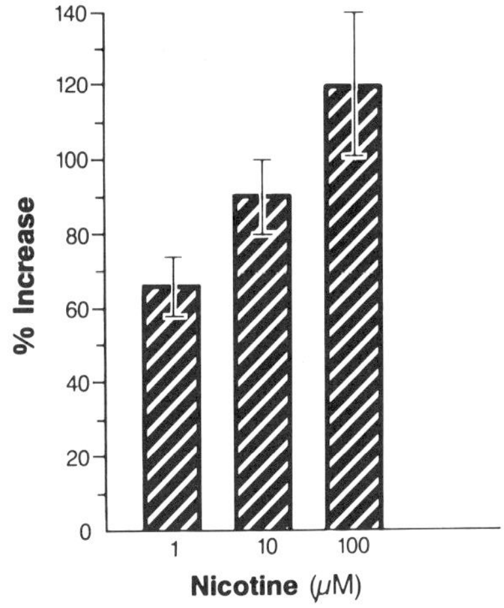

Fig. 9

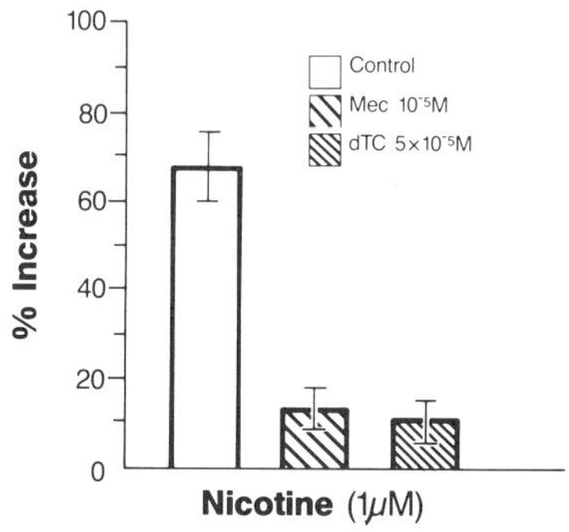

Fig. 10

Fig. 9. The effect of nicotine on the release of [^{3}H]dopamine newly synthesized from [^{3}H]tyrosine from slices continuously superfused with [^{3}H]tyrosine. Data are plotted as a percent age of the mean spontaneous release of [^{3}H]dopamine (see Methods). Each bar is the mean ± S.E.M. of 5-8 experiments versus concentration of nicotine in μM. All three concentrations produced a highly significant increase over the spontaneous release (P <.01 or <.001) for 1 μM or 10 and 100 μM, respectively.

Fig. 10. The effect of mecamylamine (10^{-5}M) or d-tubocurarine (5 x 10^{-5} M) on the nicotine evoked (1 μM) release of [^{3}H]tyrosine in slices continuously superfused with [^{3}H]tyrosine. Data are plotted as the percent increase of [^{3}H] dopamine over the spontaneous release. Each bar represents the mean ± S.E.M. of 4-6 experiments. Both mecamylamine and d-tubocurarine produced a significant attentuation (P <.01) of the nicotine-evoked response.

response curve is depicted by Fig. 9. In contrast to the experiments assessing the release of newly taken up [^{3}H]dopamine or endogenous dopamine, both mecamylamine and d-tubocurarine, in concentrations which did not alter the release of dopamine utilizing these two techniques, produced a marked and highly significant reduction of newly synthesized [^{3}H]dopamine released in response to 1 μM nicotine (Fig. 10). The nicotine-induced release of newly synthesized dopamine was also attenuated by removal of extracellular calcium (Fig. 11) or the presence of tetrodotoxin (Fig. 12).

A summary of the comparison between low and high concentrations of nicotine observed from our studies is depicted in Table 1.

DISCUSSION

The present experiments have shown that both nicotine and the nicotinic agonist DMPP produced a release of dopamine from rat striatal

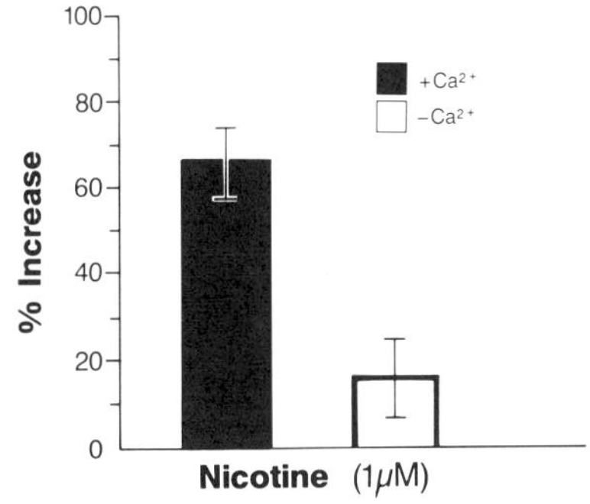

Fig. 11

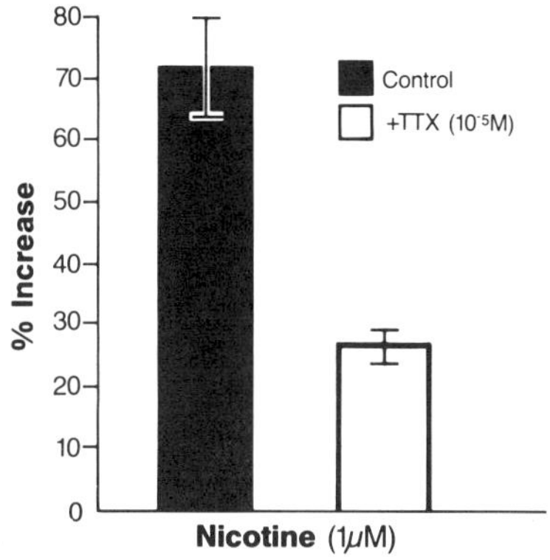

Fig. 12

Fig. 11. The effect of nicotine (1 μM) in the absence or presence of extracellular calcium in the superfusion buffer on the release of [^{3}H]dopamine newly synthesized from [^{3}H]tyrosine in slices continuously superfused with [^{3}H]tyrosine. Data are plotted as percent increase of [^{3}H]dopamine over the spontaneous release. Each bar represents the mean ± S.E.M. of 4-6 experiments. Removal of extracellular calcium resulted in a significant attenuation of the nicotine-induced response (P<.01).

Fig. 12. The effect of nicotine (1 μM) in the absence or presence of tetrodotoxin (10^{-5} M) in the superfusion buffer on the release of [^{3}H]dopamine nearly synthesized from [^{3}H]tyrosine in slices continuously superfused with [^{3}H]tyrosine. Data are plotted as percent increase of [^{3}H]dopamine over the spontaneous release. Each bar represents the mean ± S.E.M. of 4-6 experiments. The presence of tetrodotoxin resulted in a significant attenuation of the nicotine-induced release (P<.001).

slices. This is consistent with the results of numerous investigators including ourselves who have observed a similar nicotine-induced release of dopamine (4,8,9,10,11,12,13,14,15). It appears, therefore, that there is universal agreement that nicotine or nicotine-like drugs release dopamine from dopaminergic neurons in the central nervous system.

The mechanism by which nicotine exerts this effect is much less clear. As mentioned in the Introduction, some investigators have reported that the effect of nicotine or acetylcholine on dopamine release can be reduced by classical nicotinic antagonists (4,8,10,12,14,15), while others have failed to observe such a reduction by similar drugs (16,17). The importance of extracellular calcium in the nicotine-induced release of dopamine from the striatum is also controversial. Some investigators have claimed that the nicotine-induced release of dopamine requires extracellular calcium in the superfusion or incubation medium (12,16), while others have reported that such release is independent of extracellular calcium (9,17,21).

It is thought that results obtained in the present study provide an explanation for some of the discordant results reported in the

Table 1. Comparison of Low and High Concentrations of Nicotine in Releasing Dopamine From Rat Striatal Slices

Item	Low Concentration	High Concentration
Method of Assessing	•Newly Synthesized ^{3}H dopamine	•Newly taken up dopamine •Endogenous dopamine
Concentration range to see effect	1 μM-100 μM	100 μM-1000 μM
Ca^{2+} Dependent	Yes	No
TTX Dependent	Yes	No
Antagonist Block		
d-TC	Yes	No
Mec	Yes	No
Amine Pump Blocker	——	No

literature. An examination of the literature reveals that, in most cases, high concentrations of nicotine or acetylcholine were used to induce the release of dopamine from striatal tissue. The concentrations of nicotine reported in those studies were in the range of 10^{-5} to 10^{-2} M. In the present study we utilized striatal slices continuously superfused with [^{3}H]tyrosine and were able to monitor the release of newly synthesized dopamine in 2.5 min fractions of superfusate effluent. This procedure turned out to be very sensitive to the effect of nicotine in that a concentration as low as 1 μM produced a readily detectable and reproducible release of dopamine. We further observed that the release of newly synthesized dopamine by low concentrations of nicotine appears to be due to activation of nicotinic-cholinergic receptors since the releasing effect was significantly reduced by the classical nicotinic antagonists, d-tubocurarine and mecamylamine.

In contrast, much larger concentrations of nicotine (0.1 to 10 mM) were necessary to induce a release of newly taken up [^{3}H]dopamine or endogenous dopamine. The higher concentrations of nicotine also followed a concentration effect relationship but the mechanism of inducing dopamine release was different since the same antagonists (d-tubocurarine or mecamylamine) failed to antagonize the nicotinic effect.

A second major difference between the effect of low and high concentrations of nicotine observed in the present study was the dependency of extracellular calcium. We observed that the effect of the high concentrations of nicotine on dopamine release was independent of extracellular calcium in the incubation buffer while the effect of low concentrations of nicotine required extracellular calcium for the releasing effect to be seen.

A summary of the comparison between low and high concentrations of nicotine observed in the present study is depicted in Table 1, while a schema depicting the two mechanisms is presented in Fig. 13. These data are highly suggestive of the idea that nicotine can release dopamine from striatal slices by at least two mechanisms. The releasing effect of low concentrations appears to require the activation of nicotinic-cholinergic receptors, requires extracellular calcium, and is sensitive to tetrodotoxin, while the effect of high concentrations may be independent of nicotinic-cholinergic receptors, does not require extracellular calcium, and is not sensitive to tetrodotoxin. It is still possible that the effect of high concentrations of nicotine may be due to activation of an atypical nicotinic receptor. This would be consistent with the results of binding studies which suggest that there are nicotinic receptors in

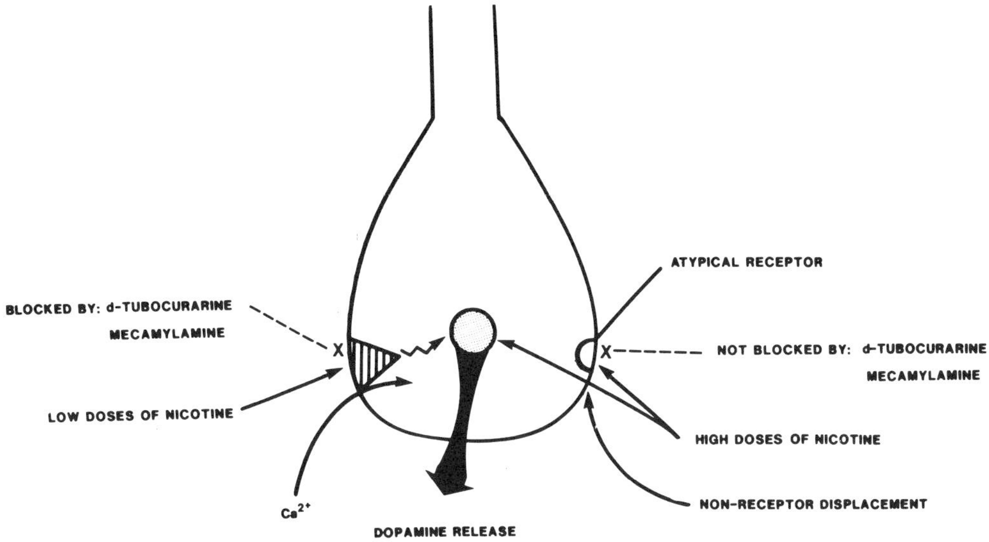

Fig. 13. Depicts a scheme showing the dual mechanism of action of nicotine in releasing dopamine from the rat striatum. Low concentrations of nicotine evoke release following activation of a nicotine receptor. The response can be blocked by classical nicotinic antagonists (d-tubocurarine; mecamylamine) and requires extracellular calcium. High concentratons release dopamine either by a displacement mechanism similar to sympathomimetic amines or following activation of an atypical nicotinic receptor. This response is not blocked or is irregularly blocked by classical nicotinic antagonists and does not require extracellular calcium.

brain that have characteristics quite different from ganglionic or neuromuscular nicotinic cholinergic receptors. This is also consistent with the conclusions of Connelly and Littleton (16) who have observed that both the naturally occurring (-) isomer and the synthetic (+) isomer of nicotine released [^{3}H]dopamine from a crude synaptosomal fraction of striatal membranes. This may also explain why some antagonists appear to attenuate the response while others do not.

Another explanation for the releasing effect of high concentrations of nicotine is that the drug may be displacing dopamine from vesicular storage sites in a manner similar to sympathomimetic amines.

The possible dual mechanism of the central action of nicotine is interesting in view of what has been observed in the peripheral nervous system. Nicotine is well known to release norepinephrine from adrenergic nerve terminals and catecholamines from the adrenal medulla (22,23,24). There has been little agreement as to the importance of action potential generation as compared to local graded depolarization as well as the role of calcium in the release of norepinephrine following activation of nicotinic cholinergic receptors. Some investigators have reported that tetrodotoxin did not prevent the nicotine-induced release of norepinephrine from various tissues (25,26,27,28) while others report that tetrodotoxin does block the release (29,30). The explanation for this discrepancy may also be a dose-related phenomenon. It has been observed that the effect of low doses of nicotine in causing a release of norepinephrine from spleen slices was blocked by tetrodotoxin and was dependent upon extracellular calcium. In contrast, the effect of high doses of nicotine was blocked by tetrodotoxin and persisted in the absence of extracellular calcium (31). The similarities between the effects on norepinephrine release from the spleen and what we now report for dopamine release are certainly of great interest.

Supported by USPHS Grants NINCDS 16215 and NIDA 02668

REFERENCES

1. Armitage, AK and Hall, GH: Further evidence relating to the mode of action of nicotine in the central nervous system. Nature 214:977-979, 1967.
2. Clarke, PBS and Kumar, R: Characterization of the locomotor stimulant action of nicotine on tolerant rats. Br. J. Pharmacol. 80:587-594, 1983.
3. Clarke, PBS, Hommer, DW, Pert, A and Skirboll, LR: Electrophysiological actions of nicotine on substantia nigra single units. Br. J. Pharmacol. 85:827-835, 1985.
4. Giorguieff, MF, LeFloch, ML, Westfall, TC, Glowinski, J and Besson, MJ: Nicotinic effects of acetylcholine on the release of newly synthesized [^{3}H]dopamine in rat striatal slices and cat caudate nucleus. Brain Res. 106:117-131, 1976.
5. Krnjevic, K: Acetylcholine receptors in vertebrate central nervous system. In (eds) Iversen, LL, Iversen, SD and Snyder, SH, Handbook of Psychopharmacology, Vol. 6, New York, 1975, 97-126.
6. Balfour, DJK: The effects of nicotine on brain neurotransmitter systems. Pharmacol. Ther. 16:269-282, 1982.
7. Chesselet, MF: Presynaptic regulation of neurotransmitter release in the brain: facts and hypothesis. Neurosci. 12:347-375, 1984.
8. Westfall, TC: Effect of nicotine and other drugs on the release of [^{3}H]norepinephrine and [^{3}H]dopamine from rat brain slices. Neuropharmacol. 13:693-700, 1974.
9. Arqueros, L, Naquira, D and Zunino, E: Nicotine-induced release of catecholamines from rat hippocampus and striatum. Biochem. Pharmacol. 27:2667-2674, 1978.
10. DeBelleroche, J and Bradford, HF: Biochemical evidence for the presence of presynaptic receptors on dopaminergic nerve terminals. Brain Res. 142:53-68, 1968.
11. Giorguieff, MF, LeFloch, ML, Glowinski, J and Besson, MJ: Involvement of cholinergic presynaptic receptors of nicotinic and muscarinic types in the control of the spontaneous release of dopamine from striatal dopaminergic terminals in the rat. J. Pharmacol. Exp. Ther. 200:535-544, 1977.
12. Giorguieff-Chesselet, MF, Kemel, ML, Wandscheer, D and Glowinski, J: Regulation of dopamine release by presynaptic nicotinic receptors in rat striatal slices: Effect of nicotine in a low concentration. Life Sci. 25:1257-1262, 1979.
13. Sakurai, Y, Takano, Y, Kohjimoto, Y, Honda, K and Kamiya, HO: Enhancement of [^{3}H]dopamine release and its [^{3}H]metabolites in rat striatum by nicotinic drugs. Brain Res. 242:99-106, 1982.
14. Westfall, TC, DiFazio, CA and Saunders, J: Local anesthetic and halothane alterations of the stimulation-induced release of [^{3}H]dopamine from rat striatal slices. Anesthesiol. 48:118-124, 1978.
15. Westfall, TC, Grant, H and Perry, H: Release of dopamine and 5-hydroxytryptamine from rat striatal slices following activation of nicotinic-cholinergic receptors. Gen. Pharmacol. 14:321-325, 1983.

16. Connelly, MS and Littleton, JM: Lack of stereoselectivity in ability of nicotine to release dopamine from rat synaptosomal preparations. J. Neurochem. 41:1297-1302, 1983.

17. Marien, M, Brien, J and Jhamandas, K: Regional release of [^{3}H]dopamine from rat brain in vitro: Effects of opioids on release induced by potassium, nicotine and α-glutamic acid. Can. J. Physiol. Pharmacol. 61:43-60, 1983.

18. Westfall, TC, Besson, MJ, Giorguieff, MF, and Glowinski, J: The role of presynaptic receptors in the release and synthesis of [^{3}H]dopamine by slices of rat striatum. Naunyn-Schmiedeberg's Arch. Pharmacol. 292:279-287, 1976.

19. Nieoullon, A, Cheramy, A and Glowinski, J: An adaptation of the push-pull cannula method to study the in vivo release of [^{3}H]dopamine synthesized from [^{3}H]tyrosine in the cat caudate nucleus: Effects of physical and pharmacological treatments. J. Neurochem. 28:819-828, 1977.

20. Voigt, MM, Wang, RY and Westfall, TC: The effects of cholecystokinin on the in vivo release of newly synthesized [^{3}H]dopamine from the nucleus accumbens of the rat. J. Neurosci. 5:2744-2749, 1985.

21. Takano, Y, Sakurai, Y, Kohjimoto, Y, Honda, K and Kamiya, HO: Presynaptic modulation of the release of dopamine from striatal synaptosomes: Differences in the effects of high K^+ stimulation, methamphetamine and nicotinic drugs. Brain Res. 279:330-334, 1983.

22. Loffelholz, K: Nicotinic drugs and postganglionic sympathetic transmission. Naunyn-Schmiedeberg's Arch Pharmacol. 267:64-73, 1970.

23. Muscholl, E: Cholinomimetic drugs and release of the adrenergic transmitter. In (eds) Schumann, HJ and Kroneberg, G, New Aspects of Storage and Release Mechanisms of Catecholamines, Berlin, 1970, 168-186.

24. Westfall, TC: Local regulation of adrenergic neurotransmission. Physiol. Rev. 57:659-728, 1977.

25. Jayasunder, S and Vohra, MM: An analysis of action of nicotinic agents on adrenergic nerve terminals in rat isolated vas deferens. Arch. Int. Pharmacodyn. 232:192-201, 1978.

26. Muramatsu, I, Fujiwara, M, Osumi, Y and Shibata, S: Vasoconstrictor and dilator actions of nicotine and electrical transmural stimulation on isolated dog cerebral arteries. Blood Vessels 15:110-118, 1978.

27. Su, C and Bevan, JA: Blockade of nicotine-induced norepinephrine released by cocaine, phenoxybenzamine and desipramine. J. Pharmacol. Exp. Ther. 175:533-540, 1970.

28. Westfall, TC and Brasted, M: The mechanism of action of nicotine on adrenergic neurons in the perfused guinea pig heart. J. Pharmacol. Exp. Ther. 182:409-418, 1972.

29. Bell, C: Differential effects of tetrodotoxin on sympathomimetic actions of nicotine and tyramine. Br. J. Pharmacol. 32:96-103, 1968.

30. Furchgott, RF, Steinsland, O and Wakade, TD: Studies on prejunctional muscarinic and nicotinic receptors. (In (eds.) Almegren, A, Carlsson, A and Engel, J, Chemical Tools in Catecholamine Research, Vol. II, North Holland, Amsterdam, 1975, 167-174.

31. Kirpekar, SM, Garcia, AG and Prat, JC: Action of nicotine on sympathetic nerve terminals. J. Pharmacol. Exp. Ther. 213:133-138, 1980.

EFFECTS OF NICOTINE AND EXPOSURE TO CIGARETTE SMOKE ON DISCRETE DOPAMINE AND NORADRENALINE NERVE TERMINAL SYSTEMS OF THE TELENCEPHALON AND DIENCEPHALON OF THE RAT: RELATIONSHIP TO REWARD MECHANISMS AND NEUROENDOCRINE FUNCTIONS AND DISTRIBUTION OF NICOTINIC BINDING SITES IN BRAIN

K. Fuxe, K. Andersson, P. Eneroth[1], A. Härfstrand
A. Nordberg[2] and L.F. Agnati[3]

Dept. of Histology, Karolinska Institutet, Stockholm, Sweden. [1]Research and Development Laboratory, Dept. of Obstetrics & Gynecology, Karolinska Hospital, Stockholm, Sweden. [2]Dept. of Pharmacology Biomedical Center, Univ. of Uppsala, Uppsala, Sweden. [3]Dept. of Human Physiology, Univ. of Modena, Modena, Italy

INTRODUCTION

The anatomy of the brain cholinergic neurons has recently been explored by immunocytochemistry using antibodies against choline acetyltransferase (CAT) (1). Large numbers of cholinergic interneurons were then demonstrated within the nucleus caudatus putamen, the nucleus accumbens, and the tuberculum olfactorium. But in the hypothalamus, only cholinergic nerve terminals were described, while cholinergic nerve cell bodies could not be identified with certainty (see *inter alia* 2, 3). Acetylcholinesterase-positive but not CAT-positive nerve cell bodies have been described within the mediobasal hypothalamus, but the origin of the cholinergic nerve terminals in the hypothalamus is presently unknown.

Both muscarinic and nicotinic cholinergic receptors are linked to the brain cholinergic synapses (4). Although the effect of nicotine has been believed, at least to some degree, to be linked to nicotinic receptors in brain, the number of nicotinic receptors was for a long time considered to be low since the snake venom α-bungarotoxin (α-BTX) was a poor binder to brain tissue. Recently, tubocurarine, nicotine, and acetylcholine ([^{3}H]ACh) have been used as nicotinic ligands in binding studies (for review see 5). These data indicate a high-affinity binding site with a similar B_{max} for all ligands and an additional low-affinity site for nicotine and tubocurarine (6,7). The regional distribution of [^{3}H]nicotine as well as [^{3}H]acetylcholine (in the presence of high concentrations of a muscarinic agonist atropine) binding sites have recently been described using brain membranes or autoradiography (7-14). The studies indicated a particularly high density of nicotinic cholinergic receptors in the cerebral cortex and in the thalamus. A relatively low density of [^{3}H]nicotine binding sites was observed within the hypothalamus. This finding is of special interest since the number of tubocrarine binding sites is high in the hypothalamus (15).

In the present article we will review the evidence of the existence of nicotine/catecholamine (CA) interactions in reward and neuroendocrine mechanisms with special reference to the neostriatum, nucleus accumbens, tuberculum olfactorium and the hypothalamus. The distribution of nicotinic cholinergic receptors in the telencephalon and diencephalon has been analyzed by means of quantitative receptor autoradiography using [^{3}H]nicotine and [^{3}H]acetylcholine as receptor radioligands, to explore an anatomical relationship to CA neurons. Results will be presented which indicate that nicotinic cholinergic receptors can also control neuropeptide systems (especially LHRH neuron systems) in the hypothalamus and the preoptic area.

MATERIALS AND METHODS

In all studies, specific pathogen-free Sprague-Dawley rats have been used. Rats were put in a Walton Horizontal Smoking Machine for exposure to cigarette smoke (16). The cigarettes were of the Kentucky Reference 1R1 type (Tobacco and Health Research Institute, University of Kentucky, Lexington, USA). Glass fiber filters have been used to remove nicotine from the smoke. The types of smoke exposure procedures are displayed in Figs. 1 and 2. In the experiments with intermittent exposure to cigarette smoke, one, two, or four cigarettes were burned with 30 min intervals during 2 h. When rats were continuously exposed to cigarette smoke, one to four cigarettes were burned during 10-40 min, respectively. The efficacy of the exposure to cigarette smoke was evaluated by measurements of serum nicotine and cotinine levels as described by Curvall et al. (17).

CA levels in individual dopamine (DA) and noradrenaline (NA) nerve terminal systems of the tel- and diencephalon were measured by quantitative histofluorimetry using formaldehyde fluorescence histochemistry. This method allows demonstration of the cellular stores of CA's (18). Changes in CA turnover and utilization were evaluated by studying the disappearence of the CA stores after tyrosine hydroxylase inhibition induced by α-methyl-p-DL-tyrosine methylester hydrochloride (αMT) injections (250 mg/kg, i.p., dissolved in saline) as recently described (18).

Serum levels of pituitary hormones and corticosterone were determined by radioimmunoassay procedures using GH, TSH, prolactin, LH, and FSH Kits kindly supplied by the National Hormone and Pituitary Program, the National Institute of Arthritis, Diabetes and Digestive and Kidney Diseases, National Institute of Health (NIADDKD). Both RP1 and RP2 standards were used. Radioimmunoassay of ACTH or β-endorphin was performed using reagents and procedures from Immunonuclear Corp. (Stillwaters, Minnesota, U.S.A.). Vasopressin or LHRH were determined

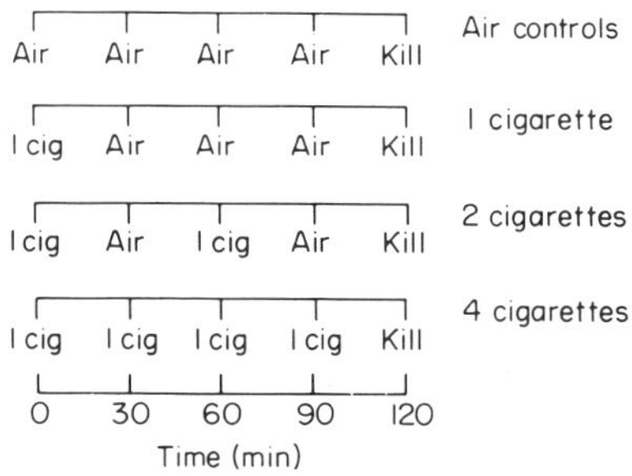

Fig. 1. Treatment schedules in the experiments involving acute intermittent exposure of male rats to cigarette smoke. "One cig" means exposure either to filtered or unfiltered cigarette smoke. The control rats were instead exposed to puffs of air according to the schedule indicated in the fig.

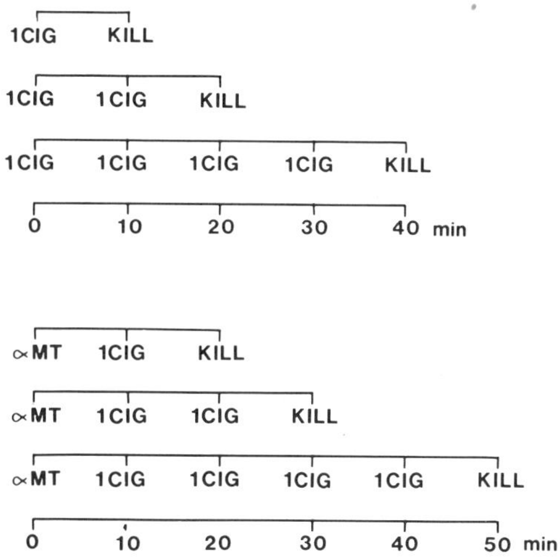

Fig. 2. Treatment schedules in the experiments involving acute continuous exposure of male rat to cigarette smoke. "One cig" means exposure either to filtered or unfiltered cigarette smoke. The control rats were instead exposed to puffs of air according to the schedule indicated in the figure. αMT (250 mg/kg, i.p.) was given 10 min before onset of exposure to air, filtered or unfiltered cigarette smoke.

using reagents and procedures from Buhlmann Laboratories (Basel, Switzerland). Measurements of serum levels of corticosterone were made as outlined by Nilsen et al. (19). For determination of substance P and CCK-like immunoreactivity, see Fuxe et al. (20).

The autoradiography techniques used in the present investigation on the distribution of [^{3}H]nicotine (NEN, U.S.A.) and of nicotinic [^{3}H]acetylcholine (Amersham, U.K.) binding sites in the tel- and diencephalon of the male rat were based on the biochemical binding procedures used to study binding in brain membranes (5,7,10). [^{3}H]nicotine (71.9 Ci/mmol) was dissolved in Tris HCl buffer (0.05 M, pH 8.0) to obtain a final concentration of 20 nM of [^{3}H]nicotine. The incubation time was 10 min at +4°C. These incubation conditions are very important; otherwise, there is a loss in [^{3}H]nicotine binding (10). Nonspecific binding was the binding in the presence of nicotine 10^{-4} M (nicotine hydrogen (+) - tartrate BDH Chemicals Ltd., Poole, England). For details on the [^{3}H]acetylcholine binding procedure, see Table 1. The quantitative autoradiographical analysis was performed using the procedures of Agnati et al., (21), Benfenati et al. (22), and Fuxe et al. (23).

AUTORADIOGRAPHICAL STUDIES ON THE DISTRIBUTION OF [^{3}H]NICOTINE AND NICOTINIC [^{3}H]ACETYLCHOLINE BINDING SITES IN THE TELENCEPHALON AND DIENCEPHALON OF THE MALE RAT

The distribution of [^{3}H]nicotine binding sites in the telencephalon is shown in Fig. 3 at bregma level +1.2 mm. A 100-μM nicotine solution containing [^{3}H]nicotine was used to define unspecific binding. The degree of specific labeling by [^{3}H]nicotine in the neostriatum, nucleus accumbens, and tuberculum olfactorium was of a similar magnitude with a fairly uniform labeling throughout all three regions (Fig. 3). A specific labeling is also observed, especially in layers II-III in the frontoparietal cortex and the anterior cingulate cortex. The labeling in

TABLE 1. Quantitative autoradiography of nicotine [^{3}H]acetylcholine binding sites in the tel- and diencephalon of the male rat. [^{3}H]Acetylcholine (86 Ci/mmol) was used in 50 mM Tris HCl (pH 7.4) buffer containing 1 mM $MgCl_2$, 2 mM $CaCl_2$, and 120 mM NaCl. Preincubation with paraoxon 10^{-4} M and atropine ($1.5 \cdot 10^{-6}$ M) at 4°C for 60 min. Incubation with 100 mM [^{3}H]acetylcholine was performed for 40' at 4°C. Unspecific binding was defined as the binding in the presence of 10^{-4} carbachol. Means ± S.E.M.

BRAIN REGION	BINDING OF ^{3}H-ACETYLCHOLINE (100 nM) (fmol/mg protein)
Bregma +1.2 mm	
ANTERIOR CINGULATE CORTEX	46±7
FRONTOPARIETAL CORTEX MOTOR AREA	30.6
FRONTOPARIETAL CORTEX SOMATOSENSORY AREA	21.9
NUCLEUS CAUDATUS PUTAMEN	31.5
TUBERCULUM OLFACTORIUM	15.8
NUCLEUS ACCUMBENS	26.1
SEPTUM	23.5
Bregma -2.3 mm	**BINDING of ^{3}H-ACETYLCHOLINE (100 nM) (fmol/mg protein)**
POSTERIOR CINGULATE CORTEX	27.8±10.5
FRONTOPARIETAL CORTEX MOTOR AREA	23.0±7.0
FRONTOPARIETAL CORTEX SOMATOSENSORY AREA	20.5±7.5
ENDOPIRIFORM CORTEX	17.1
HIPPOCAMPUS DORSAL PART	13.0
NUCLEUS LATERO-DORSALIS THALAMI	68.5±35.5
NUCLEUS MEDIODORSALIS THALAMI	52.0±19.0
MEDIAL HABENULAR NUCLEUS	57.0±15.0
VENTROMEDIAL AND VENTROLATERAL THALAMIC NUCLEI	48.0±18.0
NUCLEUS GELATINOSUS THALAMI	60.5±26.5
REUNIENS AND RHOMBOID THALAMIC NUCLEI	50.0±19.0
CENTRAL MEDIAL THALAMIC NUCLEUS	42.7
AMYGDALA	11.3
PARA- AND PERIVENTRICULAR HYPOTHALAMIC NUCLEI	16.0±5.0

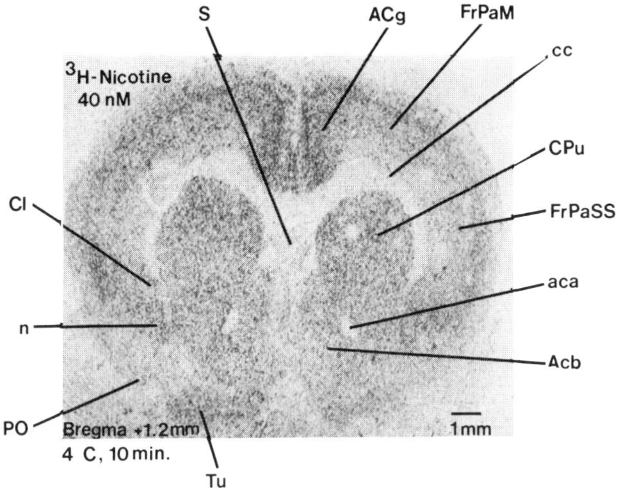

Fig. 3. Autoradiography of [^{3}H]nicotine binding sites in the coronal section (14 μm) of the rat telencephalon at Bregma level +1.2 mm using tritium-sensitive film ([^{3}H]Ultrofilm, LKB, Sweden). The concentration used was 40 nM. The [^{3}H]nicotine (NEN, U.S.A.) had a specific activity of 71.9 Ci/mmol. The experiments were performed during optimal equilibrium conditions using a Tris-HCl buffer (0.05 M, pH = 7.5). Incubation time was 10 min at +4°C. Abbreviations used: ACg = anterior cingulate cortex, FrPaM = frontoparietal cortex, motor area; FrPaSS = frontoparietal cortex somatosensory area; cc = corpus callosum; CPu = caudate putamen; aca = anterior part of the anterior commisure; Acb = accumbens nucleus; Tu = Olfactory tubercle; En = endopiriform nucleus; Cl = claustrum; S = septum. The bar indicates 1000 μm.

the piriform cortex and in the septum is substantially less than in the striatum and the frontoparietal cortex.

In Fig. 4, the distribution of [^{3}H]nicotine binding sites is shown at bregma level -2.3 mm. The autoradiogram demonstrates a very high degree of specific labeling in various thalamic nuclei such as the laterodorsal thalamic nucleus, the mediodorsal thalamic nucleus, the medial habenular nucleus, nucleus gelatinosus thalami, the rhomboid and the reuniens thalamic nuclei, and in the centromedial thalamic nucleus. A moderate degree of specific labeling was also observed in the frontoparietal cortex and in the posterior cingulate cortex, while a low degree of specific labeling is observed in the hippocampus and in the amygdaloid cortex. A uniform low degree of specific labeling of the mediobasal hypothalamus (the arcuate nucleus and the median eminence) and of the remaining areas of the hypothalamus was observed.

The distribution of the nicotinic [^{3}H]acetylcholine binding sites at bregma levels +1.2 mm and -2.3 mm are shown in Fig. 5 and Table 1. The distribution of nicotinic [^{3}H]acetylcholine binding sites mimics closely that of [^{3}H]nicotine binding sites. These findings confirm previous studies (9,13) and strengthen the suggestion that the [^{3}H]nicotine binding sites reveal cholinergic synapses in the brain.

The results indicate that in the thalamus a high affinity nicotinic cholinergic receptor dominates, while within the hypothalamus a low affinity nicotinic cholinergic receptor may play a major role, most of

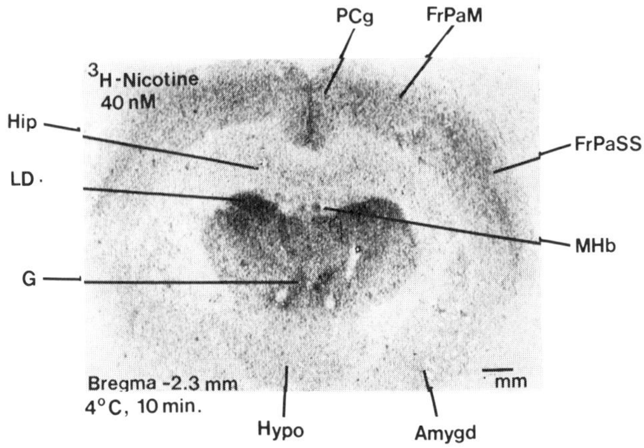

Fig. 4. Autoradiography of [^{3}H]nicotine binding sites in the coronal section (14 μm) of the rat diencephalon at Bregma level -2.3 mm. For experimental details see text to Fig. 3. Abbreviations used: PCg = posterior cingulate cortex; FrPaM = frontoparietal cortex; motor area; FrPaSS = frontoparietal cortex somatosensory area; Hip = hippocampus; LD = laterodorsal thalamic nucleus; MHb = medial habenular nucleus of the thalamus; G = gelatinosus nucleus; H = hypothalamus; Amygd = amygdala. The bar indicates 1000 μm.

which cannot be visualized by autoradiography due to rapid dissociation. It must still be considered that the low density of high affinity hypothalamic [^{3}H]nicotine binding sites can play an important role in the regulation of neuroendocrine function. Furthermore, the high density of [^{3}H]tubocurarine binding sites in the hypothalamus reported by Nordberg et al. (7) may indicate the existence also of nicotinic cholinergic synapses of the neuromuscular type. The [^{3}H]tubocurarine binding sites probably label the low affinity type of nicotinic cholinergic receptors (6,10). Alpha-bungarotoxin binding sites (BTX), which have been related to neuromuscular nicotinic receptors, also show a relatively high number of binding sites in the hypothalamus compared to other brain regions (9,10).

EFFECTS OF ACUTE SINGLE DOSE ADMINISTRATION OF NICOTINE AND EXPOSURE TO CIGARETTE SMOKE ON DISCRETE DA NERVE TERMINAL SYSTEMS OF THE FOREBRAIN

The subcortical forebrain DA systems are divided mainly into the mesostriatal and the mesolimbic systems. The mesolimbic DA nerve terminals consist of both diffuse and dotted types of nerve terminals. The dotted type of DA nerve terminal also stores cholecystokinin (CCK)-like immunoreactivity, while the diffuse type of DA nerve terminal lacks CCK-like immunoreactivity (24).

Effects of Nicotine

In our earlier work it was found that acute single injections (i.p.) of nicotine induced a selective reduction of DA levels and increases in DA utilization in discrete mesostriatal and mesolimbic DA systems (25). Thus, only the DA nerve terminals in the medial caudate and the diffuse types of DA nerve terminals within the nucleus accumbens and tuberculum olfactorium were affected. Nicotine probably partially induces an

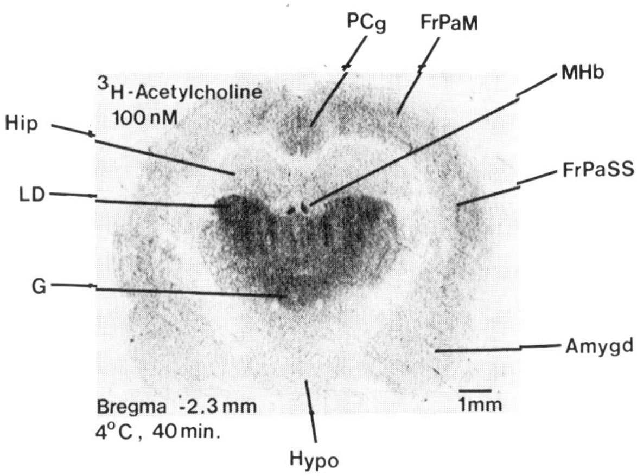

Fig. 5. Autoradiography of [^{3}H]acetylcholine binding sites in the coronal section (14 μm) of the rat diencephalon at Bregma level -2.3 mm using a tritium-sensitive film ([^{3}H]Ultrofilm, LKB, Sweden). The concentration used was 100 nM. The [^{3}H]acetylcholine had a specific activity of 86 Ci/mmol. The experiments were performed during optimal equilibrium conditions using 50 mM Tris buffer (pH = 7.4) containing 120 mM NaCl, 5 mM KCl, 2 mM $CaCl_2$, 1 mM $MgCl_2$. 1.5 μM atropine sulphate was added to the buffer. The sections were preincubated for 10 min with the Tris buffer containing atropine as well as for 10 min in the Tris buffer containing atropine and paraoxan (5 mM) at +4°C. The incubation time was 40 min at +4°C. For abbreviations, see Fig. 4. The bar indicates 1000 μM.

enhanced DA release by an action at cholinergic receptors of the nicotine type located on the DA cell bodies of the ventral tegmental area and of the substantia nigra leading to an increase in the firing rate of the ascending DA systems (26). An activation of nicotinic cholinergic presynaptic receptors located on the DA nerve terminal systems of the forebrain have also been shown to enhance DA release (27). Taken together these results indicate that cholinergic receptors of the nicotine type regulate DA release predominantly in the diffuse types of mesolimbic and in the medial and caudal mesostriatal DA systems. Roberts et al. (28) have correlated the enhanced DA receptor activity in nucleus accumbens with cocaine-induced euphoria and dependence. It therefore seems possible that DA-induced release by nicotine in the anterior nucleus accumbens may also be involved in nicotine-induced euphoria and dependence.

Intraventricular (i.v.t.) injections and intravenous (i.v.) infusion of nicotine via activation of central nicotinic cholinergic receptors were found to reduce forebrain DA stores and to increase forebrain DA utilization in a dose-dependent fashion. However, this action occurred in the vast majority of the striatal, accumbens, and tuberculum olfactorium DA nerve terminal systems (29). The DA levels and utilization in the anteromedial frontal cortex were not affected in these experiments indicating that mesolimbocortical DA neurons were not clearly involved. It is generally believed that reward mechanisms are associated with changes in the mesolimbic DA systems (30). The rewarding actions of nicotine may mainly involve DA systems in the anterior part of the nucleus accumbens (25,29).

Effects of Exposure to Cigarette Smoke

Acute intermittent exposure to cigarette smoke results in dose-dependent increases in serum nicotine and cotinine levels (Fig. 6). Exposure to air or to filtered cigarette smoke does not lead to any changes in the serum nicotine and cotinine levels. The tyrosine inhibitor αMT does not influence cigarette smoke-induced increases in serum nicotine and cotinine levels. Following acute intermittent exposure to cigarette smoke from 4 cigarettes (the highest dose), a small reduction of DA levels was observed only in the DA nerve terminal systems in the posterior part of the nucleus accumbens, where the dotted DA/CCK-positive nerve terminals are found (31). Thus, unlike acute i.p. or i.v.t. injections of nicotine, acute intermittent exposure to cigarette smoke does not affect the DA levels in the mesostriatal and mesolimbic DA systems. But acute intermittent exposure to cigarette smoke induces a dose-related increase in DA utilization in the diffuse types of DA nerve terminals in the nucleus accumbens and tuberculum olfactorium (Fig. 7) (31). All the other DA nerve terminal systems were unaffected. In summary, our data show that nicotine predominantly affects the diffuse types of DA nerve terminals in the subcortical limbic forebrain. It may be speculated that smoking-induced euphoria and dependence is related, at least in part, to enhancement of DA utilization and release in the diffuse and CCK negative DA nerve terminal systems of the nucleus accumbens and tuberculum olfactorium.

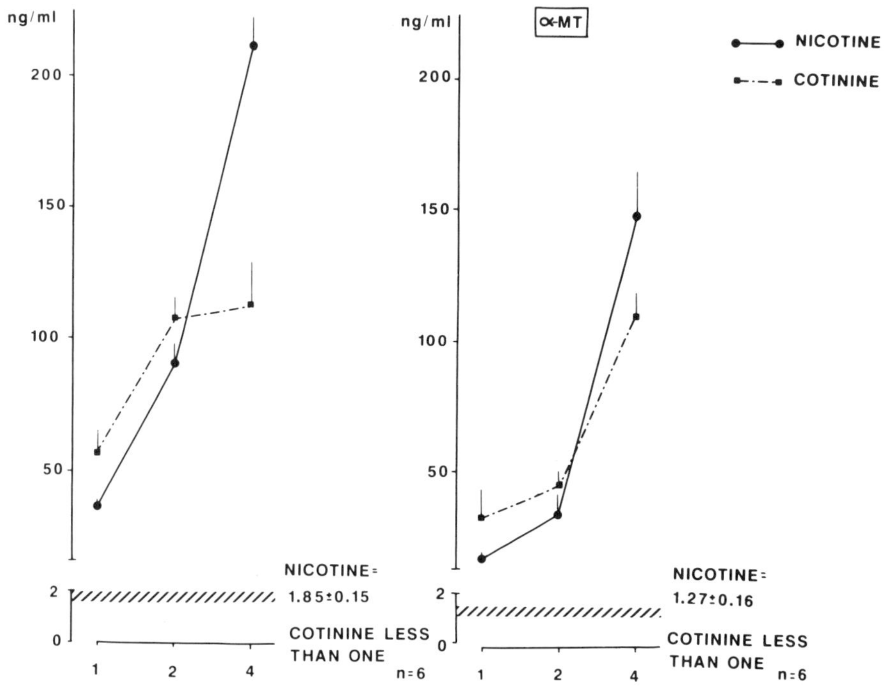

Fig. 6. The effects of acute intermittent exposure to unfiltered cigarette smoke on serum nicotine and cotinine levels of male rats. For exposure procedure see Fig. 1. Means ± s.e.m. are shown after exposure to one, two or four cigarettes. The hatched area indicates the levels in the unexposed animals (means ± s.e.m.). Data are taken from (14).

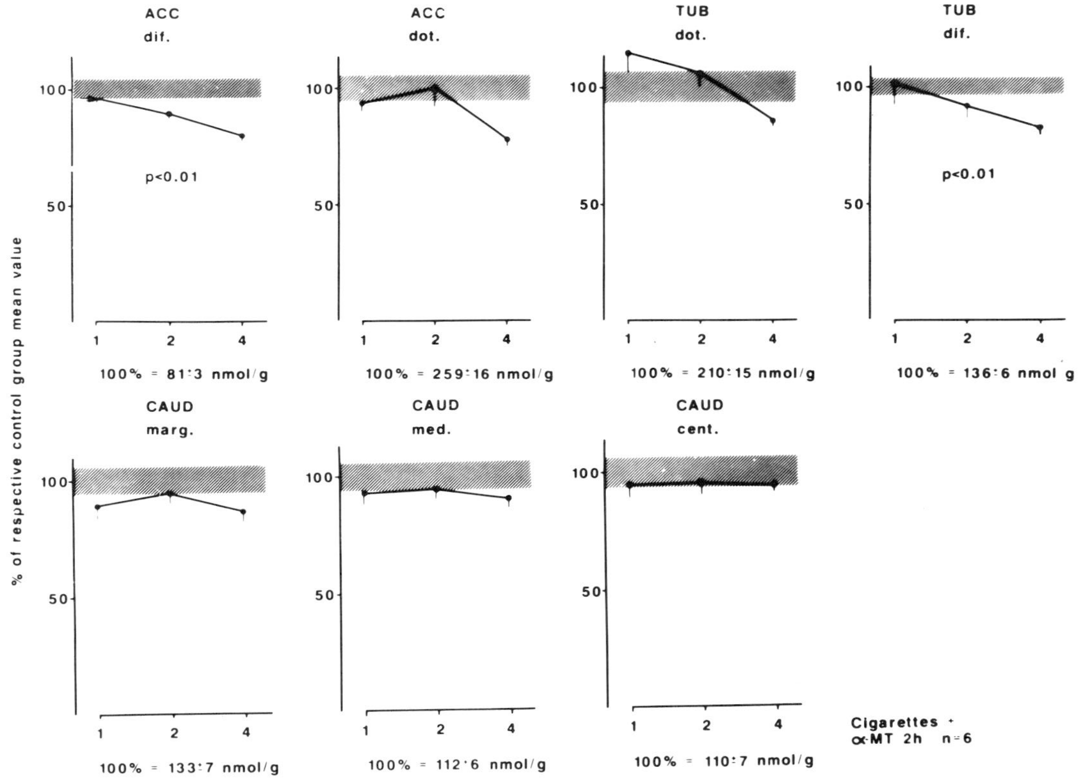

Fig. 7. Effects of acute intermittent exposure to cigarette smoke on the αMT induced DA fluorescence disappearance in various forebrain DA nerve terminal systems of the normal male rat. For exposure procedures see fig. 1. All groups were given αMT (250 mg/kg, i.p.) immediately before the onset of exposure. Means ±s.e.m. are shown in percent of air-+ αMT-treated group mean value. N = 6 rats in all groups except in the "air controls = αMT" group and the "four cig + αMT" group where n = 12. Below the respective 100% value the absolute concentrations of DA are given in nmol/g of tissue wet weight. Statistical analysis according to the Jonckheere-Terpstra test for ordered alternatives.

TABLE 2. Effect of acute intermittent exposure to cigarette smoke using a Walton smoke exposure machine on DA fluorescence in various forebrain DA nerve terminal systems of the normal male rat

	CAUD marg.	CAUD med.	CAUD cent.	ACC diff.	ACC dot.	TUB dot	TUB dif.
Air control	100±3	100±4	100±3	100±4	100±3	100±4	100±3
	(196±12)	(164±6)	(166±10)	(126±6)	(381±24)	(251±13)	(165±9)
1 cigarette	98±10	95±9	106±6	108+8	107±6	118±13	113±10
2 cigarettes	109±4	113±7	110±1	107±6	113±5 *	113±5	105±6
4 cigarettes	102±3	100±4	103±3	97±3	82±4	107±7	98±5

The animals were adjusted to the exposure chambers 15 min/day during 4 days before the experiment.The animals were exposed to air or the smoke of 1,2 or 4 cigarettes (8 puffs/cig) according to the following schedule. 1. Air exposure 4 times during 8 min every 30 min. 2. The smoke of 1 cigarette at the beginning of the experiment and afterwards 3 times air exposure every 30 min. 3. The smoke of 1 cigarette at the beginning of the experiment and after 1 h. At 30 min and 90 min the rats were exposed to air. 4. The smoke of 1 cigarette 4 times every 30 min. Means ± s.e.m. are shown in per cent of air treated group mean value n. = 6 rats in all groups except in the air control group (n = 12) and the 4 cigarette group (n = 11, all areas ecert in ACC dot., when n = 6). Below respective 100 per cent value the absolute concentrations of DA are given in nmol/g of tissue (within parenthesis). Statistical analysis according to Wilcoxon test; one-way classification, comparing all possible pairs of treatments. * = $p<0.05$; CAUD marg. = marginal zone of the caudate nucleus; CAUD med. = medial part of the caudate nucleus; CAUD cent. = central part of the caudate nucleus; ACC dif. = diffuse type of DA fluorescence in the anterior nucleus accumbens; ACC dot. = dotted type of DA fluorescence in the posterior nucleus accumbens; TUB dot. = dotted type of DA fluorescence in the medial posterior tuberculum olfactorium; TUB dif. = diffuse type of DA fluorescence in the lateral posterior tuberculum olfactorium.

EFFECTS OF NICOTINE AND EXPOSURE TO CIGARETTE SMOKE ON NEUROENDOCRINE FUNCTION AND ON HYPOTHALAMIC CATECHOLAMINES AND NEUROPEPTIDES

In the earlier work, high doses of nicotine were used to study the influence on DA, NA, and adrenaline (A) levels and utilization in the hypothalamus (see Fig. 8), (32). Nicotine given twice i.p. in a dose of 2 mg/kg and with a 30 min time interval markedly enhanced the depletion of the NA, A, and DA stores induced by the tyrosine hydroxylase inhibitor αMT. This action of nicotine on hypothalamic CA's was counteracted by pretreatment with mecamylamine in a s.c. dose of 1 mg/kg. These results indicate that nicotinic receptors of the ganglionic type are involved in the regulation of DA, NA, and A utilization and release within the hypothalamus. These receptors are probably located, at least in part, on the CA nerve terminals within the hypothalamus (14). Cholinergic receptors of the nicotine type may, however, also be located on the NA and A nerve cell bodies of the lower brain stem and also on the DA nerve cell bodies of the hypothalamus. We will now report how nicotine and exposure to cigarette smoke can affect discrete hypothalamic DA and NA nerve terminal systems as well as the neuropeptide systems and how these actions relate to neuroendocrine function.

Short-Term Nicotine Actions on Neuroendocrine Function and on Hypothalamic CA's and Neuropeptides

Intravenous injections of nicotine can induce very rapid increases in ACTH, vasopressin, and prolactin serum levels (33). These increases were noted within minutes following the i.v. injection of nicotine (0.3 mg/kg). The changes were associated with a rapid decrease in NA levels in the subependymal layer (SEL) of the median eminence and in the posterior periventricular hypothalamic area (PV II). The DA levels in the medial (MPZ) and lateral (LPZ) palisade zones of the median eminence were unaffected. A preferential increase following i.p. nicotine administration of amine turnover in discrete NA nerve terminal systems was also found to be associated with increases in serum prolactin concentrations (34). In these experiments nicotine increased NA turnover in the SEL and PV II. It is generally accepted that NA exerts a facilitatory action on prolactin release (35). Rat prolactin given i.v.

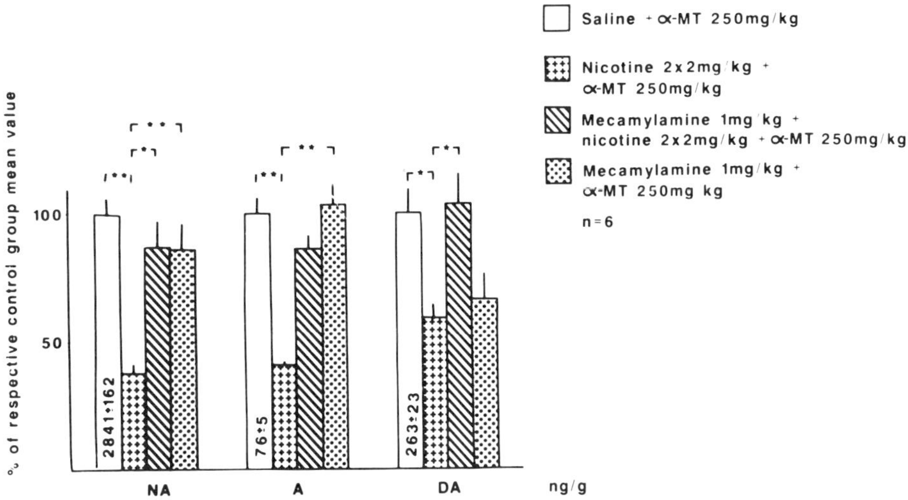

Fig. 8. Reduction by mecamylamine of nicotine-induced increases in the αMT-induced depletion of NA, A, and DA stores in the hypothalamus of the normal male rat. Nicotine (2 mg/kg, s.c.) was given two times with 30 min intervals, the first dose given immediately before αMT (250 mg/kg, i.p., 1 h before decapitation). Mecamylamine (1 mg/kg, s.c.) was given immediately before the first nicotine or saline injection. Means ± s.e.m. are given in percent of the αMT group mean value; n = 6. Statistical analysis according to Wilcoxon: one-way classification comparing all possible pairs of treatments. * = $p<0.05$; ** = $p<0.01$. NA 100% = 2841 ± 162 ng/g tissue; a 100% = 76 ± 5 ng/g of tissue; DA 100% = 263 ± 23 ng/g of tissue. Data taken from Andersson et al. (32).

has been demonstrated to affect the NA nerve terminal system in the PV II (36). The rapid rise in prolactin concentrations following nicotine administration may thus in part be secondary to the enhancement of NA release in the PV II (33).

The increase in ACTH plasma levels following nicotine administration was likely caused by the activation of CRF neurons in view of the probable inhibitory function of the SEL NA nerve terminals in the control of CRF secretion (37). Thus, reduced NA levels were observed following nicotine treatment, and CRF and ACTH administration has been shown to increase NA utilization in this system (32,33,37,38).

Acute i.p. injection of nicotine (2 mg/kg) within 5 min produced marked increases in prolactin, β-endorphin, LH, and corticosterone blood levels, while serum TSH levels were reduced (unpublished data) (Fig. 9). In the same rats it was found that nicotine induced increases in LHRH immunoreactivity in the median eminence and especially in the medial preoptic nucleus, leaving the arcuate nucleus unaffected (Fig. 10). These results are very interesting since they indicate that the rapid rise of LHRH immunoreactivity in the nerve terminals and in the cell body region of the LHRH immunoreactive neurons innervating the median eminence are associated with a rise of LH secretion. Thus, in spite of the fact that LHRH release is increased within minutes, the LHRH levels are themselves increased. These results can be explained by a nicotine-induced activation of the LHRH immunoreactive neurons leading to an enhanced processing of the precursor peptide. By this mechanism, a rapid

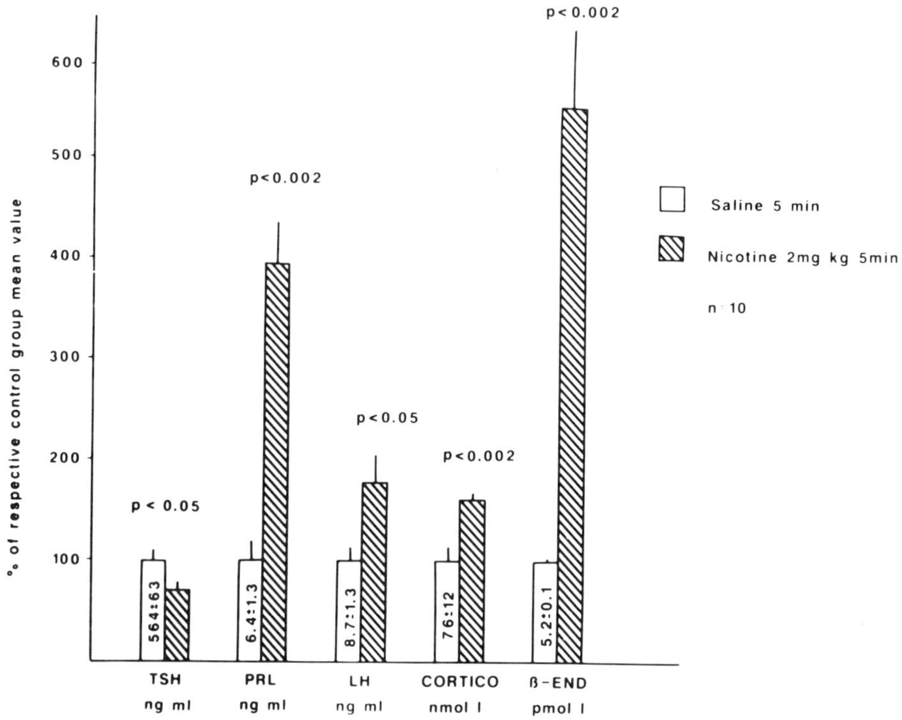

Fig. 9. Effects of an acute i.p. injection of nicotine on the serum levels of TSH, LH, prolactin, corticosterone and β-endorphin. Means ± s.e.m. are shown. Mann-Whitney U-test.

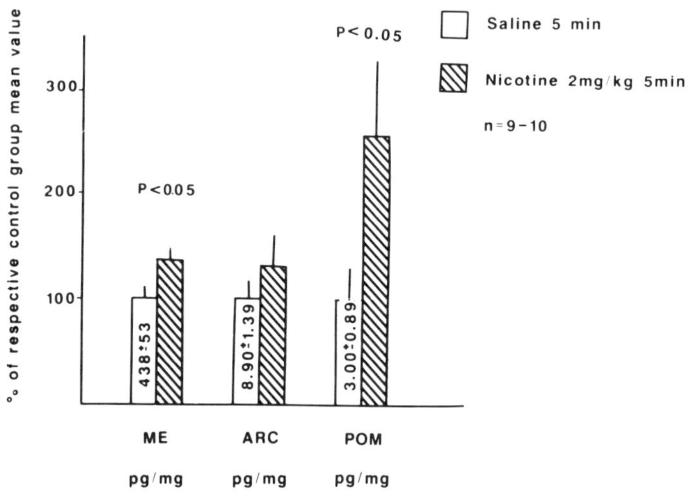

Fig. 10. The effects of acute nicotine treatment on LHRH immunoreactivity in various brain regions of the normal male rat. The animals were given saline or nicotine (2.0 mg/kg), i.p., 5 min before decapitation. Means ± s.e.m. are given in pg/mg wet weight of tissue. N = 9-10. Mann-Whitney U-test. ME = median eminence; ARC = arcuate nucleus; POM = medial preoptic area.

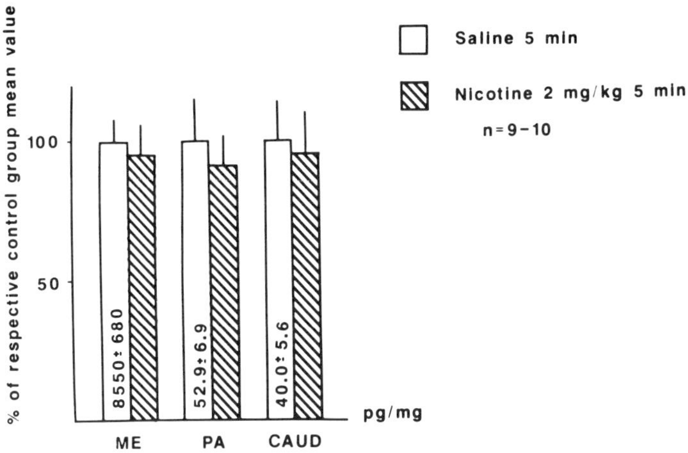

Fig. 11. Effects of acute treatment with nicotine on somatostatin immunoreactivity in various brain regions of the normal male rat. The animals were given saline or nicotine (2.0 mg/kg), i.p., 5 min before decapitation. Means ± s.e.m. are given in pg/mg wet weight of tissue. N = 9-10. Mann-Whitney U-test. ME = median eminence; PA = paraventricular hypothalamic nucleus; CAUD = caudate nucleus.

enhancement of LHRH immunoreactive compounds can be observed despite a concomitant prompt LHRH release. This illustrates the important role of processing of precursor peptides by cleavage enzymes in the control of synaptic function in the peptide neurons. These cleavage enzymes may be activated by increased impulse traffic in the neurons (39). In contrast to LHRH, the somatostatin-like and met-enkephalin-like immunoreactivity in the median eminence, in the paraventricular hypothalamic area and in the nucleus caudatus (see Fig. 11, 12), were unaffected by this high i.p. dose of nicotine, suggesting that neuropeptides are not generally affected by short-term nicotine treatment.

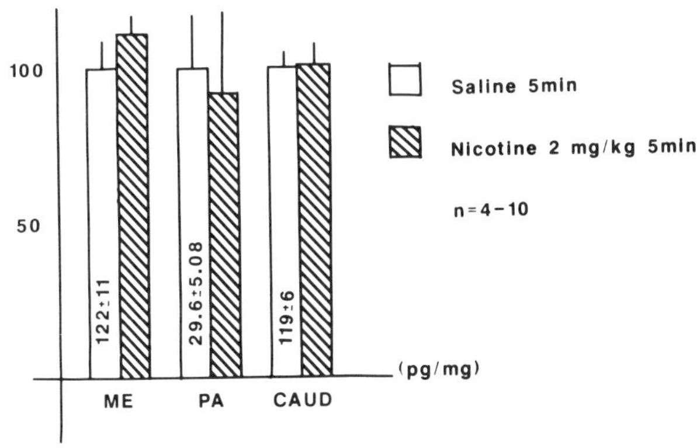

Fig. 12. Effects of acute nicotine treatment on met-enkephalin immunoreactivity in various brain regions of the normal male rat. The animals were given saline or nicotine (2.0 mg/kg), i.p., 5 min before decapitation. Means ± s.e.m. are given in pg/mg wet weight. N = 4-10. Mann-Whitney U-test. ME = median eminence; PA = paraventricular hypothalamic nucleus; CAUD = caudate nucleus.

Effects of Acute Central and Peripheral Administration of Nicotine on Hypothalamic CA's and on Neuroendocrine Function in the Male Rat

One hour after nicotine administration (i.v.t. injection or i.v. infusion), reduced serum levels of TSH, prolactin, and LH were observed (40). In contrast, i.v.t. injections of nicotine increased serum FSH levels. In these experiments increases in DA and NA utilization were observed in the median eminence and in various hypothalamic NA nerve terminal systems. These results suggest that the short-term facilitatory noradrenergic influence on prolactin, LH, and TSH release (see above) has been overridden by the activation of a relatively long-term inhibitory dopaminergic mechanism in the median eminence.

Effects of Acute Repeated Treatment with Nicotine on Neuroendocrine Function and on Various Hypothalamic CA Nerve Terminal Systems in the Normal Male Rat, in the Castrated Male Rat, and in the Castrated Female Rat

A number of studies indicate that the tubero-infundibular DA neurons innervating the MPZ release DA as a prolactin inhibitory factor (PIF), while the tubero-infundibular DA neurons innervating the LPZ are inter

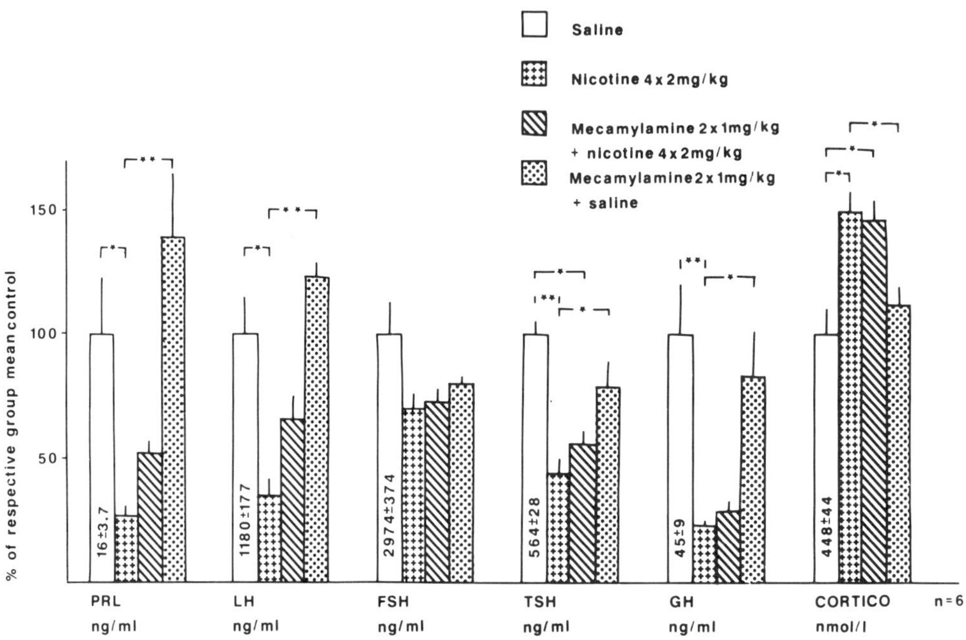

Fig. 13. Effects of mecamylamine on nicotine-induced changes in prolactin, LH, FSH, TSH, GH and corticosterone serum levels of the ovariectomized rat. Means ± s.e.m. in percent of saline-treated groups means. Mean ± s.e.m. in mecamylamine + nicotine- and mecamylamine + saline-treated groups have been adjusted for the effects on the levels of the various hormone produced by mecamylamine alone 1 mg/kg (1 h), then taken in percent of saline-treated group mean value (1). N = 6. The absolute concentrations of the respective serum hormone levels of the saline group are shown in ng/ml or nmol/l in the columns. Statistical analysis according to Wilcoxon test: one-way classification, comparing all possible pairs of treatments. *= p<0.05; **p<0.01. Data from (1).

alia involved in the inhibition of the release of LHRH from the median eminence (36,41,42). Following acute repeated (4 x 2 mg/kg, i.p., 30 min interval) treatment with nicotine in normal male rats and in castrated animals, only inhibitory effects have been observed on the secretion of prolactin, LH and TSH (Fig. 13) (43,44). However, increases in serum corticosterone levels have been obtained (40,45,46). The serum FSH levels have not shown a consistent pattern, but in most instances a small reduction in FSH serum levels has been noted.

The neuroendocrine effects of high repeated doses of nicotine have always been associated with widespread reductions in CA levels and increases in CA turnover in discrete hypothalamic NA nerve terminal systems as well as in the DA nerve terminal systems of the median eminence (MPZ and LPZp see Figs. 14, 15). Both in the normal male rat and the castrated male and female rat the inhibitory effects of nicotine on LH and prolactin secretion may be mediated in part via activation of the tubero-infundibular DA neurons innervating the LPZ and MPZ, respectively. Many of the effects of nicotine on hypothalamic CA's and on neuroendocrine function were found to be, in part, counteracted by pretreatment with the ganglion blocking agent mecamylamine, suggesting an involvement of nicotinic cholinergic receptors of the ganglionic type in the regulation of hypothalamic DA and NA nerve terminal systems and in the control of the secretion of pituitary hormones (see Figs. 14,15). In the experiments with mecamylamine in the ovariectomized rat (46), some new, interesting observations were made. Mecamylamine could partly counteract the inhibitory effects of nicotine on prolactin, LH, and TSH serum concentrations, but neither the nicotine-induced reduction in growth hormone serum levels nor the nicotine-induced increase in corticosterone serum levels could be counteracted (46). These results open up the possibility that different types of nicotinic cholinergic receptors may exist within the hypothalamus in the control of neuroendocrine function. It may be speculated that the nicotinic cholinergic receptors controlling CRF neuronal activity and GH secretion may be controlled predominantly by nicotinic cholinergic receptors of the neuromuscular type. The nicotine-induced reduction in NA levels and increases in NA turnover in peri- and paraventricular hypothalamic NA nerve terminal systems were counteracted to various degrees by mecamylamine pretreatment. One conclusion that can be drawn is that there exist different types of nicotinic cholinergic receptors involved in the regulation of DA and NA release within the hypothalamus and the median eminence (Fig. 15) (46). In addition, these experiments clearly suggest that inhibitory systems other than the tuberoinfundibular DA neurons in the MPZ and the LPZ must be involved in mediating the inhibitory effects of nicotine on prolactin, LH, and TSH secretion. Thus, the experiments show that the median eminence CA's were unaffected by mecamylamine pretreatment, while the inhibitory effects of nicotine on prolactin, LH, and TSH secretion were partly counteracted by mecamylamine.

Furthermore, in comparing the effects of nicotine on CA levels in the castrated male rat with the corresponding effects in the ovariectomized rat, an interesting sex difference was discovered within the hypothalamus and especially in the MPZ (see Figs. 14, 15). Repeated high doses of nicotine induced a larger depletion of NA and DA levels in many hypothalamic CA nerve terminal systems in the castrated male rat compared with the ovariectomized rat. Thus, it seems as if the compensatory activation of CA synthesis to restore CA levels in response to nicotine-induced CA release is considerably less pronounced in the castrated male rat than in the castrated female rat (45,46). Another interesting difference is the failure of nicotine in repeated high doses to significantly reduce LH secretion in the castrated male rat (45). The latter finding does not seem to be related to a differential sensitivity

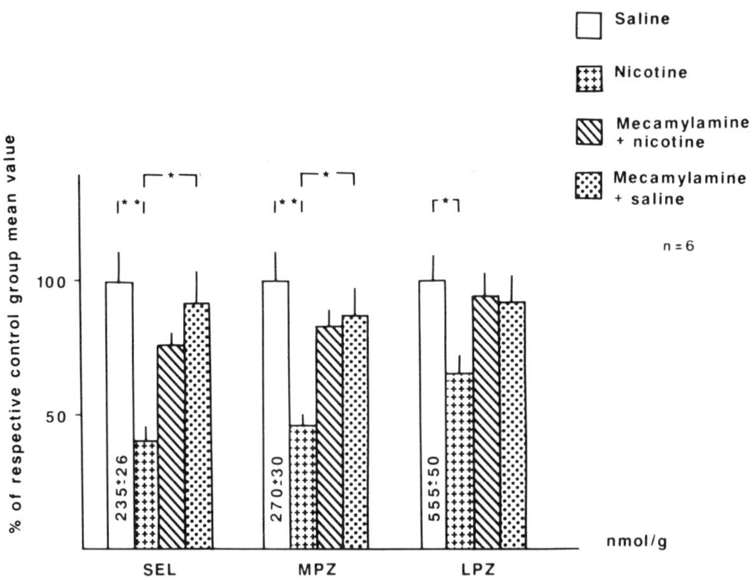

Fig. 14. Effects of mecamylamine on nicotine-induced changes in DA and NA fluorescence in various median eminence regions of the castrated male rat. Castration was performed three months before killing. Nicotine (2 mg/kg, s.c.) was given 4 times with 30 min intervals, the animals being killed 30 min after the last injection. Mecamylamine (5 mg/kg, s.c.) was given twice, 1 h before the first injection of nicotine or saline and 1 h before killing, immediately before the third injection of nicotine or saline. Means ± s.e.m. in percent of saline-treated group means. Means ± s.e.m. in mecamylamine + nicotine- and mecamylamine + saline-treated groups have been adjusted for the effects on CA levels produced by mecamylamine alone 5 mg/kg, 1 h, and then taken in percent of saline-treated group mean value (see Andersson et al. 1982). N = 6. The absolute concentrations for the control group are given in nmol/g of tissue wet weight in the columns. Statistical analysis according to Wilcoxon test: one-way classification comparing all possible pairs of treatments. * = $p<0.05$; ** = $p<0.01$. Abbreviations used: SEL = subependymal layer of the median eminence; MPZ = medial palisade zone of the median eminence; LPZ = lateral palisade zone of the median eminence. Data from (10).

to nicotine treatment of the inhibitory LPZ dopaminergic mechanism regulating LHRH secretion. A marked lowering of LH secretion had been observed after this type of nicotine treatment in the castrated female rat previously (44).

Studies on the Ability of a D1 Receptor Antagonist to Counteract the Inhibitory Effects of Nicotine on Neuroendocrine Function

During recent years we have obtained evidence that the DA receptor in the median eminence which inhibits the secretion of LHRH may be of the D1 type (47). Thus, autoradiography experiments with the selective D1 receptor antagonist [^{3}H]spiperone could not label the median eminence, while the unselective DA antagonist [^{3}H]flupenthixol resulted in labeling of this area (47). We have therefore studied the possible modulation of the nicotine-induced changes in neuroendocrine function by

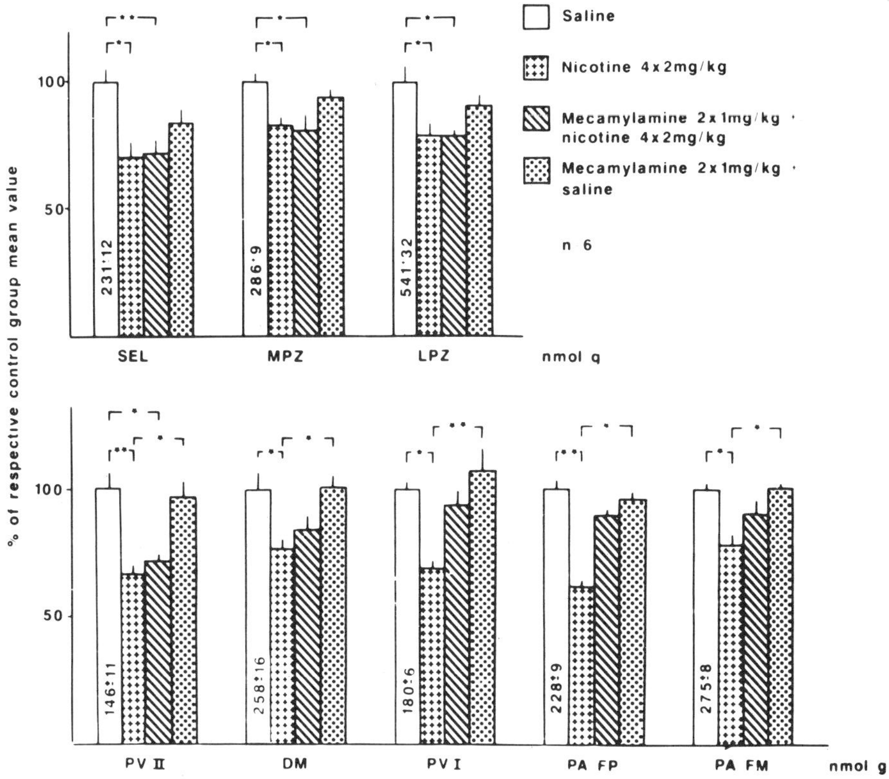

Fig. 15. Effects of mecamylamine on the nicotine-induced changes in DA and NA fluorescence in various hypothalamic regions of the ovariectomized rat. Ovariectomy was performed one month before killing. Nicotine (2 mg/kg, s.c.) was given 4 times with 30 min intervals, the animals being killed 30 min after the last injection. Mecamylamine (1 mg/kg, s.c.) was given twice, 1 h before the first injection of nicotine or saline and 1 h before killing, immediately before the third injection of nicotine or saline. Means ± s.e.m. in percent of saline-treated group means. Mecamylamine + nicotine- and mecamylamine + saline-treated groups have been adjusted for the effects on CA levels produced by mecamylamine alone 1 mg/kg, (1 h) and then taken in percent of saline-treated group mean value (11). N = 6. The absolute concentrations are presented as nmol/g of tissue wet weight in the saline group and shown in the open columns. Statistical analysis according to Wilcoxon test: one-way classification comparing all possible pairs of treatments. * = $p<0.05$; ** = $p<0.01$. SEL = subependymal layer of the median eminence; MPZ = medial palisade zone of the median eminence; LPZ = lateral palisade zone of the median eminence; PV II = posterior periventricular hypothalamic region; DM = dorsomedial hypothalamic nucleus; PV I = anterior periventricular hypothalamic region; PA FP = parvocellular part of the paraventricular hypothalamic nucleus; PA FM = magnocellular part of the paraventricular hypothalamic nucleus.

pretreatment with the D1 receptor antagonist SCH 23390 (48). As seen in Fig. 16, the D1 receptor antagonist could inhibit the nicotine-induced reduction of LH serum levels (49). In contrast, the inhibitory effects of nicotine on serum prolactin and TSH concentrations were not at all antagonized. Likewise, the D1 antagonist did not block the nicotine-induced enhancement in corticosterone serum levels. The counteraction by SCH 23390 of the lowering of LH serum levels by nicotine was dose related (Fig. 16). Obviously the D1 receptor is involved in

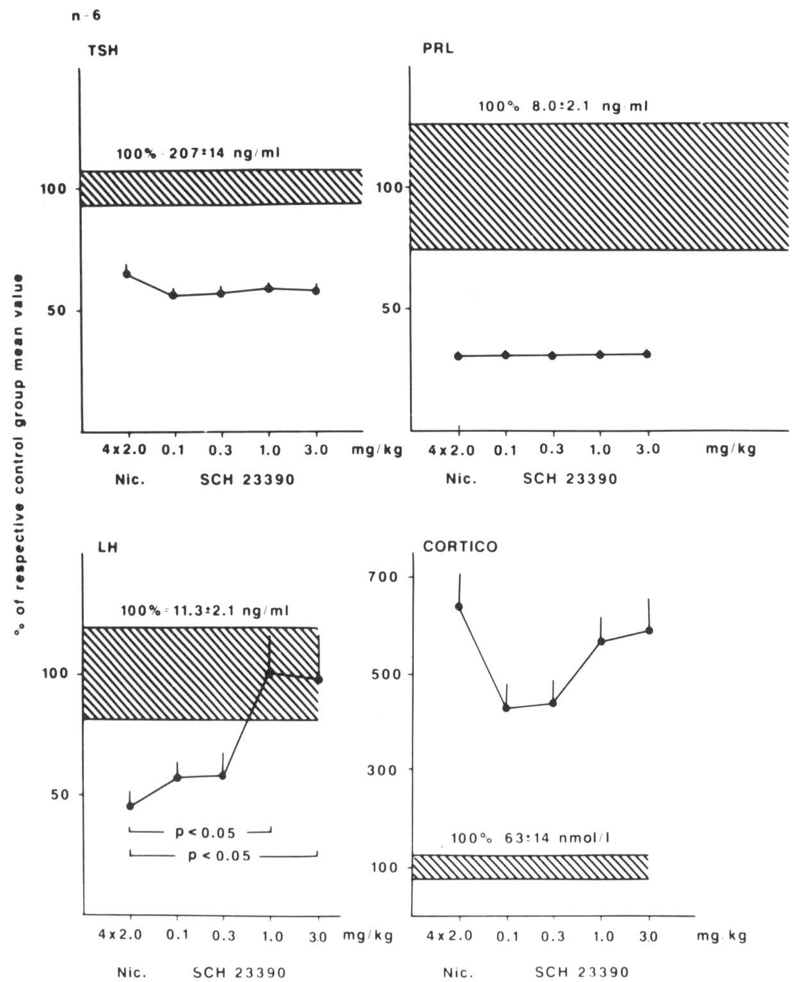

Fig. 16. Dose-effects of the D1-antagonist SCH 23390 (gift from Scheering AG, Berlin, W.G.) on nicotine-induced changes in serum levels of TSH, prolactin, LH, and corticosterone in the normal male rat. Saline or nicotine (2 mg/kg, i.p.) was given 4 times with 30 min intervals, the animals being decapitated 30 min after the last injection. Saline or various doses of SCH 23390 (0.1-3.0 mg.kg, i.p.) was given 5 min before the onset of the first repetitive saline or nicotine injection. Means ± s.e.m. are given in % of control-treated group mean values (hatched area represents the means ± s.e.m. of the control group). N = 6. Statistical analysis according to the Wilcoxon test; one-way classification, comparing all possible pairs of treatments.

mediating the inhibitory effects of nicotine on LH secretion. It has recently been shown that LHRH given i.v. to the hypophysectomized male rat can increase the DA utilization in the LPZ (41). Taken together the results support the hypothesis that the tubero-infundibular DA neurons in the LPZ are involved in inhibiting LH secretion and that the D1 receptor is linked to this system.

The lack of effect of the D1 receptor antagonist in counteracting the inhibitory effects of nicotine on prolactin secretion was expected, since the DA receptor on the prolactin-containing pituitary gland cell is of the D2 type. Thus, the results are compatible with the hypothesis that nicotine induces a reduction of prolactin secretion, mainly by releasing DA from the medial palisade zone (MPZ) into the portal vessels (36,42).

The absence of effect of SCH 23390 on the nicotine-induced reduction in serum TSH levels have been interpreted to mean that the D1 receptors do not participate in the regulation of TSH secretion.

Effects of Acute Intermittent Exposure to Cigarette Smoke on Neuroendocrine Function and on Hypothalamic CA's

Like the results obtained with repeated injections of nicotine (see above), acute intermittent exposure to cigarette smoke resulted in a marked lowering of serum LH, prolactin and TSH levels as well as increases in serum corticosterone levels (16). Filtered cigarette smoke had no such effect. Furthermore, as seen in Fig. 17, mecamylamine pretreatment abolished the inhibitory effects of the exposure to cigarette smoke on TSH, LH and prolactin secretion (50). These results show that cigarette smoke acts via its nicotine content. It is of substantial interest that mecamylamine counteracted only in part the increase in corticosterone serum levels produced by acute intermittent exposure to cigarette smoke. These results are therefore in line with the view that the nicotinic cholinergic receptor controlling the secretion of ACTH is different from those inhibiting the TSH, LH, and prolactin secretion. It is possible that this type of receptor is related to the neuromuscular type of nicotinic cholinergic receptor. This hypothesis is currently being tested by i.v.t. administration of tubocurarine.

Mecamylamine prevented the ability of acute intermittent exposure to cigarette smoke to increase CA turnover in various hypothalamic DA and NA nerve terminal systems (see Fig. 18). The results reinforce the finding that the nicotine content of cigarette smoke mediates the actions on the CA systems. Furthermore, these results indicate that the nicotinic cholinergic receptors involved are of the ganglionic type.

Effects of Acute Continuous Exposure to Cigarette Smoke on Neuroendocrine Function and Hypothalamic CA Nerve Terminal Systems of the Male Rat

In these experiments, one, two, or four cigarettes were burned without any time interval in between the cigarettes (Fig. 2), (51). The exposure time for each cigarette was 10 min. Control animals were exposed to air or to filtered cigarette smoke. It must be emphasized that in order to avoid stress, the rats were adjusted to the exposure chambers for a total of 3 days prior to the experiment. Efficacy of the exposure to cigarette smoke was tested by measurements of serum nicotine levels. The nicotine levels obtained were of similar magnitude as those obtained following acute intermittent exposure (see Fig. 6).

In the neuroendocrine experiments, we were particularly interested in analyzing whether a 10 min continuous exposure to cigarette smoke might lead to increases in prolactin and LH serum levels, since acute injection

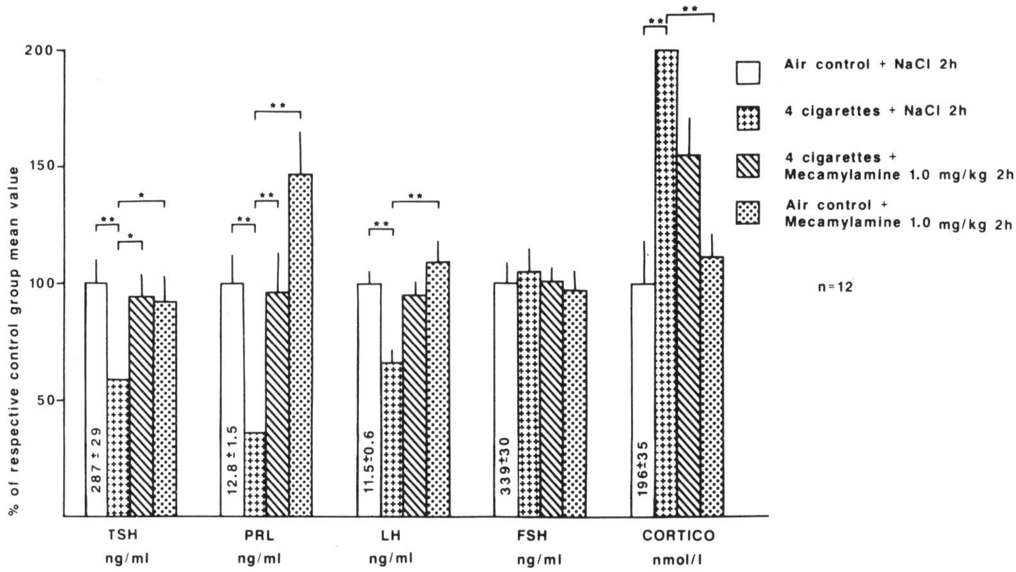

Fig. 17. Effects of mecamylamine and acute intermittent exposure to cigarette smoke on serum levels of TSH, prolactin, LH, FSH, and corticosterone in the normal male rat. For exposure procedure see Fig. 1. Means ± s.e.m. are shown in percent of the air-treated group mean value. N = 12 in all groups. The serum levels of respective hormones are given in the open columns representing the saline-treated group. Statistical analysis according to the Wilcoxon test: one-way classification, comparing all possible pairs of treatments. * = $p<0.05$; ** = $p<0.01$.

of nicotine gave these effects within 2 to 5 min. Unlike acute intermittent exposure, acute continuous exposure to cigarette smoke produced no significant changes in serum LH levels (Fig. 19). But a trend for enhanced LH serum levels was observed after continuous exposure to smoke from one cigarette. It is interesting to note that during the 10 min continuous exposure to the smoke from 1 cigarette, no significant increase was observed in serum LH or prolactin levels, whereas following acute injection of nicotine, such effects were obtained within 2 to 5 min. However, a trend for enhanced LH serum concentrations was observed, but decreased serum prolactin levels were found. In an overall analysis of the effects following acute continuous exposure to smoke from one to four cigarettes, a reduction of TSH and prolactin serum levels was noted, as well as an enhancement of ACTH plasma concentrations (Fig. 19). The effects of acute continuous exposure to cigarette smoke on the CA levels in the various hypothalamic DA and NA nerve terminal systems are shown in Fig. 20. A depletion of CA's was observed in all the CA systems in the hypothalamus. The results from the CA utilization experiments are shown in Fig. 21 and the rate constants are reported in Table 3. A significant enhancement of CA utilization following continuous exposure to cigarette smoke was observed only within the hypothalamic NA nerve terminal systems, while no changes were induced in DA utilization in the MPZ and LPZ. In contrast, acute intermittent exposure to cigarette smoke resulted in marked increases in DA utilization in the median eminence (16). One possible explanation for these findings is that upon acute continuous exposure to cigarette smoke, a depolarization-induced blockade may develop in the tubero-infundibular DA neurons during certain time periods after

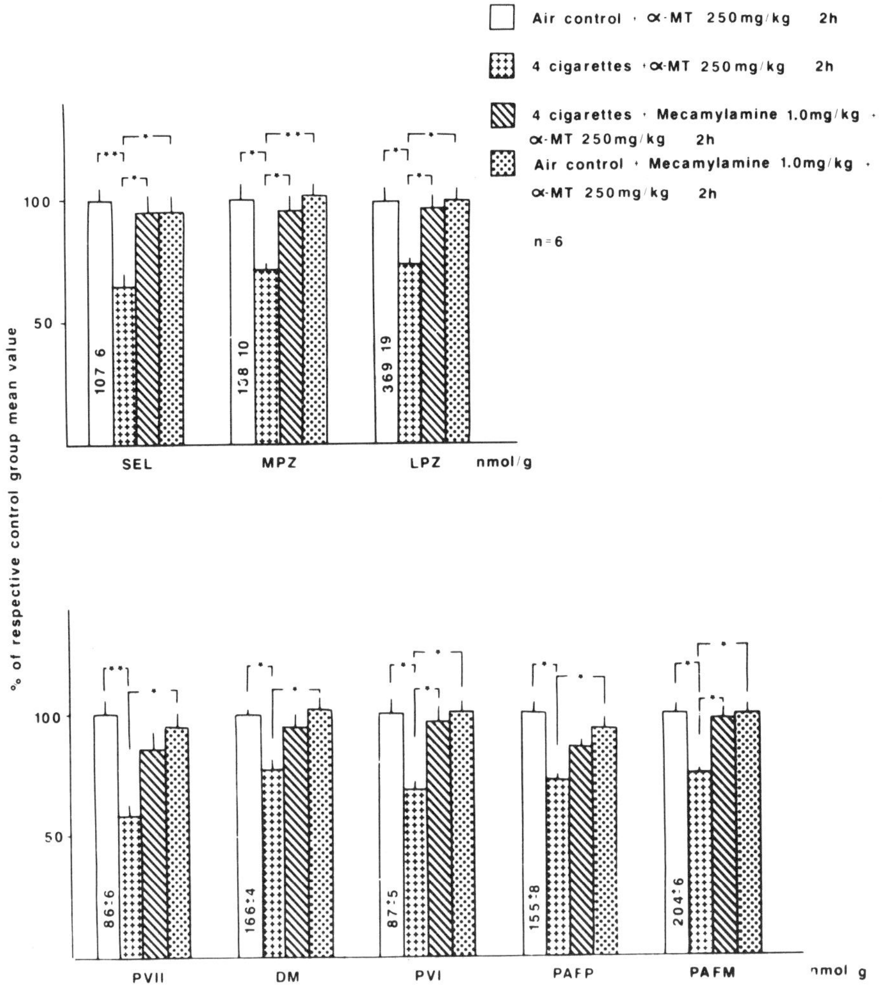

Fig. 18. Effects of mecamylamine and acute intermittent exposure to cigarette smoke on DA and NA fluorescence in discrete hypothalamic regions of the normal male rat. For exposure procedure see Fig. 1. Means ± s.e.m. are shown in percent of air-treated group mean value. N = 6 in all groups. The absolute concentrations of CA in the respective areas are given in nmol/g of tissue wet weight in the open columns representing the saline control group. Statistical analysis according to the Wilcoxon test: one-way classification comparing all possible pairs of treatments. * = p<0.05; ** = p<0.01. Abbreviations used, see text to Fig. 15.

continuous exposure to smoke from two and especially four cigarettes. Also, a small functional pool of DA may become depleted. When correlating the results obtained on serum hormone levels and on median eminence CA levels, it must be remembered that there is a time lag between hypothalamic CA events and establishment of altered serum hormone levels. Such factors may also be involved in the failure of acute continuous exposure to cigarette smoke to produce a reduction in LH serum levels. However, a depletion of DA stores in the MPZ and LPZ (probably reflecting DA release) does take place, which may be sufficient to contribute to the observed lowering of TSH and prolactin secretion.

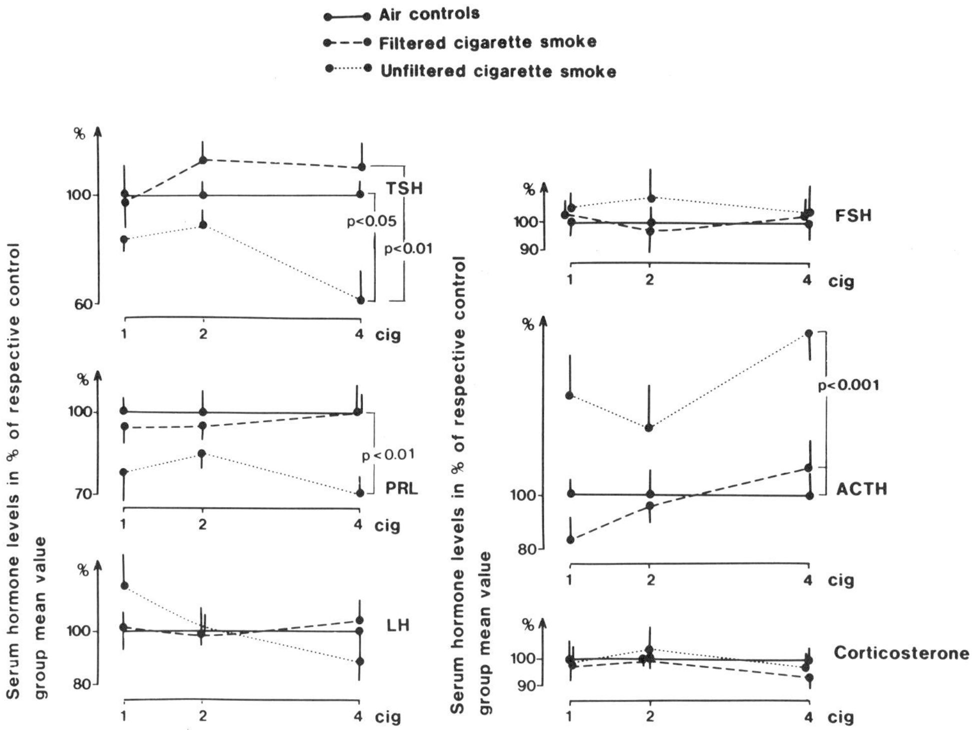

Fig. 19. Effects of acute continuous exposure to cigarette smoke on TSH, prolactin, LH, FSH, ACTH, and corticosterone serum levels in the normal male rat. For exposure procedures see Fig. 2. Means ± s.e.m. are shown in percent of respective air-exposed control group mean values. N = 9. For statistical analysis a parametrical ANOVA testing procedure was used. The absolute concentrations (100%) of respective hormone in serum were as follows: Air 10 minutes (one cigarette): TSH = 294±32 ng/ml, prolactin = 70±9 ng/ml; LH = 21±1.3 ng/ml; FSH = 371±26 ng/ml; ACTH = 155±18 pg/ml; corticosterone = 403±28 nmol/l. Air 20 minutes (two cigarettes): TSH = 316±35 ng/ml; prolactin = 48±4.8 ng/ml; LH = 11±1.8 ng/ml; FSH = 336±17 ng/ml; ACTH = 130±9 pg/ml; corticosterone = 456±23 nmol/l. Air 40 minutes (four cigarettes): TSH = 354±25 ng/ml; prolactin = 65±9.1 ng/ml; LH = 8.3±0.8 ng/ml; FSH = 394±24 ng/ml; ACTH = 139±12 pg/ml; corticosterone = 474±33 nmol/l.

Effects of Acute Continuous Exposure to Cigarette Smoke on Neuropeptide Neurons in the Hypothalamus and in the Medial Preoptic Area

By means of radioimmunoassay procedures, CCK and substance P-like immunoreactivity have been measured within the median eminence, the arcuate nucleus, the paraventricular hypothalamic nucleus, and the medial preoptic area (20). Acute continuous exposure to cigarette smoke (two cigarettes) produced increases in CCK-like immunoreactivity in the paraventricular hypothalamic region, as well as reductions in CCK-like immunoreactivity in the median eminence. This treatment also produced increases in the concentrations of CCK and substance P-like immunoreactivity in the medial preoptic area. The results suggest that CCK and substance P immunoreactive neuron systems are controlled by nicotinic cholinergic receptors. It seems possible that the influence of nicotine and exposure to cigarette smoke on LH secretion and sexual

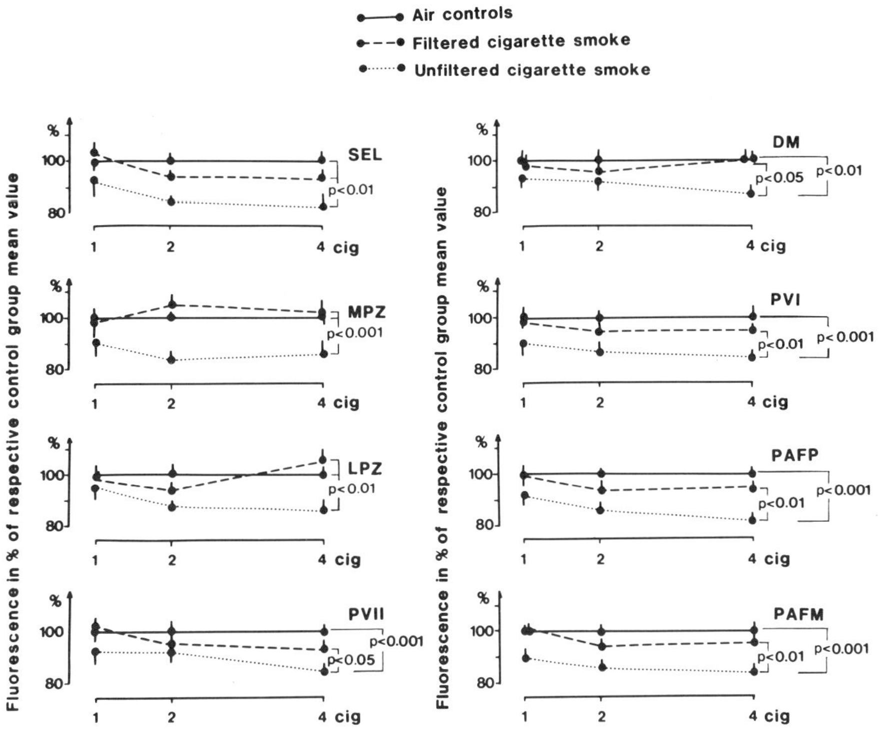

Fig. 20. Effects of acute continuous exposure to cigarette smoke on CA fluorescence in various hypothalamic CA nerve terminal regions in the normal male rat. For exposure procedures see Fig. 2. Means ± s.e.m. are shown in percent of respective air-exposed group mean values. N = 9 in all groups. For statistical analysis a parametrical ANOVA testing procedure was used. For abbreviations see text to Fig. 15. The absolute concentrations of catecholamines expressed in nmol/g of tissue wet weight representing 100% in the respective regions were as follows: Air 10 minutes (one cigarette), SEL = 158±6, MPZ = 218±7, LPZ = 605±24, PV II = 109±4, DM = 251±10, PV I = 136±5, PA FP = 213±6, PA FM = 289±9. Air 20 minutes (two cigarettes); SEL = 152±6; MPZ = 214±6; LPZ = 602±24; PV II = 111±4; DM = 243±7; PV I = 127±4; PA FP = 204±6; PA FM = 275±8. Air 40 minutes (four cigarettes); SEL = 158±6; MPZ = 212±6; LPZ = 569±17; PV II = 113±2; DM = 243±7; PV I = 130±5; PA FP = 205±4; PA FM = 275±8.

behavior (52) may to some extent involve substance P- and CCK-immunoreactive neurons in the medial preoptic area in view of the present results (53,54).

The ability of nicotine to produce a rapid dose-dependent reduction of CCK levels in the median eminence may suggest an enhanced CCK release, and an involvement of CCK mechanisms in cigarette smoke-induced changes in anterior pituitary hormone secretion.

Oxytocin and CCK-like immunoreactivity are costored in a substantial number of magnocellular neurons belonging to the hypothalamic paraventricular nucleus (55,56). The ability of acute continuous exposure to cigarette smoke to increase CCK-like immunoreactivity in this area may reflect an increase in the synthesis of CCK-like peptides in these neurons.

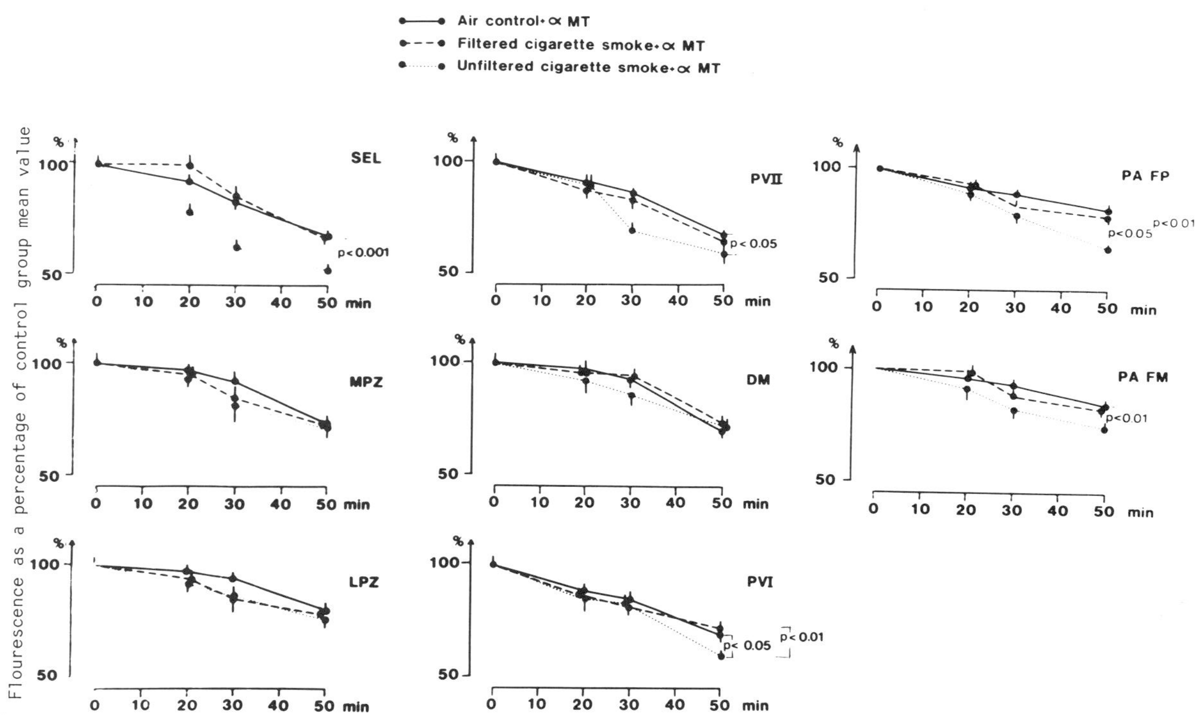

Fig. 21. Effects of acute continuous exposure to cigarette smoke on the αMT-induced amine fluorescence disappearance in discrete hypothalamic CA nerve terminal regions in the normal male rat. For exposure procedure see Fig. 2. In addition, the rats were given αMT (250 mg/kg, i.p.) 10 min before onset of exposure. Means ± s.e.m. are shown in % of air-treated group mean values. N = 9 in all groups except for unfiltered cigarette smoke-exposure groups where N = 6. For statistical analysis a parametical ANOVA testing procedure was used. The absolute concentrations of catecholamines expressed in nmol/g of tissue wet weight representing 100% in the air control + αMT-treated group in the respective regions were as follows: SEL = 147±7; MPZ = 212±6; LPZ = 589±18; PV II = 103±4; DM = 244±10; PV I = 120±5; PA FP = 193±6; PA FM = 278±8. For abbreviations see text to Fig. 15.

TABLE 3. Pool-size, half-life, rate constant, and turnover rate in discrete hypothalamic CA nerve terminal networks of the male rat following exposure to air or to continuous cigarette smoke with filter or without filter. For exposure procedure see Fig. 2. For abbreviations see text to Fig. 15.

	Pool size nmol/g w.wt means±s.e.m.	T1/2 min	Rate constant min^{-1} ±95% confidence intervall	Turnover rate nmol/g w.wt/min
Air control				
SEL	158±4	90	-0.0078±0.0023	1.232
MPZ	218±7	118	-0.0058±0.0023	1.264
LPZ	605±24	153	-0.0046±0.0021	2.783
Continuous cigarette smoke with filter				
SEL		81	-0.0085±0.0027	1.343
MPZ		110	-0.0062±0.0025	1.352
LPZ		137	-0.0051±0.0026	3.086
Continuous cigarette smoke without filter				
SEL		52	-0.0131±0.0025	2.070
MPZ		99	-0.0069±0.0032	1.504
LPZ		120	-0.0058±0.0026	3.509

There also exist parvocellular CCK immunoreactive nerve cells in the paraventricular hypothalamic nucleus projecting into the median eminence, probably representing CRF neurons (55,57). Consequently, the nicotine-induced reduction of CCK stores in the median eminence may represent release of CCK from CCK/CRF nerve terminals belonging to the parvocellular CCK/CRF immunoreactive neurons of the paraventricular hypothalamic nucleus. Thus, the increased CCK levels in this region following acute continuous exposure to cigarette smoke may also reflect a nicotine-induced increase in CCK synthesis in parvocellular paraventricular cells. Thus, the ability of nicotine to induce release of ACTH, may be related to an activation of CCK/CRF neurons. The release of CCK peptides concomitant with CRF may serve, for example, to modulate the action of CRF on ACTH secretion (55). The exact neuroendocrine role of CCK remains to be demonstrated, but it has been reported that CCK-8 given i.v.t. can modulate the secretion of LH, prolactin, growth hormone, and TSH (54,58,59,60).

When discussing the functional role of increased CCK levels in the paraventricular hypothalamic nucleus upon exposure to cigarette smoke, it must also be considered that the CCK-like immunoreactivity is also located in afferents or local circuit neurons. Injections of CCK into the paraventricular hypothalamic nucleus have been reported to reduce food intake in rats (61,62). It must therefore be considered that the nicotine-induced increase in CCK-like immunoreactivity in this area may in part mediate the anti-appetite properties of nicotine (61). Thus, CCK peptides may have multiple functional roles in the paraventricular hypothalamic nucleus.

Effects of Chronic Exposure to Cigarette Smoke on Neuroendocrine Function and on Hypothalamic CA Nerve Terminal Systems

The results are summarized in Figs. 22 and 23 (63). It is shown that chronic exposure to cigarette smoke over a period of 9 days does not lead to tolerance development with regard to the ability of acute intermittent exposure to cigarette smoke to produce a reduction in the serum levels of prolactin, LH, and FSH. Instead, a tolerance developed with regard to the ability of acute intermittent exposure to cigarette smoke to increase ACTH and corticosterone secretion. Thus, these results again indicate a difference between the nicotinic cholinergic receptors involved in the control of ACTH secretion and those involved in the control of prolactin, LH, and FSH secretions. Interestingly, it was also found that no tolerance developed with regard to the ability of acute intermittent exposure to cigarette smoke to increase DA utilization in the MPZ and LPZ or to increase NA utilization in the periventricular NA nerve terminal

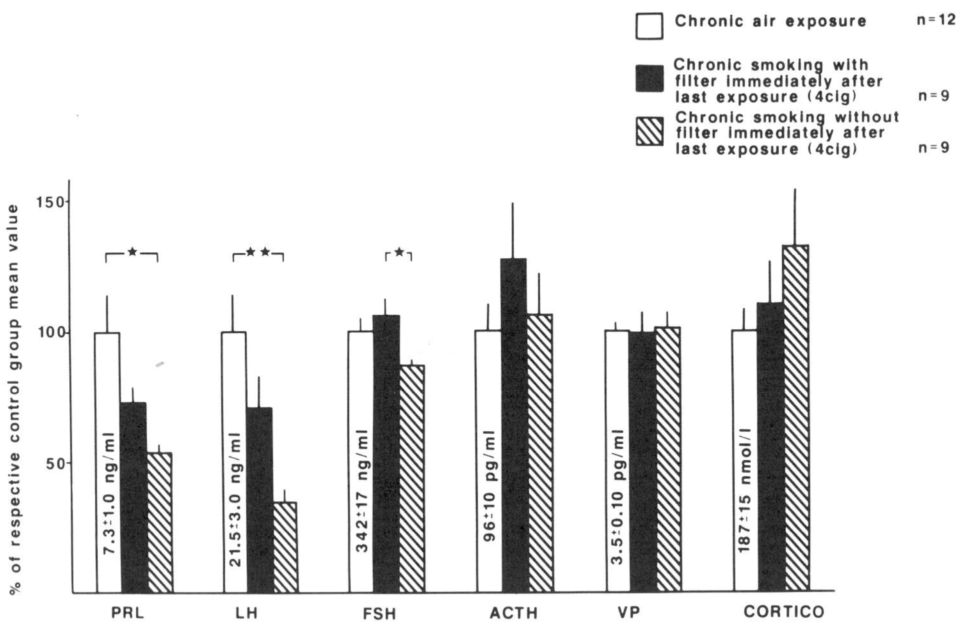

Fig. 22. Effects of chronic exposure to filtered or unfiltered cigarette smoke on prolactin, LH, FSH, ACTH, vasopressin, and corticosterone levels in the male rat peripheral blood. Male rats were exposed to air (controls) or filtered smoke (glass fiber filters) or unfiltered smoke from two cigarettes each morning between 9:00 – 11:00 during 5 plus 4 days with a smoke-free period of 2 days between the two periods. In the morning of day 12 the animals were exposed 4 times to air or the filtered or unfiltered smoke from one cigarette every 30' during 2 h according to the procedure shown in Fig. 1. The animals were decapitated immediately after the end of exposure. Means ± s.e.m. are expressed in percent of means obtained in the respective control groups. N = 12 in air-treated groups. N = 9 in all other groups. The absolute hormone levels are presented within the open columns. Statistical analysis according to the Wilcoxon test, one-way classification, comparing all possible pairs of treatments. * = $p<0.05$; ** = $p<0.01$. Data from (4).

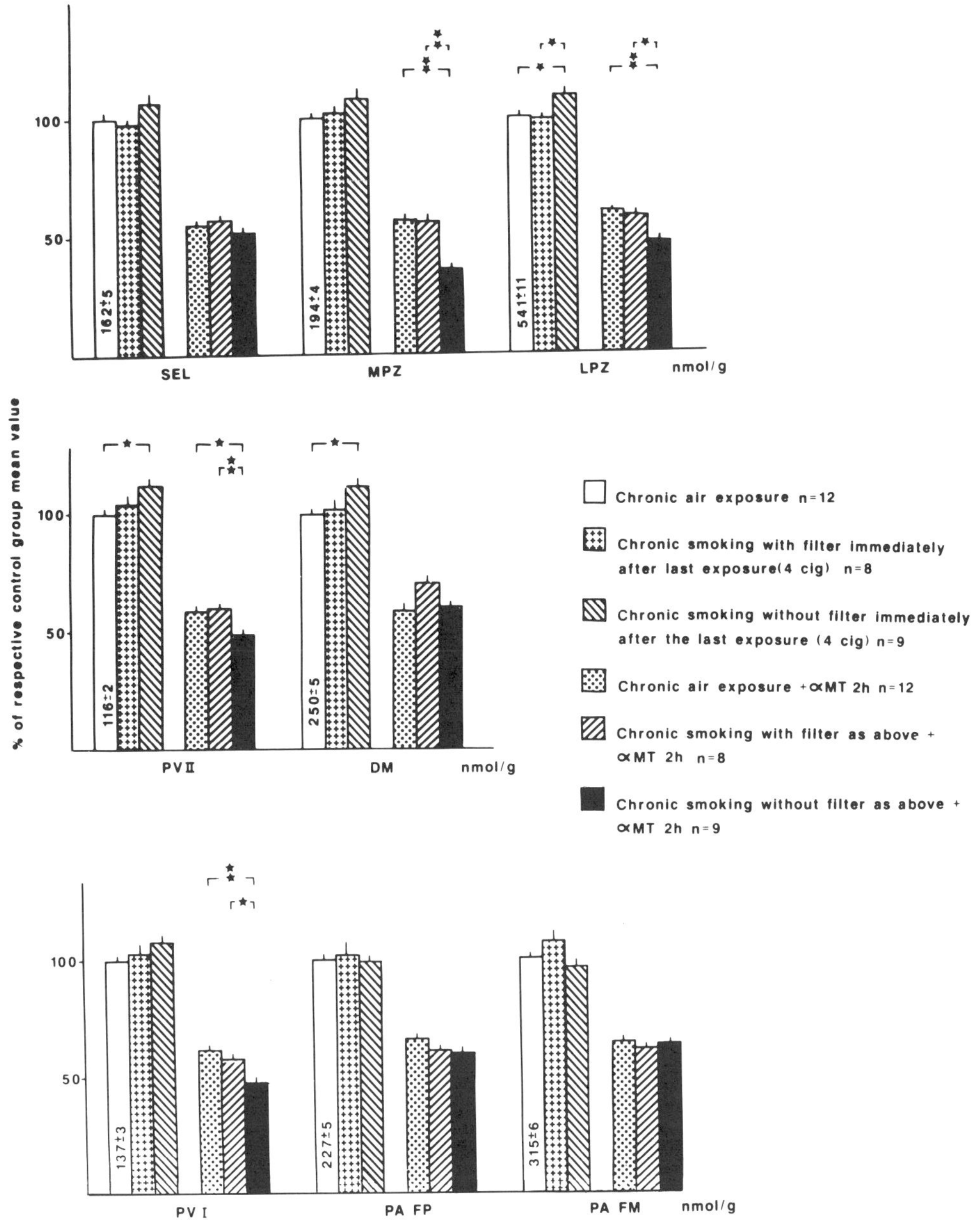

Fig. 23. Effects of chronic exposure to filtered or unfiltered cigarette smoke on CA fluorescence and αMT-induced CA fluorescence disappearance in various hypothalamic DA and NA nerve terminals of the male rat. For experimental procedure see text to Fig. 22. Means ± s.e.m. are expressed in percent of the mean value obtained in the respective control groups. N = 11-12 in air treated groups. N = 8 in the groups exposed to filtered smoke, n = 9 in the groups exposed to unfiltered smoke. Within the open columns (control group), the absolute concentrations of CA's have been expressed in nmol/g tissue wet weight. Statistical analysis according to the Wilcoxon test, one-way classification, comparing all possible pairs of treatments. * = $p<0.05$; ** = $p<0.01$. For abbreviations see text to Fig. 15. Data from (4).

systems. However, tolerance did develop with regard to the abilities of cigarette smoke to induce increases in NA utilization in the subependymal, the dorsomedial and the paraventricular hypothalamic NA nerve terminal systems. Thus, at least the nicotinic cholinergic receptors located on these NA nerve terminals may have become desensitized upon chronic exposure to cigarette smoke.

Furthermore, tolerance developed with regard to the ability of acute intermittent exposure to cigarette smoke to reduce CA levels in the various hypothalamic CA nerve terminal systems. Instead, small increases in DA and NA levels were observed in discrete DA and NA nerve terminal systems. These results are probably related to an activation of CA synthesis induced by the chronic exposure to cigarette smoke produced, for example, by increased tyrosine hydroxylase activity.

Effects of a 48-Hour Withdrawal Following a 10-Day Exposure to Chronic Cigarette Smoke on Hypothalamic CA's and on Blood Levels of Anterior Pituitary Hormones

The results are summarized in Figs. 24, 25. CA utilization in almost all regions was not significantly different from that in the control group which was exposed to air alone. An exception was the CA utilization in the MPZ. CA levels in any of the CA nerve terminal systems were unaltered as compared to those in the control group with the exception of those in the magnocellular part of the paraventricular hypothalamic nucleus where a small increase in CA levels was observed (data not shown, 64). These results indicate that the increase in CA utilization within the MPZ is maintained, even 48 hours following a withdrawal. As seen in Fig. 25, an enhancement of CA utilization was associated with a significant reduction of prolactin serum levels, which to a minor extent is also present following tyrosine hydroxylase inhibition (Fig. 24). DA released from the MPZ is probably a prolactin inhibitory factor (34,42); thus it seems likely that the reduction of prolactin serum levels was caused by this mechanism. The mechanism of the maintained activation of the DA neuronal systems in the MPA is presently unknown. However, it may be related to the humoral regulation by prolactin of the DA nerve terminals in the MPZ (34). Thus, the prolactin receptors controlling DA utilization in the MPZ may become more responsive so that the basal state of DA utilization in the MPZ is increased, leading to maintained low prolactin levels after withdrawal. Thus, prolactin receptor supersensitivity in the median eminence may develop in response to the maintained low prolactin serum levels caused by the chronic exposure to cigarette smoke.

SUMMARY

1. Nicotinic receptors have been visualized in brain using quantitative autoradiographic techniques. The regional distribution of [^{3}H]nicotine- and [^{3}H]acetylcholine-labeled nicotinic binding sites in the brain was very similar to those of a high number of binding sites in the thalamus, a moderate number in the cerebral cortex, and a low number in the hypothalamus.

2. The releasing effects of nicotine and exposure to cigarette smoke on CA's in the DA nerve terminals of the forebrain and in the DA and NA nerve terminals of the hypothalamus and the median eminence have been shown (Figs. 26-28).

3. The activation of the limbic subcortical DA nerve terminal systems and of the LHRH immunoreactive nerve terminals in limbic structures may be related to the rewarding actions of nicotine and smoking.

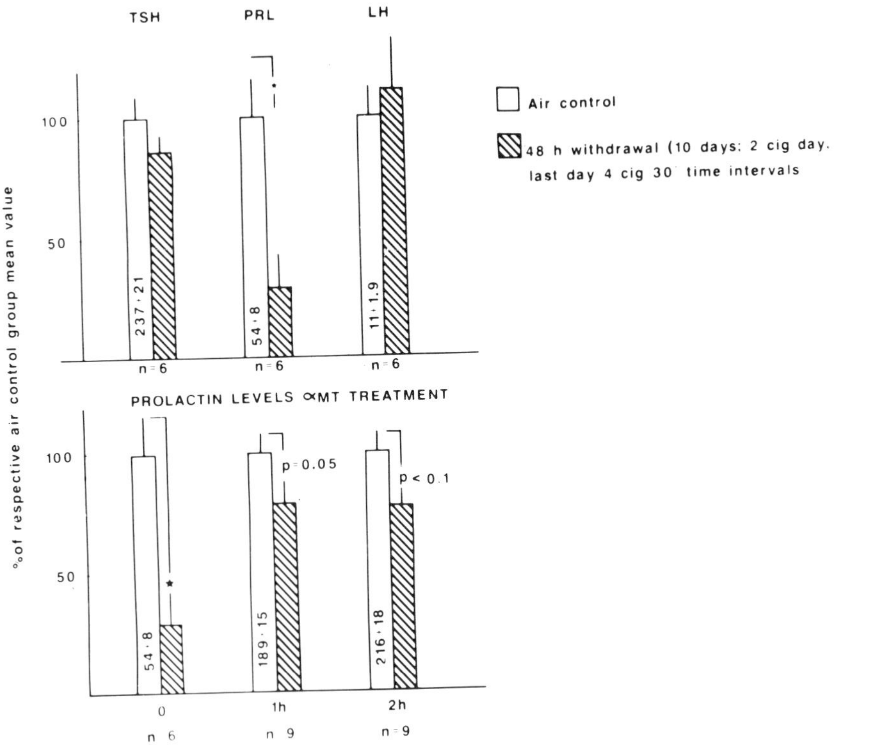

Fig. 24. Effects of a 48-h withdrawal from chronic exposure to cigarette smoke on TSH, prolactin, and LH serum levels of the normal male rat. The animals were exposed to air or to the smoke from two cigarettes each morning for 10 days. In the morning of day 11 the animals were exposed 4 times to air or to the smoke from one cigarette every 30 min for 2 h according to the procedure shown in Fig. 1. During the following 48 h the animals were kept free from exposure. The animals were then treated with saline or αMT (250 mg/kg, i.p.) for 1 h or 2 h before decapitation. Means ± s.e.m. are given in percent of respective control group mean values. The absolute values in the control groups are given in pg/ml. N = 6 in saline treated groups. N = 9 in αMT treated groups. Statistical analysis according to Mann-Whitney U-test. * = $p<0.05$.

4. The actions of nicotine and exposure to cigarette smoke on LH, FSH, prolactin, and TSH secretion are also summarized. Following an initial and marked activation of prolactin, LHRH, and LH secretion, nicotine and exposure to cigarette smoke induce a marked lowering of LH, prolactin, and TSH secretion and a small reduction of FSH secretion. The inhibitory effects of nicotine and exposure to cigarette smoke on gonadotrophin and TSH secretion may be related, at least in part, to activation of the tubero-infundibular DA neurons. This action of nicotine may be exerted both at the soma-dendritic level and at the level of the nerve terminals of the tubero-infundibular DA neurons. The cholinergic nicotinic receptors are probably of the ganglionic type as evidenced from experiments with

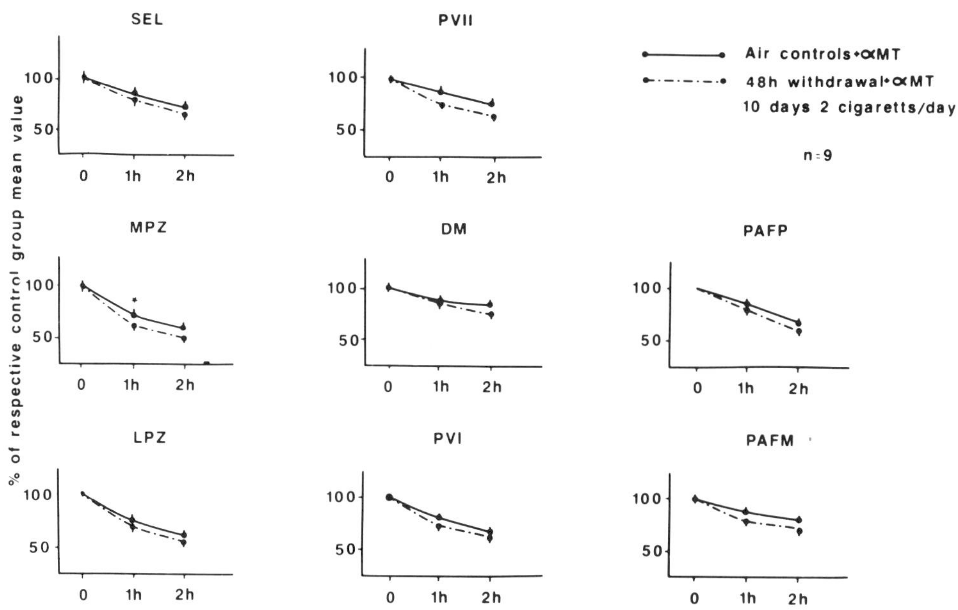

Fig. 25. Effects of 48-h withdrawal following chronic exposure to cigarette smoke on αMT-induced CA fluorescence disappearance in discrete hypothalamic DA and NA nerve terminal regions of the male rat. For experimental procedure see text to Fig. 24. Means ± s.e.m. are given in percent of respective saline-treated control group mean values. N = 6 in saline-treated groups. N = 9 in αMT-treated groups. Statistical analysis according to Mann-Whitney U-test. * = $p<0.05$. The respective absolute concentrations of CA's representing 100% were as follows (nmol/g of tissue wet weight): SEL = 220±9 ; MPZ = 300±17; LPZ = 675±39; PV II = 134±4; DM = 298±10; PV I = 166±10; PA FP = 267±10; PA FM = 357±13.

mecamylamine. It seems possible that the initial increase in prolactin and LH secretion induced by nicotine could be related, at least in part, to a rapid and preferential release of NA in discrete hypothalamic and preoptic areas. The results also give evidence for an enhanced processing of LHRH from precursors in relation to nicotine-induced increases in LH secretion.

5. The ability of nicotine to increase ACTH and vasopressin secretion may involve activation of nicotinic cholinergic receptors in the paraventricular hypothalamic nucleus, where inter alia the CCK/CRF immunoreactive nerve cell bodies are located, as well as vasopressin immunoreactive cell bodies. This nicotinic cholinergic receptor appears to be different from that controlling gonadotrophin and TSH secretion in view of its relative resistance to mecamylamine induced blockade and its more rapid desensitization. Enhancement of NA release in the paraventricular hypothalamic nucleus may contribute to the enhancement of CRF release in the median eminence by nicotine.

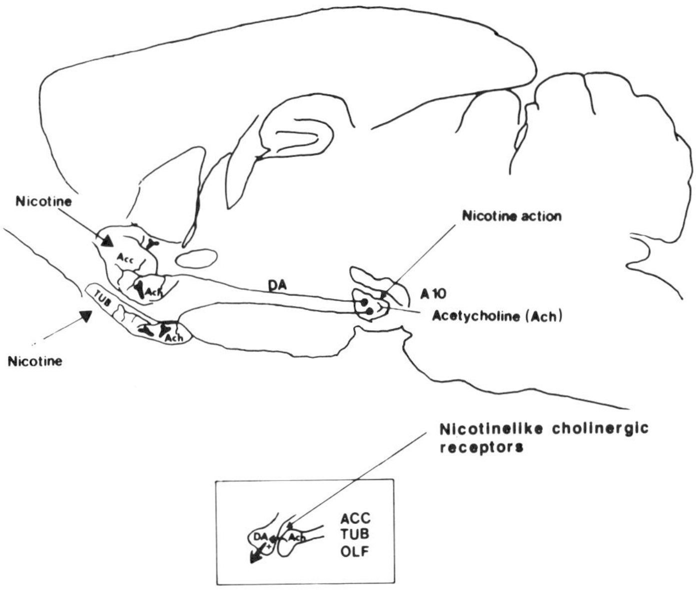

	STRIATUM		ACCUMBENS		TUBERCULUM	
	levels	turnover	levels	turnover	levels	turnover
NICOTINE	∽	↑	↓	↑	↓	↑
SMOKING	∽	∽	↓	↑	∽	↑

Fig. 26. Schematic illustration of the actions of nicotine on the mesolimbic DA neurons and an overall summary of the actions of nicotine and exposure to cigarette smoke on DA levels and utilization in the striatum, nucleus accumbens, and tuberculum olfactorium.

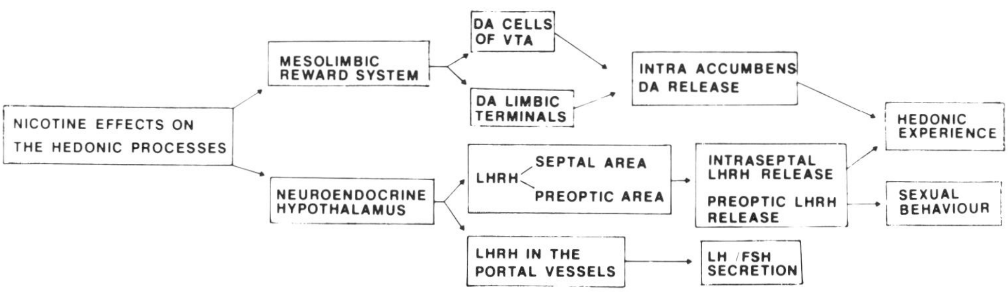

Fig. 27. Possible neurochemical mechanisms underlying the effects of nicotine on hedonic processes. The possible involvement of limbic, DA and LHRH release is indicated. Preoptic LHRH release may possibly mediate the ability of nicotine to enhance sexual behavior.

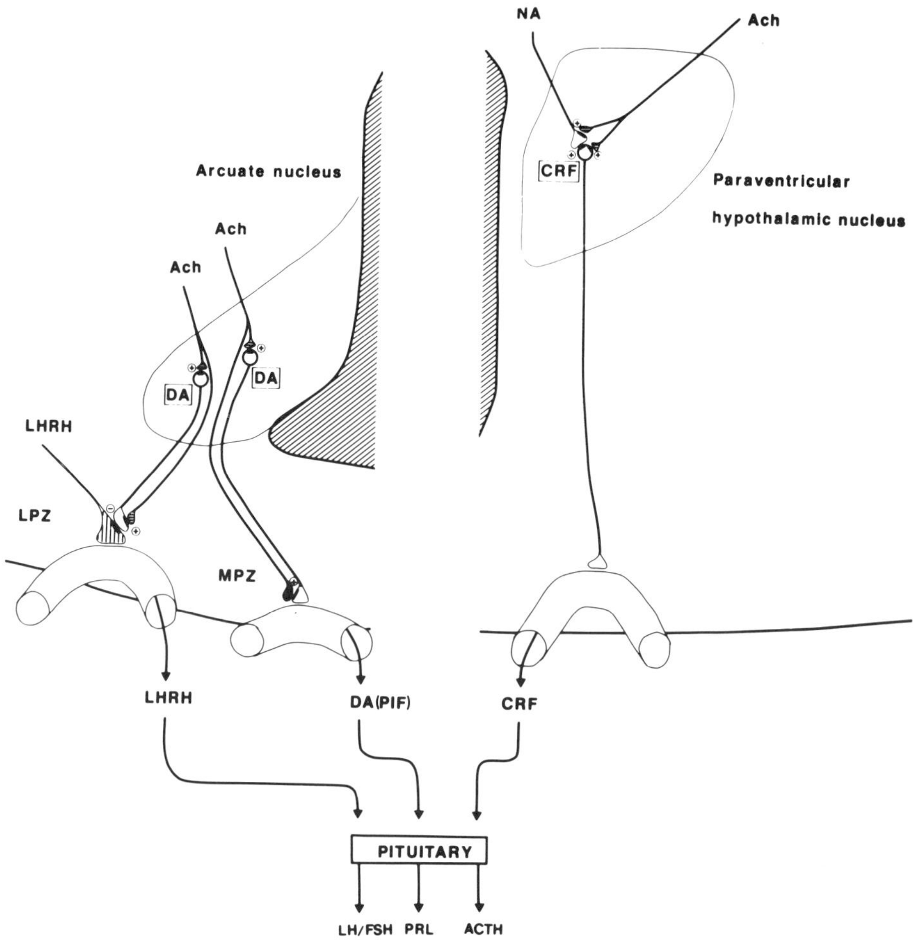

	CA TURNOVER SEL	MPZ	LPZ	NA TURNOVER PAFP	PAFM	LH	FSH	PRL	ACTH	SEXUAL BEHAVIOUR
NICOTINE	↑	↑	↑	↑	↑	(↑)↓	↓	↓↓	↑	↑↑
SMOKING	↑	↑	↑	↑	↑	(↑)↓	↓	↓	↑	↑ ?

Fig. 28. An overall summary of the effects of nicotine and exposure to cigarette smoke on median eminence and paraventricular CA levels and utilization as well as neuroendocrine function. In the upper part of the figure the possible presence of cholinergic synapses in the paraventricular nucleus, in the arcuate nucleus, in the external layer of the median eminence and in the SEL is indicated. These synapses may in part operate via nicotinic cholinergic receptors. These receptors are in part located on the DA and NA nerve terminals within these areas but may also control the tubero-infundibular DA neurons at the arcuate level. The effects of nicotine on the lateral and medial tubero-infundibular DA systems may at least in part mediate inhibitory effects of nicotine on LH and prolactin secretion, respectively. The ability of nicotine to enhance ACTH secretion is probably related to an activation of CRF secretion. This involves in part an activation of a facilitatory noradrenergic influence exerted mainly at the paraventricular hypothalamic level. It seems likely that the action of nicotine on the tubero-infundibular DA neurons clearly involves an action at the DA cell body level or in the network regulating these neurons in view of the latency in the onset of the inhibitory action of nicotine on LHRH secretion.

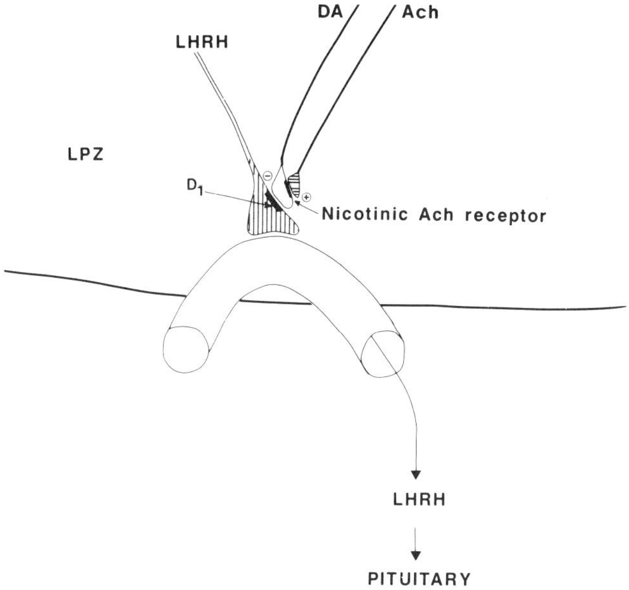

Fig. 29. Illustration of the possible mechanism mediating the nicotine-induced reduction of LH secretion. Nicotine release DA from the LPZ DA terminals to act on inhibitory DA receptors located on the LHRH terminals in the LPZ.

6. Many substance P and CCK immunoreactive neurons in the hypothalamus and the preoptic area have been shown to be counteracted by nicotinic cholinergic receptors. The increase in CCK-like immunoreactivity produced by nicotine and exposure to cigarette smoke in the paraventricular hypothalamic nucleus may inter alia contribute to the anti-satiety action of nicotine.

7. The D1 receptor antagonist SCH 23390 selectively counteracts the nicotine-induced reduction of LH secretion in the normal male rat. These results give evidence that nicotine may inhibit LHRH release via releasing DA from median eminence DA terminals onto a D1 receptor located on the LHRH nerve terminals (Fig. 29).

8. Studies on a 48-h withdrawal from chronic exposure to cigarette smoke show a maintained increase in DA turnover in the MPZ associated with a maintained lowering of prolactin secretion. These results may be explained by the development of prolactin receptor supersensitivity in the median eminence.

ACKNOWLEDGEMENTS

This work has been supported by a grant (1223) from the Council of Tobacco Research and by a grant from Svenska Tobaks AB, Stockholm, Sweden. We gratefully acknowledge the excellent technical assistance of Mrs. Eva Andersson, Mrs. Brita Blidemo, Mrs. Inger Feldt, Mrs. Ulla Britt Finnman, Mrs. Lotta Frosell, Mrs. Marita Johansson, Mrs. Siv Nilsson, Mrs. Birgitta Nyberg, Miss Barbro Tinner, and Mrs. Gaby Astrom. For excellent secretarial assistance we are grateful to Ms. Rose Marie Gustafsson. We also gratefully acknowledge the determinations of serum nicotine and cotinine by Mrs. Eva Kazemi-Wala.

REFERENCES

1. Cuello, A.C., and Sofroniew, M.V.: The anatomy of the CNS cholinergic neurons. TINS, March 74-78, 1984.
2. Satoh, K., Armstrong, D.M. and Fibiger, H.C.: A comparison of the distribution of central cholinergic neurons as demonstrated by acetylcholinesterase pharmacohistochemistry and CAT immunocytochemistry. Brain Res. Bull., 11:693-720, 1983.
3. Rodriguez-Sierra, J.F. and Morley B.J.: Evidence that cell bodies in the arcuate nucleus of the hypothalamus are not cholinergic Neuroendocrinology, in press, 1986.
4. Pepeu, G.: Brain acetylcholine: An inventory of our knowledge on the 50th anniversary of its discovery. TIPS Reviews, October 416-418, 1983.
5. Larsson, C.: Nicotine receptors in the central nervous system: Methodological and functional aspects. Acta Universitas Upsaliensis, abstracts of Uppsala disserition from the Faculty of Pharmacy 104:4-45, 1985.
6. Larsson, C. and Nordberg, A.: Studies of nicotine-like binding sites in brain. In: Neurotransmitters and their receptors, eds. V.Z. Littauer, Y. Ondai, J. Silman and V.I. Vogel, pp. 297-301, John Wiley & Sons, 1980.
7. Nordberg, A., Larsson, C., Falkeborn, Y., Lundberg, P.-A. and Premysl, S.: Development of nicotine-like binding sites in discrete areas of mouse brain findings using different nicotinic ligands. In: Developmental Neuroscience: Physiol. Pharmacol. and Clinical Aspects eds. F. Caciagli, E. Giacobini and R. Paoletti. Elsevier Science Publ. B.V. pp. 101-106, 1984.
8. Clarke, P.B., Pert, C.B. and Pert, A.: Autoradiographic distribution of nicotine receptors in rat brain. Brain Res., 323:390-395, 1984.
9. Clarke, P.B.S., Schwartz, R.D., Paul, S.M. Pert, C.B. and Pert, A.: Nicotinic binding in rat brain: Autoradiographic comparison of [^{3}H]nicotine and [^{125}I]α-bungarotoxin. J. Neurosci. 5:1307-1315, 1985.
10. Larsson, C. and Nordberg, A.: Comparative analysis of nicotine-like ligand receptor interaction in rodent brain homogenate. J. Neurochem. 45:24-31, 1985.
11. London, E.D., Waller, S.B. and Wamsley, J.K.: Autoradiographic localization of [^{3}H] nicotine binding sites in the rat brain. Neurosci. Lett. 53:179-184, 1985.
12. Morley, B.J., Farley, G.R. and Javel, E.: Nicotinic acetylcholine receptors in mammalian brain. Trends Pharmacol. Sci., 4:225-227, 1983.
13. Rainbow, T.C., Schwartz, R.D., Parsons, B. and Kellar, K.J.: Quantitative autoradiography of nicotinic [^{3}H]acetylcholine binding sites in rat brain. Neurosci. Lett., 50:193-196, 1984.
14. Schwartz, R.D., Lehmann, J. and Kellar, K.J.: Presynaptic nicotinic cholinergic receptors labeled by [^{3}H]acetylcholine on catecholamine and serotonin axons in brain. J. Neurochem., 42:1495-1498, 1984.
15. Nordberg, A. and Larsson, C.: Studies of muscarinic and nicotinic binding sites in brain. Acta Physiol. Scand. Suppl. 479:19-23, 1980.
16. Andersson, K., Fuxe, K., Eneroth, P., Mascagni, F. and Agnati, L.F.: Effects of acute intermittent exposure to cigarette smoke on catecholamine levels and turnover in various types of hypothalamic DA and NA nerve terminal systems as well as on the secretion of adenohypophyseal hormones and corticosterone. Acta Physiol. Scand. 124:277-285, 1985.

17. Curvall, M., Kazemi-Vala, E. and Enzell, C.R.: Simultaneous determination of nicotine and cotinine in plasma using capillary column gas chromatography with nitrogen-sensitive detection. J. Chromotogr., 232:283-293, 1982.
18. Andersson, K., Fuxe, K. and Agnati, L.F.: Determinations of catecholamine half-lives and turnover rates in discrete catecholamine nerve terminal systems of the hypothalamus, the preoptic region and the forebrain by quantitative histofluorimetry. Acta Physiol. Scand. 123:411-426, 1985.
19. Nilsen, O.G., Toftgard, R. and Eneroth, P.: Effects of acrylonitrile on rat liver cytochrome P-450 benzo(a)pyrene metabolism and serum hormone levels. Toxicol. Lett., 6:399-404, 1980.
20. Fuxe, K., Siegel, R.A., Andersson, K., Eneroth, P., Mascagni, F. and Agnati, L.F.: Acute continuous exposure to cigarette smoke produces discrete changes in cholecystokinin and substance P levels in the hypothalamus and preoptic area of the male rat. Acta Physiol. Scand., 125:437-443, 1985.
21. Agnati, L.F., Fuxe, K, Benfenati, F., Calza, L. and Fabbri, L.: Computer-assisted morphometry in immunocytochemistry and receptor autoradiography: impact on neuroendocrine control. In: Computers in Endocrinology eds. D. Rodbard and G. Forti, pp. 171-185, Raven Press, New York, 1984.
22. Benfenati, F., Agnati, L.F. and Fuxe, K.: Quantitative autoradiography of central neurotransmitter receptors: Methodological and statistical aspects with special reference to computer-assisted image analysis. Acta Physiol. Scand., in press, 1986.
23. Fuxe, K., Calza, L., Benfenati, F., Zini, I. and Agnati, L.F.: Quantitative autoradiographic localization of [^{3}H]imipramine binding sites in the brain of the rat: relationship to ascending 5-hydroxytryptamine neuron systems. Proc. Natl. Acad. Sci., USA 80:3836-3840, 1983.
24. Hökfelt, T., Johansson, O., Ljungdahl, A., Lundberg, J.M. and Schultzberg, M.: Peptidergic neurons. Nature, (London), 248:515, 1980.
25. Andersson, K., Fuxe, K. and Agnati, L.F.: Effects of single injections of nicotine on the ascending dopamine pathways in the rat. Acta Physiol. Scand. 112:345-347, 1981.
26. Lichtensteiger, W, Felix, D., Hefti, F. and Schlumpf, M.: Effects of nicotine on dopamine neurons of adult and prenatal mammals and of invertebrates. In Electrophysiological effects of nicotine, eds. A. Remond and C. Izard, pp. 15-30, Elsevier/North Holland Biomedical Press, Amsterdam, 1979.
27. Giorguieff, M.F., Le Floc'h, M.L., Westfall, T.C., Glowinski, J. and Besson, M.J.: Nicotinic effect of acetylcholine on the release of newly synthesized [^{3}H] dopamine in rat striatal slices and cat caudate nucleus. Brain Res., 106:117-131, 1976.
28. Roberts, D.C.S., Corcoran, M.E. and Fibiger, H.C.: Recovery of cocaine self-administration after 6-OHDA lesion of the nuc. accumbens correlates with residual dopamine levels. In: Catecholamines: Basic and clinical frontiers, eds. E. Usdin, I.J. Kopin & Barchas, Vol. 2 pp. 1774-1776, Pergamon Press, New York, 1979.
29. Andersson, K., Fuxe, K., Agnati, L.F. and Eneroth, P.: Effects of acute central and peripheral administration of nicotine on ascending dopamine pathways in the male rat brain. Evidence for nicotine induced increases of dopamine turnover in various telencephalic dopamine nerve terminal systems. Med. Biol., 59:170-176, 1981.
30. Usdin, E., Carlsson, A., Dahlstrom, A. and Engel, J.: Catecholamines Part C: Neuropharmacology and central nervous system - therapeutic aspects. In: Neurology and Neurobiology Vol. 8C. Alan R. Liss, Inc., New York, 1984.

31. Fuxe, K., Andersson, K., Harfstrand, A. and Agnati, L.F.: Increases in dopamine utilization in certain limbic dopamine terminal populations after a short period of intermittent exposure of male rats to cigarette smoke. J. Neural Transm. 67:15-29, 1986.
32. Andersson, K., Fuxe, K., Eneroth, P., Gustafsson, J.-A. and Agnati, L.F.: Mecamylamine induced blockade of nicotine induced inhibition of gonadrotrophin and TSH secretion and of nicotine induced increases of catecholamine turnover in the rat hypothalamus. Acta Physiol. Scand., Suppl. 479:27-29, 1980.
33. Andersson, K., Siegel, R., Fuxe, K. and Eneroth, P.: Intravenous injections of nicotine induced very rapid and discrete reductions of hypothalamic catecholamine levels associated with increases of ACTH, vasopressin and prolactin secretion. Acta Physiol. Scand. 118:35-40, 1983.
34. Andersson, K., Eneroth, P. and Agnati, L.F.: Nicotine-induced increases of noradrenaline turnover in discrete noradrenaline nerve terminal systems of the hypothalamus and the median eminence of the rat and their relationship to changes in the secretion of adenohypophyseal hormones. Acta Physiol. Scand., 113:227-231, 1981.
35. McCann, S.M., Krulich, L., Ojeda, S.R., Negro-Vilar, A and Vijayan, E.: Neurotransmitters in the control of anterior pituitary function. In: Central regulation of the endocrine system, eds. K. Fuxe, T. Hokfelt, and R. Luft, pp. 329-347, Plenum Press, New York and London, 1979.
36. Andersson, K., Fuxe, K., Eneroth, P., Nyberg, P. and Roos, P.: Rat prolactin and hypothalamic catecholamine nerve terminal systems. Evidence for rapid and discrete increases in dopamine and noradrenaline turnover in the hypophysectomized male rat. Eur. J. Pharmacol. 76:261-265, 1981.
37. Andersson, K., Agnati, L.F., Fuxe, K., Eneroth, P., Harfstrand and Benfenati, F.: Corticotropin-releasing factor increases noradrenaline turnover in the median eminence and reduces noradrenaline turnover in the paraventricular region of the hypophysectomized male rat. Acta Physiol. Scand. 120:621-624, 1984.
38. Andersson, K., Fuxe, K., Eneroth, P., Blake, C.A., Agnati, L.F. and Gustafsson, J.-A.: Effects of androgenic and adrenocortical steroids on hypothalamic and preoptic catecholamine nerve terminals and on the secretion of anterior pituitary hormones. In: Steroid Hormone Regulation of the Brain, eds. K. Fuxe, J.-A. Gustafsson, L. Wetterberg, pp. 117-133, Pergamon Press, Oxford, 1981.
39. Gainer, H.: Current topics in neurobiology. In: Peptides in Neurobiology, pp. 1-464, Plenum Press, New York, 1977.
40. Andersson, K., Fuxe, K., Eneroth, P. and Agnati, L.F.: Effects of acute central and peripheral administration of nicotine on hypothalamic catecholamine nerve terminal systems and on the secretion of adenohypophyseal hormones in the male rat. Med. Biol., 60:98-111, 1982.
41. Andersson, K., Fuxe, K., Agnati, L.F., Eneroth, P. and Camurri, M.: Luteinizing hormone-releasing hormone increases dopamine turnover in the lateral palisade zone of the median eminence and reduces noradrenaline turnover in the nuc. preopticus medialis of the hypophysectomized male rat. Neurosci. Lett., 45:253-258, 1984.
42. Fuxe, K., Andersson, K., Lofstrom, A., Hokfelt, T., Ferland, L., Agnati, L.F., Perez de la Mora, M. and Schwarcz, R.: Neurotransmitter mechanisms in the control of secretion of hormones from the anterior pituitary. In: Central Regulation of the Endocrine System, eds. K. Fuxe, T. Hokfelt & R. Luft, pp. 349-380, Plenum Press, New York, 1979.

43. Fuxe, K. Agnati, L.F., Eneroth, P., Gustafsson, J.-A., Hokfelt, T., Lofstrom, A., Skett, B. and Skett, P.: The effect of nicotine on central catecholamine neurons and gonadotropin secretion. I. Studies in the male rat. Med. Biol., 55:148-157, 1977.
44. Eneroth, P., Fuxe, K., Gustafsson, J.-A., Hokfelt, T., Lofstrom, A., Skett, P. and Agnati, L.F.: The effect of nicotine on central catecholamine neurons and gonadotropin secretion. II. Inhibitory influence of nicotine on LH, FSH and prolactin secretion in the ovariectomized female rat and its relation to regional changes in dopamine and noradrenaline levels and turnover. Med. Biol., 55:158-166, 1977.
45. Andersson, K. Fuxe, K., Eneroth, P. and Agnati, L.F.: Involvement of cholinergic nicotine-like receptors as modulators of amine turnover in various types of hypothalamic dopamine and noradrenaline nerve terminal systems and of prolactin, LH, FSH and TSH secretion in the castrated male rat. Acta Physiol. Scand. 116:41-50, 1982.
46. Andersson, K., Fuxe, K., Eneroth P. and Agnati, L.F.: Differential effects of mecamylamine on the nicotine induced changes in amine levels and turnover in hypothalamic dopamine and noradrenaline nerve terminal systems and in the secretion of adenohypophyseal hormones in the castrated female rat. Evidence for involvement of cholinergic nicotine-like receptors. Acta Physiol. Scand., 120:489-498, 1984.
47. Fuxe, K., Agnati, L.F., Benfenati, F., Andersson, K., Camurri, M. and Zoli, M.: Evidence for the existence of a dopamine receptor of the D1 type in the rat median eminence. Neurosci. Lett., 43:185-190, 1983.
48. Iorio, L.C., Barnett, A., Leitz, F.H., Houser, V.P. and Korduba, C.A.: SCH 23390 a potential benzazepine antipsychotic with unique interactions on dopaminergic systems. J. Pharmacol. Exp. Ther. 255: 539, 1983.
49. Andersson, K., Fuxe, K., Eneroth, and Harfstrand, A.: The effects of a D1 dopamine receptor blocking drug on the neuroendocrine actions of nicotine treatment and of acute intermittent exposure to cigarette smoke in the male rat. Relationship to median eminence catecholamines. Acta Physiol. Scand., in press, 1986.
50. Andersson, K.: Mecamylamine pretreatment counteracts cigarette smoke induced changes in hypothalamic catecholamine neuron systems and in anterior pituitary function. Acta Physiol. Scand. 125:445-452, 1985.
51. Andersson, K., Fuxe, K., Eneroth, P. and Agnati, L.F. Effects of acute continuous exposure of the male rat to cigarette smoke on amine levels and utilization in discrete hypothalamic catecholamine nerve terminal systems and on neuroendocrine function. Possible presence of depolarization induced blockade in catecholamine nerve terminals. Acta Physiol. Scand., in press, 1986.
52. Fuxe, K., Everitt, B.J. and Hokfelt, T.: Enhancement of sexual behavior in the female rat by nicotine. Pharmacol. Biochem. Behav. 7:147-151, 1977.
53. Vijayan, E. and McCann, S.M.: In vivo and in vitro effects of substance P and neurotensin on gonadotropin and prolactin release. Endocrinology, 105:64-68, 1979.
54. Vijayan, E., Samson, W.K. and McCann, S.M.: In vivo and in vitro effects of cholecystokinin on gonadotropin, prolactin, growth hormone and thyrotropin release in the rat. Brain Res. 172:295-302, 1979.
55. Kiss, J.Z.: Anatomical studies of cholecystokinin in neurons and pathways involved in neuroendocrine regulation. In: Neuronal Cholecystokinin, eds. J.J. Vanderhaeghen & J.N. Crawley, Annals of the New York, Academy of Sciences, Vol. 448, pp. 144-151, 1985.

56. Vanderhaeghen, J.J., Lotstra, F. and Gilles, C.: Gastrin(s) and cholecystokinin(s) in central nervous system and pituitary: relationship with dopamine, oxytocin and alpha-MSH containing cells and with limbic nigrostriatal systems. Horm. Res., 12:182-183, 1980.
57. Mezey, E., Reisine, T.D., Skirboll, L. Beinfeld, M. and Kiss, J.Z.: Cholecystokinin in the medial parvocellular subdivision of the paraventricular nucleus. Co-existence with corticotropin-releasing hormone. In: Neuronal Cholecystokinin, eds. J.J. Vanderhaeghen & J.N. Crawley. Annals of the New York Academy of Sciences, Vol. 448:152-156, 1985.
58. Fuxe, K., Andersson, K., Agnati, L.F., Eneroth, P., Locatelli, V., Caviocchioli, L., Mascagni, F., Tatemoto, K. and Mutt, V.: The influence of cholecystokinin peptides and PYY on the amine turnover in discrete hypothalamic dopamine and noradrenaline nerve terminal systems and possible relationship to neuroendocrine function. Inserm, Vol. 110: 65-86, 1982.
59. Fuxe, K., Andersson, K., Eneroth, P., Siegel, R.A. and Agnati, L.F.: Immobilization stress-induced changes in discrete hypothalamic catecholamine levels and turnover, their modulation by nicotine and relationship to neuroendocrine function. Acta Physiol. Scand., 117:421-426, 1983.
60. Fuxe, K., Agnati, L.F., Vanderhaeghen, J.-J., Tatemoto, K., Andersson, K., Eneroth, P., Härfstrand, A., von Euler, G., Toni, R., Goldstein, M. and Mutt, V.: Cholecystokinin neuron systems and their interactions with the presynaptic features of the dopamine neuron systems. A morphometric and neurochemical analysis involving studies on the action of cholecystokinin-8 and Cholecystokinin-58. In: Neuronal Cholecystokinin, eds. J.J. Vanderhaeghen & J.N. Crawley, Ann. N.Y. Acad. Sci., 448:231-254, 1985.
61. Baile, C.A. and Della-Fera, M.A.: Central Cholecystokinin System and the control of feeding. In: Neuronal Cholecystokinin, eds. J.J. Vanderhaeghen & J.N. Crawley, Annals of the New York, Academy of Sciences, Vol. 448, pp. 424-430, 1985.
62. Faris, P.L., Scallett, A.C., Olney, J.W., Della-Fera, M.A. and Baile, C.A.: Behavioral and immunohistochemical analysis of the function of cholecystokinin in the hypothalamic paraventricular nucleus. Soc. Neurosci. Abstr., 9:184, 1983.
63. Andersson, K., Eneroth, P., Fuxe, K., Mascagni, F. and Agnati, L.F.: Effects of chronic exposure to cigarette smoke on amine levels and turnover in various hypothalamic catecholamine nerve terminal systems and on the secretion of pituitary hormones in the male rat. Neuroendocrinology, 41:462-466, 1985.
64. Andersson, K., Fuxe, K., Eneroth, P. and Härfstrand, A.: Effects of withdrawal from chronic exposure to cigarette smoke on amine levels and turnover in various hypothalamic catecholamine nerve terminal systems and on the secretion of pituitary hormones in the male rat. Acta Physiol. Scand., in press, 1986.

NICOTINIC REGULATION OF SYMPATHOADRENAL CATECHOLAMINE SECRETION: CROSS-TOLERANCE TO STRESS

Glen R. Van Loon, Judith A. Kiritsy-Roy, Laura V. Brown and Frances A. Bobbitt

Department of Medicine, University of Kentucky

VA Medical Center, Lexington, KY 40511

INTRODUCTION

The effects of nicotine on the regulation of sympathoadrenal secretion have long been recognized (1,2). Recent demonstration of endogenous opioid peptides colocalized with catecholamines in the adrenal medulla and in sympathetic nerve terminals has provided new impetus for the further study of both the physiologic regulation of sympathoadrenal secretion by nicotinic cholinergic receptors and the pharmacologic effects of this alkaloid (3,4). In addition to actions on adrenal chromaffin cells and sympathetic ganglia (5), it is thought that nicotine may stimulate sympathoadrenal secretion of catecholamines and opioid peptides by acting in brain to increase central sympathetic outflow. However, the latter issue has not been established, and we have initiated a series of studies comparing plasma catecholamine and cardiovascular responses to peripheral (intraarterial; ia) versus central (intraventricular; ivt) administration of nicotine.

Repeated administration of nicotine rapidly induces a state of tolerance in which there is a reduction or loss of response to some of the pharmacologic effects of the drug (6,7). Tolerance with respect to sympathoadrenal secretion has also been suggested (8). However, the sites at which this tolerance is mediated remain undefined. In these studies, we have compared catecholamine and cardiovascular responses not only to initial exposure but also to the repeated administration of nicotine by both systemic and central routes, and we have examined carefully the rate of development of tolerance with respect to the effects of nicotine on sympathoadrenal secretion.

METHODS

Studies were carried out in conscious rats and in anesthetized dogs. Adult, male Sprague-Dawley rats were prepared for cardiovascular measurements and blood withdrawal by surgically implanting a polyethylene (PE50) cannula in the left internal carotid artery under Nembutal anesthesia. For experiments with brain nicotine injections, a stainless steel guide cannula was also implanted over the right lateral ventricle and secured to the skull with dental acrylic and screws. Blood samples for plasma catecholamine determinations were taken in 0.5 ml volumes from the arterial cannula as described elsewhere (9). Blood pressure and heart rate were monitored continuously using a Statham pressure transducer connected to a Buxco Cardiovascular Analyzer. Drugs were injected intraventricularly through a prefilled microinjector which was inserted into the chronically implanted guide cannula 30 min before an experiment. Rats were stressed by restraining in a soft, flexible holder for 2 min.

Adult male mongrel dogs were anesthetized with Nembutal and maintained on a respirator throughout the experiment. The lumboadrenal vein was cannulated as described previously (10) for collection of adrenal blood samples and measurements of adrenal secretion rates of catecholamines. Cannulae were also inserted into the femoral artery for blood pressure recording and peripheral blood sampling and into the femoral vein for drug administration. Data were analyzed statistically by analysis of variance, analysis of covariance, or by paired or unpaired t-tests where appropriate.

RESULTS

Acute Effects of Systemic Nicotine on Sympathoadrenal Catecholamine Secretion and on Cardiovascular Parameters in Conscious Rats

The systemic administration of nicotine by intraarterial (ia) injection produced dose-related increases in plasma concentrations of NE and EPI (Table 1). Plasma catecholamines reached maximal levels at 2 min and returned to basal concentrations by about 10 min after drug injection. In a subsequent experiment we found that the nicotine-induced increases in plasma NE and EPI were almost completely prevented by the prior administration of the ganglionic blocking agent, hexamethonium 5 mg/kg ia (data not shown).

Systemic nicotine also increased systolic, diastolic, and mean arterial blood pressure, and simultaneously produced a large dose-related bradycardia (Table 1).

Tolerance to Systemic Nicotine

Repetition of systemic 0.1 mg/kg doses of nicotine once each day for 7 days produced plasma NE and EPI responses on the seventh day which were comparable to the initial responses (Table 2). That is, with this dose

TABLE 1. Sympathoadrenal and cardiovascular responses[a] to systemic nicotine in rats

Dose (mg/kg)	n	NE (nM)	EPI (nM)	SP[b] (mmHg)	DP[b] (mmHg)	MAP[b] (mmHg)	HR[b] (bpm)
0	40	0.2±0.1	0.1±0.1	2±2	1±1	1±1	5±5
0.1	30	10.2±1.6*	8.4±1.2*	43±6*	18±3*	27±3*	-31±14
0.5	20	17.4±2.2*+	26.8±2.9*+	51±6*	16±7*	27±6*	-182±32*+

[a]Responses represent the mean ± S.E.M. change at 2 min after nicotine compared to pretreatment (baseline) values.
[b]SP: Systolic pressure, DP: Diastolic pressure, MAP: Mean arterial pressure, HR: Heart rate
*P<0.01 compared to vehicle-injected control (Dose =0)
+P<0.01 compared to both control and 0.1 mg/kg nicotine groups

and interexposure intervals of 24 hr, tolerance did not develop within 1 week. Since tolerance or tachyphylaxis to the sympathomimetic effects of nicotine has been suggested, we further investigated the effects of varying dose, interdose interval, and number of exposures of nicotine on the rate of development of tachyphylaxis.

After three ia injections of nicotine 0.1 mg/kg at 70 min intervals, there was no significant decrease in the plasma NE or EPI responses. However, three similar injections of nicotine at 30 min intervals did result in a state of tachyphylaxis, in which the plasma NE and EPI responses to the third nicotine treatment were reduced to about one sixth of the responses observed in naive rats (Table 2). Complete tolerance to the catecholamine-releasing effect of nicotine was achieved when the dose was raised from 0.1 to 0.5 mg/kg, the number of exposures increased from 3 to 15, and the interexposure interval decreased to 15 min.

The cardiovascular responses differed considerably from the catecholamine responses with regard to development of tolerance. Partial tolerance to nicotine-induced increases in systolic and/or mean arterial pressure was observed when 0.1 mg/kg of the drug was injected at either 70 or 30 min intervals. However, after 15 injections of nicotine 0.5 mg/kg at 15 min intervals, the drug elicited similar increases in systolic, diastolic, and mean pressure when compared with its effects in naive rats. The heart rate response showed tolerance, but not until the most extreme of the dosage regimens was used. Bradycardic responses showed no tolerance to nicotine 0.1 mg/kg when dosage intervals of 24 hr, 70 min or 30 min were used, but when doses of nicotine 0.5 mg/kg were repeated 15 times at 15 min intervals, the large bradycardia response of -182 ± 32 bpm was lost completely (-19 ± 14 bpm). Clearly, the degree of tolerance to systemically administered nicotine depends upon the parameter measured, the dose of the drug, the number of treatments, and the time interval between exposures.

Sympathoadrenal Responses to Stress after Repeated Exposure to Nicotine in Rats

The catecholamine responses described above represent the responses to nicotine in a basal or resting state. The effects of nicotine on stress-induced catecholamine secretion were studied in response to 2 min

TABLE 2. Development of tolerance to systemic nicotine

Dose (mg/kg)	Interval	Number of Nicotine Injections	NE	EPI	Response to Final Nicotine Injection as % of Initial Response[a] SP	DP	MP	HR
0.1	24 hr	7	116±36	134±59	76±17	87±36	81±23	59±66
0.1	70 min	3	74±40	56±22	48±17*	69±28	57±21	32±36
0.1	30 min	3	14±4**	20±3**	51±17*	59±25	54±17*	119±38
0.5	15 min	15	5±1**	5±2**	76±13	70±20	76±13	11±8**

[a]Initial response is the mean of all responses to nicotine (0.1 or 0.5 mg/kg) pooled from all experiments.
*P<0.05 **P<0.01 indicating a statistically significant reduction in response to nicotine after repeated administrations.

restraint stress administered 30-70 min after the last dose of nicotine using a time interval similar to that between drug exposures.

Restraint stress in vehicle-treated control rats produced increases in plasma concentrations of both NE (1.9±0.2nM, n=38, P<0.0001) and EPI (6.2±0.9, n=38, P<0.0001). Treatment of rats with repeated injections of nicotine prior to restraint stress modified the stress-induced increase in plasma catecholamine levels. As shown in Table 3, both the plasma NE and EPI responses to restraint stress were blunted after 15 exposures to nicotine 0.5 mg/kg at 15 min intervals when compared with stress responses in vehicle-treated control animals. The lower dose of nicotine (0.1 mg/kg) given at longer time intervals did not significantly alter the plasma NE response to restraint but did attenuate the plasma EPI response, when the drug was injected at 30- and 70-min intervals. Exposure to nicotine 0.1 mg/kg at a longer time interval (24 hr) for one week had no significant effect on the plasma catecholamine responses to stress. Thus, decreasing the interval between nicotine treatments resulted in a progressive decline in the sympathoadrenal response to restraint stress, suggesting that animals made tolerant to nicotine with respect to sympathoadrenal stimulation became cross-tolerant to restraint stress.

Effects of Acute and Repeated Intracerebral Administration of Nicotine on Sympathoadrenal Catecholamine Secretion and on Cardiovascular Parameters

Intraventricular (ivt) administration of 30 nmoles or about 5 ug of nicotine produced a rapid increase in plasma EPI concentration which peaked at 2 min and returned to basal level by 15 min; plasma NE concentration was unaffected. The plasma catecholamine responses to ivt nicotine were dose-related (Table 4).

An ivt dose of 120 nmoles of nicotine produced a plasma EPI response more than 50% larger than that produced by 150 nmoles (0.1 mg/kg) of ia nicotine (compare Tables 1 and 4). Lower doses (30-60 nmol) produced smaller incremental effects on plasma EPI. Plasma NE increased only in response to the largest ivt dose (120 nmol) examined, although this dose produced an 8-fold greater increase in plasma EPI than NE.

Ivt administration of either hexamethonium or mecamylamine, 120 nmoles, prevented the plasma EPI response to ivt nicotine 120 nmoles administered 10 min later; in this study, nicotine ivt alone did not increase plasma NE significantly.

TABLE 3. Cardiovascular and sympathoadrenal responses to restraint stress after pretreatment with systemic nicotine

Dose	Interval	Number of Nicotine Injections	Stress Response in Nicotine Group as % of Response in Saline Groups NE	EPI
0.1	24 hr	7	123±37	71±17
0.1	70 min	3	167±34	43±16**
0.1	30 min	3	73±19	47±16*
0.5	15 min	15	37±10**	23±6**

*P<0.05 **P<0.01 indicating a significant reduction in the stress response of nicotine-treated animals compared to the stress response of rats treated similarly with saline.

TABLE 4. Sympathoadrenal and cardiovascular responses[a] to ivt nicotine in rats

Dose (nmol)	N	NE (nM)	EPI (nM)	MAP (mmHg)	HR (bpm)
0	5	0.6±0.1	-0.2±0.3	2±4	0±2
30	11	0.3±0.1	3.8±0.6*	11±14	12±3
60	6	0.1±0.3	3.4±0.3*	-5±9	3±5
120	5	1.7±0.6*+	13.0±3.2*+	8±14	9±8

[a]Responses represent the mean ± S.E.M. change at 2 min after nicotine ivt compared to pretreatment (baseline) values.
*P<0.05 compared to control (Dose = 0)
+P<0.05 compared to other doses of nicotine

When nicotine, 120 nmoles, was administered ivt repeatedly at 24 hr intervals, the plasma EPI response to the second injection was markedly attenuated (48.3 ± 15.8% of first response, P<0.05) and even lower in response to the seventh daily dose (13.1 ± 3.5% of first response, P<0.05). The plasma NE response was also absent with subsequent exposures after the first injection of nicotine. The development of sustained tolerance lasting 24 hours after a single intracerebral nicotine injection contrasts sharply with the rate of development and duration of tolerance after systemic nicotine.

These ivt doses of nicotine did not alter blood pressure or heart rate after either initial or repeated administration (Table 4).

Effects of Acute Single and Repeated Systemic Administration of Nicotine in Anesthetized Dogs

Nicotine was administered intravenously (iv) in two bolus doses of 40 ug/kg separated by 60 min, and an infusion of 5 ug/kg/min was administered during the interbolus interval.

Peripheral plasma concentration of EPI increased to a peak at 2 min after the initial bolus of nicotine, but in spite of the continued low-dose infusion, plasma EPI fell at 5 min and remained at the basal level 10, 20, and 60 min after the initial injection (Fig. 1). The transient plasma EPI response to the second bolus of nicotine given 60 min after the first was equal and parallel to the response after the first bolus. A small, transient increase in peripheral plasma NE concentration was also detected after each bolus injection of nicotine.

The initial bolus of nicotine significantly increased the adrenal NE and EPI secretion rates (Fig. 2). In contrast to the peripheral plasma EPI responses where no tachyphylaxis was observed, we did observe blunted responses of adrenal NE and EPI secretion rates to the second bolus of nicotine when compared with the responses to the first bolus.

Nicotine increased systolic, diastolic and mean arterial pressure and heart rate (Fig. 3). Also, tachyphylaxis was observed for blood pressure and heart rate responses to the second bolus of nicotine.

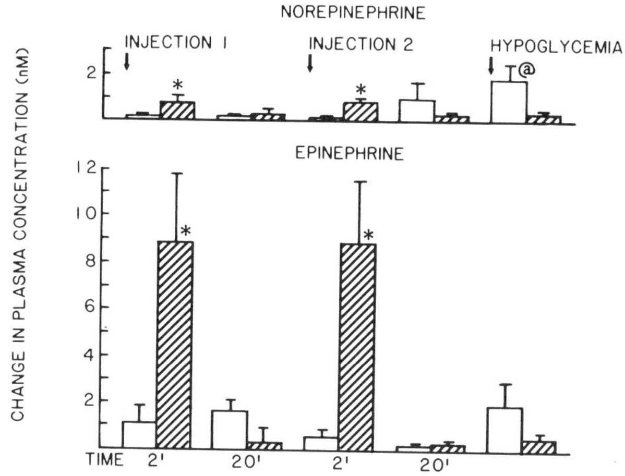

Fig. 1. Plasma catecholamine responses to nicotine in pentobarbital-anesthetized dogs. Two bolus injections of nicotine (40 ug/kg, n=10, hatched bars) or saline (n=5, open bars) were administered 60 min apart and blood samples were withdrawn from the femoral vein at 2 min and 20 min after each injection. Sixty minutes after the second bolus, insulin was injected and blood was sampled 15, 30, 45, and 60 min later. Data represent the mean ± S.E.M. change in plasma norepinephrine (upper panel) and epinephrine (lower panel) concentration from pretreatment levels. *P<0.05 compared with the saline control groups. @ P<0.05 indicating a significant increase in plasma concentration in response to insulin-induced hypoglycemia.

Sympathoadrenal Responses to Insulin-Induced Hypoglycemia after Administration of Nicotine

Administration of insulin 0.16-0.18 units/kg in vehicle-treated control dogs decreased plasma glucose from 5.8±0.2 to 1.9±0.1 mM (P<0.001) at 45 min. This hypoglycemia increased adrenal secretion rates of both EPI and NE (P<0.05), with a much greater EPI response (Fig. 2).

Since we anticipated that the nicotine-induced increase in plasma EPI would produce some resistance to the hypoglycemic effects of insulin, we administered a larger dose (0.20 units/kg) of insulin to nicotine-treated dogs in order to achieve the same degree of hypoglycemia. Plasma glucose in nicotine-treated dogs decreased from 5.8±0.3 to 2.1±0.1 mM (P<0.001) at 45 min. This represented identical decrement and nadir of plasma glucose to that produced in control dogs, confirming similar degrees of hypoglycemic stress in the two groups. Prior exposure to nicotine inhibited the adrenal NE and EPI secretion rate responses to hypoglycemia (Fig. 2).

In contrast to the cross-tolerance of hypoglycemia-induced adrenal catecholamine secretion observed after nicotine, hypoglycemia-induced tachycardia was not significantly modified by prior exposure to nicotine (Fig. 3).

DISCUSSION

Nicotine administered either systemically in the rat and dog or intracerebroventricularly in rats increases plasma catecholamine

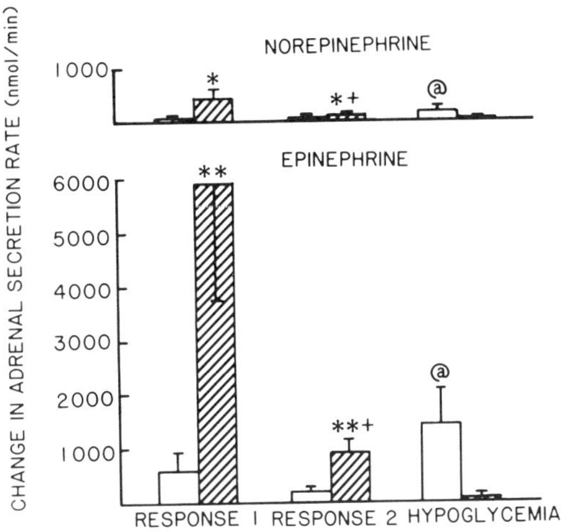

Fig. 2. Adrenal secretion rates of norepinephrine (upper panel) and epinephrine (lower panel) in response to nicotine in pentobarbital-anesthetized dogs. Nicotine (hatched bars) or saline (open bars) was administered and hypoglycemia induced as described in the legend to Fig. 1. Blood was sampled from the adrenolumbar vein for measurement of adrenal catecholamine secretion rates. Data represent the mean ± S.E.M. change in secretion rate from the baseline (pretreatment) rate to the average response over the peak hypoglycemic epoch (30–60 min post-insulin). $^{*}P<0.05$ $^{**}P<0.025$ compared with saline control group. $^{+}P<0.05$ compared with the initial response to nicotine. @ $P<0.05$ indicating a significant increase in secretion rate in response to hypoglycemia.

concentrations. Systemic administration of nicotine in rats produced increases in plasma NE about equal to those in plasma EPI. This finding is similar to other recent reports in conscious dogs (8) and rats (11) in which plasma NE and EPI levels increased in response to nicotine given intravenously. Earlier reports using less specific methodology described nicotine-induced increases in plasma EPI and NE in anesthetized cats (12) and increases in urinary EPI and its major O-methylated metabolite, metanephrine, in conscious rats (13). Elevated plasma (14,15) and urinary (16) catecholamines have also been reported after tobacco smoking. More recently, Dominiak et al. (11) found significantly higher levels of plasma NE and EPI in anesthetized rats compared to conscious rats at rest as well as in response to nicotine. These data differ somewhat from our findings in pentobarbital-anesthetized dogs in which systemic nicotine produced a marked increase in plasma EPI but only a small increase in plasma NE. Pentobarbital anesthesia has been reported to decrease basal NE spillover rate into plasma by more than 90 percent (17) and this may have accounted largely for the very small plasma NE response produced by nicotine in our experiment. The equal peripheral plasma EPI response to the second nicotine bolus in spite of blunting of the adrenal venous EPI response raises the possibility that nicotine may decrease the clearance rate of catecholamines.

The source of the plasma NE released in response to nicotine could be either from sympathetic nerve endings and overflow into plasma in addition to release from adrenal medulla (18) or alternatively from proportional adrenomedullary NE and EPI secretion without concomitant release of NE from sympathetic nerves (19). It seems likely, however, that systemic nicotine does increase plasma NE in part by an action at sympathetic

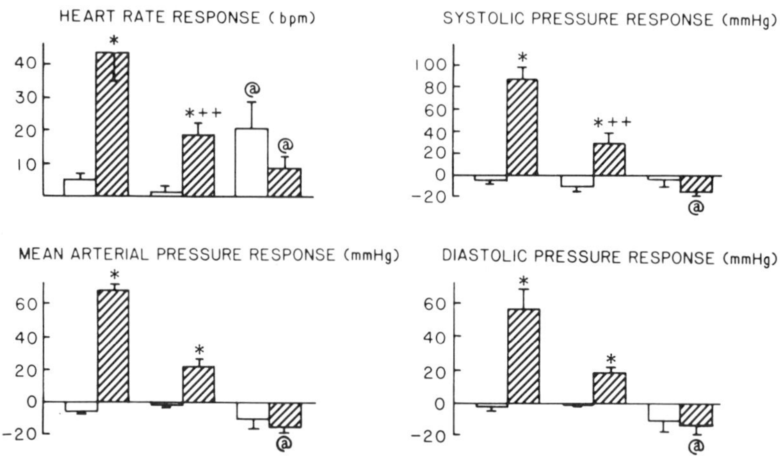

Fig. 3. Cardiovascular responses to nicotine in pentobarbital-anesthetized dogs. Nicotine (hatched bars) or saline (open bars) was administered and hypoglycemia induced as described in the legend to Fig. 1. Blood pressure was recorded from a femoral arterial catheter connected to a Statham pressure transducer, and heart rate was obtained with a tachometer triggered by the pulse wave. The peak cardiovascular responses after treatment are given as the differences from pretreatment levels. Data represent the mean ± S.E.M. for 4 saline- and 7 nicotine-treated dogs. $^{*}P<0.05$ compared to saline control. $^{++}P<0.025$ compared with the initial response to nicotine. @ $P<0.05$ indicating a significant response to hypoglycemia.

ganglia with subsequent NE release from sympathetic nerves and in part by an action at adrenal chromaffin cells.

In contrast to our findings with systemic nicotine, intracerebral administration of nicotine appeared to preferentially activate central sympathetic outflow to the adrenal medulla, since plasma EPI levels increased much more than those of NE in response to ivt nicotine. It would appear reasonable to suggest that in contrast to rather generalized sympathetic activation by actions of systemically administered nicotine at nicotinic receptors in adrenal medulla, sympathetic ganglia and even in brain, intracerebrally administered nicotine stimulates brain nicotinic receptors which predominantly activate sympathetic outflow to the adrenal medulla.

Systemic administration of nicotine increased systolic, diastolic, and mean arterial blood pressure in both conscious rats and pentobarbital-anesthetized dogs in our studies. Associated with these pressor effects of nicotine, we found bradycardia in conscious rats and tachycardia in anesthetized dogs. Variable effects of nicotine on cardiovascular parameters have been reported previously (20-22). Because of the complex pharmacological effects of nicotine on sympathetic and parasympathetic ganglia, central nervous system, adrenal medulla, and chemoreceptor and baroreceptor reflex mechanisms, the observed cardiovascular responses to nicotine may vary substantially, depending on the dose used, and the species and state of consciousness of the animal. Mandel et al. (23) described a triphasic cardiovascular response to intravenous nicotine in the conscious dog which consisted of initial bradycardia due to vagal ganglionic stimulation followed by tachycardia from catecholamine release and, finally, reflex bradycardia. The reflex bradycardia observed in that study in conscious dogs and in our studies in conscious rats may have been masked in our dog experiments by the presence of anesthesia since barbiturates have been shown to depress baroreflex mechanisms (24).

Intracerebroventricular nicotine in the doses used in our study had little effect on mean arterial pressure and heart rate. Hoffman (25) reported similarly that injection of low doses (0.1-1.0 ug) of nicotine into the third cerebral ventricle did not alter blood pressure in unanesthetized, unrestrained rats. On the other hand, anesthetized rats responded to intracisternal nicotine (1-3 ug) with an initial pressor response followed by sustained hypotension; a small bradycardia was also noted in that study (26). It is unclear whether these differences relate to effects of anesthesia or to different brain sites of action.

The effects of both systemic and ivt nicotine were blocked by the classical nicotinic antagonists, hexamethonium and mecamylamine. These data suggest that the nicotinic receptor mediating the actions of nicotine in brain is similar to the peripheral ganglionic nicotinic receptor. Others have suggested that behavioral effects of nicotine are mediated through a brain receptor with characteristics similar to the ganglionic receptor (27, 28).

Acute ivt injection of nicotine produced a prolonged state of tolerance with respect to sympathoadrenal stimulation. A single nicotine treatment markedly altered the plasma EPI response to a second nicotine injection given 24 hr later. Facile development of long-lasting tolerance has also been reported for other brain effects of nicotine. Treatment with nicotine once daily for 3 days shifted the dose-response curve for nicotine-induced depression of locomotor activity (29). Progressive loss of the prostration-immobilization syndrome was observed with 3 daily ivt nicotine injections, so that by the third day of treatment the response was only 10 percent of that observed on the first day (30).

In contrast to the ease of development of tolerance to ivt nicotine, no reduction in catecholamine secretion was noted when nicotine injections were given ia daily for 7 days or when three ia nicotine treatments were given at 70-min intervals. Tachyphylaxis to systemic nicotine was observed only when the interval between treatments was reduced to at least 30 min.

It is of interest that the time course for development of tachyphylaxis to other effects of systemically administered nicotine is quite similar to that for nicotine-induced catecholamine secretion. Thus, nicotine-induced tremor (31), respiratory stimulation, and electroencephalographic as well as behavioral arousal (32) were attenuated when the interval between injections of nicotine was 30 min or less. Kovacic and Robinson (33) observed tachyphylaxis of EPI secretion from the isolated dog adrenal with repeated injections of nicotine at 100-sec intervals. It is not likely that the reduced EPI secretion was due to depletion of adrenal EPI since responses to other secretagogues were potentiated at the time of maximum nicotine tolerance. Westfall and Brase (34) observed tolerance to daily injections of nicotine with respect to urinary catecholamine excretion; they attributed this tolerance to increased monoamine oxidase and catechol-O-methyl transferase activities and consequently enhanced EPI metabolism. No changes in adrenal medullary catecholamine content were detected in their study.

It does appear that brain nicotinic receptors or postreceptor mechanisms in brain undergo desensitization to nicotine by a different mechanism from the peripheral nicotinic receptor or peripheral postreceptor mechanisms. An increase or up-regulation in brain nicotinic binding has been reported after chronic nicotine treatment (35, 36).

However, no changes in K_D or Bmax for DL-[^{3}H]nicotine binding were detected after a single injection of the drug (35). The mechanism of tolerance to a single ivt injection of nicotine with respect to stimulation of catecholamine secretion and the difference between the development of tolerance to nicotine administered in brain versus periphery remains unclear.

Even within the sympathoadrenal system, tolerance to nicotine does not develop and is not manifested similarly for all parameters. For example, tachyphylaxis to nicotine was noted for its sympathoadrenal catecholamine-releasing effects at a time when it was not present for its pressor effects. This indicates that the effect of nicotine on blood pressure is only partially mediated by circulating catecholamines. Nicotinic receptor stimulation at sympathetic nerve endings in the heart (37) and vasculature (38) causes norepinephrine release resulting in vasoconstriction and increased cardiac contractility, which probably represents a more important mechanism of the pressor effect of nicotine. A minor role may also be provided by nicotine-induced vasopressin secretion (39).

The antianxiety effect of chronic cigarette smoking has been attributed to nicotine (40). However, the mechanism of this effect of nicotine has not been established. It is possible that chronic exposure to tobacco smoke alters physiological stress responses thereby reducing stress-evoked anxiety. Since stress-induced sympathoadrenal catecholamine secretion is mediated through nicotinic receptors, it seems reasonable to expect that tolerance of the sympathoadrenal responses to nicotine might result also in inhibition of the plasma catecholamine responses to stress.

We tested this hypothesis in two animal models. Conscious rats were stressed by restraint and anesthetized dogs were exposed to the stress of insulin-induced hypoglycemia after receiving repeated treatments with nicotine. Rats rendered tolerant to nicotine by repeated ia injections of the drug at 15-30 min intervals demonstrated cross-tolerance to the plasma catecholamine responses to restraint stress. In contrast, an injection schedule of nicotine that did not induce tachyphylaxis also did not alter stress-induced catecholamine secretion. Similar to the data in conscious rats, when anesthetized dogs showed tachyphylaxis to the catecholamine-releasing effects of nicotine, they also showed blunting of the sympathoadrenal catecholamine responses to hypoglycemia. These findings could be explained by a nicotine-induced depletion of adrenal catecholamines. Although large quantities of catecholamines appear to be present in the adrenal after such stimulation by nicotine, a readily releasable pool of catecholamines may have been depleted. Alternatively, changes in sensitivity of adrenal, ganglionic, and brain nicotinic receptors or postreceptor mechanisms by repeated injections of drug might alter responses not only to nicotine but also to other stimuli acting through nicotinic receptors.

Such a cross-tolerance between nicotine and stress is not, however, a universal finding for all parameters affected by nicotine. For example, Cam and Bassett (41) reported lack of cross-tolerance between nicotine and footshock stress with respect to stimulation of corticosterone secretion. Animals tolerant to the corticosterone-releasing effect of daily nicotine demonstrated no tolerance to stress-induced corticosterone secretion and vice versa. Thus, effects of nicotine on the sympathoadrenal system are regulated differently from effects on the hypothalamic-pituitary-adrenocortical axis.

Tolerance to the catecholamine-releasing and bradycardic effects of nicotine may be mediated by different mechanisms or at least at different sites, since the tolerance of the catecholamine-releasing effects but not of the bradycardia effect showed cross-tolerance to hypoglycemic stress.

Our data do suggest that when subjects are tolerant to nicotine with respect to stimulation of sympathoadrenal catecholamine secretion, as might be the case in cigarette smokers, they may also fail to show normal catecholamine responses to stress.

SUMMARY

Nicotine is a potent autonomic stimulant that elicits pronounced sympathetic nervous system activation at several sites along the neuraxis. Recently, stereospecific binding sites for nicotine have been demonstrated in the central nervous system, but it has not been clear whether nicotine can act in the brain to stimulate sympathetic outflow. Additionally, the effect of nicotine on stress-induced sympathetic activation has not been reported. We studied acute and chronic cardiovascular and sympathoadrenal effects of nicotine after systemic and intracerebral administration, as well as the effect of nicotine on physiologic responses to stress.

Nicotine rapidly increased plasma concentrations of norepinephrine and epinephrine after either systemic or intraventricular brain injections, but a pressor effect occurred only with peripheral nicotine. Also, intracerebral nicotine stimulated predominantly epinephrine secretion consistent with a selective effect on central sympathetic outflow to the adrenal medulla. All of these responses were antagonized by hexamethonium. Tolerance developed to the effects of nicotine to stimulate catecholamine secretion, but the rate of tolerance development differed substantially with the two routes of administration. Injections given more than one hour apart did not attenuate systemic nicotine-induced plasma catecholamine increases, whereas a reduction in the response to intracerebral nicotine was observed 24 hours after a single treatment. Animals rendered tolerant to nicotine by systemic injections demonstrated cross-tolerance to the sympathoadrenal stimulant effect of stress. This cross-tolerance was found with restraint stress in rats and with hypoglycemic stress in dogs.

The following conclusions were drawn from the results of these studies:

1. Sympathoadrenal stimulation is produced by an action of nicotine in brain as well as periphery. Activation of brain nicotinic receptors selectively stimulates sympathetic outflow to adrenal medulla.

2. Brain nicotinic receptors involved in sympathetic outflow may differ from peripheral nicotinic cholinoceptors since the rate of tolerance development to intracerebral nicotine differs considerably from that to systemic nicotine.

3. Cross-tolerance between nicotine and stress with respect to sympathoadrenal stimulation may relate to depletion by nicotine of a readily releasable pool of catecholamines or may indicate desensitization of a common nicotinic cholinergic receptor mechanism involved in catecholamine secretion.

ACKNOWLEDGEMENTS

These studies were supported by grants from the Tobacco and Health Research Institute of the University of Kentucky and from the Veterans Administration.

REFERENCES

1. Stewart, G.N. and Rogoff, J.M. The action of drugs on the output of epinephrine from the adrenals. III. Nicotine. J. Pharmacol. Exp. Ther. 13: 183-241 (1919).
2. Woods, L.F., Richardson, J.A., Richardson, A.K. and Bozeman, R.F. Plasma concentrations of epinephrine and arterenol following the actions of various agents on the adrenals. J. Pharmacol. Exp. Ther. 116: 351-355 (1956).
3. Viveros, O.H., Diliberto, E.J., Hazum, E. and Chang, K.J. Opiate-like materials in the adrenal medulla: evidence for storage and secretion with catecholamines. Mol. Pharmacol. 16: 1101-1108 (1979).
4. Livett, B.G., Day, R., Elde, R.P. and Howe, P.R.C. Costorage of enkephalin and adrenaline in bovine adrenal medulla, Neuroscience 7: 1323-1332 (1982).
5. Chauchard, A.B. and Chauchard, P. Actions localizees aux centres medullaires et leur retentissement peripherique sur le nerf dans le domaine splanchnique. C.R. Soc. Biol. (Paris) 136: 382-383 (1942).
6. Thienis, C.H. Nicotine tolerance in the white rat. Proc. Soc. Exp. Biol., 26: 845-846 (1929).
7. Clark, M.S.G., Rand, M.J. and Vanov, S. Comparison of pharmacological activity of nicotine and related alkaloids occurring in cigarette smoke. Arch. Int. Pharmacodyn. 156: 363-379 (1965).
8. Holtz, J., Sommer, O. and Bassenge, E. Development of specific tolerance to nicotine infusions in dogs on chronic nicotine treatment. Klin. Wochenschr. 623 (Suppl II): 51-57 (1984).
9. Van Loon, G.R., Appel, N.M. and Ho, D., β-endorphin-induced increases in plasma epinephrine, norepinephrine and dopamine in rats: inhibition of adrenomedullary responses by intracisternal somatostatin, Brain Res. 212: 207-214 (1981).
10. Hume, D.M. and Nelson, D.H. Adrenal cortical function in surgical shock, Surg. Forum, 5: 568-575 (1955).
11. Dominiak, P., Kees, F. and Grobecker, H. Changes in peripheral and central catecholaminergic and serotoninergic neurons of rats after acute and subacute administration of nicotine. Klin. Wochenschr. 62 (Suppl II): 76-80 (1984).
12. Armitage, A.K. and Milton, A.S. The release of adrenaline by nicotine from the adrenal medulla, in: Tobacco Alkaloids and Related Compounds, U.S. von Euler (ed.), Pergamon Press, New York, pp. 205-214 (1965).
13. Westfall, T.C. Effect of nicotine and nicotine analogues on tissue and urinary catecholamines in the rat. Acta Physiol. Scand. 63: 77-83 (1965).
14. Kashemsant, U., Smulyan, H. and Eich, R.H. Changes in blood catecholamines with smoking. Circulation 30 (Suppl. II):19 (1965).
15. Cryer, P.E., Haymond, M.W., Santiago, J.V. and Shah, D.A. Norepinephrine and epinephrine release and adrenergic mediation of smoking-associated hemodynamic and metabolic events. New Engl J. Med. 295: 573-577 (1976).
16. Westfall, T.C. and Watts, D.T. Catecholamine excretion in smokers and non-smokers. J. Appl. Physiol. 19: 40-42 (1964).

17. Best, J.D., Taborsky, G.J. Jr., Flatness, D.E. and Halter, J.B. Effect of pentobarbital anesthesia on plasma norepinephrine kinetics in dogs. Endocrinology 115: 853-857 (1984).
18. Malmejac, J., Schlotterer, M., Clostre, F. and Roux, A. Sur l'importance des secretions medullosurrenales dans la production de l'hypertension par la nicotine, C.R. Soc. Biol. (Paris) 162: 1276-1280 (1968).
19. Miele, G. The nicotinic stimulation of the cat adrenal medulla. Arch. Int. Pharmacodyn. 179: 343-351 (1969).
20. Sutton, I. and Isaac, P.F. The cardiovascular actions of nicotine in the rat. Arch. Int. Pharmacodyn. 202: 5-16 (1973).
21. Dominiak, P., Fuchs, G., von Toth, S. and Grobecker, H. Effects of nicotine and its major metabolites on blood pressure in anesthetized rats. Klin. Wochenschr. 63: 90-92 (1985).
22. Papacostas, C.A. and Reed, J.P. Influence of β-receptor blockade on certain cardiovascular actions of nicotine. Arch. Int. Pharmacodyn. Ther. 164: 167-172 (1966).
23. Mandel, W.J., Laks, M., Hayakawa, H., Obayashi, K. and McCullen, A. Cardiovascular effects of nicotine in the conscious dog: modification by changes in autonomic tone. Am. J. Cardiol. 32: 947-955 (1973).
24. Cox, R.H. and Bagshaw, R.J. Influence of anesthesia on the response to carotid hypotension in dogs. Am. J. Physiol. 237: H424-H432 (1979).
25. Hoffman, W.E. Central cholinergic receptors in cardiovascular and antidiuretic effects in rats. Clin. Exp. Pharmacol. Physiol. 6: 373-380 (1979).
26. Kubo, T. and Misu, Y. Cardiovascular responses to intracisternal administration of nicotine in rats. Can. J. Physiol. Pharmacol. 59: 615-617 (1980).
27. Clarke, P.B.S. and Kumar, R. Characterization of the locomotor stimulant action of nicotine in tolerant rats. Brit. J. Pharmacol. 80: 587-594 (1983).
28. Romano, C., Goldstein, A. and Jewell, N.P. Characterization of the receptor mediating the nicotine discriminative stimulus. Psychopharmacology 74: 310-315 (1981).
29. Stolerman, I.P., Bunker, P., and Jarvik, M.E. Nicotine tolerance in rats: Role of dose and dose interval. Psychopharmacology 34: 317-324 (1974).
30. Abood, L.G., Lowy, K. and Booth, H. Acute and chronic effects of nicotine in rats and evidence for a noncholinergic site of action. In: Cigarette Smoking as a Dependence Process, N.A. Krasnegor (ed.), NIDA Research Monograph No. 23, DHEW Pub. No. 79-800, pp. 136-149 (1979).
31. Holmstedt, B. and Lundgren, G. Arecoline, nicotine, and related compounds. Tremorgenic activity and effect upon brain acetylcholine. Ann. N.Y. Acad. Sci. 142: 126-142 (1967).
32. Yamamoto, K. and Domino, E.F. Nicotine-induced EEG and behavorial arousal. Int. J. Neuropharmacol. 4: 359-373 (1965).
33. Kovacic, B. and Robinson, R.L. Facilitation of drug-induced release of adrenal catecholamines during nicotine blockade. Fed. Proc. 27: 601 (1968).
34. Westfall, T.C. and Brase, D.A. Studies on the mechanism of tolerance to nicotine-induced elevations of urinary catecholamines. Biochem. Pharmacol. 20: 1627-1635 (1971).
35. Marks, M.J., Burch, J.B. and Collins, A.C. Effects of chronic nicotine infusion on tolerance development and nicotinic receptors. J. Pharmacol. Exp. Ther. 226: 817-825 (1983).
36. Schwartz, R.D. and Kellar, K.J. Nicotinic cholinergic receptor binding sites in the brain: regulation _in vivo_. Science 220: 214-216 (1983).

37. Westfall, T.C. and Brasted, M. The mechanism of action of nicotine on adrenergic neurons in the perfused guinea-pig heart. J. Pharmacol. Exp. Ther. 182:409-418 (1972).
38. Su, C. and Bevan, J.A. Blockade of the nicotine-induced norepinephrine release by cocaine, phenoxybenzamine and desipramine. J. Pharmacol. Exp. Ther. 175: 533-540 (1970).
39. Bisset, G.W. and Walker, J.M. The effects of nicotine, hexamethonium and ethanol on the secretion of the antidiuretic and oxytocic hormones of the rat. Br. J. Pharmacol. Chemother. 12: 461-467 (1957).
40. Finnegan, J.K., Larson, P.S. and Haag, H.B. The role of nicotine in the cigarette habit. Science 102: 94-96 (1945).
41. Cam, G.R. and Bassett, J.R. Effect of prolonged exposure to nicotine and stress on the pituitary-adrenocortical response: the possibility of cross-adaptation. Pharmacol. Biochem. Behav. 20: 221-226 (1984).

CENTRAL NICOTINIC REGULATION OF ARTERIAL BLOOD PRESSURE

T. Kubo

Department of Pharmacology
Yokohama City University School of Medicine
Urafunecho, Minami-ku, Yokohama 232, Japan

Effects of nicotine on the cardiovascular control of the central nervous system have been studied for many years by numerous investigators (1). Intraventricular or intracisternal administration of nicotine produced an increase or a decrease in blood pressure and heart rate in many animal species including the cat (2-6), dog (7), rat (8,9), and goat (10). Further, some investigators have studied cardiovascular responses to nicotine applied at discrete regions of the brain, including the posterior hypothalamus (11), ventral surface of the medulla oblongata (12-14), and lateral medullary reticular pressor areas (11) of the cat, and the nucleus ambiguus of the dog (15). These central actions of nicotine were abolished by hexamethonium or mecamylamine (4,7-12,15), suggesting the involvement of a central nicotinic cholinergic mechanism in the actions. Recently, nicotinic acetylcholine receptors have been detected on rodent (16-19) and human (20) brain membranes. However, exact central nicotinic mechanisms responsible for cardiovascular responses are not yet fully understood. The purpose of the present study was to localize nicotine-sensitive sites responsible for cardiovascular effects in the dorsal medulla oblongata of the rat. Central or peripheral mechanisms relating to the action of nicotine were also examined. Further, the medullary nicotine action was evaluated also in experimentally hypertensive rats.

METHODS

Male Wistar rats, weighing 250-320 g, were used. In some experiments, 5- or 15-18-week-old spontaneously hypertensive rats (SHR), age-matched Wistar-Kyoto rats (WKY), and DOC-salt hypertensive rats were used. The rats were anesthetized with urethane (1.2 g/kg, i.p.), and the femoral artery and vein were cannulated for continuous recordings of blood pressure and intravenous injections, respectively. Heart rate was computed by a cardiotachometer.

The rats were placed in a stereotaxic apparatus with the head fixed at 45°. The dorsal surface of the lower brainstem was exposed as described previously (21). Application of nicotine on the dorsal surface was made through a polyethylene tube (outer diameter, 1.3 mm) placed on the surface. Microinjections of drugs were made using a glass cannula (outer diameter, 0.1 mm) connected to a Hamilton microsyringe and a micrometer. The location of the needle tracks was controlled microscopically.

Electrolytic lesions of the nucleus tractus solitarius were produced bilaterally by passing an anodal DC current of 5 mA for 1-2 sec through monopolar Teflon-coated stainless steel electrodes (outer diameter, 0.15 mm) with a tip exposure of 0.2 mm. In some experiments, the carotid sinus and aortic baroreceptors were denervated bilaterally according to the method described by Krieger (22).

Data are reported as mean ± S.E. and statistical significance was evaluated by Student's t-test.

RESULTS AND DISCUSSION

The Dorsal Surface Areas of the Medulla Oblongata

Following intracisternal injections of nicotine, a biphasic effect consisting of an initial rise in blood pressure followed by a delayed fall was observed in urethane-anesthetized rats (9). To examine possible areas responsible for blood pressure response to nicotine on the dorsal surface of the medulla oblongata, nicotine (3 μg) was applied using a polyethylene tube placed on the surface. When the tube was placed around the area postrema (positions 1,2, and 4 shown in Fig. 1), nicotine produced a biphasic blood pressure response. The greatest changes in blood pressure were observed when the center of the tube was at the caudal tip of the area postrema (position 1).

To more precisely localize the nicotine-sensitive sites, nicotine (0.1 μg) was microinjected into several sites 0.2 mm below the dorsal surface. Nicotine usually produced hypertension followed by hypotension (Fig. 2) and bradycardia. Figure 3 shows the nicotine injection sites and the blood pressure responses. The highest sensitivity was observed following injections into sites just caudal to the area postrema. These findings are compatible with the study of Brezenoff and Jenden (23) who found only an insignificant fall in blood pressure following injections of nicotine into the floor of the 4th ventricle which lies more rostral to the nicotine-sensitive sites. Hexamethonium (2 μg), injected at the

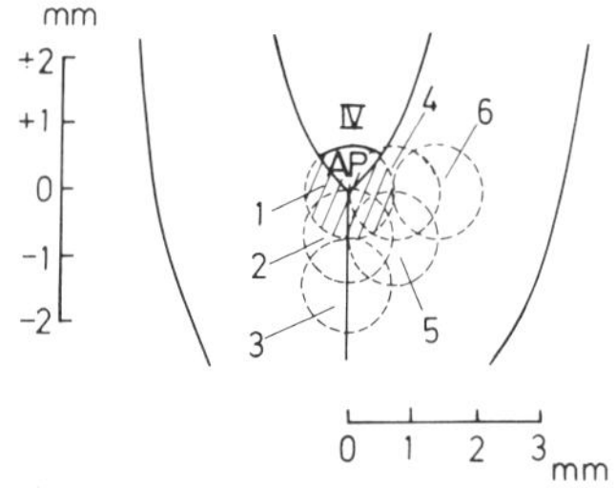

Fig. 1. Diagram of the dorsal surface of the medulla oblongata of the rat. The interrupted circles indicate the areas covered by the polyethylene tube when placed across the medulla in six positions. When applied to the striped area (position 1), the cardiovascular effect of nicotine (3 μg) was maximal. AP, area postrema; IV, fourth ventricle.

Fig. 2. The effect on arterial blood pressure of nicotine (N, 0.1 μg) injected into a site just caudal to the area postrema. The interval represents 5 min.

nicotine-sensitive sites before nicotine, abolished the blood pressure and heart rate responses to nicotine, while atropine (2 μg), similarly injected, did not alter the responses, indicating that these responses are mediated by nicotinic receptors.

The medulla oblongata contains acetylcholine, choline acetyltransferase, and acetylcholinesterase (24,25). Microinjections of physostigmine (1 and 3 μg), an acetylcholinesterase inhibitor, or choline (1 and 3 μg), a precursor of acetylcholine, into the nicotine-sensitive sites also caused a decrease in blood pressure (Fig. 4A and B) and heart

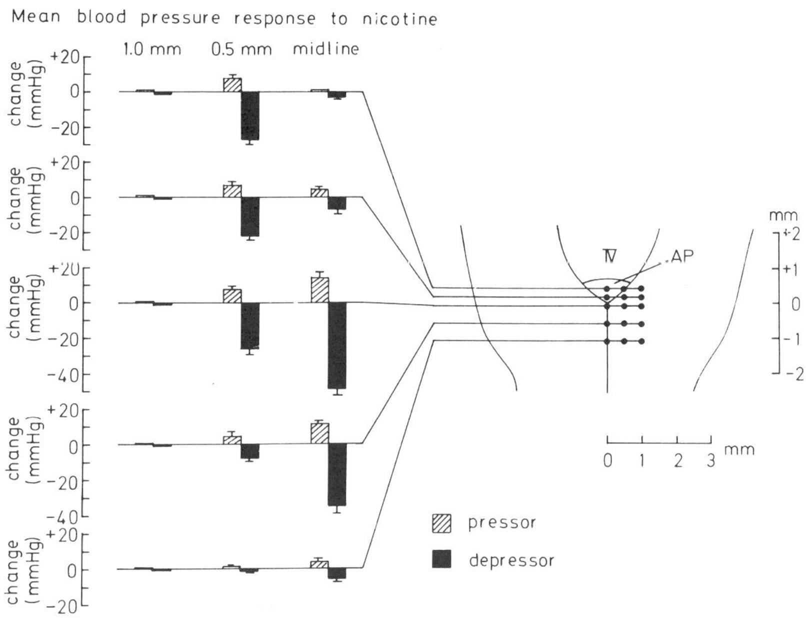

Fig. 3. Maximal increases and decreases in blood pressure following injections of nicotine (0.1 μg) into discrete medullary sites in urethane-anesthetized rats. Injection sites were located at 5 rostro-caudal levels. At each level, injections were made at a depth of 0.2 mm below the surface of the medulla and at distances lateral to the midline as indicated. Data shown are mean ± S.E. of the responses of mean blood pressure from 5-8 rats. AP, area postrema; IV, fourth ventricle.

rate. The cardiovascular responses to both agents were abolished by hexamethonium (2 μg) but not by atropine (2 μg), similarly injected. The hypotensive response to choline was enhanced by prior administration of physostigmine (0.3 μg) into the medullary site (Fig. 4C and D), while hemicholinium-3 (3 μg), similarly injected, abolished the blood pressure response to choline. Hemicholinium-3 is thought to block the high affinity uptake of choline into cholinergic neurons (26,27). Thus, these observations suggest that the depressor effect of choline is mediated by an increase in synthesis or release of acetylcholine. Further, it may be

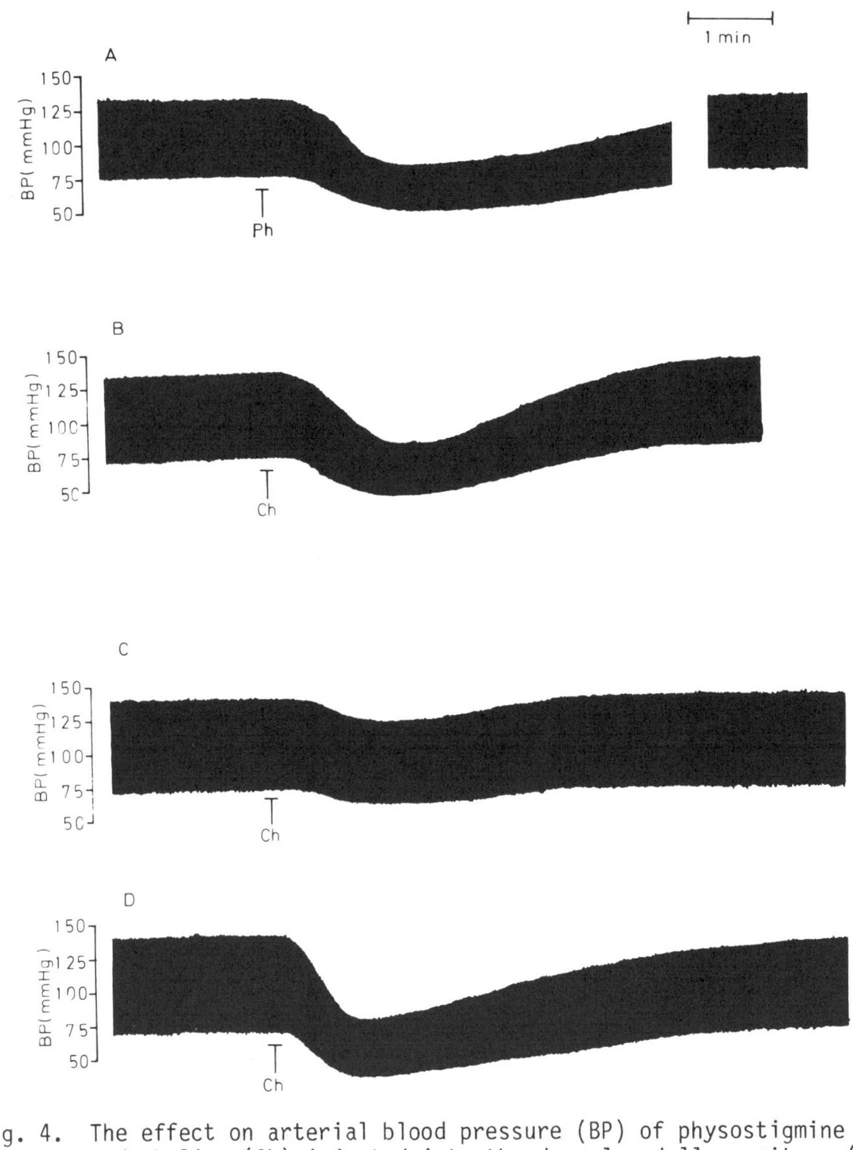

Fig. 4. The effect on arterial blood pressure (BP) of physostigmine (Ph) and choline (Ch) injected into the dorsal medullary site. (A) BP response to physostigmine (3 μg) in the rat (280 g). The interval in the tracing represents 6 min; (B) BP response to choline (3 μg) in the rat (270 g); (C) and (D) BP response to choline (1 μg) before (C) and 3 min after (D) physostigmine (0.3 μg) injected into the same site in the rat (300 g). Choline was injected at 30-min intervals.

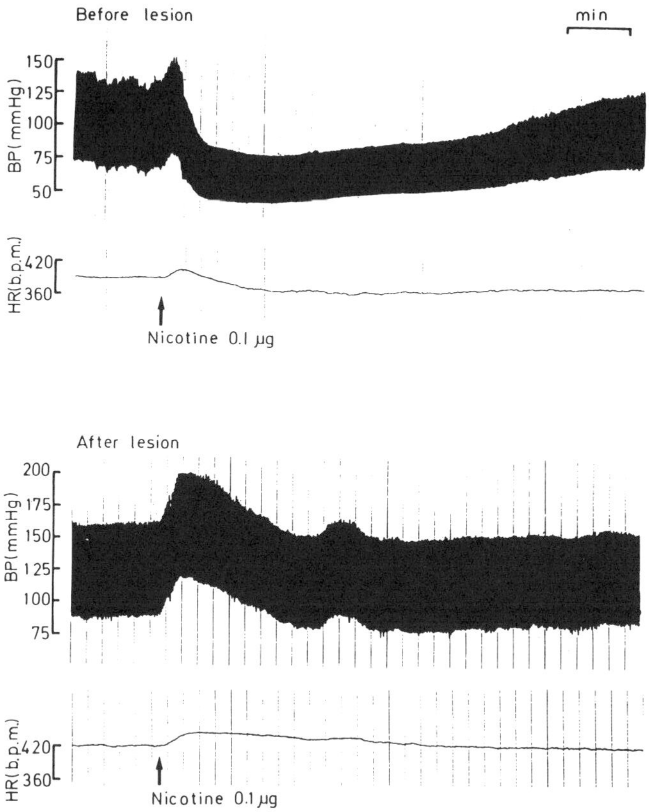

Fig. 5. The effect on arterial blood pressure (BP) and heart rate (HR) of nicotine (0.1 μg) injected into the dorsal medullary site before (upper two records) and after (lower two records) bilateral electrolytic lesions of the nucleus tractus solitarius in a urethane-anesthetized rat (310 g). Nicotine was injected at 35-min intervals.

speculated that this nicotinic cholinergic mechanism is involved in the central control of the cardiovascular system.

The nicotine-sensitive medullary area lies just above the nucleus tractus solitarius which plays an important role in the central cardiovascular regulation (28). Thus, it is possible that this nucleus is involved in the expression of the action of nicotine applied at the nicotine-sensitive site. When the nucleus tractus solitarius was electrolytically lesioned bilaterally at the level of the obex, the depressor response to nicotine (0.1 μg) was markedly inhibited (Fig. 5). In contrast, bilateral sino-aortic denervation did not affect the response. These findings indicate that the nucleus tractus solitarius is essential for the expression of the depressor action of nicotine. Further, bradycardic responses to phenylephrine (10 and 30 μg/kg, i.v.) were significantly enhanced after injections of nicotine (0.03 μg) into the nicotine-sensitive site (Fig. 6), suggesting that the medullary nicotinic mechanism can affect baroreflex responses, probably through mediation of the nucleus tractus solitarius function.

The nucleus tractus solitarius is densely innervated by adrenergic nerve terminals (24,29) and the adrenergic neurons are thought to be involved in the cardiovascular control at this nucleus (21). When

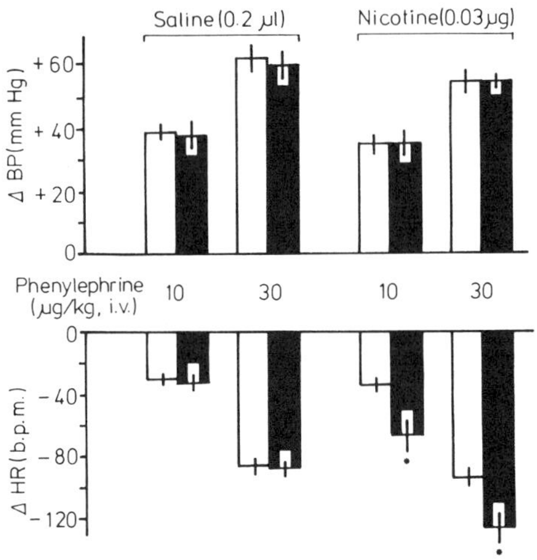

Fig. 6. Effects of nicotine (right columns) or saline (left columns) injected into the dorsal medullary site on phenylephrine-induced bradycardia in urethane-anesthetized rats. Phenylephrine (10 and 30 μg/kg) was intravenously injected before (open columns) and 3-6 min after (filled columns) nicotine (0.03 μg) or saline (0.2 μl). Values are mean ± S.E. from 5 experiments. Statistically significant P<0.01, compared with the respective before value.

phentolamine (10 nmol), an alpha adrenoceptor antagonist, was injected bilaterally into the nucleus tractus solitarius, the depressor response to physostigmine (3 μg) injected into the nicotine-sensitive site was decreased to 44% of control, suggesting that the hypotensive effect is partly mediated via the release of catecholamines in this nucleus.

The major source of catecholamine terminals in the nucleus tractus solitarius has been morphologically demonstrated to be the catecholaminergic nerve cell group A2 (29,30). Since the A2 areas are located close to the nicotine-sensitive sites, it may be possible that nicotine injected at the nicotine-sensitive sites diffuses into the A2 area and directly stimulates the A2 neurons. When microinjected directly into the A2 areas, however, nicotine (0.1 μg) produced only a small or no decrease in blood pressure (Fig. 7). Thus, it seems likely that nicotine injected at the nicotine-sensitive sites does not stimulate the A2 neurons directly but activates them indirectly via interneurons.

Methylatropine (1 mg/kg, i.v.) reduced the bradycardic effect of nicotine (0.1 μg) to 9% of control, suggesting the involvement of increased vagal tone in mediation of the bradycardic response. The blood pressure response to nicotine was abolished after spinal transection, suggesting that this response is mediated mainly by changes in sympathetic outflow. Although the dorsal area has been reported to be involved in regulation of vasopressin release (31,32), any contribution of vasopressin in the nicotine action could not be detected.

THE NUCLEUS TRACTUS SOLITARIUS

Following unilateral injections of acetylcholine (10-100 ng) into the nucleus tractus solitarius at the level of the obex (21), a dose-dependent decrease in blood pressure was observed. The response to acetylcholine

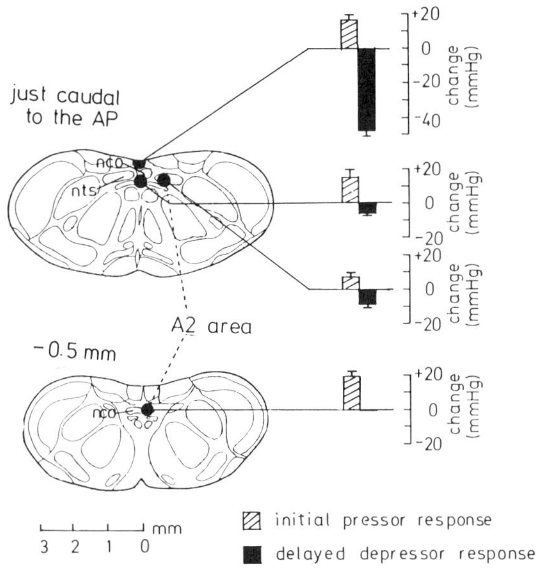

Fig. 7. Maximal increases and decreases in blood pressure following injections of nicotine (0.1 μg) into the A2 areas in urethane-anesthetized rats. Data shown are mean ± S.E. from 18–24 experiments. AP, area postrema; nco, nucleus commissuralis; nts, nucleus tractus solitarius.

(100 ng) was abolished by atropine (300 ng), suggesting the involvement of muscarinic receptors in mediation of the response. Bilateral injections of atropine (1 μg) into the nucleus tractus solitarius produced a pressor response, suggesting existence of endogenous cholinergic muscarinic tone in this nucleus of the rat. These results confirm those of Criscione et al. (33).

Microinjections of nicotine (10 ng) into the nucleus also produced a decrease in blood pressure and heart rate. The depressor response was abolished by either hexamethonium (30 ng) or atropine (30 ng), applied at the same site, while physostigmine (30 ng) similarly injected did not affect the response. Thus, no specific receptors responsible for nicotine action could be determined in this region. It seems unlikely that the nicotine action is mediated by released acetylcholine.

THE DORSAL MEDULLARY NICOTINIC MECHANISMS IN EXPERIMENTALLY HYPERTENSIVE RATS

The depressor response to nicotine (0.03 and 0.1 μg) injected into the dorsal medulla nicotine-sensitive sites was significantly smaller in 15–18-week-old hypertensive SHR than that in age-matched WKY (Fig. 8A). A similar difference was observed also between 5-week-old prehypertensive SHR and age-matched WKY (Fig. 8B). In contrast, there was no significant difference in the response between DOC-salt hypertensive and normotensive control rats (Fig. 8C). These findings indicate that the depressor reactivity to nicotine is diminished genetically in SHR. This evidence supports the hypothesis that alterations in the nicotinic cholinergic system of the brain contribute to development or maintenance of hypertension in SHR. Yamada et al. (34) have found a decrease in specific [^{3}H]nicotine binding in the lower brainstem of SHR.

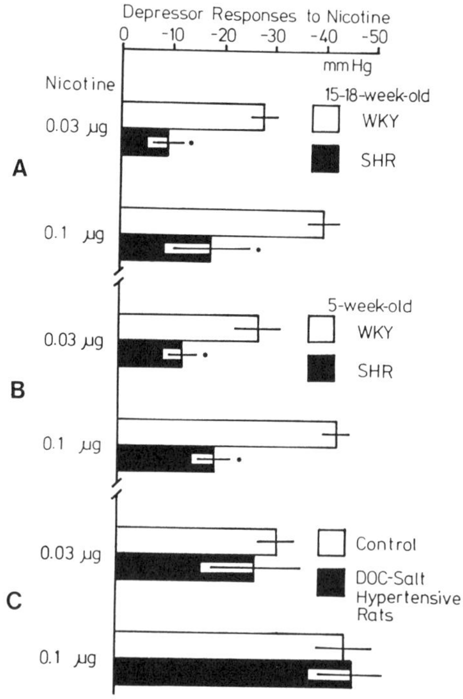

Fig. 8. Depressor responses to nicotine injected into the dorsal medullary site in 15–18-week-old (A) and 5-week-old (B) WKY and SHR, and DOC-salt hypertensive and normotensive control rats (C). Values are mean ± S.E. from 5 experiments. Statistically significant $P<0.05$, compared with WKY. Systolic blood pressure: 131 ± 3 mm Hg, 187 ± 4 mm Hg in 15–18-week-old WKY and SHR, respectively; 128 ± 3 mm Hg, 134 ± 4 mm Hg in 5-week-old WKY and SHR, respectively; 121 ± 3 mm Hg, 198 ± 7 mm Hg in control and DOC-salt hypertensive rats, respectively.

In conscious animals, central administration of nicotine usually causes a marked rise in blood pressure (7,8). Previously we have reported that the pressor response to intraventricular administration of nicotine is increased specifically in SHR (8). The enhanced pressor reactivity may be, in part, due to the diminution of the inhibitory nicotinic function in the medulla oblongata. The same tendency could possibly be found in human essential hypertension, and thus nicotine or tobacco smoking might aggravate such cardiovascular diseases.

SUMMARY AND CONCLUSION

Microinjections of nicotine into the vicinity of the area postrema of the rat elicited hypertension followed by hypotension and bradycardia. The highest sensitivity to nicotine was found at sites just caudal to the area postrema. Microinjections of physostigmine or choline into the site also caused a decrease in blood pressure and heart rate. The hypotensive effect of choline was enhanced by prior administration of physostigmine into the medullary site, while hemicholinium-3, similarly injected, abolished the response. Bilateral electrolytic lesions of the nucleus tractus solitarius markedly reduced the hypotensive and bradycardic response to nicotine. The depressor response to nicotine was decreased in both prephypertensive and hypertensive SHR as compared with age-matched WKY.

From these results, it is concluded that there exist nicotinic receptors responsible for cardiovascular effects in the dorsal medulla oblongata, and the nucleus tractus solitarius is essential for the expression of the nicotine action. Further, the results of the present study suggest the involvement of cholinergic nicotinic mechanisms of the brain in the control of the cardiovascular system. However, many studies will be necessary for elucidation of the exact mechanisms.

REFERENCES

1. Kubo, T: Brain nicotinic acetylcholine receptors and their functions in special reference to central cardiovascular regulation. Folia Pharmacol. Japon. 85: 425-433, 1985.
2. Mitchell, RA, Loeschecke, HH Severinghaus, JW, Richardson, BW and Massion, WH: Regions of respiratory chemosensitivity on the surface of medulla. Ann. N.Y. Acad. Sci. 109: 661-681, 1963.
3. Armitage, AK and Hall, GH: Effects of nicotine on the systemic blood pressure when injected into the cerebral ventricles of cats. Int. J. Neuropharmacol. 6: 143-149, 1967.
4. Armitage, AK and Hall, GH: Further evidence relating to the mode of action of nicotine in the central nervous system. Nature 214: 977-979, 1967.
5. Schaeppi, U: Effects of nicotine administration to the cat's lower brain stem upon electroencephalogram and autonomic system. Ann. N.Y. Acad. Sci. 142: 40-49, 1967.
6. Pradhan, SN, Bhattacharya, IC and Atkinson, KS: The effects of intraventricular administration of nicotine on blood pressure and some somatic reflexes. Ann. N.Y. Acad. Sci. 142: 50-66, 1967.
7. Lang, WJ and Rush, ML: Cardiovascular responses to injections of cholinomimetic drugs into the cerebral ventricles of unanaesthetized dogs. Br. J. Pharmacol. 47: 196-205, 1973.
8. Kubo, T and Misu, Y: Increased pressor responses to nicotine in spontaneously hypertensive rats. Jpn. J. Pharmacol. 31: 221-226, 1981.
9. Kubo, T and Misu, Y: Cardiovascular responses to intracisternal administration of nicotine in rats. Canadian J. Physiol. Pharmacol. 59, 615-617, 1981.
10. Vandeputte-Van Messom, G: Effect of intracerebroventricular injection of nicotine and acetylcholine on renal water and salt excretion in conscious goats. Arch. Int. Pharmacodyn. Ther. 253: 52-67, 1981.
11. Bhargava, KP, Jain, IP, Saxena, AK, Sinha, JN and Tangri, KK: Central adrenoceptors and cholinoceptors in cardiovascular control. Br. J. Pharmacol. 63: 7-15, 1978.
12. Feldberg, W and Guertzenstein, PG: Vasodepressor effects obtained by drugs acting on the ventral surface of the brain stem. J. Physiol. 258: 337-355, 1976.
13. Dev, NB and Loeschcke, HH: Topography of the respiratory and circulatory responses to acetylcholine and nicotine on the ventral surface of the medulla oblongata. Pfluegers Arch. 379: 19-27, 1979.
14. Dev, NB and Loeschcke, HH: A cholinergic mechanism involved in the respiratory chemosensitivity of the medulla oblongata in the cat. Pfluegers Arch. 379: 29-36, 1979.

15. Wu, KM and Martin, WR: An analysis of nicotine and opioid processes in the medulla oblongata and nucleus ambiguus of the dog. J. Pharmacol. Exp. Ther. 227: 302-307, 1983.
16. Yoshida, K and Imura, H: Nicotinic cholinergic receptors in brain synaptosomes. Brain Res. 172: 453-459, 1979.
17. Romano, C and Goldstein, A: Stereospecific nicotine receptors on rat brain membranes. Science 210: 647-650, 1980.
18. Morley, BJ and Kemp, GE: Characterization of a putative nicotinic acetylcholine receptor in mammalian brain. Brain Res. Reviews 3: 81-104, 1981.
19. Schwartz, RD, McGee, R Jr and Kellar, KJ: Nicotinic cholinergic receptors labeled by [^{3}H]acetylcholine in rat brain. Mol. Pharmacol. 22: 56-62, 1982.
20. Shimohama, S, Taniguchi, T, Fujiwara, M and Kameyama, M: Biochemical characterization of the nicotinic cholinergic receptors in human brain: binding of (-)-[^{3}H]nicotine. J. Neurochem. 45: 604-610, 1985.
21. Kubo, T and Misu, Y: Pharmacological characterisation of the α-adrenoceptors responsible for a decrease of blood pressure in the nucleus tractus solitarii of the rat. Naunyn-Schmiedeberg's Arch. Pharmacol. 317: 120-125, 1981.
22. Krieger, EM: Neurogenic hypertension in the rat. Circulation Res. 15: 511-521, 1964.
23. Brezenoff, HE and Jenden, DJ: Changes in arterial blood pressure after microinjections of carbachol into the medulla and 4th ventricle of the rat brain. Neuropharmacology 9: 341-348, 1970.
24. Palkovits, M and Jacobowitz, DM: Topographic atlas of catecholamine and acetylcholinesterase-containing neurons in the rat brain. II. Hindbrain (mesencephalon, rhombencephalon). J. Comp. Neurol. 157: 29-42, 1974.
25. Helke, CJ, Muth, EA and Jacobowitz, DM: Changes in central cholinergic neurons in the spontaneously hypertensive rat. Brain Res. 188: 425-436, 1980.
26. Yamamura, HI and Synder, SH: High affinity transport of choline into synaptosomes of rat brain. J. Neurochem. 21: 1355-1374, 1973.
27. Barker, LA and Mittag, TW: Comparative studies of substrates and inhibitors of choline transport and choline acetyltransferase. J. Pharmacol. Exp. Ther. 192: 86-94, 1975.
28. Miura, M and Reis, DJ: Termination and secondary projections of carotid sinus nerve in the cat brain stem. Am. J. Physiol. 217: 142-153, 1969.
29. Dahlstrom, A and Fuxe, K: Evidence for the existence of monoamine neurons in the central nervous system: II Experimentally induced changes in the interneuronal amine levels of bulbospinal neuron system. Acta Physiol Scand. Suppl. 247: 5-36, 1965.
30. Takahashi, Y, Satoh, K, Sakumoto, T, Tohyama, M and Shimizu, N: A major source of catecholamine terminals in the nucleus tractus solitarii. Brain Res. 172: 372-377, 1979.
31. Kubo, T, Amano, H, Katsumata, M and Misu, Y: Further studies on vasopressin-induced pressor responses to kainic acid injected into the nucleus tractus solitarii of the rat. Arch. Int. Pharmacodyn. Ther. 276: 73-81, 1985.
32. Sved, AF, Imaizumi, T, Talman, WT and Reis, DJ: Vasopressin contributes to hypertension caused by nucleus tractus solitarius lesions. Hypertension 7: 262-267, 1985.
33. Criscione, L, Reis, DJ and Talman, WT: Cholinergic mechanisms in the nucleus tractus solitarii and cardiovascular regulation in the rat. European J. Pharmacol. 88: 47-55, 1983.
34. Yamada, S, Ishima, T, Hayashi, M, Tomita, T and Hayashi, E: Central acetylcholine receptors in spontaneously hypertensive rats. Jpn. Heart J. 25: 857, 1984.

CONTROL OF BLOOD PRESSURE BY MUSCARINIC AND NICOTINIC RECEPTORS IN THE VENTROLATERAL MEDULLA

Hreday N. Sapru

Section of Neurological Surgery, University of Medicine and Dentistry, New Jersey Medical School, Newark, NJ 07103 and Veterans Administration Medical Center
East Orange, NJ 07019

ABSTRACT

Nicotine, one of the most widely used drugs in human society, has been implicated as a risk factor in many cardiovascular diseases. The mechanisms by which nicotine evokes cardiovascular responses are not clear. This lack of information may be partially attributed to our incomplete understanding of cardiovascular actions of this substance even under normal physiological conditions. The results of several investigators indicate that the effects of nicotine on cardiovascular function are variable. Our results indicate that the cardiovascular actions of nicotine may be variable because it evokes different responses from different regions of the brain. For example, it induces an increase and decrease in blood pressure when microinjected into the pressor and depressor areas of the ventrolateral medulla, respectively. Heart rate and resistance in major vascular beds are increased when nicotine is microinjected into the pressor area. Opposite effects on these parameters are observed when it is microinjected into the depressor area. Other investigators have shown that nicotine induces depressor and bradycardic effects when microinjected into the dorsal medulla. Thus, the net effect of nicotine on cardiovascular function depends on the accessibility of different cardiovascular neuronal pools to sufficient concentration of this substance. Specific brain cholinergic nicotinic receptors and the sympathetic nervous system are involved in these actions of nicotine. The cardiovascular actions of nicotine are comparable to those evoked by other cholinergic agonists such as carbachol and physostigmine.

INTRODUCTION

Tobacco smoke has been implicated as a risk factor in many cardiovascular diseases (1) and some of its effects have been attributed to nicotine which is one of its most active constituents (2). Nicotine is absorbed from the lungs rapidly and distributed to the central nervous system within seconds (2). Central cardiovascular actions of nicotine (3-15; Table I) and other cholinergic agonists (16) are variable. These agents may evoke qualitatively divergent responses from different sites in the central nervous system. This hypothesis was tested by microinjecting nicotine and carbachol into the different sites in the ventrolateral medulla and investigating the cardiovascular responses. Kubo and Misu (5)

Table I.

Species:Anesthesia	Route and dose	HR	BP	Ref.
Man : 0	smoking 2 cig.	↑	↑	3-4
Rat : urethane	0.1 ug, microinj. near AP	↓	↑↓	5
Cat : 0	1.0-10 ug, 4th. vent.	–	↑↓	6
: 0	0.3-3 ug, i.vertebral	–	↑↓	6
Cat : decerebrate	50 ug/ml, pledget VL med.	–	↑	7
Cat : chloralose	20-80 ug, i.vent.	0,↑,↓	↓,↑↓,↓↑	8
Cat : chloralose	4.0-50 ug, i.vent.	0,↑,↓	↓,↑↓,↓↑	9
Cat : pentobarb.	0.01-1 mg/ml, 3rd. vent.	0	↓	10
Cat : pentobarb.	vent. med. surf., 5-60 ug	–	↓	11
Cat : chlor-ureth	vent. med. surf., superfu.	–	↑	12
Dog : 0	smoking 7 cig/day x 22 mo.	0	↑	13
Dog : 0	20-60 ug i.vent.	↑	↑	14
Dog : pentobarb.	10 ug/kg, i.v.	–	↑	15

Symbols: 0 = none; ↑ = increase; ↓ = decrease; – = unknown ; ↑↓ = increase followed by decrease; ↓↑ = decrease followed by increase. Abbreviations: AP = area postrema; i.v = intravenous administration; i.vent. = intraventricular administration; perf. = perfusion; superfu. = superfusion; VL med. = ventrolateral medulla; vent. med. surf. = ventral medullary surface application by Perspex rings.

have described the actions of these agents in the nucleus tractus solitarius (NTS) which is another important site of cardiovascular regulation in the brainstem.

METHODS

Animal Model

Male Wistar rats (Royal Hart Farms, NY), weighing 300 to 350 gm and anesthetized with urethane (1 gm/kg, i.p.), were used. The results obtained in anesthetized rats were compared with those obtained in midcollicular decerebrate rats to investigate the role, if any, of rostral brain structures (e.g., hypothalamus and pituitary) and anesthetics in the observed responses. The procedure for decerebration is described elsewhere (17).

General Procedures

One of the femoral arteries and veins and the trachea were cannulated. The pulsatile arterial blood pressure (BP), the mean arterial pressure (MAP), and heart rate (HR) were monitored and recorded on a polygraph (Grass Instruments, Model 7D). Rectal temperature was monitored and maintained at 37 ± 0.5°C by a temperature controller (Bailey Instruments, NJ, Model BAT 8). The animal was ventilated artificially to maintain blood gases at normal levels throughout the experiment. These procedures are described in detail elsewhere (18).

Microinjections

The animals were fixed in a stereotaxic instrument in a supine position with the incisor bar at the level of interaural line. The larynx and the underlying muscles were excised to expose the ventral aspect of the occipital bone which was drilled to create a window about 6 mm wide and 7 mm long. The solutions of all the agents to be tested were prepared in artificial cerebrospinal fluid (CSF) (19) or normal saline. The pH of the solutions was adjusted to 7.4. The solutions were filled in multibarreled glass micropipettes pulled to a tip size of 50 μm. Each barrel was connected to a polyethylene (PE) tubing which was sealed when the barrel was not in use. The barrel from which the microinjection was to be delivered was connected via the PE tubing to a microsyringe (Hamilton, 1 μl) mounted on an infusion pump (Sage Instruments, Model 341). The microinjections (0.1 μl volume, delivered over a period of 5 sec) were made bilaterally in all experiments.

Histology

At the end of the experiment, the site of microinjection was marked by microinjecting 0.1 μl of fast green dye via one of the barrels of the micropipette. In some experiments, the microinjection site was marked by passing a constant current (3 mA) for 20 sec via a monopolar platinum-iridium electrode. The brains were quickly removed and frozen, and thin sections (12 μm) were cut in a cryostat (Hacker Instruments). The site of microinjection was ascertained histologically using the atlas of Paxinos and Watson (20).

Splanchnic Nerve Recordings

In the rat, the celiac ganglion is unpaired. Rostral to this ganglion, another ganglion (called cardiac ganglion, perhaps, because of its proximity to the cardiac end of the stomach) is located on either side. The preganglionic segment of the greater splanchnic nerve (rostral to the cardiac ganglion) was identified under an operation microscope (Carl Zeiss, OPMI-1), sectioned, desheathed at its end, and placed on a bipolar platinum-iridium electrode. A pool of warm paraffin oil was used to prevent the nerve from drying. The sympathetic activity was amplified by a band pass amplifier (Ortec 4660), displayed on an oscilloscope (Tektronix R51039N), and recorded on a tape (Hewlett Packard 3968A). The recordings were played back into a visicorder (Honeywell 1858) for making permanent records. At the end of the experiment the nerve, from which the activity was being recorded, was sectioned so that the segment on the electrode remained undisturbed. Background noise, if any, from this segment was recorded under identical conditions. This recording was fed into a Schmitt trigger to set the background noise level in each experiment. The whole-nerve activity was fed into the Schmitt trigger and the logic pulses generated were delivered into a counter (Nicolet Model 1072). Count rate analysis of these signals was performed. Interspike intervals of 20 μsec were used to ensure that all the spikes were counted. The counts were displayed on the graphics terminal (21).

Regional Blood Flow Measurement

Blood flow was monitored in the lower abdominal aorta (1 mm rostral to the iliac arteries), superior mesenteric and renal arteries. A pulsed Doppler flowmeter was used for this purpose. Miniaturized probes, with lumen diameters from 0.7 to 1.2 mm, were filled with coupling gel and placed around the above-mentioned blood vessels in the same animal. The wire leads from the probes were connected to the Doppler flowmeter. The signals were recorded as pulses or mean flow velocity of a Doppler shift (KHz) on a polygraph. Resistances in each vascular bed were calculated as described elsewhere (22).

Statistical Analyses

Paired t-test was used to assess the significance of differences between means when the animals served as their own controls. Independent t-test was used to compute the differences between means when different groups of animals were compared (23). A 2x2 factorial analysis of variance was used to test for significance of differences between treatments. The Newman-Keul's test was used to make multiple mean comparisons (24).

RESULTS

Pressor and Depressor Areas

In the pressor area of the ventrolateral medulla (VLPA), bilateral microinjections (300 ng/site) of L-glutamate monosodium evoked an increase

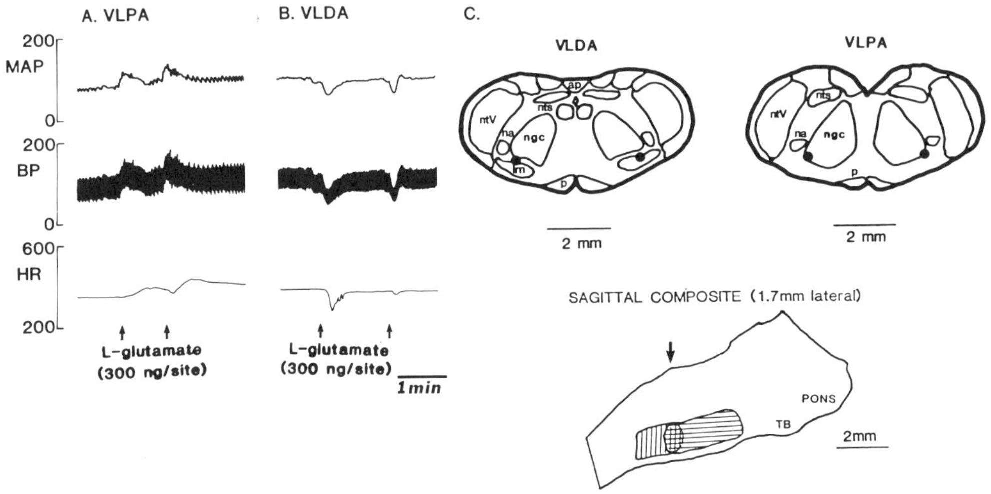

Fig. 1. Pressor (VLPA) and depressor (VLDA) areas in the ventrolateral medulla. In panels A and B: top trace is mean arterial pressure (mmHg), middle trace is pulsatile femoral arterial pressure (mmHg) and bottom trace is heart rate (beats/min). A: pressor and tachycardic responses to bilateral microinjections (300 ng/site) of L-glutamate monosodium (arrows) into the VLPA. B: depressor and bradycardic responses to bilateral microinjections of L-glutamate (300 ng/site) into VLDA. C: Diagrams showing the coronal sections of brainstem at the level (top left) 0.5 mm and 1.5 mm (top right) rostral to the calamus scriptorius. The locations of the VLDA and VLPA are shown by the filled circles. At the bottom of panel C is a diagram of a sagittal section 1.7 mm lateral to the midline; VLPA (horizontal lines) is located rostral to the obex (arrow) while VLDA (vertical lines) covers an area caudal and rostral to the obex and is located caudal to the VLPA. Abbreviations: ap, area postrema, lrn, lateral reticular nucleus, na, nucleus ambiguus; ngc, nucleus reticularis gigantocellularis; nts, nucleus tractus solitarius; ntV, nucleus tractus spinalis nervi trigemini; P, pyramidal tract; TB, trapezoid body.

in MAP (30-50 mmHg) and HR (30-50 beats/min) as shown in Fig. 1A. In the depressor area of the ventrolateral medulla (VLDA) microinjections of L-glutamate (300 ng/site) evoked a decrease in MAP (30-50 mmHg) and HR (145-155 beats/min) as shown in Fig. 1B. VLPA includes the ventrolateral aspect of the nucleus reticularis gigantocellularis and the rostral aspect of lateral reticular nucleus (nucleus reticularis paragigantocellularis). It is located 0.6-2.2 mm rostral to the obex, 1.3-2.1 mm lateral to the midline and 0.5-1.3 mm deep from the ventral surface of the medulla (Fig. 1C). VLDA includes the caudal part of nucleus reticularis gigantocellularis, the A_1 region, the nucleus reticularis ventralis and the nucleus ambiguus. It is located 0.4 mm caudal to 1.1 mm rostral to the obex, 1.4-2.1 mm lateral to the midline and 0.2 to 1.3 mm deep from the ventral surface of the medulla (Fig. 1C). Doses of L-glutamate ranging from 3 ng-3 μg elicit responses from these areas and these actions are mediated via sympathetic nervous system (25).

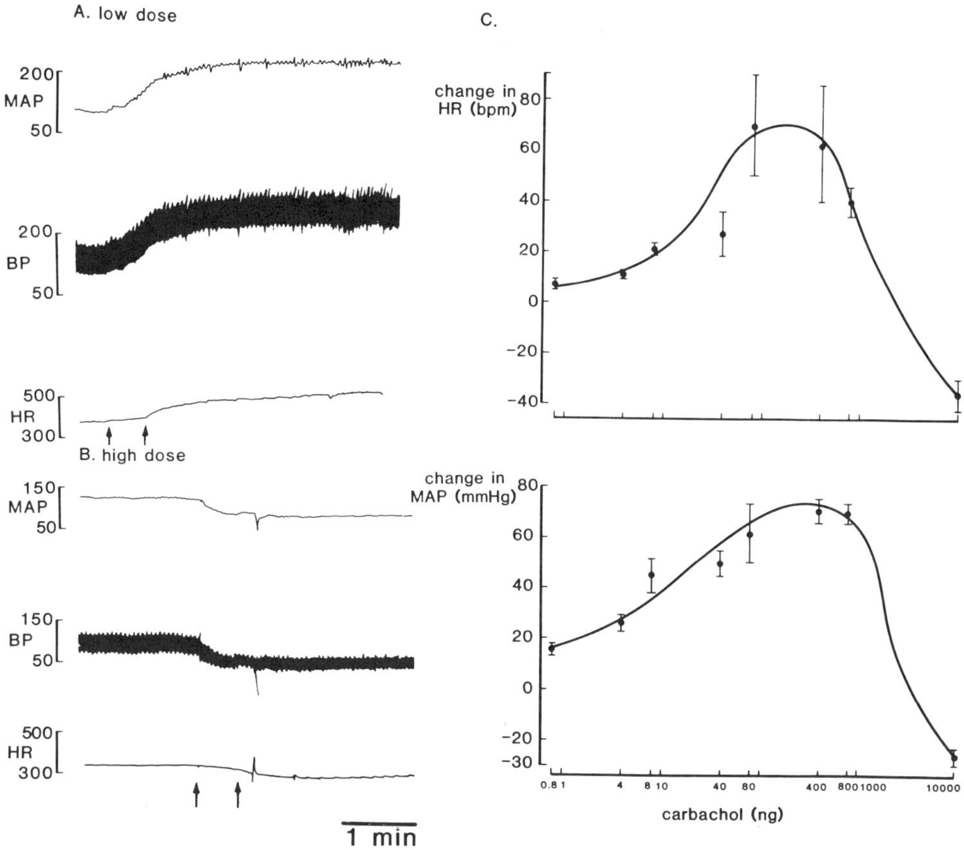

Fig. 2. Cardiovascular effects of carbachol in VLPA. In panels A and B: legends are same as in Fig. 1. A: Microinjections of low dose (100 ng/site) of carbachol (arrows) increased BP and HR. B: At a high dose (10 ug/site), carbachol decreased BP and HR. C: dose-response curves for HR (top) and BP (bottom). Doses ranging from 0.8-1000 ng/site increased BP and HR while doses of 10 μg/site decreased BP and HR. These results suggest the presence of muscarinic cholinergic receptors in VLPA; low doses activate these receptos to increase BP and HR while high doses decrease these values by depolarization blockade (with permission from Williams and Wilkins).

Effect of Cholinergic Muscarinic Agonists in VLPA

Carbachol is a stable analogue of acetylcholine. Microinjections of a small dose (100 ng/site) of this agent into the VLPA increased BP and HR (Fig. 2A) while a higher dose (10 μg/site) at the same sites caused a fall in BP and HR (Fig. 2B). Dose-response curves for these actions of carbachol are shown in Fig. 2C; low doses (0.8-1000 ng/site) increased HR and BP while higher doses (10 μg/site) caused a fall in BP and HR. The onset of effects was 5-10 sec and the duration 15-20 min.

Physostigmine, an acetylcholine esterase inhibitor that is known to increase the levels of endogenous acetylcholine (16), when microinjected into the VLPA in the dose of 200 ng/site, increased BP and HR. At a high dose of 10 μg/site, it decreased BP and HR. Doses ranging from 4-1000 ng/site increased BP and HR while higher doses (10 μg/site) caused opposite effects.

A transection of the brainstem at the level rostral to the VLPA did not alter the pressor and tachycardic responses to smaller doses of carbachol or physostigmine. The increase in BP and HR caused by the microinjections of physostigmine and carbachol into the VLPA produced increases in sympathetic nerve activity (SNA) that could be prevented as well as reversed (Fig. 3C) by intravenous administration of atropine sulfate (0.5-1 mg/kg). It is probable that the increases in BP and HR induced by these drugs were mediated via this increase in SNA.

Microinjections of atropine sulfate in the doses of 0.4 μg to 15 μg into the VLPA area caused a dose-related decrease in BP and HR. These results suggest that acetylcholine may be involved in the tonic regulation of blood pressure at the VLPA.

Nicotinic Receptors in VLPA

Bilateral microinjections of nicotine (10 ng/site) into the VLPA caused an increase in BP and HR within 10-15 sec and the effects lasted for 5-10 min (Fig. 4A). The interval between different doses was 45-60 min to avoid the possibility of development of tachyphylaxis to nicotine. Doses ranging from 1-100 ng/site caused pressor and tachycardic responses in VLPA (Table IIA). Microinjections of hexamethonium (2 μg/site), a ganglion blocker, into the VLPA prevented as well as reversed the pressor and tachycardic responses to nicotine (Figs. 5A and B). Microinjections of nicotine (100 ng/site) into the VLPA increased the resistances of the superior mesenteric, renal, and hindquarter vascular beds (Table IIA). The responses evoked by nicotine in the VLPA were mediated via the sympathetic nervous system because spinal transection at the C_1 level abolished these responses (Figs. 6A and B).

Nicotinic Receptors in VLDA

Bilateral microinjections of nicotine (10 ng/site) into the VLDA induced a decrease in BP and HR (Fig. 4B) within 10-15 sec and the responses lasted for 3-10 min. Dose-response relationships for BP and HR are shown in Table IIB. These responses were mediated via specific nicotinic receptors because microinjections of hexamethonium at the same sites prevented as well as blocked them (Figs. 5C and D). Resistances in the mesenteric, renal, and hindquarter vascular beds were decreased by microinjections of nicotine into the VLDA (Table IIB). Spinal transection at the C_1 level abolished the depressor and bradycardic responses indicating the involvement of the sympathetic nervous system (Figs. 6C and D).

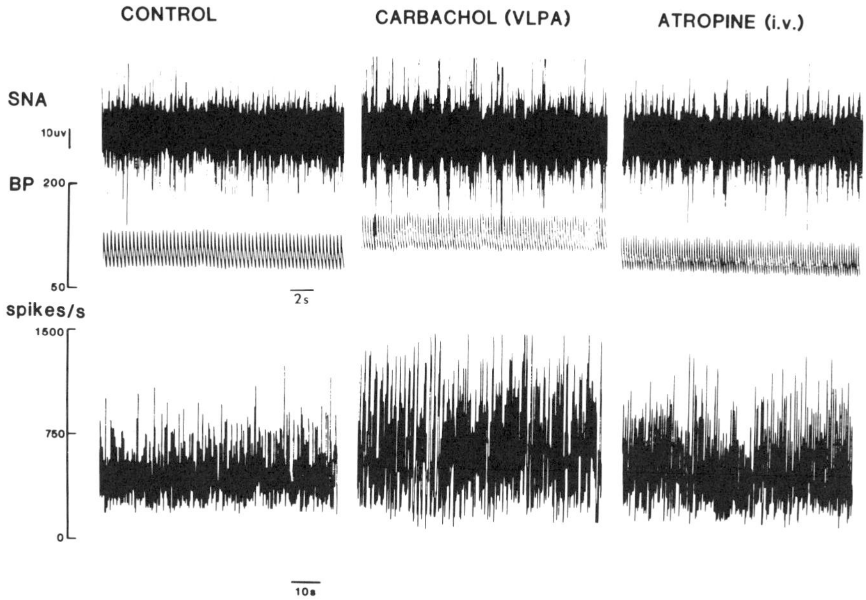

Fig. 3. In each panel: top trace is preganglionic greater splanchnic nerve activity (SNA), middle trace is BP and bottom trace is count-rate analysis of splanchnic nerve activity (spikes/sec). Left panel: Control values for BP and SNA. Middle panel: Microinjections of carbachol (40 ng/site) into the VLPA increased SNA (top and bottom traces) and BP (middle trace). Right panel: Atropine sulfate (1 mg/kg, i.v.) reversed the increase in SNA and BP induced by carbachol. These results indicate that the pressor action of carbachol is mediated via sympathetic nervous system and muscarinic receptors are involved (with permission from Williams and Wilkins).

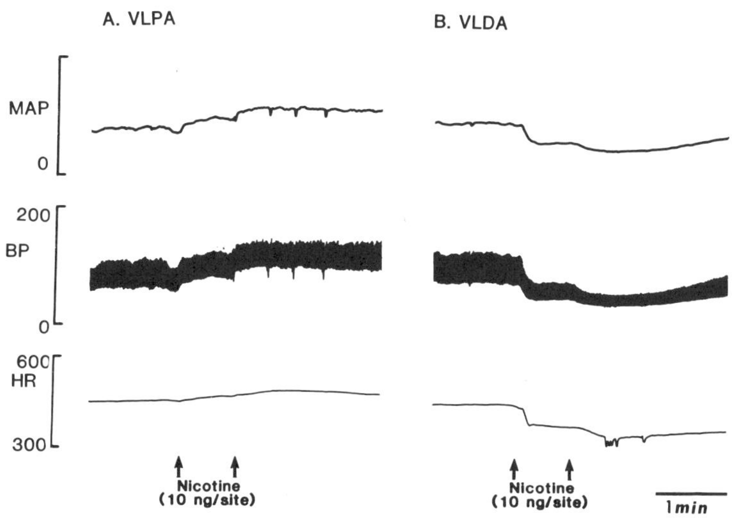

Fig. 4. Legends same as in Fig. 1. A: bilateral microinjections (arrows) of nicotine (10 ng/site) into the VLPA increased BP and HR. B: Identical microinjections into the VLDA decreased BP and HR. VLPA and VLDA were identified by microinjections of L-glutamate (not shown).

Table II.

A: VLPA

	1 ng/site	10 ng/site	100 ng/site
Increase in BP (mmHg)	18 ± 4	40 ± 6	42 ± 4
Increase in HR (bts/min)	40 ± 6	50 ± 3	51 ± 7
Increase in Resistance (% change)	ND	ND	29 ± 6 mesenteric
	ND	ND	97 ± 7 hindquarter
	ND	ND	75 ± 25 renal

B: VLDA

	1 ng/site	10 ng/site	100 ng/site
Decrease in BP (mmHg)	13 ± 2	34 ± 4	45 ± 2
Decrease in HR (bts/min)	20 ± 10	67 ± 20	125 ± 30
Decrease in resistance of: vascular beds (% change)	ND	ND	17 ± 3 mesenteric
	ND	ND	20 ± 3 hindquarter
	ND	ND	14 ± 5 renal

Abbreviations: bts/min = beats/minute; BP = blood pressure; HR = heart rate; ND = not determined; VLDA = depressor area in the ventrolateral medulla; VLPA = pressor area in the ventrolateral medulla

DISCUSSION

Wang and Ranson (26) were the first to explore systematically the cardiovascular areas in the brainstem of the cat using stereotaxic instruments and electrical stimulation. Using this technique, pressor responses were elicited from scattered points in the brainstem. These results were confirmed by Alexander (27). In many subsequent studies, similar techniques of electrical stimulation were used to characterize cardiovascular areas in the medulla (28-31). The electrical stimulation technique is not accurate because it results in activation of cell bodies, ascending or descending fibers of passage, as well as nerve terminals at the site of stimulation. Therefore, it is difficult to ascribe a particular response to a particular site. Recently, L-glutamate has been shown to stimulate only cell bodies (32). This amino acid was, therefore, used as a tool to investigate cardiovascular areas in the ventrolateral medulla; the latter has been implicated as an important site for cardiovascular regulation (33-36). Using this technique, pressor or depressor areas in the ventrolateral medulla were identified in each experiment before microinjecting nicotine or other cholinergic agonists. This procedure ensured that the responses to the agents under investigation were evoked from functionally identified neuronal pools.

Neuro-chemical studies have demonstrated the presence of cholinergic neurons in the ventrolateral medulla (37-38). Central cholinergic neurons are believed to play a role in the neural control of circulation (16). Thus, the intracerebroventricular and intravenous administration of cholinomimetic agents increases BP, and inhibition of acetycholine synthesis with hemicholinium-3 decreases BP.

In our studies, stimulation of cholinergic receptors in the VLPA with carbachol caused a dose-related increase in MAP and HR at low doses and a fall in MAP and HR at high doses. Similar results were obtained following the inhibition of acetylcholine esterase with physostigmine. These actions were mediated via an increase in sympathetic outflow. Intravenous injection of atropine sulfate, which crosses the blood-brain barrier antagonized these actions of cholinomimetic agents, indicating the involvement of specific muscarinic receptors (39). Muscarinic receptor

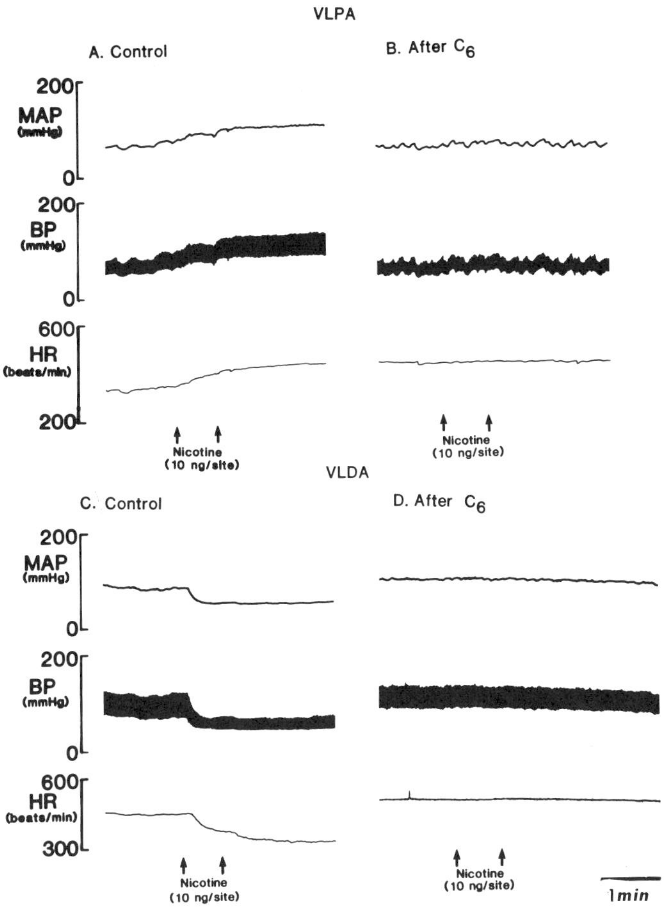

Fig. 5. Legends same as in Fig. 1. A: Control pressor and tachycardic responses to microinjections of nicotine (10 ng/site) into the VLPA. B: 50 min later hexamethonium (2 ug/site, C_6) was microinjected into the VLPA (not shown); subsequent microinjection of nicotine (10 ng/site) at the same sites failed to evoke a response. C: control depressor and bradycardic responses to microinjections of nicotine (10 ng/site) into the VLDA. D: microinjection of hexamethonium at the same sites (not shown) blocked the action of nicotine. These results indicate that specific cholinergic nicotinic receptors are mediating the responses to nicotine.

blockade with atropine sulfate microinjections in the VLPA caused a dose-related fall in MAP and HR. This observation suggests that acetylcholine may be involved in tonic regulation of BP at the level of VLPA.

Nicotine, one of the most potent constituents of cigarette smoke, is absorbed rapidly through the lungs and distributed in various tissues in the body. It crosses the blood-brain barrier and is distributed throughout the central nervous system. The results of the present study indicate that nicotine exerts different actions in different regions of the ventrolateral medulla. In the VLPA nicotine evoked a pressor and tachycardic effect while in the VLDA it exerted opposite effects. Kubo and Misu (5) have shown that nicotine evokes a depressor effect in the

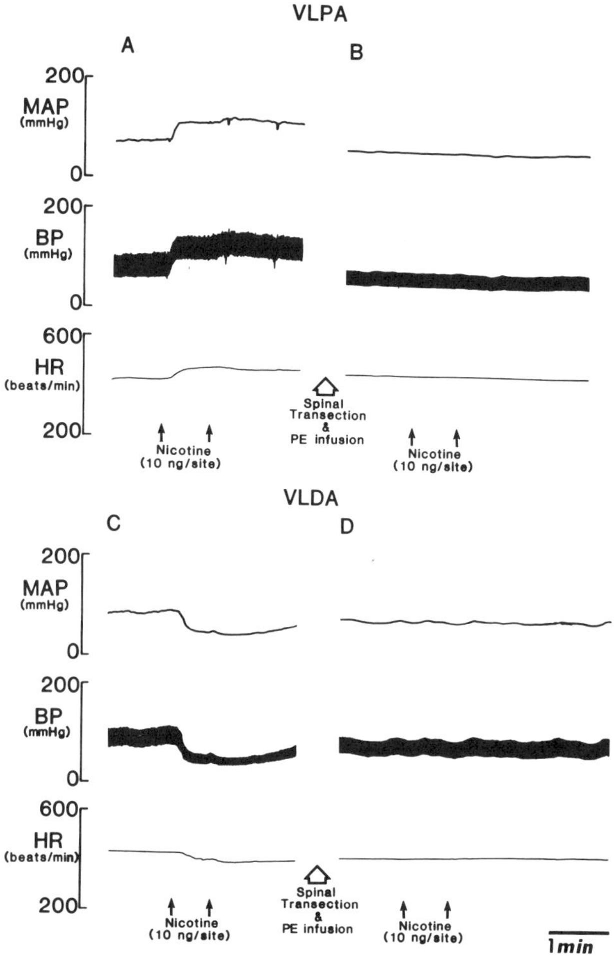

Fig. 6. Legends same as in Fig. 1. A: control pressor and tachycardic responses to microinjections of nicotine (10 ng/site) into the VLPA. B: Transection of the spinal cord at the C_1 level abolished the responses to subsequent microinjections of nicotine. C: control depressor and bradycardic responses to microinjections of nicotine into the VLDA. D: spinal cord transection abolished these responses to nicotine. The BP was restored to control levels after spinal cord transection by intravenous infusions of phenylephrine.

NTS. The responses to nicotine will, therefore, depend on the accessibility of cardiovascular neuronal pools to sufficient concentrations of nicotine in the blood or CSF. For example, if nicotine, in sufficient concentrations, reaches the NTS or VLDA first, a depressor effect is expected while the opposite response is expected if it reaches the VLPA. Hexamethonium blocked the actions of nicotine but not of L-glutamate, suggesting the involvement of specific cholinergic nicotinic receptors. Direct recordings from the greater splanchnic nerves indicated the involvement of the sympathetic nervous system.

Higher doses (10 μg) of carbachol into the VLPA decreased BP. This effect may be due to depolarization of neurons in VLPA; depolarization blockade is a well-known property of cholinergic agonists (40). The hypotensive responses to direct application of cholinergic agonists to the ventral surface of the medulla may be explained by depolarization blockade (11-12).

Previous studies (41-46) have shown that urethane does not induce any qualitative changes in cardiovascular responses to gamma-aminobutyric acid, bicuculline, and enkephalins in the ventrolateral medulla. The responses observed in decerebrate rats suggest that urethane does not modify qualitatively the responses to nicotine and carbachol. These agents evoked pressor responses from VLPA in decerebrate rats in which structures rostral to the inferior colliculus (including pituitary and hypothalamus) were removed. This observation excluded the possibility that the rise in pressure induced by nicotine and carbachol may be due to vasopressin release.

Based on the above discussion, the following scheme is presented to indicate the role of nicotinic and muscarinic receptors in blood pressure regulation (Fig. 7). The presence of depressor and pressor areas in the ventrolateral medulla has been amply demonstrated in our and other laboratories (25, 33-36, 39, 41-46). The intermediolateral column of the thoraco-lumbar cord (IML) is the site where the preganglionic sympathetic neurons are located (28-31). Anatomic studies using horseradish

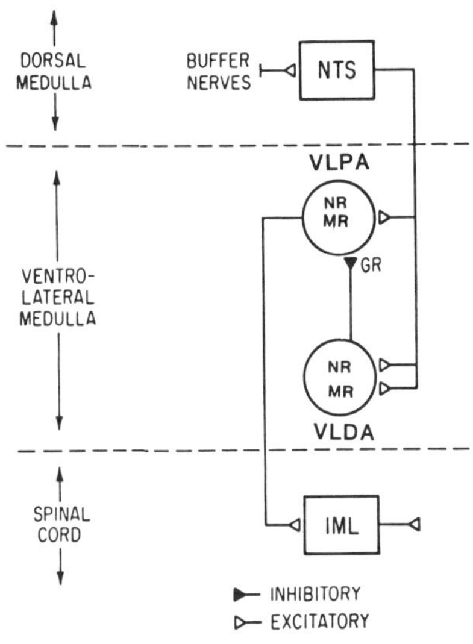

Fig. 7. A scheme showing the role of muscarinic (MR) and nicotinic (NR) cholinergic receptors in the cardiovascular responses evoked from the ventrolateral medulla. Intermediolateral column of the thoraco-lumbar cord (IML) is the site where preganglionic sympathetic neurons (PSN) are located. Pressor area (VLPA), but not the depressor area (VLDA), of the ventrolateral medulla projects to IML. The responses evoked from VLDA are mediated via VLPA. NR and MR agonists activate the neurons in the VLPA which results in excitation of the PSN in IML. Sympathetic nerve activity (SNA), BP and HR are, therefore, increased. VLDA has an inhibitory input, probably involving gamma aminobutyric acid receptors (GR), to the VLPA. Activation of NR and MR in the VLDA increases the inhibition on the VLPA neurons which results in a decrease in the activity of the PSN in IML and reduction in SNA, BP and HR.

peroxidase indicate that there are projections from the pressor area of the ventrolateral medulla to the IML. No such projections have been identified from the caudal depressor area in the ventrolateral medulla (47-48). Thus, the responses evoked from the depressor area are probably mediated via the pressor area. This hypothesis was supported by our pharmacological experiments (45-46). In these experiments, pressor and depressor areas were identified bilaterally in the same animal. The pressor area was then inhibited by microinjections of muscimol (a gamma-aminobutyric acid receptor agonist) which resulted in a fall in blood pressure. Blood pressure was restored to control levels by intravenous injections of phenylephrine (an alpha adrenergic receptor agonist). Muscimol was immediately injected into the depressor area; however, the expected rise in blood pressure did not occur, suggesting that the responses evoked from the depressor area are mediated via the pressor area.

The results presented in this paper suggest that muscarinic and nicotinic receptors are present in the depressor and the pressor areas of the ventrolateral medulla. Iontophoretic studies have shown that nicotine and carbachol excite neurons (49). Activation of nicotinic and muscarinic receptors in the pressor area causes a rise in blood pressure and heart rate while activation of these receptors in the depressor area causes a fall in blood pressure and heart rate. It has been suggested that the depressor area has an inhibitory input to the pressor area (41-42). It follows that activation of neurons in the VLDA by nicotine or carbachol would exert more inhibition on the neurons in the VLPA which would result in a fall in blood pressure and heart rate.

ACKNOWLEDGEMENTS

This paper embodies the work done in this laboratory with my colleagues, Dr. Robert N. Willette and Dr. Susan Punnen. The research work presented in this paper was supported by the following grants: N.I.H. (HL24347), American Heart Association (New Jersey affiliate) and East Orange Veterans Administration Center, NJ.

REFERENCES

1. Surgeon General: Smoking and health. U.S. Department of Health, Education and Welfare, publication no. (PHS) 79-50066, U.S. Govt. Printing Office, Washington D.C., 1979.
2. Schievelbein, H: Nicotine resorption and fate. In (ed.) Balfour, DJK, International encyclopedia of pharmacology and therapeutics, section 114, Nicotine and the smoking habit, Pergamon Press, New York, pp. 1-15, 1984.
3. Koch, A, Hoffmann, K, Steck, W, Horsch, A, Hengen, N, Morl, H, Harenberg, J, Spohr, U and Weber, E: Acute cardiovascular reactions after cigarette smoking. Atherosclerosis 35: 67-75, 1980.
4. Armitage, AK: Blood levels of nicotine and cotinine attained during smoking. In: The workshop on nicotine, Nov. 11-13, 1974, Stockholm, Sweden.
5. Kubo, T and Misu, Y: Changes in arterial blood pressure after microinjections of nicotine into the dorsal area of medulla oblongata of the rat. Neuropharmacology 20: 521-524, 1981.
6. Schaeppi, U: Effects of nicotine administration to the cat's lower brainstem upon electroencephalogram and autonomic nervous system. Ann. N.Y. Acad. Sci. 142: 40-49, 1967.
7. Cozine, RA and Ngai, SH: Medullary surface chemoreceptors and regulation of respiration in the cat. J. Appl. Physiol. 22: 117-121, 1967.

8. Armitage, AK and Hall, GH: Effects of nicotine on systemic blood pressure when injected into the cerebral ventricles of cats. Int. J. Neuropharmacol. 6: 143-149, 1967.
9. Pradhan, SN, Bhattacharya, IC and Atkinson, KS: The effects of intraventricular administration of nicotine on blood pressure and some somatic reflexes. Ann. N.Y. Acad, Sci. 142: 50-66, 1967.
10. McCarthy, LE and Borison, HL: Separation of effects of CO_2 and nicotine on ventilation and blood pressure. Resp. Physiol. 15: 321-330, 1972.
11. Feldberg, W and Guertzenstein, PG: Vasodepressor effects obtained by drugs acting on the ventral surface of the brainstem. J. Physiol. (Lond) 258: 337-355, 1976.
12. Dev, NB, and Loeschcke, HH: Topography of the respiratory and circulatory responses to acetylcholine and nicotine on the ventral surface of the medulla oblongata. Pflugers Arch. 379: 19-27, 1979.
13. Ahmed, SS, Moschos, CB, Lyons, MM, Oldewurtel, HA, Columbus, RJ and Regan, TJ: Cardiovascular effects of long-term cigarette smoking and nicotine administration. Am. J. Cardiol. 37: 33-40, 1976.
14. Lang, WJ, and Rush, ML: Cardiovascular responses to injections of cholinomimetic drugs into the cerebral ventricles of unanesthetized dogs. Brit. J. Pharmac. 47: 196-205, 1973.
15. Domino, E.F. A role of the central nervous system in the cardiovascular actions of nicotine. Arch. Int. Pharmacodyn. Ther. 179: 169-179, 1969.
16. Brezenoff, HE and Giuliano, R: Cardiovascular control by cholinergic mechanisms in the central nervous system. Annu. Rev. Pharmacol. Toxicol. 22: 341-381, 1982.
17. Sapru, HN and Krieger, AJ: Procedure for the decerebration of the rat. Brain Res. Bull., 3: 675-679, 1978.
18. Sapru, HN, Willette, RN and Krieger, AJ: Stimulation of pulmonary J receptors by an enkephalin-analog. J. Pharmacol. Exp. Ther. 217: 228-234, 1981.
19. Merlis, JK: The effect of changes in the calcium content of the cerebrospinal fluid on spinal reflex activity in the dog. Am. J. Physiol. 131: 67-72, 1940.
20. Paxinos, G and Watson, C: The rat brain in stereotaxic coordinates. Academic Press: New York, 1982.
21. Sapru, HN, Gonzales, ER and Krieger, AJ: Greater splanchnic nerve activity in the rat. Brain Res. Bull., 8: 267-272, 1982.
22. Haywood, JR, Schaffer, RA, Fastenow, C, Fink, GD and Brody, MJ: Regional blood flow measurements with pulsed Doppler flow-meter in conscious rat. Am. J. Physiol. 241: H273-278, 1981.
23. Croxton, FE: Elementary statistics with applications in medicine and biological sciences. New York: Dover Publications pp. 235-239, 1959.
24. Zar, JH: Biostatistical analysis. Prentice Hall: Englewood, NJ, pp. 151-155, 1974.
25. Willette, RN, Barcas, PP, Drieger, AJ and Sapru, HN: Vasopressor and depressor areas in the rat medulla. Neuropharmacology, 22: 1971-1079, 1983.
26. Wang, SC and Ranson, SW: Autonomic responses to electrical stimulation of the lower brainstem. J. Comp. Neurol. 71: 437-455, 1939.
27. Alexander, RS: Tonic and reflex functions of medullary sympathetic cardiovascular centers. J. Neurophysiol. 9: 205-217, 1946.
28. Galosy, RA and Clarke, LS: Neurophysiology and neuropharmacology of cardiovascular regulation and stress. Neuroscience and Biobehav. Rev. 5: 137-175, 1981.
29. Spyer, KM: Neural organization and control of the baroreceptor reflex. Rev. Physiol. Biochem. and Pharmcaol. 88: 23-125, 1981.
30. Calaresu, FR, Faiers, AA, and Mogenson, GJ: Central neural regulation of heart and blood vessels in mammals. Prog. Neurobiol. 5: 1-35,1975.

31. Wang, SC: Physiology and Pharmacology of the Brain Stem. Futura Publishing Co., New York, pp. 5-80, 1980.
32. Krnjevic, K. and Phillis, JW: Iontophoretic studies of neurons in mammalian cerebral cortex. J. Physiol. (Lond.). 165: 274-304, 1963.
33. Reis, DJ: The brain and hypertension: reflections on 35 years of inquiry into the neurobiology of the circulation. Circulation 70: (III-31)-(III-45), 1984.
34. Reis, DJ, Ross, CA, Ruggiero, DA, Granata, AR, and Joh, TH: Role of adrenaline neurons of ventrolateral medulla (the C_1 group) in the tonic and phasic control of arterial pressure. In (ed.) Slater, IH, Clinical and experimental hypertension, A6: 221-241, 1984.
35. Dampney, RAL, Goodchild, AK and Tan, E: Identification of cardiovascular cell groups in the brain stem. In (ed.) Slater, IH, Clinical and experimental hypertension, A6: 205-220, 1984.
36. Blessing, WW, Sved, AF and Reis, DJ: Arterial pressure and plasma vasopressin: regulation by neurons in the caudal ventrolateral medulla of the rabbit. In (ed.) Slater, IH, Clinical and experimental hypertension, A6: 149-156, 1984.
37. Willenberg, IM, Dermietzel, R, Leibstein, AG and Effenberger, M: Mapping of cholinoceptive (nicotinoceptive) neurons in the lower brainstem: with special reference to the ventral surface of the medulla. J. Auton. Nerv. Syst. 14: 287-298, 1984.
38. Kimura, H, McGeer, PL, and Peng, JH: Choline acetyltransferase-containing neurons in the rat brain. In (ed.) Bjorklund, Hokfelt, T and Kuhar, MJ. Handbook of chemical neuroanatomy, Vol. 3, Elsevier, New York, pp. 51-67, 1984.
39. Willette, RN, Punnen, S, Krieger, AJ, and Sapru, HN: Cardiovascular control by cholinergic mechanisms in the rostral ventrolateral medulla. J. Pharmacol. Exp. Ther. 231: 475-463, 1984.
40. Remond, A, and Izard, D: Electrophysiological effects of nicotine, Elsevier, New York, 1979.
41. Willette, RN, Krieger, AJ, Barcas, PP and Sapru, HN: Medullary gamma-aminobutyric acid receptors and the regulation of blood pressure in the rat. J. Pharmacol. Exp. Ther. 226: 893-899, 1983.
42. Willette, RN, Barcas, PP, Krieger, AJ, and Sapru, HN: Endogenous GABAergic mechanisms in the ventrolateral medulla and the regulation of blood pressure. J. Pharmacol. Exp. Ther. 230: 34-39, 1984.
43. Willette, RN, Punnen, S, Krieger, AJ, and Sapru, HN: Hypertensive response following opiate receptor stimulation in the caudal ventrolateral medulla. Neuropharmacology, 23: 401-406, 1984.
44. Punnen, S, Willette, RN, Krieger, AJ, and Sapru, HN: Cardiovascular response to enkephalin microinjections in the pressor area of the ventrolateral medulla. Neuropharmacology, 23:939-946, 1984.
45. Punnen, S, and Sapru, HN: Blockade of cholinergic receptors in C1 area abolishes hypertensive response to opiates in the A1 area of the ventrolateral medulla. Brain Res. 336: 180-186, 1985.
46. Willette, RN, Punnen, S, Krieger, AJ, and Sapru, HN: Interdependence of rostral and caudal ventrolateral medullary areas in the control of blood pressure. Brain Res. 321: 169-174, 1984.
47. Amendt, K, Czachurski, J, Dembowsky, K, and Seller, H: Bulbospinal projections to the intermediolateral cell column; a neuroanatomical study. J. Auton. Nerv. Syst. 1: 103-117, 1979.
48. Ross, CA, Ruggiero, DA, Park, DH, Joh, TH, Sved, AF, Fernandez-Pardal, J, Saavedra, JM and Reis, DJ: Tonic vasomotor control by the rostral ventrolateral medulla: effect of electrical or chemical stimulation of the area containing C1 adrenergic neurons on arterial pressure, heart rate and plasma catecholamines and vasopressin. J. Neurosci. 4: 474-494-1984.
49. Bradley, PB and Wolstencroft, JH: Effects of acetylcholine, nicotine and muscarine on brainstem neurons. Ann. N.Y. Acad. Sci. 142: 15-20, 1967.

STEREOSELECTIVITY OF NICOTINE'S CENTRAL EFFECTS AND ITS RELATIONSHIP TO PAIN AND CARDIOVASCULAR FUNCTION

Billy R. Martin

Department of Pharmacology and Toxicology
Medical College of Virginia
Virginia Commonwealth University
Richmond, VA 23298

INTRODUCTION

Nicotine produces a wide range of effects on the central nervous system which include convulsions (1), prostration (2), antinociception (3-11), release of antidiuretic hormone (12,13), tremors (13,14), as well as alterations in behavior (15,16), motor performance (3, 10, 17-19), gastric acid secretion (20-22), the cardiovascular system (7,23-30), learning (31), etc. The mechanisms by which nicotine produces these myriad effects have not been fully elucidated. Numerous studies have been carried out in an effort to characterize the effects of nicotine on central neurotransmitter functionality, as reviewed by Aceto and Martin (32) as well as contributors in this volume.

As a result of nicotine's actions on receptors in the neuromuscular junction and autonomic ganglia, it is logical to assume that nicotine's central effects are also receptor mediated. Considerable in vitro binding data have been generated over the past few years which support the notion that there are specific nicotine receptors in the central nervous system (33-44). Indeed, there is strong evidence for more than one nicotine receptor subtype. It now becomes important to establish a relationship between these nicotine binding sites and the pharmacological effects of nicotine. The implication of receptor involvement in nicotine's central effects has been supported by antagonism studies with ganglionic blockers as well as determination of structure-activity relationships and stereoselectivity (45-56). S-(-)-Nicotine is the naturally occurring form of nicotine (Fig. 1). In general, the stereoselectivity of nicotine, as summarized in Table 1, is less than that of other chiral agents which interact with central receptors. The stereoselectivity of most nicotine effects, both central and peripheral, occurs within a range of 5-25, with

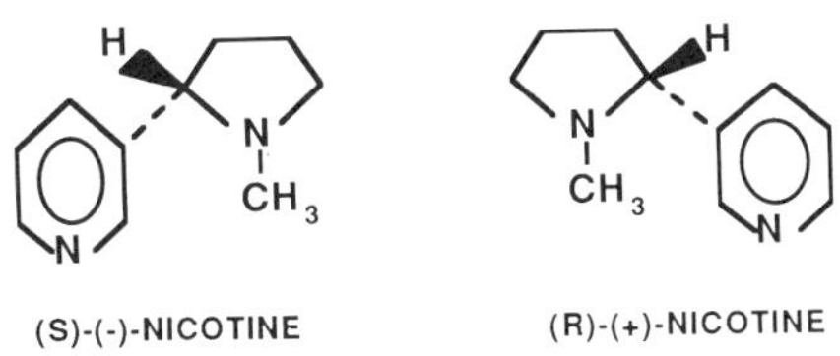

Fig. 1. Structures of the stereoisomers of nicotine

Table 1. Stereoselectivity of Nicotine's Central and Peripheral Effects

Test	P.R.*	Species	Reference
Central Effects			
Lethality	8	Mouse	45
"	7	Mouse	46
"	5	Guinea pig	45
"	1	Guinea pig	47
"	3	Rat	45
"	1	Rat	47
Convulsions	28	Rat	48
Convulsions	6	Rat	49
Prostration	45	Rat	2
Conditioned Avoidance	8	Rat	49
Conditioned taste aversion	4	Rat	50
Discrimination	9	Rat	51
"	25-50	Rat	52
Open-field	6	Rat	10
Rotarod	15	Rat	10
Locomotor activity	16	Rat	53
Dopamine release	1	Rat	54
Peripheral Effects			
Ileum Contracture	53	Guinea Pig	46
Ileum Contracture	42	Guinea Pig	55
Diaphragm blockade	1	Rat	55
Diaphragm blockade	1	Rat	56
Ganglionic blockade	14	Cat	55
Ganglionic blockade	5	Cat	56
Ganglionic stimulation	5	Cat	56
Artery contracture	>10	Rabbit	56
Anterior tibialis	3	Cat	55
Biventer contracture	5	Chick	55
Rectus contracture	10	Frog	55
AChE inhibition	1	Ox	55

* Potency ratio defined as the dose or concentration of (+)-nicotine divided by that of (-)-nicotine which produces comparable effects. Potency ratios were estimated when ED50 or LD50 were not provided.

the major exceptions being guinea-pig-ileum contracture and the prostration syndrome. The primary objective of this article is to discuss the stereoselectivity of nicotine's central effects, particularly as it relates to alterations in pain perception and cardiovascular function.

ANTINOCICEPTIVE ACTIVITY

The antinociceptive activity of nicotine has been studied in numerous laboratories using a diversity of analgesic techniques. It is not too surprising that there is a lack of unanimity among these studies. Nicotine has been reported to lack antinociceptive activity in mouse electroshock (62,63), mouse hot plate (62), rat-flinch jump (17), and in monkey tests (64,65). It has also been reported to cause both hyposensitivity and hypersensitivity in the rat tail pinch (5). However, in most studies nicotine has been found to be an effective antinociceptive agent. A summary of these data, presented in Table 2, indicate a wide range of nicotine potencies depending upon the test, species, route of administration, etc. It appears that rather high doses of nicotine are required to produce antinociception when it is given i.p. (4,59), while somewhat lower doses given s.c. are effective (3,5,6,9,58). As might be expected, nicotine is much more effective when administered i.v. (7,10,60) or i.v.t. (4,5,7). Obviously, a complete characterization is necessary for establishing credence in nicotine's antinociceptive effects. The question arises as to whether nicotine produces antinociception through a specific mechanism in the central nervous system or whether it is merely producing a general stimulation which may result in a nonspecific

Table 2. Evidence for Nicotine Antinociceptive Activity

Species	Test	Route	ED50*	Reference
Cat	Distend bladder	IP	0.1+	57
Rabbit	Tooth stim.	SC	2.5+	58
Rat	Flinch jump	IP	>1.0**	59
Rat	Hot plate	IP	5.0+	4
Rat	Tail flick	IVT	0.15+	4
Rat	Tail flick	IVT	0.025+	5
Rat	Tail flick	SC	0.5+	5
Rat	Tail flick	SC	0.7	9
Rat	Tail flick	SC	0.7	6
Mouse	Hot plate	SC	2.5+	58
Mouse	Hot plate	SC	2.5+	3
Mouse	Hot plate	IP	7.5	4
Mouse	Hot plate	SC	2.2	6
Mouse	Tail flick	SC	5.2	6
Mouse	Tail flick	SC	1.8	8
Mouse	Tail flick	SC	2.0	9
Mouse	Tail flick	IV	0.3+	60
Mouse	Tail flick	IV	0.2+	10
Mouse	PPQ stretching	SC	1.3	6
Mouse	Tail shock	SC	6.5	6
Dogs	Skin twitch	IV	0.8+	7
Dogs	Skin twitch	IVT	0.2+	7

* mg/kg

\+ Effective dose; ED50 was not determined.

** Not active at 1.0 mg/kg.

disruption of normal neuronal function. We have attempted to address this issue by determining the specificity and stereoselectivity of nicotine's antinociceptive activity and by investigating its site of action.[1]

Antagonism Studies

Antinociceptive activity of nicotine has been characterized in both mice and rats in our laboratory using tail-flick, p-phenylquinone-stretching and hot-plate tests (6,8,9). However, the tail-flick procedure has been used most extensively due to the fact that it is highly predictive of clinically useful analgesics. Following s.c. administration of nicotine, there is a rapid onset of antinociception in mice and rats which has a relatively short duration of action (20-30 min). The ED50's for nicotine in the tail-flick procedure were found to be 1.25 and 3 mg/kg in rats and mice, respectively, following s.c. administration (9). Characterization of this antinociception included the use of the antagonists described in Table 3. It is clear from these results that mecamylamine antagonized nicotine's effects in both rats and

[1]Procedures used to evaluate the ability of a drug to alter an animal's response to a noxious stimulus may be measuring the ability of the drug to block either the perception of the pain or the animal's reaction to the stimulus. It is for this reason that these effects are termed antinociceptive rather than analgesic. However, it should be pointed out that these tests are highly predictive of clinically useful analgesics.

mice. Hexamethonium produced partial blockade that was not dose responsive and occurred only after high doses. Inability of hexamethonium, a quaternary nicotinic blocker, to attenuate nicotine's antinociception may be due to either its lack of brain penetrability or to its lack of interaction with the nicotine receptor involved with antinociception. It should be pointed out that Asghar and Roth (66) have shown that hexamethonium does enter the brain to a small degree. Possible differences between hexamethonium and mecamylamine could best be determined by comparing their antagonistic activities after i.v.t. administration. Other investigators (4,5,58) have also found that mecamylamine, but not hexamethonium, blocks nicotine antinociception.

The involvement of muscarinic receptors in nicotine's antinociceptive activity is presently controversial. It has been known for some time that pain perception is associated with the central cholinergic nervous system. Several investigators have found that muscarinic agonists will produce antinociception in a variety of tests (67-71). Pedigo et al. (72) demonstrated that i.v.t. administration of acetylcholine resulted in attenuation of the tail-flick response. It is therefore attractive to postulate that nicotine produces antinociception by stimulating release of acetylcholine. Sahley and Berntson (5) found that scopolamine, but not scopolamine methylnitrate, would antagonize nicotine's effects in the rat-tail-flick procedure which supports this notion. On the other hand, physostigmine has been reported to have no effect (58) or to actually decrease (4) nicotine's antinociceptive activity. Phan et al. (4) also reported that the antimuscarinic agent, N(4-diethylamino-2-butynyl)-succinimide, was unable to antagonize nicotine's effects in the mouse-hot-plate assay. The lack of involvement of muscarinic receptors is further supported by the data presented in Table 3 which show that neither scopolamine nor atropine diminished nicotine's effects in mice and rats.

There have been a number of studies which suggest that nicotine is capable of interacting with opioid systems. Naloxone has been reported to attenuate smoke intake in chronic smokers (73), while mecamylamine exacerbates opioid withdrawal in rats (74). Kumakura et al. (75) have shown that opioids can modify nicotinic receptors in cultured adrenal

Table 3. Antagonism of Nicotine's Antinociceptive Activity

	Rats*		Mice+	
Antagonist	Dose**	%Inhibition	Dose**	%Inhibition
Mecamylamine	0.1	50	0.2	50
Hexamethonium	5.0	47	5.0	34
Atropine	10.0	0	10.0	0
Scopolamine	1.0	0	---	---
Naloxone	10.0	0	0.2	50
Yohimbine	3.3	50	1.0	50

*Rats were pretreated s.c. with the antagonist 10 min (20 min for atropine) before s.c. administration of 1.25 mg nicotine/kg body weight. They were tested 2 min after nicotine.
+Mice were pretreated s.c. with the antagonist 10 min before the s.c. administration of 3.0 mg nicotine/kg body weight. They were tested 5 min after nicotine.
**mg/kg

chromaffin cells. Mattila et al. (58) reported that some cross tolerance developed between morphine and nicotine (mouse, hot plate) but data were not provided. The results in Table 3 show that naloxone doses up to 10 mg/kg failed to block nicotine in the rat tail-flick procedure. These data as well as those of Sahley and Berntson (5) suggest that nicotine is acting independently of the opioid system in rats. However, a different situation was found in mice where naloxone effectively blocked nicotine in the tail-flick procedure at doses comparable to those which antagonize morphine. Given the fact that naloxone blocked the antinociceptive effect of i.v.t.-administered acetylcholine (72) and that opioid antagonists will reverse opioid-induced inhibition of acetylcholine release from both peripheral tissues and brain (76-78), the mechanisms of action of nicotine and acetylcholine in mice may be intertwined. However, a difficulty arises with atropine which blocked acetylcholine-induced antinociception (72) and not that of nicotine (9).

Stereoselectivity

In order to be able to fully characterize the pharmacological stereoselectivity of nicotine, it was necessary to have sufficient quantities of the unnatural (+)-isomer. A straight-forward procedure was developed which allowed for the preparation of large quantities of (±)-nicotine (61). The racemate could then be resolved into the enantiomers with d- and l-tartaric acid as described previously (46).

The stereoselectivity of nicotine's antinociceptive activity, summarized in Table 4, is dependent upon the test as well as the species. One consistent finding was the greater potency of the (-)-isomer as compared to the (+)-isomer. There was modest stereoselectivity in most of the mouse tests and somewhat higher stereoselectivity in the rat tail-flick procedure. Actually, some complications arose in establishing the potency of (+)-nicotine in the rat-tail-flick test due to its toxicity (10). While the stereoselectivity of nicotine spans a rather broad range for those effects summarized in Table 1, the stereoselectivity of nicotine's antinociceptive effects, except for possibly the mouse tail shock, is well within this range.

Structure-Activity Relationship

Despite the fact that a systematic evaluation of the structural requirements for nicotine's antinociceptive effects is lacking, a few nicotine analogs have been investigated which provide some insight into the importance of nicotine's structural integrity. 1,1-Dimethyl-4-phenylpiperazine (DMPP) iodide is a ganglionic stimulant which has been shown to produce nicotine-like central effects (4,58) as well as to compete with [^{3}H]nicotine binding to brain tissue (79). It was found to be equiactive with nicotine in the rat tail-flick procedure following

Table 4. Stereoselectivity of Nicotine's Antinociceptive Effects

Species	Test	Potency ratio	Reference
Mouse	Tail flick	5	6
Mouse	PPQ	5	6
Mouse	Hot plate	11	6
Mouse	Tail shock	2	6
Rat	Tail flick	29	10

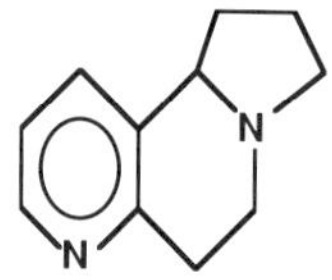

Fig. 2. Structure of the bridged-nicotine analog (BN) 1,2,3,5,6,10b-hexahydropyrido[2,3g]indolizine.

i.v.t. administration and its effects were blocked by mecamylamine (4). However, Mattila et al. (58) found DMPP to be less potent than nicotine in the rabbit-tooth method after s.c. administration. It may be that DMPP did not readily penetrate the brain following s.c. administration due to the fact that it is a quaternary compound.

Recently, the stereoisomers of the bridged-nicotine analog, 1,2,3,5,6,10b-hexahydropyrido[2,3g]indolizine (Fig. 2), were compared to (-)-nicotine in the mouse-tail-flick procedure following i.v.administration. Doses up to 5 mg/kg of both isomers were found to be devoid of antinociceptive activity, whereas doses of 0.1-0.3 mg/kg of (-)-nicotine were active (60). However, these bridged-nicotine isomers were also less active than (-)-nicotine in other pharmacological tests (60,80).

The methiodides of nicotine (structures in Table 5) have been used extensively as probes for distinguishing central and peripheral effects of nicotine (49,81-83) and have been utilized to a lesser extent for studying structure-activity relationships. The loss of central activity after quaternization of nicotine could be due either to a decrease in the penetrability of the blood-brain barrier or to an alteration in the pharmacological activity or a combination of both. In studying the peripheral actions of the methiodides, Gillis and Lewis (84) found that nicotine pyrrolidine methiodide had potent nicotine-like properties, nicotine bis methiodide possessed only slight nicotinic activity and nicotine pyridine methiodide was devoid of nicotine-like effects as measured in guinea-pig ileum, frog rectus abdominis muscle and rat phrenic diaphragm preparations as well as pressor effects in cats. Barlow and Dobson (85) evaluated the methiodides for pressor activity in cats and observed that nicotine pyrrolidine methiodide was as active as nicotine, whereas the other two methiodides were at least 10 times less potent. These findings were in agreement with those of Larson and Haig (86) which showed that nicotine and nicotine pyrrolidine methiodide were equally toxic in mice and rats and in producing pressor effects in dogs. These studies suggest that quaternizing the pyrrolidine moiety has little effect on nicotine's peripheral effects while quaternization of the pyridine nitrogen attenuates activity. Studies were conducted in our laboratories for the purpose of evaluating the antinociceptive activity of nicotine methiodides in both mice and rats after s.c. and i.v.t. administration (8). A summary of these findings are presented in Table 5. The pyrrolidine methiodide was considerably less active than nicotine in the mouse and rat tail-flick procedures following s.c. treatment. It was not possible to determine the potency of the pyrrolidine methiodide due to its peripheral toxicity. The other two methiodides lacked both antinociceptive and tremorgenic activity even after high doses. However, the pyrrolidine methiodide was approximately half as potent as nicotine in blocking PPQ-induced stretching in mice, an action which may be both centrally and peripherally mediated. Again, the pyridine and bis methiodides were without effect. Following i.v.t. administration, the pyrrolidine methiodide was actually more potent than nicotine in the mouse tail-flick test, while the pyridine methiodide was somewhat less potent

Table 5. Antinociceptive Activity of the Methiodide Derivatives of (-)-Nicotine

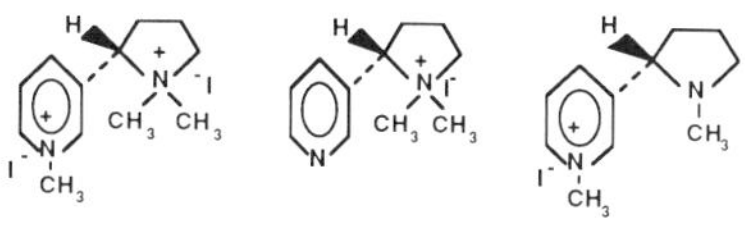

Methiodide analog	Route	Test	Potency relative to (-)-nicotine Mice	Rats
Pyrrolidine	SC	Tail flick	< 0.50	< 0.50
Pyridine	SC	Tail flick	< 0.01	< 0.08
Bis	SC	Tail flick	< 0.03	< 0.08
Pyrrolidine	IVT	Tail flick	3.89	-----
Pyridine	IVT	Tail flick	≈ 1.00	-----
Bis	IVT	Tail flick	6.90	-----
Pyrrolidine	SC	PPQ*	0.40	-----
Pyridine	SC	PPQ	0.004	-----
Bis	SC	PPQ	0.009	-----

* *p*-Phenylquinone (PPQ)-induced stretching

than nicotine. One of the most surprising findings was the high potency of the bis methiodide when it was injected i.v.t. The results with the methiodides provide further evidence that structural modifications are important with regard to nicotine's antinociceptive activity. In addition, they lend additional support to the notion that nicotine's antinociceptive effects are centrally mediated.

Site of Action

Evidence has been presented in the foregoing discussion which is consistent with nicotine producing its antinociceptive effects by interacting with specific receptors. Due to the fact that the origin, expression, and perception of pain may involve both central and peripheral components, it is important to establish the site of nicotine's action. The results with the methiodides suggest that nicotine's actions are of central origin. In addition, Sahley and Berntson (5) have shown that nicotine is effective in producing antinociception following i.v.t. doses that are not effective when administered s.c.

Studies have also been carried out in an effort to determine which areas of the brain may be involved in nicotine's antinociceptive effects (11). The potency of nicotine in the rat tail-flick procedure was compared after injections into the common carotid arteries (which serve predominately the forebrain), vertebral arteries (which primarily serve the hindbrain), subarachnoid space (at the high lumbar level, L_1-L_2), intravenously and s.c. Some of the results from these studies are summarized in Table 6. Doses of 1 mg/kg of nicotine produced 30% and 88% antinociception after s.c. and i.v. administration, respectively.

Table 6. Site of (-)-Nicotine's Antinociceptive Action in the Rat

Route	N	Dose*	%MPE+
Subcutaneous	6	500	0 ± 0
	6	1000	30 ± 13
Intravenous	6	1000	88 ± 16
Intraventricular	5	100	7 ± 3
	7	200	7 ± 4
	4	300	100 ± 0
Common carotid artery	4	40	-5 ± 4
	4	200	17 ± 13
	4	400	100 ± 0
Vertebral artery	3	25	0 ± 0
	5	50	60 ± 18
	1	250	100 ± --
Subarachnoid space	5	5	40 ± 25
	4	10	100 ± 0
	5	25	100 ± 0
	3	40	100 ± 0

*μg/kg
+Percent maximum possible effect (%MPE) in the tail flick procedure was determined 2 min after all routes of administration, except for subarachnoid injections which were determined after 1 min.

Nicotine produced complete antinociception following i.v.t. injections of 300 μg/kg, a dose which is ineffective when given peripherally. However, i.v.t. doses of 100 and 200 μg/kg were without effect. These results contrast those obtained by Sahley and Berntson (5), who found nicotine to be effective in the rat tail-flick procedure after i.v.t. doses of 25 and 50 μg, doses which are probably close to 100 and 200 μg/kg. Administration of nicotine into the common carotid artery produced results similar to those following i.v.t. injection. However, nicotine was more potent via the vertebral artery than after administration into either the ventricles or the common carotid artery, which suggests that the hindbrain is involved in mediating antinociception. Clearly, nicotine was most effective when administered directly into the subarachnoid space. Doses as low as 5 and 10 μg/kg produced 40% and 100% antinociception, respectively.

The observations that the methiodides exhibit antinociceptive activity only after central administration and that nicotine is more effective after central administration as compared to peripheral injections supports a central site of action. The mechanism, as well as the primary site of action, of nicotine remains elusive. At least in the rat, nicotine's effects appear to be mediated primarily via a cholinergic mechanism in that its effects are blocked by mecamylamine. The effectiveness of nicotine as an analgesic agent when applied directly to the spinal cord and the fact that appreciable quantities of nicotine are present in the spinal cord during antinociception, regardless of the route of administration (11), implicate the involvement of the spinal cord. It has been established that the spinal cord contains both muscarinic and nicotinic cholinergic receptors (87,88) and it would therefore seem reasonable to conclude that nicotine is stimulating the nicotinic cholinergic receptor in the spinal cord to block the tail-flick response. Nicotinic cholinergic innervation of the Renshaw cell results in neuronal inhibition via the recurrent collaterals as well as inhibition of motor neurons. It is therefore possible that nicotine's stimulation of the Renshaw cell or a similar inhibitory system could result in blocking either the perception or reaction to painful thermal stimuli. However, antinociceptive effects most likely do not arise from a single action of nicotine, such as blocking the reaction to the pain stimulus, due to the fact that nicotine is effective in blocking different stimuli in numerous tests (6). Also, nicotine's effects in the mouse-tail-flick procedure are blocked by naloxone, which suggests an opioid mechanism.

CARDIOVASCULAR EFFECTS

Despite the considerable interest in the cardiovascular effects of nicotine, the underlying mechanisms have not been clearly defined. There is evidence that nicotine alters cardiovascular function through several peripheral mechanisms which include stimulation of sympathetic ganglia and the adrenal medulla (see chapter by Van Loon) and activation of chemoreceptors of the aortic and carotid bodies. Central mechanisms are also thought to contribute as evidenced by numerous reports that cardiovascular changes result from central administration of nicotine at doses that are ineffective when administered peripherally (23,26,30); see also chaters by Van Loon, Kubo and Sapru. Further insight may be gained by identifying nicotinic receptors that may be involved in these cardiovascular effects.

Stereoselectivity

The ability of (+)-and (-)-nicotine to increase blood pressure and decrease heart rate in laboratory animals is summarized in Table 7. There is a consensus in all of the studies that (-)-nicotine is more effective than (+)-nicotine in producing these cardiovascular effects. The stereoselectivity of the hypertensive effects is somewhat greater in the rat than in the other species. In general, this stereoselectivity is consistent with that for most of the other pharmacological effects of nicotine as summarized in Table 1.

Table 7. Stereoselectivity of Nicotine's Cardiovascular Effects

Species	Potency Ratio	Reference
	Increased Blood Pressure	
Kitten	6	55
Dog	4	49
Rat	14	55
Rat	17	46
Rat	17	21
	Bradycardia	
Rat	17	46

Table 8. Comparison of the Cardiovascular Effects of the Stereoisomers of Bridged Nicotine to those of (–)-Nicotine in the Anesthetized Rat*

Dose (mg/kg)	N	Systolic	Diastolic	Heart Rate
		(-)-Nicotine		
0.01	6	2 ± 1	2 ± 1	-2 ± 2
0.03	8	3 ± 2	2 ± 2	2 ± 4
0.10	6	48 ± 7	31 ± 7	-53 ± 29
0.30	6	70 ± 4	42 ± 5	-66 ± 34
		(+)-Bridged Nicotine		
1.0	8	6 ± 1	-6 ± 1	-3 ± 5
3.0	9	-5 ± 2	-5 ± 2	3 ± 4
10.0	8	3 ± 2	4 ± 2	-5 ± 3
30.0	5	14 ± 4	14 ± 3	-4 ± 5
100.0	9	42 ± 3	44 ± 3	0 ± 4
		(-)-Bridged Nicotine		
0.3	7	0 ± 2	0 ± 2	-8 ± 3
1.0	7	-7 ± 1	-6 ± 1	-3 ± 3
3.0	8	-16 ± 2	-11 ± 2	-11 ± 5
10.0	8	-52 ± 3	-52 ± 4	-49 ± 10

*Results expressed as means ± S.E.M. for change in blood pressure (mm Hg) and heart rate (beats/min) following i.v. administration.

Bridged-Nicotine Analogs

Studies were carried out to determine whether the imposition of conformation restrictions on nicotine would enhance stereoselectivity. The bridged-nicotine analogs that were described above were also evaluated for their effects on blood pressure and heart rate in anesthetized rats (60). It is evident from the results in Table 8 that bridging the pyridine and pyrrolidine rings dramatically alters the cardiovascular effects of nicotine. At an i.v. dose of 0.3 mg/kg, (–)-nicotine increased systolic and diastolic blood pressure and decreased heart rate. The (+)-isomer of bridged nicotine produced similar hypertensive effects but was more than a 100 times less potent. Heart rate was essentially unaffected. The (–)-isomer of bridged nicotine differed from (+)-bridged nicotine and from (–)-nicotine in that it produced a decrease in both blood pressure and heart rate. It was also demonstrated that atropine would block these cardiovascular effects of (–)-bridged nicotine.

Unfortunately, stereoselectivity of the bridged nicotine isomers was not greater than that of the nicotine stereoisomers with regard to cardiovascular effects. However, these studies demonstrated that this structural modification resulted in an attenuation of cardiovascular effects, much like that of other nicotine effects. It appears that the (+)-isomer of bridged nicotine may be interacting with the same receptor as (–)-nicotine but that it has lesser affinity or intrinsic activity. On the other hand, the (–)-isomer of bridged nicotine appears to be interacting with a receptor distinct from the nicotinic receptor. Blockade of its cardiovascular effects with atropine demonstrated that muscarinic receptors are involved either directly or indirectly.

SUMMARY

The pharmacological profile of (+)-nicotine is similar to that of the natural isomer in that it appears to produce all of the same effects. However, there are slight differences in potency regarding the different effects. To a considerable extent, these differences could be attributed to the vagaries expected from comparison of data from different laboratories using various experimental techniques. It appears that (+)-nicotine is approximately 8-15 times less active than (-)-nicotine in producing most of the centrally mediated effects, such as motor performance and drug discrimination. The question remains as to whether or not all of these effects are receptor mediated, particularly when the relatively low stereoselectivity is taken into account. Of course, most nicotine receptor binding studies also exhibit low stereoselectivity and there is ample information to suggest multiple nicotine binding sites (79), see also chapter by Sloan and Martin. Additional evidence must be forthcoming before an association can be established between the central effects of nicotine and its binding sites in the brain.

In light of the fact that nicotine produces diverse antinociceptive effects in animal tests, an important question arises as to whether cigarette smoking has any relevance to pain perception. At present, there is only limited direct evidence that cigarette smoking either attenuates or enhances pain perception (89). It may be that nicotine exerts antinociceptive effects only under selective conditions that are not achieved by smoking. The importance of nicotine may prove to be its use as a tool for unraveling the complexities of pain mechanisms rather than its alteration in pain perception in smokers.

The mechanisms responsible for smoking-induced changes in cardiovascular function have not been fully characterized, although it is quite likely that both central and peripheral nicotine receptors are associated with these effects. The administration of the unnatural (+)-isomer of nicotine to laboratory animals results in hypertension and bradycardia, a pharmacological profile similar to that produced by (-)-nicotine. The modest stereoselectivity of the cardiovascular effects are comparable to those of other nicotine effects. Additional investigations will be necessary to determine the extent to which peripheral and central receptors contribute to nicotine's cardiovascular effects.

ACKNOWLEDGEMENTS

The following individuals contributed to the research that was conducted in our laboratories: M. D. Aceto, H. Awaya, W. L. Dewey, J. L. Egle, T.-C. Fu, J. L. Kachur, E. L. May, H. L. Tripathi, I. M. Uwaydah and W. C. Vincek. The research was supported by grants from the Council for Tobacco Research (No. 1157) and from the USPHS (DA-00490, DA-07027 and DA-02384).

REFERENCES

1. Aceto, MD, Bentley, HC and Dembinski, JR: Effects of ganglion blocking agents on nicotine extensor convulsions and lethality in mice. Br. J. Pharmacol. 37:104-111, 1969.

2. Abood, LG, Lowry, K, Tometsko, A and Booth, H: Electrophysiological, behavior, and chemical evidence for a noncholinergic, stereospecific site for nicotine in rat brain. J. Neurosci. Res. 3:327-333, 1978.
3. Mansner, R: Relation between some central effects of nicotine and its brain levels in the mouse. Ann. Med. Exp. Biol. Fenn. 50:205-212, 1972.
4. Phan, DV, Doda, M., Bite, A and Gyorgy, L: Antinociceptive activity of nicotine. Acta Physiol. Acad. Sci. Hung. 44:85-93, 1973.
5. Sahley, TL and Berntson, GG: Antinociceptive effects of central and systemic administration of nicotine in the rat. Psychopharmacology. 65:279-283, 1979.
6. Aceto, MD, Tripathi, HL, May, EL, Jacobson, AE, and Martin, BR: Antinociceptive effects of the optically pure stereoisomers of nicotine. Pharmacologist 22:302, 1980.
7. Kamerling, SG, Wettstein, JG, Sloan, JW, Su, T-P and Martin, WR: Interaction between nicotine and endogenous opioid mechanisms in the unanesthetized dog. Pharmacol. Biochem. Behav. 17:733-740, 1982.
8. Aceto, MD, Awaya, H, Martin, BR and May, EL: Antinociceptive action of nicotine and its methiodide derivatives in mice and rats. Brit. J. Pharmacol. 79:868-876, 1983.
9. Tripathi, HL, Martin, BR and Aceto, MD: Nicotine-induced antinociception in rats and mice: correlation with nicotine brain levels. J. Pharmacol. Exp. Ther. 221:91-96, 1982.
10. Martin, BR, Tripathi, HL, Aceto, MD and May, EL: Relationship of the biodisposition of the stereoisomers of nicotine in the central nervous system to their pharmacological actions. J. Pharmacol. Exp. Ther. 226:157-163, 1983.
11. Aceto, MD, Bagley, RS, Dewey, WL, Fu, T-C and Martin, BR: Spinal cord as a major site for nicotine's antinociceptive action in the rat. Neuropharmacol. (in press).
12. Jones, JJ and Lee, J: The value of rats with hereditary hypothalamic diabetes insipidus for the bioassay of vasopressin. J. Endocrinol. 37:335-344, 1967.
13. Mansner, R and Mattila, MJ: Nicotine induced tremor and antidiuresis and brain nicotine levels in the rat. Med. Biol. 53:169-176, 1975.
14. Holmstedt, B and Lundgren, G: Arecoline, nicotine and related compounds. Tremorgenic activity and effect upon brain acetylcholine. Ann. N.Y. Acad. Sci. 142:126-142, 1967.
15. Hendry, JS and Rosecrans, JA: Effects of nicotine on conditioned and unconditioned behaviors in experimental animals. In (ed.) Balfour, DJK, Nicotine and the Tobacco Smoking Habit, International Encyclopedia of Pharmacology and Therapeutics, Pergamon Press, New York, 1984, 75-100.
16. Rosecrans, JA and Chance, WT: The discriminative stimulus properties of N- and M-cholinergic receptor stimulants. In (eds.) Ho, BT, Richards, DW, III, and Chute, DL, Drug Discrimination and State Dependent Learning, Academic Press, New York, 1977, 119-130.
17. Rodgers, RJ: Effects of nicotine, mecamylamine, and hexamethonium on shock-induced fighting, pain reactivity, and locomotor behaviour in rats. Psychopharmacology. 66:93-98, 1979.
18. Clarke, PBS and Kumar, R: The effects of nicotine on locomotor activity in nontolerant and tolerant rats. Br. J. Pharmacol. 78:329-337, 1983.
19. Clarke, PBS and Kumar, R: Characterization of the locomotor stimulant action of nicotine in tolerant rats. Br. J. Pharmacol. 80:587-594, 1983.
20. Osumi, Y, Ishikawa, T, Nagasaka, Y and Fujiwara, M: Central effect of nicotine on gastric acid secretion in rats. Eur. J. Pharmacol. 68:409-416, 1980.

21. Ishikawa, T, Osumi, Y, Fujiwara, M and Nagata, M: Possible roles of central cholinergic nicotinic mechanisms in regulation of gastric functions. Eur. J. Pharmacol. 80:331-336, 1982.
22. Nagata, M, Ishikawa, T and Osumi, Y: Effect of d-nicotine on the l-nicotine-induced increase in gastric acid secretion in rats. Jpn. J. Pharmacol. 37:215-217, 1985.
23. Armitage, AK and Hall, GH: Effects of nicotine on the systemic blood pressure when injected into the cerebral ventricles of cats. Int. J. Neuropharmacol. 6:143-149, 1967.
24. Armitage, AK and Hall, GH: Further evidence relating to the mode of action of nicotine in the central nervous system. Nature (London) 214:977-979, 1967.
25. Armitage, AK, Hall, GH, Milton, AS and Morrison, CF: Effects of nicotine injected into and perfused through the cerebral ventricles of the cat. Ann. N.Y. Acad. Sci. 142:27-39, 1967.
26. Pradhan, SN, Bhattacharya, IC and Atkinson, KS: The effects of intraventricular administration of nicotine on blood pressure and some somatic reflexes. Ann. N. Y. Acad. Sci. 142:50-66, 1967.
27. Schaeppi, U: Effects of nicotine administration to the cat's lower brain stem upon electroencephalogram and autonomic nervous system. Ann. N.Y. Acad. Sci. 142:40-49, 1967.
28. Schaeppi, U: Nicotine treatment of selected areas of the cat brain: effects upon EEG and autonomic system. Int. J. Neuropharmacol. 7:207-220, 1968.
29. McCarthy, LE and Borison, HL: Separation of effects of CO_2 and nicotine on ventilation and blood pressure. Resp. Physiol. 15:321-330, 1972.
30. Lang, WJ and Rush, ML: Cardiovascular responses to injections of cholinomimetic drugs into the cerebral ventricles of unanesthetized dogs. Br. J. Pharmacol. 47:196-205, 1973.
31. Larson, PS, Haag, HB and Silvette, H, Tobacco, Experimental and Clinical Studies, Williams and Wilkins, Baltimore, Maryland. Also Suppls. I (1968), II (1971) and III (1975).
32. Aceto, MD and Martin, BR: Central actions of nicotine. Med. Res. Rev. 2:43-62, 1982.
33. Abood, LG, Lowry, K, Tometsko, A and Booth, M: Electrophysiological behavior, and chemical evidence for a noncholinergic stereospecific site for nicotine in rat brain. J. Neurosci. Res. 3:327-333, 1978.
34. Yoshida, K and Imura, H: Nicotinic cholinergic receptors in brain synaptosomes. Brain Res. 172:453-459, 1979.
35. Vincek, WC, Martin, BR, Aceto, MD and Bowman, ER: Synthesis and preliminary binding studies of 4,4-ditritio-(-)-nicotine of high specific activity. J. Med. Chem. 23:960-962, 1980.
36. Vincek, WC, Martin, BR, Aceto, MD, Tripathi, HL, May, EL and Harris, LS: Synthesis of 4,4-ditritio-(+)-nicotine: comparative binding and distribution studies with natural enantiomer. J. Pharm. Sci. 70:1292-1293, 1981.
37. Abood, LG, Grassi, S and Costanza, M: Binding of optically pure (-)-[^{3}H]nicotine to rat brain membranes (FEBS Lett. 157:147-149, 1983.
38. Romano, C and Goldstein, A: Stereospecific nicotine receptors on rat brain membranes. Science (Wash.) 210:647-649, 1980.
39. Costa, LG and Murphy, SD: [^{3}H]Nicotine binding in rat brain: alteration after chronic acetylcholinesterase inhibition. J. Pharmacol. Exp. Ther. 226:392-397, 1983.
40. Marks, MJ and Collins, AC: Characterization of nicotine binding in mouse brain and comparison with the binding of alpha-bungarotoxin and quinuclidinyl benzilate. Mol. Pharmacol. 22:554-564, 1982.
41. Sloan, JW, Martin, WR and Todd, GD: Multiple nicotine binding sites in rat brain P2 fraction. Psychopharmacol. Bull. 19:402-408, 1983.

42. Sershen, H, Reith, MEA, Lajtha, A and Gennaro, J, Jr: Noncholinergic, saturable binding of (+)-[^{3}H]-nicotine to mouse brain. J. Recep. Res. 2:1-15, 1981.
43. Schwartz, RD, McGee, R, Jr and Kellar, KJ: Nicotinic cholinergic receptors labeled by [^{3}H]acetylcholine in rat brain. Mol. Pharmacol. 22:56-62, 1982.
44. Abood, LG, Latham, W, and Grassi, S: Isolation of a nicotine binding site from rat brain by affinity chromatography. Proc. Natl. Acad. Sci. U.S.A. 80:3536-3539, 1983.
45. Match, DI and Davis, ME: Toxicity of alpha- and beta-nicotines and nornicotine. J. Pharmacol. Exp. Ther. 50:93-99, 1934.
46. Aceto, MD, Martin, BR, Uwaydah, IM, May, EL, Harris, LS, Izazola-Conde, C, Dewey, WL and Bradshaw, TJ: Optically pure (+)-nicotine from (±)-nicotine and biological comparisons with (-)-nicotine. J. Med. Chem. 22:174-177, 1979.
47. Hicks, CS and Sinclair, DA: Toxicities of the optical isomers of nicotine and nornicotine. Aust. J. Exp. Bio. Med. Sci. 25:83-86, 1947.
48. Caulfield, MP and Higgins, GA: Mediation of nicotine-induced convulsions by central nicotine receptors of the 'C6' type. Neuropharmacology 22:347-351, 1983.
49. Domino, EF: Some comparative pharmacological actions of (-)-nicotine, its optical isomer, and related compounds. In (ed.) Von Euler, US, Tobacco Alkaloids and Related Compounds, Pergamon Press, Oxford, 303-313, 1965.
50. Kumar, R, Pratt, JA and Stolerman, IP: Characteristics of conditioned taste aversion produced by nicotine in rats. Br. J. Pharmacol. 79:245-253, 1983.
51. Meltzer, LT, Rosecrans, JA, Aceto, MD and Harris, LS: Discriminative stimulus properties of the optical isomers of nicotine. Psychopharmacology 68:283-286, 1980.
52. Romano, C, Goldstein, A and Jewell, MP: Characterization of the receptor mediating the nicotine discriminative stimulus. Psychopharmacology 74:310-315, 1981.
53. Clark, PBS and Kumar R: Characterization of the locomotor stimulant action of nicotine in tolerant rats. Br. J. Pharmacol. 80:587-594, 1983.
54. Connelly, MS and Littleton, JM: Lack of stereoselectivity in ability of nicotine to release dopamine from rat synaptosomal preparations. J. Neurochem. 41:1297-1302, 1983.
55. Barlow, RB and Hamilton, JT: The stereospecificity of nicotine. Br. J. Pharmacol. 25:206-212, 1965.
56. Ikushima, S, Muramatsu, I, Sakakibara, Y, Yokotani, K and Fujiwara, M: The effects of d-nicotine and l-isomer on nicotinic receptors. J. Pharmacol. Exp. Ther. 222:463-470, 1982.
57. Davis, L, Pollock, LJ, Stone, TT: Visceral pain. Surg. Gynecol. Obstet. 55:418-426, 1932.
58. Mattila, MJ, Ahtee, L, Saarnivaara, L: The analgesic and sedative effects of nicotine in white mice, rabbits and golden hamsters. Ann. Med. Exp. Fenn. 46:78-84, 1968.
59. Rodgers, RJ: Effects of nicotine, mecamylamine, and hexamethonium on shock-induced fighting, pain reactivity, and locomotor behaviour in rats. Psychopharmacology 66:93-98, 1979.
60. Kachur, JF, May, EL, Awaya, H, Egle, JL, Jr., Aceto, MD and Martin, BR: Pharmacological effects of 1,2,3,5,6,10b-hexahydropyrido[2,3g]-indolizine, a bridged-nicotine analog. Life Sci. 38:323-330, 1986.
61. Bowman, ER, McKennis, H, Jr. and Martin, BR: A convenient method for the preparation of racemic nicotine. Synth. Comm. 12:871-879, 1982.
62. Kumat, UG, Pradham, RJ and Sheth, UK: Potentiation of a non-narcotic analgesic, dipyrone, by cholinomimetic drugs. Psychopharmacology 23:180-186, 1972.

63. Oelssner, W and Andreas, K: Die Hemmung von Schmerzreaktionen der Maus durch Cholinomimetika und deren antagonistische Beeinflussung. Acta Biol. Med. Germ. 22:369-385, 1969.
64. Pert, A: The cholinergic system and nociception in the primate: interactions with morphine. Psychopharmacology 44:131-137, 1975.
65. Pert, A and Maxey, G: Asymmetrical cross-tolerance between morphine and scopolamine induced antinociception in the primate: differential sites of action. Psychopharmacology 44:139-145, 1975.
66. Asghar, K and Roth, LJ: Entry and distribution of hexamethonium in the central nervous system. Biochem. Pharmacol. 20:2787-2795, 1971.
67. Flodmark, S and Wramner, T: The analgesic action of morphine, eserine, and prostigmine studied by a modified Hardy-Wolff-Goodell method. Acta Physiol. Scand. 9:88-96, 1945.
68. Chen, G: The anti-tremorine effect of some drugs as determined by Hoffner's method of testing analgesia in mice. J. Pharmacol. Exp. Ther. 124:73-76, 1958.
69. Herz, A: Actions of arecoline on the central nervous system. Naunyn-Schmiedebergs Arch. Exp. Pathol. Pharmakol. 242:414-420, 1962.
70. Harris, LS, Dewey, WL and Howes, JF: The tail-flick test, cholinergic mechanisms. Fed. Proc. 27:753, 1968.
71. Ireson, JD: A comparison of the antinociceptive actions of cholinomimetic and morphine-like drugs. Br. J. Pharmacol. 40:92-101, 1970.
72. Pedigo, NW, Dewey, WL and Harris, LS: Determination and characterization of the antinociceptive activity of intraventricularly administered acetylcholine in mice. J. Pharmacol. Exp. Ther. 193:845-852, 1975.
73. Karras, A and Kane, JM: Naloxone reduces cigarette smoking. Life Sci. 27:1541-1545, 1980.
74. Pinsky, CR, Frederickson, CA and Vazquez, AJ: Morphine withdrawal syndrome responses to cholinergic antagonists and to a partial cholinergic agonist. Nature 242:59-60, 1973.
75. Kumakura, K, Karoum, F, Guideotta, A and Costa, E: Modification of nicotinic receptors by opiate receptor agonists in cultured adrenal chromaffin cells. Nature 283:489-492, 1980.
76. Waterfield, AA and Kosterlitz, HW: Stereospecific increase by narcotic antagonists of evoked acetylcholine output in guinea-pig ileum. Life Sci. 16:1787-1792, 1975.
77. Domino, EF, Vasko, MR and Wilson, AE: Mixed depressant and stimulant actions of morphine and their relationship to brain acetylcholine. Life Sci. 18:361-376, 1978.
78. Jhamandas, K and Sutak, M: Action of enkephalin analogues and morphine on brain acetylcholine release: differential reversal by naloxone and an opiate pentapeptide. Br. J. Pharmacol. 71:201-210, 1980.
79. Martin, BR: Nicotine receptors in the central nervous system. In (ed.) Conn, P., The Receptors, Academic Press (in press).
80. Rondahl, L: Synthetic analogues of nicotine. VIII. Synthesis and biological testing of two tetrahydro-5-quinolinamines. Acta Pharm. Suec. 17:288-291, 1980.
81. Geller, I, Hartmann, R and Blum, K: Effects of nicotine, nicotine monomethiodide, lobeline, chlordiazepoxide, meprobamate and caffeine on a discrimination task in laboratory rats. Psychopharmacology 20:355-365, 1971.
82. Schechter, MD and Rosecrans, JA: Nicotine as a discriminative cue in rats: inability of related drugs to produce a nicotine-like cueing effect. Psychopharmacology 27:379-387, 1972.
83. Thompson, JH, Angulo, M, Choi, L, Roch, M and Jenden, DJ: The chronic effects of nicotine monomethiodide on gastic secretion in pylorus-ligated rats. Experientia 28:1176-1177, 1972.

84. Gillis, CN and Lewis, JJ: The pharmacology of nicotine monomethiodide. J. Pharm. Pharmacol. 8:46-54, 1956.
85. Barlow, RB and Dobson, NA: Nicotine monomethiodide. J. Pharm. Pharmacol. 7:27-34, 1955.
86. Larson, PS and Haig, HB: Studies on the fate of nicotine in the body. III. On the pharmacology of some methylated and demethylated derivatives of nicotine. J. Pharmacol. Exp. Ther. 77:343-349, 1943.
87. Headley, PM, Lodge, D and Biscoe, TJ: Acetylcholine receptors on Renshaw cell of the rat. Eur. J. Pharmacol. 30:252-259, 1975.
88. Kayaalp, SO and Neff, NH: Regional distribution of cholinergic muscarinic receptors in spinal cord. Brain Res. 196:426-436, 1980.
89. Mousa, SA, Aloyo, VJ and Van Loon, GR: Tobacco smoke and nicotine alter pain sensitivity in rats. Soc. Neurosci. 1:104, 1984.

NICOTINIC RECEPTORS IN THE MOTOR ENDPLATE AND IN SYMPATHETIC GANGLIA

Robert L. Volle

Department of Pharmacology
College of Medicine
University of Kentucky
Lexington, Kentucky

The first important account of the actions of a drug on autonomic ganglia was given in a series of papers by Langley (1, 2) in which the stimulating and blocking actions of nicotine on sympathetic ganglia were described. By applying nicotine to the surface of the ganglia it was possible to demonstrate that nicotine first imitated and then suppressed the responses of various end-organs to electrical stimulation of preganglionic nerves. The application of this experimental technique to a wide variety of sites in the peripheral nervous system made it possible for Langley to outline the general pattern of autonomic innervation of organs.

The nicotine-like activities of choline and several esters of choline, particularly acetylcholine, were described in detail by Dale in 1914. Dale (3) recognized what Boehm (4) had demonstrated previously, that the nitrous acid ester of choline evoked a pressor response in animals treated with doses of atropine that were adequate to prevent depressor activity of the ester. The observation was extended by Dale to include choline and acetylcholine (ACh). He made the additional important observation that the pressor responses of atropinized animals to the choline derivatives were abolished by the prior administration of nicotine. It should be pointed out also that in this paper, Dale made the distinction between the "nicotinic" and "muscarinic" activities of ACh; the former an excitatory and blocking action in autonomic ganglia and the latter an excitatory action of peripheral parasympathetic postganglionic junctions. Moreover, special attention was directed to the highly active but unstable actions of ACh. This latter property of the ester led Dale to propose that ACh was rapidly hydrolyzed in the blood by an esterase.

Ubiquitous Distribution on Nicotinic Receptors

Both conducting and transmitting processes in neurons are sensitive to the actions of nicotine (Table 1). Vagal C fibers, adrenergic axons, and sensory neurons are depolarized by nicotine and related drugs. Conduction in C fibers is blocked, norepinephine is released from adrenergic nerve endings, and sensory neurons are stimulated by nicotine. Antagonists, such as d-tubocurarine (dTC) or hexamethonium (C_6), prevent these actions of nicotine without affecting neuronal conduction or neuronal responses to physiologic stimulation. At some sites, e.g., neuromuscular junction, autonomic ganglia, adrenal medulla, and selected central

Table 1. Responses to nicotine and physiological stimulation after acetylcholine receptor blockade by hexamethonium (C_6).

	CONTROL RESPONSES		AFTER C_6	
Tissue	Nicotine	Physiol. Stim	Nicotine	Physiol. Stim
C fibers*	depol.	NAP+	none	NAP+
Sensory N. ending	excit.	excit.	none	excit.
Adrenegic N.	norepinephrine release	norepinephrine release	none	norepinephrine release
Adrenal medulla	norepinephrine release	norepinephrine release	none	none
Ganglia	depol.	EPSP**	none	none
Neuromuscular	depol.	EPP***	none	none

* Vagus nerve
\+ Nerve action potential (NAP)
** Excitatory Postsynaptic Potential
*** End plate potential

neurons, nicotine imitates ACh, the transmitter substance. At these sites, dTC or C_6 prevents or reverses the actions of nicotine and blocks cholinergic transmission.

Neuromuscular Block by Nicotine and Lobeline

Nicotine depolarizes excitable tissues by a mechanism that is not sustained in the continued presence of nicotine and is followed by a restoration of the membrane potential to normal values or, in some cases, by an increase in membrane potential. Depolarization by nicotine of skeletal muscle is associated with a block of transmission both during depolarization and after repolarization takes place (Fig. 1, Row 3). Neuromuscular block by dTC does not involve a change in membrane potential (Fig. 1, Row 1).

In skeletal muscle, depolarization block by nicotine is due probably to sodium inactivation in the muscle cell membrane. Under these conditions, regenerative ionic events in the muscle membrane are difficult to initiate. It can be demonstrated that direct electrical stimulation of skeletal muscle fibers is blocked during depolarization caused by nicotine. However, in the continued presence of nicotine the muscle response to electricial stimulation but not to motor nerve stimulation is restored when the muscle membrane repolarizes. The muscle response to nerve stimulation is depressed by nicotine both during depolarization and repolarization (Fig. 1, Row 3).

Repolarization block by nicotine and related drugs is due to desensitization of the end plate acetylcholine receptor (5). Desensitization consists, first, of the formation of a nicotine-receptor complex (NR) that changes slowly to a new conformational state (NR'); second, to dissociation of NR' resulting in an unoccupied receptor of the new conformation (R'); and finally, in the slow reversion of (R') to its natural state (R). The receptor (R') is said to be desensitized.

It is clear also that the repolarization block by nicotine does not depend upon prior depolarization of the end plate. Conditions that

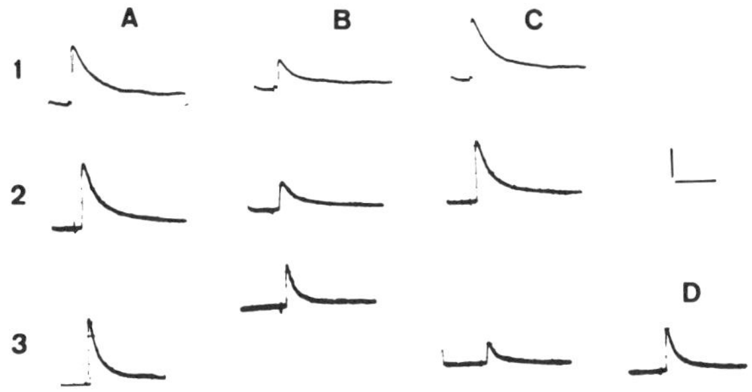

Fig. 1. Effects of d-tubocurarine (Row 1), lobeline (Row 2), and nicotine (Row 3) on endplate potentials in frog muscle. Row 1: A, control; B, 15 minutes in d-tubocurarine; C, washout. Row 2: A, control; B, 15 minutes in lobeline; C, washout. Row 3: A, control; B, 20 minutes in nicotine; C, 80 minutes in nicotine; D, washout. Calibration, 8 mV vertical and 15 msec horizontal. In contrast to the situation with d-tubocurarine and lobeline, block of the end plate potential with nicotine occurs both during depolarization of the end plate (Row 3B) and at a time when the end plate has repolarized (Row 3C).

prevent nicotine-induced depolarization of the end plate do not prevent repolarization blockade from occurring. The combination of nicotine with the receptor to form (NR) and, subsequently, (NR') does not require ion conductance in the receptor channel. The result, however, is the transformation of (R) to a nonconducting state (R'), and, subsequently, to blocked transmission and reduced sensitivity to activation by nicotinic drugs (e.g., succinylcholine).

Some insight into this process is provided by a comparison of neuromuscular blockade by dTC, nicotine, and lobeline (Fig. 1). Since Edmunds (1904) first compared lobeline with nicotine, it has been accepted that they differ only quantitatively. Lobeline, like nicotine, depolarizes and blocks ganglionic transmission, blocks conduction in vagal C fibers, stimulates peripheral sensory receptors, and depolarizes central neurons. At frog neuromuscular junctions, lobeline blocks transmission but, unlike nicotine, does not depolarize the end plate (Fig. 1, Row 2). Whereas the block of end plate cholinergic receptors by dTC is competitive, the block induced by lobeline is not (Fig. 2). Similarly, the repolarization block caused by nicotine is noncompetitive (6).

At the end plate, lobeline blocks neuromuscular transmission by a dual action on the receptor. It blocks the ACh recognition site and, in addition, blocks ion flow through the ACh receptor (7). The depression by lobeline of miniature end plate currents (m.e.p.c.) and end plate current (e.p.c.) is due mainly to a block of the ACh receptor. The shortening of the m.e.p.c. and the e.p.c. time course by lobeline can be explained by an action on the receptor-activated ionic channel. It is probable that lobeline accelerates e.p.c. decay by blocking open end plate channels (7).

Ganglion Block by Nicotine

Ganglionic blockade by drugs related to nicotine occurs in two phases (Fig. 3). The first phase coincides with the ganglionic depolarization and firing evoked by the drugs; the second phase occurs after the depolarization and firing have subsided.

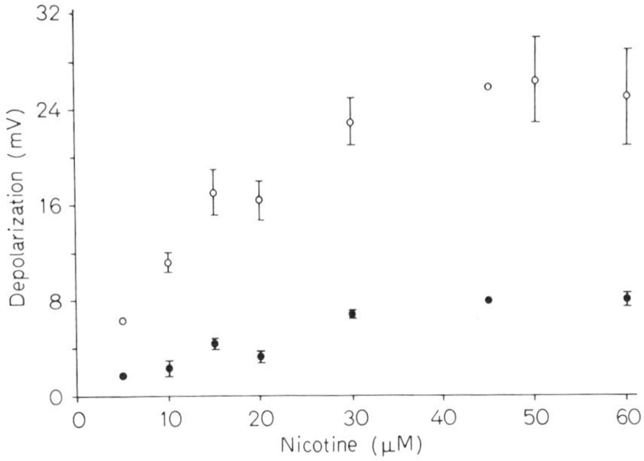

Fig. 2. Depolarization by nicotine of control (○) and lobeline-treated (●) paired frog sartorius muscles. Lobeline block of nicotine depolarization was not antagonized by increasing concentration of nicotine. Similar results were obtained using nicotine-treated muscles, i.e., nicotine did not depolarize nicotine-treated muscle when membrane potential had returned to control values.

Two lines of evidence show that the second phase of blockade by nicotine-like drugs is unlike that produced by dTC or C_6. First, the second phase of the blockade is accompanied by an increase in the demarcation potential recorded between the surface of the ganglion and the crushed end of the postganglionic nerve. The alterations in the contour of the ganglionic afterpotentials are consistent with the presence of hyperpolarization. Second, pretreatment of the ganglion with ouabain prevents the second phase of the blockade and the attending hyperpolarization but has no effect on ganglionic blockade by drugs related to dTc or C_6(8).

Ganglionic hyperpolarization is due to extrusion of sodium ions and may or may not be related to the persistent ganglion block. It is likely,

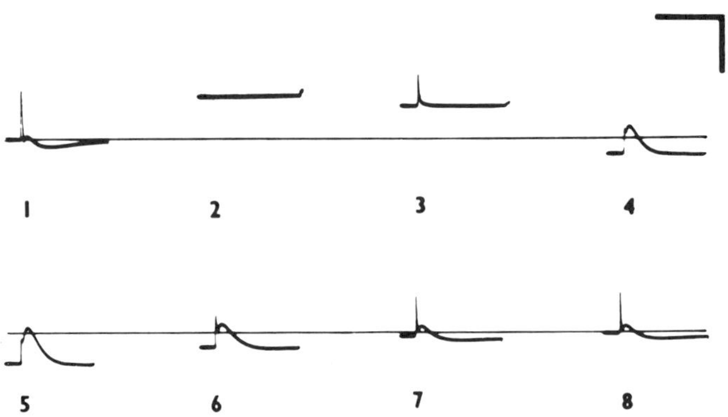

Fig. 3. Ganglion block by DMPP, a nicotinic drug. Ganglionic action potentials were evoked by preganglionic stimulation at a rate of 0.3Hz. Tracings 2 through 8 were recorded 10, 30, 40, 60, 120, 180, and 240 seconds after the drug was given. Displacement above the line signifies depolarization. Considerable hyperpolarization indicated by displacement below the line and altered contour of ganglionic action potential.

but not established, that the second phase of blockade results from receptor desensitization. If this is so, then the effects of ouabain on transmission block and receptor desensitization are of interest.

Nicotinic, Muscarinic, and Other Receptors in Autonomic Ganglia

Transmission in sympathetic ganglia is expressed as a complex pattern of excitatory and inhibitory potentials (Table 2). The ACh receptor in ganglia is present in two forms, nicotinic and muscarinic. Moreover, there is some evidence of peptidergic transmission. The actions of nicotine and related drugs on ganglionic transmission need to be viewed against this complex background.

Ganglionic firing evoked by acetylcholine (ACh) occurs in two distinctive phases (9). The first phase has a rapid onset, is sensitive to blockade by C_6, and can be activated by nicotine, tetramethylammonium (TMA) and dimethylphenylpiperazinium (DMPP) ions. This phase of excitation is attributed to the activation of ganglionic nicotinic sites. The second phase in the response to ACh has a delayed onset, is sensitive to blockade by small doses of atropine, and can be activated by muscarine, methacholine, and related drugs. The late occurring response to ACh is attributed to the activation of ganglionic muscarinic sites.

This classification of cholinoceptive sites depends upon the findings that: (a) atropine has no effect on firing caused by nicotinic drugs or causes ganglionic transmission block; (b) atropine completely blocks the firing caused by the muscarinic drugs; and (c) C_6 depresses ganglionic transmission or firing evoked by nicotinic drugs but has no effect on the firing evoked by muscarinic drugs.

In view of the foregoing, it is of some interest that several studies have demonstrated an interaction between nicotinic and muscarinic ganglionic receptors. In superior cervical ganglia, small doses of nicotine cause a marked increase in the firing that results from the activation of muscarinic receptors (10).

Cyclic Nucleotide Accumulation in a Sympathetic Ganglion

Isolated rat ganglia respond to preganglionic stimulation with a frequency-dependent increase in ganglion levels of both cAMP and cGMP. When ganglia are stimulated for 60 s at 10 Hz, cGMP levels increase from 0.1-0.3 $pmol \cdot mg^{-1}$ wet weight to about 1.0-1.5 $pmol \cdot mg^{-1}$. For cAMP, preganglionic nerve stimulation at 10 Hz increases levels from about 3-5 $pmol \cdot mg^{-1}$ to 8-9 $pmol \cdot mg^{-1}$. The cGMP and cAMP responses to

Table 2. Nicotinic, muscarinic and polypeptide receptors in sympathetic ganglia

Receptor	Transmitter	Synaptic Potential	Agonist	Antagonist
Nicotinic	ACh	EPSP*	Nicotine	Hexamethonium
Muscarinic	ACh	sIPSP*	Methacholine	Gallamine
Muscarinic	ACh	sEPSP*	Methacholine	Pirenzepine
Peptidergic	LHRH+	lsEPSP*		

* Excitatory postsynaptic potential, slow inhibitory postsynaptic potential, slow EPSP, and late, slow EPSP.
\+ Luteinizing Hormone Releasing Hormone

preganglionic stimulation require Ca^{2+} and are resistant to block by hexamethonium and atropine (11).

Thus, synaptic activation by preganglionic stimulation cause cyclic nucleotide content to increase in the face of muscarinic and nicotinic receptor blockade by atropine and C_6, respectively, and in the face of adrenergic receptor blockade by phentolamine and practolol (11). This makes it unlikely that the activation of adenyl and guanyl cyclase involves cholinergic or adrenergic transmitters in the rat ganglion.

Of the catecholamines, isoproterenol is the most potent activator of ganglionic adenyl cyclase, whereas dopamine is without effect on cAMP levels in rat ganglia. Bethanechol, a muscarinic receptor agonist, has no effect on ganglion cAMP. In some experiments, but not others, bethanechol causes an increase in ganglion cGMP. Ganglion depolarization during nicotinic receptor activation by DMPP affects neither cAMP nor cGMP. Therefore, like their antagonist counterparts, cholinergic and adrenergic receptor agonists have actions that are not consistent with the notion that cholinergic or adrenergic transmission in ganglia activate adenyl or guanyl cyclase.

Of the polypeptides tested to date, none has much effect on cGMP content (Table 3). Since the finding of noncholinergic and nonadrenergic accumulation of cAMP in rat ganglia is recent (11), not all polypeptides and modulators have been tested for their effects on cAMP accumulation. The finding that VIP has a powerful effect on cAMP accumulation in sympathetic ganglia is encouraging (12).

Conclusions

Nicotine has long-lasting effects on synaptic transmission that result from the combination of receptor desensitization and metabolic changes in cellular activity. Moreover, it is clear that the nicotinic receptor has different properties at different sites. Studies of the peripheral nervous system responses to nicotine have revealed much about the details of nicotine's actions on synapses and junctions. It will be important to conduct similar studies of receptor desensitization, ion channel activation and block, and altered metabolic responses produced by nicotine in central neurons.

Table 3. Cyclic Nucleotide Content of Isolated Rat Superior Cervical Ganglia Treated with Putative Transmitter Substances and Agonists (11)

Agonist		cGMP		cAMP	
		% of Control			
Isoproterenol	$(10^{-6}M)$	100± 7	(8*)	950±100	(5)
Dopamine	$(10^{-3}M)$	100± 6	(8)	116± 8	(5)
Bethanechol	$(10^{-4}M)$	250±10	(6)	112± 7	(8)
DMPP	$(10^{-4}M)$	110±20	(6)	95± 7	(3)
Substance P	$(10^{-4}M)$	135±30	(3)	-	
LHRH	$(10^{-4}M)$	76±18	(3)	-	
Met-enkephalin	$(5x10^{-5}M)$	100± 5	(10)	-	
VIP	$(5x10^{-6}M)$	190±25	(3)	800±75	(3)

*Mean ± S.E. Number of ganglia in parentheses.

REFERENCES

1. Langley, JN: On the union of cranial autonomic (visceral) fibres with the nerve cells of the superior cervical ganglion. J. Physiol. 23:240-270, 1898.
2. Langley, JN and Dickinson, WL: On the local paralysis of peripheral ganglia, and on the connection of different classes of nerve fibres with them. Proc. R. Soc. Lond. 46:423-431, 1889.
3. Dale, HH: The action of certain esters and ethers of choline, and their relation to muscarine. J. Pharmacol. 6:174-190, 1914.
4. Boehm, R: Uber das Vorkommen und die Wirkungen des Cholins und die Wirkungen des kunstlichen Muscarins. Arch. Exp. Path. Pharm. 19:87-100, 1885.
5. Thesleff, S: The mode of neuromuscular block caused by acetylcholine, nicotine, decamethonium and succinylcholine. Acta. Physiol. Scand. 34:218-231, 1955.
6. Steinberg, MI and Volle, RL: A comparison of lobeline and nicotine at the frog neuromuscular junction. Naunyn Schmiedebergs Arch. Pharmacol. 276:16-31, 1972.
7. Lambert, JJ, Volle, RL, and Henderson, EG: An attempt to distinguish between the interactions of tubocurarine and lobeline on the acetylcholine and its associated ionic channel. Proc. Natl. Acad. Sci. 77:5003-5007, 1980.
8. Jaramillo, J and Volle, RL: Effects of lithium on ganglionic hyperpolarization and blockade by drugs related to nicotine. J. Pharmacol. Exp. Ther. 164:166-175, 1968.
9. Takeshige, C and Volle, RL: Biomodal response of sympathetic ganglia to acetylcholine following eserine or repetitive preganglionic stimulation. J. Pharmacol. Exp. Ther. 138:66-73, 1962.
10. Gebber, GL: Dissociation of depolarization and ganglionic blockade induced by nicotine. J. Pharmacol. Exp. Ther. 160:124-134, 1968.
11. Volle, RL, Quenzer, LF and Patterson, BA: The regulation of cyclic nucleotides in a sympathetic ganglion. J. Auton. Nerv. Syst. 6:65-72, 1982.
12. Volle, RL and Patterson, BA: The regulation of cAMP accumulation in a rat sympathetic ganglion: effects of vasoactive intestinal polypeptide. J. Neurochem. 39:1195-1197, 1982.

NICOTINIC AND MUSCARINIC REGULATION OF ADRENAL CATECHOLAMINE SECRETION

Arun R. Wakade

Department of Pharmacology
State University of New York
Downstate Medical Center
450 Clarkson Avenue
Brooklyn, NY 11203

INTRODUCTION

The adrenal medulla has served as a very useful experimental tool to investigate a wide variety of noradrenergic mechanisms. The adrenal medulla is innervated by splanchnic neurons originating in the spinal cord, and about three synaptic contacts are made by splanchnic nerve endings on one chromaffin cell (1). Splanchnic neurons have been regarded as predominantly cholinergic in nature and contain acetylcholine as a principal neurotransmitter (2). The well-known sequence of events is that an increase in the activity of presynaptic splanchnic nerves causes the release of acetylcholine which, in turn, activates postsynaptic cholinergic receptors to trigger the secretion of medullary hormones.

It was widely held that the nicotine receptors were primarily involved in the secretion of catecholamines, and the muscarine receptors had only a minor role, if any, in the secretion of adrenal catecholamines. The physiological significance of muscarine receptors of the adrenal medulla, therefore, was considered to be unimportant (3-5). This conclusion was derived from the observation that secretory response to injected acetylcholine was almost completely abolished by treatment of the adrenal medullary cells with nicotinic antagonists such as hexamethonium (6,7). However, recent studies from our laboratory on the isolated perfused adrenal gland of the rat have provided evidence for the existence of muscarine receptors on chromaffin cells, and these receptors, when activated by appropriate agonists, lead to the secretion of catecholamines (8). The adrenal gland of the guinea pig is endowed with muscarine receptors that secrete catecholamines when stimulated with acetylcholine (9). Excitatory function of nicotine and muscarine receptors has allowed us to utilize the perfused adrenal gland of the rat to investigate a wide spectrum of interesting problems such as the subtypes of muscarine receptors, utilization of external vs. internal calcium ions, and the effects of agents that influence second messengers (i.e., cyclic-AMP, protein kinase C) on the secretion evoked by activation of nicotine and muscarine receptors. This communication briefly describes some of the findings.

METHODS

The preparation used to study the secretion of catecholamines from the isolated adrenal gland of the rat is shown in Fig. 1 and is described in detail in Reference 31. The adrenal gland was perfused at 37°C with Krebs-bicarbonate solution at a rate of 0.35 ml/min via renal and adrenal veins. The perfusate that escaped from a slit made in the adrenal cortex, was used to measure catecholamines by the fluorometric method (10). In some experiments, catecholamines in the perfusate were analyzed by high-pressure liquid chromatography to estimate separately epinephrine and norepinephrine (11). The stimulation of the adrenal medulla was achieved either by injecting different agents in the perfusion stream via a 3-way stopcock or by adding agents to the perfusion medium at a desired concentration. The secretion of catecholamines (ng), in most instances, is expressed as "net secretion" (total secretion minus spontaneous secretion within an equivalent time period).

RESULTS AND DISCUSSION

The effect of nicotine, muscarine, and acetylcholine on the secretion of catecholamines is shown in Fig. 2. Increasing concentrations of nicotine produced a sharp increase in the secretion of catecholamine, which reached a peak level with a 3-μg dose. Further increases in nicotine concentration produced a decline in the secretion. On the other hand, increasing concentrations of muscarine up to 300 μg produced a gradual increase in the secretion. The effects of acetylcholine were parallel to those of muscarine. In the case of muscarine and acetylcholine, the secretory response continued to rise with higher concentrations without reaching a maximum value, as was the case with nicotine.

Table 1 shows the proportions of norepinephrine and epinephrine in the perfusates of nonstimulated and stimulated glands. Epinephrine constitutes 70-80% and norepinephrine 20-30% of the total catecholamines present in the perfusate of the unstimulated adrenal glands. After

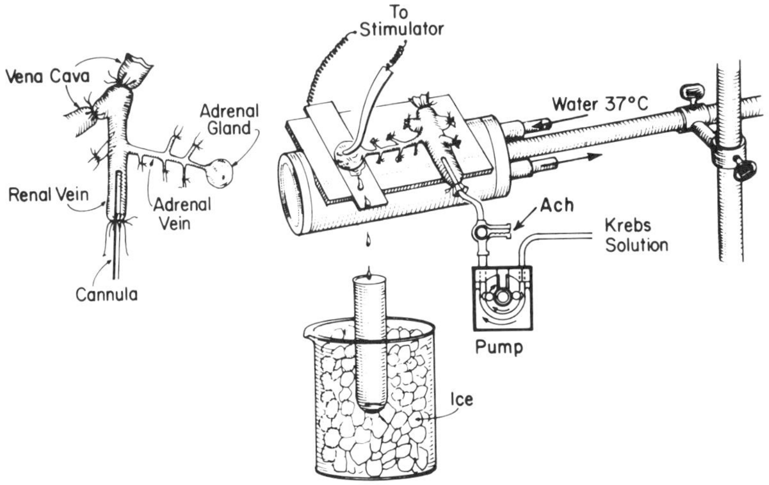

Fig. 1. Schematic drawing of the technique used for retrograde perfusion of the left adrenal gland of the rat. Reproduced from Wakade (1981), with permission (Ref. 31).

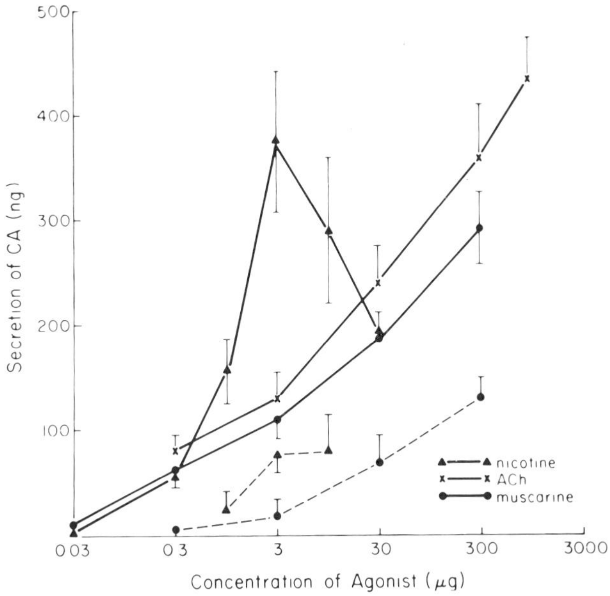

Fig. 2. Secretion of catecholamines in response to nicotine, muscarine, and acetylcholine. ●—● 0.5 μM atropine plus muscarine; ▲—▲ 0.3 mM hexamethonium plus nicotine. Reproduced from Wakade and Wakade (1983), with permission (Ref. 8).

stimulation with muscarine or nicotine, the ratio of epinephrine to norepinephrine remains in the neighborhood of 3 to 4. These observations suggest that stimulation of both types of receptors recruit essentially the same proportion of epinephrine- and norepinephrine- containing storage vesicles. Other researchers have reported that stimulation of chromaffin cells by different secretagogues produces differential secretion of epinephrine and norepinephrine (6,12-15).

Since secretion declined with an increasing concentration of nicotine but not muscarine, a possibility was considered that nicotine receptors

Table 1. Secretion of epinephrine and norepinephrine after stimulation with different agents.

Stimulating agent	No. of experiments	Epinephrine (ng)	Norepinephrine (ng)	EPI/NE
Krebs solution (50 μl)	25	4.8 ± 0.002	1.3 ± 0.001	3.69 ± 0.002
Nicotine (3 μg)	6	345 ± 38	100 ± 28	3.45 ± 0.29
Muscarine (300 μg)	6	323 ± 72	68 ± 2	4.72 ± 0.59

Fifteen min after the beginning of perfusion of the adrenal gland, secretion of epinephrine (EPI) and norepinephrine (NE) was evoked by injecting one of the above test substances into the perfusion stream in a volume of 50 μl. Perfusate was collected for 6-8 min.

undergo a phenomenon of desensitization whereas muscarine receptors do not. This problem was approached by perfusion with a fixed concentration of nicotine and muscarine. Results of such experiments are shown in Fig. 3. Nicotine produced a brisk secretory response within the first 5 min, began to fade immediately, and reached essentially the prestimulation level within 30 min of continuous perfusion with nicotine-Krebs solution. In contrast, muscarine-evoked secretion of CA remained almost at the initial peak level even after continuous perfusion with this agent for as long as 40 min. Furthermore, it was demonstrated that once the secretory process was inactivated by continuous perfusion of the adrenal medulla with nicotine, muscarine was still able to evoke the normal amounts of secretion. These findings clearly show that nicotine receptors of the rat adrenal medulla are desensitized by continued exposure to nicotine, but activation of muscarine receptors does not produce inactivation of the secretion. Thus, muscarine receptors may be of considerable physiological importance in the secretion of medullary hormones when the nicotinic component has become nonfunctional.

To establish the specificity of receptors involved in the secretion of catecholamines, some of the well-established antagonists of nicotine and muscarine receptors were utilized along with the newly introduced antagonists and agonists of muscarine receptors. Figure 4 shows that a classical nicotine receptor antagonist, mecamylamine (300 μM), produced over 90% inhibition of catecholamine secretion evoked by nicotine but had no effect on the secretion evoked by muscarine. The same figure shows that 1 mM tetraethylammonium also selectively and almost completely blocked the secretion evoked by nicotine but not muscarine. The inhibitory effects of mecamylamine and tetraethylammonium were fully reversible. Perfusion of the same adrenal gland with 3.5 μM atropine blocked muscarine-evoked secretion without affecting the secretion evoked by nicotine. These results clearly show that nicotine and muscarine

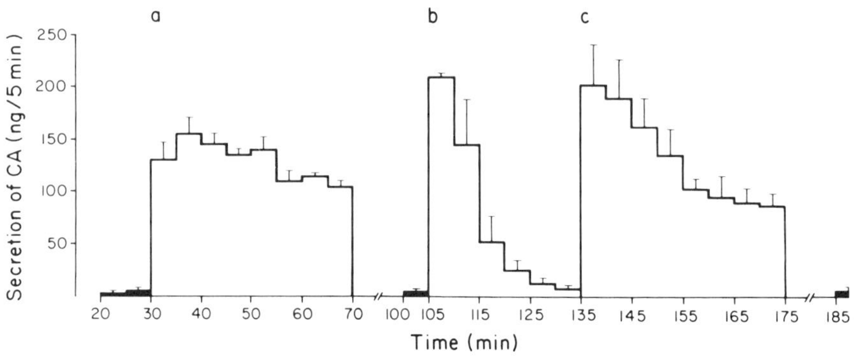

Fig. 3. Lack of desensitization of catecholamine secretion evoked by continuous stimulation with muscarine. After collecting 2 samples of perfusate in the Krebs solution (solid columns), 100 μM muscarine was introduced into the perfusion medium for up to 40 min (a). Thirty min after the washout of muscarine, one 5-min perfusate was collected in Krebs solution, and then nicotine (10 μM) was introduced into the perfusion medium for 30 min (b). The medium was then switched over to nicotine plus muscarine for 40 min. Perfusates were collected every 5 min. Ten min after washout of nicotine plus muscarine, one 5-min perfusate was collected to determine the spontaneous secretion of catecholamines. All of the columns represent a mean of 6 experiments. Vertical lines show S.E. of Mean.

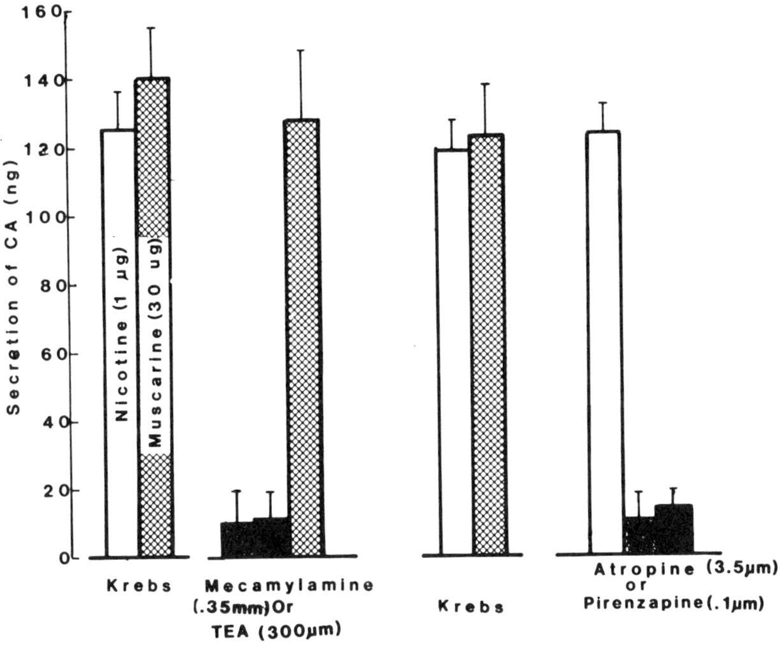

Fig. 4. Specificity of nicotine and muscarine receptor antagonists in the secretion of catecholamines evoked by injecting a slug of either nicotine or muscarine. After evoking the secretion in normal medium (Krebs), the adrenal gland was perfused with either 300 μM mecamylamine (first solid column) or 1 mM tetraethylammonium (second solid column) for 15 min, and in the presence of mecamylamine (or tetraethylammonium) the adrenal glands were exposed to nicotine and muscarine. Thirty min after washout, the glands were perfused with either 3.5 μM atropine (first solid column) or 0.1 μM pirenzepine (second solid column) for 15 min, and then the secretion was evoked by nicotine and muscarine. In control experiments, agonists were tested at 15-min intervals without the addition of antagonists. Each column is a mean of 7 experiments.

receptors of the adrenal medulla are pharmacologically distinct and react to antagonists in an expected manner.

More recently, muscarine receptors have been subdivided into two classes, namely, M_1 and M_2 types (16,17). This classification is principally based on the differential effects of newly synthesized agonist and antagonist on various neuroeffector organs, including the neuronal tissues. Fig. 4 shows that 0.1 μM pirenzepine, a selective antagonist of M_1 receptors (18,19), blocked the secretion evoked by muscarine but had no effect on the secretion evoked by nicotine. Almost a parallel shift to the right of concentration-secretion curves after pirenzepine indicated the competitive nature of antagonism between pirenzepine and muscarine (Fig. 5). From these studies it appears that the rat adrenal medulla contains mainly M_1-subtype of muscarine receptors that are involved in regulating catecholamines secretion. This suggestion was further tested by using a specific agonist of M_1 receptors, McN-A-343 (20,21). Perfusion of the adrenal gland with increasing concentrations of McN-A-343, starting from 3 to 300 μM, caused a small increase in the secretion of catecholamines (Fig. 6). Oxotremorine was also an extremely

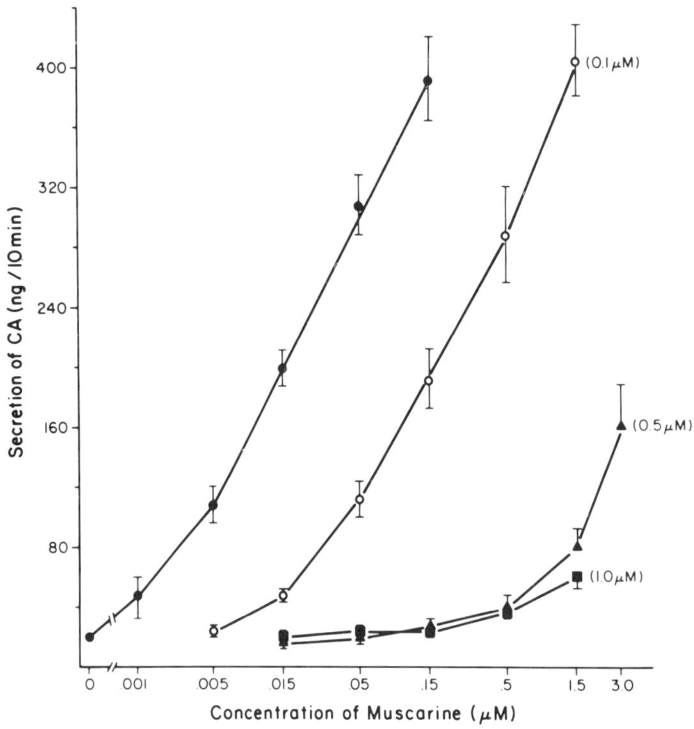

Fig. 5. Competitive antagonism between muscarine and pirenżepine. After determining the concentration-secretion curve with muscarine in Krebs solution (●), the medium was changed to increasing concentrations of pirenzepine (o, ▲ and □), as shown, and in its presence the adrenal gland was perfused with increasing concentrations of muscarine. Each concentration of pirenzepine was tested in separate adrenal gland. Each point is a mean of 3 to 5 observations. Reproduced from Wakade et al. (1986), with permission (Ref. 33).

weak agonist in evoking the secretion of catecholamines. The same figure shows the powerful and full agonistic activity of muscarine. Because of the weak agonistic property of McN-A-343 and oxotremorine, a possibility was considered that these agents may be acting as partial agonists in this test preparation. As shown in Fig. 7, oxotremorine, in a concentration-dependent manner, reduced the secretion of catecholamines evoked by muscarine but had no effect on the secretion evoked by nicotine. On the other hand, as low as 3μM McN-A-343 significantly inhibited the secretion evoked by nicotine, and the effect increased with an increase in the concentration of McN-A-343 (Fig. 8). Thus, it was possible to almost abolish nicotine-evoked secretion at 100 to 300 μM. McN-A-343 also reduced the secretion evoked by muscarine; the effect was significant at 30 μM and reached a maximum at 300 μM. This is the first agent that effectively blocked the secretion evoked by both nicotine and muscarine. The inhibitory effect on both types of receptors was reversible.

The obligatory role of calcium in the secretion of catecholamines has been well established (22,23). In the case of the rat adrenal gland, it was demonstrated that secretion evoked by stimulation of splanchnic nerves was more sensitive to the changes in the concentration of calcium in the perfusion medium than that evoked by injected acetylcholine (Fig. 9).

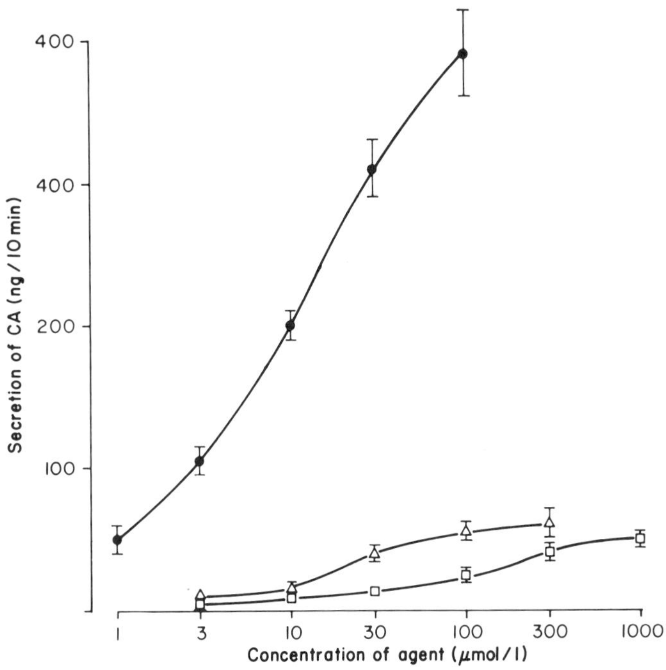

Fig. 6. Comparison of effects of muscarine, McN-A-343 and oxotremorine on the secretion of catecholamines. Increasing concentrations of muscarine (●), oxotremorine (▵) and McN-A-343 (□) were added to the perfusion medium as shown, and catecholamines secreted within a 10-min period were plotted against the concentration of an agent. Each agent was tested in separate adrenal gland. Each point represents a mean of 3 to 12 observations. Reproduced from Wakade et al. (1986), with permission (Ref. 33).

Although it was initially difficult to understand this observation in light of the stringent necessity of calcium ions in the secretory process evoked by acetylcholine (22), it has now become clear that the difference in our results and those of others (24) is due to the participation of muscarine receptors in the secretion and their differential mode of calcium utilization (see below) in the secretory process.

Figure 10 shows that after perfusion of the adrenal gland with calcium-free Krebs solution for 1 hr, nicotine did not evoke the secretion but muscarine was still effective in evoking the secretion of catecholamines. The secretory response by muscarine in zero mM calcium was 50% of that obtained in 2.5 mM calcium medium. However, addition of 1 mM EGTA to Ca-free Krebs solution resulted in almost complete loss of secretory response. From these and other unpublished results, we have concluded that activation of muscarine receptors mobilizes calcium ions mostly from intracellular sites. On the other hand, stimulation of nicotine receptors causes influx of mostly extracellular calcium ions. Consistent with such a conclusion, it was found that the calcium channel-blocker, verapamil, significanlty lowered the secretion of catecholamines evoked from the rat adrenal gland by nicotine but had no effect on the secretion evoked by muscarine (Schneider and Wakade, unpublished results). Undoubtedly, this represents an interesting situation where calcium ions from two different sources are being used upon activation of membrane receptors for regulation of catecholamine secretion from two different sources.

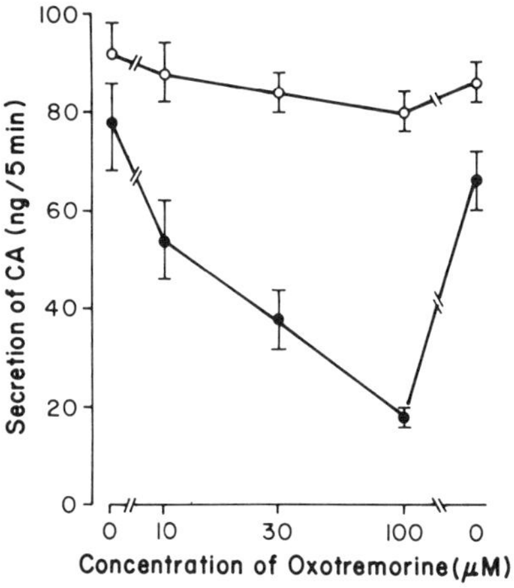

Fig. 7. Antimuscarinic action of oxotremorine. Secretion of catecholamines in response to muscarine (•) and nicotine (o) was tested first in Krebs solution and then in increasing concentrations of oxotremorine, as shown. Each concentration remained in contact with the adrenal medulla for at least 20 min prior to introduction of nicotine or muscarine. Each point is a mean of 5 experiments. Reproduced from Wakade et al. (1986), with permission (Ref. 33).

Extracellular messages are converted to a final physiological response by intracellular messengers such as cyclic-AMP and protein kinase C (25,26). Because nicotine and muscarine utilize calcium ions from different sources, it was of special interest to know whether these receptors are linked to the secretory apparatus by a common or different intracellular messenger.

The role of protein kinase C and cyclic-AMP was investigated by using pharmacological tools such as phorbol ester and forskolin (27-30), respectively. Fig. 11 shows that secretion evoked by nicotine, carbamylcholine, and excess K was enhanced about 2-fold by phorbol 12,13-dibutyrate. However, muscarine-evoked secretion of catecholamines, which was approximately of the same magnitude as that evoked by other agents, remained unaffected by phorbol 12,13-dibutyrate. Although not shown, higher concentrations of phorbol did not have any effect on muscarine-evoked secretion. As discussed above, since nicotine and muscarine utilize calcium ions from different sources for evoking secretion, it was decided to study the effects of phorbol in different concentrations of calcium ions. When calcium concentration of the perfusion medium was reduced to zero mM by adding 1 mM EGTA to calcium-free Krebs solution, muscarine and nicotine did not evoke the secretion, nor did phorbol have any effect (Fig. 12). In 0.1 mM calcium medium both agents were able to evoke the secretion. However, nicotine caused only about 20%, whereas muscarine caused 60% of the secretion obtained in the normal calcium medium (2.5 mM). In 0.1 mM calcium-Krebs solution containing phorbol, secretion evoked by muscarine was unaffected, but that evoked by nicotine was increased over 3-fold. From these observations, we have suggested that protein kinase C plays a crucial role in the exocytotic secretion of catecholamines by controlling the intracellular concentrations of calcium ions.

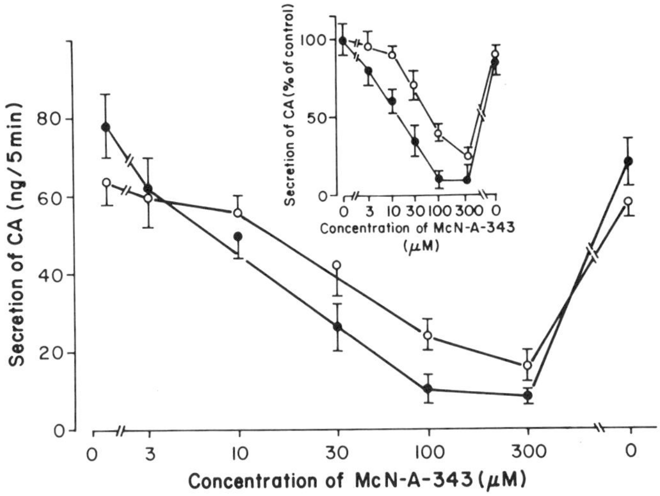

Fig. 8. Antinicotinic and antimuscarinic actions of McN-A-343. Secretion of catecholamines was first evoked in Krebs solution by nicotine (●) and muscarine (o), and then in increasing concentrations of McN-A-343, as shown. Each concentration of McN-A-343 remained in contact with the adrenal medulla for at least 20 min prior to introduction of nicotine or muscarine. After testing the final concentration, the perfusion medium was switched to Krebs solution and 30 min later the secretion was evoked by nicotine and muscarine. The inset shows the relationship between concentration of McN-A-343 and percent inhibition of catecholamine secretion evoked by nicotine (●) and muscarine (o). Each point is a mean of 4 experiments. Reproduced from Wakade et al. (1986) with permission (Ref. 33).

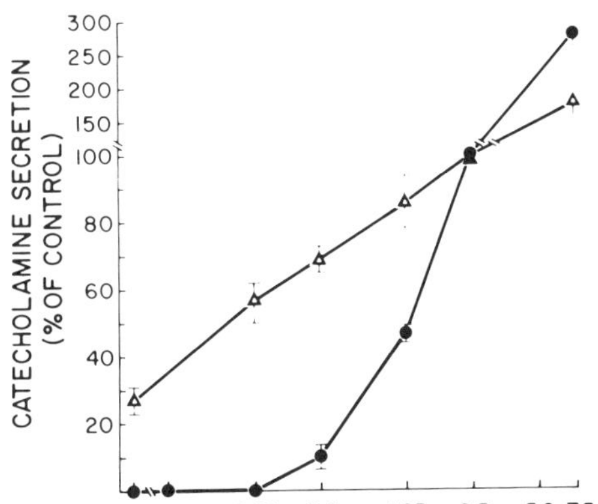

Fig. 9. Relationship between external concentration of calcium ions and secretion of catecholamines evoked by stimulation of splanchnic nerves (●) and acetylcholine (Δ). Reproduced from Wakade (1981), with permission (Ref. 31).

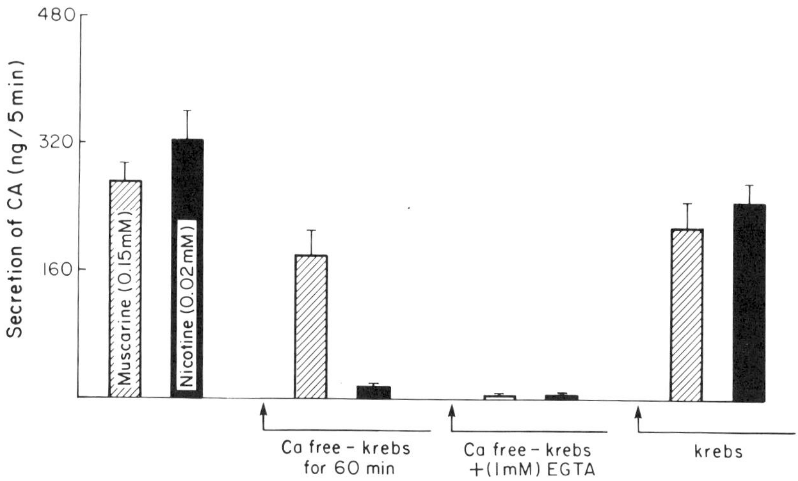

Fig. 10. Differential need of calcium in the secretion of catecholamines evoked by nicotine and muscarine. After evoking the secretion of catecholamines in Krebs solution by muscarine and nicotine, the perfusion medium was changed over to calcium-free Krebs solution for 60 min, and then to 1 mM EGTA, calcium-free Krebs solution for 15 min, and finally to Krebs solution for 30 min. Secretion was evoked in each medium by nicotine and muscarine. Each column is a mean of 3 experiments.

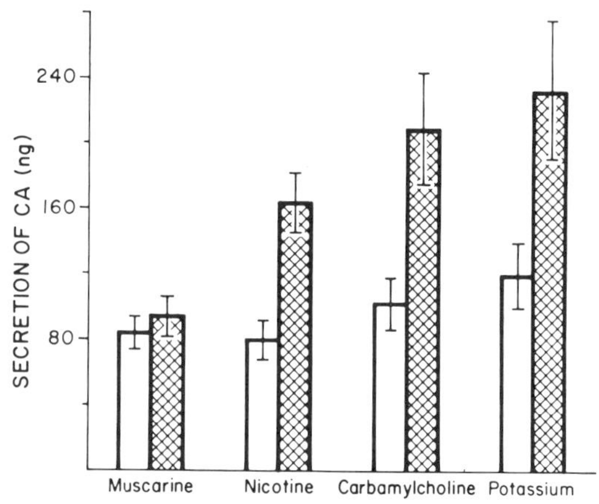

Fig. 11. Effects of phorbol ester on the secretion of catecholamines evoked by various agents. Adrenal gland was perfused with Krebs solution and then secretion was evoked by injecting muscarine (30 μg) and nicotine (0.3 μg) (open columns). The perfusion medium was then changed to Krebs solution containing 30 nM phorbol ester for 15 min, and in its presence muscarine and nicotine were injected (shaded columns), as described above. The identical protocol was carried out in another series of adrenal glands, except that nicotine was replaced with either carbamylcholine (4 μg) or K (40 μg) before (open columns) and after (shaded columns) 30 nM phorbol ester. Each column is a mean of 5 experiments, except column represented by muscarine, which is a mean of 10 experiments. Reproduced from Wakade et al. (1986), with permission (Ref. 32).

Fig. 13 schematically represents our hypothesis. Accordingly, protein kinase C, when activated by diacylglycerol or its substitute, phorbol esters (27-29), controls the intracellular concentration of calcium ions which enter the cell via calcium channels opened by activation of nicotine receptors or changes in the membrane potential. Calcium ions mobilized within intracellular sites by activation of muscarine receptors are not under the control of protein kinase C. Some experimental evidence to support the above hypothesis was collected by measuring the accumulation of ^{45}Ca in the adrenal medulla. The results are summarized in Table 2. Secretion of catecholamines evoked by nicotine, carbamylcholine, and excess K was associated with a substantial increase in the accumulation of ^{45}Ca. However, muscarine-evoked secretion of catecholamine, which was of the same magnitude as that by other agents, was not associated by increase in the accumulation of ^{45}Ca. Furthermore, phorbol facilitated the accumulation of ^{45}Ca, initiated by nicotinic agonists and high K, but phorbol had no effect on the accumulation of ^{45}Ca. These results provide clear evidence that nicotine mobilizes external calcium ions, whereas muscarine has no effect on this source of calcium, and the overall data are consistent with the proposal that nicotine receptor-linked calcium channels (and voltage-sensitive calcium channels) are regulated by protein kinase C.

Unlike differential effects of activation of protein kinase C on the secretion evoked by nicotine and muscarine, activation of adenylate cyclase by forskolin produced almost a 2-fold increase in the secretion of catecholamines evoked by nicotine, as well as muscarine (Fig. 14). The same figure shows that addition of phorbol ester on top of forskolin

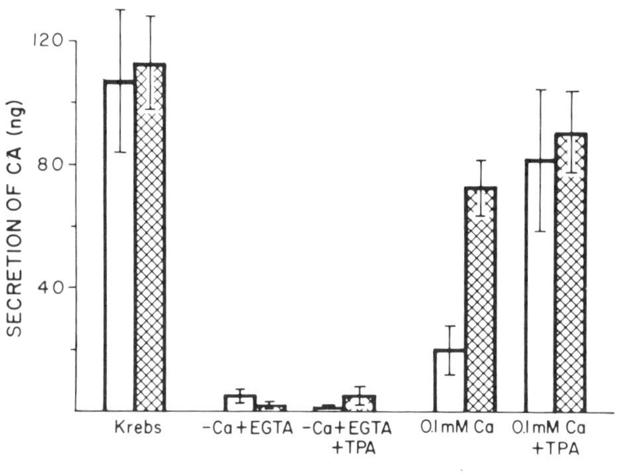

Fig. 12. The role of the calcium ions in the facilitation of secretion of catecholamines evoked by nicotine and muscarine. Secretion of catecholamines was evoked by nicotine (open column) and muscarine (shaded column), as described in Fig. 11, in Krebs solution, (2.5 mM calcium), calcium-free, plus 1 mM EGTA-Krebs solution, and 0.1 mM calcium-Krebs solution, before and after addition of 30 nM 12-O-tetradecanoylphorbol 13-acetate (TPA) to various types of solutions. The adrenal gland was perfused with each type of solution for 15 to 30 min prior to collection of samples. The entire protocol was carried out in the same adrenal gland. Each column represents a mean of 4 experiments. Vertical lines represent S.E. of mean. Reproduced from Wakade et al. (1986), with permission (Ref. 32).

TABLE 2. EFFECTS OF DIFFERENT STIMULATORY AGENTS ON THE ACCUMULATION OF ^{45}Ca, AND SECRETION OF CATECHOLAMINES IN ABSENCE AND PRESENCE OF PHORBOL ESTER

Stimulating Agent	Secretion of Catecholamines (ng/mg) Krebs	Secretion of Catecholamines (ng/mg) Phorbol Ester[c] (30 nM)	Accumulation of ^{45}Ca (pg/mg)[a] Krebs	Accumulation of ^{45}Ca (pg/mg)[a] Phorbol Ester[c] (30 nM)
None (control)	2 ± 0.4	2 ± 0.4	110 ± 20	123 ± 25
	(N.S.)		(N.S.)	
Muscarine (30 μg/ml)	38 ± 4.5	41 ± 5.1	126 ± 11	124 ± 28
	(N.S.		(N.S.)	
Carbamylcholine[b] (4 μg/ml)	39 ± 6.3	68 ± 7.4	303 ± 34	489 ± 32
	($p < 0.001$)		($p < 0.001$)	
Nicotine (3 μg/ml)	29 ± 0.2	67 ± 8.2	343 ± 21	463 ± 26
	($p < 0.001$)		($p < 0.001$)	
Excess K (1.5 mg/ml)	48 ± 6.5	91 ± 17.4	303 ± 25	507 ± 48
	($p < 0.001$)		($p < 0.001$)	

[a]Thirty min after perfusion of the adrenal gland with Krebs solution, the medium was changed to Krebs solution containing 0.5 μg/ml ^{45}Ca (as $^{45}CaCl$ specific activity 23.69 mci/mg, New England Nuclear) for 10 min. The gland was perfused for an additional 30 min with Krebs solution to wash out most of the extracellular ^{45}Ca. Within such time period the radioactivity in wash fluid approaches a very low and steady-state level (about 2% of that in first 5-min period).

[b]Identical protocol was carried out in other series of adrenal glands, except that 10 min perfusion with ^{45}Ca-Krebs solution contained various types of secretagogues. Perfusates collected for 10 min during ^{45}Ca exposure and stimulatory agents were assayed for catecholamines to insure the expected effects of these agents on catecholamines secretion.

[c]In still another series, the above (a & b) protocol was carried out, except that prior to exposure to ^{45}Ca medium the adrenal gland was perfused with Krebs solution containing 30 nM phorbol ester. Phorbol ester was also present during 10-min perfusion with ^{45}Ca medium alone, or that containing stimulatory agent. At the end of 30-min wash period, the adrenal gland was removed, the adrenal medulla was separated, weighed, homogenized in 1 ml of 0.05N perchloric acid, the homogenate was centrifuged, and the supernatant (0.5 ml) was counted in a liquid scintillation counter (Beckman, Model LS7000). Total counts after correction for dilution and quenching were converted to pg and expressed as pg/mg of the wet weight of the adrenal medulla. Each value represents a mean of 4 to 5 experiments.

N.S.: Statistically not significant; other p values show comparison between the data obtained in Krebs solution and phorbol ester-Krebs solution. Reproduced from Wakade et al. (1986), with permission (Ref. 32).

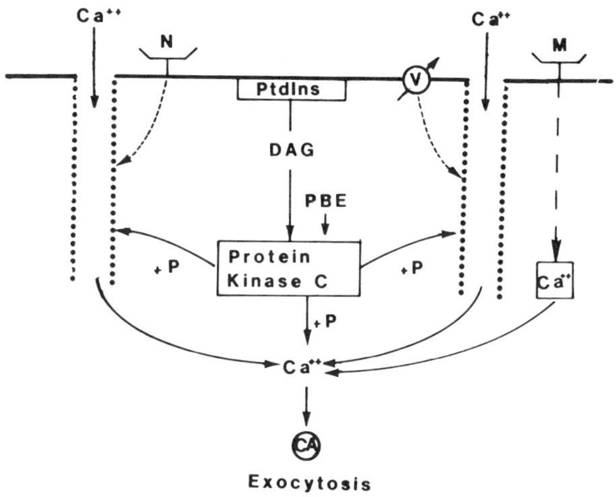

Fig. 13. Schematic representation of the role of protein kinase C in controlling the activity of calcium channels. PtdIns, polyphosphoinositides; DAG, diacylglycerol; PBE, phorbol esters; +P, phosphorylation; V, depolarization of membrane; N, nicotinic receptors; M, muscarinic receptors; CA, catecholamines.

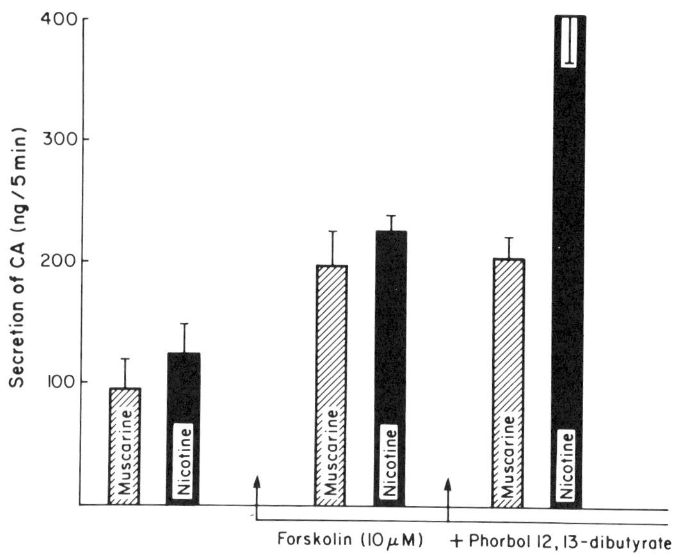

Fig. 14. Interactions between phorbol ester and forskolin on the secretion of catecholamines evoked by nicotine (10 μM) and muscarine (100 μM). After evoking the secretion in normal medium by muscarine and nicotine, the perfusion medium was changed over to 10 μM forskolin for 15 min and secretion was evoked with nicotine and muscarine. Finally, the medium contained forskolin plus phorbol ester (30 nM) for 15 min prior to collection of samples. The entire protocol was carried out in the same adrenal gland. Each column represents a mean of 4 experiments. Vertical lines show S.E. of mean.

produced still additional facilitation of secretion evoked only by nicotine, but muscarine-evoked secretion was not affected by phorbol.

SUMMARY

Nicotine and muscarine receptors are linked to the secretory apparatus of the rat chromaffin cells, and their stimulation with appropriate agonists causes secretion of epinephrine and norepinephrine. Atropine and pirenzepine block muscarine-evoked (whereas tetraethylammonium and mecamylamine block nicotine-evoked) secretion of catecholamines. McN-A-343, classified as a specific agonist of M_1-muscarine receptors of autonomic ganglia, is a powerful antagonist of nicotine and muscarine receptors of the rat adrenal medulla. Secretory response is inactivated after 30 min of continuous exposure to nicotine. However, 60 min of continuous exposure to muscarine does not produce inactivation of catecholamine secretion. Increase in the intracellular concentration of Ca^{++} ions obtained after stimulation with nicotine is due mostly to the influx of extracellular Ca^{++} ions. Stimulation with muscarine produces an increase in intracellular Ca^{++} ions by mobilization of mostly intracellular Ca^{++} ions. Activation of protein kinase C by phorbol ester potentiates the secretion evoked by nicotine but not muscarine. Activation of adenylate cyclase by forskolin potentiates secretion evoked by nicotine and muscarine. Simultaneous analysis of cyclic nucleotides, protein kinase C, and ^{45}Ca accumulation will be of considerable importance in understanding the physiological role of these constituents in the process of exocytosis.

REFERENCES

1. Coupland, R: Electron microscopic observations on the structure of the rat adrenal medulla. J. Anat. 99:231-254, 1965.
2. Feldberg, W. Minz, B and Tsudzimura, H: The mechanism of the nervous discharge of adrenaline. J. Physiol. 81:286-304, 1934.
3. Douglas, WW: Secretomotor control of adrenal medullary secretion: synaptic, membrane, and ionic events in stimulus-secretion coupling. In (eds.) Creep, RO and Astwood, EB, Handbook of Physiology: Endocrinology, pp. 367-388, 1975.
4. Guidotti, A and Costa, E: A role for nicotinic receptors in the regulation of the adenylate cyclase of adrenal medulla. J. Pharmac. Exp. Ther. 189:665-675, 1974.
5. Wilson, SP and Kirshner, N: The acetylcholine receptor of the adrenal medulla. J. Neurochem. 28:687-695, 1977.
6. Douglas, WW and Poisner, AM: Preferential release of adrenaline from the adrenal medulla by muscarine and pilocarpine. Nature 208:1102-1103, 1965.
7. Lee, FL and Trendelenburg, U: Muscarinic transmission of preganglionic impulses to the adrenal medulla of the cat. J. Pharmac. Exp. Ther. 158:73-79, 1967.
8. Wakade, AR and Wakade, TD: Contribution of nicotinic and muscarinic receptors in the secretion of catecholamines evoked by endogenous and exogenous acetylcholine. Neuroscience 10:973-978, 1983.
9. Role, LW and Perlman, RL: Both nicotinic and muscarinic receptors mediate catecholamine secretion by isolated guinea-pig chromaffin cells. Neuroscience 10:979-985, 1983.
10. Anton, AH and Sayre, DF: A study of the factors affecting the alumina oxidetrihydroxyindole procedure for the analysis of catecholamines. J. Pharmac. Exp. Ther. 138:360-375, 1962.
11. Kissinger, PT: Amperometric and coulometric detectors for high-performance liquid chromatography. Analyt. Chem. 49:447A-456A, 1977.

12. Hagen, P: The storage and release of catecholamines. Pharmac. Rev. 11:361-373, 1959.
13. Mirkin, B: Factors influencing the selective secretion of adrenal medullary hormones. J. Pharmac. Exp. Ther. 132:218-225, 1961.
14. Rubin, RP and Miele, E: A study of the differential secretion of epinephrine and norepinephrine from the perfused cat adrenal gland. J. Pharmac. Exp. Ther. 164:115-121, 1968.
15. Silver, M.: The output of adrenaline and noradrenaline from the adrenal medulla of the calf. J. Physiol. 152:14-29, 1960.
16. Caulfield, MP and Straughan, DW: Muscarinic receptors revisited. Trends Neurosci. 6:73-75, 1983.
17. Birdsall, NJM and Hulme, EC: Muscarinic receptor subclass. Trends. Pharmacol. Sci. 4:459-463, 1983.
18. Hammer R, Berrie, CP, Birdsall, NJM, Burgen ASV and Hulme, EC: Pirenzepine distinguishes between different subclasses of muscarinic receptors. Nature 283:90-92, 1980.
19. Hammer, R: Subclasses of muscarinic receptors and pirenzepine. Further experimental evidence. Scand. J. Gastroenterol. 17, Suppl. 72:59-67, 1982.
20. Roszkowski, AP: An unusual type of sympathetic ganglionic stimulant. J. Pharmac. Exp. Ther. 132:156-170, 1961.
21. Hammer, R and Giachetti, A: Muscarinic receptor subtypes: M_1 and M_2 biochemical and functional characterization. Life Sci. 31:2991-2998, 1982.
22. Douglas, WW: Stimulus-secretion coupling: The concept and clues from chromaffin and other cells. Br. J. Pharmac. 34:453-474, 1968.
23. Rubin, RP: The role of calcium in the release of neurotransmitter substances and hormones. Pharmacol. Rev. 22:389-428, 1970.
24. Douglas, WW and Rubin, RP: The role of calcium in the secretory response of the adrenal medulla to acetylcholine. J. Physiol. 159:40-57, 1961.
25. Nestler, EJ, Walaas, S. I. and Greengard, P: Neuronal phosphoproteins: physiological and clinical implications. Science 225:1357-1364, 1984.
26. Nishizuka, Y: The role of protein kinase C in cell surface signal transduction and tumor promotion. Nature 308:693-698, 1984.
27. Castagna, M, Takai, Y, Kaibuchi, K Sano, K., Kikkawa, U and Nishizuka, Y: Direct activation of calcium-activated, phospholipid dependent protein kinase by tumor promoting phorbol esters. J. Biol. Chem. 257:7847-7851, 1982.
28. Niedel, JE, Kuhn, LJ and Vandenbark, GR: Phorbol diester receptor copurifies with protein kinase C. Proc. Natl. Acad. Sci. USA 80:36-40, 1983.
29. Seamon, KB, Padgett, W and Daly, JW: Forskolin: unique diterpene activator of adenylate cyclase in membranes and intact cells. Proc. Natl. Acad. Sci USA 78:3363-3367, 1981.
30. Daly, JW, Padgett, W, and Seamon, KB: Activation of cyclic AMP-generating systems in brain membranes and slices by the diterpene forskolin: augmentation of receptor mediated responses. J. Neurochem. 38:532-544, 1982.
31. Wakade, AR: Studies on secretion of catecholamines evoked by acetylcholine or transmural stimulation of the rat adrenal gland. J. Physiol. 313:463-480, 1981.
32. Wakade, A.R., Malhotra, R.K., and Wakade, T.D. Phorbol ester facilitates ^{45}Ca accumulation and catecholamine secretion by nicotine and excess K, but not by muscarine in rat adrenal medulla. Nature 321:698-700, 1986.
33. Wakade, A.R., Kahn, R., Malhotra, R.K., Wakade, C. and Wakade, T.D. McN-A-343, a specific agonist of M, muscarinic receptors, exerts antinicotinic and antimuscarinic effects in the rat adrenal medulla Life Sci. 39:2073-2080, 1986.

NICOTINIC REGULATION OF ADRENOMEDULLARY OPIOID PEPTIDE SYNTHESIS AND SECRETION: A MODEL TO STUDY MONOAMINE NEUROPEPTIDE COTRANSMISSION

O. Humberto Viveros*, Christopher D. Unsworth*+,
Tomoyuki Kanamatsu‡¶, Jau-Shyong Hong‡,
and Emanuel J. Diliberto, Jr.*

*Department of Medicinal Biochemistry
The Wellcome Research Laboratories
Research Triangle Park, NC 27709
‡Laboratory of Behavioral and Neurological Toxicology
National Institute of Environmental Health Sciences
Research Triangle Park, NC 27709

ABSTRACT

The adrenal medulla chromaffin cell synthesizes and stores in various subcellular compartments a number of proteins, peptides, nucleotides, and other small molecules to be secreted by Ca^{2+}-dependent, nicotine receptor-mediated mechanism(s). Of these proteins and peptides, the enkephalins and other proenkephalin-derived opioid peptides are found in all species examined in substantial amounts, where they are costored with the catecholamines in the chromaffin vesicles. Splanchnic nerve stimulation, nicotine, and other secretagogues induce the cosecretion of these opioid peptides with the amines and other soluble components of these vesicles by the process of exocytosis. Regulatory mechanisms triggered by activation of nicotinic receptors, depolarization, and catecholamine depletion that involve cAMP-dependent and -independent mechanisms control the synthesis of enkephalins at the transcriptional, translational, and peptide processing levels. These mechanisms allow for rapid recovery of the opioid peptide content after secretion and for long-term modulation of the relative proportions and amounts in which catecholamines and enkephalins are costored and cosecreted. Opioid peptides secreted from the adrenal medulla reach ubiquitous opiate receptors throughout the organism and may modulate a number of important systemic functions including behavioral responses to stress. Enkephalins and norepinephrine also coexist in postganglionic sympathetic neurons, and some of the effects of nicotine administration may result from peripheral opiatergic responses through its powerful activation of the sympathetic system. The costorage and cosecretion of opioid peptides and catecholamines is only one of a growing number of examples of coexistence of multiple chemical messengers in single neurons or endocrine cells. This new principle of cotransmission is drastically changing our concepts

+Present address: Howard Hughes Medical Institute, Massachusetts General Hospital, Boston, MA 02114
¶Present address: Second Department of Physiology, Toho University School of Medicine, Tokyo, Japan

and understanding of synaptic and endocrine function. Fast and slow dynamic changes in the ratios in which cotransmitters are stored and released and, thus, coact at the effector sites illustrate an unsuspected degree of synaptic plasticity. The exploration of the short- and long-term effects of chronic use of tobacco needs to take into consideration not only the effect of nicotine on classical transmitters but particularly how the biochemistry and function of these central and peripheral cotransmitter systems are being modified.

TABLE 1. Multiple Exportable Products of Adrenomedullary Chromaffin Cells Secreted by a Nicotine-Receptor-Mediated Calcium-Dependent Release

Subcellular Origin	Product
Chromaffin vesicle	Catecholamines
	Adenine Nucleotides
	Vesicular ascorbic acid
	Chromogranins
	Dopamine β-hydroxylase
	Enkephalins and proenkephalin-derived peptides
	Dynorphin and prodynorphin-derived peptides
	Vasoactive intestinal polypeptide (VIP)
	Somatostatin
	Neurotensin
	Neuropeptide Y (NPY)
	Proopiomelanocortin-related peptides
	Oxytocin
	Arginine-vasopressin
	Other neuropeptides
	Carboxypeptidase H (enkephalin-converting enzyme)
	Other vesicular soluble components
Lysosomes	Acid deoxyribonuclease
	Acid ribonuclease
	Cathepsin
	β-glucuronidase
	Acid phosphatase
	Phospholipases A_1 and A_2
Endoplasmic reticulum	Soluble acetylcholinesterase
Cytosol	Cytosolic ascorbic acid
	Newly taken α-aminoisobutyric acid

(For references, see 2, 5, 9, 43 and 47)

(For references see 2, 5, 9, 43, and 47)

INTRODUCTION

During the past 20 years, it has become increasingly evident that the catecholamines (CA) epinephrine and norepinephrine are not the only bioactive products stored and secreted from the chromaffin vesicle during stimulation of the adrenal medulla. It has also become apparent that the CA-containing chromaffin vesicle is not the only subcellular compartment of the chromaffin cell that responds to nicotine receptor activation with secretion of its content to the extracellular space (Table 1). This costorage and cosecretion of multiple products from the adrenal medulla is just one more example of the now common finding of cotransmission by multiple chemical messengers that are released from endocrine cells and neurons (for reviews, see references 1-4). Of the different classes of products stored and secreted from the adrenal medulla, the CA, protein/peptides, and nucleotides are most abundant (5). Met- and leu-enkephalin and other enkephalin-extended opioid peptides (OP) were the first neuropeptides to be discovered in the adrenal medulla (6-8) and, in general, are contained in higher amounts and have a wider interspecies distribution than other neuropeptides so far found in this tissue (9). Extensive studies of the adrenomedullary OP have resulted in the cloning, identification, and sequencing of the pre-proenkephalin A mRNA (10-12). This polyprotein precursor contains four copies of met-enkephalin, one leu-enkephalin, and one met-enkephalin-arg^{6}-gly^{7}-leu^{8} flanked by paired basic amino acids and one C-terminal met-enkephalin-arg^{6}-phe^{7} preceded by a pair of basic amino acids. Processing of proenkephalin is incomplete in the adrenal medulla resulting in a predominance of intermediate size enkephalin-containing peptides. Some of these extended enkephalins have biological activity *in vivo*, and *in vitro* show high-affinity binding to opiate receptors on disrupted cell membrane preparations. This activity *in vitro* is the basis of a highly sensitive radio-receptor assay. Nevertheless, most of these larger OP are precursors to the shorter enkephalins and their activity is cryptic (COP): they become fully active on a receptor assay only after the enkephalin pentapeptides are released by a partial digestion with trypsin and carboxypeptidase B. The penta-, hepta-, octapeptides and other OP that have high affinity for the receptor binding assay without requiring digestion have been called native OP (NOP).

The ubiquitous nature of intercellular communication by the simultaneous secretion of multiple chemical transmitters (cotransmission) has become a challenge to a number of established dogmas in synaptic biology and endocrinology. The adrenal medulla chromaffin cell has been particularly useful as a model system to study some of the fundamental characteristics of costorage and cosecretion of multiple chemical messengers. In many of these studies, nicotine has been extensively used because of its potency and specificity in promoting adrenomedullary secretion by activation of nicotinic receptors. *In vivo*, nicotine acts not only by a direct stimulation of the cholinergic nicotinic receptor but also indirectly through CNS stimulation and reflexly by activation of arterial chemoreceptors (see papers by Van Loon and by Wakade in this volume and reference 13). In this paper we will review some of the effects of nicotine on OP/CA cosecretion, the role of nicotinic receptor activation on the regulation of OP biosynthesis, and finally we will briefly discuss some of the functional implications of cotransmission.

SUBCELLULAR LOCALIZATION OF OPIOID PEPTIDES IN CHROMAFFIN CELLS AND THEIR COSECRETION WITH CATECHOLAMINES ON ACTIVATION OF NICOTINIC RECEPTORS

Opioid peptides from dog adrenal medulla, bovine chromaffin cells in culture, and human pheochromocytoma cosediment with CA and other chromaffin vesicle markers on differential and density gradient centrifugations (7,14,15), suggesting that OP are stored in this particular organelle. The costorage of OP and CA has been confirmed by immunoelectronmicroscopy (16) and by release studies where the molar OP/CA ratios of secreted and stored materials are identical under allexperimental circumstances, as would be expected from secretion by exocytosis (see later).

Nicotine induces a Ca^{2+}-dependent, dose-related cosecretion of OP and CA from the perfused dog adrenal gland (7,17), from bovine adrenal medulla chromaffin cells in primary culture (18), and from cultured human pheochromocytoma cells as shown in Fig. 1. The cosecretion of OP and CA induced by activation of nicotinic receptors is suppressed by the corresponding pharmacological antagonists and by CA^{2+}-channel blockers (17,18). Other medullary secretagogues like veratridine, Ba^{2+}, the Ca^{2+}-inophore ionomycin, and depolarization by elevated extracellular K^+ also produce stoichiometric release of OP and CA (18). These secretagogues also stimulate the proportional release of COP from cultured chromaffin cells and from perfused adrenals (18,19).

Because of the relatively slow turnover rate of the chromaffin vesicles in the adrenal medulla (20), it is possible to study adrenomedullary secretion in the intact unanesthetized animal by measuring the decline in tissue levels of OP and CA after a brief period of intense stress that increases splanchnic stimulation. Figure 2 shows the proportional decrease in epinephrine, norepinephrine, and NOP in the rat adrenal medulla at the end of a two-hour period of immobilization stress. Direct measurements of increased levels of enkephalins and other related OP in the adrenal vein during splanchnic stimulation *in vivo* have also been reported (21,22). The activation of the rat adrenal medulla by stress can be prevented by splanchnic denervation (23) or by the combined administration of a ganglionic blocker and atropine as shown in Fig. 3 (in the rat adrenal medulla, both the nicotinic and the muscarinic receptors are coupled to the secretory response).

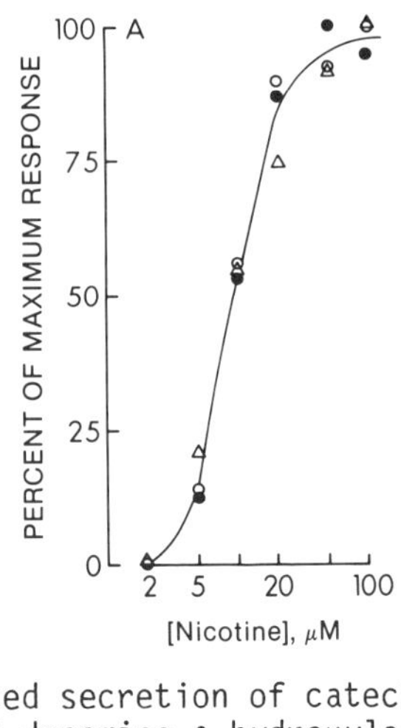

Fig. 1. Nicotine-stimulated secretion of catecholamines (○), opioid peptides (●), and dopamine β-hydroxylase (Δ) from human pheochromocytoma cells in culture. The cells were exposed to the indicated concentrations of nicotine for 10 min. Maximum secretion was approximately 30 to 50% of initial content. See References 14, 15, and 18 for methodological details.

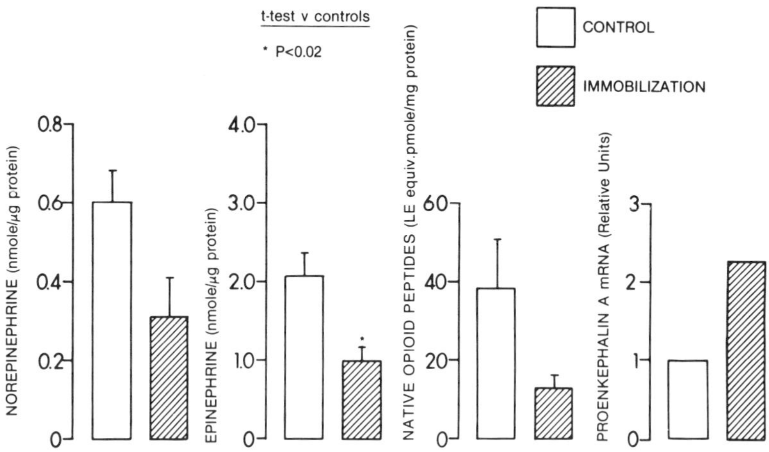

Fig. 2. The effect of 2 hr of immobilization stress on the catecholamine, opioid peptide, and proenkephalin mRNA levels in the rat adrenal medulla. Rats were sacrificed after 2 hr of immobilization and the adrenal medullae freed of cortex under a dissecting microscope in the cold. Individual medullae pairs were immediately frozen in liquid nitrogen. Individual pairs were homogenized in 0.5 M acetic acid for OP assay (29), and aliquots were treated with perchloric acid (final concentration, 0.5 M) for protein and CA determination (29). Proenkephalin mRNA was measured by hybridization with a rat cDNA probe kindly donated by Dr. S. Sabol (48) as described (49). For each mRNA determination, total RNA was extracted from eight pairs of pooled frozen glands. The mRNA values are the means of determinations on two sets of pooled medullae per point. For catecholamines and opioid peptides, N = 5.

THE REGULATION OF ENKEPHALIN BIOSYNTHESIS THROUGH CHOLINERGIC RECEPTOR ACTIVATION AND CATECHOLAMINE LEVELS

It has been almost two decades since the first demonstrations that the adrenal medulla and adrenergic neurons respond to increased cholinergic stimulation and CA depletion with long-term homeostatic changes including an increase in the rate of biosynthesis of storage vesicles and of the CA synthesis enzymes dopamine β-hydroxylase and tyrosine hydroxylase (20,24). Since we now know that increased adrenomedullary stimulation results in the secretion of OP, it was of interest to study whether there were also homeostatic mechanisms for the maintenance of stores and secretion of OP under conditions of increased demand.

Early studies on the acute and long-term changes in NOP levels in the adrenal medulla of guinea pigs that were exposed to six hours of hypoglycemia or that had been injected with reserpine for two successive days showed a 2- to 3-fold increase in OP content three to six days after the initial treatment (17). Furthermore, the increase in OP induced by reserpine was totally blocked by repeated injections of cycloheximide, suggesting that these increases in OP content could, at least in part, be a consequence of increased protein synthesis.

A more detailed study of the effects that increased splanchnic discharge by stress has on the OP system of the rat adrenal medulla can be observed in Figs. 2, 3, and 4 (25). Marked increases in proenkephalin mRNA levels are already observed after two hours of immobilization stress

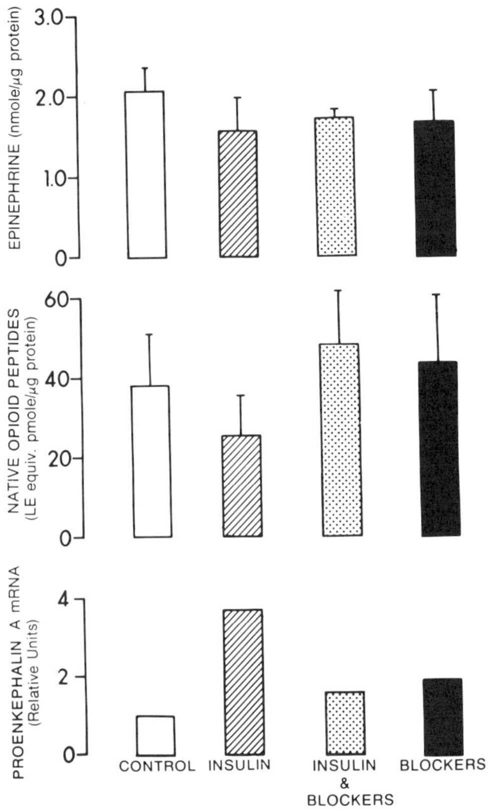

Fig. 3. Epinephrine, opioid peptides, and proenkephalin mRNA levels after a brief period of increased splanchnic firing induced by insulin hypoglycemia: blockade by cholinergic antagonists. Fasted rats were injected with insulin (10 U/kg, i.p.); 30 min later the hypoglycemia was terminated by administration of 2 ml of 40% sucrose by stomach tube and 1 ml of 20% dextrose i.p. Animals were sacrificed 135 min after the insulin injection. To block the splanchnic-adrenomedullary synapse, rats were treated with chlorisondamine (5 mg/kg, i.p.) and atropine (1 mg/kg, i.p.) 15 min prior to and again 45 min after the injection of insulin. For assay procedures, see legend to Fig. 2.

(Fig. 2) or insulin hypoglycemia (Fig. 4). No increases in mRNA were observed when hypoglycemic rats were sacrificed at 30, 60, or 90 min after insulin administration. Nevertheless, 30 min of hypoglycemic shock were enough to produce a 3.5-fold increase in specific proenkephalin A mRNA if the animals were not sacrificed until 100 min later (Fig. 3).This lag time may be related to transcription and maturation of the nuclear proenkephalin mRNA.

When the hypoglycemic stress is stopped after two hours (Fig. 4), the mRNA levels continue to increase for the next 24 hours and then decline with a half-life of approximately four days. If the assumption is made that the half-life of proenkephalin mRNA is the same in the unstimulated gland, a complete blockade of mRNA degradation induced by stress would account for not more than a doubling of mRNA levels in 24 hours as compared with the 15-fold increase observed after hypoglycemia. This result strongly suggests that the effect of increased neurogenic stimulation on proenkephalin mRNA is a transcriptional event, probably a

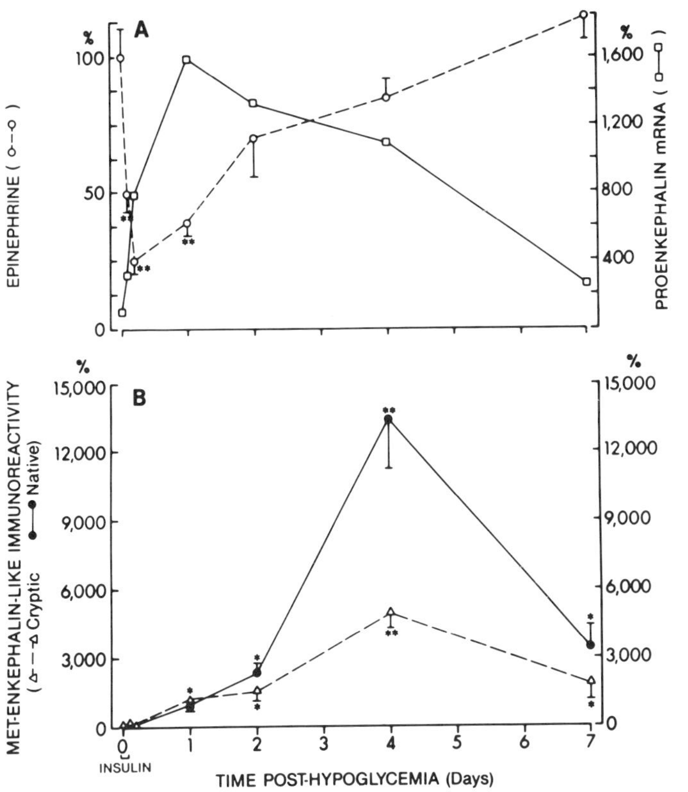

Fig. 4. Long-term effects of a 2-hr period of increased splanchnic firing on the levels of epinephrine, met-enkephalin immunoreactivity, and proenkephalin mRNA in the rat adrenal medulla. After 2 hr of insulin (10 U/kg, i.p.) hypoglycemia, rats were recovered by administration of sucrose and food ad libitum and sacrificed at the different times indicated in the abscissa. Assay procedures are as in Fig. 2, except that to measure opioid peptides the radioreceptor assay was replaced by a highly specific radioimmunoassay (49) for met-enkephalin. Native and cryptic opioid peptides are measured without or with previous partial digestion with trypsin and carboxypeptidase B, respectively. Unstimulated values were: epinephrine, 1.97 ± 0.21 μmole/μg protein; native, 0.9 ± 0.3 pmole/mg protein; cryptic, 12.1 pmole/mg protein (mRNA is expressed in relative densitometric units per μg of total RNA). For epinephrine and opioid peptides, N = 8; *p <0.02, **p <0.005. Each mRNA value obtained from a pool of eight medulla pairs. The experiment was replicated once with identical results.

derepression of the proenkephalin gene. Adrenomedullary COP levels do not increase significantly during the first five hours after insulin injection, but then show a rapid rise between 5 and 24 hours to attain a peak value 50-fold above control by the fourth day after insulin. After a similar delay, the NOP also start to increase initially at a rate lower than COP, but later they reach a 135-fold increase above the unstimulated levels. Although there may be other possible explanations for this late massive increase in NOP, it may well represent an increase in the rate of proenkephalin processing in addition to the increased rate of *de novo* biosynthesis. At seven days post-stress, there still is an 18-fold increase in COP and 34-fold in NOP above pre-stimulation values.

At this time, the CA content has increased only by 27% above control such that the OP/CA ratio is greatly increased after this short period of adrenomedullary stimulation. As shown on Fig. 3, cholinergic antagonists completely prevent the increase in proenkephalin mRNA.

Increased splanchnic nerve firing, thereby activating post-synaptic nicotinic receptors, initiates a large number of events in the chromaffin cells: depolarization by increased Na^+ and Ca^{2+} conductance, secretion of CA and other vesicular components, activation of adenylate cyclase, and possibly many others. The role of some of these changes in the regulation of OP biosynthesis has been further explored in chromaffin cells in primary culture. Depletion of cellular CA by exposure of the cultures to reserpine for two or more days results in increases in NOP levels that are partially blocked by cycloheximide or by actinomycin D (26). This treatment also increases the incorporation of radioactive amino acids into met- and leu-enkephalin. Since the ratio of met- to leu-enkephalin was also increased, it was concluded that reserpine accelerated both the rate of de novo biosynthesis of OP as well as its rate of processing to the shorter polypeptides (27).

In general, all treatments that deplete CA without stimulating cell secretion or cAMP accumulation, as seen with inhibitors of CA uptake and of CA biosynthesis, induce increases in NOP and total OP, but no change or a decrease in COP (29). These results are again suggestive of a combined increase in the rate of proenkephalin de novo synthesis, plus an increase in the rate of processing. As shown in Figs. 5 and 6, the addition of tetrabenazine to bovine chromaffin cell cultures produces a marked increase in the levels of proenkephalin mRNA and in the degree of precursor processing. Three days after exposure of the cultures to tetrabenazine, there are large increases in the levels of met- and

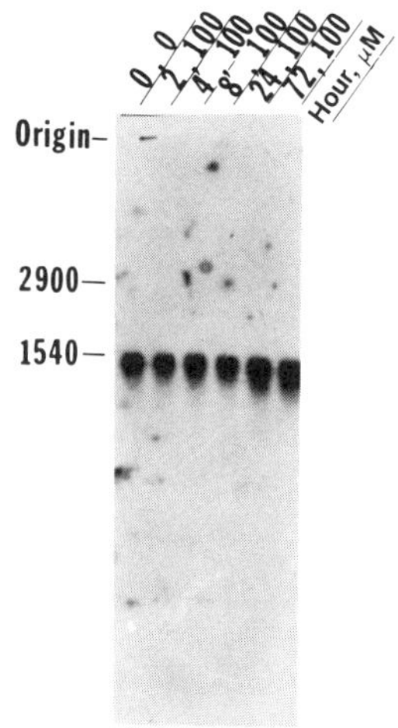

Fig. 5. Effect of depleting catecholamines by tetrabenazine treatment of bovine adrenal chromaffin cells in primary culture on proenkephalin mRNA. Cultures were exposed to tetrabenazine (100 µM), and after the indicated times, the medium was removed and plates frozen for later analysis of specific proenkephalin mRNA by Northern blot hybridization (49) with a bovine cDNA probe (12). Note that the cDNA hybridized with a single mRNA species of approximately 1.4 kbases, in accord with previous measurements of bovine adrenal medulla proenkephalin mRNA (10-12).

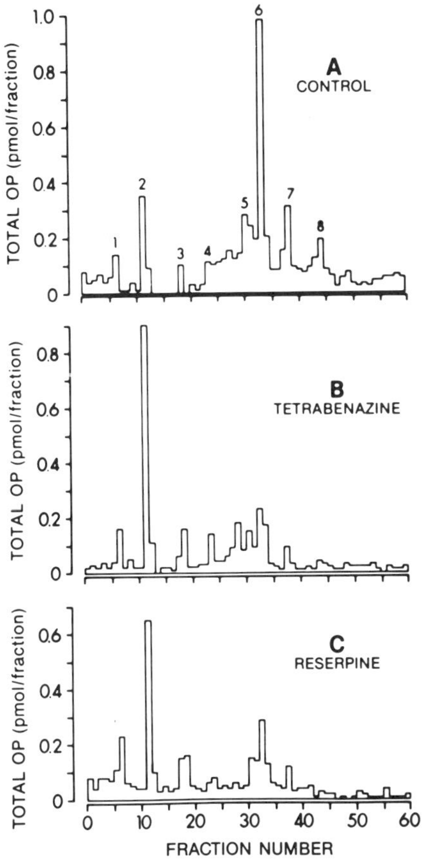

Fig. 6. Catecholamine depletion by tetrabenazine and reserpine increases the processing of opioid peptides in bovine chromaffin cells in culture. After three days of exposure of the cultures to tetrabenazine (100 µM) or reserpine (0.2 µM), cells were extracted with 0.5 M acetic acid. After removing precipitated proteins, the acid extracts were dried, redissolved in HPLC buffer A, applied to a reverse phase HPLC column (µBondapak, Waters Association) and fractionated with a nonlinear gradient from 16% acetonitrile, 5% tetrahydrofuran and 0.05% trifluoroacetic acid (Buffer A) to 60% acetonitrile, 5% tetrahydrofuran and 0.05% trifluoroacetic acid (Buffer B) (Diliberto, Unsworth and Viveros, in preparation). Each fraction was partially digested with trypsin and carboxypeptidase B previous to the radioreceptor assay to measure total opioid peptides eluted from the column. The different opioid peptides have been tentatively characterized by coelution with synthetic standards: peak 1, met-enkephalin sulfoxide and/or met-enkephalin-arg^6; peak 2, met-enkephalin; peak 3, leu-enkephalin; peak 4, met-enkephalin-arg^6-phe^7; peak 5, peptide B and/or amidorphin; peak 6, peptide E; peak 7, peptide F; and peak 8, unknown. Met-enkephalin-arg^6-gly^7-leu^8 has a typical retention time of 21.3 min. Note the marked decrease in the intermediate-size, opioid peptides with longer retention times (peaks 5-8) and increase in the enkephalin peaks (peaks 1-3) after tetrabenazine treatment.

leu-enkephalin with a corresponding decrease in the intermediate size OP that tentatively were characterized as peptides B, E, and F (Fig. 6). Cultures treated with reserpine at the minimum doses that produce nearly maximal increases in NOP show a similar shift in intermediate size to small OP. Nevertheless, we have been unable to show changes in proenkephalin mRNA after this dose of reserpine. Higher concentrations of the drug have been reported to decrease the levels of proenkephalin mRNA (30,31). Thus, drugs that deplete CA may, in addition to increasing processing and, at least in some cases possibly increasing transcription, accelerate OP biosynthesis by increasing the efficiency of translation.

Nicotine treatment increases the content of OP in human pheochromocytes in culture (32) and produces a rapid rise in proenkephalin mRNA in bovine chromaffin cell cultures that is followed by a delayed increase in the levels of met-enkephalin (33). Activation of the nicotinic receptor not only causes CA secretion and eventually produces intracellular CA depletion, but also depolarizes the cells and increases the content of cAMP (33,34). Depolarization of the cultures by high external K^+ also produces an enhancement of proenkephalin gene transcription and OP synthesis. The effect of high K^+, like the effect of nicotine, is dependent on the concentration of external Ca^{2+} (35).

Increased cellular levels of cAMP produced by addition of 8-bromo-cAMP, forskolin, or inhibitors of cyclic nucleotide phosphodiesterase result in proportional increases of both NOP and COP in the cultures. These results suggest that <u>de novo</u> synthesis, rather than processing, is primarily activated by cAMP (15,30). Increased cAMP also induces significant increases in the levels of proenkephalin mRNA (30,36). The effects of cAMP on OP levels and proenkephalin mRNA appear to be independent of changes in cell CA content (29) and do not require extracellular Ca^{2+} (33).

These studies on the regulation of proenkephalin gene transcription, translation, and processing in chromaffin cells in culture have dissected a number of factors that can regulate enkephalin biosynthesis. All of these factors probably are part of the homeostasis of adrenomedullary enkephalin synthesis, storage, and secretion. Nevertheless, in spite of the similarities observed in the regulation of enkephalin biosynthesis <u>in vivo</u> and in culture, there are also some important quantitative and qualitative differences. The magnitude of the changes of NOP, COP, and mRNA in culture is severalfold lower and of slower development when compared to the changes seen after increased neurogenic stimulation <u>in vivo</u>. Of greater importance is the failure of CA depletion or increased levels of cAMP to increase the levels of soluble and membrane-bound dopamine β-hydroxylase in the cultures (15), which are readily seen <u>in vivo</u> after reserpine or neurogenic stimulation (7,37). This dissociation in the response of two intravesicular proteins (dopamine β-hydroxylase versus proenkephalin) indicates that there may be independent regulation on the biogenesis of the different components of the catecholamine storage vesicle. Furthermore, the lack of increase of the membrane-bound dopamine β-hydroxylase, an integral protein of the vesicle membrane, suggests that there is no increase in the biogenesis of new vesicles. The newly synthesized OP would have to be accommodated in preexisting chromaffin vesicles. This, in turn, would require an intracellular system to channel the newly synthesized OP from the endoplasmic reticulum-Golgi complex to mature vesicles. Further studies are required to solve this issue and find the missing factors for regulation of enkephalin and dopamine β-hydroxylase biosynthesis.

PHYSIOLOGICAL ROLE OF ADRENAL MEDULLA OPIOID PEPTIDES

There has been a progressive awareness that peripheral administration of enkephalins can produce a large variety of direct and indirect effects on all major body systems. The plethora of recent articles in this area of research cannot be summarized here (partial lists of recent publications in this subject can be found in references 2, 9, and 38). Nevertheless, these studies indicate the potential for varied effects of the OP secreted from the adrenal medulla under situations of stress or in response to the administration of drugs like nicotine that directly and indirectly activate adrenomedullary secretion. We will limit this discussion to some functional implications of a variable CA/OP ratio.

As has been described above, neurogenic stimulation *in vivo* and a number of treatments in culture produce marked and long-lasting increases in the cell levels of OP with no change or a decrease in CA content. It seems important to determine if these dynamic changes in the ratio of stored OP/CA have any functional consequences. Since OP and CA are in the same secretory compartment, the chromaffin vesicle, and the entire vesicle content is released in a quantal fashion by exocytosis (39), it is expected that the ratio of OP/CA in the extracellular medium would be identical to the ratio inside the vesicles despite the changes in OP/CA ratios resulting from the recent functional history of the chromaffin cells. Furthermore, if the treatments that change the OP/CA ratio do not simultaneously modify the coupling of cell stimulus to secretion, the same stimulus applied before or after the change in ratio should result in

TABLE 2. Proportional Secretion of Catecholamines and Opioid Peptides from Chromaffin Cells Treated with Reserpine or Tetrabenazine

Experiment	Addition of Culture Medium	Catecholamines/Culture		Opioid Peptides/Culture	
		Cell Content	Nicotine-evoked Secretion	Cell Content	Nicotine-evoked Secretion
		nmol		pmol	
I	None	55.3 ± 1.4	12.8 ± 0.8	24.2 ± 1.2	6.9 ± 0.3
	Reserpine	9.4 ± 0.2	1.8 ± 0.1	48.2 ± 1.7	14.6 ± 0.6
	Ratio	0.17	0.14	1.99	2.12
II	None	112.0 ± 1.2	26.4 ± 1.9	27.7 ± 1.9	7.7 ± 0.9
	Tetrabenazine	20.5 ± 0.9	4.7 ± 0.4	72.5 ± 2.1	19.6 ± 1.9
	Ratio	0.18	0.18	2.62	2.55

Chromaffin cells were used three days after the indicated additions of 100 μM reserpine of 100 μM tetrabenazine to the culture medium. Cells were stimulated with 20 μM nicotine for 10 min. (Experiment I) or 10 μM nicotine for 20 min. (Experiment II). Cell contents represent the amounts of catecholamines and opioid peptides present in unstimulated cells after the incubation period. Nicotine-evoked secretion values are the losses in cell content evoked by the exposure of nicotine. The data are expressed as mean ± S.E.M.; n = 3 for Experiment I and n = 6 for Experiment II. The ratio indicates the quotient of the drug-treated value over the untreated value. For methodological detail, see Reference 18.

marked differences in the absolute amounts of each cotransmitter being released. Indeed, Table 2 shows precisely these expected results with the OP and CA secretion from cultures that had been previously exposed to reserpine and tetrabenazine when compared with sister cultures not exposed to those treatments but stimulated with identical concentrations of nicotine. Thus the amount of CA and OP secreted is directly proportionalto their individual levels inside the cell. Lewis et al. (40) described in the rat an opioid-mediated, stress-induced analgesia that is dependent on a functional adrenal medulla. When animals were treated with reserpine two days before testing, the latency of the response to the noxious stimuli more than doubled, the content of OP in the reserpine-treated, unstressed animals increased 2.5-fold and the amount of OP secreted (estimated by the difference in content of stressed and unstressed animals) was increased in the same proportion (41). A non-opioid, stress-analgesia paradigm that is independent of the adrenal medulla showed a decrease in response latency in these same animals, demonstrating that the increased opioid-dependent response was not due to a decrease in the general responsiveness of the rats because of the reserpine treatment. Furthermore, since the CA content of the glands of these animals had been reduced to 25% of control, this experiment also indicates that the opioid stress-analgesia is not positively correlated to the adrenal medulla CA content and secretion. These experiments not only illustrate that secretion of endogenous OP from the adrenal medulla may have survival value under conditions of intense stress, but exemplify the high degree of plasticity of this cotransmitter system. Thus the quantity and quality of the cotransmitter mixture to be secreted in response to a future challenge (nicotine, stress) are determined by the integrated cell response to a current or previous stimulus.

COSECRETION OF MULTIPLE CHEMICAL MESSENGERS: A NEW PRINCIPLE OF INTERCELLULAR COMMUNICATION

Although the origins of the idea of costorage and cosecretion of two neurotransmitters can be traced back for at least 30 years (42), it is only recently that the discovery of transmitter coexistence has changed from a rare phenomenon to a commonplace occurrence. It is now believed that cotransmission may be an intrinsic characteristic of every neuron and endocrine cell (1-4). Multiple transmitter coexistence is also found quite frequently and becomes extreme in the adrenal medullary chromaffin cells (Table 1). In spite of our extensive knowledge of the biochemistry of the adrenal medulla, the number of characterized proteins, peptides and other export products contained in the chromaffin vesicle and in other secretory compartments constitutes only a small fraction of the total products secreted. Nicotine and splanchnic stimulation induce secretion not only from the chromaffin vesicles but also from at least three other subcellular compartments. That adrenomedullary secretion of yet unknown product(s) may be as important as secretion of CA and OP is undoubtedly possible. An extensive review of the characteristics of secretion from compartments other than the chromaffin vesicle is out of the scope of this article (the interested reader can consult references 38, 43, and 44). The most important general difference in the characteristics of secretion of multiple products from one particular compartment, as compared with secretion from different compartments, is that in the latter situation, secretion from each compartment can be independently regulated. As an example, the ratio of secreted ascorbate to secreted CA varies markedly from one instant to the next when cultures are stimulated with nicotine or other secretagogues (43). Secretion from multiple compartments is also found in neurons: neuropeptides seem to be generally associated with large synaptic vesicles, classical transmitters are usually stored in small vesicles or sometimes in both classes of storage organelles. Opioid peptides in sympathetic postganglionic neurons are stored in large dense-cored vesicles, and norepinephrine is stored in both the large and small dense-cored vesicles (45,46). This rapid change in the relative proportions of secreted cotransmitters gives an added dimension to chemical synaptic plasticity that is complementary to the long-term plasticity produced by changes in the ratio of stored cotransmitters that we have previously discussed.

The emerging concepts of coexistence, cosecretion, and coaction of multiple messengers, as illustrated by the complex arrangement of secreting compartments and interactions observed in the chromaffin cell, are forcing neurobiologists and endocrinologists to drastically revise some of the fundamental principles of chemical intercellular communication (synaptic transmission). In particular, the high degree of short- and long-term, pre- and postsynaptic chemical plasticity characteristic of cotransmitter systems makes the concept that synapses are just unidirectional relays in hard-wired circuits obsolete. The semantics of the chemical language for intercellular communication is vastly enriched by cotransmission. Use of a single chemical transmitter reduces intercellular language to a one-word vocabulary (either open or close an ion channel, activate or inactivate an enzyme, etc.). The neuronal and endocrine language becomes expanded to phrases and sentences in proportion to the number of multiple transmitters that are cosecreted from a single neuron or endocrine cell. The ability of secretory cells to rapidly and/or slowly change the mixture of chemical words further increases the complexity of intercellular communication. Long-term changes in the ratios of the stored and secreted cotransmitters may form the substrate for some simple forms of learning and memory. Interactions of multiple transmitters with multiple receptors at the target cells seem to confer even more complexity and plasticity to the system. Postsynaptic interactions may account for increased specificity and sensitivity in the reception of a signal by a target cell; these may be particularly important in endocrine or paracrine communication. Within this theoretical context, the anatomical connectivity of adult neurons into complex, more or less hard-wired networks can be compared to the hardware of a mainframe computer. Functionally, synapses with the large degree of biochemical plasticity conferred by the use of multiple chemical transmitters would operate analogously to a software that can be autoprogrammed based on previous experience and on immediate functional demands.

Nicotine and other psychotropic drugs probably exert their effects by modifying the software rather than by changing the neuronal network. Our current knowledge of the effects of nicotine at the central and peripheral nervous system is limited to a few classical transmitters and neuronal groups. It can be expected that analysis of the effects of nicotine on other transmitter and cotransmitter systems may give a closer insight into how and why this alkaloid exerts its behavioral effects and becomes a determinant factor in initiating and establishing the use of tobacco.

REFERENCES

1. Hökfelt, T, Johansson, O, Ljungdahl, Å, Lundberg, JM and Schultzberg, M: Peptidergic neurones. Nature (Lond.) 284:515-521, 1980.
2. Viveros, OH and Wilson, SP: The adrenal chromaffin cell as a model to study the co-secretion of enkephalins and catecholamines. J. Auton. Nerv. Syst. 7:41-58, 1983.
3. Cuello, AC (ed.): Co-transmission, MacMillan, London, 1982.
4. Chan-Palay, V and Palay, SL (eds.): Coexistence of Neuroactive Substances in Neurons, John Wiley & Sons, New York, 1984.
5. Winkler, H and Westhead, E: The molecular organization of adrenal chromaffin granules. Neuroscience 5:1803-1823, 1980.

6. Schultzberg, M, Lundberg, JM, Hökfelt, T, Terenius, L, Brandt, J, Elde, RP and Goldstein, M: Enkephalin-like immunoreactivity in gland cells and nerve terminals of the adrenal medulla. Neuroscience 3:1169-1186, 1978.
7. Viveros, OH, Diliberto, EJ, Jr., Hazum, E and Chang, K-J: Opiate-like materials in the adrenal medulla: evidence for storage and secretion with catecholamines. Mol. Pharmacol. 16:1101-1108, 1979.
8. Yang, H-YT, Hexum, T and Costa, E: Opioid peptides in adrenal gland. Life Sci. 27:1119-1125, 1980.
9. Unsworth, CD and Viveros, OH: Neuropeptides of the adrenal medulla. In (ed.) Rosenheck, K, Stimulus-Secretion Coupling in Chromaffin Cells, CRC Press (in press).
10. Noda, M, Furutani, Y, Takahashi, H, Toyosato, M, Hirose, T, Inayama, S, Nakanishi, S and Numa, S: Cloning and sequence analysis of cDNA for bovine adrenal preproenkephalin. Nature (Lond.) 295:202-206, 1982.
11. Gubler, U, Seeburg, P, Hoffman, BJ, Gage, LP and Udenfriend, S: Molecular cloning establishes proenkephalin as precursor of enkephalin-containing peptides. Nature (Lond.) 295:206-208, 1982.
12. Comb, M, Seeburg, PH, Adelman, J, Eiden, L and Herbert, E: Primary structure of the human Met- and Leu-enkephalin precursor and its mRNA. Nature (Lond.) 295:663-666, 1982.
13. Armitage, AK: Effects of nicotine and tobacco smoke on blood pressure and release of catecholamines from the adrenal glands. Brit. J. Pharmacol. 25:515-526, 1965.
14. Viveros, OH, Wilson, SP, Cubeddu, LX and Kirshner, N: Opioid peptides in human adrenal medulla and pheochromocytoma. In (eds.) Ehrenpreis, S and Sicuteri, F, Degradation of Endogenous Opioids: Its Relevance in Human Pathology and Therapy, Raven Press, New York, 1983, pp. 11-24.
15. Wilson, SP, Viveros, OH and Kirshner, N: Relationship between the regulation of enkephalin-containing peptide and dopamine β-hydroxylase levels in cultured adrenal chromaffin cells. J. Neurochem. 45:1363-1370, 1985.
16. Varndell, IM, Tapia, FJ, De Mey, J, Rush, RA, Bloom, SR and Polak, J: Electron immunocytochemical localization of enkephalin-like material in catecholamine-containing cells of the carotid body, the adrenal medulla and pheocyromocytomas of man and other mammals. J. Histochem. and Cytochem. 30:682-690, 1982.
17. Viveros, OH, Diliberto, EJ, Jr., Hazum, E and Chang, K-J: Enkephalins as possible adrenomedullary hormones: storage, secretion, and regulation of synthesis. In (eds.) Costa, E and Trabucchi, M, Neural Peptides and Neuronal Communication, Raven Press, New York, 1980 pp. 191-204.
18. Wilson, SP, Chang, K-J and Viveros, OH: Proportional secretion of opioid peptides and catecholamines from adrenal chromaffin cells in culture. J. Neurosci. 2:1150-1156, 1982.
19. Kilpatrick, DL, Lewis, RV, Stein, S and Udenfriend, S: Release of enkephalins and enkephalin-containing polypeptides from perfused beef adrenal glands. Proc. Natl. Acad. Sci. (U.S.A.) 77:7473-7475, 1980.
20. Viveros, OH, Arqueros, L, Connett, RJ and Kirshner, N: Mechanism of secretion from the adrenal medulla. IV. The fate of the storage vesicles following insulin and reserpine administration. Mol. Pharmacol. 5:69-82, 1969.
21. Ryder, SW and Eng, J: Radioimmunoassay of leucine-enkephalin-like substance in human and canine plasma. J. Clin. Endocrinol. Metab. 52:367-369, 1981.
22. Hexum, TD, Hanbauer, I, Govoni, S, Yang, H-YT and Costa, E: Secretion of enkephalin-like peptides from canine adrenal gland following splanchnic nerve stimulation. Neuropeptides 1:137-142, 1980.

23. Viveros, OH, Lee, C-L, Abou-Donia, MM, Nixon, JC and Nichol, CA: Biopterin cofactor biosynthesis: Independent regulation of GTP cyclohydrolase in adrenal medulla and cortex. Science 213:349-350, 1981.
24. Thoenen, H, Mueller, RA and Axelrod, J: Tran-synaptic induction of adrenal tyrosine hydroxylase. J. Pharmacol. Exp. Therap. 169:249-254, 1969.
25. Kanamatsu, T, Unsworth, CD, Diliberto, EJ, Viveros, OH and Hong, J-S: Insulin-induced hypoglycemia alters the levels of proenkephalin A mRNA and proenkephalin-related peptides. Soc. Neurosci. Abst. 11:563, 1985.
26. Wilson, SP, Abou-Donia, MM, Chang, K-J and Viveros, OH: Reserpine increases opiate-like peptide content and tyrosine hydroxylase activity in adrenal medullary chromaffin cells in culture. Neuroscience 6:71-79, 1981.
27. Wilson, SP, Chang, K-J and Viveros, OH: Synthesis of enkephalins by adrenal medullary chromaffin cells: Reserpine increases incorporation of radiolabeled amino acids. Proc. Natl. Acad. Sci. (U.S.A.) 77:4364-4368, 1980.
28. Abou-Donia, MM, Wilson, SP, Zimmerman, TP, Nichol, CA and Viveros, OH: Regulation of guanosine triphosphate cyclohydrolase and tetrahydrobiopterin levels and the role of the cofactor in tyrosine hydroxylation in primary cultures of adrenomedullary cells. J. Neurochem. 46:1190-1199, 1985.
29. Wilson, SP, Unsworth, CD and Viveros, OH: Regulation of opioid peptide synthesis and processing in adrenal chromaffin cells by catecholamines and cyclic adenosine 3':5'-monophosphate. J. Neuroscience 4:2993-3001, 1984.
30. Eiden, LE, Giraud, P, Affolter, H-A, Herbert, E and Hotchkiss, AJ: Alternative modes of enkephalin biosynthesis regulation by reserpine and cyclic AMP in cultured chromaffin cells. Proc. Natl. Acad. Sci. (U.S.A.) 81:3949-3953, 1984.
31. Moccheti, I, Guidotti, A, Schwartz, JP and Costa, E: Reserpine changes the dynamic state of enkephalin stores in rat striatum and adrenal medulla by different mechanisms. J. Neuroscience 5:3379-3385, 1985.
32. Wilson, JP, Chang, K-J and Viveros, OH: Opioid peptide synthesis in bovine and human adrenal chromaffin cells. Peptides 2 (suppl. 1):83-88, 1981.
33. Eiden, LE, Giraud, P, Dave, JR, Hotchkiss, AJ and Affolter, H-U: Nicotinic receptor stimulation activates enkephalin release and biosynthesis in adrenal chromaffin cells. Nature (Lond.) 312:661-663, 1984.
34. Guidotti, A and Costa, E: A role for nicotinic receptors in the regulation of the adenylate cyclase of adrenal medulla. J. Pharmacol. Exp. Therap. 189:665-675, 1974.
35. Siegel, RE, Eiden, LE and Affolter, H-U: Elevated potassium stimulates enkephalin biosynthesis in bovine chromaffin cells. Neuropeptides 6:543-552, 1985.
36. Quach, TT, Tang, F, Kageyama, H, Mocchetti, I, Guidotti, A, Meek, JL, Costa, E and Schwartz, JP: Enkephalin biosynthesis in the adrenal medulla, modulation of proenkephalin mRNA content of cultured chromaffin cells by 8-bromo-adenosine 3',5'-monophosphate. Molec. Pharmacol. 26:255-260, 1984.
37. Viveros, OH, Arqueros, L and Kirshner, N: Mechanism of secretion from the adrenal medulla: VI. Effect of reserpine on the dopamine β-hydroxylase and catecholamine content and on the buoyant density of adrenal storage vesicles. Molec. Pharmacol. 7:434-443, 1971.

38. Viveros, OH, Daniels, AJ and Diliberto, EJ, Jr.: Cosecretion of catecholamines, opioid peptides, ascorbate, and other secretory products from multiple compartments within adrenomedullary chromaffin cells. In (eds.) Chan-Palay, V and Palay, SL, Coexistence of Neuroactive Substances in Neurons, John Wiley & Sons, New York, 1984, pp. 305-323.
39. Viveros, OH, Arqueros, L and Kirshner, N: Quantal secretion from adrenal medulla: All-or-none release of storage vesicle content. Science 165:911-913, 1969.
40. Lewis, JW, Tordoff, MG, Sherman, JE and Liebeskind, JC: Adrenal medullary enkephalin-like peptides may mediate opioid stress analgesia. Science 217:557-559, 1982.
41. Lewis, JW, Tordoff, JC, Liebeskind, JC and Viveros, OH: Evidence for adrenal medullary opioid involvement in stress analgesia. Soc. Neurosci. Abstr. 8:778, 1982.
42. Abrahams, VC, Koelle, GB and Smart, P: Histochemical demonstration of cholinesterases in the hypothalamus of the dog. J. Physiol. (Lond.) 139:137-144, 1957.
43. Daniels, AJ, Dean, G, Viveros, OH and Diliberto, EJ, Jr.: Secretion of newly taken up ascorbic acid by adrenomedullary chromaffin cells originates from a compartment different from the catecholamine storage vesicle. Molec. Pharmacol. 23:437-444, 1983.
44. Viveros, OH, Diliberto, EJ, Jr. and Daniels, AJ: Biochemical and functional evidence for the cosecretion of multiple messengers from single and multiple compartments. Fed. Proc. 42:2923-2928, 1983.
45. Wilson, SP, Klein, RL, Chang, K-J, Gasparis, MS, Viveros, OH and Yang, W-H: Are opioid peptides co-transmitters in noradrenergic vesicles of sympathetic nerves? Nature (Lond.) 288:707-709, 1980.
46. Klein, RL, Wilson, SP, Dzielak, DJ, Yang, W-H and Viveros, OH: Opioid peptides and noradrenaline co-exist in large dense-cored vesicles from sympathetic nerves. Neuroscience 7:2255-2261, 1982.
47. Hook, VYH and Eiden, LE: (Met)enkephalin and carboxypeptidase processing enzyme are co-released from chromaffin cells by cholinergic stimulation. Biochem. Biophys. Res. Comm. 128:563-570, 1985.
48. Yoshikawa, K, Hong, J-H and Sabol, SL: Electroconvulsive shock increases preproenkephalin messenger RNA abundance in rat hypothalamus. Proc. Natl. Acad. Sci. (U.S.A.) 82:589-593, 1985.
49. Kanamatsu, T, Unsworth, CD, Diliberto, EJ, Jr., Viveros, OH and Hong, J-S: Reflex splanchnic nerve stimulation increases levels of proenkephalin A mRNA and proenkephalin A related peptides in the rat adrenal medulla. Proc. Natl. Acad. Sci. (U.S.A.) 83:9245-9249, 1986.

METABOLISM, PHARMACOKINETICS, AND PHARMACODYNAMICS OF NICOTINE IN MAN

Neal L. Benowitz and Peyton Jacob III

Medical Service
San Francisco General Hospital Medical Center
and the Department of Medicine
and Langley Porter Psychiatric Institute
University of California, San Francisco
San Francisco, California 94143

INTRODUCTION

Why do people habitually consume tobacco and how does tobacco use influence the health of the habitual user? Complete answers to these questions are many years away. But a beginning is an understanding of the human pharmacology of nicotine, which is the primary pharmacologically active substance in tobacco. Central issues include: (1) What are the pathways and rate of metabolism? (2) How does elimination rate influence effects of nicotine or smoking behavior? (3) What is the time course of nicotine in the body during regular tobacco use? (4) What is the relationship between dose, concentration, and effects? (5) How does tolerance influence effects of nicotine? We will review a number of studies conducted in our laboratory examining these issues.

METABOLISM OF NICOTINE

Understanding pathways of metabolism of nicotine is important in determining if metabolites contribute to actions of nicotine and in understanding factors that influence rate of metabolism in different people.

Nicotine is extensively metabolized, primarily in the liver, but also to a small extent in the lung and kidney (1,2). Renal excretion depends on urinary pH and urine flow and accounts for from 2-35% of total elimination (3-5). The major metabolites of nicotine are cotinine and nicotine-N-oxide (Fig. 1). Cotinine is formed in a two-step process, the first of which involves oxidation of the 5-position of the pyrrolidine ring in a cytochrome P450-mediated process to nicotine-1'(5')-iminium ion (6,7). The latter is further metabolized by a cytoplasmic aldehyde oxidase to cotinine. Cotinine is also extensively metabolized, with only about 17% excreted unchanged in the urine (4,8).

Nicotine-1'-N-oxide is a minor metabolite of nicotine. Oxidation of the nitrogen atom of the pyrrolidine ring is dependent upon a microsomal flavoprotein system (1). After intravenous injection, 100% of nicotine-N-oxide is excreted unchanged in the urine (9). After oral

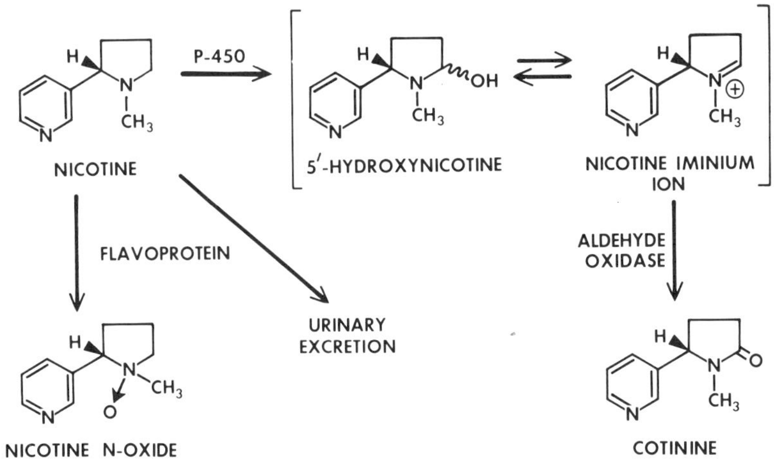

Fig. 1. Major pathways of nicotine metabolism (from Benowitz et al. (4) reprinted with permission).

administration, only 30% is recovered unchanged; the remainder is recovered as nicotine and its metabolites. After rectal administration, less than 10% is recovered as nicotine oxide. These findings indicate extensive reduction of nicotine oxide back to nicotine within the gastrointestinal tract, a phenomenon believed to be related to bacterial action.

Quantitative aspects of the conversion of nicotine to its metabolites have not been well defined. Studies of urinary excretion of cotinine in urine collected for 24 hours after intravenous injection of nicotine indicate a range from less than 10% in nonsmokers to an average of 25% in smokers (10). Another study, comparing 24 hour urinary excretion of cotinine to nicotine content of cigarette butts after smoking, indicated 46% recovery as cotinine (11). However, both of these studies underestimate the conversion of nicotine to cotinine because the urine collection period, 24 hours, was too short. Cotinine has a half-life averaging 20 hours (4), so that in 24 hours only a little more than half of cotinine will be recovered. Urine collection for at least 72 hours is necessary to recover more than 90% of cotinine in most subjects. In addition, since only 17% of cotinine is excreted unchanged (4), urinary recovery underestimates cotinine generation rate.

To investigate quantitative aspects of conversion of nicotine to its metabolites, we measured urinary excretion of nicotine and its metabolites, cotinine, nicotine-N-oxide, and cotinine-N-oxide, in 24-hour urine collections in habitual smokers (Fig. 2). Assuming steady state, the rate of excretion of metabolites reflects generation rate. One hundred percent of nicotine-N-oxide but only 17% of cotinine is excreted unchanged in the urine. Based on the ratio for cotinine to nicotine-N-oxide of 2.9, the relative generation rate of cotinine compared to nicotine-N-oxide is 17 to 1. We and Beckett et al. (9) have found that about 4% of nicotine is excreted as nicotine-N-oxide. Therefore, about 70% of nicotine appears to be converted to cotinine. Our current estimate of quantitative disposition of nicotine is shown in Fig. 3. Of note is the 17% of nicotine that is not accounted for by metabolism to cotinine or nicotine-N-oxide. Other metabolites, such as nicotine isomethonium ion and nornicotine, which may have pharmacological activity, need to be investigated as candidates for the missing 17%.

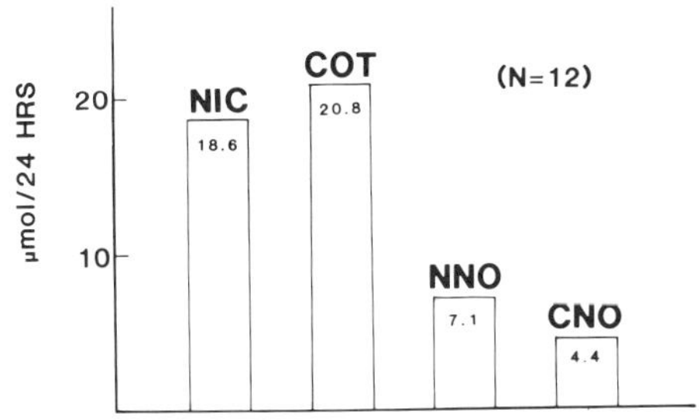

Fig. 2. Excretion of nicotine and metabolites in 24 hour urine while smoking cigarettes ad libitum. Values are averages for 12 subjects. NIC = nicotine, COT = cotinine; NNO = nicotine N-oxide, CNO = cotinine N-oxide.

Cotinine is known to accumulate to higher levels and persist in the body for a much longer time than nicotine. *In vivo* and *in vitro* animal studies indicated that cotinine might have activity in high doses (12,13,14). To examine the possibility that cotinine has pharmacologic activity in man, we infused cotinine by vein into habitual smokers who had been abstinent from smoking for two days (4). Blood cotinine concentration rose to levels (average 387 ng/ml) seen in heavy smokers. No cardiovascular changes were observed. Subjectively, there was reduction of anxiety (consistent with termination of any infusion) but no reports of stimulation, euphoria, or sedation. At this time there is no evidence that cotinine, at levels to which smokers are usually exposed, exerts any pharmacologic effect.

PHARMACOKINETICS

Absorption of Nicotine

Nicotine is distilled from burning tobacco and is carried proximally on tar droplets (0.1-0.4 microns) that are inhaled. Absorption of nicotine across biological membranes depends on pH (15,16). Nicotine is a weak base with a pKa of 7.9. Consequently, in acidic environments it is ionized and does not rapidly cross membranes. The pH of smoke from the flue-cured tobaccos, which are found in most cigarettes, is acidic (pH 5.5). At this pH, the nicotine is primarily ionized. As a consequence, there is little buccal absorption of nicotine from cigarette smoke, even when it is held in the mouth (17). The pH of smoke from air-cured tobaccos, such as in pipes, cigars, and in a few European cigarettes, is

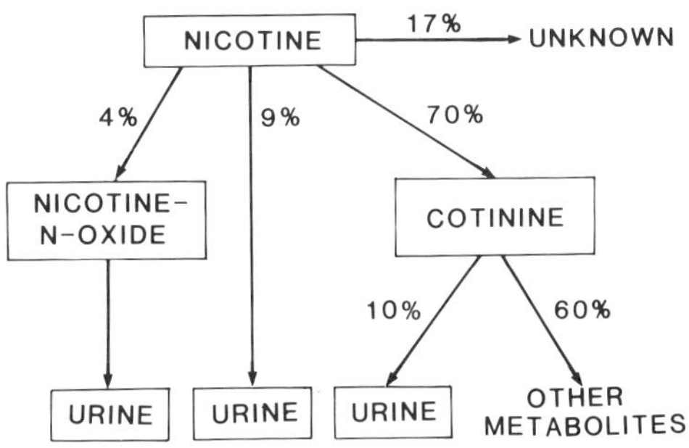

Fig. 3. Quantitative disposition of nicotine in habitual smokers.

alkaline (pH 8.5), and nicotine from these products is well absorbed through the mouth (18,19). Chewing tobacco, snuff, and nicotine gum are buffered to alkaline pH to facilitate nicotine absorption.

When tobacco smoke reaches the small airways and alveoli of the lung, the nicotine is rapidly absorbed independent of pH of the smoke. Armitage and co-workers (20), measuring exhalation of radiolabeled nicotine, found that 82-92% of nicotine in mainstream smoke was absorbed by four habitual smokers, 29% by another habitual smoker who was presumed to be a noninhaler, and 30-66% by three nonsmokers (who were instructed to smoke as deeply as possible). The rapid absorption of nicotine from cigarette smoke through the lung is presumably because of the huge surface area of the alveoli and small airways and dissolution of nicotine into fluid of pH in the human physiologic range, which facilitates transfer across cell membranes.

Distribution of Nicotine

After absorption, nicotine enters the bloodstream where, at pH 7.4, it is about 69% ionized and 31% un-ionized. Binding to plasma proteins is less than 5% (21). The drug is distributed extensively to body tissues with a steady state volume of distribution averaging 2.6 times body weight (Table 1). The pattern of tissue uptake cannot be studied in humans, but we have, by measuring concentrations of nicotine in various tissues after infusion of nicotine to steady state, examined tissue uptake in rabbits (Table 2). Spleen, liver, lungs, and brain have high and adipose tissue has relatively low affinity for nicotine.

After rapid intravenous injection, nicotine blood concentration falls rapidly due to tissue uptake of the drug. During this phase, blood concentrations are quite high while tissue concentrations would be expected to still be low. Distribution to brain and heart should be rapid because of relatively high rates of perfusion. Thus, direct effects on these organs are expected within 1 or 2 min. Distribution to muscle is by comparison predicted to be much slower, but muscle represents a storage reservoir of greater magnitude (Table 2). The consequence of uptake into muscle is that the blood concentration of drug continues to decline at a rate faster than can be explained by metabolism for 20 or 30 min after administration. Thereafter, the blood concentration declines much more slowly, the rate being determined by rate of metabolism and rate of distribution out of tissues.

Smoking represents an exposure similar to that of rapid intravenous injection except that the point of entry into the circulation is through the pulmonary rather than the systemic venous circulation. As a consequence of delivery into the lung, the lag time between smoking and entry into the brain is shorter than that after intravenous injection. Thus, after smoking, nicotine will enter the brain quickly, but then brain

TABLE 1. Pharmacokinetics of Nicotine

Half-life	120 minutes
Volume of Distribution	180 liters
Total Clearance	1300 ml/min
Renal Clearance (Acid Urine)	200 ml/min
Nonrenal Clearance	1100 ml/min

Average values, based on data from Benowitz et al. (18).

TABLE 2. Distribution of Nicotine

Tissue	Rabbit Tissue:Blood Ratio (R)	Human Simulation Mass (M) (kg)	Human Simulation MxR	% Total Body Nicotine
Blood	1.0	5.4	5.4	4.8
Brain	3.0	1.5	4.5	4.0
Heart	2.0	0.3	0.6	0.5
Muscle	2.0	30.0	60.0	53.1
Adipose	0.5	10.0	5.0	4.4
Kidney	21.6	0.3	6.5	5.7
Liver	3.7	1.7	6.3	5.6
Lung	3.7	1.0	3.7	3.2
GI Tissue	3.5	2.0	7.0	6.2
Spleen	9.3	1.5	14.0	12.4
	Totals	53.7	113.0	

$V_D = \Sigma MxR/\Sigma M = 2.1$ L/kg

Tissue to blood nicotine concentration ratio in rabbits based on measurements after 24 hr constant infusion of nicotine. Human simulation based on typical organ mass and partition ratios observed in rabbits. V_D = volume of distribution predicted by simulation.

levels will decline rapidly as nicotine is distributed to other tissues. The distribution half-life (about 8 min) rather than the elimination half-life (about 2 hours) determines the time course of central nervous system effects after smoking.

Rate of Elimination of Nicotine From the Body

Metabolism. The rate of metabolism of nicotine can be determined by measuring blood levels after administration of a known dose of nicotine. We studied habitual cigarette smokers given intravenous infusions of nicotine for 30-60 min (21). Total and renal clearance were computed directly, and the nonrenal or metabolic clearance computed as the difference between total and renal clearance. Nonrenal clearance averaged 1089 ml/min (Table 1). This represents about 70% of liver blood flow. Assuming most nicotine is metabolized by the liver (data in animals indicate only a small contribution by the lung), this means about 70% of the drug is extracted from the blood in each pass through the liver. Based on an extraction ratio of 70%, one would predict an oral bioavailability of 30%, which is similar to that reported by Jenner et al. (22).

An implication of the high degree of hepatic extraction is that clearance of nicotine should be dependent upon liver blood flow. That is, about 70% of whatever volume of blood the circulation brings to the liver should be cleared of nicotine. Thus, physiologic events such as meals, posture, exercise, or other drugs which perturb hepatic blood flow are predicted to have significant effects on the rate of nicotine metabolism. The biological significance of this prediction remains to be determined.

The level of any drug in the body during repeated dosing depends on the rate of intake and the rate of elimination of the drug. There is considerable variability in rate of elimination of nicotine among individuals (21). In pharmacokinetic terms, clearance is a number which predicts steady state blood level of a drug achieved at a particular level of intake. The relationship is expressed mathematically as $Css = R/Cl$, where Css is the steady state blood concentration, R is the rate of intake, and Cl is clearance. Since there may be fourfold interindividual differences in clearance of nicotine, a person with an extremely high clearance will have blood and body nicotine levels one-fourth as high as a person consuming the same amount of nicotine but having a low clearance. If people smoke to achieve a particular body level of nicotine, then clearance could influence self-determined rate of cigarette smoking. Differences in clearance among people could result in different exposures to and risks from other combustion products.

Renal Excretion. Nicotine is excreted by glomerular filtration and tubular secretion, with variable reabsorption depending on urinary pH (Table 2). In acid urine, nicotine is mostly ionized and tubular reabsorption is minimized; renal clearance may be as high as 600 ml/min (urinary pH 4.4), depending on urine flow rate (4,5). In alkaline urine, a larger fraction of nicotine is un-ionized, allowing net tubular reabsorption with renal clearances as low as 17 ml/min (urine pH 7.0). With uncontrolled urine pH, renal clearance averages about 100 ml/min, accounting for the elimination of about 10% of the daily intake of nicotine. Because urinary pH is a determinant of elimination rate and, hence, body levels of nicotine, it has been suggested that urine pH may influence self-determined nicotine intake (23). This issue will be addressed in a later section of this paper.

NICOTINE BLOOD LEVELS DURING TOBACCO USE

Blood or plasma nicotine concentrations measured in the afternoon in smokers generally range from 10-50 ng/ml, although we have observed one patient with a level of over 100 ng/ml. The increment in blood nicotine concentration after smoking a single cigarette ranges from 5 to 30 ng/ml, depending on how the cigarette is smoked (20,24.25). Blood levels peak at the end of smoking a cigarette and decline rapidly over the next 20 min due to tissue distribution.

Peak blood levels of nicotine are similar for cigar smokers and users of snuff and chewing tobacco, although the rate of rise of nicotine is slower (18,26-29). Pipe smokers, particularly those who have previously smoked cigarettes, may have blood levels of nicotine as high as cigarette smokers (26,30). Others who are pipe smokers, who have not previously smoked cigarettes, tend to have relatively low nicotine levels. Blood levels of nicotine while chewing 4-mg gum are similar to those observed during cigarette smoking, while blood levels while chewing 2-mg gum are less than seen in most smokers (31).

The earliest published studies of the elimination kinetics of nicotine reported half-lives of 20-40 min (20,25). In those studies, blood levels were followed only for 30 to 60 min, which is not long enough to determine the terminal half-life. Thus, half-lives were based on blood levels which included the distribution phase. When blood levels are followed for several hours after the end of nicotine infusion, there is a log linear decline phase with a half-life of about 2 hours.

The half-life of a drug is useful in predicting the rate of accumulation of that drug in the body with repetitive doses, and the time course of decline after cessation of dosing. Based on a half-life of two hours for nicotine, one would predict accumulation over 6-8 hours (3-4 half-lives) of regular smoking, and persistence of significant nicotine levels for 6–8 hours after cessation of smoking. If a smoker smokes until bedtime, significant levels should persist all night. Studies of blood levels in regular cigarette smokers confirm these predictions (Fig. 4) (32). Peaks and troughs follow each cigarette, but as the day progresses, trough levels rise and the influence of peak levels becomes less important. Thus, nicotine is not a drug to which people are exposed intermittently and which is eliminated rapidly from the body. To the contrary, smoking represents a multiple dosing situation with considerable accumulation while smoking and persistent levels for 24 hours of each day.

INFLUENCE OF ELIMINATION RATE ON REGULATION OF NICOTINE INTAKE DURING CIGARETTE SMOKING

There is considerable evidence that smokers adjust their smoking behavior to try to regulate or maintain a particular level of nicotine in the body (33,34). We have applied techniques for measuring daily intake of nicotine (35) to study the influence of elimination on nicotine intake.

Rate of renal elimination of nicotine was manipulated by administration of ammonium chloride or sodium bicarbonate to acidify or alkalinize the urine, respectively (36). Compared with daily excretion during placebo treatment (3.9 mg nicotine per day), acid loading increased (to 12 mg per day) and alkaline loading decreased (to 0.9 mg per day)

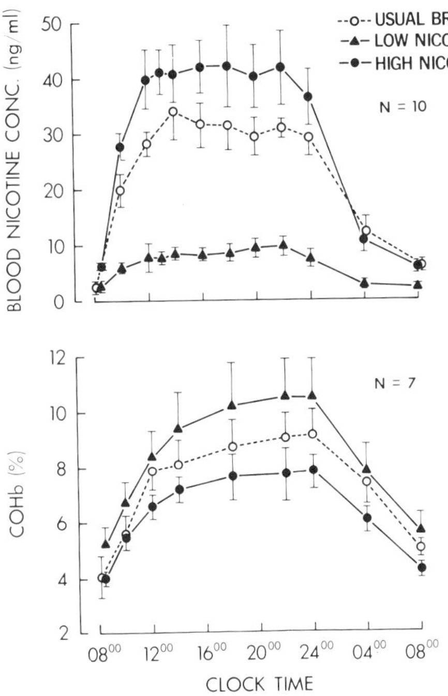

Fig. 4. Blood nicotine and carboxyhemoglobin concentrations in subjects smoking high-nicotine (2.5 mg) and low-nicotine (0.4 mg) Kentucky Reference cigarettes and their usual brand (average nicotine yield 1.2 mg) of cigarettes. Subjects smoked on a fixed schedule of one cigarette every half-hour from 8:30 A.M. to 11:00 P.M. for a total of 30 cigarettes per day. Blood samples were collected just before the next scheduled cigarette (figure taken from Benowitz et al. (29), reprinted with permission).

daily excretion of nicotine. Average blood nicotine concentrations were similar in placebo and bicarbonate treatment conditions, but were 15% lower during ammonium chloride treatment. Daily intake of nicotine was 18% higher during acid loading, indicating compensation for increased urinary loss. The compensatory increase in nicotine consumption was only partial, replacing about half the excess urinary nicotine loss.

Caution should be exercised in applying these findings to usual smoking situations. Our studies were performed under conditions of extreme urinary acidification or alkalinization, so that the changes in renal clearance would be maximized. Even with extreme differences in urinary pH, differences in overall nicotine elimination rate and smoking behavior were modest. This is because renal excretion is a minor pathway for elimination of nicotine; most is metabolized. Smaller changes in urinary pH, such as occur spontaneously throughout the day or that might be related to stressful events, would not be expected to substantially influence nicotine elimination or smoking behavior. Alkali therapy has been advocated as a way to slow the rate of nicotine elimination and to reduce smoking (37). Our data show that there is relatively little excretion of nicotine in baseline conditions (averaging 4 mg per day) and a decrease to 0.9 mg per day with alkali treatment has very little effect on body levels or smoking behavior. Thus, bicarbonate therapy would be expected to have little, if any, effect on cigarette consumption. This prediction is supported by other experimental studies (38).

The primary route of nicotine elimination is metabolism. In preliminary studies, we have found a significant correlation between metabolic clearance and daily intake of nicotine in smokers smoking their customary brands. There was no correlation between average nicotine blood levels and clearance. These observations support the idea that rapid metabolizers generally consume more nicotine (smoke more) than slow metabolizers, while maintaining similar nicotine blood levels and, presumably, similar effects.

PHARMACODYNAMICS

General Considerations

The basic pharmacology of nicotine is extremely complex. A detailed review of this pharmacology is beyond the scope of this paper. Several fundamental pharmacologic issues are relevant to the human pharmacology of nicotine. First, the actions of nicotine in the intact organism are complex and difficult to predict from *in vitro* and animal studies. This is because nicotine can have effects on multiple neuroeffector systems and can affect many or all body organs, the effects of which depend on the prevalent autonomic tone of that particular organ. Effects on various organs may have opposing actions, the net effect depending upon the interaction of the two. For example, it is known that nicotine influences heart rate. Based on *in vitro* and animal studies, we know that nicotine can influence heart rate by: action on peripheral chemoreceptors (via the central nervous system), direct actions on the medullary center of the brain, direct release of neuronal norepinephrine, facilitation of catecholamine release in response to neural activation, direct peripheral ganglionic stimulation with resultant cardiac sympathetic or parasympathetic (or both) nerve stimulation, release of epinephrine from the adrenal, vagal responses to nicotine-induced stimulation of the emetic chemoreceptor trigger zone in the medulla, and/or baroreceptor-mediated reflex responses to nicotine-induced blood pressure changes (39,40). It is obvious why it is difficult to predict from basic pharmacologic studies what the influence of nicotine on heart rate will be in the intact human.

A second basic pharmacologic issue is the nature of the dose-response relationship for nicotine. In classical pharmacology, nicotine is commonly discussed as an example of a drug which in low doses causes ganglionic stimulation and in high doses causes ganglionic blockade (following brief stimulation) (39). Dose-response characteristics *in vivo* are often biphasic as well, although the mechanisms are far more complex. For example, at very low doses, similar to those seen during cigarette smoking, cardiovascular effects appear to be mediated by the central nervous system, either via activation of chemoreceptor afferent pathways or via direct effects on the brain stem (39,40). The net result is sympathetic neural discharge with an increase in blood pressure and heart rate. At higher doses, there may be direct effects on the peripheral nervous system, such as ganglionic stimulation, with the release of adrenal catecholamines. With extremely high doses, there may be hypotension and heart rate slowing, mediated either by peripheral vagal activation or direct depressor effects mediated by effects on the brain (41,42). Before extrapolating pharmacologic observations from animals to humans, blood concentrations should be measured to ensure that the effects are being studied in a portion of the dose-response curve related to smokers.

A third pharmacologic issue of importance is development of tolerance. Smokers know that tolerance develops to some effects of smoking. Smoking the first cigarette as a teenager is commonly associated with dizziness, nausea and/or vomiting, effects to which the habitual smoker rapidly becomes tolerant. Likewise, in *in vitro* pharmacologic studies, tolerance to various effects develops rapidly, although tolerance may not be complete (43,44). Studies of the human pharmacology of nicotine necessarily take place in people who have a degree of tolerance prior to dosing with nicotine. The extent of tolerance depends on both the level of nicotine, the duration of exposure to a given level, and the rate of increase of nicotine exposure to a particular organ. In designing pharmacologic studies in man, it is essential to consider recent exposure to nicotine and the degree of tolerance that may be present at the time nicotine is administered. Individual differences in degree of tolerance may be important determinants of individual differences in pharmacologic responses observed in experimental studies.

In light of the above issues, it is important to consider not only the actions but also dose-response characteristics and tolerance, as well as pharmacokinetics and metabolism, in interpreting studies of nicotine effects in man.

Cardiovascular Effects

The cardiovascular effects of cigarette smoking have been extensively studied. Smoking a cigarette activates the sympathetic nervous system and results, in healthy people, in an increase in heart rate and blood pressure (45), cardiac stroke volume and output (46), and coronary blood flow (47). Cigarette smoke or nicotine causes peripheral vascular changes, including cutaneous vasoconstriction (48), associated with a decrease in skin temperature, systemic venoconstriction (49), and increased muscle blood flow (50). Smoking results in increased circulating concentrations of norepinephrine, consistent with neural adrenergic stimulation, and epinephrine, indicating adrenal medullary stimulation (45). Circulating free fatty acids, glycerol, and lactate concentrations increase. Cardiovascular and metabolic effects are prevented by combined alpha and beta adrenergic blockade, indicating that the cardiovascular effects of cigarette smoking are mediated by activation of the sympathetic nervous system.

Nicotine appears to be the substance in cigarette smoke responsible for activation of the sympathetic nervous system. Evidence for this comes from studies comparing the effects of nicotine-containing versus denicotinized cigarettes, comparing cigarettes of differing nicotine yield, and studying the effects of nicotine per se, either as intravenous nicotine or nicotine gum (21,51-53).

Pharmacodynamics

Although there have been many studies of the actions of smoking or nicotine per se, until recently little has been known about the relationship between nicotine blood concentrations and cardiovascular effects. Such considerations are important in relating knowledge of nicotine blood levels throughout the day while smoking to the expected cardiovascular consequences. We studied the effects of intravenous bolus injections of nicotine given in a manner to simulate cigarette puffing (5). Injections were given every 30 min for 3 hours. Blood levels and cardiovascular effects were similar to those observed after cigarette smoking (Fig. 5). Heart rate and blood pressure increased sharply after the first series of nicotine injections. Values remained above baseline, but there was little increment with successive injections, despite rising and falling nicotine blood levels. In contrast, skin temperature fell progressively during the period of nicotine dosing, gradually returning toward baseline at the end of the study. These data indicated rapid development of tolerance to heart rate and blood pressure responses, but tolerance was not complete in that heart rate and blood pressure remained above baseline.

To clarify these pharmacodynamic observations, a second group of subjects was studied with continuous infusion of nicotine. Heart rate and systolic blood pressure increased mostly in the first 5 to 10 min after infusion (Fig. 6), despite the progressive increase in nicotine blood levels over 30 min. Skin temperature declined during the infusion and increased gradually after cessation of the infusion. Analysis of the blood concentration-response characteristics by hysteresis plots (Fig. 7) illustrates that heart rate increases at relatively low blood nicotine concentrations and plateaus despite rising levels. Heart rate was lower for a given blood nicotine concentration in the decline phase, indicating development of tolerance. Skin temperature declined and rose in association with changes in blood nicotine concentrations, showing no evidence of tolerance.

These pharmacodynamic observations have potential clinical significance. That low concentrations of nicotine increase heart rate to a maximum suggests that heart rate will increase most with the first few cigarettes of the day but subsequently will not vary in relation to the amount of nicotine consumed. That only partial tolerance developed to heart rate acceleration due to nicotine suggests that effects on heart rate may persist as long as significant levels of nicotine persist, including overnight. These predictions were tested in a study during which healthy volunteers smoked either high or low yield research cigarettes, or abstained from smoking (54). The research cigarettes were nonfilter cigarettes which were high or low yield because they contained more or less nicotine. Thus, full compensation for the low yield cigarette was impossible. Resultant nicotine blood levels were fourfold different (Fig. 4). As predicted, heart rate, assessed by continuous ambulatory EKG monitoring, increased in the morning, more on smoking than nonsmoking days, and the increase occurred with the first few cigarettes of the day (Fig. 8). Subsequently, heart rate followed a normal circadian pattern but was always higher when smoking was compared with abstinence. Only at 6-7 A.M., prior to the next day's smoking, were heart rates during

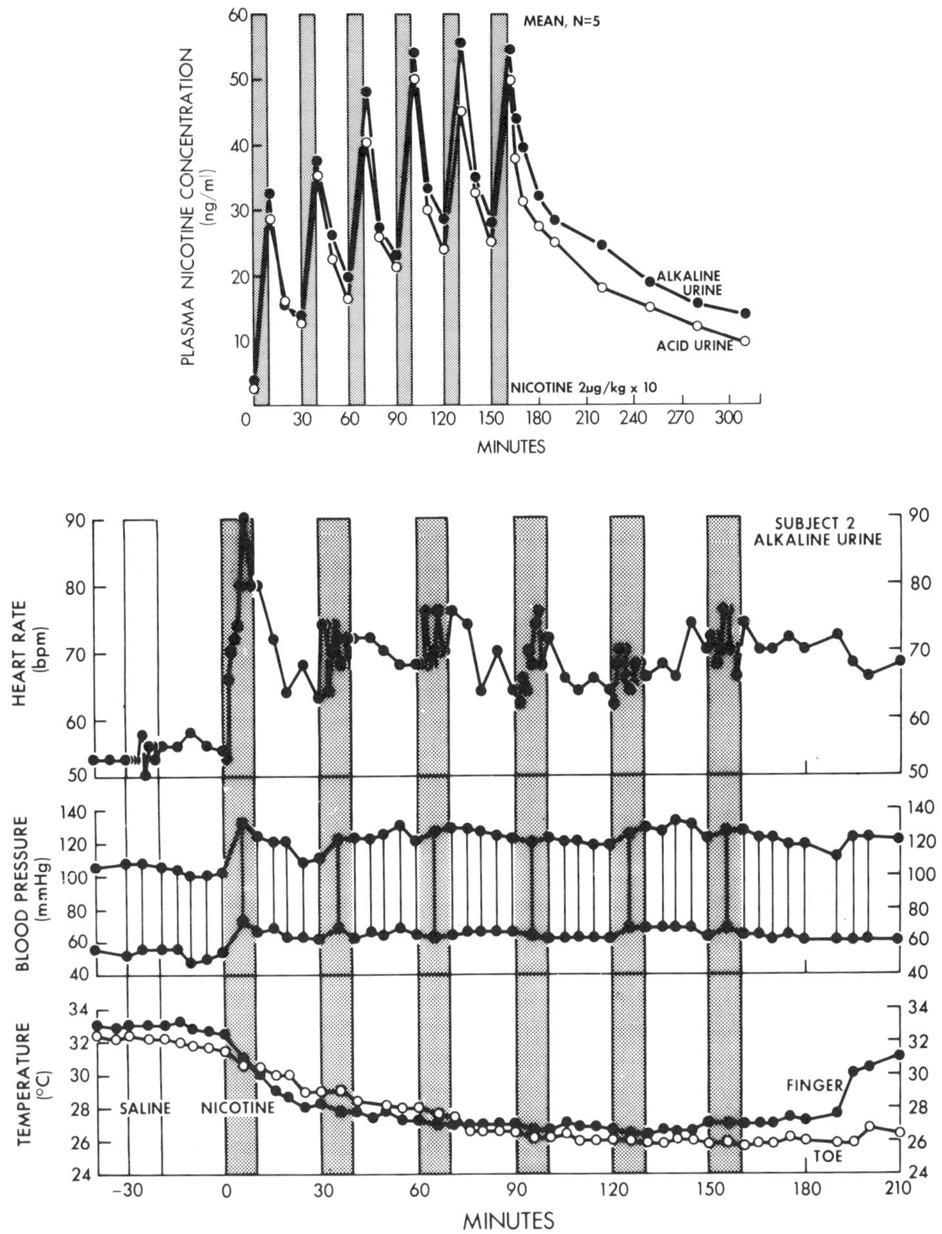

Fig. 5. Upper panel: Mean plasma nicotine concentrations for repetitive injections of nicotine with alkaline and acid urine. Hatched areas indicate 10 min intervals during which intravenous injections of nicotine, 2 μg/kg/min, were given. Data represent mean values for five subjects. (Rosenberg et al. (5), reprinted with permission).

Lower panel: Physiologic responses during repetitive series of injections of nicotine for a representative subject. Hatched areas indicate 10 min intervals during which intravenous injections of nicotine, 2 μg/kg/min, were given. Clear area indicates 10 min intervals during which intravenous injections of saline, 1 ml/min, were given. (Rosenberg et al. (5), reprinted with permission).

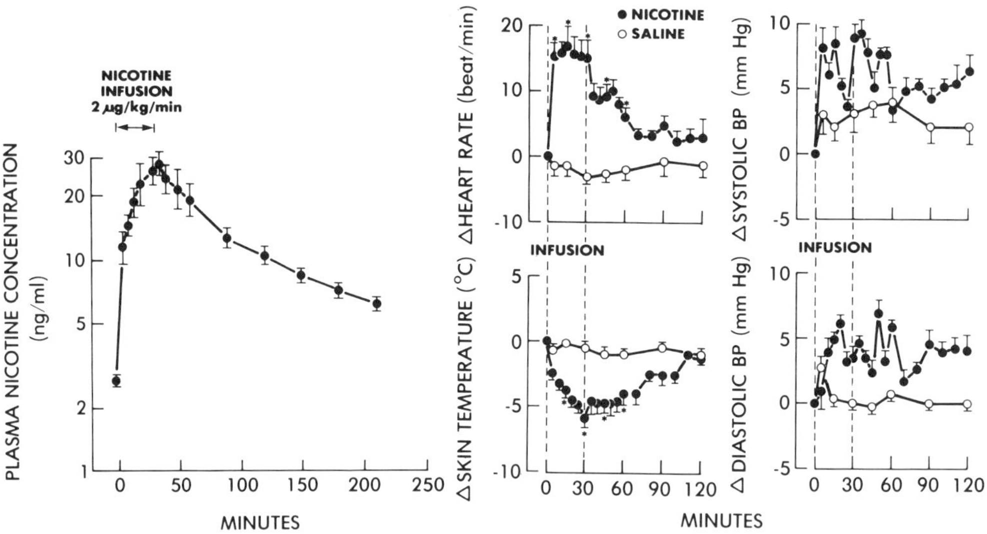

Fig. 6. Left panel: Plasma nicotine concentrations (± S.E.M.) in five subjects during and after constant infusion for 30 min (from Benowitz et al. (18), reprinted with permission).

Right panel: Cardiovascular responses to nicotine (N=5) and saline (N=5) infusions. Asterisks indicate P<0.05, comparing nicotine and saline conditions (Benowitz et al. (18), reprinted with permission).

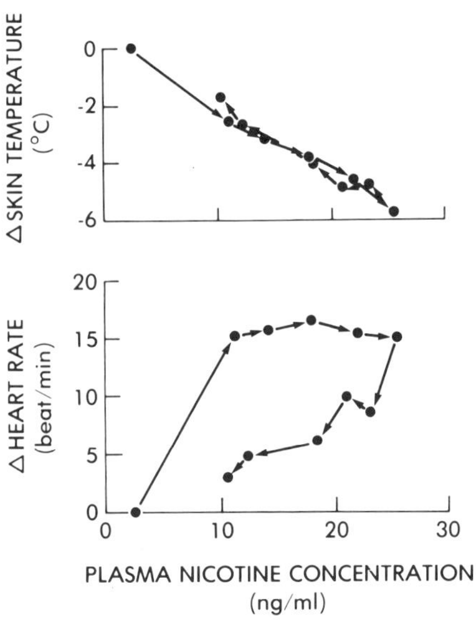

Fig. 7. Hysteresis loops for skin temperature and heart rate responses plotted against simultaneous plasma nicotine concentration (mean values N=5). Arrows indicate progression of time during and after nicotine infusion. (Benowitz et al. (18), reprinted with permission).

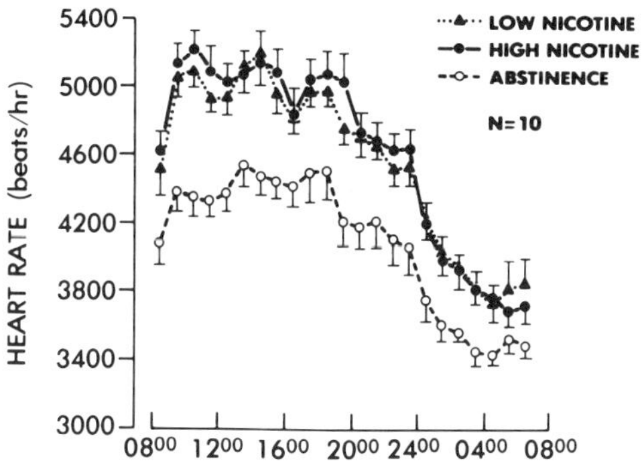

Fig. 8. Hourly HR while smoking LN or HN cigarettes and while abstaining ($\bar{X}$ ± S.E.). (Benowitz et al. (51), reprinted with permission).

smoking similar to those in abstinence. Also, as predicted, heart rate was no different when smoking low yield compared to high yield cigarettes, despite the fourfold difference in blood nicotine concentration.

Recently we have studied cardiovascular effects of nicotine gum. Nicotine blood levels were substantially lower while chewing gum than while smoking customary brands of cigarettes but were as high or higher than levels in subjects smoking low yield research cigarettes. Somewhat unexpectedly, nicotine gum did not produce heart rate acceleration throughout the day, although it did increase blood pressure to a similar magnitude as smoking. We have repeated nicotine infusion studies, but this time we maintained infusions for 3 hours. With prolonged infusion, nearly complete tolerance developed to heart rate acceleration but not to blood pressure elevation. These observations suggest that the route and time course of delivery of nicotine is an important determinant of nicotine effects. While smoking there are repeated brief, high level exposures of the brain to nicotine interspersed with periods of falling levels. It is less likely for complete tolerance to occur in this situation than in chewing nicotine gum, where sustained levels of nicotine without marked fluctuations are observed. If this hypothesis is correct and if the pharmacodynamics of central nervous system actions are similar, this may explain why smoking is more highly addictive than using chewing tobacco or nicotine gum.

The above studies indicate rapid development of tolerance to some (but not all) actions of nicotine. These studies were performed in habitual smokers after overnight abstinence. But even after overnight abstinence, significant quantities of nicotine persist in the body, and there is conceivably some persistence of tolerance. To determine if there is a longer-term tolerance and, if so, its extent, we examined cardiovascular responses to infused nicotine after overnight and after seven days tobacco abstinence. Heart rate and blood pressure responses were substantially greater after more prolonged abstinence (Fig. 9). However, within 60-90 minutes the blood concentration-effect relationship in subjects after prolonged abstinence approximated that observed after brief abstinence. Thus, we conclude that there is an element of tolerance that persists throughout the daily smoking cycle and which is lost with prolonged abstinence, but that tolerance is rapidly re-established with subsequent exposure.

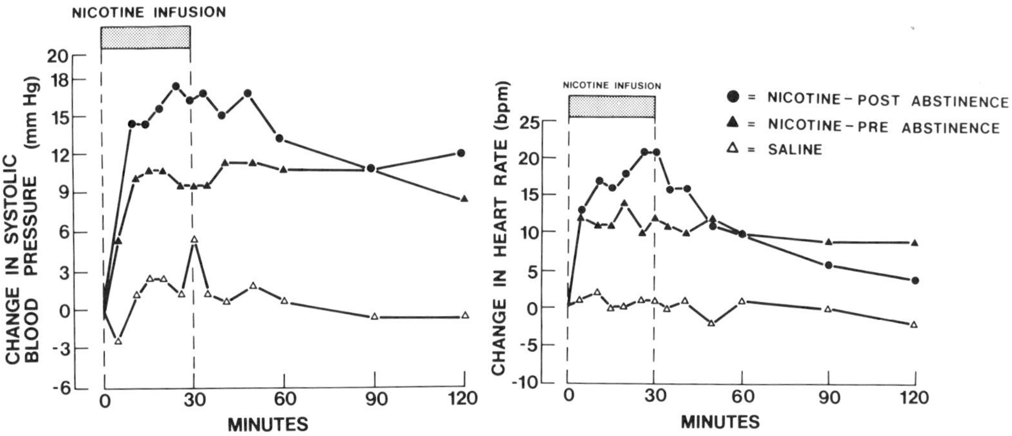

Fig. 9A/B. Cardiovascular responses to nicotine infusion (2 μg/kg/min x 30 min) after overnight compared to 7 days tobacco abstinence (mean of 7 subjects). Control saline infusions had no significant effect on cardiovascular indices. Changes in systolic blood pressure and heart rate were significantly greater when nicotine was infused after prolonged compared to brief abstinence.

SUMMARY AND CONCLUSIONS

Our studies have several implications with respect to the neurobiology of nicotine and tobacco smoking.

1. Quantitative metabolism studies account for only a fraction of the metabolites generated from nicotine. The possibility that unidentified metabolites contribute to the pharmacologic actions of nicotine remains. Further studies to identify and quantitate metabolites and evaluate their pharmacologic activity at exposure levels relevant to human smokers are needed.

2. Pharmacokinetic studies indicate that with regular use of tobacco nicotine accumulates in the body (and presumably the brain) throughout the day and persists at significant levels overnight. Thus, smoking cannot be considered an intermittent nicotine dosing with rapid elimination type exposure. Smoking results in exposure to nicotine that lasts 24 hours of each day.

3. The rate of metabolism of nicotine is highly variable among individuals. Because of the tendency for smokers to regulate nicotine levels in the body, rate of elimination may be an important determinant of smoking behavior and possibly of tobacco-smoke related health risks.

4. Pharmacodynamic studies indicate a nonlinear dose-effect relationship for some (e.g., heart rate) but not all responses. The consequence is that many of the cardiovascular effects of smoking, as observed throughout the day, are not dose dependent. This may explain in part why there is a poor correlation between changing yields of cigarettes over the past 30 years and the smoking-related risk of coronary heart disease.

5. Substantial and rapid development of tolerance to subjective effects (primarily stimulation) and heart rate acceleration occurs during nicotine infusion. Some tolerance reverses after overnight abstinence, but a degree of tolerance persists for days. Short- and long-term tolerances need to be considered in planning and interpreting studies of the pharmacology of nicotine.

6. The route and rate of administration may have an important influence on the effects of and extent of development of tolerance to nicotine. Such considerations may explain differences in subjective and cardiovascular effects of cigarette smoking and nicotine gum chewing. The smoking-rapid brain-uptake process may explain in part the highly addictive nature of cigarette smoking.

REFERENCES

1. Gorrod JW and Jenner P: The metabolism of tobacco alkaloids. In "Essays in Toxicology," Vol. 6, Academic Press, New York, 1975, 35-78.
2. Turner DM, Armitage AK, Briant RH and Dollery CT: Metabolism of nicotine by the isolated perfused dog lung. Xenobiotica 5:539-551, 1975.
3. Beckett AH, Rowland M and Triggs EB: Significance of smoking in investigations of urinary excretion rates of amines in man. Nature 207:200-201, 1965.
4. Benowitz NL, Kuyt F, Jacob P III, Jones RT and Osman A-L: Cotinine disposition and effects. Clin. Pharmacol. Ther. 309:139-142, 1983.
5. Rosenberg J, Benowitz NL, Jacob P III and Wilson KM: Disposition kinetics and effects of intravenous nicotine. Clin. Pharmacol. Ther. 28:516-522, 1980.
6. Brandange S and Lindblom L: The enzyme "aldehyde oxidase" is an iminium oxidase. Reaction with nicotine delta-1-(5')iminium ion, Biochem. Biophys. Res. Commun. 91:991-996, 1979.
7. Gorrod JW and Hibberd AR: The metabolism of nicotine-delta-1'(5')-iminium ion, in vivo and in vitro. Eur. J. Drug Metab. Pharmacokin. 7:293-298, 1982.
8. Bowman ER and McKennis H Jr.: Studies on the metabolism of (-)-cotinine in the human. J. Pharmacol. Exp. Ther. 135:306-311, 1962.
9. Beckett AH, Gorrod JW and Jenner P: The analysis of nicotine-1'-N-oxide in urine, in the presence of nicotine and cotinine, and its application to the study of in vivo nicotine metabolism in man. J. Pharm. Pharmacol. 23:55S-61S, 1971.
10. Beckett AH, Gorrod JW and Jenner P: The effect of smoking on nicotine metabolism in vivo in man. J. Pharm. Pharmacol 23:62S-67S, 1971.
11. Schievelbein H: Nicotine. Resorption and fate. Pharmac. Ther. 18:233-248, 1982.
12. Borzelleca JF, Bowman, ER and McKennis H: Studies on the respiratory and cardiovascular effects of (-)-cotinine. J. Pharmacol. Exp. Ther. 137:313-318, 1962.
13. Kim KS, Borzelleca JF, Bowman ER and McKennis H: Effects of some nicotine metabolites and related compounds on isolated smooth muscle. J. Pharmacol. Exp. Ther. 161:59-69, 1968.
14. Fuxe K, Everitt BJ and Hokfelt T: On the action of nicotine and cotinine on central 5-hydroxytryptamine neurons. Pharmacol. Biochem. Behav. 10:671-677, 1979.
15. Armitage AK and Turner DM: Absorption of nicotine in cigarette and cigar smoke through the oral mucosa. Nature 226:1231-1232, 1970.

16. Schievelbein H, Eberhardt R, Loschenkohl K, Rahlfs V and Bedall FK: Absorption of nicotine through the oral mucosa. I. Measurement of nicotine concentration in the blood after application of nicotine and total particulate matter. Agents and Actions 3/4:254-258, 1973.
17. Gori GB, Benowitz NL and Lynch CJ: Mouth and deep airways absorption of nicotine in cigarette smokers. Clin. Pharmacol. Ther. (Submitted, 1985).
18. Armitage A, Dollery C, Houseman T, Kohner E, Lewis PJ and Turner D: Absorption of nicotine from small cigars. Clin. Pharmacol. Ther. 23:143-150, 1978.
19. Russell MAH, Raw M and Jarvis MJ: Clinical use of nicotine chewing gum. Br. Med. J. 280:1599-1602, 1980.
20. Armitage AK, Dollery Ct, George CF, Houseman TH, Lewis PJ and Turner DM: Absorption and metabolism of nicotine from cigarettes. Br. Med. J. 4:313-316, 1975.
21. Benowitz NL, Jacob III P, Jones RT and Rosenberg J: Interindividual variability in the metabolism and cardiovascular effects of nicotine in man. J. Pharmacol. Exp. Ther. 221:368-372, 1982.
22. Jenner P, Gorrod JW and Beckett AH: The absorption of nicotine-1'-N-oxide and its reduction in the gastrointestinal tract in man. Xenobiotica 3:341-349, 1973.
23. Schachter S: Pharmacological and psychological determinants of smoking. Ann. Int. Med. 88:104-114, 1978.
24. Herning RI, Jones RT, Benowitz NL and Mines AH: How a cigarette is smoked determines nicotine blood levels. Clin. Pharmacol. Ther. 33:84-90, 1983.
25. Isaac PF and Rand MJ: Cigarette smoking and plasma levels of nicotine. Nature 236:308-310, 1972.
26. Turner JAM, Sillett RW and McNicol MW: Effect of cigar smoking on carboxyhaemoglobin and plasma nicotine concentrations in primary pipe and cigar smokers and ex-cigarette smokers. Br. Med. J. 2:1387-1389, 1977.
27. Gritz ER, Baer-Weiss V, Benowitz NL, Van Vunakis H and Jarvik ME: Plasma nicotine and cotinine concentrations in habitual smokeless tobacco users. Clin. Pharmacol. Ther. 30:201-209, 1981.
28. Russell MAH, Jarvis MJ and Feyerabend C: A new age for snuff?, Lancet 1:474-475, 1980.
29. Russell MAH, Jarvis MJ, Devitt G and Feyerabend C: Nicotine intake by snuff users. Br. Med. J. 283:814-817, 1981.
30. McCusker K, McNabb E and Bone R: Plasma nicotine levels in pipe smokers. J. Am. Med. Assn. 248:577-578, 1982.
31. Russell MAH, Sutton SR, Feyerabend C, Cole PV and Saloojee Y: Nicotine chewing gum as a substitute for smoking. Br. Med. J. 1:1060-1064, 1977.
32. Benowitz NL, Kuyt F and Jacob P III: Circadian blood nicotine concentrations during cigarette smoking. Clin. Pharmacol. Ther. 32:758-764, 1982.
33. Russell MAH: Tobacco smoking and nicotine dependence. In Research Advances in Alcohol and Drug Problems, Vol. 3, John Wiley. New York, 1976, 1-48.
34. Gritz ER: Smoking behavior and tobacco abuse. In (ed.) Mello, N, Advances in Substance Abuse, Vol. 1, JAI Press, Greenwich, CN, 1980, 91-158.
35. Benowitz NL and Jacob III P: Daily intake of nicotine during cigarette smoking. Clin. Pharmacol. Ther. 35:499-504, 1984.
36. Benowitz NL and Jacob III P: Nicotine renal excretion rate influences nicotine intake during cigarette smoking. J. Pharmacol. Exp. Ther. 234:153-155, 1985.
37. Fix AJ, Daughton, D, Kass I, Smith JL, Wickiser A, Golden CJ and Wass AR: Urinary alkalinization and smoking cessation. J. Clin. Psychol. 39:617-623, 1983.

38. Cherek DR, Lowe WC and Friedman TT: Effects of ammonium chloride on urinary pH and cigarette smoking behavior. Clin. Pharmacol. Ther. 29:762-770, 1981.
39. Comroe JH: The pharmacological actions of nicotine. Ann. N.Y. Acad. Sci. 90:48-51, 1960.
40. Su C: Actions of nicotine and smoking on circulation. Pharmac. Ther. 17:129-141, 1982.
41. Ingenito AJ, Barrett JP and Procita L: Direct central and reflexly mediated effects of nicotine on the peripheral circulation. Eur. J. Pharmacol. 17:375-385, 1972.
42. Porsius AJ and Van Zwieten PA: The central actions of nicotine on blood pressure and heart rate after administration via the left vertebral artery of anaesthetized cats. Arzheim Forsch. 28:1628-1631, 1978.
43. Löffelholz K: Autoinhibition of nicotinic release of noradrenaline from postganglionic sympathetic nerves. Naunyn-Schmiedbergs. Arch. Pharmak. Bd. 267:49-63, 1970.
44. Steinsland OS and Furchgott RF: Desensitization of the adrenergic neurons of the isolated rabbit ear artery to nicotinic agonists. J. Pharmacol. Exp. Ther. 193:138-148, 1975.
45. Cryer PE, Haymond MW, Santigao JV and Shah SD: Norepinephrine and epinephrine release and adrenergic mediation of smoking-associated hemodynamic and metabolic events. N. Engl. J. Med. 295:573-577, 1976.
46. Irving DW and Yamamoto T: Cigarette smoking and cardiac output. Br. Heart J. 25:126-132, 1963.
47. Bargeron LM, Ehmke D, Gonlubol F, Castellanos A, Siegel A and Bing RJ: Effect of cigarette smoking on coronary blood flow and myocardial metabolism. Circulation 15:251-257, 1957.
48. Freund J and Ward C: The acute effect of cigarette smoking on the digital circulation in health and disease. Ann. N.Y. Acad. Sci. 90:85-101, 1960.
49. Eckstein JW and Horsley AW: Responses of the peripheral veins in man to the intravenous administration of nicotine. Ann. N.Y. Acad. Sci. 90:133-137, 1960.
50. Rottenstein H, Peirce G, Russ E, Felder D and Montgomery H: Influence of nicotine on the blood flow of resting skeletal muscle and of the digits in normal subjects. Ann. N.Y. Acad. Sci. 90:102-113, 1960.
51. Spohr U, Hofmann K, Steck W, Harenberg J, Walter E, Hengen N, Augustin J, Morl H, Koch A, Horsch A and Weber E: Evaluation of smoking-induced effects on sympathetic, hemodynamic and metabolic variables with respect to plasma nicotine and COHb levels. Atherosclerosis 33:271-283, 1979.
52. Aronow WS, Dendinger BS and Rokaw SN: Heart rate and carbon monoxide level after smoking high, low and non-nicotine cigarettes. Ann. Intern. Med. 74:697-702, 1971.
53. Nyberg G, Panfilov V, Sivertsson R and Wilhelmsen L: Cardiovascular effects of nicotine chewing gum in healthy nonsmokers. Eur. J. Clin. Pharmacol. 23:303-307, 1982.
54. Benowitz NL, Kuyt F and Jacob III P: Influence of nicotine on cardiovascular and hormonal effects of cigarette smoking. Clin. Pharmacol. Ther. 36:74-81, 1984.

METABOLISM OF NICOTINE IN SMOKERS AND NONSMOKERS*

D.W. Sepkovic and N.J. Haley

American Health Foundation
Division of Nutrition and Endocrinology
Naylor Dana Institute
Valhalla, N.Y. 10595

ABSTRACT

While a number of secondary reinforcers may exert some influence on cigarette smoking behavior, the major factor responsible for tobacco habituation is believed to be nicotine. Two studies on the metabolism of the N-oxidative metabolites have revealed substantial *in vivo* reduction of these compounds to nicotine after their chronic and subchronic administration to rats. Nicotine-N,N'-dioxide was also shown to be back-reduced to the parent alkaloid as evidenced by high levels of nicotine and cotinine in the plasma and urine of treated rats. In another study, the rate of elimination of cotinine in smokers and passively exposed nonsmokers was measured. Ten smokers quit smoking and the elimination of cotinine in plasma and urine was measured for eight days after cessation. Four nonsmokers were exposed to a sidestream smoke-polluted environment and the rate of cotinine disappearance was assessed. Cotinine elimination in the plasma of the passively exposed nonsmokers took at least twice as long as that of chronic cigarette smokers. The rate of elimination from the urine was also significantly slower in nonsmokers when compared with smokers. A new nicotine delivery method, a nicotine aerosol rod, was examined under standard FTC machine smoking conditions. In a pilot study, the uptake of nicotine from this nicotine aerosol rod was also measured in smokers and nonsmokers. The results indicated that per puff deliveries of nicotine were too small to be assessed by radioimmunoassay and that this method of nicotine delivery did not result in smoker satisfaction. More work is necessary if alternate methods of nicotine delivery are to be accepted by the population which continues to smoke.

INTRODUCTION

Exposure to tobacco smoke constituents can occur through the active intake of mainstream smoke, the passive intake of sidestream smoke, or the transfer of tobacco smoke constituents by the maternal bloodstream to the developing fetus (1,2,3). Exposure can be evaluated by physiological or

*Supported by National Cancer Institute grants P01 CA322617-04, P01 CA32261702 and P01 CA29580.

biochemical means as well as by self-reported daily consumption of tobacco products in active smokers (4). However, the actual uptake and body burden of tobacco components can be quantitated only by direct measures of the constituents or their metabolites.

The measurement of nicotine is specific for tobacco but, for *in vivo* measurements, is limited by a short biological half-life. Assessment of its terminal metabolite cotinine has proven more useful in appraising active and passive uptake of tobacco smoke compounds (5,6,7).

In this article, we present a series of investigations on nicotine uptake, absorption, and metabolism, beginning with two studies on the less prominent N-oxidative metabolites of nicotine and then proceeding to several studies on the metabolism of nicotine in active smokers as well as passively exposed adults and neonates. Finally, some data are presented on a nicotine aerosol delivery method that is currently being marketed in the U.S. for use by smokers and others who desire nicotine.

Studies on the N-oxidative Metabolites of Nicotine in Rats

It is well established that the major pathway of nicotine metabolism to cotinine is by α-hydroxylation (8,9). However, substantial quantities of nicotine are metabolized to nicotine-N'-oxides by N-oxidation (10,11). While less than 1/3 of absorbed nicotine is biotransformed by this pathway, it is nonetheless important for two reasons. First, the fact that nicotine-N'-oxides are rapidly back-converted to nicotine both *in vivo* and *in vitro* (12,13) means that these metabolites represent a *nicotine reserve* that is involved in reinforcing nicotine habituation. Secondly, these compounds are easily nitrosated to form the powerful tobacco specific carcinogens N'-nitrosonornicotine (NNN) and 4-(methylnitrosoamino)-1-(3-pyridyl)-1-butanone (NNK). Klemish and Stadler in 1979 demonstrated that the *in vitro* formation of NNN is 3 times higher from nicotine-N'-oxides than from nicotine itself (14).

In a recent study that we conducted on the chronic effects of the metabolites of nicotine, male Fischer rats were divided into one control and 3 experimental groups (15). Each treatment group received either

TABLE 1. Nicotine and Cotinine Concentrations in Rats Administered Cotinine and Nicotine-N-Oxides

	Serum Levels		Urine Levels	
	Nicotine ng/ml	Cotinine ng/ml	Nicotine ng/per mg creatinine	Cotinine μg/per mg creatinine
Cotinine (N = 29)	ND*	23206 ±1300	76. ±5.	>650
Nicotine-N-Oxide (100% trans) (N = 33)	119 ±12	995 ±125	6150 ±880	18.7 ±2.0
Nicotine-N-Oxide (64% trans-36% cis) (N = 32)	123 ±17	605 ±144	3300 ±269	14.3 ±1.9

*ND = Not detected
From Sepkovic, et al. (15)

0.02% pure *trans* nicotine-N'-oxide, 0.02% of a diastereomeric N'-oxide mixture, or 0.1% cotinine in drinking water for 78 weeks. Plasma and urinary nicotine and cotinine levels were measured in the various treatment groups (Table 1). Substantial concentrations of plasma and urinary cotinine were observed in both N'-oxide treatment groups which were checked to be the result of back-conversion of these metabolites to nicotine and subsequent α-hydroxylation of nicotine to cotinine via 5'-hydroxycotinine. Pure *trans*-N'-oxide readily back-converts to nicotine while the diastereomeric mixture is less readily reduced *in vivo*. Body weights of the animals were significantly reduced in both of the N'-oxide treatment groups (Fig. 1). These findings prompted us to conduct a short-term study administering purified preparations of *trans*- or *cis*-nicotine-N'-oxide or nicotine-N,N'-dioxide in rats for three weeks and then measuring the back-conversion of these metabolites to nicotine (16). The metabolites were given in drinking water in the same concentrations as in the previous chronic study. Blood samples were obtained weekly for three weeks. After seven days of metabolite administration, nicotine levels in the plasma of the *trans*-nicotine-N-'-oxide group rose to approximately twice the concentrations observed in the other groups (Fig. 2). After two weeks, plasma nicotine levels declined for both nicotine-N'-oxide diasteromers and were approximately equal. Nicotine concentrations in these groups remained similar and continued to decline in week three. Plasma nicotine concentrations above 200 ng/ml were observed during week one of the study in nicotine-N,N'-dioxide-treated animals. Mean plasma nicotine concentrations decreased to approximately 100 ng/ml by week two and continued to decrease during week three. Plasma cotinine concentrations were similar for both *cis*-and *trans*-nicotine-N'-oxides during week one (Fig. 3). However, cotinine levels were increased in the *trans*-N'-oxide group relative to *cis*-N'-oxide-treated rats in week two. At week three, both groups began to exhibit decreases in plasma cotinine concentrations. Plasma cotinine concentrations in nicotine dioxide-treated rats rose less dramatically when compared with the N'-oxide groups, but significant amounts of cotinine were present in the serum of these animals. The results of this short-term study confirm that both isomers of nicotine-N'-oxide are readily back-converted *in vivo* from the nicotine-N'-oxide to its nicotine congener. Nicotine-N,N'-dioxide was also shown to be reduced to nicotine. This seldom-studied compound comprises approximately 5% of the N-oxidation products of nicotine metabolism and its back-conversion has not been previously reported. These findings and those of other workers suggest the involvement of the liver and/or other tissues in the reduction of N-oxidative metabolic products of nicotine (12,13). Increases of cotinine in the urine of all three treatment groups reflect decreases in the blood as well as concomitant induction of α-hydroxylation hepatic microsomal enzymes.

These findings have relevance to smoking and health issues since nicotine has the chemical potential to act as a precursor for the formation of the tobacco specific nitrosamines (17,18). This is especially important since age-related shifts in nicotine metabolism from cotinine to nicotine-N'-oxides have been reported (19). The increased availability of nicotine from the *in vivo* reduction of nicotine-N'-oxides and from nicotine-N,N'-dioxides provides the substrate for increased nitrosation. Also, the capacity of the nicotine-N'-oxides to be directly nitrosated to form NNN (14) emphasizes the possibility that individuals who undergo shifts in nicotine metabolism, either caused by aging or by metabolic alteration, might be at increased risk for tobacco-related diseases.

Cotinine Elimination in Smokers and Nonsmokers

A number of epidemiological studies have indicated an association of environmental smoke exposure with tobacco-related diseases (20-23).

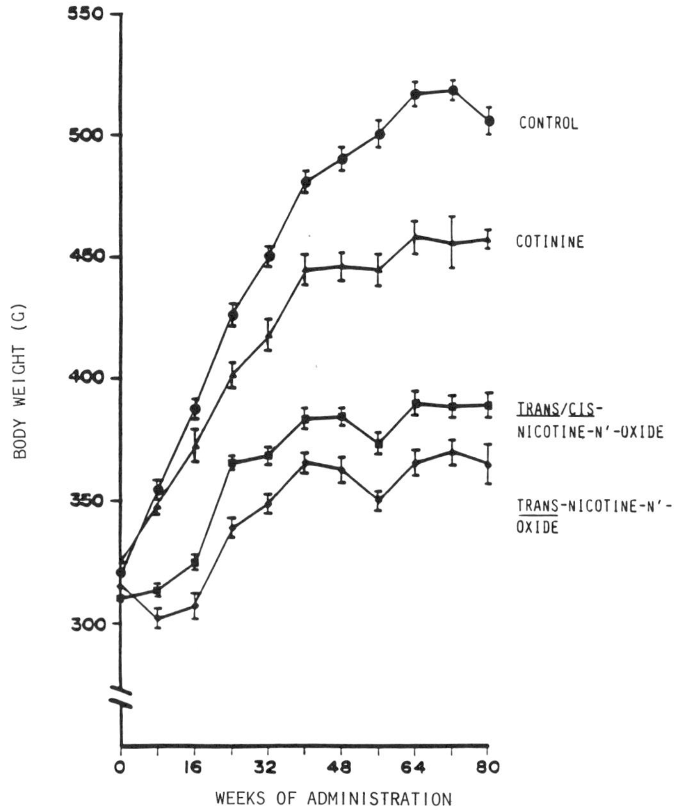

Fig. 1. Body weights from rats administered nicotine metabolites for 78 weeks. Each point equals the mean ± SEM. From Sepkovic, et al. (15).

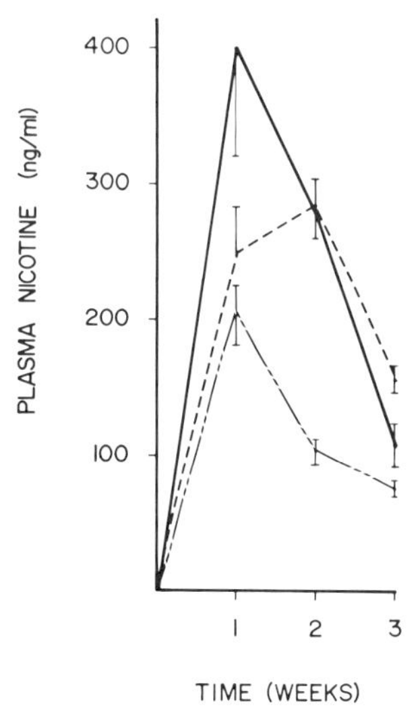

Fig. 2. Plasma nicotine concentrations in rats receiving trans-nicotine-N'-oxide (——), cis-nicotine-N'-oxide (----) or nicotine-N,N'-oxide (—-—-) for 3 weeks. Each point equals the mean ± the SEM of 20 animals. From Sepkovic, et al. (16).

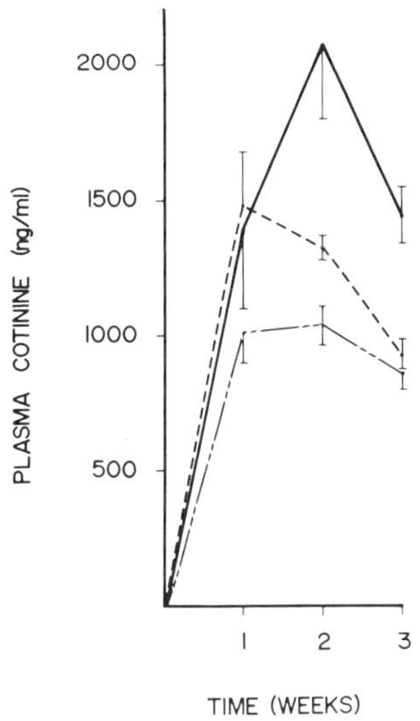

Fig. 3. Plasma cotinine concentrations in rats receiving trans-nicotine-N'-oxide (——), cis-nicotine-N'-oxide (----) or nicotine-N, N'-dioxide (—-—) for 3 weeks. Each point equals the mean ± the SEM of 20 animals. From Sepkovic, et al. (16).

Environmental tobacco smoke pollutants are primarily those products of tobacco combustion that emit as sidestream smoke from the smoldering product in between puff-drawing (Table 2). The composition of tobacco sidestream smoke differs significantly from that of mainstream smoke in that, before dilution with air, the former is actually enriched in some toxic gas phase components including carbon monoxide, formaldehyde, nitrogen oxides, and acetone (Table 3). N-nitroso compounds, benzo(a)pyrene, and heavy metals are also more abundant in sidestream than in mainstream smoke before dilution (24) (Table 4). Elevated concentrations of such harmful chemical compounds in heavily polluted environments may constitute a health risk to passively exposed individuals. In comparison with tobacco smokers, nonsmokers thus exposed may actually be subjected to a disproportionate increase in body burden of toxic substances since the hepatic microsomal enzyme systems that detoxify tobacco compounds in smokers may not be fully induced. By measuring cotinine in plasma and in urine, we have estimated the rate of elimination

TABLE 2. Comparisons of Mainstream and Sidestream Smoke of Cigarettes

Parameters	MS	SS/MS
Peak Temperature During Formation (°C)	900	600
Particle Sizes (μm)	0.1 - 1.0	0.01 - 0.1
Median Diameter	0.4	
Smoke Dilution (Vol. %) (10 mm from burning cone)		
Carbon Monoxide	3 - 5	1
Carbon Dioxide	8 - 11	2
Oxygen	12 - 16	16 - 20
Hydrogen	15 - 3	0.5

From Hoffmann, et al. (24)

TABLE 3. Distribution of Compounds in the Gas Phase in Cigarette Mainstream Smoke (MS) and Sidestream Smoke (SS) - Nonfilter Cigarettes

Gas Phase	MS	SS/MS
Carbon Monoxide	10 - 23 mg	2.5 - 4.7
Carbon Dioxide	20 - 60 mg	8 - 11
Formaldehyde	70 - 100 μg	0.1 - -50(?)
Acrolein	60 - 100 μg	8 - 15
Acetone	100 - 250 μg	2 - 5
Pyridine	20 - 40 μg	10 - 20
3-Vinylpyridine	15 - 30 μg	20 - 40
Hydrogen Cyanide	400 - 500 μg	0.1 - 0.25
Nitrogen Oxides	100 - 600 μg	4 - 10
Ammonia	50 - 130 μg	40 - 130
N-Nitrosodimethylamine	10 - 40 μg	20 - 100
N-Nitrosopyrrolidine	6 - 30 μg	6 - 30

From Hoffmann, et al. (24)

of this end product of nicotine metabolism in both smokers and in passively exposed nonsmokers (25). Ten volunteer smokers selected for the study smoked their customary cigarettes ad libitum for 5 days and reported to the clinic each day at 9:30 a.m. for blood and saliva sampling. At 12:00 midnight, on day 5, the subjects quit smoking. Sampling on the following two days provided two blood samples and four saliva samples per day. Blood and saliva samples were taken once daily from day 8 through day 12 when the study was terminated. Total urine voids were collected daily throughout the study period.

TABLE 4. Distribution of Compounds in the Particulate Phase in Cigarette Mainstream Smoke (MS) and Sidestream Smoke (SS) Nonfilter Cigarettes

Particulate Phase	MS	SS/MS
Particulate Matter	15 - 40 mg	1.3 - 1.9
Nicotine	0 - 2.3 mg	2.6 - 3.3
Phenol	60 - 120 μg	2.0 - 3.0
Catechol	100 - 280 μg	0.6 - 0.9
Aniline	360 ng	30
2-Toluidine	160 ng	19
2-Naphthylamine	1.7 ng	30
Benz(a)anthracene	20 - 700 ng	2 - 4
Benzo(a)pyrene	20 - 40 ng	2.5 - 3.5
N'-Nitrosonornicotine	200 - 3,000 ng	0.5 - 3
NNK	100 - 1,000 ng	1 - 4
N-Nitrosodiethanolamine	20 - 70 ng	1.2
Nickel	20 - 80 ng	13 - 30
Polonium-210	0.03 - 0.5 pCi	

From Hoffmann, et al. (24)

In this same study, four nonsmokers, 25-35 years of age, were exposed to the sidestream smoke-polluted environment generated by four Kentucky 1R1 Reference cigarettes twice daily for 80 minutes each time in a test laboratory. This exposure schedule was maintained for four consecutive days. The laboratory consisted of a bare room of 16 m^3 with 6 air changes per hour (26). An indwelling catheter was inserted before exposure permitting continuous blood sampling. Saliva and urine samples were also taken from each subject after each exposure period at a different location which was free of smoke pollutants. Monitoring continued for 7 days. Cotinine elimination in the plasma of the passively exposed nonsmokers took at least twice as long as that of the chronic cigarette smokers (Fig. 4). The rate of cotinine disappearance from the urine had also significantly slowed down in the nonsmokers compared with smokers (Fig. 5).

These findings have significance in several ways. First, the slower clearance in nonsmokers of a terminal nicotine metabolite commonly used to measure exposure to tobacco smoke could result in a misinterpretation of "cigarette equivalents" which some researchers use to calculate passive exposure (27). The prolonged elimination of cotinine shown by passive smokers precludes an extrapolation to "cigarette equivalents of smoke uptake" from a single measurement of urinary cotinine.

Secondly, the prolonged elimination and likely slower metabolism of nicotine in nonsmokers would suggest that passively exposed individuals also carry a body burden of other toxic tobacco components for longer periods than do long-term cigarette smokers. The extended residence time of nicotine (and possibly of other tobacco alkaloids) increases also the probability of the endogenous formation of carcinogenic tobacco-specific N-nitrosamines (28). Thus, this phenomenon of prolonged elimination should not be disregarded in the analysis of relative risk for tobacco-related cancers.

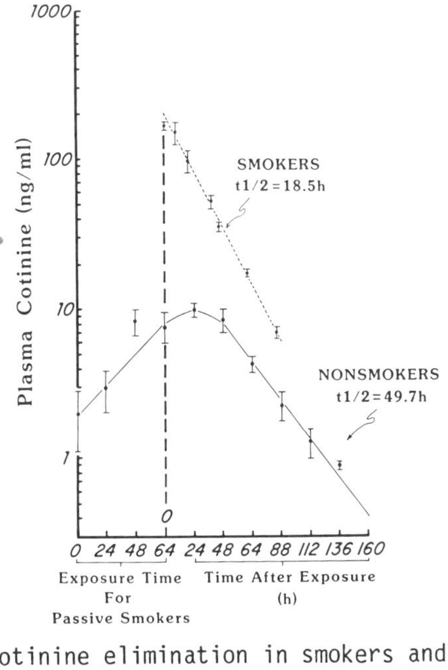

Fig. 4. Plasma cotinine elimination in smokers and passively exposed nonsmokers. The computer program used for pharmacokinetic analysis was provided by Johnston and Wollard (33). From Sepkovic et al. (25).

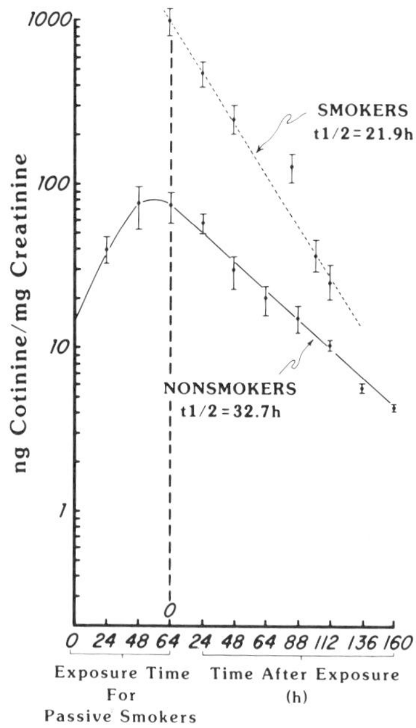

Fig. 5. Urinary cotinine elimination in smokers and in passively exposed nonsmokers. Urinary cotinine concentrations are normalized by creatinine. The computer program for pharmacokinetic analysis was provided by Johnston and Wollard (33). From Sepkovic et al. (25).

Cotinine Concentrations in Passively Exposed Neonates

An important field study was conducted to measure passive exposure to environmental tobacco smoke in infants from mothers who smoke. The study was conducted with the collaboration of Drs. Robert Greenberg, Ruth Etzel,and Frank Loda, from the University of North Carolina at Chapel Hill (29). Urine and saliva was collected from infants one to three months of age who were not breast fed. Based on self-report of the primary caretaker, infants were classified as exposed or nonexposed. Cotinine measures clearly differentiated these groups (Fig. 6). There was a dose-response relationship between the daily number of cigarettes smoked by the mother and infant excretion of cotinine (Fig. 7). While the levels of cotinine observed in adult and child passive smokers are very low, the study indicates that infants, including neonates who are cared for in households that include smokers, absorb tobacco smoke constituents and excrete nicotine in addition to metabolizing nicotine to form cotinine. These results conform to those of other workers (30).

Upon determining that neonates were able to take up and metabolize nicotine from an environment that was tobacco smoke polluted, it then became important to explore whether the fetus of a smoking mother is exposed to nicotine. In collaboration with the Chapel Hill group we selected 11 neonates of smoking mothers and 12 neonates of nonsmoking mothers who were delivered at North Carolina Memorial Hospital (3). To be eligible for the study, infants had to be isolated from exposure to all sources of tobacco smoke products after birth, including breast feeding. On the day after delivery, all the mothers of the eligible babies were asked whether they had smoked in the 24 hours just before delivery. Urine was collected from each group and analyzed for cotinine (Fig. 8). Significant concentrations of urinary cotinine were observed in the

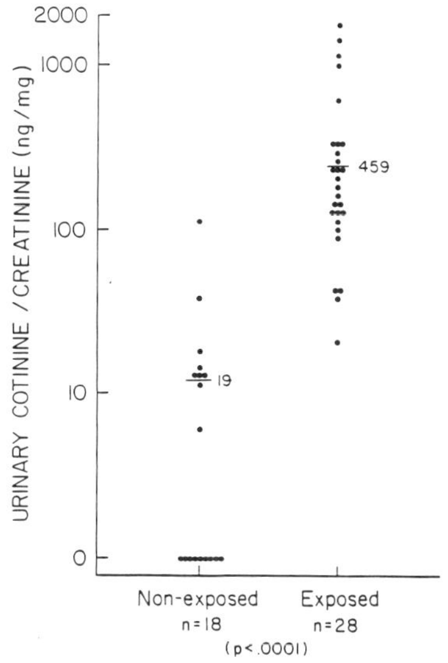

Fig. 6. Urinary cotinine concentrations in infants exposed and not exposed to tobacco smoke. Cotinine concentrations were normalized over urinary creatinine. From Greenberg, et al. (29).

neonates of smoking mothers when compared with the neonates of nonsmoking mothers. Interestingly, the half-elimination time of cotinine in newborns was discovered to be two to three times longer than in adults (68 hours).

Commercial Aerosol Rods as Nicotine Supplements

The acquisition of nicotine represents the primary motivation for cigarette smoking. During the act of smoking, a wide variety of toxic substances are inhaled along with nicotine. These vapor phase and particulate phase compounds have been implicated in a number of pulmonary disease states. The Surgeon General reports that cigarette smoking is the major cause of chronic obstructive lung disease in the United States for both men and women. The mortality ratios for chronic obstructive lung

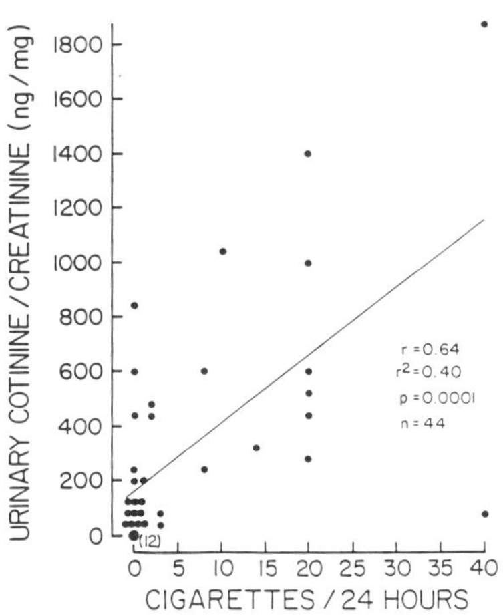

Fig. 7. Maternal cigarette smoking versus infant urinary cotinine excretion from Greenberg, et al. (29).

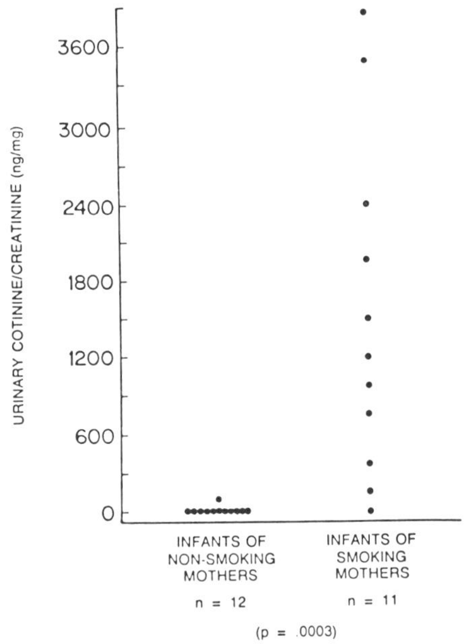

Fig. 8. Urinary concentrations of cotinine during the first day of life in neonates of smoking and nonsmoking mothers. From Etzel, et al. (3).

disease in smokers are as large as or larger than for lung cancer, the disease most people associate with cigarette smoking (31).

The elimination of such toxic tobacco smoke compounds would significantly decrease the health risk associated with cigarette smoking. The possibility exists that nicotine supplementation by inhalation of nicotine as an aerosol could alter smoking behavior patterns and reduce the uptake of more hazardous tobacco smoke compounds. For this reason, we

TABLE 5. Nicotine Yield of Nicotine Aerosol Rod Under Standard Machine Smoked Conditions*

Puffs	Average Nicotine/Puff (μg)	Total Nicotine (μg)
1 - 10	0.3	3
11 - 20	1.8	18
21 - 30	3.5	35
31 - 40	4.4	44
41 - 50	5.8	58
51 - 60	6.4	64
	Total	222

*Smoked on a Hamburg Heiner Borgwaldt single port piston smoking machine (1 puff/min, 35 ml volume of 2 sec duration). Nicotine was trapped on a Cambridge filter pad and determined by gas chromatography. From Sepkovic et al. (32).

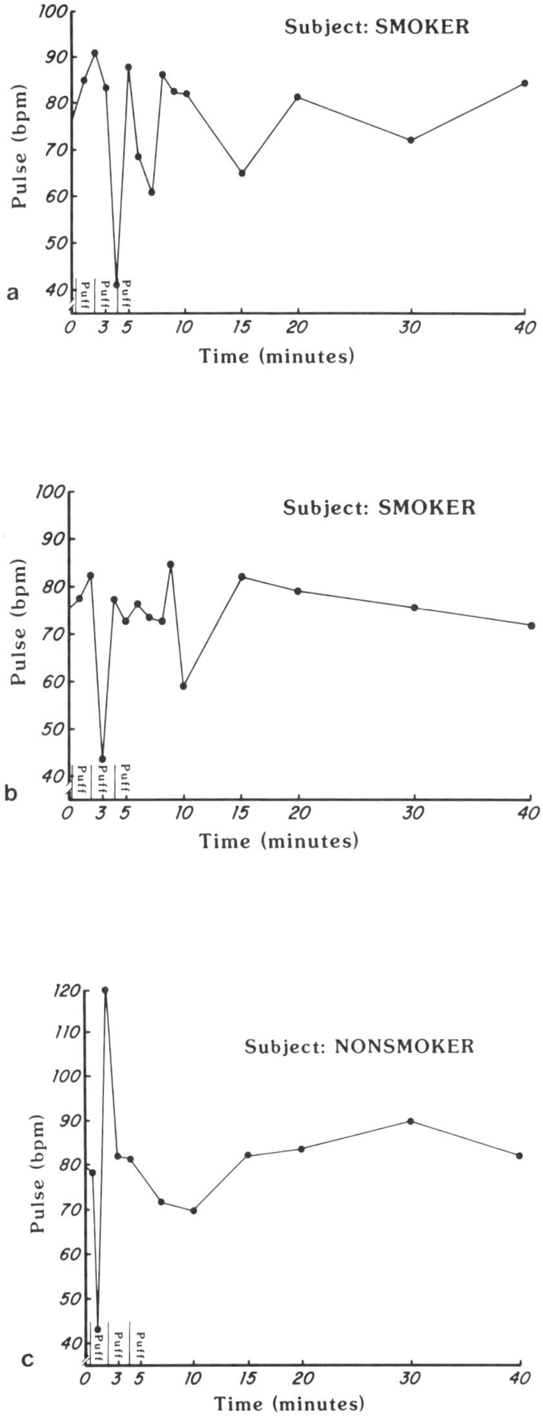

Fig. 9a. Pulse rate in a moderate cigarette smoker (plasma cotinine 223 ng/ml) upon using a nicotine aerosol rod.

b. Pulse rate of a light cigarette smoker (plasma cotinine 116 ng/ml) after using a nicotine aerosol rod.

c. Pulse rate of a nonsmoker after using a nicotine aerosol rod.

became interested in the commercial nicotine aerosol rod as a substitute source of nicotine for cigarette smokers.

The nicotine aerosol rod is currently being test-marketed in Texas. The per puff delivery approximates nicotine concentrations delivered by conventional cigarettes according to the manufacturer.

The aerosol rods were first examined under standard machine smoking conditions with the following results (Table 5) (32). The average nicotine per puff increased with the number of puffs taken from 0.3 μg/puff at 10 puffs to 6.4 μg/puff after 50 puffs.

In order to measure the uptake of nicotine in users of the aerosol rod, we designed the following study. Four smokers and three nonsmokers were asked to abstain for at least 12 hours prior to reporting to each session. After they were comfortably seated, an indwelling catheter was inserted into the antecubital vein of the right forearm and an automatic blood pressure cuff was placed on the left arm. Baseline measurements of blood pressure and pulse were also taken and physiological measurements were continued throughout each session. Puffs of the aerosols were inhaled every 2 minutes for 6 minutes.

In two of the four smokers, an immediate drop in pulse was noted after the second puff (Figs. 9a and b). In one nonsmoker, heart rate decreased after the first puff and then immediately showed a dramatic increase (Fig. 9c). In the other subjects, no change in heart rate was observed. The plasma obtained from smoking and nonsmoking users was analyzed for nicotine and cotinine with the following results (Table 6). No nicotine was observed in the plasma or urine of smokers or nonsmokers after using the aerosol rod.

The amount of nicotine delivered per puff was the equivalent of a hypothetical 0.01-mg nicotine content cigarette, which would provide per puff deliveries too small for nicotine absorbed to be assessed in the plasma. Only after a total of 60 puffs would this rod approximate the nicotine delivery of a 0.2-mg nicotine content cigarette. Furthermore, the rate of absorption of nicotine in aerosol form may not be rate-equivalent to tobacco smoke absorption.

Product satisfaction was also assessed in this pilot study. All subjects commented on a burning sensation following each puff and the nonsmokers complained of sore throats. The physiological effects elicited by the nicotine aerosol rod are somewhat more serious in nature. The heart rate depression in two of our four smokers and the depression followed by the compensatory increase in heart rate in one of three nonsmokers suggests that a more in-depth study with a larger volunteer population is necessary to identify the causative agent responsible for the fluctuations in heart rate.

TABLE 6. Nicotine and Cotinine in the Plasma of Smoking and Nonsmoking (Nicotine Aerosol Rod) Users

	Plasma Nicotine	Plasma Cotinine (ng/ml)	Urinary Cotinine
Smokers	ND	197 ± 174 S.D.	---
Nonsmokers	ND	ND	ND

ND = Not detected

These investigations into the metabolism of nicotine and the disposition of its metabolites were conducted with the overall goal of more closely identifying the health risk associated with cigarette smoking. Since the acquisition of nicotine is a dependence process and represents the primary motivation for cigarette smoking, any attempt to reduce the hazardous ancillary compounds absorbed along with nicotine would be beneficial to all habituated smokers. The ultimate goal of tobacco-related research should be the elimination of smoking behavior in existing smokers and the prevention of inception of the habit by nonsmokers. However, the health of 52 million people in the United States who continue to smoke cannot be ignored.

REFERENCES

1. Hopkins, R, Wood, LW and Sinclair, NM: Evaluation of methods to estimate cigareete smoke uptake. Clin. Pharmacol. Ther. 36:788-794, 1984.
2. Hoffmann, D, Haley, NJ, Adams, JD and Brunnemann, KD: Tobacco sidestream smoke: Uptake by nonsmokers. Prev. Med. 13:608-617, 1984.
3. Etzel, RA, Greenberg, RA, Haley, NJ and Loda, FA. Urinary cotinine excretion in neonates exposed to tobacco smoke products in utero. J. Pediatr. 107:146-148, 1985.
4. Ashton, H, Stepney, T, Teleford, R and Thompson, JW: Cardiovascular and behavioral responses to smoking. In (ed.) Greenhalgh, RM, Smoking and Arterial Disease, Pitman Medical Press, London, 1981, p. 258.
5. Isaac, PF and Rand, MJ: Cigarette smoking and plasma levels of nicotine. Nature 236:308, 1972.
6. Hill, P and Marquardt, H: Plasma and urine changes after smoking different brands of cigarettes. Clin. Pharmacol. 27:652-655, 1980.
7. Sepkovic, DW, Haley, NJ: Biomedical applications of cotinine concentrations in biological fluids. Am. J. Public Health 75:663-664, 1985.
8. McKennis, H, Turnbull, LB, Bowman, ER: L-(3-pyridyl) 7-methyl-amino-butyric acid as a urinary metabolite of nicotine. J. Am. Chem. Soc. 79:6342, 1957.
9. Gorrod, JW and Jenner, P: The metabolism of tobacco alkaloids. Essays Toxicol. 6:35-78, 1975.
10. Gorrod, JW, Jenner, P, Keysell, GR and Mikhael, BR: Oxidative metabolism of nicotine by cigarette smokers with cancer of the urinary bladder. J. Natl. Cancer Inst. 52:1421-1424, 1974.
11. Booth, J and Boyland, E: The metabolism of nicotine into two optically-active stereoisomers of nicotine-1'-oxide by animal tissues in vitro and by cigarette smokers. Biochem. Pharmacol. 19:733-742, 1970.
12. Booth, J and Boyland, E: Enzymatic oxidation of (-) nicotine by guinea pig tissue in vitro. Biochem. Pharmacol. 20:407-412, 1971.
13. Dajani, RM, Gorrod, JW and Beckett, AH: Hepatic and extrahepatic reduction of nicotine-1'-N-oxide in rats. Biochem. J. 130:88, 1972.
14. Klemish, HJ and Stadler, L: Untersuchungen Zur Bildung Von N'-nitroso-nornikotin auc nikotin-N'-oxid. Talanta 23:614-616, 1979.
15. Sepkovic, DW, Haley, NJ, Axelrad, CM and LaVoie, EJ: Thyroid hormone concentrations in rats after chronic nicotine metabolite administration. Proc. Soc. Exp. Biol. Med. 177:412-416, 1984.
16. Sepkovic, DW, Haley, NJ, Axelrad, CM, Shigematsu, A and LaVoie, EJ: Short-term studies on the in vivo metabolism of N'-oxides of nicotine in rats. J. Toxicol. Environ. Health 18:205-214, 1986.
17. Hoffmann, D and Adams, JD: Carcinogenic tobacco-specific N-nitrosamines in snuff and in the saliva of snuff dippers. Cancer Res. 41:4305-4308, 1981.

18. Hoffmann, D, LaVoie, EJ and Hecht, SS: Nicotine: A precursor for carcinogens. Cancer Lett. 26:67-75, 1985.
19. Klein, AE and Gorrod, JW: Age as a factor in the metabolism of nicotine. Eur. J. Drug Metab. Pharmacokinet. 1:51-58, 1978.
20. Correa, P, Pickle, LW, Fontham, L, Lin T and Haenszel W: Passive smoking and lung cancer. Lancet 2:595-597, 1983.
21. Hirayama, T: Passive smoking and lung cancer: Consistency of association. Lancet 2:1425-1426, 1983.
22. Garfinkel, L, Auerbach, O and Joubert, L: Involuntary smoking and lung cancer: a case control study. J. Natl. Cancer Inst. 75:463-469, 1985.
23. Preston-Martin S, Yu MC, Benton B, Henderson, BE: N-Nitroso compounds and childhood brain tumors: a case control study, Cancer Res. 42:5240-5245, 1982.
24. Hoffmann, D, Haley, NJ, Brunnemann, KD, Adams, JD and Wynder, EL: Cigarette Sidestream Smoke: Formation analysis and model studies on the uptake by nonsmokers. U.S., Japan Meeting "New Etiology of Lung Cancer," Honolulu, Hawaii, March 21-23, 1983.
25. Sepkovic DW, Haley, NJ, Hoffmann D. Elimination from the body of tobacco products by smokers and passive smokers. J. Amer. Med. Assoc. 256:863, 1986.
26. Hoffmann, D, Haley, NJ, Adams, JD and Brunnemann, KD: Tobacco sidestream smoke: uptake by nonsmokers. Prev. Med. 13:608-617, 1984.
27. Repace, JL and Lowrey, AH: A quantitative estimate of nonsmokers lung cancer risk from passive smoking. Environ. Int. 11:3-22, 1985.
28. Hoffmann, D and Hecht, SS: Nicotine-derived N-nitrosamines and tobacco-related cancer: current status and future directions. Cancer Res. 45:935-944, 1985.
29. Greenberg, RA, Haley, NJ, Etzel, RA and Loda, FA: Nicotine and cotinine in urine and saliva. New Engl. J. Med. 310:1075, 1984.
30. Luck, W and Heinz, N: Nicotine and cotinine concentrations in serum and urine of infants exposed via passive smoking or milk from smoking mothers. J. Pediat. 107:816-820, 1985.
31. U.S. Department of Health and Human Services, 1984. The Health Consequences of Smoking: Chronic Obstructive Lung Disease. DHHS publication No. (PHS) 84-50205.
32. Sepkovic DW, Colosimo, SG, Axelrad, CM, Adams, JD, and Haley, NJ. The delivery and uptake of nicotine from an aerosol rod. Amer. J. Publ. Health 76:1343, 1986.
33. Johnston, A and Woollard, RC: STRIPE: an interactive computer program for the analysis of drug pharmacokinetics. J. Pharmacol. Methods 9:193-199, 1983.

DOES SMOKING DECREASE EATING AND EATING INCREASE SMOKING?

Murray E. Jarvik

Department of Pharmacology
and
Department of Psychiatry
The Neuropsychiatric Institute and Hospital
School of Medicine
University of California, Los Angeles
Los Angeles, California 90024
and
Veterans Administration Medical Center West Los Angeles
Brentwood Division
Los Angeles, California 90073

In the 1920s there used to be a popular cigarette ad that read "Reach for a Lucky instead of a sweet." The implication was that smoking would keep weight down, reduce food intake, and substitute for a dessert. The purpose of this paper is to examine those implications and to explore the mechanisms underlying the slogan.

Most smokers would agree that smoking helps them stay slim by curbing their appetite. They also find that eating a meal stimulates the desire for a cigarette. Research on the weight-reducing property of smoking has been reviewed (1,2) but is not yet fully explained. Even more mysterious is the commonly known fact that eating a meal tends to stimulate smoking. Very little research has been done on the enhanced desirability of the postprandial cigarette (3,4).

There are two possible explanations of weight loss that could account for the nicotine effect. First, energy utilization or metabolism might be increased by smoking. Second, food intake (and especially the intake of certain types of food) may be curbed by smoking. Grunberg (5) has evidence that sweet foods are particularly vulnerable to the anorectic action of smoking and nicotine, both in rats and in man (See Fig. 1).

Before trying to answer the question in the title we will define some of the terms in the paper. Cigarette smoking is an acquired behavior involving the burning of and the inhaling of smoke from a paper-wrapped cylinder of specially prepared tobacco. The seeking of a substance, nicotine, plays a dominant role in the behavior. Eating is another substance-seeking behavior which, however, is largely innate. The substance in this case is food, which represents a complex configuration of substances, only some of which are nutritious. Both smoking and eating involve (a) the active search for chemicals, (b) rituals associated with their procurement and use, (c) ingestion into the mouth, and (d) sensations, perceptions, and pleasure associated with the behavior.

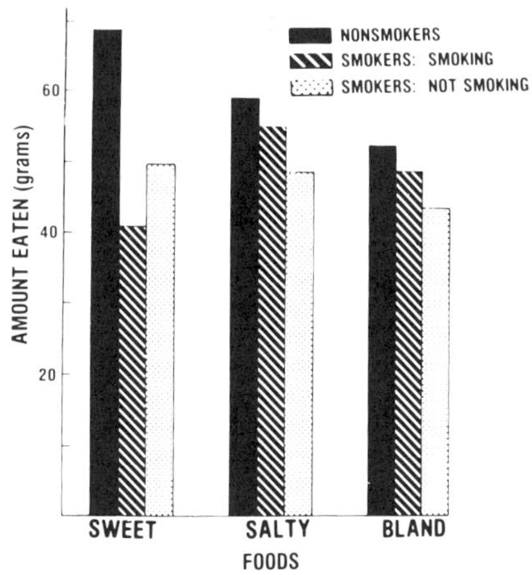

Fig. 1. Average amount of food eaten within each taste class. This indicates that smokers not smoking ate more sweets than smokers smoking. Nonsmokers consumed significantly more sweets than smokers. This accounted for most of the appetite suppressant effect of smoking.

(Source: Grunberg NE. The effects of nicotine and cigarette smoking on food consumption and taste preference. Addictive Behaviors, 7:317-331, 1982).

Appetite usually describes a perception antecedent to food intake. It indicates a craving or hunger that can be satisfied by eating or satiety. Appetite, in a broader sense, need not involve food and may also be applied to anticipation of any form of reinforcement including drugs such as nicotine for which smokers evince a craving. "Taste" for cigarettes is sometimes substituted for craving for or satisfaction with nicotine.

Eating is one of the most intensively studied forms of reinforcement (6,7,8), and its relationship to smoking promises to yield valuable information about both habits. A great deal is known about anatomical and chemical determinants of eating or feeding (9,10).

Since the preservation of the individual and the species are so important, it is not surprising that eating and sex provide most of the sources of primary reinforcement for animals and humans. Other motives (7) are subservient to these. Life for most subhuman species consists of three drives: eating, not being eaten, and reproduction. Play, curiosity, and species specific behaviors are really ancillary to the preservation of the species and its immortal DNA. Drugs may be considered surrogate incentives for the enjoyment of food, safety, and sex. Even stomachics, stimulants, and aphrodisiacs are used to help satisfy these three drives.

Drugs self-administered for pleasure ("recreation") may be considered an exotic form of food in which reinforcement mechanisms developed for eating have been usurped. In fact, the three most commonly used drugs - alcohol, caffeine, and nicotine - are sold in grocery stores. Although nicotine is the focus of this paper, I thought it important to stress the teleological context in which its use was developed.

A good deal of the structure and function of the brain and the body has been devoted to the pursuit of food. The olfactory and gustatory systems and the teeth are almost exclusively devoted to the pursuit, screening, and processing of food. The amount of space taken up by the gastrointestinal system including the liver and pancreas is substantial.

In the evolution of the nervous system, pathways and centers associated with the primitive function of eating have always played a prominent role. Central nervous system structures in the brain of the fish or frog or alligator that control feeding, including vagal, hypothalamic, and limbic nucleii, still exist in the mammalian brain almost unchanged. Feeding systems in the rat brain are illustrated in Fig. 2. The development of the forebrain accompanied development of structures surrounding the gastrointestinal tract such as arms and legs, sensory organs and speech apparatus. This can be seen both in phylogeny (evolution) and ontogeny (embryology). These enable us to enjoy a dinner party as an intellectual experience rather than a pure experience in gluttony.

Rudyard Kipling (11) wrote, "A woman is only a woman but a good cigar is a smoke," indicating the relative importance he attached to both (The Betrothed, Stanza 25). Why do smokers enjoy a good cigar after dinner, or much more commonly nowadays, a cigarette? The postprandial cigarette is usually considered the favorite cigarette of the day and the one which is hardest to relinquish. There are several possible reasons for its exceptional reinforcing properties. (1) First, eating may cause a lowering of blood nicotine levels due to increased hepatic metabolism or sequestering in the stomach or in postprandial lipids which accumulate in the blood. (2) Second, meals often produce sedation which nicotine stimulation can overcome. (3) Third, nicotine may counteract the parasympathetic dominance produced by eating. (4) Fourth, satiety

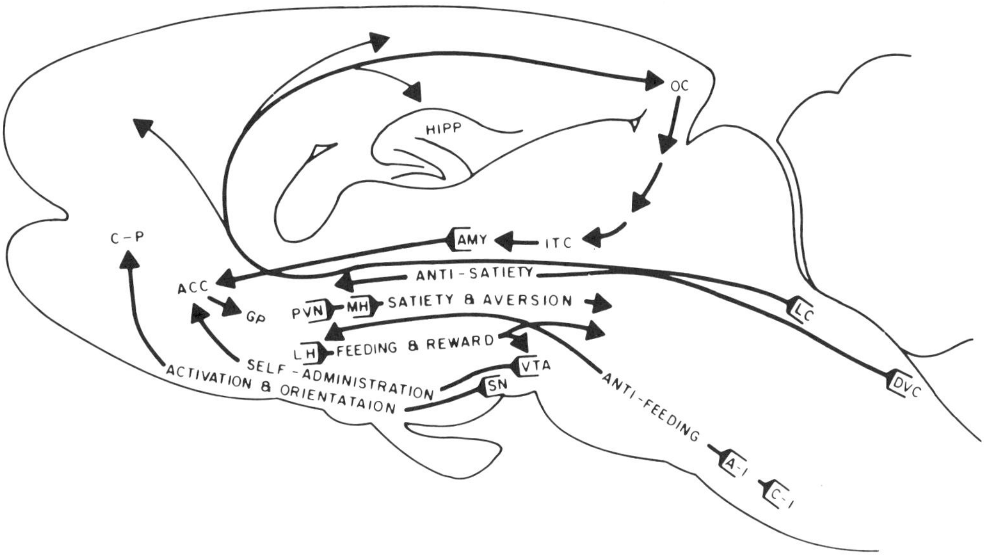

Fig. 2. Diagrammatic representation of the feeding system in the rat brain. The major regions, pathways and functions associated with feeding are shown. Some of the important centers in the hypothalamus are paraventricular (PVN), medial (MH) and lateral (LH) hypothalamic nuclei. Analogous regions associated with smoking have not yet been fully defined, although London et al. (1985) have begun to identify such areas. (Source: For a further explanation of abbreviations see Hoebel BG. Neurotransmitters in the control of feeding and its rewards. In: AJ Stunkard and E Stellar (Eds.) Eating and Its Disorders. New York: Raven Press, pp. 15-38, 1984).

hormones such as cholecystokinin released from the gastrointestinal tract and brain by food may induce a feeling of fullness which is potentiated or modulated by nicotine. (5) Fifth, eating initiates a positive feedback reinforcement system which encourages smoking and other rewarding activities (the "Cracker Jack" effect).

Recently we have instituted studies on the remarkable craving and satisfaction for cigarette smoking induced by food, the postprandial effect. It may be stronger than other motives for smoking that we have examined including stress, alcohol, and caffeine (12,13,14,15).

We conducted an experiment giving different types of meals at dinner time (6 PM) to determine their effect on craving for and enjoyment of a cigarette. In a within-subjects design, on three separate occasions 12 subjects were given either: 1) a meal of their choice (steak, chicken, fish, or vegetarian); 2) a liquid meal of equal caloric value (Sustacal); or 3) water. The results indicate that the meal of their choice had the greatest statistically significant effect, the liquid meal the next greatest, and the water the least effect in increasing craving for and satisfaction from the postprandial cigarette. We are currently exploring reasons for these effects (Fig. 3).

Our first hypothesis presumes that eating may lower nicotine levels in the blood and brain. Nicotine levels have been thought to control

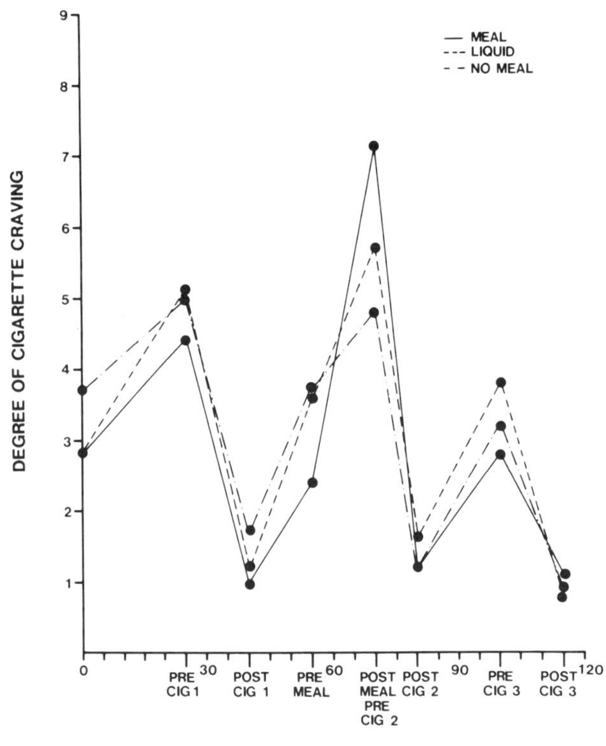

Fig. 3. Effects of two types of 6 PM meal upon cigarette craving. The self-selected meal gave rise to significantly greater craving that the liquid (Sustacal) meal. Water raised the level of craving only to the degree expected from nicotine deprivation. (Source: Jarvik ME. Unpublished data).

smoking, and avoidance of withdrawal is a powerful drug-seeking incentive. However, compensation by smoking for changes in blood nicotine levels (titration) is far from perfect (16,17,18,19). On the other hand, peripheral actions of cigarettes and smoke possibly not related to nicotine have been demonstrated to influence smoking (13). Nevertheless it is possible that eating, as well as stress, may reduce nicotine levels. Schachter et al. (20) and Schachter (21) have indicated that stress may reduce nicotine levels by enhancing renal excretion through lowering pH. On the other hand, Benowitz (16) has shown that blood nicotine levels are controlled only to a minor extent by changes in blood pH. Similarly, eating produces an alkaline tide because of gastric acid production. Although this may lower renal nicotine excretion, it may increase gastric sequestration of nicotine. In fact, Schmiterlow et al. (22) demonstrated that nicotine was concentrated in the stomach as well as the brain. Thus far, no one has demonstrated whether eating causes a change in nicotine concentration in the blood or brain. Another possibility is that eating may stimulate blood flow to the liver, thereby increasing the possibility of metabolism of nicotine. If so, the levels of cotinine should rise and a direct test of this hypothesis is possible. A third less likely possibility is that lipemia following a meal may sequester nicotine in the lipid compartments in the blood and prevent its passage into the brain. We plan to compare postprandial lipemia induced by high and low fat meals with postprandial cigarette craving.

Second, there is a strong possibility that eating may produce somnolence and smokers may take advantage of the stimulating properties of nicotine to increase alertness. This is obviously situationally determined, since desired levels of arousal depend upon the activities in which the smoker wishes to be involved. For example, after lunch it is more important to be aroused for an American academic examination than for a Spanish siesta.

The waking brain, so aptly named by Magoun (23), represents the optimal state of arousal seen during the course of a day. Although arousal is a term fraught with vagueness, it describes a vital function of the brain. Arousing centers in the brain were first demonstrated by Moruzzi and Magoun (24) in the reticular activating center, and the behavioral implications were investigated by Lindsley (25). Routtenberg (26) proposed a second limbic activating center and since then other centers have been localized. For example, the apparent pacemaker of the brain for diurnal or circadian rhythms seems to be in the suprachiasmatic nucleus of the hypothalamus.

Hunger and the search for food as well as its ingestion depend upon alertness. As a general rule, one may say that arousal precedes meals but drowsiness follows meals. Discrete meals depend on the existence of intermittent food supplies and are characterized by refractory periods (satiety) following feeding (27). Such behavior exists in very primitive creatures (e.g., Venus fly trap, hydra), whereas continuous feeders such as clams and barnacles may not utilize meals. We may speculate that a possible teleological explanation for the existence of diurnal variations in arousal lies in the eating habits of organisms as well as in the influence of night and day.

Animals who must eat almost constantly must keep active for most of the day or night. On the other hand, the evolution of circumscribed meals, particularly evident in carnivores, left intervening time for rest. During rest periods, animals that can do so (not fishes or elephants) generally retreat to a place of safety, a den or burrow, and sleep until hunger activates them for the next meal. Humans also have these periods of drowsiness following a meal or during the night. It was required that people stay alert at times that they might not ordinarily do so. They discovered that stimulant drugs could counteract primitive somnolent influences and they could spend their waking hours in activities other than eating. For hundreds of years nicotine and caffeine have probably been used partly for this purpose, and newer stimulants such as amphetamine and cocaine are also taken for animation.

At the behavioral level, smoking and eating clearly interact and arousal may play a key role. Smoking (at least in some doses) increases arousal, and in appropriate circumstances alertness may be rewarding. Meade and Wald (28) have shown that office workers consume most of their cigarettes at work (study done before restrictions on smoking became popular). Demands of the workplace require alertness. Smokers frequently report that smoking improves their ability to think and concentrate (29). A variety of studies have indicated smoking or nicotine improves performance on reaction time (30,31), visual surveillance (32), mental efficiency (33), and rapid information processing. Wesnes and Warburton (34) concluded on the basis of reviewing many studies that smoking facilitates performance, especially on monotonous tasks, and improves speed and accuracy of information processing.

Nicotine is clearly a psychoactive drug, but there is still disagreement about whether nicotine should be classified as a stimulant (such as amphetamine or caffeine) or a depressant (such as diazepam), or whether it has dose-related stimulant and depressant properties (35,36). There is much objective evidence that it has stimulant properties (37,38), but much less that it has depressant properties (39). However, smokers often report a calming effect from smoking (40,41,42), and we have shown a stress-reducing action under certain circumstances (Jarvik et al. (43), in preparation).

Arousal is one mechanism connecting eating and smoking. Could nicotine have amphetamine-like effects by influencing the same mechanisms as amphetamine? Like amphetamine it might produce both stimulation and anorexia. Nicotine stimulation may counteract the sedative effects of eating. Amino acids such as tryptophan enter the brain (44) and are transformed into 5-hydroxytryptamine (serotonin) which has sedative effects. Spring et al. (45) have shown that a carbohydrate meal, paradoxically enough, has a more sedating effect than other types, probably because it releases insulin, which selectively diverts amino acids other than tryptophan to peripheral tissues and away from the brain. This favors the entry of tryptophan into the brain with resulting sedation.

Third, high sympathetic tone may sometimes be rewarding as in a competition or exercise. Smoking a cigarette produces enough nicotine to cause the release of norepinephrine and epinephrine from the adrenal medulla (46,47,64). This may counteract some of the unwanted parasympathetic dominance which has been shown to be aversive by Janowsky et al. (48). The actions of catecholamines include tachycardia, gluconeogenesis, and increased muscle blood flow, all of which may be rewarding under certain circumstances. However, it is more likely that the reinforcing effects of nicotine are primarily centrally mediated (49,50, in press); (21,51). Suppression of centrally mediated sympathetic activity by clonidine has been reported to diminish smoking (52).

The fourth hypothesis presupposes that satiety, which is produced by eating, stimulates smoking. The mechanisms may involve interaction between satiety hormones such as cholecystokinin and other neurotransmitter mechanisms, particularly in the brain. For decades, amines, such as dopamine (DA), norepinephrine (NE), epinephrine (Epi), and serotonin (5-HT), have been known to be involved in feeding and satiety, especially in the lateral and medial hypothalamus. Recent work indicates that hormones such as dynorphin and other opioids may play a key role in stimulating eating and perhaps also in stimulating smoking (6,53,54). Dozens of other peptides such as arginine vasopressin (AVP) and corticotrophin releasing factor (CRF) influence feeding and probably smoking, as well as stress and cognitive functioning (55). Many brain regions (e.g., hypothalamus, reticular system, nucleus of tractus solitarius, ventral tegmental area, nucleus accumbens, and others have been studied for their involvement in both smoking (56,57,54) and eating (58,7). Also, eating and smoking appear to be related by mechanisms subsuming reward, arousal, and emotion (59,54, in press).

Only recently has the distribution of nicotinic receptors been mapped in the brain (60; and see papers by Kellar, Westfall in this volume). Surprisingly, there are few receptors in the hypothalamus. However, the effects of the nicotine cascade might easily influence the hypothalamus. As early as 1945, Pickford (61) showed that nicotine had a strong influence on the supraoptic nucleus of the hypothalamus, which secretes AVP. Nicotine has also been shown to increase catecholamine synthesis and turnover in the hypothalamus and other parts of the brain (62,63,65,66,67). There may be nicotinic receptive cells in the locus ceruleus which, in turn, may stimulate the release of norepinephrine from the cells they occupy (49).

More recently Wilkins et al. (57) have shown that growth hormone, prolactin, and cortisol are released by nicotine. This implicates corticotrophic releasing factor (CRF) and adrenocorticotrophic hormone (ACTH) which may mediate the stimulating effect of smoking on the hypothalamic-pituitary-adrenal axis, but the anatomical connections have not yet been worked out. However, Pomerleau and associates have not been able to demonstrate the release of ACTH by nicotine (53).

Pomerleau has, nevertheless, implicated opioids in the action of nicotine. Opioids are also involved in eating as well as in smoking. Naloxone, an opioid antagonist, reduces feeding both in animals and humans. Karras and Kane (68) reported that naloxone caused a reduction in smoking. This study has not yet been replicated, although two laboratories have attempted to do so. Palmer and Berens (69) reported significant dose-response relationship between nicotine and beta-endorphin in plasma. Pomerleau et al. (53) proposed that some of the effects of nicotine may be mediated by beta-endorphin, which has opiate-like actions, including antinociceptive and anxiolytic properties. They suggested that sedation or calming may result from the ability of nicotine to release central beta-endorphins and produce an opiate-like calming effect. This presumably occurs only at high doses, whereas at low doses the catecholaminergic effect predominates. Perhaps clonidine reduces smoking because it diminishes the desired stimulating effect of nicotine. Risch et al. (70) reported that physostigmine, which nonselectively increases both nicotinic and muscarinic activity, also produces elevations of plasma beta-endorphin and cortisol.

Tobin et al. (71) showed that nicotine initially increases respiration, but following smoking respiration was significantly depressed, implying an opiate-like effect of endorphins coming into play. Naloxone given before smoking had no effect on the respiratory stimulation, but diminished the respiratory depression that followed smoking. This is further indirect evidence for the involvement of beta-endorphin in the action of nicotine.

The fifth, or "Cracker Jack" hypothesis, implies that cigarette smoking is reinforcing in its own right, and that interacting the positive reinforcing effects of eating and smoking may produce a synergistic or at least an additive rewarding action. As long ago as 1929, Bayer showed that satiated chickens could be induced to eat by allowing them to watch hungry chickens eat (72). Two reinforcements given simultaneously are not antagonistic but rather additive so long as a ceiling or aversive level is not reached. Henningfield and Griffiths (73) found that amphetamine, a stimulant and highly reinforcing drug, significantly increased smoking even though both drugs presumably produced a similar type of reinforcement.

The reinforcing properties of nicotine have been reviewed in many publications (e.g., 35,74,75,76,77,36,78, and 79). At least ten of the contributors to this volume have examined the reinforcing role of nicotine. Today, it is the consensus that CNS actions of nicotine are a determining factor in tobacco use, especially cigarette smoking. The major reasons why people smoke are to avoid withdrawal, to increase arousal and attention, and to reduce anxiety, aggression, and appetite. Although most investigators agree that nicotine underlies the smoking habit, the mechanism of reinforcement is not clear. There is an important relationship between eating and smoking which must still be explored.

In summary, smoking may decrease appetite for food and particularly for sweet foods by an action on the liver, on glucoreceptors in the brain, or by an effect on catecholamines similar to amphetamine. It also decreases weight by increasing oxygen utilization or metabolism. Eating may stimulate the desire to smoke by reducing nicotine levels in the blood, by counteracting the sedation produced by a meal, or by potentiating the rewarding effect of a meal by releasing endogenous opioids, catecholamines or other substances involved in the mediation of reward. Until recently, investigators have tended to ignore the postprandial influence. A greater understanding of the reward mechanisms controlling smoking and eating may result from further study of this phenomenon.

REFERENCES

1. Grunberg NE. Nicotine, cigarette smoking, and body weight. British Journal of Addictions, 80:369-377, 1985.
2. Wack J and Rodin J. Smoking and its effects on body weight and the system of caloric regulation. American Journal of Clinical Nutrition, 35:366-380, 1982.
3. McKennell AC. Smoking motivation factors. British Journal of Social and Clinical Psychology, 9:8-22, 1970.
4. Shiffman S, Read L, Maltese R, Rapkin D, and Jarvik M. Preventing relapse in ex-smokers: a self-management approach. In: Relapse Prevention (GA Marlatt and JR Gordon, Eds.) New York: Guilford Press, pp. 472-520, 1985.
5. Grunberg NE. The effects of nicotine and cigarette smoking on food consumption and taste preference. Addictive Behaviors, 7:317-331, 1982.
6. Morley JE and Levine AS. The pharmacology of eating behavior. In: Annual Review of Pharmacology and Toxicology, 25:127-146, 1985.
7. Stellar JR and Stellar E. The Neurobiology of Motivation and Reward. New York: Springer-Verlag, 1985.
8. Silverstone T, Ed. Drugs and Appetite. London: Academic Press, 1982.
9. Hoebel BG. Neurotransmitters in the control of feeding and its rewards. In: AJ Stunkard and E Stellar, (Eds.) Eating and its disorders. New York: Raven Press, pp. 15-38, 1984.
10. Woods JC and Leibowitz S. Hypothalamic sites sensitive to morphine and naloxone: effects of feeding behavior. Psychopharmacology Biochemistry and Behavior, 23:431-438, 1985.

11. Kipling R, The Betrothed.
12. Mintz J, Boyd G, Rose JE, Charuvastra VC and Jarvik ME. Alcohol increases cigarette smoking: a laboratory demonstration. Addictive Behaviors, 10:203-207, 1985.
13. Rose JE, Ananda S, and Jarvik ME. Cigarette smoking during anxiety-provoking and monotonous tasks. Addictive Behaviors, 8:353-359, 1983.
14. Rose JE, Zinser MC and Jarvik ME. Caffeine augments cigarette smoking (in press).
15. Shiffman SM and Jarvik ME. Cigarette smoking, physiological arousal, and emotional response: Nesbitt's paradox re-examined. Addictive Behaviors, 9:95-98, 1984.
16. Benowitz N. Clinical Pharmacology of Nicotine. In: Annual Review of Medicine, 37:21-32, 1986.
17. Gritz ER, Baer-Weiss V, and Jarvik ME. Titration of nicotine intake with full-length and half-length cigarettes. Clinical Pharmacology and Therapeutics, 20:552-556, 1976.
18. Kumar R, Cooke EC, Lader MH, and Russell MAH. Is tobacco smoking a form of nicotine dependence? In: RE Thornton (Ed.) Smoking Behaviour: Physiological and Psychological Influences. Edinburgh: Churchill Livingstone, pp. 244-258, 1978.
19. Russell MAH. Cigarette smoking: a dependence on high-nicotine boli. Drug Metabolism Reviews, 8:29-57, 1978.
20. Schachter S, Kozlowski LT, and Silverstein B. Effects of urinary pH on cigarette smoking. Journal of Experimental Psychology: General, 106:13-19, 1977.
21. Schachter S. Pharmacological and psychological determinants of smoking. In: RE Thornton (Ed.) Smoking Behaviour: Physiological and Psychological Influences. Edinburgh: Churchill Livingstone, pp. 208-228, 1978.
22. Schmiterlow CG, Hansson E, Andersson G, Appelgren LE, and Hoffmann PC. Distribution of nicotine in the central nervous system. Annals of the New York Academy of Science, 142(Art 1):2-14, 1967.
23. Magoun HW. The Waking Brain. Springfield, Illinois: Charles C. Thomas, 1963.
24. Moruzzi G, and Magoun HW. Brain stem reticular formation and activation of the EEG. Electroencephalography and Clinical Neurophysiology, 1:455-473, 1949.
25. Lindsley DB. Electrophysiology of the visual system and its relation to perceptual phenomena. In: MAB Brazier (Ed.) Brain and Behavior, Vol I. Washington DC: American Institute of Biological Sciences, 1961.
26. Routtenberg A. The two-arousal hypothesis: reticular formation and limbic system. Psychological Reviews, 75:51-80, 1968.
27. Smith GP. The physiology of the meal. In: T Silverstone (Ed.) Drugs and Appetite. London: Academic Press, pp. 1-21, 1982.
28. Meade TW and Wald NJ. Cigarette smoking patterns during the working day. British Journal of Prevention and Social Medicine, 31:5-29, 1977.
29. Wesnes K and Warburton DM. Smoking, nicotine and human performance. In: DJK Balfour (Ed.) Nicotine and the Smoking Habit. Section 114, International Encyclopedia of Pharmacology and Therapeutics. Oxford: Pergamon Press, pp. 133-152, 1984.
30. Frankenhaeuser M, Myrsten AL, Post B, and Johansson G. Behavioural and physiological effects of cigarette smoking in a monotonous situation. Psychopharmacologia, 22:1-7, 1971.
31. Myrsten AL and Andersson K. Effects of cigarette smoking on human performance. In: RE Thornton (Ed.) Smoking Behaviour: Physiological and Psychological Influences. Edinburgh: Churchill Livingstone, pp. 156-167, 1978.

32. Tarriere C and Hartemann F. Investigation into the effects of tobacco smoke on a visual vigilance task. In: Proceedings of the Second Annual Congress of Ergonomics, Dortmund. Supplement to Ergonomics, 525-530, 1964.
33. Williams DG. Effects of cigarette smoking on immediate memory and performance in different kinds of smokers. British Journal of Psychology, 71:83-90, 1980.
34. Wesnes K and Warburton DM. Effects of smoking on rapid information processing. Neuropsychobiology, 9:223-229, 1983.
35. Ashton H and Stepney R. Smoking: Psychology and Pharmacology. London: Tavistock, 1982.
36. Mangan GL and Golding JF. The Psychopharmacology of Smoking. Cambridge: Cambridge University Press, 1984.
37. Domino EF. Behavioral electrophysiological, endocrine and skeletal muscle actions of nicotine and tobacco smoking. In: A Remond and C Izard (Eds.) Electrophysiological Effects of Nicotine. Amsterdam: Elsevier/North-Holland Biomedical Press, 1979.
38. Ulett JA and Itil TM. Quantitative electroencephalogram in smoking and smoking deprivation. Science, 164:969-970, 1969.
39. Ashton H, Marsh VR, Millman JE, Rawlins MD, Telford R and Thompson JW. The use of event-related slow potentials of the brain as a means to analyse the effects of cigarette smoking and nicotine in humans. In: RE Thornton (Ed.) Smoking Behaviour: Physiological and Psychological Influences. Edinburgh: Churchill Livingstone, pp. 54-68, 1978.
40. Russell MAH, Peto J, and Patel UA. The classification of smoking by factorial structure of motives. Journal of the Royal Statistical Society, A, 137:313-333, 1974.
41. Pomerleau OF and Pomerleau CS. Neuroregulators and the reinforcement of smoking: Towards a biobehavioral explanation. Neuroscience and Biobehavioral Reviews, 8:503-514, 1984.
42. Pomerleau OF, Turk DC and Fertig JB. The effects of cigarette smoking on pain and anxiety. Addictive Behaviors, 9:265-271, 1984.
43. Jarvik ME, Rose JE, Herskovic JE, and Sadeghpour M. Anxiolytic effects of cigarette smoking on four types of experimental stress (in preparation).
44. Wurtman JJ and Wurtman RJ. Studies on the appetite for carbohydrates in rats and humans. Journal of Psychiatric Research, 17:213-221, 1982/1983.
45. Spring B, Maller O, Wurtman J, Digman L, and Cozolino L. Effects of protein and carbohydrate meals on mood and performance interactions with sex and age. Journal of Psychiatric Research, 17(2):155-167, 1982/83.
46. Kershbaum A, Pappajohn DJ, Bellet S, Hirabayashi M, and Shafiiha H. Effect of smoking and nicotine on adrenocortical secretion. Journal of the American Medical Association, 203:275-278, 1968.
47. Westfall TC and Watts DT. Catecholamine excretion in smokers and nonsmokers. Journal of Applied Physiology, 19:40-42, 1964.
48. Janowsky DS, El-Yousef MK, Davis JM and Sekerke HJ, Antagonistic effects of physostigmine and methylphenidate in man. American Journal of Psychiatry, 130:1370, 1973.
49. Carruthers M. Modification of the noradrenaline related effects of smoking by beta-blockade. Psychological Medicine, 6:251-256, 1976.
50. Nemeth-Coslett R, Henningfield JE, Griffiths RR, and O'Keeffe MK. Effects of mecamylamine on cigarette smoking and subjective responses Psychopharmacology, 88:420-425, 1986.
51. Stolerman IP, Goldfarb T, Fink R, and Jarvik ME. Influencing cigarette smoking with nicotine antagonists. Psychopharmacologia, 28:247-259, 1973.

52. Glassman A. Cigarette craving, smoking withdrawal and clonidine. Science, 226:864-866, 1984.
53. Pomerleau OF, Fertig J, Seyler LE, and Jaffe JH. Neuroendocrine reactivity to nicotine in smokers. Psychopharmacology, 83:61-67, 1983.
54. Wise RA, Jenck F and Raptis L. Morphine potentiates feeding via the opiate reinforcement mechanism. In: LS Harris (Ed.). Problems of Drug Dependence 1985. National Institute on Drug Abuse Research Monograph Series, Washington, DC: Department of Health and Human Services, (in press) 1985.
55. Flood JF, Landry DW, and Jarvik ME. Cholinergic receptor interactions and their effects on long-term memory processing. Brain Research, 215:177-185, 1981.
56. Goodman LS and Gilman A (Eds.). The Pharmacological Basis of Therapeutics (5th Edition). New York: Macmillan, 1975.
57. Wilkins JF, Carlson HE, Van Vunakis H, Hill MA, Gritz ER, and Jarvik ME. Nicotine from cigarette smoking increases circulating levels of cortisol, growth hormone, and prolactin in male chronic smokers. Psychopharmacology, 78:305-308, 1982.
58. Hoebel BG and Novin D. The Neural Basis of Feeding and Reward. Brunswick, Maine: Haer Institute, 1982.
59. Stein L. The chemistry of positive reinforcement. In: M Zuckerman (Ed.) Biological Bases of Sensation Seeking, Impulsivity and Anxiety. Hillsdale, NJ: L. Erlbaum Associates, p. 151-175, 1983.
60. London ED, Connolly RJ, Szikszay M, and Wamsley JK. Distribution of cerebral metabolic effects of nicotine in the rat. European Journal of Pharmacology, 110:391-392, 1985.
61. Pickford M. Pituitary antidiuretic hormone. Physiological Reviews, 25:573, 1945.
62. Andersson K, Fuxe K, Eneroth P, and Agnati LF. Effects of acute central and peripheral administration of nicotine on hypothalamic catecholamine nerve terminal systems and on the secretion of adenohypophyseal hormones in the male rat. Medical Biology 60:98-111, 1982.
63. Balfour DJ. The effects of nicotine on brain neurotransmitter systems. Pharmacology and Therapeutics, 16:269-282, 1982.
64. Cryer PE, Haymond MW, Santiago JV, and Shah SD. Norepinephrine and epinephrine release and adrenergic mediation of smoking-associated hemodynamic and metabolic events. New England Journal of Medicine, 295:573-577, 1976.
65. Goodman FR. Effects of nicotine on distribution and release of C-norepinephrine and C-dopamine in rat brain striatum and hypothalamic slices. Neuropharmacology, 3:1025-1032, 1974.
66. Westfall TC. Effect of nicotine and other drugs on the release of ^{3}H-norepinephrine and ^{3}H-dopamine from rat brain slices. Neuropharmacology, 13:693-700, 1974.
67. Yoshida K, Kato Y, and Imura K. Nicotine-induced release of noradrenaline from hypothalamic synaptosomes. Brain Research, 182:361-368, 1980.
68. Karras A and Kane J. Naloxone reduces cigarette smoking. Life Sciences, 27:1541-1545, 1980.
69. Palmer RF and Berens A. Double blind study of the effects of naloxone on the pleasure of cigarette smoking. Federation Proceedings, 42:654, 1983.
70. Risch SC, Cohen RM, Janowsky DS, Kalin NH, and Murphy DL. Mood and behavioral effects of physostigmine on humans are accompanied by elevations of plasma beta-endorphin and cortisol. Science, 207:1545-1546, 1980.
71. Tobin MJ, Jenouri BS, and Sackner MA. Effect of naloxone on change in breathing pattern with smoking. A hypothesis on the Addictive Nature of Cigarette Smoking. Chest, 82:530-537, 1982.

72. Bayer E. Beitrage aur Zweikomponententheorie des Hungers. Zeitzchrift fur Psychologie, 112:1-54, 1929. In: FA Moss (Ed.) Comparative Psychology (Rev. Ed.) New York: Prentice-Hall, Inc., p. 72 1942.
73. Henningfield JE and Griffiths RR. Cigarette smoking and subjective response: effects of d-amphetamine. Clinical Pharmacology and Therapeutics, 30:497-505, 1981.
74. Dunn WL, Jr. (Ed.) Smoking Behavior: Motives and Incentives. Washington, D.C.: V.H. Winston & Sons, 1973.
75. Henningfield JE and Jasinski DR. Human pharmacology of nicotine. Psychopharmacology Bulletin, 19:413-415, 1983.
76. Hunt WA (Ed.) Learning Mechanisms in Smoking. Chicago: Aldine 1970.
77. Jarvik ME. The role of nicotine in the smoking habit. In: Learning Mechanisms in Smoking. WA Hunt (Ed.) Chicago: Aldine, pp. 155-190, 1970.
78. Remond A, and Izard C. Electrophysiological Effects of Nicotine. Amsterdam: Elsevier/North-Holland Academic Press, 1979.
79. Thornton RE, Ed. Smoking Behaviour: Physiological and Psychological Influences. Edinburgh: Churchill Livingstone, 1978.

THE BIOCHEMICAL AND FUNCTIONAL NATURE OF THE NICOTINE RECEPTOR IN RAT BRAIN

L.G. Abood, A. Maiti, S. Grassi, and K.S. Salles

Center for Brain Research
University of Rochester School of Medicine and Dentistry
Rochester, New York 14642

ABSTRACT

An investigation has been undertaken of the biochemical nature and neuroanatomical localization of the sites involved in nicotine's action on the brain. The present report describes studies aimed at 1) the characterization and purification of the nicotine receptor in rat brain; 2) elucidating the psychopharmacologic nature of the receptor using various nicotine analogues and other psychotropic agents; and 3) determining the particular brain regions directly involved in the behavioral effects (e.g., prostration and seizures) of nicotine. The nicotine receptor, which has been purified by affinity chromatography and consists of a major protein with a molecular weight of 58 kdalton and minor components ranging from 47 to 83 kdaltons, does not appear to resemble the nicotinic cholinergic receptor purified from the electric fish. The functional significance of the nicotine receptor has been confirmed by correlating the psychotropic potency of various nicotine analogues with their affinity for the receptor. Evidence is presented for the existence of multiple forms of both the nicotine and nicotinic cholinergic receptors by comparing the binding and psychotropic characteristics of some carbamate esters, nicotine analogues, and various unrelated drugs acting as antagonists to nicotine. In order to determine its neuroanatomical site of psychotropic action, nicotine was injected into cannulae chronically implanted into various brain regions of conscious, freely moving rats. It appears that the primary locus is the vestibular system, particularly that component linked to the cerebellar nodule. Lesioning the cerebellar nodule or inhibiting it with a local anesthetic abolishes the nicotine-induced prostration and increases the number of nicotine and muscarinic cholinergic receptors in hypothalamus and caudate nucleus. It is proposed that the vestibular system, which plays a major role in the processing of sensory-motor information may be involved in the reinforcing and other affective effects of nicotine in man.

INTRODUCTION

In recent years, it has become apparent that the action of nicotine on the central nervous sytem is complex and cannot be accounted for on the basis of a single cholinergic receptor, such as that for acetylcholine

(ACh) at the neuromuscular junction (1,2). Evidence for this conclusion derives from electrophysiologic, pharmacologic, and receptor binding studies. Electrophysiologic studies have shown that the action of nicotine applied microiontophoretically to single neurons in the caudate nucleus could not be mimicked by ACh; whereas, even large doses of nicotine were unable to elicit responses in single pyramidal neurons of the pericruciate cortex obtained with iontophoretically applied nicotine (3,4). Nicotine, administered intraventricularly (i.v.c.) to rats, produces a characteristic prostration which cannot be mimicked by ACh, α-bungarotoxin, and other cholinergic agents (1). Finally, [^{3}H]nicotine binding studies with brain membranes demonstrate that nicotinic cholinergic agents such as α-bungarotoxin, ACh, and hexamethonium (4-7) exhibit low-affinity binding.

On the other hand, receptor binding studies with [^{3}H]nicotine and [^{3}H]ACh reveal a similarity in the rank order of affinity of a number of nicotinic cholinergic agents. Furthermore, autoradiographic studies of rat brain reveal a parallelism in the neuroanatomical distribution of [^{3}H]nicotine and [^{3}H]ACh binding sites, which is distinct from that of [^{125}I]α-bungarotoxin (8). Although mecamylamine is able to antagonize both the action of acetylcholine on ganglionic neurons and the nicotine-induced prostration, it has an extremely low affinity for nicotine receptors (1,6).

In an effort to help resolve the issue concerning the mechanisms of action of nicotine in the central nervous system, a study was undertaken comparing the pharmacologic actions of a variety of agents and their [^{3}H]nicotine binding properties. One basis for selecting candidate agents for the study derived from the observation that carbamylcholine, a classical cholinergic agent, had a relatively high affinity for the nicotine receptor.

Relation Between IC_{50} for Nicotine Analogues and Behavioral Activity

A plot of the log IC_{50} values for a series of N'-substituted nicotine analogues at the high affinity nicotine binding site and their ability to induce prostration following administration i.v.c. revealed an excellent correlation (Fig. 1). With increasing length of N'-alkyl substituent from methyl to butyl, there was a correlative decrease in psychotropic potency with increasing IC_{50} values. Quaternization of the pyrrolidine (N') or pyridine of nicotine virtually abolished psychotropic activity and reduced the binding affinity over three orders of magnitude. Removal of the N'-methyl diminished both binding affinity and psychotropic potency. Psychotropic potency is expressed as the "-log act," where act. (activity) is the reciprocal of the potency relative to nicotine = 1. The correlation of the IC_{50} values with the psychotropic potency was the same for the low affinity (+)-[^{3}H]nicotine binding sites.

Antagonism of Prostration and [^{3}H]Nicotine Binding by Pharmacologically Unrelated Agents

Various pharmacologically unrelated agents were tested for their ability to compete for [^{3}H]nicotine binding and to antagonize prostration induced by nicotine i.v.c. (Table 1). The most potent agent was methylcarbamylcholine, which had about 1/5 the IC_{50} and psychotropic efficacy of nicotine and was over a magnitude more potent than carbamylcholine or dimethylaminoethyl methylcarbamate. Although mecamylamine and hexamethonium were effective antagonists to the nicotine-induced prostration, they had very low affinities for the nicotine receptor. Bungarotoxin, the most potent nicotinic cholinergic antagonist had no antagonistic effect on nicotine-induced prostration and

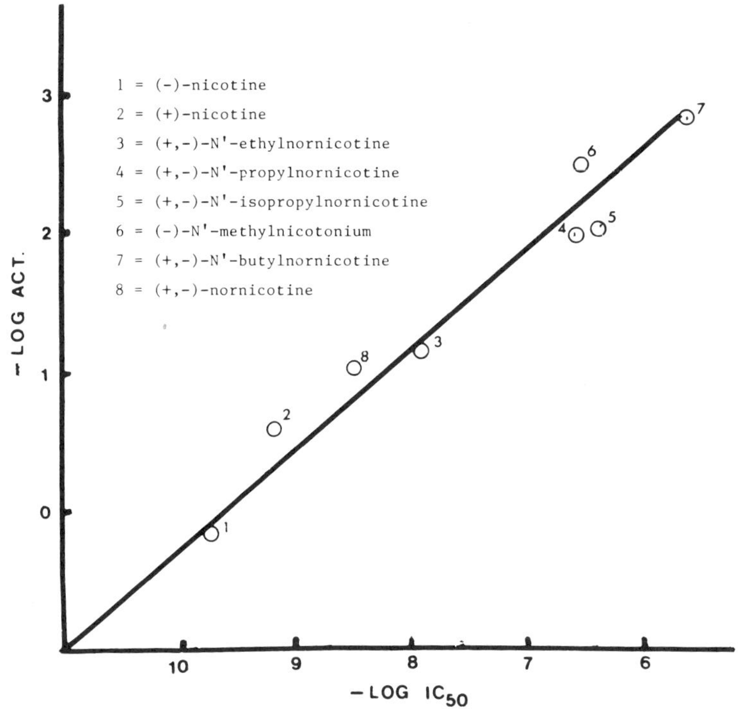

Fig. 1. Relationship of binding affinity of nicotine analogues to psychotropic potency. The abscissa referes to the IC_{50} for [^{3}H]nicotine binding to rat brain membranes. The ordinate expresses the -log of the relative potency of an analogue in producing prostration in rats i.v.c.; (-)-nicotine itself having a value of 1. The IC_{50} of the experimental drug is the numerator.

extremely low affinity for the nicotine receptor. Pyrilamine, an antihistaminic, exhibited some affinity for the receptor and antagonism of prostration. Although the local anesthetics, procaine and cocaine, antagonized the nicotine-induced prostration, only procaine exhibited affinity for the receptor.

It is apparent from the data that no simple relationship exists between the receptor affinity of an agent and its ability to antagonize the psychotropic action of nicotine. Although binding affinity is correlated with psychotropic potency, the latter can occur in the absence of affinity for the nicotine-binding site. The most likely explanation for this divergence is that the agents act on neural networks connected to the primary sites for nicotine. The fact that carbamate esters, particularly methylcarbamylcholine, display high affinity and psychotropic potency suggests that the receptors involved are nicotinic cholinergic in nature.

There is, however, a significant difference between the carbamate esters and the series of nicotine analogues: quaternization of the pyrrolidine N in nicotine diminishes affinity and psychotropic potency by three orders of magnitude, whereas quaternization of the aminoalcohol N of the carbamate esters increases activity 25-fold. One interpretation of the results is that the configurational aspects of the quarternized carbamate esters and nicotine are such that steric hindrance only

TABLE 1. Effect of Various Agents on [^{3}H]Nicotine Binding and Prostration

Agent	[^{3}H]Nicotine Binding IC_{50} M	Prostration EC_{50} nmoles
(-)-nicotine	2×10^{-9}	20
carbamylcholine	5×10^{-7}	1000
*DEAE methylcarbamate	2×10^{-7}	1000
methylcarbamylcholine	8×10^{-9}	100
hexamethonium	$> 1 \times 10^{-4}$	antagonism 50
mecamylamine	$> 1 \times 10^{-4}$	antagonism 40
α-bungarotoxin	$> 1 \times 10^{-4}$	no effect
pyrilamine	5×10^{-5}	antagonism 500
procaine	8×10^{-5}	antagonism
cocaine	$> 1 \times 10^{-3}$	antagonism

[^{3}H]nicotine binding was measured on rat brain membranes. Agents were administered into the rat's fourth ventricle to measure prostration or their ability to block the nicotine-induced prostration (1).
*DEAE = dimethylaminoethyl

restricts accessibility of N'-nicotinium to the receptor. Another explanation is that the receptors involved exist as multiple forms. Evidence for the latter derives from the observation that N'-nicotinium is even more active at peripheral nicotinic cholinergic receptors than nicotine. An understanding of the relationship of the nicotine and nicotinic cholinergic binding sites must await the elucidation of the chemical and functional nature of the multiple forms of the receptors involved.

Locus of Action of Nicotine-induced Prostration

In an effort to determine the neuroanatomical locus of nicotine associated with the prostration response, nicotine was administered through stereotactically implanted cannulae into various brain areas of conscious, freely moving rats (1). The only brain regions that appeared to be primarily involved were those immediately adjacent to the fourth ventricles, regions such as the vestibular nuclei and lobule X of the cerebellum (cerebellar nodule). Among the areas that were clearly not directly involved were the dentate, pontine, fastigial and caudate nuclei, as well as the nucleus in tractus solitarious, in addition to other areas of the brain stem, midbrain, cerebellum, and cerebral cortex (1). One area that appeared to be directly involved was the lateral vestibular nucleus and, to a lesser extent, the medial vestibular nucleus; however, since such areas were adjacent to the fourth ventricle, the possibility existed that nicotine entered the ventricle and acted on other surrounding regions, such as the cerebellar nodule. Subsequent studies revealed that the administration of nicotine into the cerebellar nodule did produce prostration and seizures. In an effort to substantiate the role of the cerebellar nodule as a site for nicotine's action, lesioning studies were undertaken employing the neurotoxin, kainic acid (KA).

Behavioral Effects of Kainic Acid (KA) Given into the Cerebellar Nodule

Kainic acid has been widely used as a selective neuronal lesioning agent in the mammalian cerebellum (9) causing severe loss of Purkinje stellate and Golgi neurons with a relative sparing of granule cells, which are thought to be glutaminergic. Stereotaxic injections of KA into various areas of the brain have been extensively studied for the assessment of neurotransmitter function within specific brain areas, including the cerebellum of different species (10). Fastigial influences on postural tonus has been studied by KA lesions of the fastigial nuclei unilaterally and bilaterally (11). In a previous study, it was postulated that the cerebellar nodule, in its role of regulating autonomic activity, may be responsive to neurotransmitters and other substances present in the spinal fluid in the fourth ventricles (12).

Although several studies have reported on the motor behavior characteristics of KA-induced cerebellar lesioned rats, which included severe postural abnormality, circular movements, and barrel rotation, the lesions in the cerebellum were very extensive. To determine whether the cerebellar nodule was essential for the nicotine-induced prostration, cannulae were chronically implanted into the cerebellar nodule and the lateral ventricle. The placement of cannulae was verified by administering a dye marker and performing histology (Fig. 2). A single dose of 8 ng KA in 1 μl vol was administered into the nodule and the animals examined for short-term and long-term alterations in behavioral locomotor activity.

Immediately following the administration of 6 ng KA or greater into the cerebellar nodule, the rats displayed violent running fits, circling, barrel rotation, and myoclonic seizures. When quiescent, the rats were generally prostrate or exhibited scoliosis with limbs in the extended position. Other symptoms included ataxia, marked hyperexcitability, stereotypic head movements, and marked aggression. Most of the abnormal

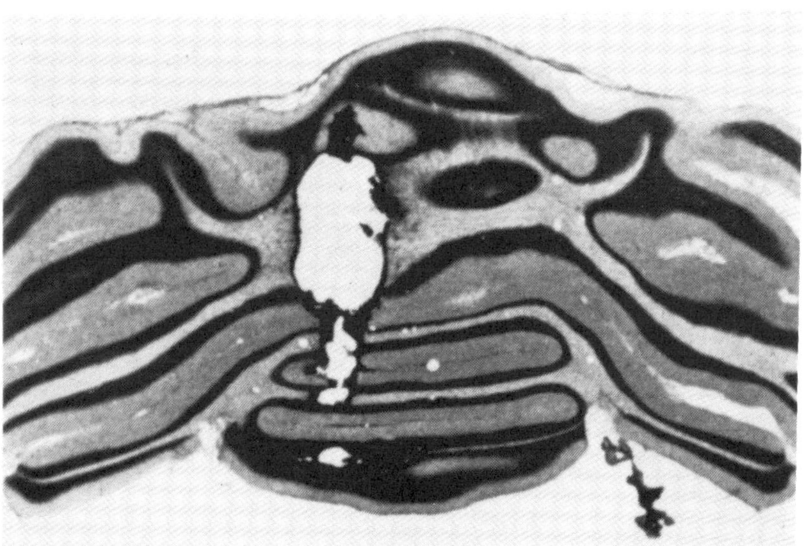

Fig. 2. A transverse section of the rat brain showing the cannula tract terminating in the cerebellar nodule. India ink (1 μl) was administered as a marker immediately before sacrifice. Staining is with cresyl violet.

motor effects were absent 8 hr after KA; however, those animals with the most severe responses displayed hyperactivity, dystonia, and aggressive behavior as long as 4 days after KA. During the period 2-3 days after KA, the animals were extremely hyperirritable, pugnacious, difficult to handle, and exhibited marked startle responses with running fits. When nicotine was administered i.v.c. 24 hr after KA lesion, it failed to produce prostration or any observable motor effects. It can be concluded that the cerebellar nodule is a primary site for nicotine's action.

Effect of Kainate Lesioning on [^{3}H]Nicotine Binding in Various Brain Areas

An increase in [^{3}H]nicotine binding ranging from 27-42% occurred in the remaining nodular area and surrounding cerebellar region 4-8 days following the administration of 4 or 12 ng of KA into the nodule; while the caudate showed a 51% and the hypothalamus a 20% increase (Table 2). After 12 days following the administration of 12 ng KA, the increase in [^{3}H]nicotine binding was 40% in the nodule, 10% in the surrounding area, 87% in the caudate, and 74% in the hypothalamus. At a dose of 12 ng KA, the increase in both the caudate and hypothalamus was 90%.

Effect of Kainate Lesioning on [^{3}H]QNB Binding in Various Brain Areas

An increase in [^{3}H]QNB binding ranging from 93-100% occurred in both the remaining nodular area and surrounding cerebellar region 4-14 days following the administration of 4 ng into the cerebellar nodule, while the caudate showed a 20% and the hypothalamus a 39% increase (Table 3). At doses of 8-12 ng of KA, the increase occurring 8-12 days later ranged from 136-153% in the nodule and surrounding cerebellar region, while the caudate showed a 23% and the hypothalamus a 60% increase. Comparable increases were seen in the surrounding cerebellar and hypothalamus following the administration of 12 ng KA, while the [^{3}H]-QNB binding in the caudate increased further to 42%.

[^{3}H]-Binding and Behavioral Effects After Chronic Exposure of Rats to Nicotine

In an effort to determine if chronic administration of nicotine to rats leads to tolerance, rats were administered 12 μM/kg of nicotine twice

TABLE 2. Effect of Kainate Lesioning on [^{3}H]Nicotine Binding to Various Regions of Rat Brain

	Nodular area moles x 10^{14}		Surrounding Cerebellum moles x 10^{14}		Caudate moles x 10^{14}		Hypothalamus moles x 10^{14}	
	Control	Kainate	Control	Kainate	Control	Kainate	Control	Kainate
4 ng KA 4 days	2.5	3.3 (32%)	2.2	2.8 (27%)	3.3	5.0 (51%)	4.5	5.5 (22%)
12 ng KA 8 days	2.3	2.8 (22%)	M	M	3.6	6.8 (89%)	4.5	8.5 (91%)

The results are an average of 3-4 separate experiments done in triplicate; the coefficient of variation of each value being no greater than 12%.
M = missing

TABLE 3. Effect of Kainate Lesioning on [^{3}H]QNB Binding to Various Regions of Rat Brain

	Nodular area moles x 10^{12}		Surrounding Cerebellum moles x 10^{12}		Caudate moles x 10^{12}		Hypothalamus moles x 10^{12}	
	Control	Kainate	Control	Kainate	Control	Kainate	Control	Kainate
4 ng KA 4 days	0.32	0.62 (93%)	0.52	0.98 (88%)	2.85	3.42 (20%)	2.30	3.20 (39%)
12 ng KA 8 days	M	M	0.58	1.58 (172%)	3.10	4.40 (42%)	2.90	4.84 (67%)

The results are an average of 3 separate experiments; the coefficient of variation of each value being less than 12%.

daily for 2 weeks and tested both for their response to nicotine administered intraventricularly and possible changes in [^{3}H]nicotine binding (Tables 4 and 5). When rats were administered 5-10 nM of nicotine into the fourth ventricle 12 hr after the last injection of nicotine, they exhibited the same degree of prostration as the control group (13). Similarly, if nicotine is infused continuously into the lateral ventricles by means of an Alzet minipump at a rate of 50 nM 1 hr for 10 days, their response to nicotine given intraventricularly on the last day was indistinguishable from the controls.

A measurement of [^{3}H]nicotine binding to membranes prepared from the brains of the rats chronically treated with nicotine revealed no differences in either the K_d or the B_{max} of [^{3}H]nicotine binding as determined by Scatchard analysis (Table 5). The studies appear to be at

TABLE 4. Effect of Various Dosage Regimens and Routes of Administration on Prostration Following Nicotine Administration into Fourth Ventricles of Rats

Dosage Schedule	Incidence of Prostration Syndrome
5 nM nicotine twice daily into fourth ventricles for 4 days	All rats responded day 1; 60% responded day 2; 30% responded day 4; 90% responded fully after 3 days without nicotine.
Alzet minipump (50 nM/hr) into lateral ventricle for 10 days	Normal response to 5 nM of nicotine into fourth ventricle on last day of infusion.
Acute injection of 12 μM/kg of twice daily for 14 days.	Most rats responded to 5 nM ip of nicotine into fourth ventricles 12 hr after last injection. Response was slightly less than in controls.

This table was reproduced from (13).

TABLE 5. Effect of Chronic Nicotine Treatment on B_{max} and K_d of $[^3H](-)$-Nicotine Binding to Rat Brain Membranes

Conditions	B_{max} M/mg x 10^{14}		K_d x 10^{10}	
	High	Low	High	Low
Normal	0.5 ± 0.1	2.9 ± 0.4	1.0 ± 0.3	3.5 ± 0.2
Chronic (twice daily)	0.4 ± 0.1	2.7 ± 0.2	0.8 ± 0.2	3.3 ± 0.3
Chronic (infusion)	0.4 ± 0.1	2.8 ± 0.3	1.1 ± 0.2	3.5 ± 0.3

Values are expressed as mean ± sd with n=3-4 rat brains. Data were derived from Scatchard plots. This table was reproduced from (13).

variance with those of Marks et al. (14) reporting up-regulation of nicotine binding as well as behavioral tolerance in Dba mice after chronic nicotine infusion. One possible explanation for the difference may be that Dba mice appear to be more sensitive to the effects of nicotine compared with other inbred strains of mice (14).

Purification of the Nicotine Receptor from Rat Brain

In a previous study, it was reported that the nicotine receptor could be purified about a 1000-fold from a Triton-solubilized extract of rat brain membranes by affinity chromatography (15). SDS gel electrophoresis revealed that the purified material consisted of a major protein with a Mr of about 56,000 and minor components ranging from 47 to 83 kdaltons. The fact that specific activity of the purified material was less than 10% of the theoretical could be attributed to either partial denaturation during preparation or inadequacy of the methods (gel filtration or polyethylene glycol precipitation) used to determine $[^3H]$nicotine binding to low concentrations of receptor protein. In an effort to determine whether either methodological factors or denaturation were responsible, alternate procedures were employed for both the purification and receptor assay.

After establishing that detergents such as CHAPS and deoxycholate were even less useful than Triton X-100, it was found that the receptor could be solubilized by exposing rat brain membranes to 0.1% solution of pancreatic trypsin for 2 hr at 37°C. The solubilized material was then passed through an affinity column prepared by conjugating 6-hydroxymethyl nicotine to epoxy sepharose (15). Previously, 6-β-hydroxyethyl nicotine was used to prepare the affinity gel, the 6-hydroxyethyl derivative having the advantage that it was less difficult to synthesize (16). A gel electrophoretic pattern of the trypsin-solubilized material after affinity chromatography revealed the presence of bands corresponding to molecular weights of 75, 58, and 47 kdaltons. The gel pattern was similar to that obtained with the Triton-solubilized material.

In order to more accurately measure $[^3H]$nicotine binding to the purified receptor, a new method was developed which involved the conjugation of the receptor proteins to heat denatured rat brain membranes via disulfide linkages utilizing the conjugating agent 4,4'-dithiopyridine. With this procedure, it was possible to demonstrate a 15,000-fold purification of the nicotine receptor of the trypsin-solubilized material and a 12,000-fold purification of the

Triton-solubilized preparation. The purified material yielded K_d values similar to membranes (1.0×10^{-9} M) and exhibited the same relative affinity to various nicotine analogues.

As discussed elsewhere (15), the gel electrophoretic pattern of the nicotine receptor bears little resemblance to that of the _Torpedo_ nicotinic cholinergic receptor consisting of subunits with Mr values of 40, 50, 57, and 66 kdaltons. It has been reported that a tryptic fragment of the _Torpedo_ 66 kdalton subunit which has a Mr of 47 kdalton may be the recognition site for the local anesthetic trimethisoquin (17). It remains to be determined if the _Torpedo_ fragment is similar to the 47 kdalton component obtained from rat brain membranes exposed to trypsin.

Homeostasis and the Vestibular Cerebellum

The vestibular system is a phylogenetically and ontogenetically primitive neural system whose function is to maintain equilibrium and spatial orientation as the organism attempts to adapt to kinematic changes in its environment. In addition to relaying information from visual, auditory, and somatosensory systems, the various vestibular nuclei, via connections to the cerebellum, reticular formation, and spinal cord, serve to integrate a multitude of viscero-somatic and affective stimuli prior to final destination to effector organs. A variety of homeostatic mechanisms, such as stress-adaption, posture, and blood glucose, involve the vestibular cerebellum by way of its connection to cortical, nigrostriatal, hypothalamic, and pontine-medullary pathways.

One of our most common experiences involving the vestibular system is the startle reaction which results from a sudden loss of equilibrium or spatial orientation. The startle reaction can be either adaptive or maladaptive, depending upon the ability of the vestibular complex to integrate and reorder the diverse sensory-motor and perceptual stimuli. Among the maladaptive reactions are vertigo, nausea, vomiting, and ataxia, as well as sensations of discomfort, apprehension, fright, and even panic. Improvement in the adaptive or accommodative efficiency of the vestibular apparatus to unnatural spatialkinematic stimuli and sensory input involves several mechanisms of vestibular habituation. Among these mechanisms are the ability to control eye, head, and body movements in response to vestibular challenges. Another involves the interplay of suppressive and switching mechanisms that allows the attenuation and rearrangement of adversive sensory-motor and perceptual stimuli for better control of motion and the affective response. The degree of effectiveness of such adaptive mechanisms, which facilitate adjustment to novel and otherwise maladaptive sensory-motor inputs, may be an important factor in one's sense of well-being.

There is a considerable body of evidence to suggest that learning disability and emotional disturbances in children are frequently associated with vestibular dysfunction as indicated by poor postural reflex and oculomotor control, inadequate equilibrium reactions, and hyponystagmus (see [18] for review). The vestibular system can be regarded as an information processing system integrating sensory-perceptual data essential for the coordination of body movements and maintaining spatial orientation; and its dysfunction could interfere with sensory-perceptual processing. Vestibular stimulation, involving such procedures as accelaratory movements in a prone and supine position or suspension in a gravitational field, has been used to treat children with learning disability (19).

Although smoking is known to produce such vestibular effects as vertigo and postural nystagmus in man, there has been no detailed study of the mechanisms involved nor the relationship of such effects to nicotine

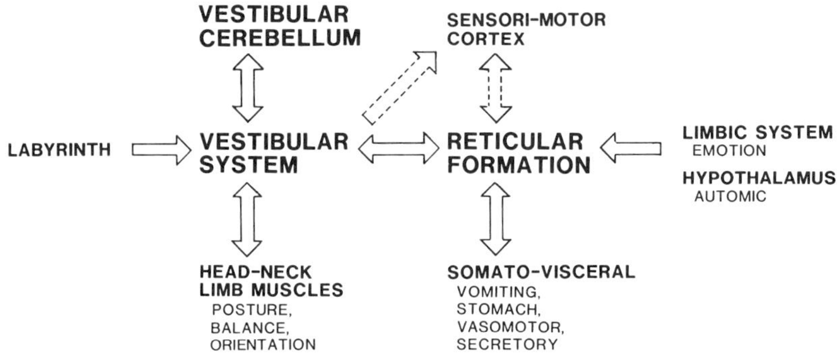

Fig. 3. Neuroanatomical and function relationship of the vestibular system. The scheme describes the connections from the vestibular system to the cerebellum, spinal cord, and reticular formation. Connections to the limbic system and sensory-motor cortex (involving the locus ceruleus with its projections to the cerebellum, hypothalamus, midbrain, and cerebral cortex) are less well defined. Some of the functional parameters relevant to the action of nicotine are indicated.

itself. It is apparent, however, from innumerable animal studies that nicotine exerts a wide variety of other sensory-motor disturbances, such as emesis, ataxia, tremors, rigidity, paralysis, and seizures (Figure 3). An understanding of the neurobehavioral mechanisms of nicotine's action is germane to the question of why people smoke. To what extent the vestibular system may be involved in nicotine's behavioral effects, particularly positive reinforcement, remains to be determined.

REFERENCES

1. Abood, LG, Reynolds, DT, Booth, H and Bidlack, JM: Sites and mechanisms of nicotine's action in the brain. Neurosci. Behav. Rev. 5: 479-486, 1981.
2. Sloan, JW, Martin, WR and Todd, GD: Multiple nicotine binding sites in rat brain P_2 fraction. Psychopharmacol. Bull. 19: 402-408, 1983.
3. McLennan, H and York, DH: Cholinergic mechanisms in the caudate nucleus. J. Physiol. Lond. 187: 163-175, 1966.
4. Dreifus, JJ, Kelley, JS and Krnjevic, R: Membrane actions of ACh and nicotine in the cerebral cortex. Fourth Int. Congress on Pharmacol. Abst. Basel: S. Karger, 1969, p 220.
5. Sloan, JW, Todd, GD and Martin, WR: Nature of nicotine binding in rat brain P_2 fraction. Pharmacol. Biochem. Behav. 20: 899-909, 1984.
6. Marks, MJ and Collins, AC: Characterization of nicotine binding in mouse brain and comparison with the binding of α-bungarotoxin and quinuclidinyl benzilate. Mol. Pharmacol. 22: 554-564, 1982.
7. Sershen, H, Reith, MEA, Lajtha, A and Gennaro, J: Noncholinergic, saturable binding of [^{3}H]nicotine to mouse brain. J. Receptor Res. 2: 1-15, 1981.
8. Clarke, PBS, Schwartz, Rd, Paul, SM, Pert, CB and Pert, A: Nicotinic binding in rat brain. Audioradiographic comparison of [^{3}H]-acetylcholine, [^{3}H]nicotine and [^{125}I]α-bungarotoxin. J. Neurosci. 5: 1307-1315, 1985.

9. Herndon, RM, Coyle, JT and Addicks, E: Ultrastructural analysis of kainic acid lesion to cerebellar cortex. Neuroscience 5: 1015-1026, 1980.
10. Foster, GA and Roberts, PJ.: Morphological and biochemical changes in the cerebellum induced by kainic acid in vivo. J. Neurochem. 34: 1191-1200, 1980.
11. Imperato, A, Nicolett, F, Diana, M, Scapagnini, U and DiChiara, G: Fastigial influences on postural tonus as studied by kainate lesions and by local infusion of GABAergic drugs in the rat. Brain Res. 295: 51-63, 1984.
12. Maiti, AK: Vestibulo-cerebellum modulates nigrostriatal dopamine synthesis: clinical implications in Parkinsonism. Neuroendocrin. Lett. 4: 196, 1982.
13. Abood, LG, Grassi, S, Costanzo, M and Junig, J: Behavioral and biochemical studies in rats after chronic exposure to nicotine. In (ed.) Sharp, CW, Mechanisms of Tolerance and Dependence, NIDA Monograph 54, pp 348-355, 1984.
14. Marks, MJ, Burch, JB and Collins AC: Effects of chronic nicotine infusion on tolerance development and cholinergic receptors. J. Pharmacol. Exp. Ther. 226: 806-816, 1983.
15. Abood, LG, Latham, W and Grassi, S: Isolation of a nicotine binding site from rat brain by affinity chromatography. Proc. Natl. Acad Sci. USA 80: 3536-3539, 1983.
16. Itokawa, H, Toshikazu, I, Haruta, R and Kameyama S: Radical methylation and radical hydroxymethylation of nicotine and quinine. Chem. Bull. 197: 295-297, 1978.
17. Wennogle, LP, Oswald, R, Saitoh, T and Changeux, JP: Dissection of the 66,00-dalton subunit of the acetylcholine receptor. Biochemistry 20: 2492-2497, 1981.
18. Morrison, DC: Neurobehavioral and Perceptual Dysfunction in Learning Disabled Children. C.J. Hogrefe, Lewiston, NY, 1985.
19. Bhatara, V, Clark, D, Arnold, L, Gunsett, R and Smeltzer, D: Hyperkinesis treated by vestibular stimulation: An exploratory study. Biol. Psychiat. 3: 269-279, 1981.

BRAIN AND MUSCLE NICOTINIC RECEPTORS: COMPLEX HOMOLOGOUS PROTEINS CARRYING MULTIPLE BINDING SITES

Bianca M. Conti-Tronconi* and Michael A. Raftery**

*Department of Biochemistry
University of Minnesota
1479 Gortner Ave.
St. Paul, MN 55108
**Division of Chemistry and Chemical Engineering
California Institute of Technology
Pasadena, CA 91125

INTRODUCTION

Elucidation of the molecular structure and function of multisubunit membrane proteins capable of complex functions, like the receptors for neurotransmitters, requires their purification in relatively large amounts, both to obtain direct information on their physicochemical and functional characteristics and to permit isolation and identification of their precursor genes. This is a formidable task to perform on any brain receptor, including the one(s) for acetylcholine and nicotine, because of the agonizingly low amounts present and the difficult isolation of active, intact form of these intrinsic membrane proteins in the presence of large amounts of endogenous proteases. A lot is known about the nicotinic acetylcholine receptor (AChR) present in peripheral tissues (see below), such as the electric organs of *Torpedo* and *Electrophorus*, and the muscle of different species, including *mammals* and, *hopefully* in the near future, man.

The fact that the AChR can be purified readily in useful amounts from the electric organs of *Torpedo* species has allowed detailed studies of its structure, which *ultimately* led to the cloning of all the genes encoding its subunits as well as to sophisticated investigation of its function, both in its native, membrane-bound form and after reconstitution of the purified AChR in artificial membranes (reviewed in 1 and 2). Elucidation of the structure of *Torpedo* AChR opened the avenue to successful studies on the structure of *muscle* AChRs (see below). Recently, isolation techniques developed during the studies of peripheral AChR have been successfully used to gain insights into the structure of brain AChRs. One would expect that in the near future the information now available on brain AChR will be instrumental in identification and cloning of the corresponding genes and their expression *in vitro* and that this will allow elucidation of some mechanisms by which brain AChR functions.

Nicotinic Acetylcholine Receptor from Torpedo Electroplax.

The acetylcholine receptor from *Torpedo electroplax* can be isolated in pure form either in its native membrane-bound state or by affinity chromatography after solubilization. This is due to dense packing of the AChR molecules in the postsynaptic membrane to the exclusion of other intrinsic membrane components. Purified membrane fragments can reseal, forming closed, right-side-out vesicles, which can be used for functional and structural studies. Due to the ease with which this receptor can be purified, both its structure and its function have been investigated to a remarkable degree of sophistication in comparison with other receptor systems.

The major physicochemical properties of Torpedo californica AChR are summarized in Fig. 1. AChRs from electric tissue of other Torpedo species have very similar properties (reviewed in 1).

STRUCTURE

SUBUNIT COMPOSITION	$\alpha_2\beta\gamma\delta$
SUBUNIT MOLECULAR WEIGHT	40, 50, 60, 65 x 10^3 DALTONS (EXPERIMENTAL)
	50, 54, 56, 58 x 10^3 DALTONS (CALCULATED)
AcChR MOLECULAR WEIGHT	270 ± 30 x 10^3 DALTONS (EXPERIMENTAL)
	267.5 x 10^3 DALTONS (CALCULATED)

PHYSICAL PROPERTIES

S VALUE	9 S
	13.7 S AS DIMER
STOKES RADIUS	72 Å
pI	4.9
SPECIFIC ACTIVITY	ONE α-BUTX PER 110 ± x 10^3 DALTONS

SEQUENCE HOMOLOGY AT N-TERMINI OF ACETYLCHOLINE RECEPTOR SUBUNITS

2 4 6 8 10 12 14 16 18 20 22 24 26 28 30 32 34 36 38 40 42 44 46 48 50 52 54 56

40K S E H E T R L V A N L L E N Y N K V I R P V E H H T H F V D I T V G L Q L I Q L I S V D E V N Q I V E T N V

50K S V M E D T L L S V L F E T Y N P K V R P A Q T V G D K V T V R V G L T L T N L L I L N E K I E E M R T N V

60K E N E E G R L I E K L L G D Y D K R I I P A K T L D H I I D V T L K L T L T N L I S L N E M E E A L T T N V

65K V N E E E R L I N D L L I V N K Y N K H V R P V K H N N E V V N I A L S L T L S N L I S L K E T D E T L T S N V

CARBOHYDRATE	~75 RESIDUES/MOLECULE	OF AcChR
O-SUBSTITUTED SER	~22 RESIDUES/MOLECULE	
O-SUBSTITUTED THR	~23 RESIDUES/MOLECULE	
PHOSPHOSERINE	~7 RESIDUES/MOLECULE	

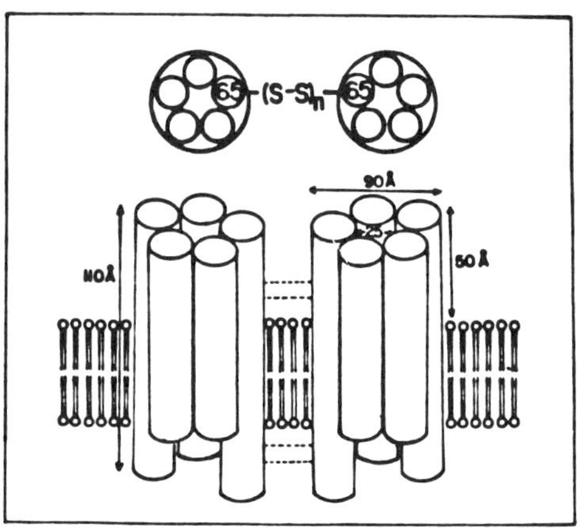

Fig. 1. Physico-chemical properties of purified acetylcholine receptor from Torpedo californica (see Ref. 1).

Torpedo AChR is formed by four different proteins which in *T. californica* have, upon SDS gel electrophoresis, apparent molecular weights of 40,000, 50,000, 60,000 and 65,000, commonly referred to as α, β, γ and δ, respectively (reviewed in 1) (Fig. 2). The true molecular weights of the AChR subunits, calculated from the recently published cDNA sequences (see below), are 50,116 (α), 53,681 (β), 56,279 (γ) and 57,565 (δ), before glycosylation or other post-translational processing. This multisubunit pattern was first described by Raftery and co-workers (3) and confirmed shortly thereafter by Weill and co-workers (4). This complex structure was challenged for many years by reports of pure AChR having simpler subunit patterns, but these erroneous reports are now known to have been the result of proteolytic artifacts (reviewed in 1). The observed subunit pattern of *Torpedo* AChR is very sensitive to proteolysis, and the AChR can be cleaved or nicked to the extent that only the α chain retains its integrity, without significant effect on functional and structural characteristics, including binding capacity for α neurotoxins, antibodies, or cholinergic ligands (5,6). This indicates that extensive proteolytic cleavage of peptide bonds results in minimal fragmentation until the protein is denatured in SDS and indicates a strong, intimate interaction between the AChR subunits.

The debate regarding the subunit composition of the AChR was resolved by the demonstration that the four peptides present in pure AChR preparations are highly homologous proteins (7; Fig. 1). It was further demonstrated by simultaneous quantitative sequencing of the mixture of polypeptides obtained by SDS denaturation of purified, intact AChR that the four subunits are present in the AChR molecule with a stoichiometry of $\alpha_2\beta\gamma\delta$ (Table 1) (8).

The complete amino acid sequence of the precursors of all *Torpedo* subunits has been deduced from corresponding nucleic acid clones obtained by recombinant DNA technology (9-15). This has shown that the subunits have homologous sequences throughout their length. The extensive homology

Table 1. AChR Subunit Stoichiometry

(a) Torpedo californica electric organ

Subunit	Residues			Triton-solubilized AcChR[a] preparation 1	Triton-solubilized AcChR[a] preparation 2	Membrane-bound AcChR[a] preparation 3	Membrane-bound AcChR[a] preparation 4	Average
α	Ala-9	Asn-10	Asn-14	1.93 ± 0.13	1.92 ± 0.14	1.96 ± 0.04	2.05 ± 0.16	1.97 ± 0.12
β	Ser-9	Val-10	Thr-14	1.02 ± 0.08	1.07 ± 0.09	1.03 ± 0.04	1.02 ± 0.01	1.03 ± 0.06
γ	Glu-9	Lys-10	Asp-14	1.00 ± 0.10	1.02 ± 0.21	1.01 ± 0.03	1.00 ± 0.07	1.01 ± 0.10
δ	Asn-9	Asp-10	Val-14	1.04 ± 0.07	1.00 ± 0.13	1.01 ± 0.08	0.93 ± 0.08	0.99 ± 0.09

(b) Electrophorus electricus electric organ

Subunit	Residues	Preparation 1	Preparation 2	Average
α_1	Val-8, Gly-14	1.90 ± 0.19	1.96 ± 0.18	1.93 ± 0.10
β_1	Met-8, Ala-14	1.02 ± 0.02	0.99 ± 0.01	1.01 ± 0.02
γ_1	Ala-8, Tyr-14	1.10 ± 0.27	1.04 ± 0.12	1.07 ± 0.20
δ_1	Ile-8, Glu-14	1.02 ± 0.07	1.00 ± 0.05	1.04 ± 0.06

(c) Mammalian muscle (fetal calf)

Subunit	Residue (cycle 8)	Ratio
α	Val-8	2.16
β	Leu-8	0.95
x	Gln-8	0.92
y	Ile-8	0.98

[a]Values are means ± S.E.M.

of the four subunits argues for a shared ancestry and allows the generation of a pseudosymmetric complex (7) similar to that observed for many complex proteins (16).

The calculated M_r from the cDNA sequences of all four T. californica AChR subunits leads to an M_r of 270,000 for the intact AChR, in excellent agreement with our experimental value of 270,000 ± 30,000 for a preparation consisting of essentially all 9S AChR (17).

Since the AChR formed by the four subunits contains both the binding sites for cholinergic ligands and the cation gating unit (see below), it must span the postsynaptic membrane. A direct demonstration that the AChR molecule protrudes from both sides of the postsynaptic membrane has come from morphological studies of binding of the anti-AChR antibodies to the outside and inside surfaces of Torpedo postsynaptic membrane fragments. By examining replicas of intact and sheared membrane vesicles (18), we found that both surfaces were labeled by antibodies, in agreement with the results obtained by thin sectioning by Tarrab-Hazdai et al. (19).

To determine which of the AChR subunits is a transmembrane protein, we studied the effect of proteases acting either inside or outside membrane vesicles containing all four AChR peptides (20,2,5). All subunits were susceptible to tryptic degradation from either side of the membrane; when trypsin was added outside, all four peptides disappeared at the same rate, whereas when the enzyme was acting from within the vesicles, they were degraded at a rate proportional to their molecular weight, the α-chain being most resistant. Therefore, it may be concluded that all the AChR subunits span the membrane and that they protrude outside to the same extent and inside to an extent approximately proportional to their M_r (scheme of Fig. 1).

All four of AChRs' homologous subunits are in contact with the hydrocarbon core of the lipid bilayer (2,21) and it is therefore possible that all the subunits interact with the surrounding membrane in a related fashion.

So far it has been possible to correlate subunit composition and full physiological functionality for Torpedo AChR only (see below). Because the Torpedo AChR complex contains both the binding site for acetylcholine and the cation gating unit (22,23,24, see below) it has been debated which subunit forms the cation channel. In this respect, the fact that all four subunits of Torpedo AChR are transmembrane proteins (20,5), that they are structurally related (7,14), arranged in a pseudosymmetrical fashion (7) to form a cylindrical structure containing an indentation in the center (25,26) argues in favor of the possibility that more than one subunit - possibly most or even all of them - participate in forming a central cation channel.

Non-Equivalence of the Two α-Subunits

Each α-subunit has a high affinity binding site for cholinergic ligands (see below) but the two sites are not equivalent. Bromoacetylcholine can be used to label one or both α-subunits while maleimidobenzyltrimethoylammonium (MBTA) labels only one (reviewed in 1). The non-equivalence of these two sites can be explained by their different microenvironments since each α-chain must be flanked by different other subunits. In addition, the two α-subunits could be chemically different, in spite of their identical primary structure. The α-subunits of Torpedo AChR have been purified by preparative SDS gel electrophoresis and after proteolysis using V8 protease two prominent peptides of M_r 17,000 and 19,000 were present in similar amounts (27,28). Only the 19K peptide can be labeled by MBTA under mild reducing conditions (28) and only the 17K peptide can be stained for carbohydrate (27,28). We have determined the amino terminal sequence of all the peptides obtained after V8 digestion and found that the 17K and 19K peptides had the same NH_2-terminal sequence starting at residue 47 (27). Since even extensive proteolysis does not convert the 19K peptide to the 17K peptide (27,28), these results indicate that the two α-chains differ in their extent of glycosylation. The binding site on the less glycosylated α-subunit is likely to be more accessible to reducing and affinity labeling reagents, thus explaining the non-equivalence of high affinity ligand binding to the two subunits.

Similar peptides have been demonstrated for mammalian muscle α subunits (29). Evidence has been presented that the α subunit of muscle AChR has a single N-asparagine-linked oligosaccharide chain (29). The only possible site of N-glycosidic linkage in the α subunit of _T. californica_ AChR is the asparagine residue at position 141 in the unique sequence Asn-Cys-Thr (10), which is contained in both peptides discussed above and is more heavily glycosylated in one of them (peptide 17K). It seems therefore that the two mature α subunits differ in the extent of glycosylation and that the presence of sugars makes the disulfide bridge close to the high-affinity binding site less accessible to reducing agents and cholinergic affinity labels. This possibility is supported by the fact that the cysteine residue at position 142 is one of the two most likely candidates for being labeled by bromoacetylcholine and MBTA (10,30,31). If so, the presence of a sugar moiety on Asn-141 of one subunit could hinder the access to this cysteine and could explain the different susceptibility of the two α subunits to affinity labeling.

The nonequivalence of the two α subunits both in _Torpedo_ and in mammalian muscle is supported also by results obtained in studies of the binding of antibodies. It has been demonstrated that anti-_Torpedo_ AChR antibodies can inhibit a maximum of half of the α-bungarotoxin (α-Btx) binding sites (32), which are two for each AChR molecule and are believed to be at least partially formed by the two α subunits (33,34). More recently, similar findings have been obtained for antibody binding to mammalian muscle AChR, and it has been demonstrated that the inhibiting antibody directed was against a carbohydrate antigen, since its binding could be inhibited by sugars (35). All of these results strongly support the notion that the two α subunits differ in the extent of their glycosylation.

The fact that one of the two mature α subunits is more heavily glycosylated raises interesting questions about the synthesis, assembly, and intracellular transport of such subunits. In this respect, the recent observation (36) that each subunit of _Torpedo_ AChR can form homopolymers during biosynthesis and transport may ultimately lead to understanding how the selective glycosylation we discuss here is achieved.

Is Torpedo AChR a Good Model for Other Nicotinic AChRs?

Study of the AChR from sources other than _Torpedo_ electric organ has been hampered by difficulties in obtaining suitable amounts of intact AChR, due to the much lower AChR content and to high levels of protease activity. Similarities in the pharmacology, morphology, antigenicity, and physical properties as well as the frequent presence upon SDS gel electrophoresis of complex polypeptide patterns (reviewed in 1), reminiscent of the subunit pattern of _Torpedo_ AChR, suggested the likelihood of close structural and functional similarities between the AChRs from different sources. To prove the existence and the extent of such similarities between AChRs from different sources is crucial because it would justify the use of analogy from _Torpedo_ AChR to structural and functional characteristics of AChRs from other species and other tissues, such as muscle and brain.

Torpedo (a marine elasmobranch) and Electrophorus (a freshwater teleost) are highly diverged species whose evolution arose separately from the primordial vertebrate stock (~ 400 million years ago) and accordingly the presence of electric organs in these two species is due to convergent evolution. Sufficient AChR can be isolated from Electrophorus electric organ and muscle to conduct structural analysis of their constituent subunits. The purified AChR from Electrophorus electric organ contains four main polypeptides in the same M_r range as Torpedo AChR (Fig. 2). For each of the four peptides, the amino terminal amino acid sequence was determined (37) (Fig. 3) and the four subunits were found to have distinct but homologous sequences with the degree of identity between pairs of subunits ranging between 47.5% and 37.5%. Conservative substitutions tend to further increase the degree of similarity among subunits. Furthermore the sequences are homologous (up to 62.5% identity) to those of Torpedo AChR subunits of comparable M_r. Simultaneous quantitative sequencing of the peptides present in preparations of intact Electrophorus AChR was used to determine the subunit stoichoimetry (37). Molar ratios of 2:1:1:1 were obtained for the subunits of M_r 41,000 (α), 50,000 (β), 55,000 (γ) and 62,000 (δ), respectively (Table 1).

Upon SDS gel electrophoresis of AChR from Electrophorus muscle, four main polypeptides were again observed whose molecular weights were similar but not identical to the molecular weights of the four peptides present in the electric organ AChR (38,39). Each of these peptides has been isolated and their amino terminal sequences have been determined (39). The AChRs from muscle were found to be composed of four homologous proteins of apparent molecular weight 40,500 (α), 50,000 (β), 56,000 (γ) and 63,000 (δ), respectively. No difference was found between the sequenced segments of corresponding subunits from muscle and from electric organ AChR. The small difference in apparent M_r upon SDS gel electrophoresis between corresponding subunits could be due to different degrees of glycosylation rather than to differences in the actual M_r of the polypeptides. Therefore AChRs from different peripheral tissues of Electrophorus have identical primary structure and they are pentameric complexes of four different subunits, one of which (α) is present in two copies in the complex. The subunits are structurally related, thus allowing the formation of a pseudosymmetric supramolecular complex from four different polypeptides, conforming to the generality that complex protein systems are often constructed from identical or related subunits (16). From the apparent M_r of the four subunits and their 2:1:1:1 stoichiometry, a M_r of 249,000 can be calculated for the Electrophorus AChR complex; this value fits with experimental determinations obtained with different approaches (reviewed in 1) and is consistent with the size of Electrophorus AChR as determined by electron microscopy (reviewed in 1).

The high degree of amino acid sequence homology between Torpedo and Electrophorus AChRs demonstrates that the receptor molecule has been highly conserved throughout animal evolution and it suggests that the genes encoding each of the four subunits originated from a single ancestral gene. A genealogical tree showing the evolutionary pathway by which the four contemporary subunits of both Electrophorus and Torpedo AChRs can be generated from a single ancestor sequence via minimum nucleotide substitution is shown in Fig. 3.

The shared ancestry and the high degree of conservation of the AChRs from distant species suggests that all AChRs have a structure similar to that of the receptors from Torpedo and Electrophorus. The remarkable correspondence between subunits of similar M_r (Fig. 3) raises the

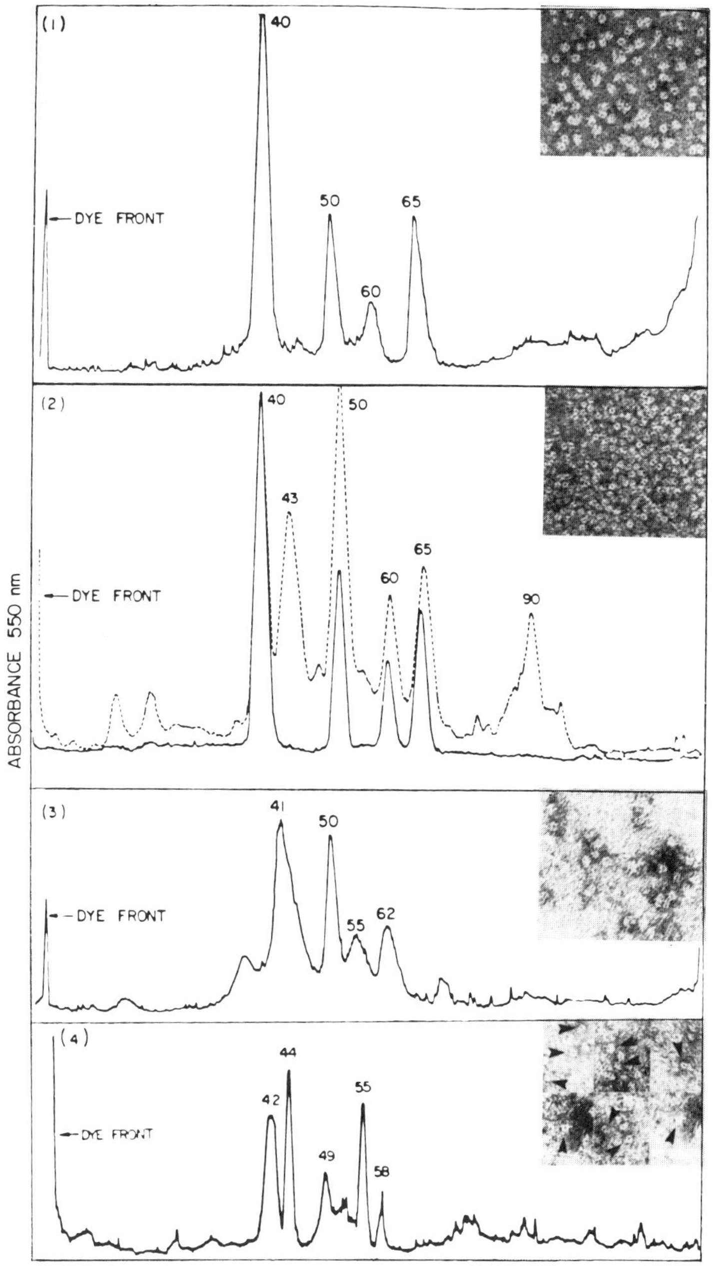

Fig. 2. SDS gel electrophoresis scans of T. californica AChR (1). Electrophorus AChR (3), and fetal calf AChR (4). Lane 2, SDS-gel scans of purified AChR-rich membrane fragments before (---) and after (——) alkali extraction (88,89). All the extrinsic membrane proteins are removed by this treatment and the AChR subunits are the only polypeptides left. All gels were stained with Coomassie blue. Number shown are $M_r \times 10^{-3}$. (Insets) Electromicrographs of these AChRs (negative staining 1,3,4 250,000x; 2, 125,000x).

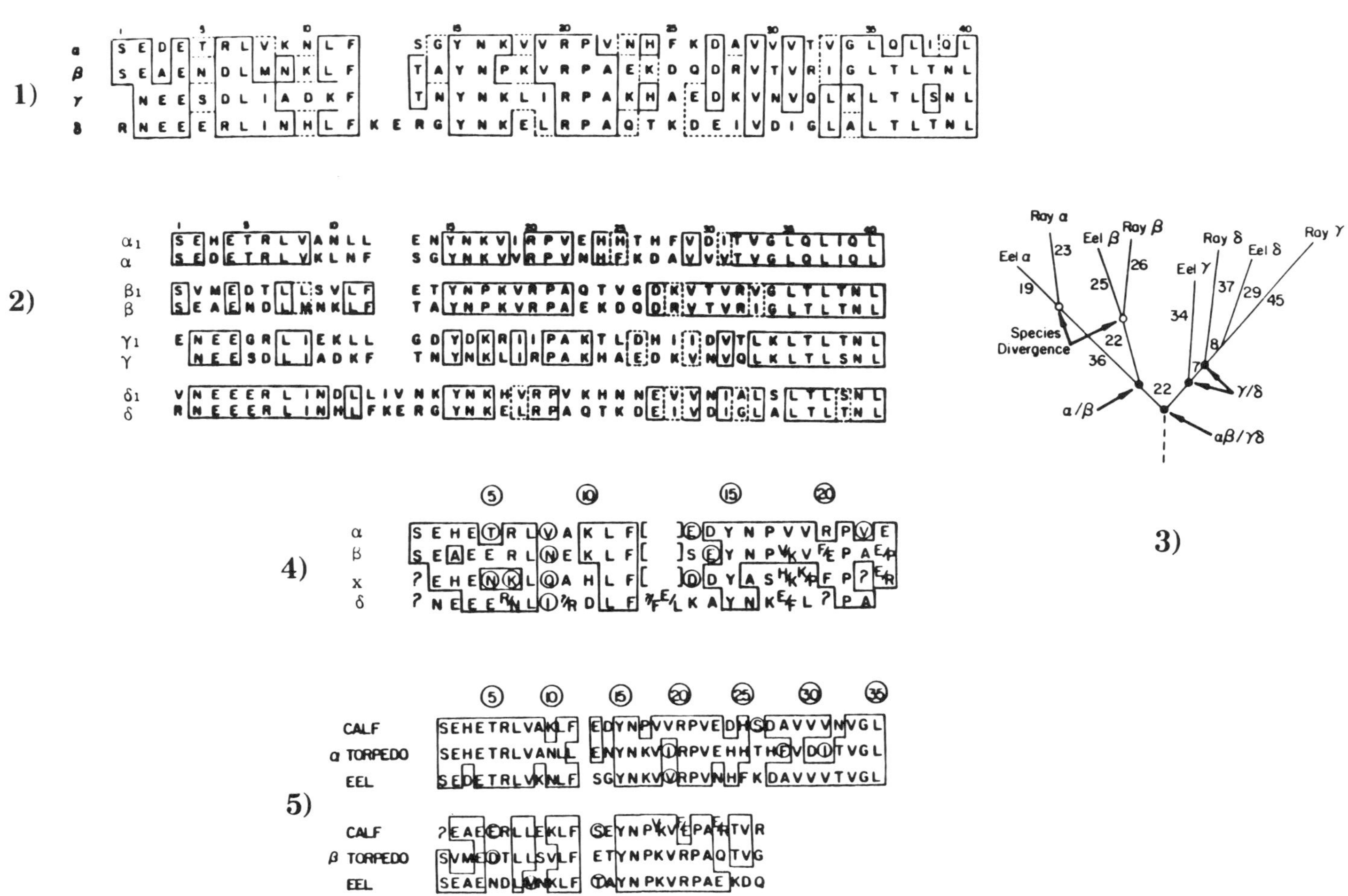
1)
α S E D E T R L V K N L F S G Y N K V V R P V N H F K D A V V V T V G L Q L I Q L
β S E A E N D L M N K L F T A Y N P K V R P A E K D Q D R V T V R I G L T L T N L
γ N E E S D L I A D K F T N Y N K L I R P A K H A E D K V N V Q L K L T L S N L
δ R N E E E R L I N H L F K E R G Y N K E L R P A Q T K D E I V D I G L A L T L T N L
2)
α1 S E H E T R L V A N L L E N Y N K V I R P V E H H T H F V D I T V G L Q L I Q L
α S E D E T R L V K L N F S G Y N K V V R P V N H F K D A V V V T V G L Q L I Q L
β1 S V M E D T L L S V L F E T Y N P K V R P A Q T V G D K V T V R V G L T L T N L
β S E A E N D L M N K L F T A Y N P K V R P A E K D Q D R V T V R I G L T L T N L
γ1 E N E E G R L I E K L L G D Y D K R I I P A K T L D H I I D V T L K L T L T N L
γ N E E S D L I A D K F T N Y N K L I R P A K H A E D K V N V Q L K L T L S N L
δ1 V N E E E R L I N D L L I V N K Y N K H V R P V K H N N E V V N I A L S L T L S N L
δ R N E E E R L I N H L F K E R G Y N K E L R P A Q T K D E I V D I G L A L T L T N L
3)
Ray α
Eel α
Eel β
Ray β
Ray δ
Eel δ
Eel γ
Ray γ
Species Divergence
α/β
γ/δ
αβ/γδ
4)
α
β
x
δ
5)
CALF
α TORPEDO
EEL
SEHETRLVAKLF EDYNPVVRPVEDHSDAVVVNVGL
SEHETRLVANLL ENYNKVIRPVEHHTHFVDITVGL
SEDETRLVKNLF SGYNKVVRPVNHFKDAVVVTVGL
CALF
β TORPEDO
EEL
SVMEDTLLSVLF ETYNPKVRPAQTVG
SEAENDLMNKLF TAYNPKVRPAEKDQ

Fig. 3. Aminoterminal sequences of the four subunits of *Electrophorus* (1) and fetal calf (4) AChR. In 2, corresponding subunits of *Torpedo* (α, β, γ, δ) and *Electrophorus* (α, β, γ, δ) are compared. In 5, the α- and β-subunits from fetal calf, *Torpedo* and *Electrophorus* AChR are compared. In 3, a phylogenetic tree generated from the aminoterminal sequence data of the four AChR subunit types from *T. californica* (α_1, β_1, γ_1, δ_1) and *E. electricus* (α_2, β_2, γ_2, δ_2) by using the best-fit matrix method. Each branch length represents the "accepted point mutations" (PAMs) per 100 amino acid residues that occurred in generating the contemporary subunits of both *Torpedo* and *Electrophorus* AChRs.

possibility that each subunit evolved to perform a crucial, independent function (see below).

Molecular Structure of Nicotinic AChR from Mammalian Muscle

To elucidate the molecular structure of mammalian muscle AChR, we purified this receptor from fetal calf muscle. The purified solubilized AChR had physical properties similar to the *Torpedo* AChR monomers (40,41). Negatively stained preparations of purified receptor show the presence of rosette-like structures, with a diameter of 95 Å and an electron-dense central pit very similar to those found in *Torpedo* and *Electrophorus* AChR (Fig. 1). Upon SDS gel electrophoresis, the purified AChR resolved into five major polypeptides having relative molecular weights of 42,000, 44,000, 49,000, 55,000, and 58,000 (Fig. 1). The peptide of M_r 44,000 is actin since it contains 3-methyl-histidine and binds antiactin antibodies (42), and the peptide of M_r 42,000 is labeled by [^{3}H]bromoacetylcholine (42).

Each polypeptide of fetal calf AChR was submitted to aminoterminal sequence analysis (42). Four homologous sequences were obtained (Fig. 3), and these were present in native AChR complexes in a stoichiometry of 2:1:1:1 (Table 1), which demonstrates that mammalian muscle nicotinic receptor is also a pentameric complex composed of two equivalent and three pseudoequivalent subunits.

Elucidation of the aminoterminal sequence of bovine muscle AChR facilitated the identification and cloning of the corresponding genes, which are now fully sequenced for the α, β, γ, and δ subunits of bovine AChR (43-46), as well as for a novel subunit called ε (47) which mostly resembles the γ subunit, and which could be expressed instead of this subunit at later stages of embryonic development (47). The sequences of the α and δ subunits of muscle AChR from mouse have been published (48,49) as well as the α and γ subunits from human muscle AChR (43,50), and they are all highly homologous proteins which belong to the same gene family. A comparison of the sequences of the α subunit from different AChRs is reported in Fig. 4.

Molecular Structure of the Nicotinic Acetylcholine Receptor from Brain

Some regions of the vertebrate brain contain a nicotinic AChR with pharmacological characteristics similar to but not identical to those of muscle AChR (51,52). In addition, some brain areas and peripheral ganglia contain high-affinity binding sites for α-bungarotoxin and similar snake toxins, which are known to bind to peripheral AChR. However, the identity of the neuronal toxin-binding component with a nicotinic AChR has been disputed (for reviews, see Refs. 53 and 54). In avian and amphibian optic lobe (53), human medullablastoma cells (55), and some sympathetic ganglionic sites (54,56-60) there is evidence in favor of this identity, even though α-neurotoxin binding does not always block receptor function. From avian optic lobe, where α-Btx blocks AChR function (53), the α-Btx binding protein has been purified (61). This protein displays the ligand-binding characteristics of a nicotinic receptor and it can be covalently labeled by the affinity reagent bromoacetylcholine in a manner similar to muscle and electric organ (61). It also shows distinct immunological crossreactivity with an antiserum to mammalian muscle AChR (61) and with some monoclonal antibodies raised against chicken muscle AChR (62,63). Identification of this brain α-Btx-binding protein as a true AChR and resolution of the question as to whether it is encoded by genes of the family described above for the nonneuronal AChRs requires information on the nature and amino acid sequence of its subunits. We have found evidence that the central nervous system (CNS) α-Btx-binding

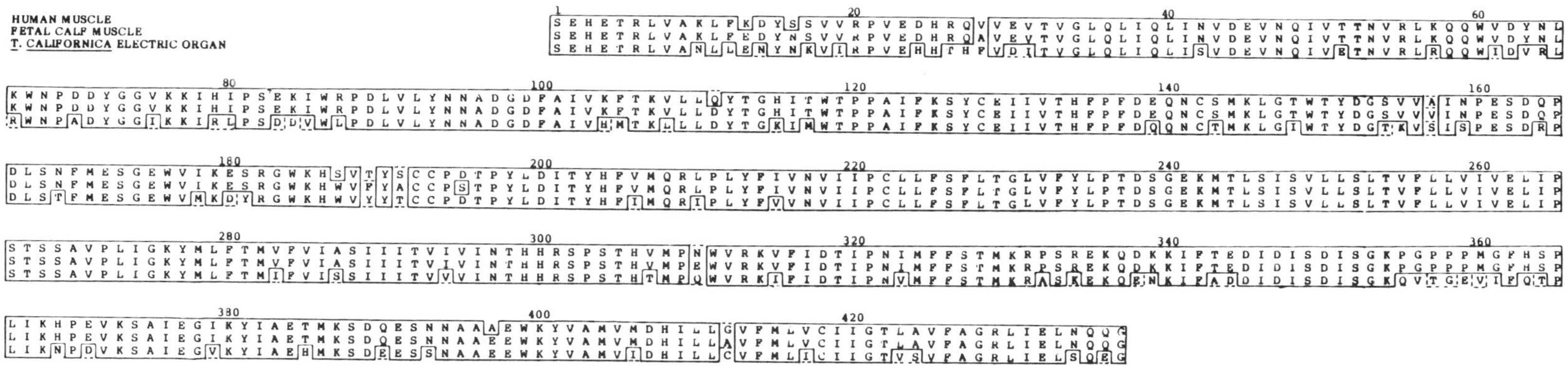

Fig. 4. Alignment of the amino acid sequences of the human (top), calf (middle) and *T. californica* (bottom) AChR α-subunits. The sequences are shown by the one-letter amino acid notation. Sets of identical residues (enclosed with solid lines) and sets of residues considered to be favored amino acid substitutions (enclosed with dotted lines). The *T. californica* sequence has been taken from Ref. 10, and the human and calf sequences from Ref. 43.

protein is a nicotinic AChR structurally homologous to peripheral AChR (64). We used for this study avian brain because it contains α-Btx binding in two well defined areas, i.e., in the optic lobe, where α-Btx blocks the cholinergic responses, and in the cortex, where on the contrary α-Btx, although it binds in a competitive fashion with cholinergic ligands, does not seem to interfere with any cholinergic function.

As a preliminary study, we purified the AChR from avian muscle (65,66). As for the AChRs from other receptor species a multi-subunit pattern was observed. The amino terminal sequence of the peptides present were found to be homologous to the corresponding subunits of nicotinic receptor from other species (65,66), and these four homologous subunits were present in the AChR molecule in the ratio $\alpha_2\beta\gamma\delta$ (66). The amino terminal sequence of the α-subunit is reported in Fig. 5.

We then proceeded to isolate the α-Btx binding component(s) from the two areas of the avian brain discussed above, i.e., optic lobe and cortex, using affinity chromatography and α-Btx as a ligand (64).

The purified α-Btx-binding component from optic lobe showed high-affinity binding of $[^{125}I]$α-Btx with specific activities of 4000-6000 nmol/g of protein in different preparations. The component purified from the cortex had similar α-Btx-binding characteristics and specific activity. In sucrose density gradients the optic lobe AChR sedimented mostly as a peak with a sedimentation coefficient of 10.5 S. For the AChRs from chick or other muscles (1,62) this value is about 9 S, and the molecular size of the optic lobe AChR is indeed significantly larger, as was shown by hydrodynamic determination with ^2H_2O correction for bound detergent (E. A. Barnard, personal communication). Upon SDS gel electrophoresis the α-Btx-binding proteins from either chick optic lobe or brain had very similar peptide compositions. Different preparations consistently contained four or five major components, whose molecular weights ranged between 48,000 and 72,000. All the peptides contained in purified optic lobe AChR were isolated and submitted to amino-terminal amino acid sequencing. The lowest molecular weight component gave a readily identifiable single sequence, reported in Fig. 5. Comparison of this sequence with the known sequences of <u>Torpedo</u>, <u>Electrophorus</u>, bovine and human peripheral AChR subunits (see above), and with the sequence of chick muscle AChR subunits (65,66,67) revealed that the optic lobe sequence is, although different, highly homologous to the subunits of the other AChRs, the highest degree of homology being with the subunits, and among these with the α subunit of <u>Torpedo</u> (Fig. 5). This would indicate that the divergence of peripheral and central nicotinic receptors happened very early during vertebrate evolution, as indicated also by the phylogenetic tree depicted in Fig. 5. Amino-terminal amino acid sequence analysis of the other subunits isolated from optic lobe AChR did not yield any signal above a consistently high background, indicating that these subunits had blocked amino termini. To determine whether this blockage was caused by the isolation of the subunits from SDS gels, intact optic lobe AChR preparations were submitted to simultaneous amino-terminal analysis of all the sequences present in such preparations. Only one signal was detected, which corresponded to the sequence reported above for the lightest component of optic lobe AChR, and it was therefore concluded that amino-terminus blockage pre-existed subunit isolation.

Due to the very low amounts of AChR present in the rest of the chick brain the isolation of individual subunits for sequencing was not feasible. Intact brain preparations were therefore submitted to simultaneous amino-terminal analysis as described above. Only one signal

1

BRAIN	XEFETKLYKELLKNYNPLEXPVAXD
RAY	SEHETRLVANLLENYNKVIRPVEHH
EEL	SEDETRLVKLNFSGYNKVVRPVNHF
CALF	SEHETRLVAKLFEDYNSVVRPVEDH
CHICK	XEHETRLVDDLFRDYSKVVRPVENH

2

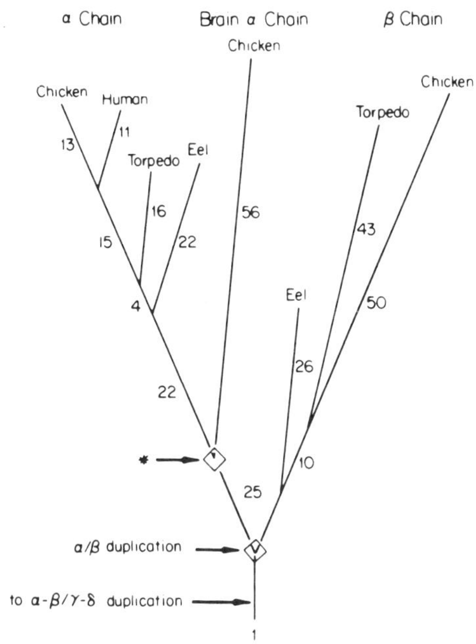

Fig. 5. (1) Comparison of the amino-terminal amino acid sequence (standard one-letter code) obtained for the M_r 48,000 subunit of brain AChR (top) (64) and the amino-terminal sequences of known α subunits of peripheral AChR from *Torpedo* electroplax (7,10), *Electrophorus* electroplax and muscle (37), calf muscle (43), and chick muscle (65). (2) A phylogenetic tree generated from the amino-terminal sequence data of the known α and β subunits of peripheral AChRs (7,10,37,42,43,65) and chicken brain AChR (64) by using the best-fit matrix method. Each branch point represents a nodal or ancestral sequence. The numbers associated with each branch length represent the "accepted point mutations" (PAMs) per 100 amino acid residues that occurred in generating the contemporary AChR subunits. The position of the α/β duplication is arbitrarily located along a 25-PAM segment separating the ancestral sequences of modern α and β subunits. The * indicates the divergence of the α subunit of the brain AChR from the ancestor of the peripheral α subunits.

was present above a high background, which indicated the existence of other proteins with blocked amino termini. The sequence obtained was identical to that of the lowest molecular weight component of optic lobe AChR. These data, the similar gel patterns and the indistinguishable reactivities (61,63) to polyclonal and monoclonal antibodies against muscle AChR, suggest that the AChRs from the optic lobe and from other regions of the chick brain are very similar or even identical complex proteins.

In the absence of sequencing data, other methods had to be used to identify the other peptides as AChR subunits. The peptide of apparent M_r 56,000 has already been shown to be the site of specific alkylation by the affinity reagent bromoacetylcholine (61) and is therefore part of the receptor, since it must contain at least part of one of the binding sites for cholinergic ligands. In the case of peripheral AChR and under similar conditions, only the α subunit is labeled by bromoacetylcholine (reviewed in 1). However, since multiple ligand binding sites have been shown to exist, at least on _Torpedo_ AChR (69-71 and below), and since bromoacetylcholine itself can label other _Torpedo_ AChR subunits under different labeling conditions (unpublished observations), it is reasonable to conclude that in the case of the highly divergent CNS receptor a different subunit can be more easily labeled by bromoacetylcholine. Likewise [^{125}I]α-Btx can be crosslinked to the optic lobe M_r 56,000 subunit by using dimethyl suberimidate (Ref. 61 and unpublished observations). Hence, the M_r 56,000 subunit carries a high-affinity site for AChR and for α-Btx, as is known for the α subunit of the peripheral AChR (1).

Further evidence that other polypeptides are components of the AChR was obtained by virtue of their immunoreactivity (64). For this test, a monoclonal antibody, 7B2, raised against chick muscle AChR, was used. This antibody crossreacts with the α-Btx complex of the AChR purified from chick optic lobe (63) and precipitates the polypeptides of M_r 48,000, 56,000, and 69,000. Note that these peptides include the subunit of lowest molecular weight, for which the sequence data of Fig. 4 reveal a correspondence with the subunit of peripheral AChR, and the subunit of (M_r 56,000) labeled by bromoacetylcholine.

In conclusion, the CNS nicotinic receptor(s) recognized by α-Btx are composed of at least three subunits of M_r 48,000, 56,000 and 69,000. In the absence of amino terminals available for sequencing it was not possible to determine the stoichiometry of the receptor subunits, or to ascertain whether the other two polypeptides present in $NaDodSO_4$ gels of purified CNS receptors, having molecular weights of 62,000 and 72,000, represent one or more contaminants or are AChR subunits or different proteolytic products or post-translational modifications of one subunit. We can exclude the possibility that the M_r 48,000 subunit is a breakdown product of any of the others, since the amino terminus is different from theirs (not blocked) and corresponds in sequence to the true amino terminus of the peripheral AChR subunit. Actin, which is found complexed to the calf muscle AChR (40,42) and which migrates to the M_r 45,000-48,000 region and has a blocked amino terminus, was absent here, as shown by comparison with a chicken gizzard actin standard and by the lack of change of the pattern upon treatment of the AChR with anti-actin antibody.

These results establish that the CNS α-BTX-binding protein is indeed a macromolecule similar to the receptors found in muscle and electric organ. Brain and muscle AChR from the same species, although homologous, must be encoded by different genes which originated from the same ancestral gene. Since the similarity between subunits of central and peripheral receptors from the same animal is much less than between subunits of peripheral AChRs from different animals, it may be concluded that the central and peripheral nicotinic receptors diverged very early during vertebrate evolution (see Fig. 5). This divergence from a common ancestor well explains the pharmacological characteristics of these receptors, which only partially overlap. Similarly, the partial structural identity explains why in certain cases polyclonal and monoclonal antibodies raised against peripheral AChRs failed to recognize central acetylcholine receptors (52,72) and conversely antibodies against chick optic lobe receptor did not bind to peripheral AChR from chick or from *Torpedo californica* (73).

Our data do not exclude the possibility that within the nervous system different nicotinic receptors may exist (see 87). This possibility is supported by the following findings: (i) neuronal proteins exist that do not bind α-Btx but bind anti-muscle AChR antibodies as well as other α-Btx-like snake venom toxins (58,74,75); and (ii) in chick sympathetic ganglia two nicotinic receptors with slightly different pharmacological specificities exist, both of which bind α-Btx-like toxins, but only one of them is functionally blocked as a consequence of α-Btx-neurotoxin binding (56,59).

Cation Gating Function of AChR

Correlation of the known subunit structure of the AChR with its physiological function necessitates the development of quantitative methods to evaluate *in vitro* the efficiency of receptor-mediated cation transport. A rapid kinetic method was developed for this purpose to allow spectroscopic detection of monovalent cation transport (22). A water-soluble fluorophore (8-aminonaphthalene-1,3,6-trisulfonate [ANTS]) is trapped within AChR-enriched vesicles, and the quench of its fluorescence, caused by agonist-mediated inward transport of thallium (Tl) ion, is monitored in stopped-flow experiments allowing detection on a millisecond time scale. The rate of fluorescence decay depends on the number of activated AChRs, and this can be used to determine the level of ion transport mediated by a single AChR molecule. This was shown (22) to correspond to 7 x 10^6 ions per second per receptor in *Torpedo* Ringer's. This value is very close to that (~10^7) estimated for the AChR at the neuromuscular junction *in vivo* (reviewed in 76) and demonstrates that the receptor in isolated vesicles is fully functional. Preparations of membrane-bound AChR composed only of the four homologous polypeptides previously discussed transported Tl^+ as efficiently as membranes containing other protein components. Therefore the pentameric complex $\alpha_2\beta\gamma\delta$ constitutes a complete physiological receptor for postsynaptic depolarization by AcCh. This conclusion is confirmed by reconstitution studies using detergent-solubilized AChR preparations where the AChR polypeptides were the only protein components reassociated with phospholipid vesicles (23). Using a different approach, Boheim and co-workers (24) confirmed that only the four receptor proteins form the active complex.

It should be noted that the midpoints of the dose-response curves for carbamylcholine-mediated flux in native membranes (22) or in reconstituted vesicles (23) are in the millimolar range, in agreement with physiological studies (77). Similarly, values for AChR (~100 μM) and other agonists (see Table 2) are in close agreement. Such high concentrations of agonist necessary for activation of ion transport are at variance with many direct studies of ligand binding to the AChR in either the resting or desensitized state (reviewed in 1).

An important aspect of the flux response is the shape of the dose-response curves. Recently, results (Fig. 6) obtained using a different fluorophore (pyrenetetrasulfonic acid) as a substitute for ANTS clearly demonstrate that the flux response is a cooperative phenomenon with a Hill coefficient of 1.7 (2). An interesting question with respect to mechanism is whether the cooperative flux response is due to multiple ligand binding and, if so, whether such binding is positively cooperative (see below).

TABLE 2. Effect of Agonists on Fluorescence of NBD-Labelled AChR and Comparison with Kd Values from Thallium Flux Data

Agonist	Kd (Fluorescence), mM	Kd (Thallium Flux), mM[1]
SdCh[2]	0.001 ± 0.0005	0.001 ± 0.007
ACh	0.09 ± 0.02	0.057
Carb	0.96 ± 0.17	1.0
Nicotine	0.60 ± 0.21	0.13
PTA	0.47 ± 0.09	0.11
Choline	35	-

[1]Data fitted to model where two ligand binding sites must be occupied for channel to open.

$$kapp = \frac{kmax\ [L]^2}{(Kd + [L])^2}$$

[2]SdCh, suberoyldicholine; ACh, acetylcholine; Carb, carbamylcholine; PTA, phenyltrimethylammonium ion.

The Nicotinic Receptor Contains Multiple Ligand Binding Sites

The cation channel contained in the AChR molecule closes as a consequence of the binding of cholinergic agonists and antagonists (reviewed in 1). A major goal in elucidating the mechanism of AChR function has been to define the ligand binding events leading to activation of the ion channel. Both electrophysiological (78-81) and stopped-flow (2) studies indicated that the opening of the channel results from acetylcholine binding to two low-affinity sites, which operate somehow in a cooperative fashion. Ligand binding also causes the AChR to slowly become desensitized to a state characterized by inability of the channel to open and by high-affinity for ligands (82-85). Since each of the two α-subunits has a high-affinity binding site for cholinergic ligands (reviewed in 1), several models of the linear type have been proposed in which both activation and desensitization of the receptor are controlled by ligand binding to these two sites and this triggers physiological effects by sequential mechanisms involving multiple conformational changes with concomitant alterations in ligand affinity (reviewed in 1). However, the relatively early divergence of the genes encoding the AChR subunits, as well as the extremely high degree of homology between corresponding subunits from distant animal species (see above), suggests that the AChR subunits might have evolved separately to perform different functions and that each subunit could in principle have an independent ligand binding site.

Ligand Binding to AChR

The AChR complex contains binding sites for several types of ligands including agonists, cholinergic antagonists, polypeptide neurotoxins, small molecule neurotoxins, and local anesthetics (reviewed in 1). Some, but not all, of these ligands compete for the same binding sites and therefore multiple receptor-ligand associations are possible. This leads to more complex models of such interactions than generally considered.

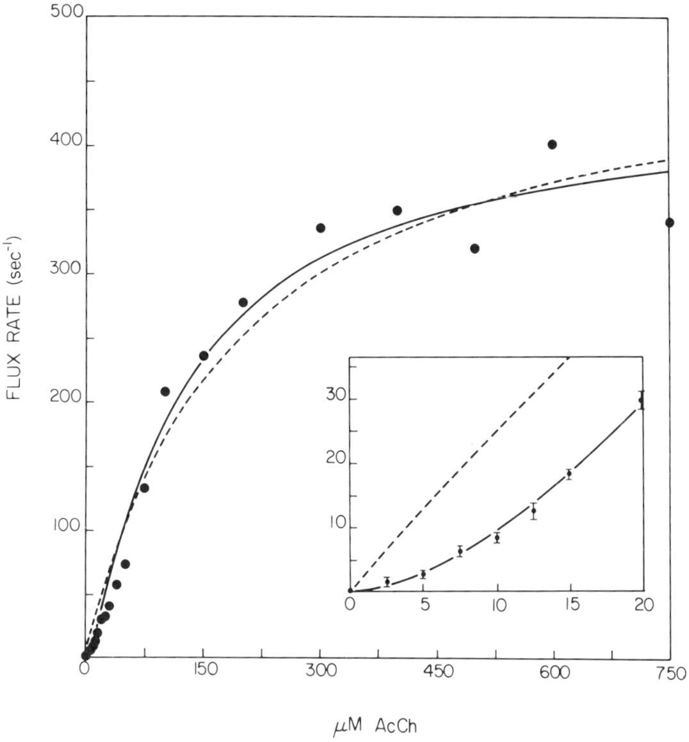

Fig. 6. Effect of AChR on kinetics of thallium influx (50 mM after mixing) monitored by fluorescence changes of pyrene tetrasulfonic acid (17 mM) trapped within AChR-enriched membrane vesicles (——) Fit to two-ligand-binding model $k_{app} = k_{max}/(1+[L]/K_c)$ using $k_{max} = 400\ s^{-1}$ and $K_d = 57\ \mu M$/ (---) fit to single-ligand-binding model $k_{app} = k_{max}[L](K_d + [L])$ with $k_{max} = 486\ s^{-1}$ and $k_d = 186\ \mu M$. (Inset) Expansion of low-concentration region and lines obtained from best-fit parameters above.

Measurements of the equilibrium binding of agonists to the AChR have shown that under these conditions in which the receptor is desensitized and the ion channel is closed, the affinity for agonists is high (Kd ~ 10 nM for ACh). Since channel opening occurs within a few milliseconds of neurotransmitter release, functionally important conformational changes must occur on rapid time scales. A variety of stopped-flow fluorescence techniques have therefore been used to investigate the interaction of agonists with the AChR under pre-equilibrium conditions that may be important in channel activation (reviewed in 1 and 2). In all early reports of such rapid kinetic studies there are two major problems in correlating any observed receptor conformation change with the open channel state: (1) apparent dissociation constants for agonist binding to the resting state of the AChR are lower than those obtained from the concentration dependence of the permeability response; and (2) no observed conformational change is fast enough to be identified with channel opening, which must occur on a millisecond time scale. It is therefore likely that these slow processes are related to desensitization or other inactivation mechanisms which in electrophysiological experiments have been shown to occur on similar time scales (77).

A high-affinity binding site for agonists has been assigned to each α-subunit since following reduction of a reactive disulphide bond near the site it can be labeled by affinity alkylating agents such as bromoacetylcholine (reviewed in 1). In some of the stopped-flow studies discussed above, the fluorescent probe IAS was used, which was shown to react with the same reduced disulfide bond near the high-affinity binding sites on the subunits as shown by the fact that after labeling of the AChR in the IAS, covalent binding of [^{3}H]BrACh was much reduced (86). In view of the proximity of the fluorophore, agonist binding to the α-subunit sites was probably measured and it is notable that all observed conformational changes were too slow to be primary events in channel activation.

It has generally been assumed that the well-characterized sites on the α-subunits are the only binding sites for agonists and that their occupancy leads in a sequential manner to both functional responses of channel opening and desensitization. Recently, however, we have demonstrated the existence of a low-affinity site(s) for agonists which is present under equilibrium conditions and which is distinct from the high affinity sites which can be labeled by bromoacetylcholine (68-71). The characteristics of agonist binding to the low-affinity site(s) suggest their involvement in channel activation.

The membrane-bound AChR was covalently labeled by the fluorescence probe 4-[(iodoacetoxy)ethylmethylamino]-7-nitro-2,1,3-benzoxadiazole (IANBD) without any impairment of function and it was shown that the fluorescence of this probe was enhanced in a saturable manner by the binding of agonists (69,70) with the same dose dependency as the cation flux response giving Kds for carbamylcholine and AChR binding of ~ 1 mM and ~ 100 μM, respectively (see Table 2). Kinetic experiments have shown that the fluorescence change is a monophasic process occurring on a rapid time scale; the rate had a simple hyperbolic dependence on agonist concentration, and the transition must therefore be a conformational change of the receptor-ligand complex, reaching a maximum rate of about 400 s^{-1} for carbamylcholine and about 600 s^{-1} for AChR (69,70).

The low-affinity site(s) is (are) distinct from the high-affinity sites on the α-subunits since even when these latter sites were maximally labelled by BrACh, the fluorescence enhancement and therefore agonist binding to the low-affinity site was unaltered (68,71). In the presence of covalently bound BrACh, no ion flux was observed. The conformational change occurring on agonist binding to the low-affinity site is therefore independent of other transitions which inhibit channel opening.

Several lines of evidence implicate the low-affinity site in channel opening and suggest that activation and desensitization are parallel pathways which are mediated by agonist binding to different sites: (1) the fluorescence enhancement is specific for agonists, is abolished by prior incubation with α-bungarotoxin, and reflects a conformational transition of the receptor-agonist complex; (2) Kd values for agonist binding correspond to those for activation (Table 2); (3) the conformational change is rapid, reaching about 400 s^{-1} for carbamylcholine and about 60 s^{-1} for ACh; (4) Q_{10} is 2.5, in agreement with electrophysiological measurements for channel opening; and (5) the binding is unaffected by desensitization, covalently bound bromoacetylcholine or prior incubation with physiologically active concentrations of curare, HTX, or local anesthetics.

Independent Pathways of Channel Activation and Desensitization

Torpedo AChR therefore appears to have two classes of agonist binding sites - those of high-affinity on the α-subunits which may be labeled by bromoacetylcholine and those of low-affinity which are revealed by NBD

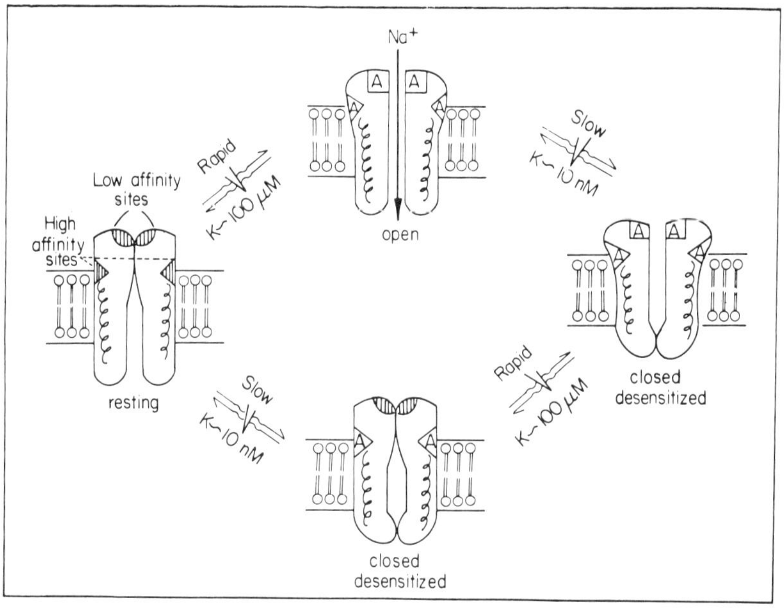

Fig. 7. Schematic representation of the ligand binding mechanism discussed in the text, in which activation and desensitization are parallel and independent mechanisms mediated by agonist binding to different sites.

fluorescence changes. The conformational change occurring on agonist binding to the low affinity site is unaffected by desensitization or by covalent labeling by bromoacetylcholine and the two pathways must therefore be independent. Such a model is illustrated pictorially in Fig. 7. In the resting state the "activation gate" is closed, but in the presence of high concentrations of agonists the low-affinity sites are occupied and the AChR undergoes a rapid conformational transition to an open channel state. Over longer time scales slow conformational transitions in another part of the molecule mediated by agonist binding to other sites (probably those of high equilibrium affinity on the α-subunits) cause the channel to close. Alternatively, at lower concentrations of agonist, occupancy of these latter sites cause slow conformational changes which close an "inactivation gate." Under these circumstances the same conformational change may be induced by agonist binding to the low-affinity sites, but this transition cannot now lead to opening of the ion channel.

CONCLUSION

The structure of the nicotinic AChR has been highly conserved during animal evolution, and in all the species and tissues studied so far, including mammals, it is a pseudosymmetric, pentameric complex of related subunits with very similar physical properties. All subunits of these nicotinic receptors were derived from a common ancestral gene, probably by way of gene duplications occurring very early in animal evolution. The likely existence of such a unique ACh-binding ancestral protein raises the possibility of a shared ancestry for many or even all proteins able to bind ACh, such as the muscarinic receptor, acetylcholinesterase, etc., whose genes would have diverged very early to perform different receptor-mediated or enzymatic functions. The relatively early divergence of the genes coding for the subunits of the nicotinic AChR, as well as their highly conserved primary structure and molecular weight, supports the possibility that the AChR subunits themselves evolved separately to perform discrete functions, such as activation, inactivation, and desensitization. The demonstration of multiple binding sites for agonists in *Torpedo* AChR provides strong support for this hypothesis.

REFERENCES

1. Conti-Tronconi, BM and Raftery, MA: The nicotinic cholinergic receptor: Correlation of molecular structure with functional properties. *Annu. Rev. Biochem.* 51:491-530, 1982.
2. Raftery, MA, Dunn, SMJ, Conti-Tronconi, BM, Middlemas, DS and Crawford, RD: The n-acetylcholine receptor: subunit structure, functional binding sites and ion transport properties. Cold Spring Harbor Symp. *Quant. Biol.* 48:21-33, 1983.
3. Raftery, MA, Vandlen, R, Michaelson, D, Bode, J, Moody, T, Chao, Y, Reed, K, Deutsch, J and Duguid, J: The biochemistry of an acetylcholine receptor. *J. Supramol. Struct.* 2:582-592, 1974.
4. Weill, CL, McNamee, MG and Karlin, A: Affinity-labeling of purified acetylcholine receptor from *Torpedo californica*. *Biochem. Biophys. Res. Commun.* 61:997, 1974.
5. Conti-Tronconi, BM, Dunn, SMJ and Raftery, MA: Functional stability of *Torpedo* acetylcholine receptor: Effects of protease treatment. *Biochemistry* 21:893-899, 1982.
6. Lindstrom, J, Gullick, W, Conti-Tronconi, B and Ellisman, M: Proteolytic nicking of the acetylcholine receptor. *Biochemistry* 19:4791-4799, 1980.
7. Raftery, MA, Hunkapiller, MW, Strader, CD and Hood, LE: Acetylcholine receptor: Complex of homologous subunits. *Science* 208:1454-1457, 1980.
8. Strader, CD, Hunkapiller, MW, Hood, LE and Raftery, MA: Determination of the subunit stoichiometry of the acetylcholine receptor from *Torpedo californica* by amino terminal sequence analysis. *In* Yamamura, H, Olsen, R and Usdin, E, Amsterdam: Elsevier/North Holland, pp. 35-46, 1980.
9. Sumikawa, K, Houghton, M, Smith, JC, Bell, L, Richards, BM and Barnard, EA: The molecular cloning and characterization of cDNA coding for the α-subunit of the acetylcholine receptor. *Nucleic Acids Res.* 10:5809-5822, 1982.
10. Noda, M, Takahashi, H, Tanabe, T, Toyosato, M, Furutani, Y, Hirose, T, Asai, M, Inayama, S, Miyata, T and Numa, S: Primary structure of α-subunit precursor of *Torpedo californica* acetylcholine receptor deduced from cDNA sequence. *Nature* 299:793-797, 1982.
11. Claudio, T, Ballivet, M, Patrick, J and Heinemann, S: Nucleotide and deduced amino acid sequences of *Torpedo californica* acetylcholine receptor γ-subunit. *Proc. Natl. Acad. Sci. USA* 80:1111-1115, 1983.
12. Noda, M, Takahashi, H, Tanabe, T, Toyosato, M, Kikyotani, S, Hirose, T, Asai, M, Takashima, H, Inayama, S, Miyata, T and Numa, S: Primary structures of β- and -subunit precursors of *Torpedo californica* acetylcholine receptor deduced from cDNA sequences. *Nature* 301:251-255, 1983.
13. Ballivet, M, Patrick, J, Lee, J and Heinemann, S: Molecular cloning of cDNA coding for the γ-subunit of *Torpedo* acetylcholine receptor. *Proc. Natl. Acad. Sci. USA* 79:4466-4470, 1982.
14. Noda, M, Takahashi, H, Tanabe, T, Toyosato, M, Kikyotani, S, Furutani, Y, Hirose, T, Takashima, H, Inayama, S, Miyata, T and Numa, S: Structural homology of *Torpedo californica* acetylcholine receptor subunits. *Nature* 302:528-533, 1983.

15. Devillers-Thiery, A, Giraudat, J, Bentaboulet, M and Jean-Pierre, C: Complete mRNA coding sequence of the acetylcholine binding α-subunit of Torpedo marmorata acetylcholine receptor: A model for the transmembrane organization of the polypeptide chain. Proc. Natl. Acad. Sci. USA 80:2067-2071, 1983.
16. Matthews, BW and Bernhard, SA: Structure and symmetry of oligomeric enzymes. Annu. Rev. Biophys. Bioeng. 4:257-317, 1973.
17. Martinez-Carrion, M, Sator, V and Raftery, MA: The molecular weight of an acetylcholine receptor isolated from Torpedo californica. Biochem. Biophys. Res. Commun. 65:129, 1975.
18. Strader, CD, Revel, J-P and Raftery, MA: Demonstration of the transmembrane nature of the acetylcholine receptor by labeling with anti-receptor antibodies. J. Cell Biol. 83:499, 1979.
19. Tarrab-Hazdai, R, Geiger, B, Fuchs, S and Amsterdam, A: Localization of acetylcholine receptor in excitable membrane from the electric organ of Torpedo: Evidence for exposure of receptor antigenic sites on both sides of the membrane. Proc. Natl. Acad. Sci. USA 75:2497, 1978.
20. Strader, CD and Raftery, MA: Topographic studies of Torpedo acetylcholine receptor subunits as a transmembrane complex. Proc. Natl. Acad. Sci. USA 77:5807-5811, 1980.
21. Middlemas, DS and Raftery, MA: Exposure of acetylcholine receptor to the lipid bilayer. Biochem. Biophys. Res. Commun. 115:1075-1082, 1983.
22. Moore, H-PH and Raftery, MA: Direct spectroscopic studies of cation translocation by Torpedo acetylcholine receptor on a time scale of physiological relevance. Proc. Natl. Acad. Sci. USA 77:4509-4513, 1980.
23. Wu, WC-S, Moore, H-PH and Raftery, MA: Quantitation of cation transport by reconstituted membrane vesicles containing purified acetylcholine receptor. Proc. Natl. Acad. Sci. USA 78:775-779, 1981.
24. Boheim, G, Hanke, W, Barrantes, FJ, Eibl, H, Sakmann, B, Fells, G and Maelicke, A: Agonist activated ionic channels in acetylcholine receptor reconstituted into planar lipid bilayers. Proc. Natl. Acad. Sci. USA 78:3586, 1981.
25. Kistler, J. Stroud, RM, Klymkowsky, MW, Lalancette, RA and Fairclough, RH: Structure and function of an acetylcholine receptor. Biophys. J. 37:371-383, 1982.
26. Brisson, A and Unwin, PNT: Quaternary structure of the acetylcholine receptor. Nature 315:474-477, 1985.
27. Conti-Tronconi, BM, Hunkapiller, MW and Raftery, MA: Molecular weight and structural nonequivalence of the mature α-subunits of Torpedo californica acetylcholine receptor. Proc. Natl. Acad. Sci. USA 81:2631-2634, 1984.
28. Gullick, W, Tzartos, S and Lindstrom, J: Monoclonal antibodies as probes of acetylcholine receptor structure. 1. Peptide mapping. Biochemistry 20:2173-2180, 1981.
29. Merlie, MP, Sebbane, R, Tzartos, S and Lindstrom, J: Inhibition of glycosylation with tunicamycin blocks assembly of newly synthesized acetylcholine receptor subunits in muscle cells. J. Biol. Chem. 257:2694-2701, 1982.
30. Cahill, S and Sehmiat, J: An immunochemical approach to the identification of the MBTA binding site of the nicotinic acetylcholine receptor of Torpedo californica. Biochem. Biophys. Res. Commun. 122:602-608, 1984.
31. McCormick, DJ and Atassi, MZ: Localization and synthesis of the acetylcholine-binding site in the α-chain of the Torpedo californica acetylcholine receptor. Biochem. J. 244:995-1000, 1984.
32. Claudio, T and Raftery, MA: Inhibition of α-bungarotoxin binding to acetylcholine receptors by antisera from animals with experimental autoimmune myasthenia gravis. J. Supramol. Struct. 14:267-279, 1980.

33. Haggerty, JG and Froener, SC: Restoration of ^{125}I-α-bungarotoxin binding activity to the α-subunit of Torpedo acetylcholine receptor isolated by gel electrophoresis in sodium dodecyl sulfate. J. Biol. Chem. 256:8294-8297, 1981.
34. Gershoni, JM, Hawrat, E and Lentz, TL: Binding of the α-bungarotoxin to isolated subunit of the acetylcholine receptor of Torpedo californica: quantitative analysis with protein blots. Proc. Natl. Acad. Sci. USA 80:4973-4977, 1983.
35. Roisin, MP, Gu, Y and Hall, ZW: The specificity of a myasthenic serum for developmentally different forms of the acetylcholine receptor. Soc. Neurosci. Abstr. 168:13, 1983.
36. Anderson, DJ and Blobel, G: Identification of homo-oligomers as potential intermediates in acetylcholine receptor subunit assembly. Proc. Natl. Acad. Sci. USA 80:4359-4363, 1983.
37. Conti-Tronconi, BM, Hunkapiller, MW, Lindstrom, JM and Raftery, MA: Subunit structure of the acetylcholine receptor from Electrophorus electricus. Proc. Natl. Acad. Sci. USA 79:6489-6493, 1982.
38. Conti-Tronconi, BM, Hunkapiller, MW, Lindstrom, JM and Raftery, MA: Multisubunit structure and amino-terminal sequences of piscine muscle acetylcholine receptor. J. Recept. Res. 4:801-816, 1984.
39. Gullick, WJ and Lindstrom, J: Comparison of the subunit structure of acetylcholine receptors from muscle and electric organ of Electrophorus electricus. Biochemistry, in press.
40. Gotti, C, Conti-Tronconi, BM and Raftery, MA: Mammalian muscle acetylcholine receptor purification and characterization. Biochemistry 21:3148, 1982.
41. Einarson, B, Gullick, W, Conti-Tronconi, B, Ellisman, M and Lindstrom, J: Subunit composition of bovine muscle AChR. Biochemistry 21:5295-5302, 1982.
42. Conti-Tronconi, BM, Gotti, CM, Hunkapiller, MW and Raftery, MA: Mammalian muscle acetylcholine receptor: A supramolecular structure formed by four related proteins. Science 218:1227-1229, 1982.
43. Noda, M, Furutani, Y, Takahashi, H, Toyosato, M, Tanabe, T, Shimizu, S, Kikyotani, S, Kayano, T, Hirose, T, Inayama, S and Numa, S: Cloning and sequence analysis of calf cDNA and human genomic DNA encoding α-subunit precursor of muscle acetylcholine receptor. Nature 305:818-823, 1983.
44. Tanabe, T, Noda, M, Furutani, Y, Takai, T, Takahashi, H, Tanaka, K, Hirose, T, Inayama, S and Numa, S: Primary structure of β subunit precursor of calf muscle acetylcholine receptor deduced from cDNA sequence. Eur. J. Biochem. 144:11-17, 1984.
45. Takai, T, Noda, M, Furutani, Y, Takahashi, H, Notake, M, Shimizu, S, Kayano, T, Tanabe, T, Tanaka, K, Hirose, T, Inayama, S and Numa, S: Primary structure of γ-subunit precursor of calf-muscle acetylcholine receptor deduced from cDNA sequence. Eur. J. Biochem. 143:109-115, 1984.
46. Kubo, T, Noda, M, Takai, T, Tanabe, T, Kayano, T, Shimizu, S, Tanaka, K, Takahashi, H, Hirose, T, Inayama, S, Kikuno, R, Miyata, T and Numa, S: Primary structure of -subunit precursor of calf muscle acetylcholine receptor deduced from cDNA sequence. Eur. J. Biochem. 149:5-13, 1985.
47. Takai, T, Noda, M, Mishina, M, Shimizu, S, Furutani, Y, Kayano, T, Ikeda, T, Kubo, T, Takahashi, H, Takahashi, T, Kuno, M and Numa, S: Cloning, sequencing and expression of cDNA for a novel subunit of acetylcholine receptor from calf muscle. Nature 315:761-764, 1985.
48. LaPolla, RJ, Mayne, KM and Davidson, N: Isolation and characterization of a cDNA clone for the complete protein coding region of the α-subunit of the mouse acetylcholine receptor. Proc. Natl. Acad. Sci. USA 81:7970-7974, 1984.
49. Boulter, J, Luyten, W, Evans, K, Mason, P, Ballivet, M, Goldman, D, Stengelin, S, Martin, G, Heinemann, S and Patrick, J: Isolation of

a clone coding for the α-subunit of mouse acetylcholine receptor. J. Neurosci. 5:2545-2552, 1985.

50. Shibahara, S, Kubo, T, Perski, HJ, Takahashi, H, Noda, M and Numa, S: Cloning and sequence analysis of human genomic DNA encoding γ-subunit precursor of muscle acetylcholine receptor. Eur. J. Biochem. 146:15-22, 1985.
51. Curtis, DR and Crawford, JM: Central synaptic transmission-microelectrophoretic studies. Annu. Rev. Pharmacol. 9:209-240, 1969.
52. Morley, BJ, Dwyer, DS, Strang-Brown, PF, Bradley, RJ and Kemp, GE: Evidence that certain peripheral anti-acetylcholine receptor antibodies do not interact with brain BuTX binding sites. Brain Res. 262:109-116, 1981.
53. Oswald, RE and Freeman, JA: Alpha-bungarotoxin binding and central nervous system nicotinic acetylcholine receptors. Neuroscience 6:1-14, 1981.
54. Jacob, MH and Berg, DK: The ultrastructural localization of α-bungarotoxin binding sites in relation to synapses on chick ciliary ganglion neurons. J. Neurosci. 3:260-271, 1983.
55. Syapin, PJ, Salvaterra, PM and Engelhardt, JK: Neuronal-like features of TE671 cells: Presence of a functional nicotinic cholinergic receptor. Brain Res. 231:365-377, 1982.
56. Conti-Tronconi, BM, Gotti, C, Paggi, P and Rossi, A: Acetylcholine receptors in the ciliary ganglion and in the iris muscle of the chick: Specific binding and effect on the synaptic transmission of the neurotoxin Naja naja siamensis. Br. J. Pharmacol. 66:33-38, 1979.
57. Marshall, LM: Synaptic localization of α-bungarotoxin binding which blocks nicotinic transmission at frog sympathetic neurons. Proc. Natl. Acad. Sci. USA 78:1948-1952, 1981.
58. Quik, M and Lamarca, MV: Blockade of transmission in rat sympathetic ganglia by a toxin which co-purifies with α-bungarotoxin. Brain Res. 238:385-399, 1982.
59. Ciofi-Luzzatto, A, Conti-Tronconi, BM, Paggi, P and Rossi, A: Binding of Naja naja siamensis α-toxin to the chick ciliary ganglion: a light microscopy autoradiographic study. Neuroscience 5:313-318, 1980.
60. Toldi, J, Joo, F, Adam, G, Feher, O and Wolff, R: Inhibition of synaptic transmission in the rat superior cervical ganglion by intracarotid infusion of bungarotoxin. Brain Res. 262:323-327, 1983.
61. Norman, RI, Mehraban, F, Barnard, EA and Dolly, JO: Nicotinic acetylcholine receptor from chick optic lobe. Proc. Natl. Acad. Sci. USA 79:1321-1325, 1982.
62. Dolly, JO and Barnard, EA: Nicotinic acetylcholine receptors: An overview. Biochem. Pharmacol. 33:841-858, 1984.
63. Mehraban, R, Kemshead, JT and Dolly, JO: Properties of monoclonal antibodies to nicotinic acetylcholine receptor from chick muscle. Biochem. 138:53-61, 1984.
64. Conti-Tronconi, BM, Dunn, SMJ, Barnard, Ea, Dolly, JO, Lai, FA, Ray, N and Raftery, MA: Brain and muscle nicotinic acetylcholine receptors are different but homologous proteins. Proc. Natl. Acad. Sci. USA 82:5208-5212, 1985.
65. Barnard, EA, Beeson, D, Bilbe, G, Brown, DA, Constanti, A, Conti-Tronconi, BM, Dolly, JO, Dunn, SMJ, Mehraban, F, Richards, BM and Smart, TG: Acetylcholine and GABA receptors. Subunits of central and peripheral receptors and their encoding nucleic acids. Cold Spring Harbor Symposia on Quantitative Biology XLVIII, 1983.
66. Beeson, MW, Jackson, JF, Barnard, EA, Conti-Tronconi, BM, Dunn, SMJ, Anderson, TL, Wilderspin, AF and Bell, LD: The chicken muscle acetylcholine receptor: subunit structure and α-subunit cDNA cloning. J. Biol. Chem., in press.
67. Nef, P, Mauron, A, Stalder, R, Alliod, C and Ballivet, M: Structure, linkage and sequence of the two genes encoding the and γ subunits

of the nicotinic acetylcholine receptor. Proc. Natl. Acad. Sci. USA 81:7975-7979, 1984.

68. Conti-Tronconi, BM, Dunn, SMJ and Raftery, MA: Independent sites of low and high affinity for agonists on Torpedo californica acetylcholine receptor. Biochem. Biophys. Res. Commun. 107:123-129, 1982.
69. Dunn, SMJ and Raftery, MA: Activation and desensitization of Torpedo acetylcholine receptor: Evidence for separate binding sites. Proc. Natl. Acad. Sci. USA 79:6757-6761, 1982.
70. Dunn, SMJ and Raftery, MA: Multiple binding sites for agonists on Torpedo californica AChR. Biochemistry 21:6264-6272, 1982.
71. Dunn, SMJ, Conti-Tronconi, BM and Raftery, MA: Separation sites of low- and high-affinity for agonsits on Torpedo californica acetylcholine receptor. Biochemistry 22:2512-2518, 1983.
72. Wonnacott, S, Harrison, R and Lunt, G: Immunological cross-reactivity between the α-bungarotoxin-binding component from rat brain and nicotinic acetylcholine receptor. J. Neuroimmunol. 3:1-13, 1982.
73. Betz, H and Pfeiffer, F: Monoclonal antibodies against the α-bungarotoxin-binding protein of chick optic lobe. J. Neurosci. 4:2095-2105, 1984.
74. Jacob, MH, Berg, DK and Lindstrom, JM: Shared antigenic determinant between the Electrophorus acetylcholine receptor and a synaptic component on chicken ciliary ganglion neurons. Proc. Natl. Acad. Sci. USA 81:3223-3227, 1984.
75. Patrick, J and Stallcup, WB: Immunological distinction between acetylcholine receptor and the α-bungarotoxin-binding component on sympathetic neurons. Proc. Natl. Acad. Sci. USA 74:4689-4692, 1977.
76. Lester, HA: The response to acetylcholine. Sci. Am. 236:106-116, 118, 1977.
77. Adams, PR: Acetylcholine receptor kinetics. J. Membr. Biol. 58:161-174, 1981.
78. Dionne, VE and Stevens, CF: Voltage dependence of agonist effectiveness at the frog neuromuscular junction: Resolution of a paradox. J. Physiol. 251:245-270, 1975.
79. Adams, PR: Relaxation experiments using bath-applied suberyldicholine. J. Physiol. 268:271-289, 1977.
80. Dionne, VE, Steinbach, JM and Stevens, CF: An analysis of the dose-response relationship at voltage-clamped frog neuromuscular junctions. J. Physiol. 281:421-444, 1978.
81. Dreyer, F, Peper, R and Sterz, R: Determination of dose-response curves by quantitative ionophoresis at the frog neuromuscular junction. J. Physiol. 281:395-419, 1978.
82. Katz, B and Thesleff, S: A study of the "desensitization" produced by acetylcholine at the motor end-plate. J. Physiol. 138:63-80, 1957.
83. Rang, MP and Ritter, JM: A new kind of drug antagonism: Evidence that agonists cause a molecular change in acetylcholine receptors. Mol. Pharmacol. 5:394-411, 1969.
84. Rang, MP and Ritter, JM: On the mechanism of desensitization at cholinergic receptors. Mol. Pharmacol. 6:357-383, 1970.
85. Rang, MP and Ritter, JM: The relationship between desensitization and the metaphilic effect at cholinergic receptors. Mol. Pharmacol. 6:383-390, 1970.
86. Dunn, SMJ, Blanchard, SG and Raftery, MA: Kinetics of carbamylcholine binding to membrane-bound acetylcholine receptor monitored by fluorescence changes of a covalently bound probe. Biochemistry 19:5645-5652, 1980.
87. Schneider, M, Adee, C, Betz, H and Schmidt, J: Biochemical characterization of two nicotinic receptors from the optic lobe of the chick. J. Biol. Chem. 260:14505-14512, 1985.

88. Elliott, J, Dunn, SMJ, Blanchard, SG and Raftery, MA: Specific binding of perhydrohistrionicotoxin to *Torpedo* acetylcholine receptor. *Proc. Natl. Acad. Sci. USA* *76*:2576-2579, 1979.
89. Neubig, RR, Krodel, EK, Boyd, ND and Cohen, JB: Acetylcholine and local anesthetic binding to *Torpedo* nicotinic postsynaptic membranes after removal of nonreceptor peptides. *Proc. Natl. Acad. Sci. USA* *76*:690-694, 1979.

THE EFFECTS OF CHRONIC NICOTINE ADMINISTRATION ON BRAIN NICOTINIC RECEPTOR NUMBERS

Allan C. Collins and Michael J. Marks

School of Pharmacy and
Institute for Behavioral Genetics
University of Colorado
Boulder, Colorado 80309

ABSTRACT

Nicotine was infused via the jugular vein into DBA mice, and the effects of such treatment on brain nicotinic receptors and various behavioral and physiological responses to nicotine were determined. Nicotinic receptors were measured using [^{3}H]nicotine or [^{125}I]α-bungarotoxin (α-BTX) as the ligands. Infusion with nicotine resulted in tolerance to several of nicotine's effects. This change in drug sensitivity was accompanied by an increase in Bmax, for both brain nicotinic receptors. A dose-response analysis indicated that [^{3}H]nicotine binding increased at lower infusion doses than did [^{125}I]α-BTX binding. The maximal increase in nicotine binding was seen at a 4-mg/kg/hr infusion dose, while α-BTX binding continued to increase with nicotine dose up to a dose of 8 mg/kg/hr. Tolerance to nicotine's effects correlated best with the changes in [^{3}H]nicotine binding. Time course studies for the onset and offset of tolerance and changes in nicotinic receptors were also carried out. Alpha-BTX binding increased rapidly at the nicotine dose used and reached its new equilibrium value within 2 days. Nicotine binding reached its new equilibrium value in 4 days. Approximately 8 days were required, following termination of nicotine treatment, for nicotine binding to return to control levels, while α-BTX binding was at control levels at the earliest post-treatment test time (4 days). The rates of acquisition and loss of tolerance to nicotine paralleled the changes in brain nicotine binding. These results suggest that chronic tobacco use may lead to changes in brain nicotinic receptors. Such effects could explain not only tolerance to nicotine, but also effects of tobacco on such processes as learning and memory.

Tolerance accompanies chronic treatment with a wide variety of centrally acting drugs, including nicotine. Tolerance to nicotine's effects on locomotor activity has been studied most frequently. The intraperitoneal injection of large (approximately 1 mg/kg) doses of nicotine induces decreased locomotor activity in the rat, and chronic treatment with nicotine results in tolerance to this effect (1,2,3). Similarly, we have demonstrated that high doses of nicotine elicit decreases in locomotor activity in the mouse, and, as is the case with the rat, that chronic treatment with nicotine results in tolerance to this effect (4,5,6,7). We have also observed tolerance to several other behavioral and physiological effects of nicotine (5,6,7). This tolerance cannot be explained by differences in the metabolism of nicotine (4,5). This finding suggests that nicotine tolerance is due to pharmacodynamic rather than pharmacokinetic changes.

Our research has focused on the possible involvement of changes in brain nicotinic receptors in the development of tolerance to nicotine. Nicotinic receptors in the central nervous system have been measured several ways. Snake neurotoxins, primarily α-bungarotoxin (α-BTX), have been widely used for these measurements (8,9), but the use of α-BTX to measure nicotinic receptors in the brain has been criticized because this compound fails to inhibit neurotransmission in several preparations (10-14). Nonetheless, the sites labeled with α-BTX display many properties expected of a nicotinic, cholinergic receptor (9,15,16), and may be related to the incidence of nicotine-induced convulsions (17).

As an alternative to α-BTX, nicotine itself has been used as a ligand for nicotinic receptors (18-21). The binding properties of the site labeled with nicotine are in many ways consistent with those anticipated for a nicotinic, cholinergic receptor. However, most classical nicotinic receptor antagonists inhibit nicotine binding poorly (18,19,21), leading to the suggestion that this binding site may not be cholinergic (18,20). Under the proper conditions, acetylcholine (ACh) has been shown to bind to sites in the brain which are apparently nicotinic, cholinergic receptors (22). These binding sites also have low affinity for most nicotinic antagonists. A recent autoradiographic analysis of radiolabeled nicotine and ACh binding has revealed a strikingly similar distribution in the binding of these two agents in rat brain (23). This, plus the observation that nicotinic cholinergic agonists inhibit the binding of both compounds while nicotinic antagonists inhibit poorly (21,22), suggests that the nicotine and ACh binding sites are very similar, if not identical. The nicotine/ACh binding site is probably different from the α-BTX site (21,23), and we (21) have suggested that the nicotine binding site is a high-affinity nicotinic receptor and that the α-BTX binding site is a lower-affinity nicotinic receptor.

Our studies, as mentioned above, have been designed to assess the relationship between brain nicotinic receptors and tolerance to nicotine. Details concerning most of the studies summarized here may be found in our earlier publications (5,6,7).

METHODS

Materials

The radiolabeled compounds, L-[^{3}H]nicotine (N-methyl-[^{3}H], specific activity 71.9 Ci/mmol), α-[^{125}I]BTX (Tyr-^{125}I, initial specific activity 136.8 Ci/mmol), and L-[^{3}H]quinuclidinyl benzilate (QNB) (benzilic-4,4'-^{3}H, specific activity 30.2 Ci/mmol) were obtained from New England Nuclear Corporation (Newton, MA). [^{3}]Nicotine was repurified by thin layer chromatography and stored frozen with a 4-fold molar excess of mercaptoacetic acid.

Mice

Female mice of the DBA/2Ibg strain were used in these studies. The mice were between 60 and 90 days old at the time of surgery.

Surgery

Cannulas made of silastic tubing were implanted in the right jugular veins of the mice using the method of Barr et al. (24). The day after the surgery, the mice were transferred to individual cages and the implanted cannula was attached to tubing that was connected to a 1-ml syringe mounted on an infusion pump (Harvard Apparatus, South Natick, MA). The animals were subsequently infused with sterile saline at a flow rate of 35 µl/hr. The next day the saline infusion was continued for control animals or drug (nicotine) treatment was started.

Chronic Drug Treatment

The nicotine doses used ranged between 0.1 mg/kg/hr and 8 mg/kg/hr. For the dose-response study, changes in dose were achieved by altering the concentration of nicotine in the infusion solution. In the time course studies, a 4 mg/kg/hr dose was used for the onset of tolerance study. Drug treatment was terminated after 1, 2, 4, 8, and 12 days of treatment. In the offset study, a 7-mg/kg/hr nicotine dose was used. Animals were infused with this dose for 8 days, and tested 0, 4, 8, 12, 16, and 20 days after drug treatment had ended. Whenever a dose exceeding 2 mg/kg/hr was used, nicotine treatment was started at this level, and the dosage was increased by 1 mg/kg/hr each day until the final desired infusion dose was achieved. Animals were infused with this final infusion dose, except in the onset study, for 8-10 days.

Tolerance Tests

Tolerance was measured using a test battery consisting of the following tests: respiratory rate, acoustic startle response, Y-maze activity (both line crossings and rears), heart rate, and body temperature. Every test was run on each animal, which was possible because the tests apparently do not interact with one another (25).

With the exception of the offset study, the animals were tested 2 hr after removal from the infusion chambers. This time period is sufficient to allow for the total elimination of nicotine from the mouse (4,5,26). This same procedure was used for the day 0 mice in the offset studies. Tolerance tests were conducted using doses of nicotine ranging between 0.5 and 2.0 mg/kg, injected i.p. Saline-infused mice were also tested after the injection of each dose of nicotine and of saline. Injection volume was 0.01 ml/g. The test battery was then conducted.

Respiratory Rate

Respiration was measured 1 min after injection of the test solution using a Respiration Monitor (Columbus Instruments, Columbus, OH). Five individual readings of respiration rate were made over a 1-min time period.

Startle Response

After respiration was measured, the mouse was transferred to a Responder V Startle Reflex Monitor (Columbus Instruments, Columbus, OH). Testing was started 3 min after nicotine injection. The mouse was exposed to 10 acoustic stimuli at 10-sec intervals. Both the magnitude of the response and the delay time to that response were recorded.

Y-Maze

After completion of the startle response test, the mouse was transferred to a Y-maze. Both line crossings and rears were recorded during a 3-min test session beginning 5 min after injection.

Heart Rate

After the Y-maze test was completed, the mouse was placed in a restrainer and needle electrodes were inserted through the skin. The electrodes were connected through a preamplifier to an E & M physiograph (Narco Biosystems, Houston, TX). Heart rate was monitored for 6 sec, 9 min after injection.

Body Temperature

Body temperature was measured with a rectal thermometer (Bailey Instruments, Saddlebrook, NJ) 15 min after injection.

Tissue Preparation

After completion of the tolerance test, the mouse brain was removed, and the brain dissected into seven regions: cortex, cerebellum, hindbrain (pons-medulla), hypothalamus, hippocampus, striatum, and midbrain. The cortex was used in the studies reported here. The tissue pieces were placed in 10 vol of HEPES-buffered Ringer's solution (5) and frozen at -70°. On the day of assay, the samples were thawed and homogenized with a glass-teflon homogenizer. The particulate fraction was prepared using the method of Romano and Goldstein (19). Prior to each of the three centrifugation steps, the homogenates were incubated for 5 min at 37° to promote the dissociation of any nicotine that may have been in the tissue (5).

[^{3}H]-L-Nicotine Binding

The binding of [^{3}H]-L-nicotine was measured using a modification of the method of Romano and Goldstein (19) as described previously (21). A single concentration of radiolabeled nicotine (4.1 ± 0.1 nM) was used for the majority of these assays. In addition, the affinity of nicotine for the binding site in cortex was estimated by displacement of the [^{3}H]-L-nicotine by unlabeled L-nicotine. The displacement results were used to calculate the K_D and the B_{max} for ligand binding after conversion of the results to a form suitable for use in Scatchard plots. Blanks were determined by including 10^{-5} M nicotine in the assays.

[^{125}I]-α-BTX Binding

The binding of [^{125}I]-α-BTX was measured as described previously (21) using a ligand concentration of 0.67 ± .04 nM. Five additional concentrations of [^{125}I]-α-BTX were used when K_D and B_{max} values were calculated from Scatchard plots. Blanks were determined by including 10^{-3} M nicotine in the assays.

[^{3}H]-L-QNB Binding

The binding of [^{3}H]-L-QNB was measured using a modification of the method of Yamamura and Snyder (27) as described previously (21). A single concentration of ligand (147 ± 8 pM) was used in most of the studies, but six QNB concentrations were used when the binding parameters (K_D and B_{max}) for this brain region were determined from Scatchard plots of the data. Blanks were obtained by omitting protein from the assays.

Protein Assay

Protein was measured using the method of Lowry et al. (28) with bovine serum albumin as the standard.

RESULTS

Fig. 1 presents dose-response curves for nicotine's effects on the test battery in DBA mice. In this study, mice were infused with saline or 4 mg/kg/hr nicotine for 10 days. Mice were challenged with saline or

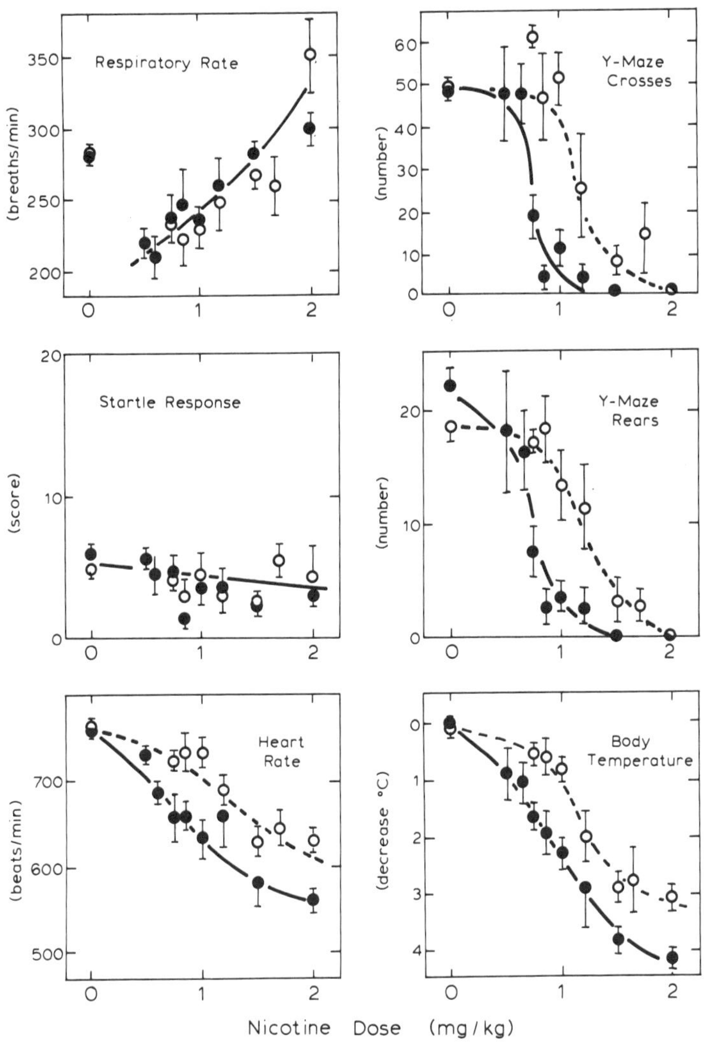

Fig. 1. Dose-Response Curves for Nicotine Effects in Control and Chronically-Treated Mice. The responses of DBA female mice chronically infused with saline (●) or 4.0 mg/kg/hr nicotine (○) were determined after administration of one of the nicotine challenge doses indicated. The responses were measured using the test battery. Results represent mean ± SEM of 5-8 animals at each point.

nicotine, injected i.p., 2 hours after termination of treatment. As can be seen in Fig. 1, low doses of nicotine depressed respiratory rate in the DBA mouse. As the nicotine challenge dose was increased, this depression in respiratory rate decreased. The dose-response curves for the saline-infused and nicotine-treated animal were identical, which indicates that tolerance does not develop to the effects of nicotine on respiratory rate. Similarly, no difference was seen between chronic saline- and nicotine-infused animals for the startle response test. DBA mice, unlike other mouse strains such as the C3H, do not exhibit an enhanced startle response when treated with nicotine (29). The results of the present study indicate that chronic nicotine treatment fails to alter the startle response either by enhancing or depressing it.

The other four tests (heart rate, Y-maze crosses, Y-maze rears, and body temperature) were affected similarly by chronic nicotine treatment. Nicotine elicited a dose-related decrease in these measures in both saline- and nicotine-infused animals. The dose-response curves for the

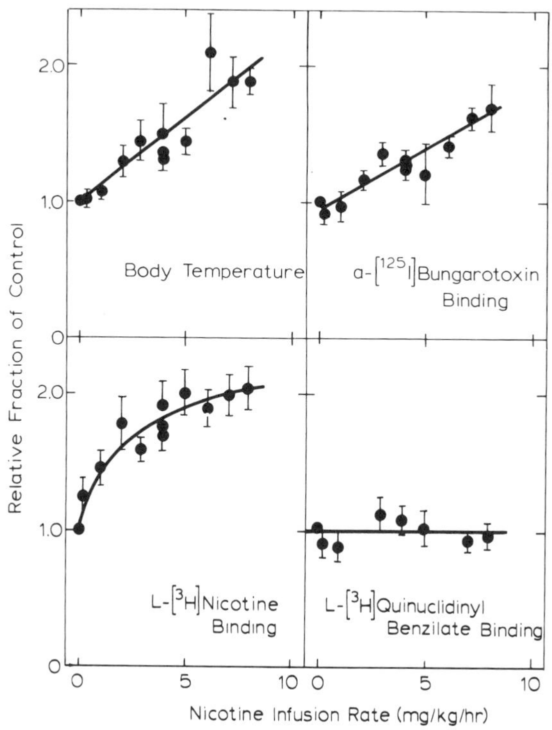

Fig. 2. Changes in Several Parameters as a Function of Nicotine Treatment Dose. DBA female mice were chronically infused with saline or one of several doses of nicotine (0.2 - 8.0 mg/kg/hr). The effects of acute i.p. administration of nicotine on body temperature was measured and the B_{max} values for $[^3H]$nicotine, $[^{125}I]\alpha$-BTX, and $[^3H]$QNB binding were determined in cerebral cortex. With the exception of control data, each point represents the mean ± SEM of separate experiments representing 7-10 individuals. The relative values for body temperature represent the estimated dose required to depress body temperature 2°C, while those for the binding measurements are the ratio of B_{max} values measured for treated animals compared to controls.

nicotine-infused animals were shifted to the right indicating that tolerance to the effects of nicotine had developed.

Fig. 2 presents the results of those studies where the effects of chronic nicotine treatment, at varying doses, on cortical [^{3}H]nicotine, [^{125}]α-BTX, and [^{3}H]QNB binding were determined. The results presented here represent, in each case, changes in ligand binding using a single ligand concentration. We have demonstrated that chronic nicotine treatment does not change the K_D values for any of these ligands (5-7). Therefore, these changes reflect changes in B_{max}. Chronic nicotine infusion resulted in a dose-related increase in the binding of both nicotinic ligands, while QNB binding was unaffected by nicotine treatment. An increase in [^{3}H]nicotine binding was seen at the lowest infusion doses used. The dose-response curve resembles a rectangular hyperbola and a maximal increase in binding was attained at the 4 mg/kg/hr infusion dose. Further increases in nicotine dose did not elicit significantly greater increases in [^{3}H]nicotine binding. Nicotine also elicited an increase in the binding of [^{125}I]α-BTX. No effects were seen at the lower infusion doses. However, as the dose of nicotine was increased, a linear increase in binding was observed. The increase in [^{125}I]α-BTX binding detected at the 8-mg/kg/hr nicotine dose was not as great as the plateau value achieved for [^{3}H]nicotine binding at the 4-mg/kg/hr dose. Therefore, it appears as though the nicotine binding site is more labile than is the α-BTX binding site in that it is affected at lower nicotine doses and increases to a greater degree. It may be, however, that further increases in α-BTX binding may be obtained if the nicotine infusion dose is increased beyond the 8-mg/kg/hr dose range.

Fig. 2 also presents the effects of chronic nicotine treatment on the body temperature test. In this study, we estimated the dose required to elicit a 2-degree decrease in body temperature at each of the infusion doses. Chronic nicotine treatment resulted in tolerance to this measure that was dose-related. As the nicotine infusion dose was increased, the response to the challenge dose decreased such that at the highest infusion doses, nearly twice as much nicotine would have been required to elicit the same hypothermic response as was obtained in control animals.

Fig. 3 presents the results of the study where the time course for the increase in nicotinic receptors and drug tolerance was measured. In this study, mice were infused with 4 mg/kg/hr nicotine for varying time periods. Chronic nicotine treatment did not alter [^{3}H]QNB binding, but was successful in eliciting changes in the binding of the two nicotinic receptor ligands. The binding of [^{125}I]α-BTX increased rapidly, and a new equilibrium value was attained in the first 2 days of infusion. The binding of [^{3}H]nicotine also increased with the new equilibrium value being attained at approximately 4 days of drug treatment. As was the case with the data reported in Fig. 2, the increase in nicotine binding was larger than the increase in α-BTX binding.

Fig. 3 also presents the time course of tolerance development. The "response" value presented in Fig. 3 represents an arithmetic mean of the change in drug response (tolerance) seen for the body temperature, Y-maze crosses, and Y-maze rears measures (see figure legend for a description of the method used to calculate this value). The tolerance to nicotine is maximal after 4 days of drug treatment and parallels the changes in [^{3}H]nicotine binding.

The results of the offset study are presented in Fig. 4. In this study, the mice were treated with a nicotine dose of 7 mg/kg/hr in order to maximize tolerance development and receptor changes. Animals were tested 2 hours after termination of drug treatment (day 0) and every 4

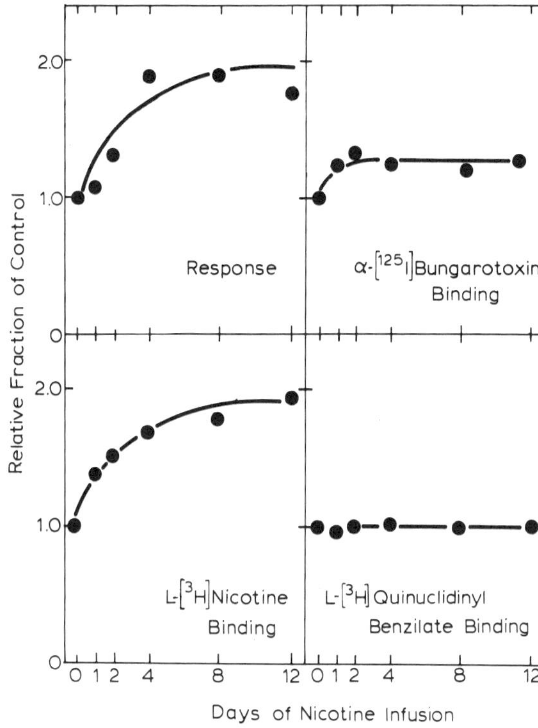

Fig. 3. Changes in Animal Responsiveness and Receptor Levels as a Function of Time of Nicotine Treatment. DBA female mice were infused with 4.0 mg/kg/hr nicotine for the times indicated, tested for nicotine effects on Y-maze crosses and rears and body temperature after administration of 0.75 mg/kg nicotine i.p., and their brains were assayed for [^{3}H]nicotine, [^{125}I]α-BTX, and [^{3}H]QNB binding. The points represent the arithmetic means for the ED_{50} values for the three tests (response) and for the binding of the three ligands in 6 brain regions. All values are expressed relative to control for ready comparison.

days thereafter. Chronic nicotine treatment, once again, was without effect on QNB binding, and QNB binding did not change during the withdrawal phase. Chronic nicotine treatment elicited increases in both nicotine and α-BTX binding as is evident at the 0 day of withdrawal time point. The binding of [^{125}I]α-BTX had returned to control levels at the earliest post-withdrawal testing time (4 days). [^{3}H]Nicotine binding also returned to control levels, but approximately 8 days were required for this process.

Fig. 4 also presents the time course for the loss of tolerance. As was the case in Fig. 3, the "response" measure is an arithmetic average of the four tests that provide evidence for tolerance to nicotine. This tolerance is lost over the withdrawal period in a fashion that most closely resembles the rate of decrease in cortical [^{3}H]nicotine binding.

DISCUSSION

The results of the studies reported here provide further evidence that chronic nicotine treatment results in the development of tolerance to some

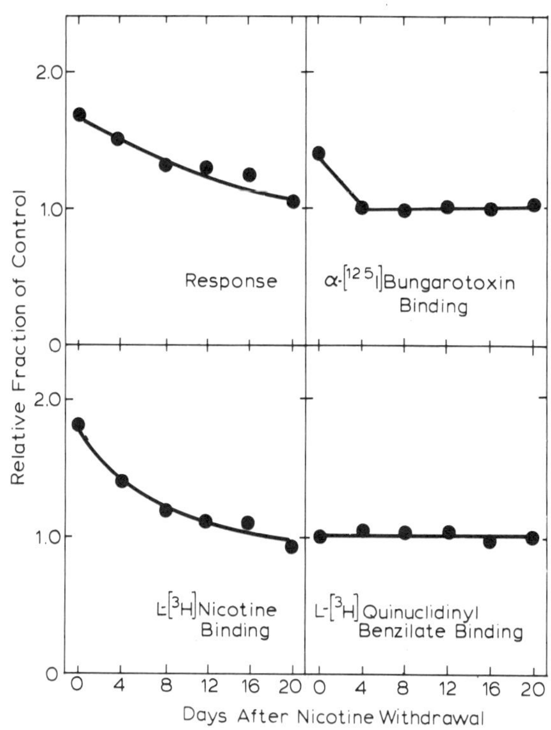

Fig. 4. Changes in Animal Responsiveness and Receptor Levels as a Function of Time After Nicotine Withdrawal. DBA female mice were chronically infused with 7.0 mg/kg/hr nicotine for 8 days after which time treatment was stopped. The effects of an acute i.p. dose of nicotine (1.5 mg/kg) on Y-maze crosses and rears and body temperature were determined at the days indicated and the binding of the three ligands were subsequently determined. Data points were determined as described in the Legend to Fig. 3.

of nicotine's effects. Interestingly, tolerance was not detected for every measure in that chronic nicotine-treated animals respond in the same fashion as do control animals in the respiration rate test. Our previous studies have not detected any change in nicotine elimination from brain, liver, or blood (4,5) following chronic treatment. Therefore, it seems reasonable to conclude that the tolerance observed in these studies reflects a change in tissue sensitivity to nicotine.

Chronic nicotine treatment elicited changes in both of the nicotinic receptors; i.e., an up-regulation of both the [^{3}H]nicotine and [^{125}I]α-BTX binding sites was detected. The nicotine binding site was affected at lower doses but the α-BTX binding site responded more quickly following initiation of treatment with an adequate dose. The time course of acquisition and loss of tolerance most closely resembled the time course for changes in the [^{3}H]nicotine binding site. However, the dose-response analysis suggested that at high doses, at least for hypothermia, changes in the α-BTX binding site may also be important in tolerance development.

The finding that chronic nicotine treatment resulted in tolerance to nicotine that was accompanied by an increase in brain nicotinic receptors is surprising. A number of studies of the effects of chronic drug treatment on receptor number have been carried out (see 30 for a review of some of this literature), and the results obtained suggest that chronic agonist treatment elicits a decrease in receptor number while chronic antagonist treatment elicits an increase in receptor number. The decrease in receptor number is accompanied by tolerance to the agonist, whereas the increase in receptor number parallels the development of supersensitivity to agonists. Based on these findings, we anticipated that chronic treatment with the nicotinic receptor agonist, nicotine, would result in a decrease in the number of brain nicotinic receptors. Clearly, the opposite result has been obtained.

Schwartz and Kellar (22,31) have also studied the effects of chronic nicotine treatment on brain nicotinic receptors. These investigators have used [^{3}H]acetylcholine (ACh) as the ligand, and they have noted that chronic injection (1 mg/kg, twice daily) in rats results in an increase in brain [^{3}H]ACh binding. This increase is maximal at 21 days of treatment. Schwartz and Kellar (22) suggested, and we agree (5), that the unexpected increase in brain [^{3}H]ACh and [^{3}H]nicotine binding may be a result of nicotine-induced desensitization of the ACh/nicotine binding site(s).

Ksir et al. (32) have reported that the chronic nicotine-induced increase in [^{3}H]ACh binding is accompanied by a supersensitivity to nicotine-induced increases in locomotor activity. This is consistent with the observations of Clarke and Kumar (3) who reported that an acute dose of nicotine induced a decrease in locomotor activity which was replaced by activation following chronic nicotine treatment. It may be that nicotine causes an increase in locomotor activity because of activation of the nicotinic receptor(s) in brain and that reduced locomotor activity results from a desensitization of the receptor. Presumably, following nicotine administration some of the nicotinic receptors are in an activated state while others are desensitized. Increased locomotor activity may be seen when a threshold value of the receptors is activated, and a decrease in locomotor activity may occur when a threshold value of the receptors is desensitized. Tolerance to the depressant effects of nicotine on locomotor activity may arise because, as a result of the up-regulation of receptors, more receptors are available for activation following injection with the same nicotine dose.

In summary, we have demonstrated that chronic nicotine treatment results in tolerance to some of nicotine's actions, and this tolerance is accompanied by an up-regulation of brain nicotinic receptors. A further understanding of nicotine tolerance will require the resolution of several problems. Among these are unequivocally determining whether the brain [^{3}H]nicotine binding site is cholinergic, whether the α-BTX binding site has a behavioral function, and ascertaining whether the brain nicotinic receptors desensitize. Future studies in our laboratory hopefully will aid in the resolution of some of these problems.

ACKNOWLEDGEMENTS

The work reported here was supported by a grant from the National Institute on Drug Abuse (DA-03194). Technical assistance on some of the studies was provided by Ms. Elena Romm and Mr. Jerry Stitzel.

REFERENCES

1. Stolerman, IP, Fink, R and Jarvik, ME: Acute and chronic tolerance to nicotine measured by activity in rats. Psychopharmacologia 30: 329-342, 1973.

2. Stolerman, IP, Bunker, P and Jarvik, ME: Nicotine tolerance in rats: Role of dose and dose interval. Psychopharmacologia 34: 317-324, 1974.
3. Clarke, PBS and Kumar, R: The effects of nicotine on locomotor activity in nontolerant and tolerant rats. Br. J. Pharmac. 78: 329-337, 1983.
4. Hatchell, PC and Collins, AC: Influences of genotype and sex on behavioral tolerance to nicotine in mice. Pharmacol. Biochem. Behav. 6: 25-30, 1977.
5. Marks, MJ, Burch, JB and Collins, AC: Effects of chronic nicotine infusion on tolerance development and nicotinic receptors. J. Pharmacol. Exp. Ther. 226: 817-825, 1983.
6. Marks, MJ and Collins, AC: Tolerance, cross-tolerance, and receptors after chronic nicotine or oxotremorine. Pharmacol. Biochem. Behav. 22: 283-291, 1985.
7. Marks, MJ, Stitzel, JA and Collins, AC: Time course study of the effects of chronic nicotine infusion on drug response and brain receptors. J. Pharmacol. Exp. Ther. 235: 619-628, 1985.
8. Morley, BJ, Kemp, GE and Salvaterra, P: α-Bungarotoxin binding sites in the CNS. Life Sci. 24: 859-872, 1979.
9. Oswald, RE and Freeman, JA: Alpha-bungarotoxin binding and central nervous system nicotinic acetylcholine receptors. Neuroscience 6: 1-14, 1981.
10. Chou, TC and Lee, CY: Effect of whole and fractionated cobra venom on sympathetic ganglion transmission. Eur. J. Pharmacol. 8: 326-330, 1969.
11. Miledi, R and Szczepaniak, AC: Effect of Dendroaspis neurotoxins on synaptic transmission in the spinal cord of the frog. Proc. R. Soc. Lond. B. Biol. Sci. 190: 267-274, 1975.
12. Duggan, AW, Hall, JG and Lee, CY: Alpha-bungarotoxin and excitation of Renshaw cells by acetylcholine. Brain Res. 107: 166-170, 1976.
13. Ko, CP, Burton, H and Bunge, RP: Synaptic transmission between rat spinal cord explants and dissociated cervical ganglion neurons in tissue culture. Brain Res. 117: 437-460, 1976.
14. Carbonetto, ST, Fambrough, DM and Muller, KJ: Nonequivalence of α-bungarotoxin receptors and acetylcholine receptors in chick sympathetic neurons. Proc. Natl. Acad. Sci. U.S.A. 75: 1016-1020, 1978.
15. Salvaterra, PM and Mahler, HR: Nicotinic acetylcholine receptor from rat brain: Solubilization, partial purification, and characterization. J. Biol. Chem. 251: 6327-6334, 1976.
16. Salvaterra, PM and Foders, RE: $[^{125}I]_2$α-Bungarotoxin and $[^3H]$quinuclidinylbenzilate binding in central nervous system of different species. J. Neurochem. 32: 1509-1517, 1979.
17. Miner, LL, Marks, MJ and Collins, AC: Classical genetic analysis of nicotine-induced seizures and nicotinic receptors. J. Pharmacol. Exp. Ther. 231: 545-554, 1984.
18. Abood, LG, Reynolds, DT and Bidlack, JM: Stereospecific ^3H-nicotine binding to intact and solubilized rat brain membranes and evidence for its noncholinergic nature. Life Sci. 27: 1307-1314, 1980.
19. Romano, C and Goldstein, A: Stereospecific nicotine receptors on rat brain membranes. Science 210: 647-650, 1980.
20. Sershen, H, Reith, MEA, Lajtha, A and Gennaro, J, Jr.: Noncholinergic saturable binding of (±)-$[^3H]$nicotine to mouse brain. J. Recept. Res. 2: 1-15, 1981.
21. Marks, MJ and Collins, AC: Characterization of nicotine binding in mouse brain and comparison with the binding of α-bungarotoxin and quinuclidinyl benzilate. Mol. Pharmacol. 22: 554-564, 1982.
22. Schwartz, RD and Kellar, KJ: Nicotinic cholinergic receptor binding sites in brain: regulation in vivo. Science 202: 214-216, 1983.

23. Clarke, PBS, Schwartz, RD, Paul, SM, Pert, CB and Pert, A: Nicotinic binding in rat brain: Autoradiographic comparison of ^{3}H-acetylcholine, ^{3}H-nicotine and ^{125}I-alpha-bungarotoxin. J. Neurosci. 5: 1307-1315, 1985.
24. Barr, JE, Holmes, DB, Ryan, LM and Sharpless, SK: Techniques for the chronic cannulation of the jugular vein in mice. Pharmacol. Biochem. Behav. 11: 115-118, 1979.
25. Marks, MJ, Romm, E, Bealer, S and Collins, AC: A test battery for measuring nicotine effects in mice. Pharmacol. Biochem. Behav. 23: 325-330, 1985.
26. Petersen, DR, Norris, KJ and Thompson, JA: A comparative study of the disposition of nicotine and its metabolites in three inbred strains of mice. Drug Metab. Dipos. 12: 725-731, 1984.
27. Yamamura, HI and Snyder, SH: Muscarinic cholinergic binding in rat brain. Proc. Natl. Acad. Sci. U.S.A. 71: 1725-1729, 1974.
28. Lowry, OH, Rosebrough, NH, Farr, AC and Randall, RJ: Protein measurement with the Folin phenol reagent. J. Biol. Chem. 193: 265-275, 1951.
29. Marks, MJ, Burch, JB and Collins, AC: Genetics of nicotine response in four inbred strains of mice. J. Pharmacol. Exp. Ther. 226: 291-302, 1983.
30. Creese, I and Sibley, DR: Receptor adaptations to centrally active drugs. Ann. Rev. Pharmacol. Toxicol. 21: 357-391, 1981.
31. Schwartz, RD and Kellar, KJ: In vivo regulation of [^{3}H]acetylcholine recognition sites in brain by nicotinic cholinergic drugs. J. Neurochem. 45: 427-433, 1985.
32. Ksir, C, Hakan, R, Hall, DP Jr., and Kellar KJ: Exposure to nicotine enhances the behavioral stimulant effect of nicotine and increases binding of [^{3}H]acetylcholine to nicotinic receptors. Neuropharmacology 24: 527-531, 1985.

MULTIPLE NICOTINIC RECEPTORS: NICOTINIC LIGANDS WITH DIFFERENT SPECIFICITIES

Jewell W. Sloan, William R. Martin and W. T. Smith

University of Kentucky, Department of Pharmacology
Research Facility No. 2, MR-104 and Department of Chemistry
Lexington, KY 40536

ABSTRACT

These studies suggest that both (-)- and (+)-nicotine are involved in complex binding site and receptor interactions which may involve positive cooperativity as well as up-regulation of binding at the high affinity sites. Early studies employing (±)-[^{3}H]nicotine as the labeled ligand suggested that (-)- and (+)-nicotine each bound to 3 rat brain membrane sites: a very high affinity site, a high-affinity site and a low-affinity site. It was also found that each of the isomers enhanced the binding of (±)-[^{3}H]nicotine in low concentrations and that (+)-nicotine was more potent and efficacious in producing this effect than (-)-nicotine. Further, both the configuration of Scatchard plots of saturation curves as well as Hill plots at low concentrations (n_H = 1.34) suggested positive cooperativity. In subsequent studies employing optically pure (-)- and (+)-[^{3}H]nicotine it was found in saturation studies that (+)-nicotine occupied only one-tenth as many binding sites as (-)-nicotine. This suggested that (+)-nicotine might be occupying a site different from most of the sites occupied by (-)-nicotine. It was further observed that (-)- and (+)-nicotine increased the binding of both (-)- and (+)-[^{3}H]nicotine. (+)-Nicotine was more effective in this regard than (-)-nicotine. In an effort to identify drugs with specificity for nicotine's binding sites, the binding characteristics of over 50 substituted pyridines, piperidines, and pyrrolidines, as well as other nicotinic drugs, were studied using (±)-[^{3}H]nicotine as the labeled ligand. Drugs were found to differ markedly both in their ability to inhibit as well as to enhance the binding of (±)-[^{3}H]nicotine. Compounds were identified which had specificity for the up-regulatory site in that they enhanced binding over a wide range of concentrations but had little ability to inhibit binding. Of these (±)-2-methylpiperidine was the most specific. Its activity in enhancing the binding of (-)-[^{3}H]nicotine resides in its (+)-isomer whereas both isomers are equally effective in inhibiting the binding of (-)- and (±)-[^{3}H]nicotine. Saturation studies with (±)-[^{3}H]-2-methylpiperidine indicate that it binds to a very high affinity binding site. These data taken together suggest that the very high affinity nicotine binding site is probably the up-regulatory site (Site I) and have identified 2 drugs with stereospecificity for this site, (+)-2-methylpiperidine and (+)-nicotine. The data further suggest that there are separate (-)- [Site II(-)] and (+)-nicotine [Site III(+)] high-affinity binding sites and a low-affinity site (Site IV).

A detailed analysis of prototypic drugs with different patterns of binding specificities indicate that they have different pharmacologic profiles and suggest that the four binding sites may have pharmacologic significance. Studies with (±)-2-methylpiperidine show that it, like low concentrations of (+)-nicotine, produces EEG synchronization after intraventricular injection into the conscious chronic spinal dog whereas higher doses of (+)-nicotine and (-)-nicotine produce desynchronization of the EEG. The prototypic nicotinic agonist (-)-cytisine, which interacts with Site I and Site II (-) but not with Site IV, presents a different pharmacologic profile from that seen with either (-)- or (+)-nicotine or (±)-2-methylpiperidine. Further, (+)-2-methylpiperidine when administered i.v. produces no changes in heart rate but antagonizes (-)-nicotine induced bradycardia in the urethane-pentobarbital anesthetized rat. Finally, arguments have been presented which suggest that the low-affinity rather than the high-affinity site may be the nicotinic cholinergic receptor.

INTRODUCTION

The purpose of this research was to determine if nicotine, in addition to interacting with classic nicotinic cholinergic receptors, might interact with other types of nicotine binding sites and receptors. Abood et al. (1) have shown that some of nicotine's central actions are not readily explained by assuming that it interacts only with nicotinic cholinergic receptors. Binding studies conducted by Yoshida and Imura (2) and by Romano and Goldstein (3) showed that (±)-[^{3}H]nicotine binds to a high and low affinity site in rat brain homogenates. These findings have been confirmed by other laboratories (4, 5, 6). This manuscript will summarize data which indicates that the binding sites of (-)-nicotine and related compounds are even more complex and that these complexities have pharmacologic significance. Table 1 summarizes four postulated nicotine binding sites which have been clearly differentiated.

METHODS AND MATERIALS

A. Binding Studies

The sources of materials, binding procedures and methods of data analysis (6, 7, 8, 9), preparation of (+)-nicotine (9) and resolution of (+)- and (-)-2-methylpiperidine (8) have been described in detail. (±)-[^{3}H]-2-Methylpiperidine, 28 Ci/mM, 99% pure as determined by HPLC and TLC, was synthetized by New England Nuclear. In brief, binding studies were conducted using a P_2 fraction prepared from the whole brains of female Sprague-Dawley rats (6-9). For competition studies, the P_2 suspension, 0.5 ml, (~ 2 mg protein) was added to each tube followed by either 0.25 ml of 50 mM, pH 7.4 Hepes (total binding) or 0.25 ml of the appropriate concentration of unlabeled ligand prepared in 50 mM Hepes, pH 7.4. [^{3}H]Nicotine, 0.25 ml (final concentration of 1.2 x 10^{-8} M for (-)- and 4 x 10^{-8} M for (+)-[^{3}H]nicotine) was then added. Each sample

Table 1. Multiple Nicotine Binding Sites

I.	UP-REGULATORY SITE (VERY HIGH AFFINITY SITE)
II.	(-)-NICOTINE HIGH AFFINITY SITE
III.	(+)-NICOTINE HIGH AFFINITY SITE
IV.	LOW AFFINITY SITE

in saturation studies contained 0.5 ml of the tissue suspension, 0.25 ml of the different concentration of the labeled ligand and 0.25 ml of either Hepes (total binding) or 1 x 10^{-2} M (-)-nicotine [nonsaturable binding for (+)- or (-)-[^{3}H]nicotine]. Nonsaturable binding for (±)-[^{3}H]-2-methypiperidine was determined in the presence of 5 x 10^{-1} M unlabeled (±)-2-methylpiperidine. All determinations were in triplicate and each experiment was repeated three to five times using different homogenate preparations. Each tube was incubated for exactly 1 hour at 4°C. After incubation, the contents of each tube were diluted with Hepes, filtered, and washed 4 times at reduced pressure using Whatman GF/C glass fiber filters soaked in poly-l-lysine, 0.1%.

Nonspecific binding for competition studies employing (±)-, (-)- or (+)-[^{3}H]nicotine was determined in the presence of 1 x 10^{-2} M (-)-nicotine. This concentration was chosen since lower concentrations produced less inhibition and higher concentrations produced no further inhibition of [^{3}H]nicotine binding.

B. Data Analysis

1. Curve Fitting: The binding curves were analyzed by iterative nonlinear curve-fitting programs, ANOVA I and ANOVA II (6) and LIGAND (10) which provide a least squares best fit of the data. K_Ds and site densities for these best fits were estimated as previously described (6-9).

2. Enhancement of Binding (Up-regulation): Up-regulation is defined as a statistically significant ($p < .05$) increase in the amounts of (-)- and (+)-[^{3}H]nicotine specifically bound. The level of up-regulation of binding varied between homogenates. In order to test the statistical significance of this up-regulation several procedures were used. The data obtained by concentrations of the unlabeled ligand which produced an enhancement of (-)-[^{3}H]nicotine binding were analyzed by a two-way analysis of variance (ANOVA) in which the between homogenates and between concentrations variance was partitioned. If the between concentrations variance was not statistically significant then each concentration of the drug for each homogenate was considered to be an independent determination, the data were pooled and the significance of the mean was determined using a t-test. A t-test was also used to assess the statistical significance of the mean enhancement for individual concentrations of the drugs. Unpaired t-tests were used to compare the pooled enhancement produced by the isomers and the racemic mix. Further evidence of up-regulation is provided by an increase in site density in the presence of the drug relative to its density in the absence of the drug as determined by the least squares best fit of the data.

RESULTS AND DISCUSSION

Scatchard analysis of both saturation and inhibition binding studies employing (±)-[^{3}H]nicotine as the labeled ligand revealed greater complexities (6) than those previously reported. Both the configuration of saturation Scatchard plots and Hill plots ($n_H = 1.34$) at low concentrations suggested positive cooperativity. As can be seen from Fig. 1 which shows Scatchard plots of inhibition data in which (±)-[^{3}H]nicotine was used as the labeled ligand, very low concentrations of (-)- and (+)-nicotine both inhibited and enhanced the binding of (±)-[^{3}H]nicotine. (+)-Nicotine was more efficacious in enhancing binding than (-)-nicotine. These data were interpreted as indicating that (-)- and (+)-nicotine were interacting with a very high affinity site (10^{-11} -10^{-13} M) and with a positive cooperativity site (Figure 1, Panels A and B). Further, Scatchard analyses also indicated the existence of both high- and low-affinity sites (Figure 1 Panels B and C).

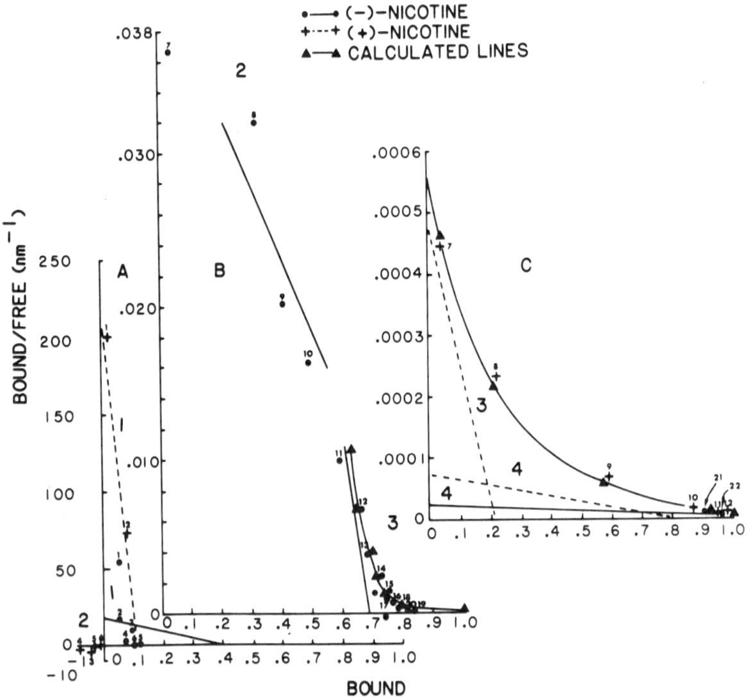

FIG. 1. Scatchard analysis of the inhibition of (±)-[^{3}H]nicotine (2.4 x 10^{-8}) binding by graded concentrations of (-)- and (+)-nicotine (10^{-12} to 10^{-2}M). Bound (shown on the abscissa of panels A, B and C) represents the combined fraction of (±)-[^{3}H]nicotine and unlabeled (-)-nicotine that is saturably bound at each point. Bound/free (the ordinate for each graph) is derived by dividing the fraction bound by the concentration of the free drug. The large numbers refer to line segments representing each binding site, whereas the small numbers refer to unlabeled ligand concentrations designated in Fig. 7 of the publication Sloan et al., 1984 (6). Panel A shows Scatchard conversions of points 1-6 for (-)-nicotine and (+)-nicotine. The solid line 1 refers to a very high affinity site for (-)-nicotine whose K_D was generated by the best fit line to data points 1-6 using ANOVA I and ANOVA II (Sloan et al., 1984). The dashed line is for a very high affinity site for (+)-nicotine whose K_D was calculated from the slope of the line between data points 1 and 2, whereas the large 2 over points 3 to 6 designates the concentrations which produce positive cooperativity. Panel B encompasses (-)-nicotine data points 7-20. The large 2 refers to data points 7-10. The line through points 8-10 represents the best fit line obtained by ANOVA I and ANOVA II. Data points 11-22 were fitted by the LIGAND program. The slopes of the lines calculated from the K_Ds are line 3, Panel B and Line 4, Panel C. Panel C for (+)-nicotine also shows data points 7-12 which were fitted with the LIGAND program. These data were best fitted by lines 3 and 4 yielding the indicated calculated line.

Subsequent evidence supported the concept that there is a nicotinic up-regulatory site. In studies employing (-)-[^{3}H]nicotine as the labeled ligand it was observed that low concentrations of (-)-nicotine (10^{-11} M) significantly increased the binding of (-)-[^{3}H]nicotine (Fig. 2). When (+)-nicotine was used to inhibit the binding of

(-)-[^{3}H]nicotine, concentrations of (+)-nicotine ranging from 10^{-14} to 10^{-8} M produced a significant degree of enhancement of the binding of (-)-nicotine (Fig. 3). When (+)-[^{3}H]nicotine was used as the labeled ligand, a very different pattern of up-regulation of binding was observed in competition studies employing (-)- and (+)-nicotine as the unlabeled ligands. The binding of (+)-[^{3}H]nicotine was enhanced in a dose-related way by both (-)- and (+)-nicotine and they were about equipotent in this regard since their dose-response lines had about the same slopes and intercepts (Figs. 4 and 5). (+)-Nicotine was a more efficacious up-regulator of (±), (-)- and (+)-[^{3}H]nicotine binding than (-)-nicotine. Both (-)- and (+)-nicotine also produced a dose-related inhibition of (+)-[^{3}H]nicotine binding. No evidence of a very high affinity site was obtained in these competition studies using either (-)- or (+)-[^{3}H]nicotine as the labeled ligand.

The unlabeled (+)-nicotine used in these studies was prepared by racemizing (-)-nicotine and the isomers were resolved by preparing the (+)- or (-)-tartrate salt and recrystallizing the two salts to constant melting points. The specific rotation of (+)-nicotine was +10° (C = 2.0 in H_2O) and that of (-)-nicotine was -8.8° (C = 2.0 in H_2O).

In the study of the ability of over 50 substituted pyridines, piperidines, and pyrrolidines as well as other nicotinic drugs to inhibit the binding of (±), (-)- and (+)-[^{3}H]nicotine (6-9), it was found that (±)-2-methylpiperidine exhibited a high degree of specificity in enhancing the binding of (+)- and (-)-[^{3}H]nicotine (Fig. 6). The isomers of 2-methylpiperidine were resolved (8). As can be seen from Fig. 7, (+)-2-methylpiperidine produced a statistically significant enhancement of (-)-[^{3}H]nicotine binding over a concentration range of 10^{-12} to 10^{-6} M, whereas (-)-2-methylpiperidine was ineffective in enhancing the binding of (-)-[^{3}H]nicotine. Both isomers, however, inhibited the binding of (-)-[^{3}H]nicotine over a concentration range of 10^{-5} to 10^{-1} M. Saturation studies have been conducted with (±)-[^{3}H]-2-methylpiperidine and these data indicate a site with a K_D of 1.8 x 10^{-8} M (Table 2). From these data the following conclusions are drawn. At the present time

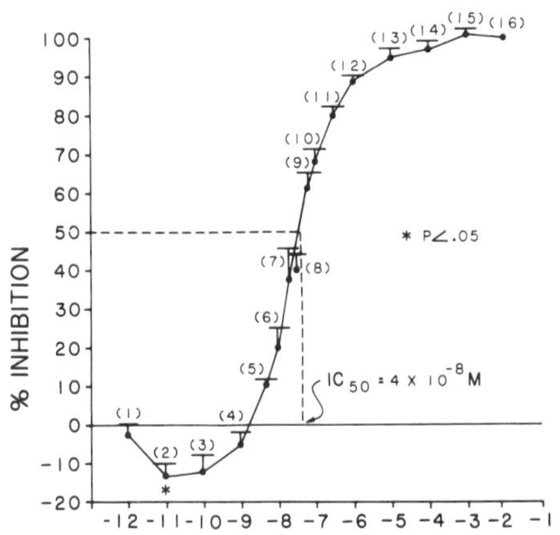

FIG. 2. The inhibition of (-)-[^{3}H]nicotine (1.2 x 10^{-8} M) by graded concentrations of (-)-nicotine. The inhibition produced by (-)-nicotine is shown as a percentage of the maximum displacement achieved by 10^{-2} M (-)-nicotine. Each point is the mean of 4 experiments with its standard error.

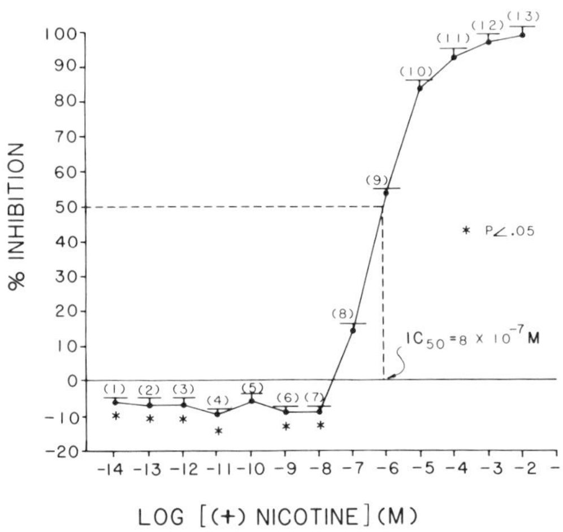

FIG. 3. The inhibition of (-)-[^{3}H]nicotine (1.2 x 10^{-8} M) binding by graded concentrations of (+)-nicotine. Each point is the mean of 4 experiments shown with its standard error.

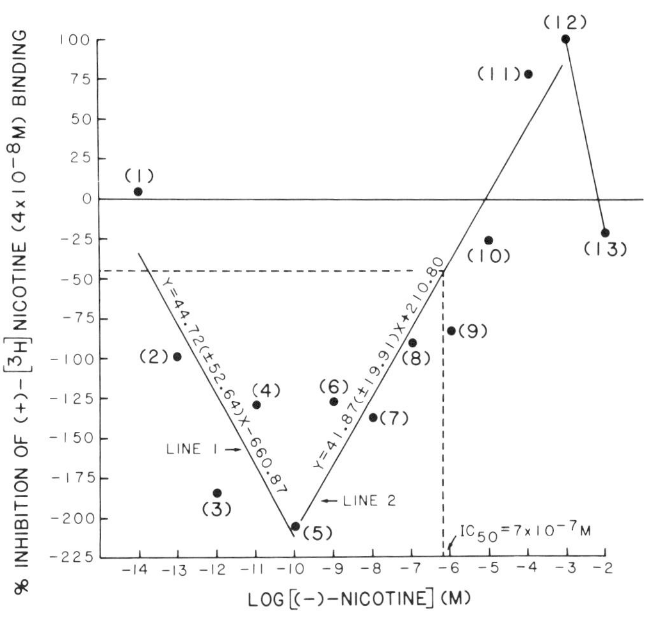

FIG. 4. The inhibition of (+)-[^{3}H]nicotine (4 x 10^{-8} M) binding by graded concentrations of unlabeled (-)-nicotine. Each point is the mean of 4 experiments. The least squares fit for lines 1 (points 1-5) and 2 (points 5-12) are shown with the equation for the fit. The slope for each line ± its 95% confidence limits was calculated by regressing the % inhibition against the log of the molar concentration of (-)-nicotine.

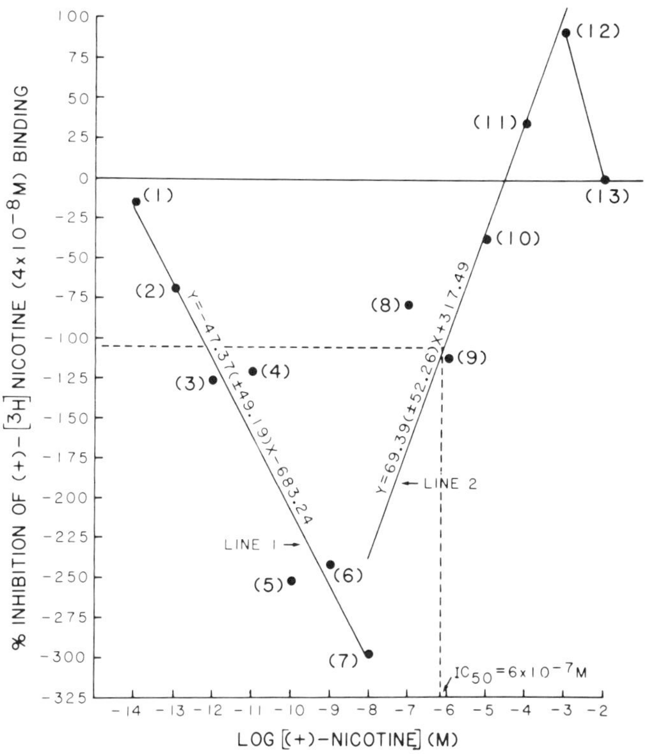

FIG. 5. The inhibition of (+)-[^{3}H]nicotine (4 x 10^{-8} M) binding by graded concentrations of unlabeled (+)-nicotine. Each point is the mean of 4 experiments. The least squares fit for lines 1 (points 1-7) and 2 (points 7-12) are shown with the equation for the fit. The slope for each line (± its 95% confidence limits) was calculated by regressing the % inhibition against the log of the molar concentration of (+)-nicotine.

it is most probable that the very high affinity site seen in studies with (±)-[^{3}H]nicotine is the up-regulatory site. Several lines of evidence would clearly indicate that the up-regulatory site can be differentiated from other binding sites. (1) A highly stereospecific ligand has been identified, (+)-2-methylpiperidine, which enhances the binding of (-)-[^{3}H]nicotine in concentrations which are 10^8-fold less than concentrations which inhibit the binding of (-)-[^{3}H]nicotine. (2) (+)-Nicotine is both more potent and more effective in up-regulating the binding of (-)-[^{3}H]nicotine than is (-)-nicotine. (3) Whereas both (+)-nicotine and (-)-nicotine are approximately equipotent in enhancing the binding of (+)-[^{3}H]nicotine, (+)-nicotine is more efficacious in this regard. These data further indicate that each of these drugs has its own unique stereospecificity.

Table 2 also summarizes the results of saturation studies conducted with (-)- and (+)-[^{3}H]nicotine. It is of particular importance that these studies indicated that there were over 10 times as many (-)-nicotine high-affinity binding sites as (+)-nicotine high-affinity binding sites. As others have found, (-)-nicotine has an affinity approximately 10 times higher for its high-affinity binding site than (+)-nicotine has for its high-affinity binding site (11). As has been shown, both (+)- and (-)-nicotine have a greater ability to up-regulate the binding of

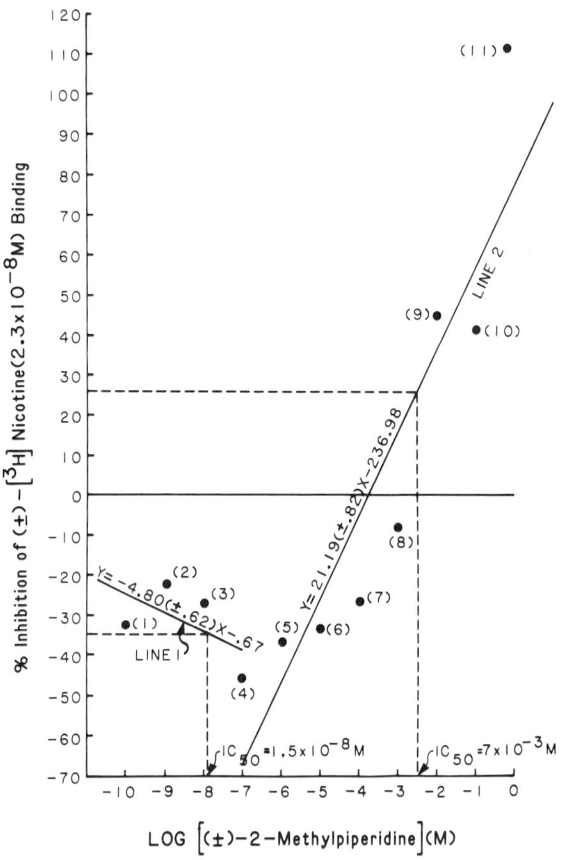

FIG. 6. Inhibition of (±)-[^{3}H]nicotine binding by graded concentrations (±)-2-methylpiperidine. Each point is the mean of four experiments. The curve has a significant quadratic component. A significant regression on dose was produced by lines 1 and 2, which are shown with their IC_{50} values, slopes ± the 95% confidence limits, and their intercepts calculated by regressing the percent inhibition against the log of the femtomolar concentration of (±)-2-methylpiperidine.

(+)-[^{3}H]nicotine than they have in up-regulating the binding of (-)-[^{3}H]nicotine. These observations suggest that there is a separate high-affinity binding site that is preferentially occupied by (+)-nicotine [III (+)] in addition to another high-affinity site that is preferentially occupied by (-)-nicotine [II (-)] (Table 1). These two pools appear to differ in their binding site density and differ with regard to their ability to be up-regulated.

Fig. 8 compares the ability of (-)-cytisine, (-)- and (+)-nicotine to enhance the binding of (±) and (-)-[^{3}H]nicotine. As can be seen in panel "A" of Fig. 8 where (±)-[^{3}H]nicotine was used as the labeled ligand, it would appear that neither (-)-cytisine nor (-)-nicotine enhanced the binding of (±)-[^{3}]nicotine whereas (+)-nicotine did. However, when (-)-[^{3}H]nicotine was employed as the labeled ligand, (-)-cytisine, (-)-nicotine, and (+)-nicotine all clearly enhanced the binding of (-)-[^{3}H]nicotine, (Panel "B", Fig. 8). Hence the binding characteristics of drugs differ depending on which labeled nicotine ligand is employed.

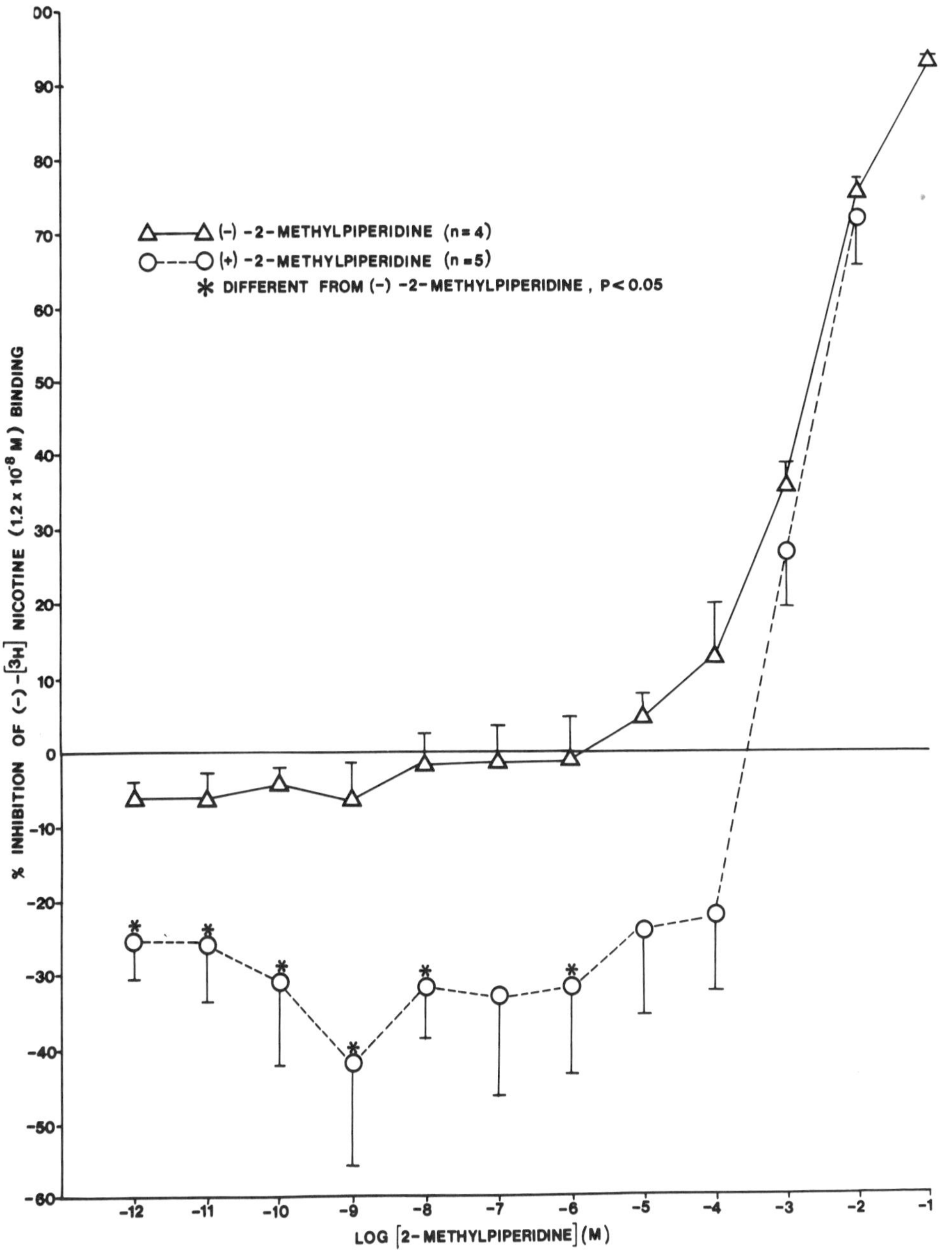

FIG. 7. The competition of (–)-[^{3}H]nicotine binding by graded concentrations of (–)- and (+)-2-methylpiperidine. Each value is the mean (±) its standard error.

Fig. 9 compares the ability of (+)-nicotine, (–)-nicotine, and (–)-cytisine to alter the binding of (+)-[^{3}H]nicotine. (+)-Nicotine has the greatest ability to enhance the binding of (+)-[^{3}H]nicotine, (–)-nicotine has a somewhat lesser ability, and (–)-cytisine has the least ability to enhance the binding of (+)-[^{3}H]nicotine. On the other hand, (–)-cytisine showed the greatest ability to inhibit the binding of (+)-[^{3}H]nicotine followed by (–)-nicotine and then (+)-nicotine. These

Table 2. A Comparison of K_Ds and Site Densities Obtained from Saturation Binding Studies.

Ligand	Site No.	K_D (M)	Site Density fm/Mg Tissue
(-)-[^{3}H]nicotine	I	5.8×10^{-10}	.11
	II (-)	5.4×10^{-8}	4.50
	IV	1.4×10^{-4}	1000.00
(+)-[^{3}H]nicotine	I	6.7×10^{-10}	0.01
	III (+)	2.0×10^{-7}	0.30
	IV	1.3×10^{-5}	62.40
(±)-[^{3}H]-2-Methylpiperidine	I	1.8×10^{-8}	0.15

data show again that the ability of these drugs to enhance binding is dissociated from their ability to inhibit binding. Further, these data show that these drugs differ in their ability to enhance binding. These findings taken together suggest that the drugs may act on the up-regulatory site (Site I) as strong or partial agonists. Table 3 compares four drugs with regard to their relative specificities based on their ability to inhibit the binding of (-)-[^{3}H]nicotine. As can be seen, all four drugs interact with the very high affinity up-regulatory site (Site I). On the other hand, (-)-nicotine and (-)-cytisine interact with Site II (-), the (-)-nicotine high-affinity site, whereas (+)-nicotine interacts with Site III (+), the (+)-nicotine high-affinity site. (+)-2- Methylpiperidine interacts only with the up-regulatory site (Site I) in concentrations that would be of pharmacologic interest. Cytisine differs from (-)-nicotine in that it interacts with the up-regulatory site (Site I) and the high-affinity site [Site II (-)] but not with the low-affinity site (Site IV).

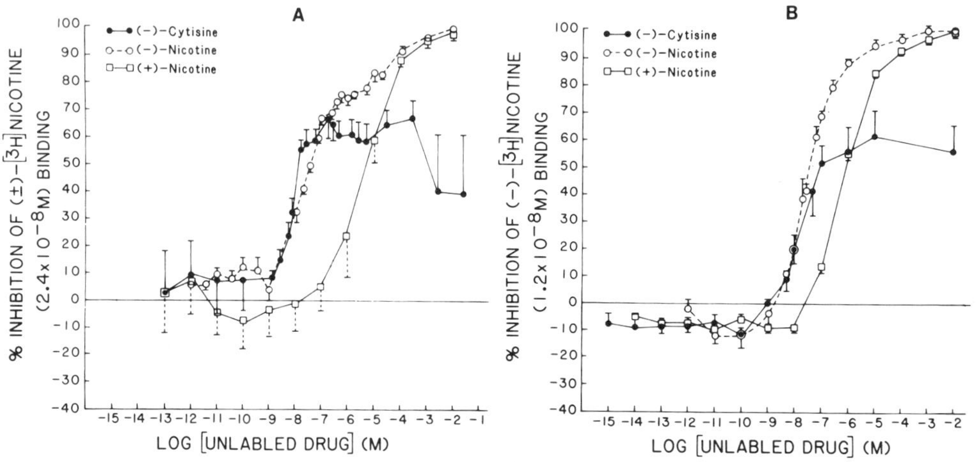

FIG. 8. Inhibition studies employing (±)-[^{3}H]nicotine (panel A) and (-)-[^{3}H]nicotine (panel B) in which (-)-cytisine, (-)-nicotine and (+)-nicotine are compared. Each point, shown with its standard error, is the mean of 4 experiments.

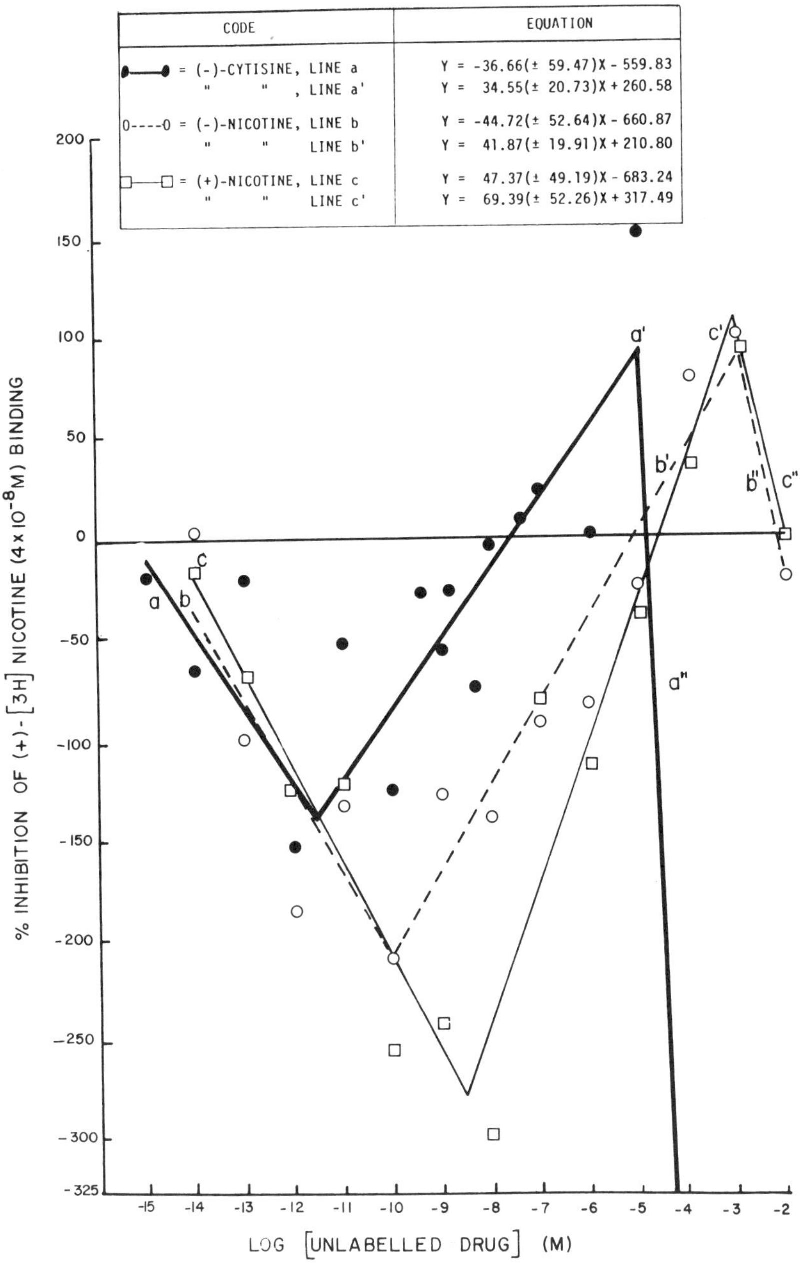

FIG. 9. A comparison of the inhibition of (+)-[^{3}H]nicotine by graded concentrations of (-)-cytisine (lines, a, a', a") with that of (-)-nicotine (lines b, b', b") and of (+)-nicotine (lines c, c', c"). Each point is the mean of 4 experiments except for (-)-cytisine which is the mean of 3 experiments. The least squares fit for lines a and a', b and b' and c and c' are shown with the equation for the fit. The slope for each line (± its 95% confidence limits) was calculated by regressing the % inhibition against the log concentration of the unlabelled drug. Slopes were not calculated for lines a", b" and c".

Table 3. Binding Parameters of Some Compounds to the Rat Brain P_2 Preparation Using (-)-[^{3}H]nicotine (1.2 x 10^{-8} M) as the Labeled Ligand.

Compound	Binding Site	K_D (M)	Site Density Moles/MG Tissue	Dose Range Producting Enhancement (M)
(-)-Nicotine	I			10^{-12} -10^{-9}
	II (-)	2.5x10^{-8}	2.8x10^{-15}	
	IV	1.5x10^{-5}	1.3x10^{-13}	
(+)-Nicotine	I			10^{-14} - 10^{-8}
	III (+)	6.7x10^{-7}	6.0x10^{-14}	
	IV	1.5x10^{-4}	1.5x10^{-12}	
(-)-Cytisine	I			10^{-15} - 10^{-10}
	II (-)	3.7x10^{-9}	8.5x10^{-16}	
(+)-2-Methylpiperidine	I			10^{-12} - 10^{-4}

Using these ligands of different specificities, studies were conducted in intact beagle-type dogs with implanted 4th ventricle cannulae. A variety of physiologic parameters were studied including skin twitch reflex latency, the parietal electroencephalogram, systolic blood pressure, heart rate, pupillary diameter, and rectal temperature (Table 4). Not only do the binding patterns of these four drugs differ but they

Table 4. The relative potencies of some compounds at the up-regulatory, high and low affinity (-)-[^{3}H]nicotine binding sites in the rat brain P_2 preparation: Comparison with their effects on some physiologic measures (i.v. infusion) in the urethane-pentobarbital anethetized rat and conscious spinal dog (4th ventricle injection).

	RAT								DOG						
	Binding Studies			Physiologic Measures+					Physiologic Measures+						
	Binding Site*													EEG	
Compound	I	II(-) or III(+)	IV	Resp. Arrest	Brady-cardia	Tachy-cardia	Systo-lic BP	Brady-pnea	Systolic BP	STRL	HR	PD	T	S	DS
(-)-Nicotine	+	1	1	1	1	1	1	1	0	↑	↓	0	↑-0	0	-↑
(+)-Nicotine	++	.04	.1	.09	.14	.09	.14	.17	↑	0-↑	↑-↓	↓-0	↑-0	↑	↑
(-)-Anabasine	0	.02	.06	.07	.04	--	.06	.06							
(-)-Cytisine	+	6.76							↓	0	↑	↓	0	0	0
(±)-2-Methyl-piperidine	+++								0	↓	0	↓	0	↑	0
(+)-2-Methyl-piperidine	+++			IE	IE	IE	IE	IE							

* I = Up-regulatory
II(-) or III(+) = High affinity site
IV = Low affinity site

+ BP = Blood pressure
STRL = Skin twitch reflex latency
HR = Heart rate
PD = Pupil diameter
T = Rectal temperature
S = EEG Synchronization
DS = EEG Desynchronization
IE = Ineffective

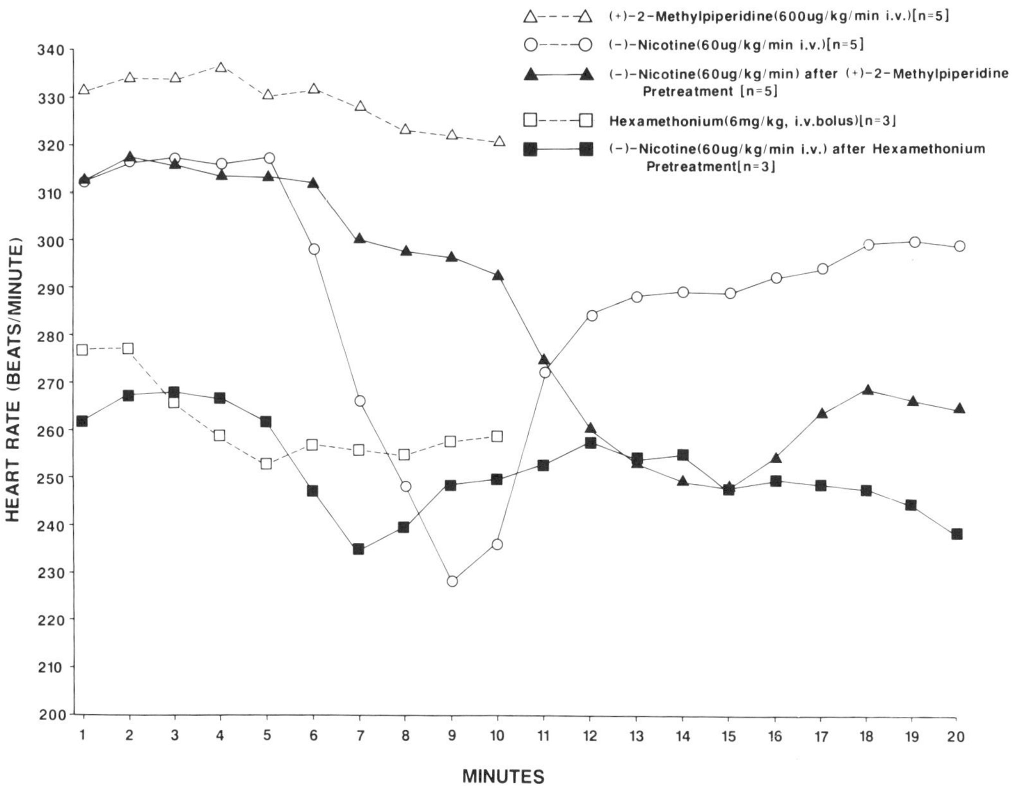

FIG. 10. The effects of (-)-nicotine, (+)-2-methylpiperidine and hexamethonium alone on heart rate in the urethane (1 g/kg/i.p.)-pentobarbital (20 mg/kg/i.p.) anethetized rat: Comparison with the effect of (-)-nicotine after (+)-2-methylpiperidine or hexamethonium pretreatment.

have different pharmacologic profiles when administered into the 4th ventricle. The most obvious relationships are that the two drugs that are the most potent and effective up-regulators tend to produce EEG synchrony [(+)-nicotine and (+)-2-methylpiperidine]. On the other hand, high doses of (-)-nicotine and (+)-nicotine which interact with Site IV cause EEG desynchronization, analgesia, and bradycardia, whereas (-)-cytisine, which interacts with the high-affinity site [II (-)] and the up-regulatory site (Site I) but not with the low-affinity site (Site IV), produces tachycardia, hypotension, and miosis. (-)-Nicotine and high doses of (+)-nicotine produced analgesia in contrast to (-)-cytisine which had no effect on the latency of the skin twitch reflex and (±)-2-methylpiperidine which produced hyperalgesia.

The effects of the i.v. infusion of these drugs were also studied in the urethane-pentobarbital anesthetized rat. (+)- and (-)-Nicotine produce a variety of pharmacologic effects including respiratory arrest, bradycardia, tachycardia, both an increase and decrease in blood pressure, bradypnea, and enhanced tidal volume. As can be seen in Table 4, if (-)-nicotine is used as a standard drug, (+)-nicotine is approximately 1/10 as potent in producing respiratory arrest, bradycardia, tachycardia, bradypnea, and in alterating blood pressure. (-)-Anabasine also produced these same effects and is basically 14-20 times less potent than (-)-nicotine. (+)-2-methylpiperidine had no effect on these variables. However, as can be seen in Fig. 10, when urethane-pentobarbital

Table 5. Endogenous Brain Chemicals Which Produce Up-Regulation of (±)-[^{3}H]nicotine (2.4 x 10^{-8} M) Binding to the Rat Brain P_2 Preparation.*

Compound	K_D (M)	Concentration (M) producing maximum enhancement	Percent enhancement (mean)
Niacinamide	3.8 x 10^{-4}	10^{-10}	17
Pipecolic Acid	0	10^{-6}	14
Histamine	3.3 x 10^{-4}	10^{-7}	13
Picolinic Acid	0	10^{-5}	10
Quinolinic Acid	1.1 x 10^{-5}	10^{-8}	8

*These values are derived from the mean of 4 experiments using different brain homogenates.

anesthetized rats were pretreated with (+)-2-methylpiperidine, both the bradycardia and the tachyphylaxis produced by (-)-nicotine were delayed and reduced. In contrast to (+)-2-methylpiperidine, hexamethonium pretreatment produced a marked decrease in heart rate. Further, both the bradycardia and the tachyphylaxis to this effect by nicotine were antagonized to a greater extent after hexamethonium pretreatment than after (+)-2-methylpiperidine pretreatment. This suggests that (+)-2-methylpiperidine and hexamethonium may antagonize nicotine's effects on the heart through different mechanisms.

As can be seen in Table 4, the potency estimates of these drugs on the various physiological parameters are most closely related to the relative potencies of these drugs at their low-affinity binding site and not to the potency estimates at the high-affinity binding site.

Drugs that are structurally related to nicotine have diverse binding characteristics on several nicotinic binding sites. Prototypic drugs with different patterns of binding specificities have been identified. An analysis of the pharmacology of these prototypic agonists would indicate that they also have different profiles of action. These data also suggest that the four binding sites identified may have pharmacologic significance. Although it is now commonly assumed that the high-affinity binding site is the nicotinic cholinergic site, arguments have been presented that the low-affinity site may be the nicotinic-cholinergic receptor. How this binding is related to the up-regulatory site and the high-affinity sites is not known. The physiologic role of these four binding sites is also unknown. The possibility that there may be endogenous ligands, for the up-regulatory site cannot be discounted, for endogenous brain compounds have been identified which interact with this binding site (Table 5).

REFERENCES

1. Abood, L.G., Lowy, K., Tometsko, A., and Booth, H. (1978): J. Neurosc. Res. 3:327.
2. Yoshida, K., and Imura, H. (1979): Brain Research 172:453.
3. Romano, C., and Goldstein, A. (1980): Science 210:647.
4. Vincek, W.C., Martin, B.R., Aceto, M.D., and Bowman, E.R. (1980): J. Med. Chem. 23:960.
5. Costa, L.G., and Murphy, S.D. (1983): J.P.E.T. 226:392.
6. Sloan, J.W., Todd, G.D., and Martin, W.R. (1984): Pharmacol. Biochem. Behav. 20:899.

7. Sloan, J.W., Martin, W.R., Hook, R., and Hernandez, J. (1985): J. Med. Chem. 28:1245.
8. Sloan, J.W., Martin, W.R., Hook R., Bostwick, M., Howell, A., and Smith, W.T. (1985): Life Sci. 37:1367.
9. Sloan, J.W., Martin, W.R., Hernandez, J., and Hook, R. (1985): Pharmacol. Biochem. Behav. 23:987.
10. Munson, P.J., and Rodbard, D. (1980): Analytical Biochemistry 107:220.
11. Abood, L.G., Grassi, S., and Noggle, H.D. (1985): Neurochem. Res. 10:259.

NICOTINIC CHOLINERGIC RECEPTOR RECOGNITION SITES IN BRAIN

Kenneth J. Kellar, Rochelle D. Schwartz
and Andrea M. Martino

Department of Pharmacology
Georgetown University
Schools of Medicine and Dentistry
3900 Reservoir Road, N.W.
Washington, D.C. 20007

Nicotinic cholinergic receptors in the mammalian central nervous system have eluded the exquisitely detailed characterization that has been achieved for the nicotinic receptors in the electric organs of fish and in the mammalian neuromuscular junction. Although radioactively labeled alpha-bungarotoxin - which has been instrumental in studies of the electric organ and muscle receptor - binds to sites in mammalian brain, the toxin itself does not appear to affect nicotinic cholinergic receptor mediated neurotransmission in the mammalian CNS, and the binding sites in brain do not appear to have certain characteristics expected of a functional nicotinic cholinergic receptor (Duggan et al., 1976a; 1976b; Carbonetto et al., 1978). The lack of a suitable probe for the CNS nicotinic cholinergic receptor has been a major obstacle to understanding the molecular pharmacology and neurobiology of these receptors, their effector mechanisms, their regulation, and their possible role in physiological and pathophysiological processes.

In the last five years, binding sites with characteristics of nicotinic cholinergic receptor recognition sites in rodent and human brain have been labeled with [^{3}H]acetylcholine ([^{3}H]ACh) (Schwartz et al., 1982; Schwartz and Kellar, 1983a; 1983b; 1985) and with [^{3}H]nicotine (Romano and Goldstein, 1980; Yoshida and Imura, 1979; Marks and Collins, 1982; Marks et al., 1983; Costa and Murphy, 1983). The pharmacological characteristics of the [^{3}H]ACh and the high-affinity [^{3}H]nicotine binding sites are similar to each other, and they differ markedly from the characteristics of alpha-bungarotoxin binding sites (Schwartz et al., 1982; Marks and Collins, 1982). In addition, the brain regional distribution of the [^{3}H]ACh and the high-affinity [^{3}H]nicotine recognition sites appear to be virtually identical, and this distribution is markedly different from that of alpha-bungarotoxin binding sites (Clarke et al., 1985; Schwartz et al., 1982). These *in vitro* binding data suggest that [^{3}H]ACh and [^{3}H]nicotine label the same site in mammalian brain - and it is probable that this site is the agonist recognition site of the elusive CNS nicotinic cholinergic receptor.

Characteristics of [^{3}H]acetylcholine nicotinic recognition sites in brain

[^{3}H]ACh binds to a site that has kinetic, pharmacological, and physico-chemical characteristics of a nicotinic cholinergic receptor recognition site in rat brain. [^{3}H]ACh is conveniently and economically synthesized from [^{3}H]choline with a specific radioactivity of approximately 80 Ci/mmol (Schwartz et al., 1982). The assay of nicotinic sites is conducted in the presence of diisopropyl fluorophosphate (DFP) to inhibit cholinesterases, and atropine (1.5 μM) to fully occupy muscarinic sites. Nonspecific binding is defined as that which occurs in the presence of 100 μM carbachol. Specific binding to nicotinic sites is defined as the difference between total binding and nonspecific binding. The dissociation constant (K_d) of [^{3}H]ACh in most brain areas examined is 6-20 nM at 0°C (Fig. 1), but it is approximately 100 nM at 25°C due to a faster rate of dissociation of the ligand from the binding site at the higher temperature (Schwartz et al., 1982).

Nicotinic cholinergic agonists such as cytisine, (-)nicotine, and carbachol have high affinity (1-100 nM) for the [^{3}H]ACh recognition site in rat cerebral cortex, and the site displays a fair degree of stereoselectivity towards nicotine, with (-)nicotine being 20 to 40 times more potent than (+)nicotine in competing for the site (Table 1). Most nicotinic antagonists have quite low affinity (20-1000 μM) for the site (Table 1); one exception, however, is the nicotinic antagonist dihydro-B-erythroidine, which has an affinity of approximately 100 nM (Table 1). Muscarinic drugs, choline, and the choline uptake inhibitor hemicholinium-3 do not compete effectively for the site (Table 1). This pharmacological profile - with agonists showing high affinity and

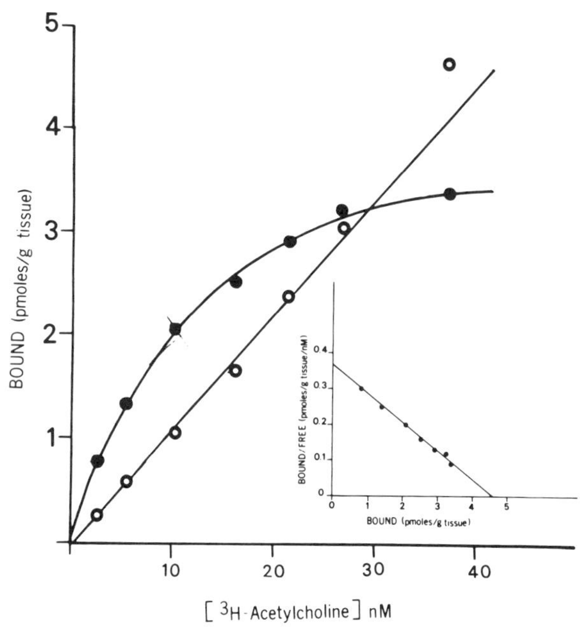

Fig. 1. Saturation analysis of [^{3}]acetylcholine ([^{3}H]ACh) binding to rat cerebral cortex homogenates. The tissue was homogenized, washed twice by centrifugation at 48,000 g for 10 min, suspended in buffer containing DFP to inhibit cholinesterases, and incubated with [^{3}H]ACh (2.5-37 nM) for 40 min at 0°C in the presence of atropine (1.5 uM) to fully occupy muscarinic sites. Nonspecific binding (O—O) was determined in the presence of 100 μM carbachol. Specific binding for (●—●) was defined as the difference between total binding and nonspecific binding. Inset: Scatchard plot of [^{3}H]ACh specific binding.

Table 1. Inhibition of [^{3}H]ACh binding in rat cerebral cortex by various drugs. Competing drugs were incubated with 10 nM [^{3}H]ACh. The inhibition constants (K_I) were derived from the IC_{50} and are the mean ± SEM of at least 3 determinations. (From Schwartz et al., 1982).

Drugs	K_I (nM
Cholinergic agonists	
cytisine	1.3 ± 0.5
(-)nicotine	6.4 ± 0.5
acetylcholine	7.6 ± 1.4
carbachol	13.4 ± 2.5
(+)nicotine	146 ± 28.5
succinylcholine	6,700 ± 1,900
piperidine	8,200 ± 2,600
decamethonium	9,700 ± 2,300
methacholine	84,300 ± 4,400
choline	134,000 ± 63,400
cotinine	175,000 ± 16,400
Cholinergic antagonists	
dihydro-B-erythroidine	111.3 ± 20.9
d-tubocurarine	28,500 ± 4400
gallamine	157,000 ± 1600
hexamethonium	182,000 ± 37,000
atropine	476,000 ± 189,000
pancuronium	556,000 ± 144,000
trimethaphan	610,000 ± 166,000
chlorisondamine	686,000 ± 187,000
mecamylamine	822,000 ± 273,000
alpha-bungarotoxin	> 20,000
Other drugs	
hemicholinium-3	80,200 ± 9,900
4-aminopyridine	284,000 ± 12,500
DFP	> 1,000,000

antagonists showing very low affinity - is also seen in studies of [^{3}H]nicotine binding in rat and mouse brain (Romano and Goldstein, 1980; Marks and Collins, 1982; Costa and Murphy, 1983; see below).

The number of [^{3}H]ACh nicotinic recognition sites varies in different areas of brain. In homogenates from 8 areas of rat brain, the thalamus, striatum and cerebral cortex had the highest number of binding sites (Schwartz et al., 1982). However, homogenate binding studies are limited to relatively gross brain areas that can be dissected relatively easily, and recent autoradiographic studies, in which much more discrete brain areas could be examined, revealed that the interpeduncular nucleus, the medial habenula, and the superficial layer of the superior colliculus have among the highest levels of [^{3}H]ACh binding in rat brain (Table 2; see also, Clarke et al., 1985).

The binding of [^{3}H]ACh to nicotinic recognition sites in brain requires intact disulfide bonds (Schwartz and Kellar, 1983a). Pretreatment of brain homogenates with dithiothreitol (DTT), which reduces disulfide bonds to sulfhydryl groups, decreases the binding of [^{3}H]ACh

Table 2. Relative density of nicotinic cholinergic [^{3}H]ACh binding sites in rat brain. Brain sections (24 μm) were incubated with [^{3}H]ACh (20 nM) and then placed on LKB Ultrofilm for 60 days to expose autoradiograms. Optical density readings of the brain areas were made and expressed as a ratio to the optical density of white matter in that section. Values are the mean ± SEM of 6 readings of specific binding per brain area from 2-3 rats.

Brain Area		Relative Density
Interpeduncular nucleus		11.45 ± 0.12
Thalamic Nuclei	laterodorsal	11.31 ± 0.06
	anteroventral	10.53 ± 0.08
	gelatinosus	9.65 ± 0.06
	anteromedial	9.40 ± 0.12
	mediodorsal	8.69 ± 0.08
	lateral posterior	7.41 ± 0.21
	anterodorsal	7.31 ± 0.11
	ventrolateral	7.31 ± 0.05
	reticular	6.98 ± 0.03
	posterior	6.41 ± 0.04
	ventroposterior lateral	6.31 ± 0.05
	paraventricular	6.06 ± 0.13
Medial Habenula		9.24 ± 0.30
Presubiculum		8.97 ± 0.03
Superior colliculus	superficial grey layer	8.83 ± 0.01
	optic nerve layer	4.40 ± 0.08
Retrosplenial cortex		7.52 ± 0.05
Lateral geniculate	dorsal	5.50 ± 0.02
	ventral	5.48 ± 0.11
Substantia nigra	pars compacta	5.30 ± 0.12
	pars reticulata	2.09 ± 0.05
Cerebral Cortex Lamina III		5.24 ± 0.14
IV		4.03 ± 0.06
I		3.95 ± 0.09
V		3.74 ± 0.14
VI		3.45 ± 0.17
II		3.12 ± 0.04
Dorsal tegmental nucleus		4.97 ± 0.12
Caudate-putamen		4.31 ± 0.10
Raphe nucleus		3.98 ± 0.08
Central grey		3.96 ± 0.12
Pontine nuclei		3.69 ± 0.07
Dentate gyrus-molecular layer		3.62 ± 0.07
Cerebellum	granule cell layer	3.49 ± 0.23
	molecular layer	2.08 ± 0.01
Cuneiform nucleus		3.49 ± 0.23
Vestibular nucleus	medial	3.36 ± 0.07
	lateral	2.04 ± 0.08
Medial geniculate		3.31 ± 0.02
Anterior pretectal area		3.22 ± 0.07
Globus pallidus		3.12 ± 0.16
Amygdala, basolateral nucleus		3.03 ± 0.06
Primary olfactory cortex		2.97 ± 0.04
Inferior colliculus		2.36 ± 0.05
Nucleus of the horizontal limb of the diagonal band		2.69 ± 0.15
Accumbens nucleus		2.60 ± 0.04
Septal nucleus		2.52 ± 0.02
Vertical limb of the diagonal band		2.46 ± 0.03
Hypothalamic nuclei		2.38 ± 0.11
Zona incerta		2.33 ± 0.07
Hippocampus (stratum oriens)		2.22 ± 0.08
Ventral tegmental area		2.08 ± 0.04
Facial nucleus		1.82 ± 0.02
Motor trigeminal nucleus		1.73 ± 0.04
Nucleus of the solitary tract		1.72 ± 0.02
Anterior olfactory nucleus		1.72 ± 0.02
Entorhinal cortex		1.60 ± 0.02

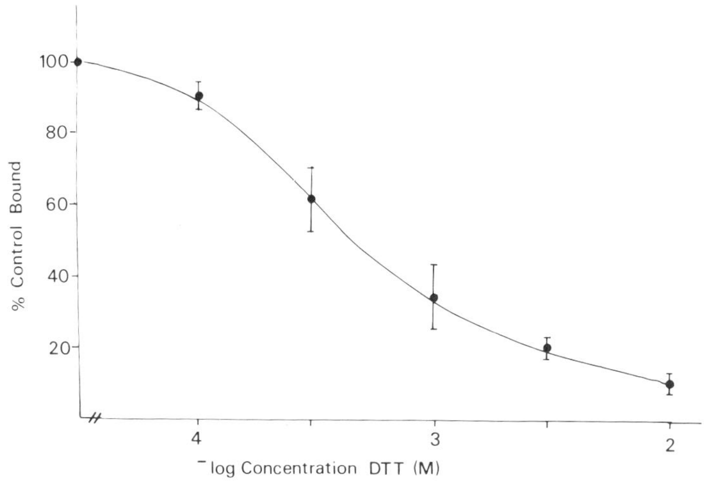

Fig. 2. Effect of reduction of disulfide bonds with dithiothreitol (DDT) on [^{3}H]ACh binding in rat cerebral cortex. Cortical homogenates were preincubated in buffer (pH 8.5) in the absence or presence of DTT (0.1-10 mM) for 30 min at 0°C. The reactions were terminated by dilution with cold buffer, and the tissues were washed twice by centrifugation at 48,000 g for 10 min. Binding of [^{3}H]ACh (10 nM) was then measured as described in Fig. 1. Each point is the mean ± SEM of 3 replicate experiments.

in a concentration-dependent manner (Fig. 2). The decrease in binding is due to an apparent decrease in the number of binding sites without alteration in the affinity of the remaining sites (Schwartz and Kellar, 1983a). Presumably, the reduction of disulfide bonds alters the receptor conformation which leads to an absolute loss in its ability to bind [^{3}H]ACh. This change is reversible, and when DTT-treated membranes are subsequently treated with dithiobis-2-nitrobenzoic acid (DTNB) to reoxidize sulfhydryl groups, the binding of [^{3}H]ACh is restored to control values (Schwartz and Kellar, 1983a).

Presynaptic Localization of [^{3}H]Acetylcholine Nicotinic Recognition Sites in Brain

Cholinergic projections to and intrinsic cholinergic neurons of the cerebral cortex, thalamus, striatum, and hypothalamus have been well-described (McGeer et al., 1971; Lewis and Shute, 1978; Armstrong et al., 1983). Nicotine induces release of catecholamines and possibly of serotonin and acetylcholine in brain (Westfall, 1974; De Belleroche and Bradford, 1978; Giorguieff-Chesselet et al., 1979; Sakurai et al., 1982; Armitage, 1968; Rowell and Winkler, 1984). To determine if [^{3}H]ACh nicotinic recognition sites are located presynaptically on certain axons in rat brain, catecholamine and serotonin axons were lesioned by intraventricular injections of the neurotoxins 6-hydroxydopamine and 5,7-dihyroxytryptamine, respectively. Binding of [^{3}H]ACh to nicotinic sites was significantly reduced in the striatum and hypothalamus 7-10 days following either lesion (Table 3). The reduced binding was due to a decrease in the density of these [^{3}H]ACh binding sites, indicating that at least some of these sites are located on catecholamine and serotonin axons in the striatum and hypothalamus (Schwartz et al., 1984). In contrast, these lesions did not alter binding of [^{3}H]ACh in the cerebral cortex or thalamus (Table 3).

Table 3. Binding of [^{3}H]ACh (10 nM) to nicotinic sites in brain regions from rats 7-10 days following intraventricular injection of 6-hydroxydopamine (6-OHDA), which lesions catecholamine axons, or 5,7-dihydroxytryptamine (5,7-DHT), which lesions serotonin axons. Values are the means ± SEM from 4-8 rats. *P<0.01 compared with control. (From Schwartz et al., 1984)

	Specific Binding (fmol/mg protein)		
Brain Area	Control	6-OHDA	5,7-DHT
Cerebral cortex	38.4 ± 2.6	36.2 ± 1.9	36.0 ± 2.1
Thalamus	53.7 ± 1.2	56.6 ± 2.1	54.0 ± 2.1
Striatum	47.3 ± 2.2	33.0 ± 3.2*	33.2 ± 1.7*
Hypothalamus	41.5 ± 3.5	16.9 ± 1.1*	23.6 ± 3.6*

A recent autoradiographic study of [^{3}H]nicotine binding has confirmed the loss of nicotinic sites in the rat striatum following lesions of dopamine axons (Clarke and Pert, 1985). In addition, this study revealed that these lesions result in a loss of nicotinic sites on mesolimbic dopamine axons in the nucleus accumbens and olfactory tubercle, and in the substantia nigra and ventral tegmental area, which contain dopamine cell bodies (Clarke and Pert, 1985).

The presence of nicotinic cholinergic recognition sites on catecholamine and serotonin axons in specific brain areas coupled with evidence of nicotine-induced release of certain neurotransmitters (see Westfall, this volume and Rowell, this volume), suggests that modulation of neurotransmitter release might be one function of these sites. Thus, the action of endogenous acetylcholine or exogenous nicotine via presynaptic nicotinic receptors on catecholamine and serotonin axons might have important effects on nigro-striatal and mesolimbic neurotransmission, and on hypothalmic neurotransmission - including the secretion of hypothalamic hormones that regulate pituitary function (c.f. Sladek and Joynt, 1979; Sharp and Beyer, 1985).

In Vivo Regulation of [^{3}H]Acetylcholine Nicotinic Recognition Sites in Brain

One of the mechanisms by which cells adjust to changes in neurotransmitter signals is to increase or decrease the number of receptors for the mediator of the signal. This adaptive up- and down-regulation is usually reciprocal to the change in stimulation and is one indication that a receptor is, in fact, functional.

Treatment of rats with the cholinesterase inhibitor DFP increases the acetylcholine concentration in brain (Russell et al., 1981) and, presumably prolongs the synaptic stimulation of cholinergic receptors. Following 10 days of treatment with DFP, there is a significant decrease in [^{3}H]ACh binding in brain (Schwartz and Kellar, 1983b; 1985). Binding in the cerebral cortex, thalamus, striatum, and hypothalamus is reduced by 20-38% (Table 4), and this reduction in binding appears to be due entirely to a decrease (down-regulation) in the number of [^{3}H]ACh recognition sites (Schwartz and Kellar, 1983b; 1985). The decrease in these sites indicates that they are responsive to the synaptic stimulation by acetylcholine *in vivo*, suggesting that they are innervated by cholinergic axons. Chronic cholinesterase inhibition also decreases the number of high-affinity [^{3}H]nicotine-binding sites in the rat forebrain (Costa and

Table 4. Effect of chronic cholinesterase inhibition with DFP on the binding of [^{3}H]ACh (10 mM) to nicotinic sites in rat brain. Rats were treated with DFP for 10 days and sacrificed 24 hours after the last injection, at which time the cholinesterase activity was inhibited by 82%. Values are the means ± SEM from 4-13 rats. $^*P< 0.05$, $^{**}P< 0.01$ compared with control. (From Schwartz and Kellar, 1985).

Brain Area	Specific Binding (pmol/g tissue) Control	DFP
Cerebral cortex	2.2 ± 0.06	1.7 ± 0.08**
Thalamus	3.2 ± 0.17	2.6 ± 0.18*
Striatum	2.2 ± 0.07	1.5 ± 0.05**
Hypothalamus	0.72 ± 0.07	0.45 ± 0.06*

Murphy, 1983), suggesting that these sites might be identical to the [^{3}H]ACh sites, or that the two recognition sites are on the same molecule. In contrast, the low affinity [^{3}H]nicotine-binding sites appear to be unaffected by cholinesterase inhibition (Costa and Murphy, 1983).

In contrast to the effects of cholinesterase inhibition, treatment of rats with nicotine increases the number of [^{3}H]ACh nicotinic recognition sites in rat brain (Schwartz and Kellar, 1983b; 1985). Binding of [^{3}H]ACh to nicotinic sites in homogenates from the cerebral cortex, thalamus, striatum, and hypothalamus is increased 20-37% following a 10-day treatment with nicotine (Table 5). This increased binding is due to an increase in the number of [^{3}H]ACh nicotinic recognition sites (Schwartz and Kellar, 1983b; 1985). Repeated administration of nicotine results in an increase in the locomotor response to nicotine which can be blocked by mecamylamine, a nicotinic antagonist (Clarke and Kumar, 1983; Ksir et al., 1985). This increased behavioral response after repeated administration of nicotine may be linked to increased nicotinic cholinergic receptors, as reflected by the increased number of [^{3}H]ACh nicotinic recognition sites (Ksir et al., 1985).

Time-course studies of the nicotine-induced increase in [^{3}H]ACh binding in rat cerebral cortex indicate that there is no change in binding

Table 5. Effect of repeated administration of nicotine on the binding of [^{3}H]ACh (10 nM) to nicotinic sites in rat brain. Rats were injected with nicotine (1 mg/kg, subcutaneously) twice daily for 10 days and sacrificed 18 hours after the last injection. Controls were injected with water. Values are the means ± SEM from 7-16 rats. $^*P< 0.01$ compared with control. (From Schwartz and Kellar, 1985).

Brain Area	Specific Binding (pmol/g tissue) Control	Nicotine
Cerebral cortex	2.0 ± 0.04	2.5 ± 0.06*
Thalamus	3.6 ± 0.11	4.6 ± 0.17*
Striatum	2.1 ± 0.05	2.5 ± 0.24*
Hypothalamus	0.67 ± 0.04	0.92 ± 0.07*

one day after a single injection of nicotine; however, following five days of nicotine treatment, binding is increased by approximately 20% (Fig. 3). The binding appears to be increased maximally after 10 days of nicotine treatment, and this increase (approximately 30%) is maintained throughout 21 days of treatment (Fig. 3). The increased binding to nicotinic sites is reversible. When rats are sacrificed seven days after the last nicotine treatment, [^{3}H]ACh binding in cortex is significantly decreased compared with that from rats sacrificed one day after the last treatment (Fig. 3); and within 21 days after the last nicotine treatment, binding returns to control values (Ksir et al., 1985).

Collins and colleagues found that chronic infusion of nicotine also increases [^{3}H]nicotine-binding sites in mouse brain (Marks et al., 1983; 1985; see also Collins, this volume). Thus, nicotine has been found to increase nicotinic cholinergic recognition sites in the two species in which it has been examined. This up-regulation of [^{3}H]ACh and [^{3}H]nicotine recognition sites during chronic administration of nicotine may be the result of a functional blockade of nicotinic receptors (Schwartz and Kellar, 1983b; 1985; Marks et al., 1983). Although nicotine is classically defined as an agonist at peripheral cholinergic receptors (Dale, 1914), and in the CNS (McLennan and Hicks, 1978; Misgeld et al., 1980; Bradley and Lucy, 1983), following exposure to nicotine there is a rapid decrease in cholinergic receptor responsivity due to depolarization blockade (Eccles, 1935; Thesleff, 1955). In frog ganglia, adequate doses of nicotine result in diminished receptor responsivity that persists beyond the depolarization phase (Ginsborg and Guerrero, 1964), possibly reflecting prolonged desensitization of the receptor. Closer to the mammalian CNS, nicotine induces a marked increase in the release of adrenocorticotropin and prolactin hormones in rats, and Sharp and Beyer (1985) found that these responses appear to desensitize for at least 6

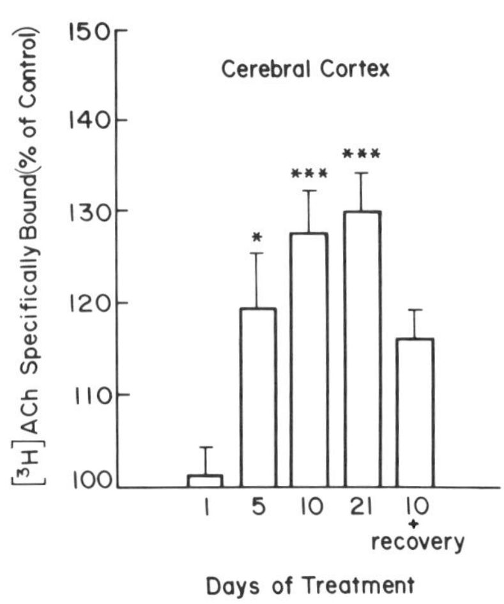

Fig. 3. Time-course of the nicotine-induced increase in [^{3}H]ACh binding sites in rat cerebral cortex. Binding of [^{3}H]ACh (10 nM) was measured in homogenates of cortices from rats that were injected with 1 mg/kg (-)nicotine twice daily for the number of days indicated. The rats were sacrificed 18 hours after the last injection except those in the "recovery" group, which were sacrificed 7 days after the last injection. Data are the mean ± SEM for 6-14 rats per group. The value for control binding was 2.2 ± 0.1 pmol/g tissue. $^*P < 0.05$; $^{***} P < 0.01$ compared with control. $^{**}P < 0.05$ compared with the 10-day treatment group that was sacrificed 18 hours after the last injection.

hours after a single injection. The duration of this apparent functional blockade of nicotinic receptors probably varies in different tissues, but it is reversible; and, in fact, increased behavioral responses to low doses of nicotine can be elicited in rats within 24 hours after the last administration (Clarke and Kumar, 1983; Ksir et al., 1985).

Thus, it is possible that repeated administration of nicotine results in a protracted functional blockade of nicotinic cholinergic receptors and that this is the signal for the cell to increase its number of receptors. This effect of nicotine could be important to the mechanisms underlying nicotine addiction in humans.

Comparison of [^{3}H]acetylcholine and [^{3}H](-)nicotine binding in brain

Nicotine binds to multiple sites, including apparent noncholinergic as well as cholinergic sites. Although the [^{3}H]ACh and high affinity [^{3}H]nicotine recognition sites in brain as measured in several laboratories are qualitatively similar, several differences have been reported - including differences in the regulation of the sites (Nordberg et al., 1985; Abood et al., 1984), which would indicate that the two recognition sites are not identical. However, it is also possible that these differences stem from differences in assay conditions used by different laboratories and, in some cases, from the use of [^{3}H]-labeled racemic nicotine, rather than the naturally occurring (-)nicotine as the binding ligand. To compare [^{3}H]ACh and high affinity [^{3}H](-)nicotine recognition sites in brain quantitatively, we measured the binding characteristics of both ligands under identical assay conditions - those described above for [^{3}H]ACh binding (Schwartz et al., 1982).

In rat and human brain assayed under these conditions, [^{3}H](-)nicotine appears to bind to a single class of sites with an affinity of 2-7 nM. In the five areas of rat brain examined, the densities of [^{3}H]ACh and [^{3}H](-)nicotine recognition sites are very similar (Table 6). The two ligands also label the same number of binding sites in normal human cerebral cortex (Table 6).

In drug competition experiments, the inhibition constants (K_I) of nicotinic agonists and antagonists for [^{3}H]ACh and [^{3}H](-)nicotine binding sites in rat cortex are not significantly different (Table 7), indicating that the affinities of these drugs for the sites labeled by the two [^{3}H]ligands are quantitatively similar.

[^{3}H]ACh binding to nicotinic recognition sites in brain is decreased by preincubation with DTT, which reduces disulfide bonds to sulfhydryl

Table 6. Binding constants of [^{3}H]ACh and [^{3}H](-)nicotine in rat brain regions and human cerebral cortex. Values were derived from Scatchard plots and are the means ± SEM of 3-7 determinations.

	B_{max} (fmol/mg protein)		K_d (nM)	
Rat Brain Region	[^{3}H]ACh	[^{3}H](-)Nicotine	[^{3}H]ACh	[^{3}H](-)Nicotine
Thalamus	91.4 ± 4.8	93.5 ± 6.3	8.6 ± 0.4	3.5 ± 0.3
Striatum	88.5 ± 8.8	80.9 ± 5.3	19.1 ± 1.7	6.5 ± 0.6
Cerebral cortex	55.4 ± 2.4	57.0 ± 4.0	10.9 ± 0.9	3.6 ± 0.4
Hypothalamus	29.5 ± 5.0	32.7 ± 4.6	7.6 ± 1.7	3.9 ± 0.4
Hippocampus	25.4 ± 6.9	23.5 ± 3.2	16.3 ± 1.5	6.7 ± 0.5
Human cortex	14.1 ± 2.7	16.1 ± 2.7	5.2 ± 0.9	1.9 ± 0.4

Table 7. Inhibition of $[^3H]$ACh and $[^3H]$(-)nicotine binding in rat cerebral cortex by cholinergic drugs. Competing drugs were incubated with 10 nM $[^3H]$ACh or 4 nM $[^3H]$(-)nicotine. K_I values are the mean ± SEM of 3-6 determinations.

Drug	K_I (nM)	
	$[^3H]$ACh	$[^3H]$(-)Nicotine
Cholinergic Agonists		
cytisine	3.8 ± 0.9	2.0 ± 0.2
(-)nicotine	5.5 ± 2.3	4.2 ± 1.3
acetylcholine	13.9 ± 5.1	11.7 ± 1.8
carbachol	93.4 ± 27.6	81.2 ± 26.2
(+)nicotine	282 ± 187	148 ± 42
Cholinergic Antagonists		
dihydro-B-erythroidine	111.9 ± 28.7	90.7 ± 9.6
d-tubocurarine	28,500 ± 9,240	17,900 ± 3880
mecamylamine	> 1,000,000	> 1,000,000

groups (Fig. 2). This effect is reversed by subsequent preincubation with dithiobis-2-nitrobenzoic acid (DTNB), which reoxidizes sulfhydryl groups (Schwartz and Kellar, 1983a). The effects of disulfide bond reduction by DTT and reoxidation of the sulfhydryl groups by DTNB were found to be quantitatively similar on $[^3H]$(-)nicotine binding and $[^3H]$ACh binding in rat cortex (Fig. 4).

Finally, the *in vivo* regulation of the $[^3H]$ACh and $[^3H]$(-)nicotine recognition sites in rat brain by repeated administration of nicotine were

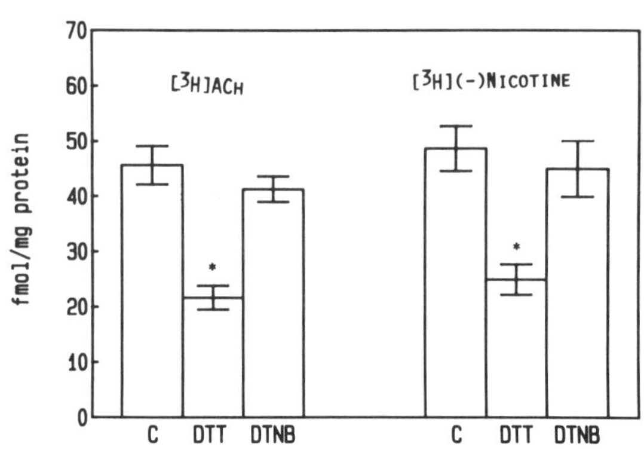

Fig. 4. Effects of reduction of disulfide bonds and reoxidation of sulfydryl groups on $[^3H]$ACh and $[^3H]$(-)nicotine binding in rat cerebral cortex. Cortical homogenates were preincubated in the absence or presence of 3 mM DTT as described in the legend of Fig. 2, washed by centrifugation, and then preincubated a second time in the absence or presence of 1 mM dithiobis-2-nitrobenzoic acid (DTNB). The homogenates were then washed again, and binding of $[^3H]$ACh (10 nM) and $[^3H]$(-)nicotine (4 nM) was measured. Preincubation with DTNB alone had no effect on the binding of either ligand. Data are the mean ± SEM from 5 experiments. *$P < 0.05$ compared to control tissues (C), which were preincubated in the absence of modifying agents.

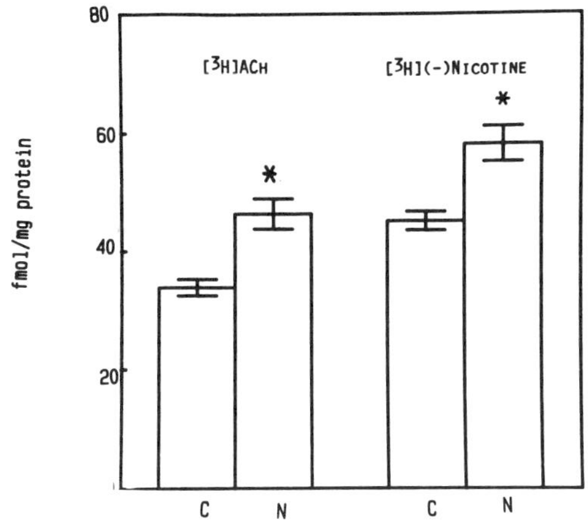

Fig. 5. _In vivo_ regulation of $[^3H]$ACh and $[^3H]$(-)nicotine recognition sites in rat cerebral cortex by chronic administration of nicotine. Rats were injected with nicotine (0.8 mg/kg, s.c.) or saline twice daily for 10 days and sacrificed 24 hours after the last injection. Binding sites were assayed using 10 nM $[^3H]$ACh or 4 nM $[^3H]$(-)nicotine. Values are the mean ± SEM of 5 saline-treated (C) and 6 nicotine-treated (N) rats. $^*P < 0.05$ compared to control.

compared. Treatment of rats with nicotine (0.8 mg/kg) for 10 days increased binding of $[^3]$ACh and $[^3H]$(-)nicotine in the cerebral cortex by approximately the same percentage (Fig. 5). Similar effects were found in the striatum.

Taken together, these quantitative comparisons argue strongly that the $[^3H]$ACh and the high affinity $[^3H]$(-)nicotine recognition sites in brain, as measured under these conditions, are identical.

CONCLUSIONS

$[^3]$ACh and $[^3H]$(-)nicotine label apparently identical recognition sites in brain which have the characteristics of nicotinic cholinergic receptors. The specific functions of these receptors in brain are not yet known; nor is the role that these receptors might have in the pharmacological effects of nicotine known. However, the location of nicotinic cholinergic recognition sites on certain catecholamine and serotonin axons and the capacity of nicotine to stimulate the release of these neurotransmitters as well as the release of acetylcholine (Balfour, 1982; Rowell and Winkler, 1984) suggest that nicotinic cholinergic receptors might have an important physiological role in mediating synaptic signals resulting in release of certain neurotransmitters and hormones. In this regard, it is worth noting that we have recently found that there is a marked reduction of nicotinic cholinergic recognition sites in the cerebral cortex from Alzheimer's patients, and we hypothesize that these sites are on the cholinergic axons that degenerate in Alzheimer's disease (Whitehouse et al., 1986).

REFERENCES

Abood, LG, Grassi, S, Costanzo, M and Junig, J: Behavioral and Biochemical studies in rats after chronic exposure to nicotine. IN NIDA Research Monograph 54 Mechanisms of tolerance and Dependence, Sharp, C Wm (ed.) 348-355, 1984.

Armitage, AK, Hall, GH and Morrison, CF: Pharmacological basis for the tobacco habit. Nature 271: 331-334, 1968.

Armstrong, DM, Saper, CB, Levey, AI, Wainer, BH and Terry, RD: Distribution of cholinergic neurons in rat brain: Demonstrated by the immunocytochemical localization of choline acetyltransferase. J. Comp. Neurol. 216: 53-68, 1983.

Balfour, DJK: The effects of nicotine on brain neurotransmitter systems. Pharmacol. Ther. 16: 269-282, 1982.

Bradley, PB and Lucy AP: Cholinoceptive properties of respiratory neurons in the rat medulla. Neuropharmacology 22: 853-858, 1983.

Carbonetto, ST, Fambrough, DM and Muller, KJ: Nonequivalence of alpha-bungarotoxin receptors and acetylcholine receptors in chick sympathetic neurons. Proc. Natl. Acad. Sci. 75: 1016-1020, 1978.

Clarke, PBS and Kumar, R: The effects of nicotine on locomotor activity in non-tolerant and tolerant rats. Brit. J. Pharmacol. 78: 329-337, 1983.

Clarke, PBS and Pert, A: Autoradiographic evidence for nicotine receptors on nigrostriatal and mesolimbic dopaminergic neurons. Brain Res. 348: 355-358, 1985.

Clarke, PBS, Schwartz, RD, Paul, SM, Pert, CB and Pert, A: Nicotinic binding in rat brain: autoradiographic comparison of [^{3}H]acetylcholine, [^{3}H]nicotine and [^{125}I]alpha-bungarotoxin. J. Neurosci. 5: 1307-1315, 1985.

Costa, LG and Murphy, SD: [^{3}H]Nicotine binding in rat brain: alteration after chronic acetylcholinesterase inhibition. J. Pharmacol. Exp. Ther. 226: 392-397, 1983.

Dale, HH: Action of certain esters and ethers of choline and their relation to muscarine. J. Pharmacol. Exp. Ther. 6: 147-190, 1914.

de Belleroche, J and Bradford, HF: Biochemical evidence for the presence of presynaptic receptors on dopaminergic nerve terminals. Brain Res. 142: 53-58, 1978.

Duggan, AW, Hall, JG, Headley, PM, Hendry, IA and Minchin, MCW: Absence of binding of alpha-bungarotoxin and cobra neurotoxin to central acetylcholine receptors - an autoradiographic study. Neurosci. Lett. 3: 123-127, 1976a.

Duggan, AW, Hall, JG and Lee, CY: Alpha-bungarotoxin, cobra neurotoxin and excitation of Renshaw cells by acetylcholine. Brain Res. 107: 166-170, 1976b.

Eccles, JC: The action potential of the superior cervical ganglion. J. Physiol. 85: 179-206, 1935.

Ginsborg, BL and Guerrero, S: On the action of depolarizing drugs on sympathetic ganglia of the frog. J. Physiol. 172: 189-206, 1964.

Giorguieff-Chesselet, MF, Kemel, ML, Wandscheer, D and Glowinski, J: Regulation of dopamine release by presynaptic nicotinic receptors in rat striatal slices: Effect of nicotine in a low concentration. Life Sci. 25: 1257-1262, 1979.

Ksir, C, Hakan, P, Hall, DP Jr and Kellar, KJ: Exposure to nicotine enhances the behavioral stimulant effect of nicotine and increases binding of [^{3}H]acetylcholine to nicotinic receptors. Neuropharmacology 24: 527-531, 1985.

Lewis, PR and Shute, CCD: Cholinergic pathways in CNS. In Handbook of Psychopharmacology vol. 9 (ed.) Iversen, LL, Iversen, SD and Snyder, SH, pp. 315-355, Plenum Press, New York, 1978.

Marks, MJ, Burch, JB and Collins, AC: Effects of chronic nicotine infusion on tolerance development and nicotinic receptors. J. Pharmacol. Exp. Ther. 226: 817-825, 1983.

Marks, MJ and Collins, AC: Characterization of nicotine binding in mouse brain and comparison with the binding of alpha-bungarotoxin and quinuclidinyl benzilate. Mol. Pharmacol. 22: 554-564, 1982.
Marks, MJ, Stitzel, JA and Collins AC: Time course study of the effects of chronic nicotine infusion on drug response and brain receptors. J. Pharmacol. Exp. Ther. 235: 619-628, 1985.
McGeer, PL, McGeer, EG, Fibiser, HC, and Wickson, V: Neostriatal choline acetylase and cholinesterase following selective brain lesions. Brain Res. 35(1): 308-14, 1971.
McKinney, M and Coyle, JT: Regulation of neocortical muscarinic receptors: effects of drug treatment and lesions. J. Neurosci. 2: 97-105, 1982.
McLennan, H and Hicks, HP: Pharmacological characterization of the excitatory cholinergic receptors of rat central neurons. Neuropharmacology 17: 329-334, 1978.
Misgeld, U, Weiler, MH and Bak, IJ: Intrinsic cholinergic excitation in the rat neostriatum: nicotinic and muscarinic receptors. Exp. Brain. Res. 39: 401-409, 1980.
Nordberg, A, Wahlstrom, G, Arnelo, U and Larsson, C: Effect of long-term nicotine treatment on [^{3}H]nicotine binding sites in the rat brain. Drug and Alcohol Depend. 16: 9-17, 1985.
Romano, C and Goldstein A: Stereospecific nicotine receptors on rat brain membranes. Science 210: 647-649, 1980.
Rowell, PP and Winkler, DL: Nicotinic stimulation of [^{3}H]acetylcholine release from mouse cerebral cortical synaptosomes. J. Neurochem. 43: 1593-1598, 1984.
Russell, RW, Carson, VG, Booth, RA and Jenden, DJ: Mechanisms of tolerance to the anticholinesterase, DFP: acetylcholine levels and dynamics in rat brain. Neuropharmacology 20: 1197-1201, 1981.
Sakurai, Y, Takano, Y, Kohjimoto, Y, Honda, K and Kamiya, H: Enhancement of ^{3}H-dopamine release and its ^{3}H-metabolites in rat striatum by nicotinic drugs. Brain Res. 242: 99-106, 1982.
Schwartz, RD and Kellar, KJ: [^{3}H]Acetylcholine binding sites in brain: effect of disulfide bond modification. Mol. Pharmacol. 24: 387-391, 1983a.
Schwartz, RD and Kellar, KJ: Nicotinic cholinergic receptor binding sites in brain: in vivo regulation. Science 220: 214-216, 1983b.
Schwartz, RD and Kellar, KJ: In vivo regulation of [^{3}H]acetylcholine recognition sites in brain by nicotinic cholinergic drugs. J. Neurochem. 45: 427-433, 1985.
Schwartz, RD, Lehmann, J and Kellar, KJ: Presynaptic nicotinic cholinergic receptors labeled by [^{3}H]acetylcholine on catecholamine and serotonin axons in brain. J. Neurochem. 42: 1495-1498, 1984.
Schwartz, RD, McGee, Jr, R and Kellar, KJ: Nicotinic cholinergic receptors labeled by [^{3}H]acetylcholine in rat brain. Mol. Pharmacol. 22: 56-62, 1982.
Sharp, B and Beyer, S: Rapid desensitization to the acute stimulatory effects of nicotine on rat plasma adrenocorticotropin and prolactin. Soc. Neurosci. Abst. 11: 657, 1985.
Sladek, CD and Joynt, RJ: Characterization of cholinergic control of vasopressin release by the organ-cultured rat hypothalamo-neurohypophyseal system. Endocrinol. 104: 659-663, 1979.
Thesleff, S: The mode of neuromuscular block caused by acetylcholine, nicotine, decamethonium and succinylcholine. Acta Physiol. 34: 218-231, 1955.
Westfall, TC: Effect of nicotine and other drugs on the release of ^{3}H-norepinephrine and ^{3}H-dopamine from rat brain slices. Neuropharmacol. 13: 693-700, 1974.
Whitehouse, PJ, Martino, AM, Antuono, PG, Lowenstein, PR, Coyle, JT, Price, DL and Kellar, KJ: Nicotinic acetylcholine binding sites in Alzheimer's disease. Brain Res. 371: 146-151, 1986.
Yoshida, K and Imura, H: Nicotinic cholinergic receptors in brain synaptosomes. Brain Res. 172: 453-459, 1979.

NICOTINIC BINDING SITES IN THE BRAIN:

PROPERTIES, REGULATION, AND PUTATIVE ENDOGENOUS LIGANDS

Henry Sershen and Abel Lajtha

Center for Neurochemistry
Nathan S. Kline Institute for Psychiatric Research
Ward's Island
New York, New York 10035

INTRODUCTION

This session has dealt with current attempts to characterize the putative receptor sites of action of nicotine in brain. Several radiolabeled ligands with cholinergic function have been used to label sites in brain; apparent similarities and differences have been observed between labeled nicotine and other labeled nicotinic agonists and antagonists. It appears that nicotine affects a number of processes and has multiple and complex effects. It is also likely, however, that nicotine has specific effects in some brain regions.

Our studies examined the characteristics of [^{3}H]nicotine binding in brain in comparison with those of other cholinergic ligands and the regulation of these sites after chronic nicotine treatment, and we attempted to isolate the endogenous substrate from brain.

[^{3}H]NICOTINE BINDING IN THE BRAIN

As shown in Table I, during the past decade a number of studies have attempted to characterize [^{3}H]nicotine binding to brain membranes. Although a binding site with affinity in the nanomolar range has been reported, in some instances the methodology made the interpretation of the results uncertain (see below). The pharmacological studies of this binding site have also been conflicting regarding its stereospecificity and cholinergic nature. At this stage it would be premature to classify the binding site as a receptor until additional kinetics and pharmacology are reported. Additionally, there are differences that are at present difficult to explain in the binding parameters of the several labeled cholinergic ligands.

The early binding studies were hindered by the lack of availability of labeled nicotine or acetylcholine of high specific activity, although binding sites for [^{3}H]nicotine and [^{3}H]acetylcholine with affinities in the low nanomolar range were reported (1). In these studies, however, no measurement was made of saturable versus nonsaturable binding. Abood et al. (2) were the first to describe a problem with [^{3}H]nicotine binding studies incorporating rapid filtration on glass fiber filters to terminate the incubation. Nicotine avidly binds to glass. Saturation

Table I. Kinetics and Characterization of Nicotinic Binding in Brain

Year	Ligand (Specific Activity)	K_d	Methodology	Result	Ref.
1974	[^{3}H]Nicotine (2.75 Ci/mmol)	8 nM	equilibrium dialysis	cholinergic	1
	[^{3}H]Acetylcholine (0.29 Ci/mmol)	23 nM	equilibrium dialysis	cholinergic	1
1978	[^{3}H]Nicotine (0.25 Ci/mmol)	10–20 nM	brain slices, glass fiber filters	noncholinergic, stereospecific	2
1979	[^{3}H]Nicotine (0.25 Ci/mmol)	2–20 nM	filtration	cholinergic, α-bungarotoxin-like	3
1980	4-4-ditrito-(–)-[^{3}H]Nicotine (4.7 Ci/mmol)	4.7 μM	centrifugation	nonstereospecific	4
	($\pm$)-[^{3}H]Nicotine (22.6 Ci/mmol)	6 nM(0.1 μM?)	centrifugation	noncholinergic, stereospecific	5
	($\pm$)[^{3}H]Nicotine (23.6 Ci/mmol)	28 nM,0.46 μM	filtration	nicotinic-cholinergic stereospecific	31
1981	($\pm$)[^{3}H]Nicotine (22.6 Ci/mmol)	60 nM,20 μM	filtration	noncholinergic, weak stereospecific	7
1982	($\pm$)[^{3}H]Nicotine (61.2 Ci/mmol)	58 nM	filtration	non α-bungarotoxin-like	8
	[^{3}H]Acetylcholine (77 Ci/mmol)	12 nM	filtration	cholinergic, stereospecific, non α-bungarotoxin-like	9
1983	(–)-[^{3}H]Nicotine (87 Ci/mmol)	0.2–2 nM	centrifugation	weakly stereospecific	10
1984	($\pm$)-[^{3}H]Nicotine ($\rightarrow$ 71.2 Ci/mmol)	5 sites	filtration	multiple pharmacology	12
1985	[^{3}H]Nicotine (high specific activity)	nanomolar & micromolar sites	filtration and centrifugation	varying high-affinity (stereospecific-cholinergic) low-affinity components (unknown pharmacology)	11,32

studies of glass fiber filters in the absence of membranes showed saturable stereospecific binding. Yoshida and Imura (3) estimated two binding sites (affinities of 2 and 20 nM) of synaptosomal membrane preparations with a pattern similar to labeled α-bungarotoxin. In this study, the specific activity of the labeled nicotine (0.25 Ci/mmol) was low, and the absence of correction for filter binding makes interpretation of data difficult.

The stereospecificity of nicotine binding has been questioned. Vincek et al. (4) detected non-stereospecific binding of 4,4-ditritio(-)-[^{3}H]nicotine. The affinity of their binding site was reported to be 4.7 μM. This lack of stereospecificity is mostly like a property of the low-affinity site they measured; subsequent binding studies that detect a high-affinity binding component show stereospecificity. Although Abood et al. (5) reported a stereospecific high-affinity binding component (6 nM K_d) utilizing nicotine of higher specific activity (22.6 Ci/mmol), their saturation experiments did not support this conclusion. IC_{50} values (where $K_i = IC_{50}/[1+(L)/K_d](6)$) of 20 nM labeled nicotine is expected to be at the saturating end of the bound-versus-free curve. Thus the reported IC_{50} value for nicotine itself at 20 nM (±)-[^{3}H]nicotine would indicate a K_d of 0.1 μM. Our studies showed two binding sites in synaptosomal membrane preparation, a high-affinity site of 60 nM and a low-affinity site of 20 μM K_d (7). Slight stereospecificity was observed in crude membranes, and increased stereospecificity is seen in synaptosomal preparations in which a greater proportion of the high-affinity component is measured.

The availability of labeled nicotine of even higher specific activity (>22 Ci/mmol) made it possible to establish the existence of a high-affinity (1-60 nM K_d) binding component. In our studies (22 Ci/mmol) with 70 nM labeled (±)-[^{3}H]nicotine, the lowest concentration that could be accurately measured, 56% of the label was bound to the high-affinity site and 44% to the low-affinity component. Marks and Collins (8) were able to label a greater percentage of the high-affinity sites with labeled nicotine of higher specific activity (61.2 Ci/mmol), even to exclude the low-affinity component by changing the incubation temperature. Under this condition stereospecificity is observed. The low-affinity component apparently binds both stereoisomers. This explains why Vincek et al. (4) did not detect stereospecificity; they measured only the low-affinity site (4.7 μM) in their binding assay.

Although nicotine is a cholinergic agonist, results have been equivocal on whether binding in brain is mainly cholinergic. The results from receptor binding studies with [^{3}H]acetylcholine (9), [^{3}H]nicotine (5,7,8), and labeled bungarotoxin (8) suggest that these ligands may bind to separate populations of receptors. The existence of noncholinergic nicotine receptors has been suggested by the psychopharmacological effects of nicotine that are not explained by its action on nicotinic sites within the brain (2). The [^{3}H]nicotine binding site from rat brain has been purified by affinity chromatography and shown to be different from the cholinergic receptor of electroplax or calf skeletal muscle (10). These studies indicated but did not prove the existence of separate noncholinergic nicotine receptors in the brain. The possibility exists that high-affinity binding sites for [^{3}H]nicotine in brain are nicotinic cholinergic receptors, and central cholinergic receptors are different from those in electric organ of fish or eel and in muscle.

We compared the binding of (±)-[^{3}H]nicotine and that of [^{3}H]acetylcholine, a ligand with the expected cholinergic profile. Regional differences were observed for the two ligands (11). Cerebellum and striatum had the lowest (±)-[^{3}H]nicotine binding, and the midbrain region the highest. [^{3}H]Acetylcholine binding was lowest in the cerebellum and highest in the midbrain and striatum. The nicotine/acetylcholine binding ratio was not constant; it was highest in the cerebellum (3.1) followed by hippocampus (2.2), and lowest in the striatum (0.9). The curve of inhibition of the binding of (±)-[^{3}H]nicotine by nicotine in the cortex was biphasic and shallow, indicating the multiplicity of the binding site. The inhibition by acetylcholine was essentially monophasic; however, only about 50% of (±)-[^{3}H]nicotine binding was inhibited at millimolar concentrations by acetylcholine. At 20 nM (±)-[^{3}H]nicotine in our experiment approximately 58% of the bound (±)-[^{3}H]nicotine represents binding to the high-affinity component and 42% to the low-affinity component. Inhibition by 50% of 20 nM (±)-[^{3}H]nicotine by acetylcholine suggests that the high-affinity component represents the cholinergic site.

The complexity of the nicotine binding site was further addressed by Sloan et al. (12), who described as many as 5 sites in rat brain P-2 preparations. There are site differences in stereospecificity, a positive cooperativity binding site, and differences in drug specificity to the multiple sites. (+)-Nicotine was more potent than (-)-nicotine in inducing positive cooperativity, whereas (-)-nicotine was 80 times more potent than (+)-nicotine in inhibiting binding at the high-affinity site. The (+)- and (-)-nicotine isomers were also about equal in inhibiting the low-affinity site. A number of cholinergic drugs besides (+)- and (-)-nicotine produced both inhibition and enhancement of (±)-[^{3}H]nicotine binding. These included lobeline, carbachol, atropine, mecamylamine, and acetylcholine, which enhanced binding. Cytisine, anabasine, choline, and cotinine inhibited (±)-[^{3}H]nicotine binding without producing positive cooperativity (12). Although the pharmacological significance of the different binding sites has not been determined, the data give some indication of the pharmacological multiplicity of the nicotine binding sites in brain. Delineating the pharmacological characteristics of these sites and their regional distribution may give insight into the mechanisms involved in nicotine effects on behavior. For instance, we have observed an effect of (-)-nicotine on brain protein metabolism. Protein metabolism is inhibited in adult and fetal brain after nicotine administration (26,33). Protein degradation measured in newborn brain slices was also inhibited by (-)-nicotine but not by (+)nicotine (26), indicating a pharmacological difference in response to the isomers. Presumably the site of action related to nicotine effects on protein metabolism is related to intracellular binding sites that are stereospecific. Recently, nicotine was shown to bind to different subcellular fractions; high binding was observed in nuclear, myelin, mitochondrial, as well as to synaptosomal fractions. Pharmacological differences and similarities have been reported for both isomers, depending on the response measured.

Autoradiographic (13,14) images showing the distribution of 10 nM [^{3}H]acetylcholine, 3.5 nM (±)-[^{3}H]nicotine, and 1.4 nM [^{125}I]α-bungarotoxin identified qualitatively similar high-affinity agonist (acetylcholine and nicotine) density patterns, different from the antagonist (α-bungarotoxin) pattern. Binding of both labeled agonists was displaced by unlabeled L-nicotine and acetylcholine. Interestingly, autoradiographic studies show little or no densities of tritiated agonist labeling in hypothalamus and hippocampus (14). We observed (±)-[^{3}H]nicotine binding in hippocampus (7,11), 2-fold higher than [^{3}H]acetylcholine, which agrees with the separate studies using [^{3}H]nicotine (8) and [^{3}H]acetylcholine (9). The differences that we observed between brain regions in the ratio of 20 nM (±)-[^{3}H]nicotine to 20 nM [^{3}H]acetylcholine binding are probably accounted for by regional differences in the properties of the high- and low-affinity binding components of [^{3}H]nicotine. For instance, the cerebellum shows three times less binding of [^{3}H]acetylcholine than the cortex, but the difference in (±)-[^{3}H]nicotine binding at 20 nM [^{3}H]nicotine between the two regions is much smaller (11). This could be due to a greater proportion of (±)-[^{3}H]nicotine (at 20 nM) bound to low-affinity sites in the cerebellum than in the cortex, hence the obscuration of a difference in the high-affinity component. In fact, the plateau in the inhibition curve occurred at a higher percentage of specific (±)-[^{3}H]nicotine binding in the cerebellum than in the cortex. The B_{max} of [^{3}H]acetylcholine is lower in the hippocampus; kinetic analysis of [^{3}H]nicotine binding has not been reported, although a single point nanomolar concentration binding study demonstrates a relatively high binding in this region (7,8,11,15). Possibly a low-affinity binding site is present in this region that is not detected by autoradiographic studies, as suggested by Clarke et al. (14). This doesn't seem to completely explain the lack of autoradiographic staining, since binding studies show considerable labeling in hippocampal membrane preparations. The regional distribution determined autoradiographically shows similarities. The differences reported tend to support the existence of noncholinergic sites of action of nicotine in brain. The fact that low-affinity component is not observed with [^{3}H]acetylcholine can reflect relevant pharmacological sites of action for nicotine. If nicotine accumulates in brain at low micromolar concentrations, these low affinity sites may have relevance.

IN VIVO REGULATION OF THE [^{3}H]NICOTINE BINDING SITE

Chronic drug treatment can change neurotransmitter receptor properties. It could be expected that exposure to an agonist nicotine would decrease (down-regulate) the receptor. The data are conflicting on the changes that occur in cholinergic receptors after chronic exposure to nicotine (Table II).

[^{3}H]Tubocurarine binding was lower in midbrain and was unchanged in cortex and hippocampus after chronic exposure to nicotine (16). The binding properties of labeled nicotine in rat brain change with age. Labeled nicotine (0.07 μM), total and saturable, increased approximately 3-fold from birth to adulthood, with near maximal binding occurring at 4 weeks. Nicotine binding is 25% higher in the offspring of rats given nicotine during gestation (17). These studies should be repeated with nicotine of higher-specific activity to examine kinetically whether the changes reflect an increase in the low- or high-affinity binding component. Most likely the low-affinity component increases more with age than the high-affinity component. At the concentration of labeled nicotine used (0.07 μM), it is likely that the increase observed in newborn rats exposed to nicotine is due to the high-affinity component, measurable because the low-affinity component is a smaller proportion of total binding. This was confirmed by Hagino and Lee (35) using high-affinity labeled nicotine. They also show age related increases in the high-affinity nicotine binding site. We did not observe an increase in nicotine binding in adult mice after exposure to injections 2 times daily of 0.4 mg/kg nicotine (17), although increases were seen if nicotine were administered by infusion (35). This might reflect the inability to measure a larger proportion of the high-affinity sites under the binding conditions of those times, but could also likely reflect pharmacokinetic differences in response to drug amounts administered. More recent studies do report an increase in nicotinic receptors ([^{3}H]acetylcholine (18,19) and (+)-[^{3}H]nicotine (20,22) after chronic exposure to nicotine. The binding studies used nicotine of higher specific activity. Exposure to cholinesterase inhibitors (diisopropylfluorophosphate), by increasing endogenous acetylcholine, decreases [^{3}H]acetylcholine binding (18). However, in organophosphate anticholinesterase disulfotan-treated rats there was significantly less [^{3}H]nicotine binding to the high-affinity component, with no change in the low-affinity component (22). Benwell and Balfour (23) did not observe increases in [^{3}H]nicotine binding after chronic nicotine (0.4 mg/kg s.c.), as in our studies (17), and suggest that the differences reported may be related to the dose, route, and schedule of nicotine administered (infusion versus repeated injections). However, Ksir et al. (24) giving low doses of nicotine similar to those used by Benwell and Balfour did observe increases in [^{3}H]acetylcholine binding and [^{3}H]nicotine (present meeting). A chronically elevated acetylcholine level in brain, by administration of cholinesterase inhibitors, decreases [^{3}H]acetylcholine and [^{3}H]nicotine binding. However, chronic nicotine, when administered to adult rodents by infusion, increases [^{3}H]nicotine binding, an effect not seen with repeated injections of nicotine under some conditions. Apparently, under certain conditions (chronic infusion) nicotine acts as an antagonist rather than an agonist. The binding site *in vitro*, however, is agonist selective, and antagonists have little affinity to the binding site. Obviously, there are similarities in the binding parameters of [^{3}H]nicotine and [^{3}H]acetylcholine; however, interesting differences exist that have not been explained.

ENDOGENOUS LIGAND FOR NICOTINE BINDING SITE

The possibility of an endogenous ligand other than acetylcholine that can bind to, and presumably modulate, the nicotine receptor is intriguing. We have observed the presence in mouse brain extracts of a substance that inhibits [^{3}H]nicotine binding (25,26). The approximate molecular weight of the inhibitory material is in the low range (<5,000 daltons) as estimated from gel filtration. There was some indication of the peptide nature of the substance in the loss of activity with acid hydrolysis; however, enzymatic degradation studies were difficult to interpret because the reagent blanks were inhibitory by themselves. The material had about equal potency in inhibiting 12-15 nM [^{3}H]nicotine or [^{3}H]acetylcholine. There was also some inhibition of [^{3}H]D-Ala-D-Leu enkephalin binding; however, the material is not likely to be enkephalin, which is eluted much later in the gel filtration profile and is inhibited by aminopeptidases. The material did not inhibit [^{3}H]QNB, [^{3}H]imipramine, [^{3}H]WB-4101, [^{3}H]spiroperidol, or [^{3}H]naloxone. Since the purification is still rather crude, it is not known whether the preparation contains more than one inhibitor with similar affinity to [^{3}H]nicotine and [^{3}H]acetylcholine binding site, or the material consists of one inhibitor inhibiting the same site, if in fact the [^{3}H]acetylcholine and [^{3}H]nicotine sites are the same. It is unlikely that the material is acetylcholine, which would be destroyed in the heat denaturation, and it would be expected to inhibit [^{3}H]QNB (muscarinic ligand). There has been contradictory evidence in support of the presence of an endogenous substrate in brain that can compete for the α-bungarotoxin binding sites (27,28). Piperidine has been suggested to be a possible substrate (2). Tryptamines and catecholamines (dopamine) react *in vivo* with carbonyls to yield β-carboline derivatives (29). β-Carboline analogues have been postulated to resemble the dopaminergic neurotoxin MPTP (29,30). There are some structural resemblances between the tobacco alkaloids, antiparkinson drugs, MPTP, and the endogenous β-carbolines (Fig. 1). If an endogenous ligand exists, it may act on a regulatory site to modulate the cholinergic receptor activity (up- and down-regulation). This action is similar to that seen with the endogenous diazepam binding inhibitor (36) which acts to regulate the high-affinity GABA binding

Table II. In Vivo Regulation of Nicotinic Receptor

Treatment	Ligand	Changes in Ligand Binding	Ref.
Nicotine (drinking fluid) Adult Rat	$[^3H]$tubocurarine	reduced	16
Nicotine (0.5 mg/kg s.c 2 x daily)	$[^3H]$nicotine		17
Offspring		increased	
Adult		unchanged	
Nicotine (0.1–1.0 mg/kg s.c 2 x daily)	$[^3H]$acetylcholine	increased	18,19,24
Diisopropylfluorophosphate		decreased	
Nicotine (infusion 1–8 mg/kg/hr)	$[^3H]$nicotine	increased	20,21
	$[^3H]$QNB	unchanged	
Disulfoton	$[^3H]$nicotine	decreased	22
	$[^3H]$QNB	decreased	
Nicotine (0.4 mg/kg s.c 1 x daily)	$[^3H]$nicotine	unchanged	23
Nicotine (infusion 175 μg/hr/rat)	$[^3H]$nicotine		
Fetus		increased	35
Adult		increased	

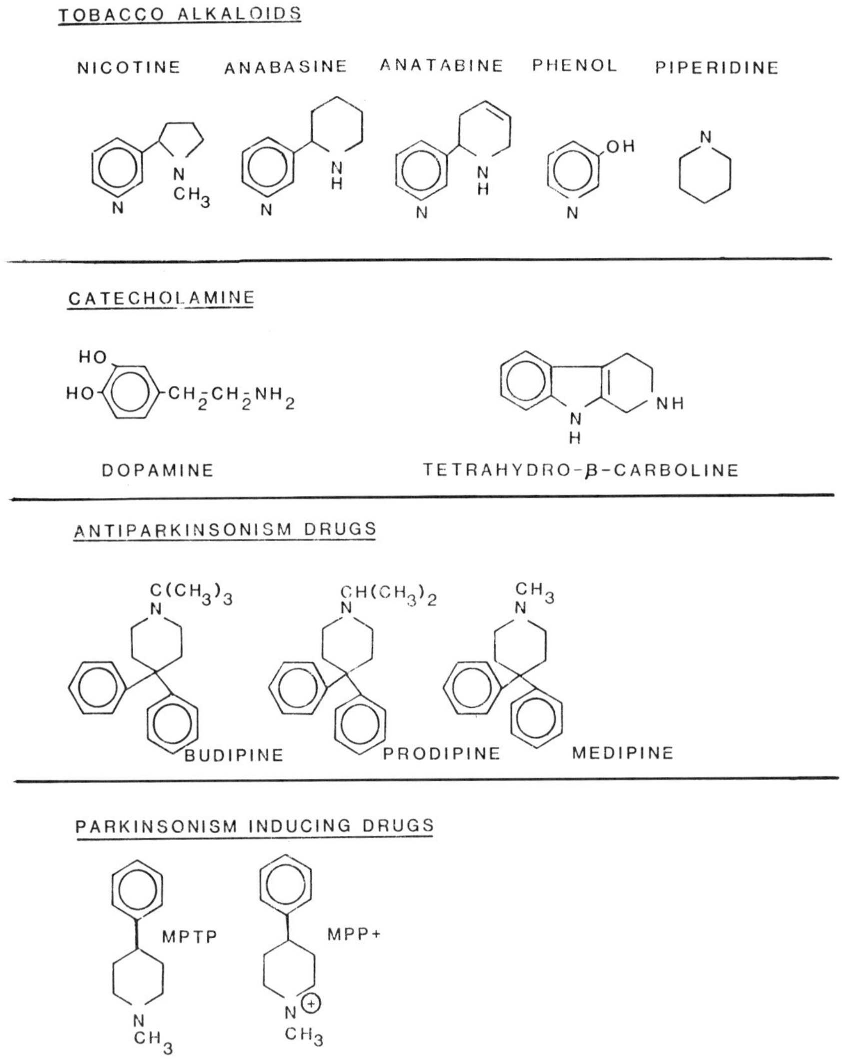

Fig. 1

site. Nicotine competes with this endogenous compound, lessening its ability to down-regulate the receptor, thereby up-regulating the recognition site. Endogenous acetylcholine acts directly on the recognition site and we observe down-regulation. However, we can at present only speculate on the existence of the endogenous ligand.

CONCLUSIONS

Several biochemical, electrophysiological, and behavioral changes have been associated with nicotine action in brain. However, the exact mechanisms and specific receptor site(s) involved are not known. We found alterations in brain protein metabolism *in vivo* (17,33) and *in vitro* (34) following nicotine administration, which *may be* mediated by *intracellular* sites of action. Of special interest is the effect of nicotine on neurotransmitters and receptors and whether specific synaptic receptor (binding) sites for nicotine exist that mediate its action on the nervous system. The possible existence of endogenous ligands for such sites and the interaction of nicotine with neurotransmitter systems in various brain areas are under investigation. Central catecholaminergic, GABAergic, enkephalinergic, and possibly noncholinergic nicotinic sites have been implicated. As many as 5 binding sites have been reported for [^{3}H]nicotine. The nanomolar high-affinity site resembles the binding site associated with [^{3}H]acetylcholine. These two ligands, however, differ in regional distribution, autoradiographic patterns of distribution, and responses to receptor changes after exposure to chronic nicotine. The low-affinity (micromolar) binding site apparently exists. Its pharmacological role is not known, but if after smoking, nicotine levels in specific brain regions reach low micromolar levels, these sites may be relevant to some of the central actions of nicotine that cannot be explained by the nicotinic-cholinergic site.

The authors would like to acknowledge Dr. Maarten E.A. Reith for his collaboration, and Audrey Hashim for technical assistance.

The work presented by the authors was supported by grants from the Council for Tobacco Research, U.S.A., Inc.

REFERENCES

1. Schleifer, LS and Eldefrawi, ME: Identification of the nicotinic and muscarinic acetylcholine receptors in subcellular fractions of mouse brain. Neuropharmacology 13:53-63, 1974.
2. Abood, LG, Lowy, A, Tometsko, A, and Booth, H: Electrophysiological, behavioral, and chemical evidence for a noncholinergic, stereospecific site for nicotine in rat brain. J. Neurosci. Res. 3:327-333, 1978.
3. Yoshida, K and Imura, H: Nicotinic cholinergic receptors in brain synaptosomes. Brain Res. 172:453-459, 1979.
4. Vincek, WC, Martin, BR, Aceto, MD, and Bowman, ER: Synthesis and preliminary binding studies of 4,4-ditritio-(-)-nicotine of high specific activity. J. Med. Chem. 23:960-962, 1980.
5. Abood, LG, Reynolds, DT, and Bidlack, JM: Stereospecific ^{3}H-nicotine binding to intact and solubilized rat brain membranes and evidence for its noncholinergic nature. Life Sci. 27:1307-1314, 1980.
6. Cheng, YC and Prusoff, WH: Relationship between the inhibition constant (K_i) and the concentration of inhibitor which causes 50 percent inhibition (IC_{50}) of an enzymatic reaction. Biochem. Pharmacol. 22:3099-3108, 1973.
7. Sershen, H, Reith, MEA, Lajtha, A, and Gennaro, J, Jr: Noncholinergic, saturable binding of (±)-[^{3}H]nicotine to mouse brain. J. Receptor Res. 2:1-15, 1981.
8. Marks, MJ and Collins, AC: Characterization of nicotine binding in mouse brain and comparison with the binding of α-bungarotoxin and quinuclidinyl benzilate. Mol. Pharmacol. 22:554-564, 1982.
9. Schwartz, RD, McGee, R, Jr, and Kellar, KJ: Nicotinic cholinergic receptors labeled by [^{3}H]acetylcholine in rat brain. Mol. Pharmacol. 22:56-62, 1982.
10. Abood, LG, Grassi, S, and Costanza, M: Binding of optically pure (-)-[^{3}H]nicotine to rat brain membranes. FEBS Lett 157:147-149, 1983.
11. Sershen, H, Reith, MEA, Hashim, A, and Lajtha, A: Comparison of [^{3}H]nicotine and [^{3}H]acetylcholine binding in mouse brain: regional distribution. Res. Commun. Chem. Path. Pharmacol. 48:345-352, 1985.

12. Sloan, JW, Todd, GD, and Martin, WR: Nature of nicotine binding to rat P_2 fraction. Pharmacol. Biochem. Behav. 20:899-909, 1984.
13. Clarke, PBS, Pert, CB, and Pert, A: Autoradiographic distribution of nicotine receptors in rat brain. Brain Res. 323:390-395, 1984.
14. Clarke, PBS, Schwartz, RD, Paul, SM, Pert, CB, and Pert, A: Nicotinic binding in rat brain: autoradiographic comparison of [^{3}H]acetylcholine, [^{3}H]nicotine, and [^{125}I]α-bungarotoxin. J. Neurosci. 5:1307-1315, 1985.
15. Abood, LG, Reynolds, DT, Booth, H, and Bidlack, JM: Sites and mechanisms for nicotine's action in the brain. Neurosci. Biobehav. Rev. 5:479-486, 1981.
16. Falkeborn, Y, Larsson, C, and Nordberg, A: Chronic nicotine exposure in rat: a behavioral and biochemical study of tolerance. Drug Alc. Depend. 8:51-60, 1981.
17. Sershen, H, Reith MEA, Banay-Schwartz, M, and Lajtha, A: Effects of prenatal administration of nicotine on amino acid pools, protein metabolism, and nicotine binding in the brain. Neurochem. Res. 7:1515-1522, 1982.
18. Schwartz, RD and Kellar, KJ: Nicotinic cholinergic receptor binding sites in the brain: regulation in vivo. Science 220:214-216, 1983.
19. Schwartz, RD and Kellar, KJ: In vivo regulation of [^{3}H]acetylcholine recognition sites in brain by nicotinic cholinergic drugs. J. Neurochem. 45:427-432, 1985.
20. Marks, MJ, Burch, JB, and Collins, AC: Effects of chronic nicotine infusion on tolerance development and nicotinic receptors. J. Pharmacol. Exp. Ther. 226:817-825, 1983.
21. Marks, MJ and Collins, AC: Tolerance, cross-tolerance, and receptors after chronic nicotine or oxotremorine. Pharmacol. Biochem. Behav. 22:283-291, 1985.
22. Costa, LG and Murphy, SD: [^{3}H]Nicotine binding in rat brain: alteration after chronic acetylcholinesterase inhibition. J. Pharmacol. Exp. Ther. 226:392-397, 1983.
23. Benwell, MEM and Balfour, DJK: Nicotine binding to brain tissue from drug-naive and nicotine-treated rats. J. Pharm. Pharmacol. 37:405-409, 1985.
24. Ksir, C, Hakan, R, Hall, DP, Jr, and Kellar, KJ: Exposure to nicotine enhances the behavioral stimulant effect of nicotine and increases binding of [^{3}H]acetylcholine to nicotine receptors. Neuropharmacology 24:527-531, 1985.
25. Sershen, H, Hashim, A, Reith, MEA, and Lajtha, A: [^{3}H]Nicotine and [^{3}H]acetylcholine binding: endogenous brain factor(s). Trans. Amer. Soc. Neurochem. 14:311, 1983.
26. Sershen, H, Reith, MEA, Hashim, A, and Lajtha, A: Endogenous material in brain inhibiting [^{3}H]nicotine and [^{3}H]acetylcholine binding. J. Neurosci. Res. 12:563-569, 1984.
27. Quik, M: Presence of an endogenous factor which inhibits binding of α-bungarotoxin 2.2 to its receptor. Brain Res. 245:57-67, 1982.
28. Connolly, JA: Is there an endogenous bungarotoxin in the vertebrate central nervous system? Brain Res. 323:307-310, 1984.
29. Testa, B, Naylor, R, Costall, B, Jenner, P, and Marsden, CD: Does an endogenous methylpyridium analogue cause Parkinson's disease. J. Pharm. Pharmacol. 37:679-680, 1985.
30. Collins, MA and Neafsey, EJ: β-Carboline analogues of N-methyl-4-phenyl-1,2,5,6-etrahydropyridine (MPTP): endogenous factors underlying idiopathic Parkinsonism? Neurosci. Lett. 55:179-184, 1985.
31. Romano, C and Goldstein, A: Stereospecific nicotine receptors on rat brain membranes. Science 210:647-650, 1980.
32. Larsson, C and Nordberg, A: Comparative analysis of nicotine-like receptor-ligand interactions in rodent brain homogenate. J. Neurochem. 45:24-31, 1985.

33. Sershen, H, Reith, MEA, Lajtha, A, and Gennaro, J, Jr.: Effect of cigarette smoke on protein synthesis in brain and liver. Neuropharmacology 20:456, 1981.
34. Sershen, H and Lajtha, A: The effect of nicotine on the metabolism of brain proteins. Neuropharmacology 18:763-766, 1979.
35. Hagino, N and Lee, JW: Effect of maternal nicotine on the development of sites for [^{3}H]nicotine binding in the fetal brain. Int. J. Develop. Neurosci. 3:567-571, 1985.
36. Costa, E and Guidotti, A: Endogenous ligands for benzodiazepine recognition sites. Biochem. Pharmacol. 34:3399-3403, 1985.

DOSE-RELATED NICOTINE EFFECTS ON CNS BINDING AND BEHAVIOR

Charles Ksir and Robert L. Hakan

University of Wyoming
Kenneth J. Kellar
Georgetown University

Repeated nicotine exposures increase both central nicotinic receptor numbers and the behavioral stimulant effect of nicotine. In the present experiments, the behavioral effects of graded doses of nicotine (0.1 to 1.6 mg/kg, s.c.) were observed in photocell activity cages both before and after the rats were given five daily injections of 0.2 mg/kg nicotine in their home cages. Comparing the behavioral response after five days of exposure to the response prior to exposure revealed an upward shift in the dose-response curve (Fig. 1).

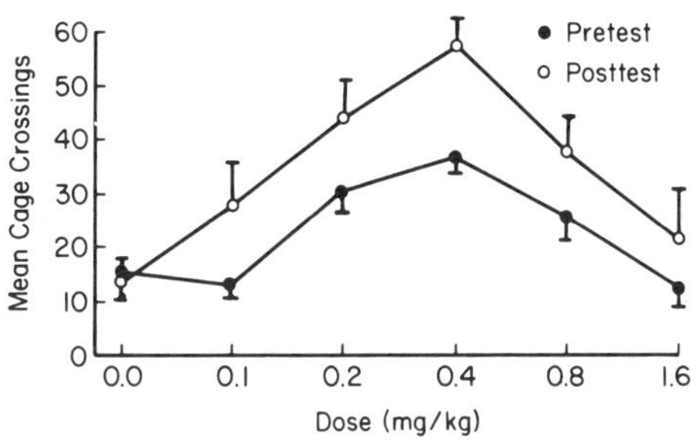

Fig. 1: Group mean response in the first 10 minutes following a s.c. injection of nicotine at the indicated test dose. Pretest: naive rats. Posttest: the same rats tested after 5 daily injections of 0.2 mg/kg nicotine.

Other rats were given five daily injections of graded low doses of nicotine (0.01 to 0.30 mg/kg). On the sixth day these rats were all tested in the photocell cages for their response to a common test dose (0.2 mg/kg). Following the test session the rats were decapitated, and the brains were removed and frozen for later assay of nicotinic cholinergic receptors. Both [^{3}H]acetylcholine and [^{3}H]nicotine were used to assay binding in the cerebral cortex and striatum. Figure 2 demonstrates that the rats exposed to 0.03 mg/kg or more for five days showed an elevated behavioral response to the test dose, and an increased nicotinic binding in the cortex. A similar pattern was found in the striatum. The binding results using a single concentration of [^{3}H]acetylcholine were highly correlated (r=0.90) with the results using [^{3}H]nicotine. Data from other rats indicate that nicotine exposure results in an increase in B_{max} and no change in Kd for either ligand.

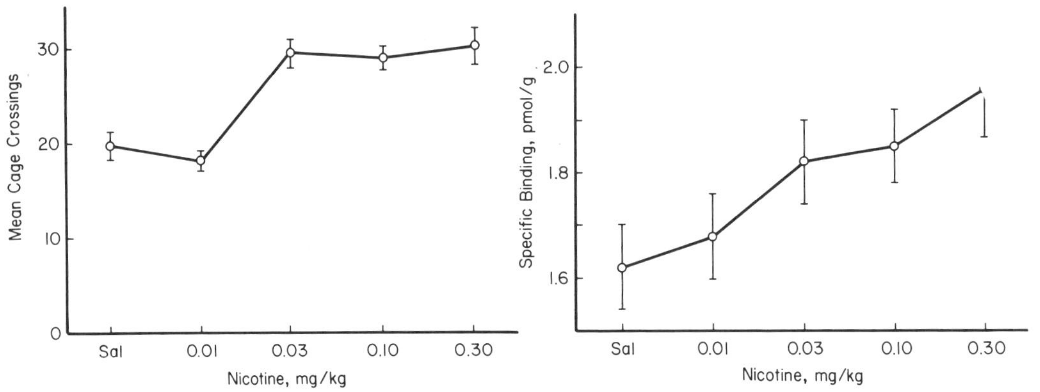

Fig. 2. Behavioral response in the first 10 minutes after 0.2 mg/kg nicotine (left panel) and binding of [^{3}H]ACh to cerebral cortex (right panel) after five days of exposure to the indicated dose of nicotine.

[^{3}H]ACETYLCHOLINE BINDING TO NICOTINIC CHOLINERGIC RECEPTORS IN BRAIN: LOCALIZATION AND REGULATION VISUALIZED BY AUTORADIOGRAPHY

Rochelle D. Schwartz[1], Barry B. Wolfe[2], Thomas C. Rainbow[2] and Kenneth J. Kellar[3].

[1]Clinical Neuroscience Branch, NIMH,
[2]Dept. of Pharmacology, University of Pennsylvania
and [3]Dept. of Pharmacology, Georgetown University

[^{3}H]Acetylcholine ([^{3}H]ACh) labels nicotinic cholinergic receptors of high affinity in brain homogenates (1). The distribution of these sites can be studied in greater detail using autoradiographic techniques. In a recent autoradiographic study, a qualitative comparison was made of [^{3}H]ACh and [^{3}H]nicotine binding to nicotinic cholinergic receptors in rat brain sections (2). In the present study, autoradiography was used to determine the relative distribution of [^{3}H]ACh binding sites throughout several levels of rat brain. Brain sections (24 μm) were thaw-mounted onto subbed slides, assayed for [^{3}H]ACh binding, and exposed to tritium-sensitive film as previously described (2,3). Optical density measurements were made of specific brain regions relative to areas of white matter in the same slice. There was a seven-fold difference between areas with the highest and lowest levels of [^{3}H]ACh binding. The greatest binding occurred in areas such as the interpeduncular nucleus, several thalamic nuclei, and medial habenula. Areas with moderate levels of binding included the superior colliculus, retrosplenial cortex, substantia nigra pars compacta, and caudate nucleus, while [^{3}H]ACh binding was virtually absent in areas such as the inferior colliculus, hippocampus (stratum oriens), and entorhinal cortex. In previous studies using brain homogenates, we have observed that repeated exposure to the cholinesterase inhibitor DFP leads to a 'down-regulation' of [^{3}H]ACh binding sites while repeated exposure to nicotine leads to an 'up-regulation' of these binding sites in brain areas such as the thalamus, cortex, caudate, and hypothalamus (4,5). In the present study, autoradiographic techniques were used in order to examine the effects of repeated nicotine administration on regional [^{3}H]ACh binding in more detail. Following 10 days of nicotine administration to rats (1 mg/kg, twice daily, ip), major increases in [^{3}H]ACh binding were observed in several brain areas. Binding was increased to the greatest extent in the anterior hypothalamus (130%), interpeduncular nucleus (96%), and substantia nigra (96%), while smaller increases were observed in the cortex (60-75%) and caudate nucleus (59%). The smallest increases in [^{3}H]ACh binding occurred in various thalamic nuclei (20-30%) and the medial habenula (30%). Thus, repeated exposure to nicotine results in a widespread up-regulation of [^{3}H]ACh binding sites in brain, however, this up-regulation occurs to varying degrees among brain regions.

1. R.D. Schwartz, R. McGee and K.J. Kellar, Mol. Pharmacol. 22, 56 (1982).
2. P.B.S. Clarke, R.D. Schwartz, S.M. Paul, C.B. Pert and A. Pert, J. Neurosci. 5, 1307 (1985) (and this symposium).
3. T.C. Rainbow, R.D. Schwartz, B. Parsons and K.J. Kellar, Neurosci. Lett. 50, 193 (1984).
4. R.D. Schwartz and K. J. Kellar, Science 220, 214 (1983).
5. R.D. Schwartz and K.J. Kellar, J. Neurochem. 45, 427 (1985).

ACUTE EFFECTS OF TOBACCO ON HUMAN BRAIN STEM EVOKED POTENTIALS

Verner J. Knott

Specialty Clinics, Royal Ottawa Hospital
Ottawa, Canada

Numerous studies employing EEG and evoked potentials to examine the acute effects of tobacco/nicotine on brain electrical activity have reported electrocortical activation to be the predominant response and have indicated that this effect is mediated via nicotine's action on the reticular formation. Noninvasive investigation of human brain stem involvement in tobacco smoking may be studied feasibily by acoustically elicited, far field, short-latency electrical potentials which reflect sequential activation of nuclei within the brain stem auditory pathway. This study examined the acute effects of tobacco smoking on these brain stem auditory-evoked potentials (BSAEP).

Twelve regular smokers were tested on two separate sessions involving sham or real smoking. On each session, BSAEPs were recorded during a baseline period and immediately after smoking. BSAEPs, recorded from Cz, were elicited by presentation of 1,000 monaural, rarefaction click stimuli. Latency and amplitudes of peak components I through V were assessed and analyzed. No significant effects were observed for latency measures or for amplitudes of the first four peaks. A significant effect was observed for peak V with tobacco resulting in larger amplitudes relative to sham smoking. Peak V reflects activity from upper pontine-lower midbrain sites and this tobacco-peak V finding is discussed in relation to arousal theories of smoking.

EFFECTS OF NICOTINE ON THE DEVELOPMENT OF AN INVERTEBRATE EMBRYONIC SYSTEM

Vicki Martin, Kathy Kolberg, and William Archer

Department of Biological Sciences
University of Notre Dame
Notre Dame, Indiana 46556

The Cnidarians have traditionally been used as model systems for the study of cell genealogies, cell lineages, cell differentiation, cell-cell interactions, establishment and maintenance of polarity, regeneration, and the factors which control and govern these processes. These animals display a simple architecture. They have a small number of cell lineages and cell types, and they exhibit exceptional morphogenetic plasticity and adaptability. Hydrozoan embryos provide good systems in which to examine the effects of toxic substances on development. These embryos have been used to validate a teratogenic risk assessment bioassay system. Adult lethal and developmentally toxic doses have been determined for a number of chemicals using the hydrozoan system, and the A/D ratios obtained are compatible with those from mammals. Once these dosages are known, one can predict with some certainty those chemicals selectively toxic to the hydrozoan embryo and of probable teratogenic risk to higher animals. The hydrozoan system is being promoted as a system that could effectively predict the teratogenic risk potential of a large number of newly developed or yet untested chemicals in a rapid and cost-effective manner. The research focuses on identifying peak exposure periods (gastrulation), length of test article exposure, and determining endpoints which may be predictive of developmental toxicity. Such endpoints include: gross morphological changes to the embryo and/or adult as assessed using light microscopy or electron microscopy, alteration in size, effect on motility, altered response to stimuli (i.e., light), and inability of embryos to form adults. Embryos were exposed to varying concentrations of nicotine at different developmental stages and fixed for either light microscopy, transmission electron microscopy, or scanning electron microscopy. Results indicate that the A/D ratio for nicotine using the hydrozoan test system is 10. Immediately upon exposure to nicotine, embryos lose their surface cilia and there is a massive discharge by glandular cells. Supportive cells and nerve cells undergo changes in cell shape. Embryos treated with nicotine and allowed to recover eventually regenerate cilia. These treated embryos display an altered gross morphology; they become negatively phototactic; and they develop into adults faster than do comparable controls. Hydrozoan embryos contain a population of multipotent stem cells, interstitial cells, which either divide to maintain the stem populations or differentiate into two terminally differentiated products: ganglionic nerve cells and nematocytes. In nicotine-treated embryos, the ratio of interstitial cells to ganglionic cells to nematocytes was significantly different when compared with comparable controls. In treated embryos an increase in the numbers of differentiated progeny was accompanied by a decrease in the stem cell population. Hence, treated embryos contain fewer stem cells. Results from our studies suggest that nicotine is a potential teratogen. This work represents the first attempt to document the effects of nicotine on the differentiation and development of a stem cell population.

EFFECTS OF NICOTINE ON THE TURNOVER RATE OF CATECHOLAMINES IN THE RAT BRAIN

T. Kubo, H. Amano, Y. Goshima, and Y. Misu

Department of Pharmacology, Yokohama City University
School of Medicine, Yokohama 232, Japan

Acute effects of _in vivo_ nicotine application on turnover rates of dopamine (DA), norepinephrine (NE), and epinephrine (EP) in the rat brain regions were examined. Turnover rates of the amines were evaluated by measuring tissue content of a metabolite of DA and NE, dihydroxyphenylacetic acid and 3-methoxy-4-hydroxyphenylethylene glycol, respectively, and/or by comparing decline of contents of the amines after blockade of synthesis. Synthesis was blocked by α-methyl-p-tyrosine 300 mg/kg, i.p., and fusaric acid 80 mg/kg, i.p., an inhibitor of tyrosine hydroxylase and dopamine-β-hydroxylase, respectively. Three kinds of the amines and the metabolites were measured using HPLC-ECD methods. Injection of nicotine (0.4–1.0 mg/kg, s.c.) increased DA turnover rate in hypothalamus, thalamus, and pons/medulla, and NE turnover rate in striatum, hypothalamus, thalamus, midbrain, pons/medulla, and cerebellum. As a parameter of turnover rate, nicotine-induced increases in content of the metabolites were more sensitively detected than nicotine-induced decreases in DA and NE content itself after blockade of synthesis. These facilitatory effects of nicotine on turnover rate of DA and NE were antagonized by pretreatment with mecamylamine 5 mg/kg, i.p., a centrally acting nicotinic antagonist, but not by hexamethonium 10 mg/kg, i.p., an antagonist poorly penetrating into the brain. Nicotine also tended to accelerate the disappearance of EP content caused by fusaric acid in hypothalamus, thalamus, midbrain, and pons/medulla. These findings demonstrate that nicotine increases the turnover rate of catecholaminergic neurons in various regions via direct activation of central nicotinic receptors.

CIGARETTE SMOKING AND PARKINSONISM

John A. Baron, Priscilla Adams, and
Hebe Quinton

Dartmouth Medical School, Hanover, New Hampshire 03756

One of the most consistent findings in the epidemiology of Parkinsonism is the apparent protective effect of cigarette smoking. At least 16 investigations have addressed this question; almost all of them have demonstrated smokers to have a lower risk than nonsmokers. In general, the results have been statistically significant, with estimated relative risks varying around 0.5 for those with a history of ever smoking compared to those who never smoked. This reduction in risk has been found among males and females, and at all ages. Several investigators have reported a pattern of decreasing risk with increasing amount smoked. Although many of the studies have limitations, in aggregate, the epidemiologic data strongly suggest that smoking could alter the risk of Parkinsonism.

Several other lines of evidence support the idea that smoking can have important effects on dopaminergic processes in the CNS. First, nicotine may ameliorate the tremor of Parkinsonism. Second, cigarette smoke or nicotine cause decreased secretion of prolactin and increased secretion of growth hormone. Both effects suggest an enhancement of CNS dopamine transmission. Finally, nicotine increases the firing of nigrostrial neurons and enhances dopamine turnover in the hypothalamus and the striatum. Thus, smoking appears to exert a central dopaminergic influence that could counter the dopamine deficiency characteristics of Parkinsonism.

Although these findings have no direct public health relevance, the effect of cigarette smoking on Parkinsonism may provide insights into the etiology and effective treatment of that syndrome.

STEREOSPECIFIC N-METHYLATION OF NICOTINE BY INTACT GUINEA PIG PULMONARY ALVEOLAR MACROPHAGES

C. Gairola[1,2], A. Al Houdi[3], C.S. Godin[3] and P.A. Crooks[2,3]

Tobacco and Health Research Institute[1], Graduate Center for Toxicology[2] and College of Pharmacy[3], University of Kentucky, Lexington, KY 40546

Pulmonary alveolar macrophages (PAMs) are primary phagocytes of the lung which interact with nicotine that is inhaled during the process of smoking. The present investigation was conducted to examine the metabolism of nicotine by PAMs. Free lung cells, obtained by bronchoalveolar lavage of male Hartley guinea pigs, were cultured in plastic petri dishes to establish macrophage monolayers. Reaction mixtures containing radiolabeled racemic nicotine or its (R-(+)- or S-(-)-isomer were placed on the monolayers and incubated at 37°C in humidified 5% CO_2/air atmosphere for 1 hr. The supernatants and cell lysates were collected separately for the analysis of nicotine metabolites by high performance cation-exchange liquid radiochromatography as described elsewhere (Cundy, K.C. and Crooks, P.A., 1984, J. Chromatog. 306:291). A single methylated metabolite, N-methylnicotinium ion, was consistently detected in cell lysates, but not in the supernatants. Oxygenated metabolites of nicotine could not be detected either in lysates or supernatants. These observations suggested that PAMs methylate nicotine intracellularly but do not carry out oxidative metabolism of nicotine under these conditions. Maintenance of PAMs in culture from 1 hr to 40 hr increased their ability to methylate [^{14}C-NCH_3]-(±)-nicotine. The conversion of nicotine by PAMs was stereospecific for the R-(+)-isomer which was converted by approximately 20% to its N-methylated metabolite. In comparison, the S-(-)-isomer of nicotine was not methylated. The macrophages also did not methylate [^{3}H]histamine under similar conditions, thus further indicating the specificity of a PAM-methyltransferase system for R-(+)-nicotine. The presence of histamine (50 μM) in the reaction mixture, however, inhibited N-methylation of R-(+)-nicotine by about 30%. The ability of PAMs from guinea pigs exposed to sidestream or mainstream cigarette smoke to methylate [^{14}C-NCH_3]-(±)-nicotine was found to be similar to that of PAMs from room and sham control animals, indicating no effect of smoke exposure. The results presented above suggest a role of PAMs in the biotransformation of R-(+)-nicotine in the respiratory tract. (Supported by KTRB grant No. 5A533 and 4C018).

BIODEGRADABLE MICROSPHERES FOR SUSTAINED DELIVERY OF α-PROTEINASE INHIBITOR

F. Rypacek and P.P. DeLuca

College of Pharmacy, University of Kentucky, Lexington, KY 40536

One of the adverse effects of smoking is related to the deterioration of α-$_1$-proteinase inhibitor activity. There is a strong evidence that deficiency in α-1-proteinase inhibitors (α-PI) may predispose persons to the development of pulmonary emphysema. Since therapeutic administration of α-PI has been suggested it is possible that such therapy would be more efficient if delivered directly to the target organ, i.e., lung in this case. A selective administration of α-PI to the lung can be achieved via inhalation or intravenously, if the protein is available in the form of spherical particles of appropriate size, i.e., in the range 1-5 μm for the delivery to the lower airways or in the range 8-20 μm for an intravenous delivery to the lung capillaries and/or precapillary vessels. The methodology for preparation of microspheres from a biodegradable hydrogel of controllable size range has been developed. Microspheres have been characterized by light and electron scanning microscopy and factors affecting the yield of incorporated protein and the rate of its release have been studied. The amount of a diffusion-released fraction of incorporated protein can represent up to 25% of the mass of microspheres and the rate of its release can be controlled through control of the matrix properties. Retention of a substantial part of the inhibitory activity of incorporated α-PI has been documented. Biodegradable hydrogel microspheres constitute an attractive system for parenteral delivery of α-PI and other biologically active proteins; however, its appropriateness for particular purposes should be considered with respect to the capacity of the system and the required amount of the protein to be delivered.

IN VITRO AND *IN VIVO* BIOMETHYLATION OF NICOTINE ENANTIOMERS

C. Steven Godin, William F. Pool,
Chukwuemeka G. Nwosu, and Peter A. Crooks

College of Pharmacy and Graduate Center for Toxicology
University of Kentucky, Lexington, KY

Both S-(-)-nicotine and R-(+)-nicotine are reported to be constituents of tobacco smoke condensate. Although numerous studies have been published on the metabolic fate of the S-(-)-enantiomer, little is known about the biotransformation of the unnatural R-(+)-isomer. In this study, the comparative metabolic fates of both optical isomers of nicotine are examined.

R-(+)-Nicotine was found to undergo *in vivo* metabolism in the guinea pig to both oxidative and quaternary N-methylated metabolites. The latter metabolic pathway affords N-methylnicotinium ion (NMN), N-methylcotininium ion (NMC), N-methylnornicotinium ion (NMNN), and the *cis*- and *trans*-diastereomers of N-methyl-N'-oxonicotinium ion (NMNO). These metabolites constitute about 15-20 percent of the total urinary metabolites of i.p. administered R-(+)-nicotine. S-(-)-Nicotine affords only oxidative urinary metabolites *in vivo* in the guinea pig, 3-hydroxycotinine being the major metabolite: no traces of methylated nicotine metabolites could be detected.

In vitro experiments have shown that 'nicotine N-methyltransferase' is widely distributed in guinea pig tissues, being particularly active in lung and spleen. The cofactor for the reaction is SAM, and the enzyme exhibits stereospecificity for the R-(+)-nicotine isomer. SAH is a feedback inhibitor in the reaction, and S(-)-nicotine potently inhibits the N-methylation of its optical antipode. Enzyme activity was located in the cytosolic cellular fraction, and the enzyme characteristics resembled those of the endogenous enzyme, histamine-N-methyltransferase. In rabbit liver, a crude enzyme preparation catalyzed the N-methylation of both S-(-)- and R-(+)-nicotine. The former substrate afforded S(-)-NMN ion as the only enzymic product, while R-(+)-nicotine methylation gave both R-(+)-NMN ion and the isomeric product, R(+)-N'-methylnicotinium ion. *In vitro* experiments with 100,000 g human liver supernatants demonstrated stereoselectivity in the N-methylation of S-(-)- and R-(+)-nicotine isomers to S-(-)-NMN and R-(+)-NMN ion, respectively. [Aided by a grant from the Tobacco and Health Research Institute, Lexington, KY.]

PERTURBATION OF S-ADENOSYLHOMOCYSTEINE AND S-ADENOSYLMETHIONINE LEVELS, AND INHIBITION OF HISTAMINE METABOLISM IN THE GUINEA PIG, BY NICOTINE

C. Steven Godin, Abdul Al Houdi, and Peter A. Crooks

College of Pharmacy and Graduate Center for Toxicology
University of Kentucky, Lexington, KY

Previous work from our laboratory has shown that both (R-(+)- and S-(-)-nicotine enantiomers are capable of interacting with endogenous N-methyltransferase systems both in vitro and in vivo. This present study was designed to determine whether nicotine was capable of perturbing SAH and SAM levels in vivo, in guinea pig lung and liver.

Groups of 5 male Hartley guinea pigs were dosed chronically with either R-(+)-nicotine or S-(-)-nicotine. Control animals received saline only. The nicotine enantiomers were administered by Alzet mini-osmotic pumps at the rate of 0.5 mg/hr for 24 days. Animals were housed in metabolic cages and urine was collected every four days. After sacrifice, lung and liver tissues were homogenized in 5% per chloric acid, centrifuged, and analyzed for SAH and SAM content by HPLC.

Guinea pig lung tissue was depleted of both SAH and SAM after chronic exposure of animals to either R-(+)- or S-(-)-nicotine enantiomers. The depletion was more pronounced with the S-(-)-enantiomer. Liver tissue levels of SAH and SAM were less effected, in fact, a significant increase in liver SAH levels was observed after exposure of animals to R-(+)-nicotine. These results indicate that chronic exposure to nicotine may perturb important endogeneous methyltransferase reactions, which could be a contributory factor in the toxicological effects produced by cigarette smoking.

In this respect it was also demonstrated that S-(-)-nicotine was a competitive inhibitor of guinea pig lung histamine N-methyltransferase (HMT) activity. (HMT is the major enzyme involved in the metabolic deactivation of histamine.) However, analysis of urine from guinea pigs that had been chronically dosed with either S-(-)- or R-(+)-nicotine enantiomers indicated no significant changes in urinary histamine or N -methylhistamine compared to control animals. [Aided by a grant from the Tobacco and Health Research Institute, Lexington, KY.]

CIGARETTE HABITUATION IN NONHUMAN PRIMATES

Walter R. Rogers and Megan L. Tuttle

Department of Bioengineering
Southwest Research Institute
San Antonio, Texas 78284
and
Department of Physiology and Medicine
Southwest Foundation for Biomedical Research
San Antonio, Texas 78284

Using operant conditioning techniques to establish and control the behavior, baboons can be trained to actively puff on cigarettes in a human-like manner. A pilot experiment was conducted to determine if baboons would become habituated to cigarette smoke.

Fourteen baboons, originally trained to smoke cigarettes for water rewards, were given gradually increasing amounts of supplementary drinking water until, in week 28 of the procedure, water was available ad lib and there was no external reward contingency for cigarette smoking. As increasing amounts of water were given, the average number of puffs per cigarette gradually decreased. However, by week 32, 4 of the animals continued to smoke, averaging between 4 and 5 puffs per cigarette, while the other subjects averaged less than 1 puff per cigarette. During the next 18 weeks, the 4 habituated animals averaged 116 puffs per day on 20 cigarettes. Mean smoke puff volume was 17.4 ml, and mean carboxyhemoglobin (COHb) level was 2.7%, a level three times greater than that of nonsmoking baboons. Two animals continued voluntary smoking for 27 months; during the last 9 months, they averaged 7.5 puffs per cigarette with a mean puff volume of 26.2 ml and mean blood COHb of 5.2%. The other 2 animals continued smoking voluntarily for 5 months, at which time the behavioral studies were terminated.

Using the 4 habituated smokers, a sequential preference experiment was conducted in which each day for a week either cigarettes or shams were placed in the mouthpiece of the SIRIC. The stimulus was changed randomly every hour, and there were four presentations per hour. The baboons made significantly ($P < .01$) more puffs per day on cigarettes (mean = 100) than on shams (mean = 23).

These results demonstrate that nonhuman primates can become habituated to cigarettes and thus could be used as an animal model to examine in detail the behavioral, pharmacologic, and neurochemical mechanisms involved in the establishment and dissolution of the cigarette smoking habit.

This research was supported by BSRG grant No. 87.68 from Southwest Foundation for Biomedical Research.

NAUSEA, CIGARETTE SMOKING, NICOTINE, AND PITUITARY HORMONE RELEASE

Mark J. Majchrzak, Ovide F. Pomerleau,
and Cynthia S. Pomerleau

Behavioral Medicine Laboratory
Department of Psychiatry
University of Michigan School of Medicine

This laboratory has previously reported increases in arginine vasopressin, neurophysin I, and beta-endorphin/beta-lipotropin in conjunction with relatively high plasma nicotine levels in subjects not experiencing nausea. The present study further explores the hormonal and neuropeptide responses provoked by nicotine from smoking. Nausea was induced in deprived smokers by having them smoke 2 high-nicotine cigarettes (2.87 mg.) in succession. Subjects who reported nausea showed a less selective release of pituitary hormones than previously observed at sub-nausea: Plasma levels of adrenocorticotropic hormone, arginine vasopressin, beta-endorphin/beta-lipotropin, growth hormone, neurophysin I, and prolactin increased over baseline; no increases were seen in follicle-stimulating hormone, luteinizing hormone, or thyroid-stimulating hormone. Nausea and significantly elevated pituitary hormone release were consistently associated with plasma nicotine levels exceeding 60 ng/ml produced by smoking high-nicotine cigarettes, but did not occur in response to low-nicotine (0.48 mg.) cigarettes (plasma nicotine levels below 10 ng/ml). Because the hormonal profile associated with nicotine-induced nausea resembles that of nausea produced by other manipulations (motion sickness, administration of other drugs), it is likely that the observed hormonal pattern at this level of stimulation is characteristic of nausea and not unique to nicotine.

THE ROLE OF TOBACCO SMOKING IN CUTANEOUS CIRCULATORY CHANGES ASSOCIATED WITH ATHEROSCLEROSIS

Daniel Richardson and Gordon Hyde

Departments of Physiology and Surgery
University of Kentucky, College of Medicine
Lexington, KY 40536

We have shown a variety of cutaneous macro- and microcirculatory disturbances distal to sclerotic lesions (Schwartz et al., J. Vas. Surg. 1, 800, 1984). Since the majority of those patients were cigarette smokers and since smoking affects the cutaneous circulation, the purpose of this on-going project is to determine if the circulatory changes characteristic of sclerotic patients are due to the disease itself or tobacco smoking. This is being done by cross-sectional comparisons of arterial (ABF) and capillary (CBF) blood flows in the cutaneous region of the toe in groups of healthy control nonsmokers (CNS); healthy control smokers (CS); smokers who have atherosclerosis in the femoral tree (AS); and nonsmokers who have femoral atherosclerosis (ANS). All groups are being matched according to age and gender. The smoker groups are also matched in accordance with smoking history. This presentation gives preliminary results of the first three groups. The rank order of ABF both at rest and at peak reactive hyperemia following a 1 min period of ischemia thus far is: CNS > CS > AS. The CS and AS values are significantly lower compared to the CNS group ($P \leq 0.10$). At the microcirculatory level resting CBF per vessel has been similar among the groups, but the AS group has had a higher capillary density such that estimated total capillary perfusion (flow per vessel x density) is higher in this group. Despite this, the AS subjects have had a lower reactive hyperemia of CBF both in terms of absolute values and percent change ($p \leq 0.05$). These preliminary results suggest that habitual tobacco smoking per se elicits changes in cutaneous arterial blood flow directionally similar to those seen in smokers who have atherosclerosis, but post-lesion microcirculatory changes seem to be a manifestation of the disease itself and are not related to tobacco smoking.

EFFECT OF CIGARETTE ROD LENGTH ON PUFF AND POST-PUFF RESPIRATION TOPOGRAPHY, RESPIRATORY CO, AND SUBJECTIVE RATINGS IN SMOKERS

Phillip P. Woodson and Roland R. Griffiths

The Johns Hopkins University School of Medicine
Department of Psychiatry and Behavioral Sciences
Division of Behavioral Biology
Behavioral Pharmacology Research Unit, D-5-West
Francis Scott Key Medical Center
4940 Eastern Avenue, Baltimore, Maryland 21224

The habitual smokers (5 males and 5 females) smoked their preferred brands of cigarettes under three different conditions separated by 30 min intervals. These three smoking conditions were presented in a counterbalanced order across a total of 10 days. Each of the three conditions involved taking eight puffs in a normal fashion at a fixed interpuff interval which varied from 45-65 sec. The three experimental conditions were: 1) Whole cigarette - one cigarette smoked progressively down the burning rod to within 5 mm of filter, 2) Full Length - one puff taken from each of eight full length cigarettes, and 3) Butt Length - one puff taken from each of eight butt length cigarettes prepared by clipping off all but 5 mm of burnable tobacco rod distal to filter.

Puff topography was characterized for each of the eight puffs in each condition by deriving puff duration (sec), puff volume (ml), peak flow (ml/sec), and latency to peak flow (sec). In a similar fashion, the first post-puff respiratory cycle was characterized by deriving inhalation volume (ml), latency to peak of inhalation volume (sec), exhalation volume (ml), and duration of this inhalation/exhalation cycle (sec). Respiratory carbon monoxide (CO) concentration (ppm) was measured using end-expiratory air after a forced expiration/inhalation maneuver both before and after each "cigarette" was smoked. Subjective ratings were measured on visual analog scales for degree of strength, heat, harshness, draw, taste, and satisfaction after each "cigarette."

Results suggest that tobacco rod length may act to at least partially control smoke exposure in nonpharmacological fashion. The precise roles played by resistance to draw, as well as nicotine, tar, and CO, need to be defined experimentally.

EFFECTS OF SMOKING TOPOGRAPHY ON NICOTINE AND CARBON MONOXIDE EXPOSURE

James P. Zacny, Maxine L. Stitzer,
Roland R. Griffiths, Fran J. Borwn,
and John E. Yingling

Johns Hopkins University School of Medicine/Key
Medical Center

Exposure to smoke constituents from a cigarette is determined by properties of the cigarette (e.g., ventilation/filtration, paper porosity) and by the manner in which the cigarette is smoked (e.g., puff number and volume). The role of several potentially important smoking behaviors in affecting biological exposure levels, though, has not been experimentally determined. The purpose of the present study was to measure acute intake of carbon monoxide (CO) and nicotine, while varying puff volume (the amount of smoky air drawn into the mouth per puff), inhalation volume (the amount of smoky air inhaled into the lungs), and breath-hold duration (the temporal period from maximum inhalation volume to exhalation onset).

Nine to 10 smokers participated in each of three experiments in which puff volume, inhalation volume, and breath-hold duration were varied. In each experiment, by using auditory feedback to aid smokers, one of these variables was manipulated while the other two were held constant. Puff volume significantly influenced CO and nicotine absorption: as puff volume increased from 15 ml to 60 ml, mean nicotine absorption after 8 puffs increased from 4.6 ng/ml to 16.2 ng/ml and mean CO absorption increased from 1 ppm to 8.7 ppm. Breath inhalation volume had no impact on either nicotine or CO absorption following intake of standard (50 ml) puffs. A similar amount of nicotine and CO was absorbed after small inhalations (20% of a smoker's vital capacity) and large inhalations (60% of a smoker's vital capacity). Breath-hold duration did not significantly affect nicotine absorption, but did affect CO absorption. As breath-hold duration increased from 0 sec to 16 sec, mean CO absorption after 8 standard (50 ml) puffs increased from 4.4 ppm to 9.3 ppm. While smokers could reduce their exposure to the particulate and gas phase of smoke by taking smaller tobacco smoke puffs, it appears inhalation volume does not influence biological exposure.

CHRONIC NICOTINE INFUSION IN RATS: EVIDENCE FOR SOME NONCHOLINERGIC ACTIONS

M.D. Aceto, S.M. Tucker, J.R. Hinson, and G.S. Ferguson

Department of Pharmacology and Toxicology
Medical College of Virginia
Virginia Commonwealth University

Recently, we reported (Aceto et al., Pharmacologist, 27, 225, 1985) that nicotine di-l-tartrate given continuously by intraperitoneal infusion (Teiger, 190, 405, 1974) at a dose of 0.1 μg/kg/sec (base) markedly suppressed drinking during the first day of infusion. In addition, when nicotine was abruptly withdrawn six days later, water consumption increased dramatically during the first 24 hrs. We confirmed these results and conducted additional studies with mecamylamine, a nicotine antagonist. When given simultaneously with nicotine at doses of 0.01 and 0.05 μg/kg/sec, mecamylamine failed to block either the initial suppression of drinking or the rebound increase during withdrawal. The vehicle and mecamylamine controls had no significant effect on drinking. In addition, the rats receiving nicotine or nicotine plus mecamylamine did not gain as much body weight as the other subjects in the study. However, four days after abrupt withdrawal was initiated, body weight was in the same range as the vehicle controls. No behavioral signs were noted either during infusion or abrupt withdrawal. These results suggest that some of nicotine's actions may not involve cholinergic mechanism.

Supported by NIDA Grant DA02384

HETEROGENEITY OF BRAIN NICOTINE RECEPTORS: A FUNCTIONAL APPROACH

S. Wonnacott, C. Rapier, and G.G. Lunt

Department of Biochemistry, University of Bath,
Bath, BA2 7AY, U.K.

To study nicotinic receptors in the brain, we have compared the biochemical and pharmacological properties of ligand binding sites with a model in which receptor function can be measured. The system is based on the nicotinic facilitation of dopamine release from striatal nerve terminals, and we have demonstrated the release of [^{3}H]dopamine from perfused synaptosomes in response to sub$_\mu$molar concentrations of nicotinic agonists (1). This nicotinic mechanism is antagonized by ganglionic antagonists and the shellfish toxin neosurugatoxin, but not α-bungarotoxin (2). Although ligand binding is generally unaffected by many antagonists, high affinity [^{3}H]nicotine binding to whole brain membranes is sensitive to dihydro-β-erythroidine and neosurugatoxin, but not α-bungarotoxin (2). However, [^{125}I]α-bungarotoxin binding is not inhibited by neosurugatoxin, although it is sensitive to higher concentrations of nicotine and dihydro-β-erythroidine.

ANTAGONIST	% INHIBITION OF NICOTINE(10^{-6}M) EVOKED DOPAMINE RELEASE	K_i VALUES (M) for LIGAND BINDING [^{3}H]NICOTINE	[^{125}I]α-BUNGAROTOXIN
Pempidine($5x10^{-6}$M)	62%	$2.6x10^{-3}$	$2.5x10^{-3}$
Mecamylamine($5x10^{-6}$M)	55%	$> 10^{-2}$	$1.2x10^{-3}$
Dihydro-β-erythroidine ($5x10^{-7}$M)	74%	$1.7x10^{-6}$	$2.7x10^{-5}$
Neosurugatoxin($5x10^{-8}$M)	54%	$1.2x10^{-6}$	$> 10^{-5}$
α-Bungarotoxin($2.5x10^{-7}$M)	1%	$> 10^{-4}$	$4.9x10^{-8}$
Histrionicotoxin($5x10^{-7}$M)	43%	ND	ND
(Nicotine)	–	$5.9x10^{-8}$	$8.9x10^{-6}$

These two ligand binding sites can be separated by affinity chromatography: the α-bungarotoxin site is retained by passage of detergent-solubilized membranes through Sepharose-α-toxin beads while the high-affinity nicotine site is quantitatively recovered in the non-bound fraction. This substantiates other reports that the ligand binding sites are not associated with the same receptor protein. A low-affinity nicotine site is present on the brain α-bungarotoxin receptor, analogous to the situation at the neuromuscular junction.

Comparison of the ligand binding properties with the nicotinic modulation of striatal dopamine release indicates that the high-affinity nicotine site may mediate these latter effects. However, preliminary studies (3) of the nicotinic autoregulation of acetylcholine release from rat hippocampus demonstrate that the action of nicotine in this preparation is antagonized by α-bungarotoxin. Thus the different classes of nicotinic ligand binding sites in the central nervous system may have different functional correlates in the brain.

References

1. Mills, A. & Wonnacott, S. (1984) Neurochem. Int. 6, 249-257
2. Rapier, C., Harrison, R., Lunt, G.G. & Wonnacott, S. (1985) Neurochem. Int. 7, 389-396
3. Moss, S. & Wonnacott, S. (1985) Biochem. Soc. Trans. 13, 164-165

Acknowledgements

This study was supported financially by the Tobacco Advisory Council.

ATTENUATION OF RAT PLASMA PROLACTIN RESPONSE TO RESTRAINT STRESS FOLLOWING ACUTE AND CHRONIC DOSING WITH NICOTINE

B. Sharp, S. Beyer, A.S. Levine, J.E. Morley, and K. McAllen

Dept. of Medicine, Hennepin County Medical Center
Minneapolis V.A.M.C., and University of Minnesota,
Mpls., MN 55415

Nicotine consumption by smoking is often associated, anecdotally with stress. Since both restraint stress and acute dosing with nicotine are known to elevate rat plasma prolactin, the interaction between nicotine and stress was studied using plasma prolactin as an end point. Groups of male Holtzmann rats, approximately 200 g, received one of several doses of nicotine sulfate or saline, intraperitoneally, either as a single injection or 5 times/day for 7 days with a single final injection on the eighth morning. One hour after the final dosing, 5 min of restraint was followed by decapitation. Plasma prolactin levels were determined by RIA and the results were analyzed by ANOVA. Prolactin levels were significantly less after stress in rats given a single injection of nicotine (N) 1.5, 2.0, and 3.0 mg (free base)/kg BW (saline 41.1 ± 3.5 ng/ml; $N_{0.75}$ 24.4 ± 4.3; $N_{1.5}$ 23.9 ± 4.1*; $N_{2.0}$ 19.9 ± 3.9***; $N_{3.0}$ 9.0 ± 1.1*** Values are mean ± SEM. *P < .05, *** p < .01 compared with saline). In nonstressed rats, a single dose of nicotine significantly suppressed prolactin levels 1 h after dosing (saline 3.4 ± 1.1; $N_{0.75}$ 1.1 ± 0.2; $N_{1.5}$ 1.2 ± 0.1; $N_{2.0}$ 1.0 ± 0.1*; $N_{3.0}$ 0.8 ± 0.1** **p < .025). Plasma prolactin levels were significantly less after stress in rats given the 7-day regimen (saline 50.2 ± 4.1; $N_{1.0}$ 48.2 ± 4.4; $N_{2.0}$ 23.6 ± 4.7**; $N_{3.0}$ 20.2 ± 3.2**). Chronic nicotine dosing did not affect prolactin levels measured in nonstressed rats 1 h after the final dose (saline 1.5 ± 0.3; $N_{1.0}$ 2.0 ± 0.4; $N_{2.0}$ 1.5 ± 0.2; $N_{3.0}$ 0.9 ± 0.1).

Tolerance to some of the acute effects of nicotine has previously been demonstrated. In contrast to the suppressed levels at 1 h, we observed that a single dose of nicotine significantly elevated prolactin levels at +15 min. (saline 2.1 ± 0.2; $N_{0.5}$ 9.8 ± 2.1*; $N_{1.0}$ 13.1 ± 3.3***; $N_{2.0}$ 13.4 ± 3.9*). However, 7-day exposure of all groups to nicotine 1.0 mg/kg eliminated the +15 min response to nicotine at all of the final doses (saline 2.7 ± 0.9; $N_{0.5}$ 3.4 ± 1.1; $N_{1.0}$ 7.8 ± 1.7; $N_{2.0}$ 8.1 ± 2.7 p = 0.1 by ANOVA).

A single dose of nicotine has a biphasic affect on basal prolactin levels and attenuates the plasma prolactin response to restraint stress. Attenuation persists with chronic exposure to nicotine in association with tolerance to the stimulatory effect of nicotine on prolactin release.

COGNITIVE, PERFORMANCE AND ELECTROPHYSIOLOGICAL EFFECTS OF NICOTINE DEPRIVATION AND NICOTINE GUM IN HUMAN VOLUNTEERS

Wallace B. Pickworth, Ronald I. Herning, Frederick R. Snyder and Jack E. Henningfield

NIDA, Addiction Research Center, Baltimore, MD 21224

Cigarette smokers often report that smoking improves their mood and concentration and that abstinence from tobacco is associated with impaired mood and concentration. In the present report, we describe the cognitive, performance, and electrophysiologic consequences of 9 days of nicotine deprivation. In a preliminary study nicotine gum was used in a substitution paradigm to assess the role of nicotine in the tobacco abstinence sequela.

The studies were performed on residential subjects who had long histories of cigarette smoking. During their first week on the ward, subjects smoked ad lib, EEG was recorded, and they trained on the Performance Assessment Battery (PAB) - five computerized tests which measure short term memory, sustained attention, logical reasoning, rapid mathematics, and a visual search task. Training continued until the subjects' performances stabilized. After the smoking phase the subjects were deprived of cigarettes for 9 days. At specified times during the deprivation, they were tested on the PAB and spontaneous EEG's were obtained. In a second study the effects of 0, 2, 4, or 8 mg nicotine gum were assessed on 12 h deprived heavy smokers. Measures of EEG, cognitive evoked potentials, mood, and performance were obtained in this study. Prior to electrophysiologic tests, subjects were occasionally pretreated with mecamylamine (10 mg) before administration of the gum.

Slowing of the EEG alpha frequency evident at 5 h of deprivation persisted throughout the abstinence period. Certain measures of cognitive performance were impaired for the duration of the abstinence such as response times in all tasks and accuracy in digit recall and arithmetic manipulation tasks. When smoking resumed, the alpha frequency and cognitive performance returned to baseline levels within 24 hrs.

In the second study, nicotine gum reversed the slowing of the EEG alpha frequency and restored the cognitive impairment. The EEG effects were most apparent in the relaxed subject. Mecamylamine pretreatment blocked the nicotine-induced reversal of the EEG alpha slowing. Latencies for the P300 auditory oddball task were lengthened with high background noise, and nicotine gum facilitated processing in this situation. This effect was not blocked by mecamylamine. Nicotine gum appears to substitute well for the electrophysiologic and cognitive effects of tobacco abstinence. The gum substituted less well for desire to smoke as measured by self-reported data.

CONTRIBUTORS

Abood, L. G.
Center for Brain Research
University of Rochester School of Medicine and Dentistry
Rochester, New York 14642

Aceto, M. D.
Department of Pharmacology and Toxicology
Medical College of Virginia
Virginia Commonwealth University
Richmond, Virginia 23284

Adams, Priscilla
Dartmouth Medical School
Hanover, New Hampshire 03756

Agnati, L. F.
Department of Human Physiology
University of Modena
Modena, Italy

Al Houdi, Abdul
College of Pharmacy
University of Kentucky
Lexington, Kentucky 40536

Amano, H.
Department of Pharmacology
Yokohama City University
School of Medicine
Yokohama 232, Japan

Andersson, K.
Department of Histology
Karolinska Institutet
Stockholm, Sweden

Archer, William
Department of Biological Sciences
University of Notre Dame
Notre Dame, Indiana 46556

Baron, John A.
Dartmouth Medical School
Hanover, New Hampshire 03756

Benowitz, Neal L.
San Francisco General Hospital and Medical Center
Building 30, Fifth Floor
101 Potrero Avenue
San Francisco, California 94110

Beyer, S.
Department of Medicine
Hennepin County Medical Center
Minneapolis V.A.M.C. and
University of Minnesotta
Minneapolis, Minnesota 55415

Bobbitt, Frances A.
Department of Medicine
University of Kentucky
Veterans Administration Medical Center
Lexington, Kentucky 40511

Brown, Fran J.
Johns Hopkins University
School of Medicine
Key Medical Center
Baltimore, Maryland 21224

Brown, Laura V.
Department of Medicine
University of Kentucky
Veterans Administration Medical Center
Lexington, Kentucky 40511

Clarke, Paul B. S.
Kinsmen Laboratory of Neurological Research
Department of Psychiatry
University of British Columbia
Vancouver, British Columbia
Canada V6T 1W5

Collins, Allan C.
School of Pharmacy
University of Colorado
Boulder, Colorado 80309

Conti-Tronconi, Bianca
Department of Biochemistry
University of Minnesota
1479 Gortner Ave.
St. Paul, Minnesota 55108

Crooks, Peter A.
Graduate Center for Toxicology
and College of Pharmacy
University of Kentucky
Lexington, Kentucky 40536

Davis, D. Layten
Tobacco and Health Research
Institute
University of Kentucky
Lexington, Kentucky 04546-0236

DeLuca, P. P.
College of Pharmacy
University of Kentucky
Lexington, Kentucky 40536

Edwin C. Williamson
Department of Pharmacology
College of Medicine
University of Kentucky and
The Tobacco and Health Research
Institute
University of Kentucky
Lexington, Kentucky 40536

Diliberto Jr., Emanuel J.
Department of Medicinal Biochemistry
The Wellcome Research Laboratories
Research Triangle Park
North Carolina 27709

Eneroth, P.
Research and Development Laboratory
Department of Obstetrics and
Gynecology
Karolinska Hospital
Stockholm, Sweden

Ferguson, G. S.
Department of Pharmacology and
Toxicology
Medical College of Virginia
Virginia Commonwealth University
Richmond, Virginia 23284

Fudala, Paul J.
Graduate Center for Toxicology
The Graduate School
University of Kentucky
Lexington, Kentucky 40536

Fuxe, Kjel
Department of Histology
Karolinska Institutet
Stockholm, Sweden

Gairola, C.
Tobacco and Health Research
Institute and
Graduate Center for Toxicology
University of Kentucky
Lexington, Kentucky 40546

Godin, C. Steven
College of Pharmacy
and
Graduate Center for Toxicology
University of Kentucky
Lexington, Kentucky 40536

Goldberg, Steven R.
National Institute on Drug Abuse
Addiction Research Center
4940 Eastern Avenue
Baltimore, Maryland 21224
and University of Maryland
School of Medicine

Goshima, Y.
Department of Pharmacology
Yokohama City University
School of Medicine
Yokohama 232, Japan

Grassi, S.
Center for Brain Research
University of Rochester School
of Medicine and Dentistry
Rochester, New York 14642

Griffiths, Roland R.
Department of Psychiatry and
Behavioral Sciences
Division of Behavioral Biology
Behavioral Pharmacology Research
Unit, D-5-West
Francis Scott Key Medical Center
4940 Eastern Avenue
Baltimore, Maryland 21224

Hakan, Robert L.
University of Wyoming
Laramie, Wyoming 82071

Haley, Nancy J.
American Health Foundation
Division of Nutrition and
Endocrinology
Naylor Dana Institute
Valhalla, New York 10595

Harfstrand, A.
Department of Pharmacology
Biomedical Center
University of Uppsala
Uppsala, Sweden

Henningfield, Jack E.
National Institute on Drug Abuse
Addiction Research Center
P. O. Box 5180
4940 Eastern Avenue
Baltimore, Maryland 21224
and The Johns Hopkins University
School of Medicine
Baltimore, Maryland 21224

Herning, Ronald I.
National Institute on Drug Abuse
Addiction Research Center
P. O. Box 5180
4940 Eastern Avenue
Baltimore, Maryland 21224

Hinson, J. R.
Department of Pharmacology and
Toxicology
Medical College of Virginia
Virginia Commonwealth University
Richmond, Virginia 23284

Hong, Jau-Shyong
Laboratory of Behavioral and
Neurological Toxicology
National Institute of Environmental
Health Sciences
Research Triangle Park
North Carolina 27709

Hyde, Gordon
Department of Surgery
College of Medicine
University of Kentucky
Lexington, Kentucky 40536

Iwamoto, Edgar T.
Department of Pharmacology
College of Medicine
University of Kentucky and
The Tobacco and Health Research
Institute
University of Kentucky
Lexington, Kentucky 40536

Jacob III, Peyton
Medical Service
San Francisco General Hospital
Medical Center/Langley Porter
Psychiatric Institute
University of California,
San Francisco
San Francisco, California 94143

Jarvik, Murray E.
Department of Pharmacology and
Department of Psychiatry
The Neuropsychiatric Institute and
Hospital, School of Medicine
University of California, Los Angeles
Los Angeles, California 90024; and
Veterans Administration Medical Center
West Los Angeles, Brentwood Division
Los Angeles, California 90073

Jasinski, Donald R.
The Johns Hopkins University
School of Medicine
Baltimore, Maryland 21224

Kanamatsu, Tomoyuki
Second Department of Physiology
Toho University School of Medicine
Tokyo, Japan

Kellar, Kenneth J.
Department of Pharmacology
Georgetown University
Schools of Medicine and Dentistry
3900 Reservoir Road, N.W.
Washington, D.C. 20007

Kiritsy-Roy, Judith A.
Department of Medicine
University of Kentucky
Veterans Administration Medical
Center
Lexington, Kentucky 40511

Knott, Verner J.
Specialty Clinics
Royal Ottawa Hospital
Ottawa, Canada

Kolberg, Kathy
Department of Biological Sciences
University of Notre Dame
Notre Dame, Indiana 46556

Ksir, Charles
University of Wyoming
Laramie, Wyoming 82071

Kubo, T.
Department of Pharmacology
Yokohama City University
School of Medicine
Urafunecho, Minami-ku
Yokohama 232, Japan

Kumar, R.
Department of Psychiatry
Institute of Psychiatry
University of London
De Crespigny Park
London SE5 8AF, England

Lajtha, Abel
Center for Neurochemistry
Nathan S. Kline Institute for Psychiatric Research
Ward's Island
New York, New York 10035

Levine, A. S.
Department of Medicine
Hennepin County Medical Center
Minneapolis V.A.M.C. and
University of Minnesotta
Minneapolis, Minnesota 55415

Lunt, G. G.
Department of Biochemistry
University of Bath
Bath, BA2 7AY, U.K.

Maiti, A.
Center for Brain Research
University of Rochester School of Medicine and Dentistry
Rochester, New York 14642

Majchrzak, Mark J.
Behavioral Medicine Laboratory
Department of Psychiatry
University of Michigan School of Medicine
Ann Arbor, Michigan 48109

Marks, Michael J.
Institute for Behavioral Genetics
University of Colorado
Boulder, Colorado 80309

Martin, Billy R.
Department of Pharmacology and Toxicology
Medical College of Virginia
Virginia Commonwealth University
Richmond, Virginia 23298

Martin, Vicki
Department of Biological Sciences
University of Notre Dame
Notre Dame, Indiana 46556

Martin, William R.
Department of Pharmacology
University of Kentucky
Lexington, Kentucky 40536

Martino, Andrea M.
Department of Pharmacology
Georgetown University
Schools of Medicine and Dentistry
3900 Reservoir Road, N.W.
Washington, D.C. 20007

McAllen, K.
Department of Medicine
Hennepin County Medical Center
Minneapolis V.A.M.C. and
University of Minnesotta
Minneapolis, Minnesota 55415

Misu, Y.
Department of Pharmacology
Yokohama City University
School of Medicine
Yokohama 232, Japan

Morley, J. E.
Department of Medicine
Hennepin County Medical Center
Minneapolis V.A.M.C. and
University of Minnesotta
Minneapolis, Minnesota 55415

Mundy, William R.
Graduate Center for Toxicology
The Graduate School
University of Kentucky
Lexington, Kentucky 40536

Nordberg, A.
Department of Pharmacology
Biomedical Center
University of UPpsala
Uppsala, Sweden

Nwosu, Chukwuemeka G.
College of Pharmacy
and
Graduate Center for Toxicology
University of Kentucky
Lexington, Kentucky 40536

Perry, Holly
Department of Pharmacology
St. Louis University
School of Medicine
St. Louis, Missouri 63104

Pert, Agu
Biological Psychiatry Branch
National Institute of Mental Health
Building 10, Room 3N212
Bethesda, Maryland 20205

Pickworth, Wallace B.
National Institute on Drug Abuse
Addiction Research Center
P. O. Box 5180
4940 Eastern Avenue
Baltimore, Maryland 21224

Pomerleau, Cynthia S.
Behavioral Medicine Laboratory
Department of Psychiatry
University of Michigan School of Medicine
Ann Arbor, Michigan 48109

Pomerleau, Ovide F.
Behavioral Medicine Laboratory
Department of Psychiatry
University of Michigan School of Medicine
Ann Arbor, Michigan 48109

Pool, William F.
College of Pharmacy
and
Graduate Center for Toxicology
University of Kentucky
Lexington, Kentucky 40536

Pratt, J. A.
Department of Pharmacology
University of Strathclyde
Scotland

Quinton, Hebe
Dartmouth Medical School
Hanover, New Hampshire 03756

Raftery, Michael A.
Division of Chemistry and Chemical Engineering
California Institute of Technology
Pasadena, CA 91125

Rainbow, Thomas C.
Department of Pharmacology
University of Pennsylvania
Philadelphia, Pennsylvania 19104

Rapier, C.
Department of Biochemistry
University of Bath
Bath, BA2 7AY, U.K.

Reavill, C.
Department of Pharmacology
and Department of Psychiatry
Institute of Psychiatry
De Crespigny Park
London SE5 8AF, England

Richardson, Daniel
Department of Physiology
College of Medicine
University of Kentucky
Lexington, Kentucky 40536

Rogers, Walter R.
Department of Bioengineering
Southwest Research Institute
San Antonio, Texas 78284

Rosecrans, John A.
Department of Pharmacology and Toxicology
School of Basic Health Sciences
Virginia Commonwealth University
Richmond, Virginia 23298

Rowell, Peter P.
Department of Pharmacology and Toxicology
University of Louisville School of Medicine
Louisville, Kentucky 40292

Russell, M. A. H.
Addiction Research Unit
Institute of Psychiatry
101, Denmark Hill
London SE5 8AF,England

Rypacek, F.
College of Pharmacy
University of Kentucky
Lexington, Kentucky 40536

Salles, K. S.
Center for Brain Research
University of Rochester School of Medicine and Dentistry
Rochester, New York 14642

Sapru, Hreday N.
Section of Neurological Surgery
University of Medicine and Dentistry
New Jersey Medical School
Newark, New Jersey 07103
and
Veterans Administration Medical Center
East Orange, New Jersey 07019

Schwartz, Rochelle D.
Department of Pharmacology
Georgetown University
Schools of Medicine and Dentistry
3900 Reservoir Road, N.W.
Washington, D.C. 20007

Sepkovic, D. W.
American Health Foundation
Division of Nutrition and
Endocrinology
Naylor Dana Institute
Valhalla, New York 10595

Sershen, Henry
Center for Neurochemistry
Nathan S. Kline Institute for
Psychiatric Research
Mabon Building
Ward's Island
New York, New York 10035

Sharp, B.
Department of Medicine
Hennepin County Medical Center
Minneapolis V.A.M.C. and
University of Minnesotta
Minneapolis, Minnesota 55415

Sloan, Jewell W.
Department of Pharmacology
Research Facility No.2
University of Kentucky
Lexington, Kentucky 40536

Smith Jr., Walter T.
Department of Chemistry
CP341 Chemistry-Physics Building
University of Kentucky
Lexington, Kentucky 40536

Snyder, Frederick R.
National Institute on Drug Abuse
Addiction Research Center
P. O. Box 5180
4940 Eastern Avenue
Baltimore, Maryland 21224

Stitzer, Maxine L.
Johns Hopkins University
School of Medicine
Key Medical Center
Baltimore, Maryland 21224

Stolerman, I. P.
Department of Pharmacology and
Department of Psychiatry
Institute of Psychiatry
De Crespigny Park
London SE5 8AF, England

Tucker, S. M.
Department of Pharmacology and
Toxicology
Medical College of Virginia
Virginia Commonwealth University
Richmond, Virginia 23284

Tuttle, Megan L.
Department of Physiology and
Medicine
Southwest Foundation for Biomedical
Research
San Antonio, Texas 78284

Unsworth, Christopher D.
Howard Hughes Medical Institute
Massachusetts General Hospital
Boston, Masssachusetts 02114

Van Loon, Glen R.
Department of Medicine
University of Kentucky
Veterans Administration Medical
Center
Lexington, Kentucky 40511

Vickery, Lillian
Department of Pharmacology
St. Louis University
School of Medicine
St. Louis, Missouri 63104

Viveros, O. Humberto
Department of Medicinal Biochemistry
The Wellcome Research Laboratories
Research Triangle Park
North Carolina 27709

Volle, Robert L.
National Board of Medical Examiners
3930 Chestnut Street
Philadelphia, Pennsylvania 19104

Wakade, Arun R.
Department of Pharmacology
State University of New York
Downstate Medical Center
450 Clarkson Avenue
Brooklyn, New York 11203

Warburton, David M.
Department of Psychology
Reading University
Reading RG6 2AL, England

Wesnes, Keith
Cognitive Drug Research, Ltd.
Opus House
72 South Street
Reading
United Kingdom RG14HW

Westfall, Thomas C.
Department of Pharmacology
St. Louis University
School of Medicine
St. Louis, Missouri 63104

Wolfe, Barry B.
Department of Pharmacology
University of Pennsylvania
Philadelphia, Pennsylvania 19104

Wonnacott, S.
Department of Biochemistry
University of Bath
Bath, BA2 7AY, U.K.

Woodson, Phillip P.
Department of Psychiatry and
Behavioral Sciences
Division of Behavioral Biology
Behavioral Pharmacology Research
Unit, D-5-West
Francis Scott Key Medical Center
4940 Eastern Avenue
Baltimore, Maryland 21224

Yingling, John E.
Johns Hopkins University
School of Medicine
Key Medical Center
Baltimore, Maryland 21224

Zacny, James P.
Johns Hopkins University
School of Medicine
Key Medical Center
Baltimore, Maryland 21224

INDEX

Acetylcholine, 115, 128, 134, 135, 153, 157, 160-162, 169, 171-177, 191, 193, 195, 196, 198, 201, 202, 219, 220, 227, 228, 231, 235, 277, 279, 282,292, 304, 305, 317, 325-327, 330, 331, 333
  binding to brain, 494-495
  binding site, nicotinic, 227
  receptor, 227, 413-437
  recognition site in brain, 467-477
Acetylcholinesterase, 129, 175, 225, 279
Acid, gastric, secretion and nicotine, 301
Activation, dopaminergic, and nicotine, 178-183
Adenosine monophosphate, (cAMP), 332, 350
S-Adenosylhomocysteine and nicotine, 500
S-Adenosylmethionine and nicotine, 500
Adenyl cyclase, 322
Adrenal gland, 344
Adrenocorticotropin, 474
Aldehyde oxidase, 357
Alzheimer's disease, 134, 158, 477
γ-Aminobutyric acid (GABA), 134
Amphetamine, 90, 119, 180-183, 195, 212-215
Amygdala, 195, 199
Anabasine, 117, 118, 463, 484
Antinociception and nicotine, 301-309
Arecoline, 126-129, 200, 202
Arteriosclerosis, 21
Atherosclerosis, 22
  and tabocco smoking, 502
Atropone, 126-129, 153, 157, 160, 170-176, 180, 197, 228, 231, 279, 282, 292-295, 304, 305, 317, 321, 322, 327, 329, 468, 484
Autoradiography, 151-167, 227-230, 494-495
Autoreceptor, 159, 198

Baboon and cigarette smoking, 500-501
Behavior alteration and nicotine, 301
Benzodiazepine, 119
Bethanechol, 322
Binding site, *see* Receptor
Binding studies of nicotine,113-124, 452-464
  inhibition, 454-458, 461
Biomethylation of nicotine enantiomer, 499
Blood pressure and nicotine, 277-286, 309, 310
  arterial, 277-286
  control of, 287-300
Brain
  areas, 194
  binding site, nocotinic, 225-262, *see also* Receptor
  neurotransmitter release, 191-208
  stem, human, and tobacco, 495
    and evoked potential, 495
  stem, cat, 294
  stimulation
    electrial (ESB), 141-144
    reward (BSR), 141-149
  substance, dependence-productiing, 1
α-Bungarotoxin, 114, 121, 153, 157-162, 171-175, 198, 225, 230, 402, 404, 440, 442, 445, 447, 467, 483-486, 505
Buprenorphine, 90

Calcium, 175, 196, 198, 212, 214, 216-222, 322, 330-335, 341-344
Carbachol, 157, 160, 161, 174, 176, 198, 200, 228, 291-293, 296, 298, 468, 476, 484
Carbamylcholine, 332 334, 336, 402, 404, *see also* Carbachol
Carbon monoxide, 503
Carboxyhemoglobin in smokers, 363

Cardiovascular effects of nicotine, 301-316, 365-370
Catecholamine, 169, 183, 191, 193, 221, 234, 243-351, 471
cosecretion, 343-345
regulation, nicotinic, 263-276
sectretion
adrenal, 325-339
sympathoadrenal, 263-276
Catechol-O-methyltransferase, 271
Central nervous system (CNS)
and nicotine, 191-208
and nicotine-binding in rat, 493-494
Cerebellum, vestibular, of rat
and behavior, 409
homeostasis, 409
startle reaction, 409
structure and function, 410
Chlordiazepoxide, 90, 119
Chlorisondamine, 114, 121, 176, 178
Cholesterol, 21
Choline, 175, 279, 280, 317, 468, 484
Choline acetyltransferase, 175, 225, 279
Cholinergic receptor, *see* Receptor, cholinergic
Cholinesterase, 128, 468, 472, 473
Chromaffin
cell, 341-345, 348-351
peptide, opioid, 343-345
vesicle, 343
Chronic obstructive lung disease (COLD), 21
Cigarette
brand-switching, 36-37, 41
compounds
health-related, 20
in smoke phase, 20
consumption and nicotine yield, 35-37
design, 18-19
filter -, 19
nicotine yields in countries, 17, 18
production trends, 16
sales, worldwide, 16
smoke chemistry, 19-21
smoking trends, U.S.A., 15-17, *see also* Cigarette smoking
tar yields in countries, 17-18
low tar, 43-46
ventilation, 19
Cigarette smoking
by baboons, habituation, 500
deprivation in human, 506-507
electroencephalogram (EEG), 507
and hormone, pituitary, 501-502
manner of, 503
Cigarette smoking (continued)
and nausea, 501-502
and nicotine, 501-502
and Parkinson's disease, 497
and performance, cognitive, 507
properties of, 503
pulse rate, 385
and rat, effects in, 225-262
trends in
countries, 15, 17
U.S.A. (1960-1983), 16
Cigar smoking, 29-30, 43
and nicotine level in blood, 30
reasons for, 192
Circulation, cutaneous
and smoking tobacco, 502
Clonidine, 94, 134
Cocaine, 182, 183, 403, 404
COLD, *see* Chronic obstructive lung disease
Communication, intercellular, 352-353
Convulsion and nicotine, 301
Corticosterone, 226, 227, 239, 243
Cosecretion
from medulla, adrenal, 343
of multiple messengers, 32
Cotinine, 200, 202, 232, 357, 376, 377, 386, 484
elimination, 377
excretion, urinary, 358, 359, 376-377, 382
formation in metabolism, 357
in plasma, 379, 381
in serum, 376-377
in urine, 358, 359, 376-377, 382
Cotransmissioners, 343, 352
Cross tolerance, 305
to stress, 263-276
Cytisine, 115-118, 179-182, 462, 463, 468, 476, 484

DDT, *see* Dithiothreitol
DEAE methylcarbamate, 404
Decamethonium, 160
Deoxycholate, 408
2-Deoxyglucose, 161
Dependence
commonality, 86
definition of, 136
neuroadaptation, 84-85
potential of nicotine, 86-95
euphoria, 87-91
psychoactivity, 87-91
physiologic, 84-85
tolerance, 84-85
treatment strategy, pharmacologic, 94-95
Desmethylimipramine, 130

Detoxification, *see* Drug abstinence
DHBE, *see* Dihydro-β-erythroidine
Diacylglycerol, 335, 337
Diazepam, 133
Diencephalon of rat, 225-262
Dihydro-β-erythroidine (DHBE), 160, 162, 176, 197, 476
5,7-Dihydroxytryptamine, 471, 472
Diisopropyl fluorophosphate, 468
Dimethylaminoethyl methylcarbamate, 402
Dimethylphenylpiperazinium (DMPP), 202, 209, 210, 212-215, 217
Disulfoton, 487
Dithiobis-2-nitrobenzoic acid (DTNB), 471
Dithiothreitol (DDT), 471
4,4-Ditritio(-)-/^{3}H/nicotine, 483
DMPP, *see* Dimethylphenylpiperazinium
Dog, 264
  hypoglycermia, insulin-induced, 268, 269
  and nicotine, 267-273
Dopamine, 110, 133, 134, 158, 161, 169-175, 193-195, 200, 201, 486, 496-497, 505
  and DMPP, 210, 212-215
  endogenous in rat, 211
  nerve terminal system, 225-262
  in neuron, 129, 169
    cholinergic, 170, 173-175,
    dopaminergic, 219
  and nicotine, 216-218
    effect, 129-131
    regulation mechanism, 209-223
  release, 173
    in neostriatum, 209-223
  striatal, 172
  in striatal slice, 210
  and tyrosine, 212
Dopamine β-hydroxylase, 344
Drug
  abstinence, 85
  abuse, 86
  dependence, *see* Dependence
  nicotinic-blockers, 114
  recreational, self-admistration, 390
    alcohol, 380
    caffeine, 380
    nicotine, 380
DTNB, *see* Dithiobis-2-nitrobenzoic acid

Eating, *see* Feeding
EEG, *see* Electroencephalogram
Efficiency, mental, 63-79
  improved by
    cigarette, 63
    mechanism, neural, 76-77
Efficiency, mental
  improvement (continued)
    nicotine, 63
Electroencephalogram (EEG)
  and nicotine, 507
Electrophorus sp., 413-437
  organ, electric, 413-437
β-Endorphin, 133, 134
Enkephalin
  biosynthesis, 345-350
  regulation of, 345-350
  see Peptide, opioid
Epinephrine, 195, 264-271, 326, 327, 343, 344, 346, 347, 496-497
Eurphoria, 87-91, 231
  and nicotine, 182
Extraversion and smoking, 55
Eyseneck Personality Inventory, 54

Fagerstrom Tolerance Questionaire, 85
Feeding, 391-395
  arousal, 394
  hunger, 393-394
  satiety, 394-395
Forskolin, 332, 335, 337
Frog muscle, 317-323
Function, cardiovascular, *see* Cardiovascular function
Fusaric acid, 497

Ganglion, sympathetic
  block by nicotine, 319-321
  nucleotide, cyclic, 321-322
  receptor
    muscarinic, 321
    nicotinic, 317-323
L-Glutamate, 291, 293, 294, 296
L-Glutamate monosodium, 290
Growth hormone, 497
Guinea pig, 499
  macrophage, 498
  and nicotine, 500

Health and tobacco
  overview, 1-13
  risks, associated, 15-23
Heart
  disease, 21
  rate and nicotine, 364-370
Hemicholinium-*3*, 176, 280, 468
Heroin, 134
Hexamethonium, 114, 120, 121, 126, 128, 160, 175, 180, 197, 264, 266, 277, 278, 292, 295, 296, 304, 318, 322, 325, 327, 402, 404, 463, 464, 497
Hippocampus, 133-134
Histamine, 464, 500
Histamine N-methyltransferase, 500

Hormone
  antidiuretic, 301
  neuroendocrine, 225-262
  pituitary, 501-502
Horseradish peroxidase, 298
Hydrogel microspheres, 498-499
6-Hydroxydopamine, 129, 130, 170, 171
6-Hydroxymethyl nicotine, 408
5-Hydroxytryptamine, 119
Hypoglycemia, insulin-induced, 268, 269, 346
Hypothalamic hormones, and nicotine, 194

ICSS, *see* Self-stimulation, intracranial
Information processing, 65-66, 71-73
Insulin, 346, 347
  and induced hypoglycemia, 268, 269, 346
Iontophoresis, 176, 177
Isoproterenol, 322

Kainic acid, 171, 177
  and behavior, 405-407
  and brain lesion, 405, 406

Learning
  and nicotine, 101-111, 301
    methods, 106-107
  theory, 102
Ligand-binding, 428-431
  mechanism, 431
  and nicotine, 481-491
  receptor, nicotinic, *see* Receptor, nicotinic
  site, 428, 504-505
  and thallium influx, 429
Lobeline, 118, 175, 200, 318-320, 484
Locomotor activity, 178, 179
Lung
  cancer, 21
  disease, chronic obstructive (COLD), 21
  and nicotine, 498

Macrophage, 498
Marijuana, 90
Mecamylamine, 90, 114, 120, 126, 128, 130, 131, 143, 144, 161, 169, 175, 176, 178, 180-182, 197, 212, 216-218, 220, 221, 225, 238, 240-245, 266, 277, 303, 306, 308, 329, 402, 404, 476, 484, 497, 504
Medulla
  adrenal, 325, 341-347
  dorsal, 283-284
Medulla (continued)
  oblongata, 278-282
  peptide, opoid, from, 351-352
  ventrolateral, 287-300
    and blood pressure, 287-300
    receptor, 287-300
Memory in rat
  and nicotine, 101-111
    methods, 106-107
  and vasopressin, 102
Met-enkephalin, 133, 134, 322, 347, 350
Methadone, 94
Methiodide, 121, 306, 307, 308
Methylatropine, 282
Methylcarbamylcholine, 402-404
4-Methylnitrosoamino-1-(3-pyridyl)-1-butanone (NNK), 376
2-Methylpiperidine, 202, 455, 458-464
α-Methyl-p-tryrsine, 183, 496
Microsphere, biodegradable, 498-499
  and alpha-proteinase inhibit -or, 498-499
  hyrogel, 498-499
Monoamine
  neuropeptide cotransmission, 341-356
  oxidase, 271
Morphine, 90, 182, 183, 305
Motor
  endplate receptor, nicotinic, 317-323
  performance and nicotine, 301
Mouse, 132-134, 439-450
  and nicotine, 132-134, 439-450
Muscarine, 326-337
Muscimol, 298
Muscle, skeletal, 318-322

Naloxone, 304, 305
Naltroxone, 84
Nausea in smokers, 501
Neostigmine, 202
Neostriatum of rat, 209-223
Neosurugatoxin, 160
Nerve terminal
  cholinergic, 225-262
  splanchnic, 325-333
Neurobiology of tobacco, 1-13
Neuron
  cholinergic, 225
  dopaminergic, 170-173
  splanchnic, 325
Neuropeptide, 234-237, 246, 341-356
Neuroticism and smoking, 55-56
Neurotoxin in smoke, 440
Neurotransmission, dopaminergic, 169-189
Neurotransmitter in CNS, 191-208
Niacinamide, 464
Nicotine, 132-134, 225-262, 357-373, 439-450, 451-465, 468-477

Nicotine (continued)
absorption, 359-360
abuse liability, 81-99
acid, gastric, secretion, 301
action, mechanism of, 134-135
activation, dopaminergic, 178-183
addiction, *see* dependence
administration, chronic, in mouse 439-450
aerosol rod as cigarette substitute, 386
analog and behavior, 402
antinoception, 301-309
aversion, 110
behavior
alteration, 301
analogs, 402
effects, 125-139
-binding, 152-163, 402-407, 413-437, 451-465, 481-494
biomethylation, 499
block, neuromuscular, 318-319
and blood, 360-364
pressure, 277-286, 309-310
and brain, *see* Brain
cardiovascular effects, 264-267, 301-316, 365-370
and catecholamine
secretion, 263-276
turnover rate, 496-497
in central nervous system (CNS), 191-208
in cnidaria, 495-496
convulsion, 301
dependence, 42, 46-47, 81-99, 125-139, 182-183
deprivation in humans, 506-507
distribution in body, 360-361
in dog, 267-273
and dopamine, 129-131, 209-223
dose-response, 104, 365, 443, 444, 493-494, 505-506
effects, 2-3, 144, 178-182, 230-252
efficency, mental, 63-79
electroencephalogram (EEG), 507
elimination from body, 361-364
embryo, invertebrate, 495
enantiomer, 456
enantiomer, biomethylation of, 499
euphoria, 182
excitability of nerve, 2-3
excretion
renal, 362
urinary, 359
ganglion block, 319-321
growth hormone release, 395
guinea pig metabolism, 500
heart rate, 364-370, 386
hormone pituitary, 501-502
hypothalamus hormones, 194

Nicotine (continued)
improvement, mental, 66-77
infusion, chronic, 504
intake regulation, 25-50
learning in rat, 54, 101-111, 301
ligand-binding, 428-430, 451-465, 481-491
locomotor behavior, 178
in lung, 498
mechanism, noncholinergic, 125-139
medulla oblongata of rat, 278-282
and memory, 54, 101-111
metabolism, 357-388
N-methylation, 498
and mood, 54
motor performance, 301
in mouse, 26, 132-134, 439-450
nausea, 501-502
neostriatum, 209-223
nerve effects, 2-3
neuroadaptation, 84-95
neuropharmacology, 114
neurotransmission, dopaminergic, 169-189
neurotransmitter rlease in CNS, 191-208
in organs, 360-361
in plasma, 378
and pain, 301-316
and peptide, opioid, 341-356
pharmacodynamics, 364-370
pharmacokinetics, 357-373
pharmacology, 192
and prolactin, 505-506
and prostration, 301, 402-404
in rat brain, 468-477, 493-494, 496-497
receptor, 301
in brain, 395, 401-450
cholinergic, 125
multiple, 451-465
in muscle, 413-437
purification, 408-409
reinforcement effect, 395-396
respiration, 395
reward, 101-111
self-administration, 84
see dependence
self-stimulation, intracranial, 141-149
stereoisomer structure, 301
steroselectivity, 301-316
stimulus effect, 113-139
stress, 265-266, 506
substitution, *see* Nicotine gum
time effect, 104-106
tolerance, 84-85, 132-134, 264-265, 365, 369, 506
transport across membrane, 359-360
tremor, 301
withdrawal syndrome, 132-134

Nicotine (continued)
  yield to smoker, 35-39
  see Cotinine, Nicotine N-oxide
Nicotine gum, 85, 95, 369, 506-507,
  see also Polacrilex
Nicotine N-methyltransferase, 499
Nicotine N-oxide, 357, 376-377
  in serum, 376-377
  in urine, 376-377
Nitrosamine in tobacco, 377
N'-Nitrosonornicotine (NNN), 376
Nodule, cerebellar, 405-406
Noradrenaline, 255-262
Norepinephrine, 130, 193-195, 200, 221, 264-271, 318, 326, 327, 343, 344, 364, 496-497
Nornicotine, 118, 160
Northern blot analysis, 348
Nucleus tractus solitarius, 281-283

Opiate use and orientation, 51
Oxotremorine, 126, 174, 175, 202, 329-332
Oxytocin, 247

Pain and nicotine, 301-316
Paraoxon, 228, 231
Pargyline, 213
Parkinson's disease, 129, 134, 497
Pempidine, 114, 121, 161, 175, 197
Pentazocine, 90
Peptide, opoid, 263, 341-356
Perhydrohistrinonicotoxin, 122
Phencyclidine, 122
Phentolamine, 282, 322
Phenylephrine, 282, 298
Pheochromocytoma, 343, 344, 350
Phorbol ester, 332-337
Physostigmine, 128, 129, 160, 174, 175, 279, 280 292, 294, 304
Picolinic acid, 464
Picrotoxin, 119
Pilocarpine, 180
Pipecolic acid, 464
Piperidine, 160, 486
Pipe smoking, 29-30, 43
Pirenzepine, 329, 330
Place-conditioning paradigm, 102-106
Pleasure, neurobiology of, 2
Polacrilex (nicotine gum), 95
Potassium, 174, 199, 200, 332, 336
Practolol, 322
Pregnancy and smoking, 382
Pre-proenkephalin A mRNA, 343, 345
Primate, nonhuman, *see* Baboon
Procaine, 403, 404
Proenkephalin, 343-347
  mRNA, 345, 346, 348, 350
Prolactin, 474, 497, 505-506
Prostration and nicotine, 402-404

α-1-Proteinase inhibitor, 498-499
Protein kinase C, 332, 335, 337
Psychoactivity, 87-91
Puff of smoke, 502-503
Pyrilamine, 403, 404

Questionnaires, 85, 90
Quinolinic acid, 464

Radioimmuneassay, 226
Radioligand, 151-152
Rat, 101-124, 141-167, 209-300, 401-411, 468-477, 493-507
Receptor
  cholinergic, 173-175, 225, 345-350, 467-479,
  muscarinic, 171, 174, 175, 225, 344
  nicotinic, 225-262, 287-300, 304, 317-324, 341-345, 401-465, 481-491, 495, 504-505
Regulation of catecholamine
  secretion, 263-276
Reserpine, 213, 215, 239-351
Restraint stress, 266, 505-506
Reward in rat and nicotine, 101-111
  mechanism, 225, 262
Ribonucleic acid, *see* RNA
nRNA, 343, 345, 347, 348, 350

Scopolamine, 304
Secretion, sympathoadrenal, 263-276
Self-administration, 84, 91-93
  see Nicotine dependence
Self-stimulation, intracranial (ICSS), 142-147
Serotonin, 130, 194-196, 394, 471, 472
Shuttle box experiments, 145-147
Single-dose questionnaire, 90
Smoke, 33-34
Smokers
  occasions (seven), 52
  types (four), 52
Smoking of cigarettes, 2, 17, 18, 21, 136, 162
  after meal, 391-392
  and appetite, 389-390
  "Craker Jack" hypothesis, 395
  and eating, 389-400
  function of, 51-61
  information processing, 65-66, 71-73
  motif for, 55, 58
  nicotine intake regulation, 25-50
  personality characteristics, 54-57
  questionnaires, 53
  regulation, 39-42
  stress, 55-58
  weight loss, 389
  withdrawl symptoms, 57
Snake neurotoxin, 440
Snuff, 30-31

Somatostatin, 237
Sparteine, 200
Startle response, 441
Stereoselectivity, 305, 310
Stimulation, electrical, 141
Stress and nicotine, 263-276
Straitum, 194, 195, 200
Stroop effect, 73
Substantia nigra, 175-178

Tail-flick test, 302-309
Telencephalon of rat, 225-262
Tetrabenazine, 348-351
Tetraethylammonium, 175, 197, 329
Tetrodotoxin, 193, 212, 214, 217, 219, 221
Thallium, 429
Tobacco
  abuse, 86
  carcinogenicity, 376-377
  dependence, 1, 4, 5, 82
  deprivation, 85
  euphoria, 87-91
  exposure, 375-376
  and health, 1-13
    risks, 15-23
  history, 15
  neurobiology, 1-13
  psychoactivity, 87-91
  puffing, 502-503
  self-administration, 84, 91-93
  smoke composition, 379-380
  smokeless, 4, 18, 43
  withdrawal, 93-94
Tolerance, 84-85, 264-265, 369, 441, 506
  see Cross-tolerance
Torpedo species, 413-437
  T. californica, 414-437
  T. electroplax, 414-437
  receptors, 413-437
Treatment for dependence, 94-95
Tremor and nicotine, 301
Tripelennamine, 90
Tryptamine, 486
Tryptophan and sedation, 394
d-Tubocurarine, 114, 160, 175, 197, 212, 216-221, 225, 230, 319, 476, 485, 487,
Tyramine, 200, 212, 213, 215
Tyrosine and dopamine release, 212, 216, 218-220
Tyrosine hydroxylase, 169, 170, 171, 226, 345

Vasopressin and memory, 102
Verapamil, 331

Y-maze, 143-145, 147, 442, 444-447
Yohimbine, 304

Zomepirac, 90